CRC Handbook of Laser Science and Technology

Volume II
Gas Lasers

Editor

Marvin J. Weber, Ph.D.

Laser Program
Lawrence Livermore National Laboratory
University of California
Livermore, California

CRC Press, Inc.
Boca Raton, Florida

Library of Congress Cataloging in Publication Data

Main entry under title:

Gas Lasers.

(Handbook of laser science and technology;
v. 2)
Bibliography: p.
Includes index.
1. Gas lasers. I. Weber, Marvin J.,
1932- . II. Series.
TA1695.G34 621.36'63 81-17087
ISBN 0-8493-3502-7 AACR2

This book represents information obtained from authentic and highly regarded sources. Reprinted material is quoted with permission, and sources are indicated. A wide variety of references are listed. Every reasonable effort has been made to give reliable data and information, but the author and the publisher cannot assume responsibility for the validity of all materials or for the consequences of their use.

Direct all inquiries to CRC Press, Inc., 2000 Corporate Blvd., N.W., Boca Raton, Florida, 33431.

International Standard Book Number 0-8493-3502-7

Library of Congress Card Number 81-17087
Printed in the United States

PREFACE

In the ten years since the publication of the *CRC Handbook of Lasers with Selected Data on Optical Technology,* the growth in the number and diversity of lasers and their applications has continued at a rapid pace. These developments have prompted an update and expansion of the original volume into the *CRC HANDBOOK SERIES OF LASER SCIENCE AND TECHNOLOGY.* The first two volumes of the series are devoted to lasers in all media. Later volumes are planned on optical materials and on laser instrumentation and operating techniques.

The object of this series is to provide a readily accessible and concise source of data in tabular and graphical form for workers in the areas of laser research and development. Volumes I and II contain extensive tables of experimental data on lasers and complete references to the original work. This is the primary emphasis. Many books and review articles describing the physics and operation of various lasers already exist. Therefore textual material is in general only that required to explain the data, to summarize the general operation and characteristics of each type of laser, and to describe important specific lasers or properties not covered elsewhere.

Of the various lasing media, gases contribute the largest number of laser transitions. This volume is devoted exclusively to gas lasers; other lasers and masers are covered in Volume I of this series. The coverage is divided into neutral atom, ion, and molecular lasers. The last group is further subdivided by whether the transition is between electronic, vibrational, or rotational states. A list of gas lasers arranged in order of wavelength is also provided.

It is a pleasure to acknowledge the efforts and cooperation of the contributors to this volume. Several chapters cover a vast body of data and required prodigious efforts to complete. The table of gas laser wavelengths was prepared with the assistance of Ann Weber and Eve Weber. The numerous suggestions and guidance of the Advisory Board throughout the preparation of this volume are greatly appreciated. Dr. Dean Hodges assisted with the initial arrangements for the preparation of the chapter on far-infrared lasers. Again, I wish to thank Marsha Baker and Pamela Woodcock of the CRC Press for their help and technical expertise in the editing of the series.

Marvin J. Weber
Danville, California
February, 1981

THE EDITOR

Marvin J. Weber is an Assistant Associate Program Leader, Basic Research, of the Laser Program at the Lawrence Livermore National Laboratory, University of California, Livermore, California.

Dr. Weber received the A.B., M.A., and Ph.D. degrees in physics from the University of California, Berkeley, in 1954, 1956 and 1959, respectively. After graduation, Dr. Weber joined the Research Division of Raytheon Company where he was a Principal Scientist and became Manager of Solid-State Lasers. In 1966, Dr. Weber was a Visiting Research Associate in the Department of Physics, Stanford University.

In 1973, Dr. Weber joined the Laser Program of the Lawrence Livermore National Laboratory. His activities have included studies of the physics, characterization, and development of optical materials for high power lasers Dr. Weber has published numerous research papers and review articles in the areas of lasers, luminescence, optical spectroscopy, and magnetic resonance in solids and holds several patents on solid-state laser materials.

Dr. Weber is a Fellow of the American Physical Society and a member of the Optical Society of America and the American Ceramics Society. He has served as a consultant for the Division of Materials Research, National Science Foundation, and is a member of the Advisory Editorial Board of the *Journal of Non-Crystalline Solids.*

ADVISORY BOARD

CONTRIBUTORS

William B. Bridges, Ph.D.
Professor of Electrical Engineering and Applied Physics
Division of Engineering and Applied Science
California Institute of Technology
Pasadena, California

Tao-Yuan Chang, Ph.D.
Member of Technical Staff
Bell Telephone Laboratories
Holmdel, New Jersey

Paul D. Coleman, Ph.D.
Professor
Department of Electrical Engineering
University of Illinois
Urbana, Illinois

Christopher C. Davis, Ph.D.
Associate Professor of Electrical Engineering
Department of Electrical Engineering
University of Maryland
College Park, Maryland

Robert S. Davis
Graduate Student
Department of Mathematics
University of Illinois at Chicago Circle
Chicago, Illinois

David J. E. Knight, D. Phil.
Head, Optical Frequencies Section
Division of Quantum Metrology
National Physical Laboratory
Teddington, Middlesex, England

Charles K. Rhodes, Ph.D.
Professor of Physics
Department of Physics
University of Illinois at Chicago Circle
Chicago, Illinois

HANDBOOK OF LASER SCIENCE AND TECHNOLOGY

VOLUME I: LASERS AND MASERS

SECTION 1: INTRODUCTION
1.1 Types and Comparisons of Laser Sources—William F. Krupke

SECTION 2: SOLID STATE LASERS
2.1 Crystalline Lasers
  2.1.1 Paramagnetic Ion Lasers—Peter F. Moulton
  2.1.2 Stoichiometric Lasers—Stephen R. Chinn
  2.1.3 Color Center Lasers—Linn F. Mollenauer
2.2 Semiconductor Lasers—Henry Kressel and Michael Ettenberg
2.3 Glass Lasers—Stanley E. Stokowski
2.4 Fiber Raman Lasers—Rogers W. Stolen and Chinlon Lin
2.5 Table of Wavelengths of Solid State Lasers—Marvin J. Weber

SECTION 3: LIQUID LASERS
3.1 Organic Dye Lasers—Richard Steppel
3.2 Inorganic Liquid Lasers
  3.2.1 Rare Earth Chelate Lasers—Harold Samelson
  3.2.2 Aprotic Liquid Lasers—Harold Samelson

SECTION 4: OTHER LASERS
4.1 Free Electron Lasers
  4.1.1 Infrared and Visible Lasers—Donald Prosnitz
  4.1.2 Millimeter and Submillimeter Lasers—Victor L. Granatstein, Robert K. Parker, and Phillip A. Sprangle
4.2 X-Ray Lasers—Raymond C. Elton

SECTION 5: MASERS
5.1 Masers—Adrian E. Popa
5.2 Maser Action in Nature—James M. Moran

SECTION 6: LASER SAFETY
6.1 Optical Radiation Hazards—David H. Sliney
6.2 Electrical Hazards from Laser Power Supplies—James K. Franks
6.3 Hazards from Associated Agents—Robin DeVore

HANDBOOK OF LASER SCIENCE AND TECHNOLOGY

VOLUME II: GAS LASERS

TABLE OF CONTENTS

SECTION 1: NEUTRAL GAS LASERS 3

SECTION 2: IONIZED GAS LASERS 171

SECTION 3: MOLECULAR GAS LASERS 271

3.1 Electronic Transition Lasers 273
3.2 Vibrational Transition Lasers 313
3.3 Far Infrared Lasers 411

SECTION 4: TABLE OF LASER WAVELENGTHS 495

INDEX 571

Section 1
Neutral Gas Lasers

1. NEUTRAL GAS LASERS

Christopher C. Davis

INTRODUCTION

Since the first report of laser action in a gas discharge in a He-Ne mixture by Javan et al.[468] in 1961, oscillation in the neutral spectra of 50 elements on over 730 identified transitions has been reported. More than 90 additional, unidentified laser lines appear likely to originate from neutral species. Figure 1.1 shows that with the exception of Groups IIIB, IVB, VIB, and the actinides, laser action has been observed in all groups of the periodic table. The development of innovative pumping schemes has helped considerably in adding 21 elements to the list of neutral gas lasers in the last decade.

TABLES OF NEUTRAL LASER TRANSITIONS

Tables 1.1.1 to 1.10.1, which follow, contain details of all the identified neutral gas laser transitions and unidentified probable neutral gas laser transitions which have been reported in the literature, to our knowledge, through early 1980. Laser transitions are arranged according to groups, in order of increasing atomic number. Wavelengths of lines *in vacuo* are given in italics.

Unless stated otherwise, and with the exception of the lanthanides, identified transitions are assigned and their calculated wavelengths, *in vacuo*, given in accordance with energy level data given by Charlotte E. Moore in "Atomic Energy Levels," Volumes I, II, and III (1949, 1952, and 1958), National Bureau of Standards (NBS), National Standard Reference Data Series, NSRDS-NBS 34, reissued 1971, (originally issued as NBS Circ. 467) U.S. Government Printing Office, Washington, D.C. 20402. Calculated wavelengths and spectral assignments for the lanthanides are taken from Martin, Zalubas, and Hagan, "Atomic Energy Levels" — The Rare-Earth Elements, NSRDS-NBS 60, 1978. However, wherever possible the latest spectroscopic literature concerning each element that postdates "Atomic Energy Levels" has been consulted. In these cases, accurately measured values of the wavelengths in air of laser transitions observed in spontaneous emission are given in the table whenever possible, together with currently accepted transition assignments. Since most modern energy level data appears to be accurate to $\lesssim 10^{-3}$ cm^{-1}, several significant digits have been retained in calculated, *in vacuo*, wavelengths. The interested reader can easily convert these figures to air values by the use of Table of Wavenumbers[582] or by direct computation from Edlen's refractive index formula.[582]

The reference underlined for each laser transition is the reference in which oscillation was first reported. Additional references are not intended to be exhaustive, but represent selected important work both on the laser system itself and also frequently on the physical processes occurring within it. Substantial numbers of literature references are listed for those laser transitions of greatest practical importance. Typical conditions under which laser action has been obtained for each transition are given. If several significantly different pumping conditions exist, these are included. However, the implication that laser action occurs only under the specific conditions indicated is not intended. For further details, the interested reader should consult the individual references. A reference number indicated with (E) contains an incorrect wavelength and/or transition assignment.

Large numbers of the transitions listed have sufficiently high gain that they will

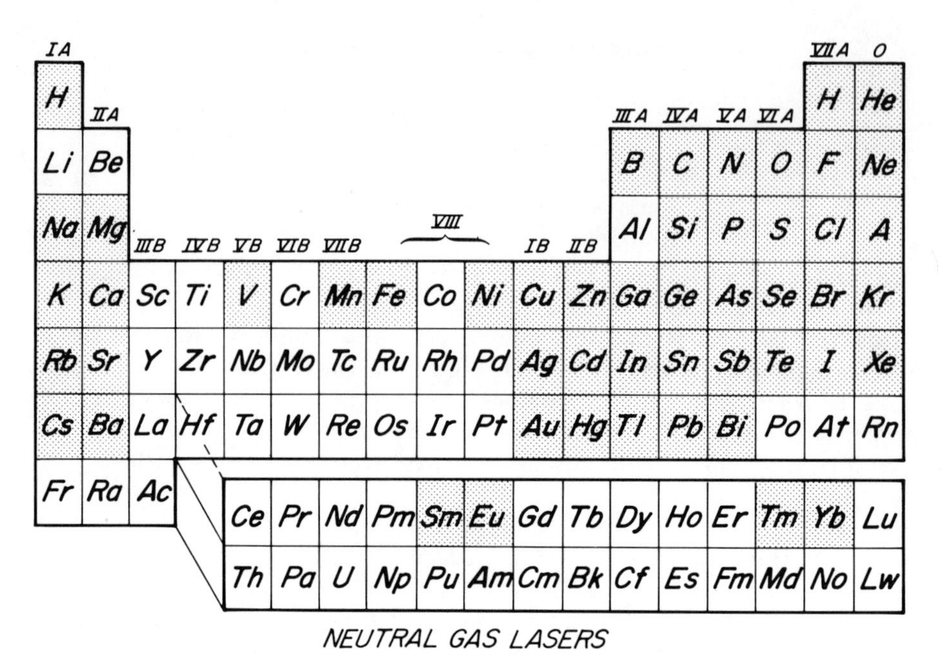

FIGURE 1.1. Periodic table of the elements showing those elements (shaded) in which laser oscillation in the neutral species has been reported.

generate coherent, directional laser output with only a single cavity reflector or with none at all. Lines which operate in this fashion generally amplify their own spontaneous emission up to the saturation intensity in a single pass. These lines which operate in an amplified spontaneous emission (ASE) mode[639-643] are designated as such to distinguish them from those few transitions which exhibit true cooperative emission (superfluorescence).

The tables of neutral gas laser transitions are arranged as follows:

GROUP IA

Table	Element
1.1.1	Hydrogen (3)
1.1.2	Sodium (4)
1.1.3	Potassium (15)
1.1.4	Rubidium (13)
1.1.5	Cesium (19)

GROUP IB

Table	Element
1.1.6	Copper (5)
1.1.7	Silver (2)
1.1.8	Gold (2)

GROUP IIA

Table	Element
1.2.1	Magnesium (6)
1.2.2	Calcium (4)
1.2.3	Strontium (4)
1.2.4	Barium (21)

GROUP IIB

Table	Element
1.2.5	Zinc (2)
1.2.6	Cadmium (17)
1.2.7	Mercury (31)

GROUP IIIA

Table	Element
1.3.1	Boron (1)
1.3.2	Gallium (5)
1.3.3	Indium (6)
1.3.4	Thallium (5)

GROUP IVA

Table	Element
1.4.1	Carbon (11)
1.4.2	Silicon (3)
1.4.3	Germanium (2)
1.4.4	Tin (4)
1.4.5	Lead (10)

GROUP VA and VB

Table	Element
1.5.1	Nitrogen (23)
1.5.2	Phosphorus (10)
1.5.3	Arsenic (15)
1.5.4	Antimony (1)
1.5.5	Bismuth (2)
1.5.6	Vanadium (2)

GROUP VIA

Table	Element
1.6.1	Oxygen (14)
1.6.2	Sulfur (9)
1.6.3	Selenium (3)
1.6.4	Tellurium (6)
1.6.5	Thulium (14)

GROUP VIIIA and VIIB

Table	Element
(See 1.1.1)	Hydrogen
1.7.1	Fluorine (18)
1.7.2	Chlorine (10)
1.7.3	Bromine (5)
1.7.4	Iodine (19)
1.7.5	Ytterbium (10)
1.7.6	Manganese (12)

GROUP VIII

Table	Element
1.8.1	Iron (7)
1.8.2	Nickel (2)
1.8.3	Samarium (8)
1.8.4	Europium (16)

GROUP 0

Table	Element
1.9.1	Helium (11)
1.9.2	Neon (200)
1.9.3	Argon (68)
1.9.4	Krypton (46)
1.9.5	Xenon (58)

Table 1.10.1
Miscellaneous and unidentified possible neutral laser transitions (45)

Note: Figures in parentheses indicate the number of laser lines in each table.

1.1A GROUP IA

Table 1.1.1
HYDROGEN[a] (FIGURE 1.2)

Wavelength (μm)	Transition assignment	Comments	Ref.
0.434046	5 → 2 (H-γ line, Balmer series)	As an impurity in a pulsed discharge in Ne at 1.5 torr; D = 25 mm; E/p = 140 V/cm torr	1
0.486132	4 → 2 (H-β line, Balmer series)	As an impurity in a pulsed discharge in Ne at 1.5 torr; D = 25 mm; E/p = 140 V/cm torr	1
1.87510	$4f^2F^o_{7/2} \rightarrow 3d^2D_{5/2}$($P_\alpha$, first member of Paschen series, strongest fine-structure component)	Pulsed; as an impurity in 3.5 torr of He; optimum H pressure 0.01 torr; D = 7mm	1,2

[a] Measured wavelengths were taken from Wiese, W. L., Smith, M. W., and Glennon, B. M., *Natl. Stand. Ref. Data Ser. Natl. Bur. Stand.*, NSRDS-NBS4, 1966.

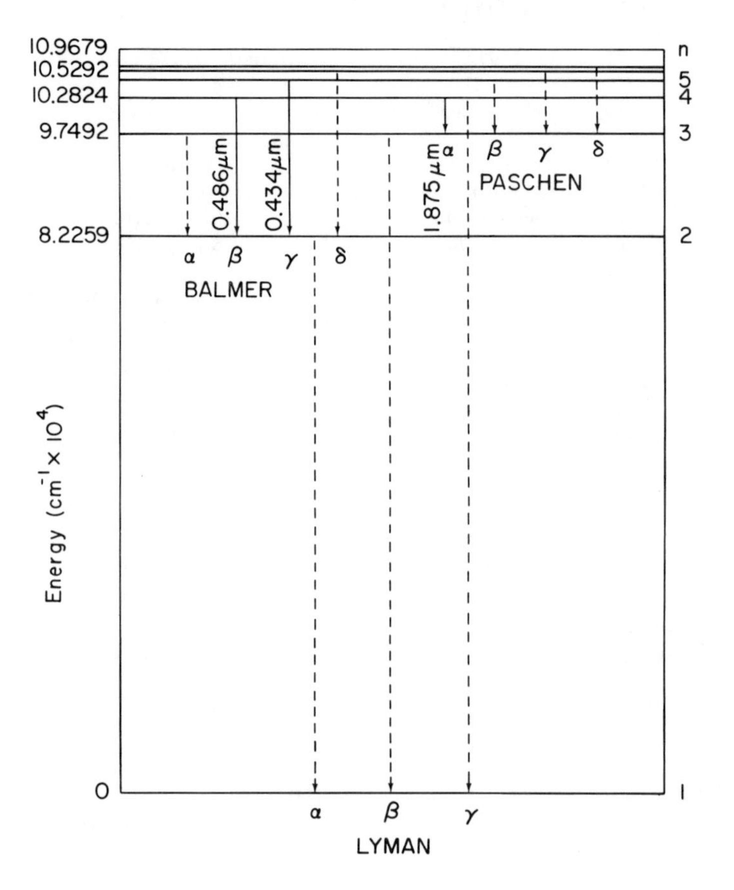

FIGURE 1.2. Partial energy level diagram of atomic H showing the reported laser transitions.

Table 1.1.2
SODIUM[a] (FIGURE 1.3)

Wavelength (μm)	Transition assignment	Comments	Ref.
0.58959236	$3p\ ^2P^o_{1/2} \rightarrow 3s\ ^2S_{1/2}$[b]	Pulsed; operates in an ASE mode following photodissociation of NaI with the fifth harmonic of a Q-switched Nd/YAG laser at 0.2128 μm; NaI in a cell at 600°C with 10 torr of Ar; NaI density $1.1 \pm 0.3 \times 10^{15}$ cm^{-3};[c] also with ArF laser pumping (193 nm) of NaI or NaBr heated in an oven at temperatures up to 1000°C, generally operated at 500—700°C which corresponds to $10^{-4} - 5 \times 10^{-1}$ torr vapor pressure; no buffer gas used	3,4
0.58899504	$3p^2P^o_{3/2} \rightarrow 3s\ ^2S_{1/2}$[b]	Pulsed under the same operating conditions as the second conditions listed above	3
1.138145	$4s\ ^2S_{1/2} \rightarrow 3p\ ^2P^o_{1/2}$[d]	Pulsed; 0.001—0.003 torr of Na with 1—10 torr of H; D = 12 mm	5-8
1.140378	$4s\ ^2S_{1/2} \rightarrow 3p\ ^2P^o_{3/2}$[d]	Pulsed; 0.001—0.003 torr of Na with 1-10 torr of H; D = 12 mm	5-8

[a] Wavelength and spectral assignments were taken from Risberg, P., *Ark. Fys.*, 10, 583—606, 1956.
[b] Resonance line.
[c] Excitation results from the reaction NaI $\xrightarrow{h\nu}$ Na(3p $^2P^o$) + I(5p^5 $^2P_{3/2}$).
[d] Selective excitation occurs via the two-body recombination reaction; $Na^+ + H^- \rightarrow$ Na(4s $^2S_{1/2}$) + H.

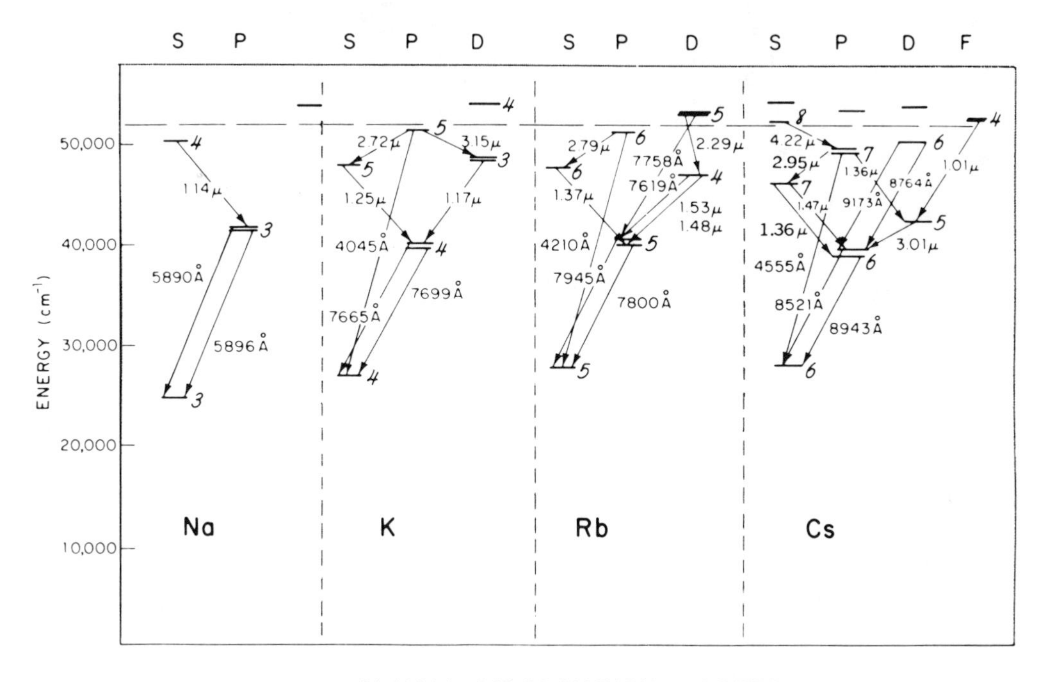

FIGURE 1.3. Laser transitions observed in the neutral alkali metals following ArF laser dissociation of alkali-iodide salts. The script numbers to the right of the levels are principal quantum numbers. (From Ehrlich D. J. and Osgood, R. M., Jr., *Appl. Phys. Lett.*, 34, 655, 1979. With permission.)

Table 1.1.3
POTASSIUM[a] (FIGURE 1.3)

Wavelength (μm)	Transition assignment	Comments	Ref.
0.4044136	5p $^2P^o_{3/2}$ → 4s $^2S_{1/2}$[b]	Pulsed; pumped with an ArF laser (193 nm) using KI or KBr heated in a cell to 500—700°C without buffer gas	3
0.40472602	5p$^2P^o_{1/2}$ → 4s $^2S_{1/2}$[b]	Pulsed; pumped with an ArF laser (193 nm) using KI or KBr heated in a cell to 500—700°C without buffer gas	3
0.7664899	4p $^2P^o_{3/2}$ → 4s $^2S_{1/2}$	Pulsed; pumped with an ArF laser (193 nm) using KI or KBr heated in a cell to 500—700°C without buffer gas	3
0.7698959	4p $^2P^o_{1/2}$ → 4s $^2S_{1/2}$	Pulsed; pumped with an ArF laser (193 nm) using KI or KBr heated in a cell to 500—700°C without buffer gas	3
1.177283	3d $^2D_{5/2}$ → 4p $^2P^o_{3/2}$	Pulsed; pumped with an ArF laser (193 nm) using KI or KBr heated in a cell to 500—700·C without buffer gas	3
1.243224	5s $^2S_{1/2}$ → 4p $^2P^o_{1/2}$	Pulsed; 0.1 torr of K with 3—5 torr of H[f]	7
1.252211	5s $^2S_{1/2}$ → 4p $^2P^o_{3/2}$	Pulsed; as for the two lines immediately above	3,7
3.1415224	5p $^2P^o_{3/2}$ → 3d $^2D_{3/2}$[c]	Pulsed; K vapor excited with a Q-switched ruby laser (694.3 nm); also as for 1.177-μm line	3,9
3.1601267	5p $^2P^o_{1/2}$ → 3d $^2D_{3/2}$[c]	Pulsed; K vapor excited with a Q-switched ruby laser (694.3 nm); also as for 1.177-μm line	3,9
6.422525	6p $^2P^o_{3/2}$ → 6s $^2S_{1/2}$[d]	Pulsed; K vapor discharge in a heat pipe at 370°C (1 torr vapor pressure) pumped with a flashlamp pumped coumarin dye laser (534.31 nm)	10,11
6.4575288	6p $^2P^o_{1/2}$ → 6s $^2S_{1/2}$[d]	Pulsed; K vapor discharge in a heat pipe at 370°C (1 torr vapor pressure) pumped with a flashlamp pumped coumarin dye laser (534.31 nm)	10,11
7.8953393	7s $^2S_{1/2}$ → 6p $^2P^o_{3/2}$	Pulsed; K vapor discharge in a heat pipe at 370°C (1 torr vapor pressure) pumped with a flashlamp pumped coumarin dye laser (534.31 nm)	10,11
9.1791962	6d $^2D_{3/2}$ → 5f$^2F^{oe}$	Pulsed; K vapor discharge in a heat pipe at 370°C (1 torr vapor pressure) pumped with a flashlamp pumped coumarin dye laser (534.3) nm)	10,11
12.568814	7p $^2P^o_{1/2}$ → 7s $^2S_{1/2}$	Pulsed; K vapor discharge in a heat pipe at 370°C (1 torr vapor pressure) pumped with a flashlamp pumped coumarin dye laser (534.31 nm)	10,11

15.968063	6d $^2D_{3/2} \rightarrow$ 7p $^2P^o_{1/2}$	Pulsed; K vapor discharge in a heat pipe at 370°C (1 torr vapor pressure) pumped with a flashlamp pumped coumarin dye laser (534.31 nm)	10,11

[a] Wavelengths and spectral assignments taken from same source as for Na
[b] Both components of this doublet probably oscillate, although this is not made clear in Reference 3.
[c] Unclear whether both components of doublet were observed in Reference 3.
[d] Unclear whether both components were observed.
[e] Weak line in competition with the *15.97-μm* transition.
[f] Tube bore apparently 12 mm.

Table 1.1.4
RUBIDIUM[a] (FIGURE 1.3)

Wavelength (μm)	Transition assignment	Comments	Ref.
0.420185[b]	6p $^2P^o_{1/2} \rightarrow$ 5s $^2S_{1/2}$	Pulsed; RbI or RbBr heated to 500—700°C in a cell without buffer gas, pumped with an ArF laser (193 nm)	3
0.421556[b]	6p $^2P^o_{3/2} \rightarrow$ 5s $^2S_{1/2}$	Pulsed; RbI or RbBr heated to 500—700°C in a cell without buffer gas, pumped with an ArF laser (193 nm)	3
0.7800268	5p $^2P^o_{3/2} \rightarrow$ 5s $^2S_{1/2}$	Pulsed; RbI or RbBr heated to 500—7—°C in a cell without buffer gas, pumped with an ArF laser (193 nm)	3
0.7947603	5p $^2P^o_{1/2} \rightarrow$ 5s $^2S_{1/2}$	Pulsed; RbI or RbBr heated to 500—700°C in a cell without buffer gas, pumped with an ArF laser (193 nm)	3
0.7621029	5d $^2D_{3/2} \rightarrow$ 5p $^2P^o_{1/2}$	Pulsed; RbI or RbBr heated to 500—700°C in a cell without buffer has, pumped with an ArF laser (193 nm)	3
0.7761570	5d $^2D_{3/2} \rightarrow$ 5p $^2P^o_{3/2}$	Pulsed; RbI or RbBr heated to 500—700°C in a cell without buffer gas, pumped with an ArF laser (193 nm)	3
1.366501	6s $^2S_{1/2} \rightarrow$ 5p $^2P^o_{3/2}$	Pulsed; RbI or RbBr heated to 500—700°C in a cell without buffer gas, pumped with an ArF laser (193 nm)	3
1.475241	4d $^2D_{3/2} \rightarrow$ 5p $^2P^o_{1/2}$	Pulsed; RbI or RbBr heated to 500—700°C in a cell without buffer gas, pumped with an ArF laser (193 nm)	3
1.528843	4d $^2D_{3/2} \rightarrow$ 5p $^2P^o_{3/2}$[c]	Pulsed; RbI or RbBr heated to 500—700°C in a cell without buffer gas, pumped with an ArF laser (193 nm)	3
1.528948	4d $^2D_{5/2} \rightarrow$ 5p $^2P^o_{3/2}$[c]	Pulsed; RbI or RbBr heated to 500—700°C in a cell without buffer gas, pumped with an ArF laser (193 nm)	3

Table 1.1.4 (continued)
RUBIDIUM[a] (FIGURE 1.3)

Wavelength (μm)	Transition assignment	Comments	Ref.
2.252965	6p $^2P^o_{3/2}$ → 4d $^2D_{5/2}$	Pulsed; Rb vapor in a cell at abot 400°C with a He buffer, pumped with a Q-switched ruby laser	9,12
2.790537	6p $^2P^o_{1/2}$ → 6s $^2S_{1/2}$	Pulsed; as for the 1.53-μm line above	3
2.293247	6p $^2P^o_{1/2}$ → 4d $^2D_{3/2}$	Pulsed; as for either the 1.53- or 2.25-μm lines above	3,9,12

[a] Unless otherwise indicated, wavelengths and spectral assignments are taken from data in Johansson, L., *Ark. Fys.*, 20, 135—146, 1961.

[b] Wavelengths taken from Meggers, W. F., Corliss, C. H., and Scribner, B. F., *Natl. Bur. Stand. U.S. Monogr.*, 145 (1), 1975. It is not clear whether one or both of these fine-structure components were observed in Reference 3.

[c] It is not clear whether one or both of these fine-structure components were observed in Reference 3.

Table 1.1.5
CESIUM[a] (FIGURE 1.3)

Wavelength (μm)	Transition assignment	Comments	Ref.
0.4555276[b]	$7p\ ^2P^o_{3/2} \rightarrow 6s\ ^2S_{1/2}$	Pulsed; CsI or CsBr in a cell at 500—700°C excited with an ArF laser (193 nm)	3
0.8521133	$6p\ ^2P^o_{3/2} \rightarrow 6s\ ^2S_{1/2}$	Pulsed; CsI or CsBr in a cell at 500—700°C excited with an ArF laser (193 nm)	3
0.87614150	$6d\ ^2D_{3/2} \rightarrow 6p\ ^2P^o_{1/2}$	Pulsed; CsI or CsBr in a cell at 500—700°C excited with an ArF laser (193 nm)	3
0.8943468[b]	$6p\ ^2P^o_{1/2} \rightarrow 6s\ ^2S_{1/2}$	Pulsed; CsI or CsBr in a cell at 500—700°C excited with an ArF laser (193 nm)	3
0.91723217	$6d\ ^2D_{5/2} \rightarrow 6p\ ^2P^o_{3/2}$	Pulsed; CsI or CsBr in a cell at 500—700°C excited with an ArF laser (193 nm)	3
1.01236025	$4f\ ^2F^o_{7/2} \rightarrow 5d\ ^2D_{5/2}$	Pulsed; CsI or CsBr in a cell at 500—700°C excited with an ArF laser (193 nm)	3
1.358831	$7s\ ^2S_{1/2} \rightarrow 6p\ ^2P^o_{1/2}$	Pulsed; CsI or CsBr in a cell at 500—700°C excited with an ArF laser (193 nm)	3
1.360257	$7p\ ^2P^o_{3/2} \rightarrow 5d\ ^2D_{5/2}$	Pulsed; Cs vapor excited by a 765.8-nm nitrobenzene Raman laser; absence of He required	9
1.375883	$7p\ ^2P^o_{1/2} \rightarrow 5d\ ^2D_{3/2}$	Pulsed; Cs vapor excited by a 765.8-nm nitrobenzene Raman laser; absence of He required; also as for the 0.4555-μm line above	3,9
1.469493	$7s\ ^2S_{1/2} \rightarrow 6p\ ^2P^o_{3/2}$	Pulsed; as for the 0.4555-μm line above	3
2.9317981	$7p\ ^2P^o_{3/2} \rightarrow 7s\ ^2S_{1/2}$	Pulsed; as for the 0.4555-μm line above	3
3.01033	$5d\ ^2D_{3/2} \rightarrow 6p\ ^2P^o_{3/2}$	Pulsed; as for the 0.4555-μm line above	3
3.0111339	$5d\ ^2D_{3/2} \rightarrow 6p\ ^2P^o_{1/2}$	Pulsed; Cs vapor excited with a Q-switched 1.06-μm laser	9
3.0961401	$7p\ ^2P^o_{1/2} \rightarrow 7s\ ^2S_{1/2}$	Pulsed; Cs vapor excited with 694.3-,765.8-, 740-900 nm or 1.06-μm laser pulses; He buffer required	9,12
3.2050778	$8p\ ^2P^o_{1/2} \rightarrow 6d\ ^2D_{3/2}$	CW; vapor pressure of Cs at 175°C optically pumped with a He lamp at 388.9 nm; D = 10 mm	13-15
3.4909363	$5d\ ^2D_{5/2} \rightarrow 6p\ ^2P^o_{3/2}$	Pulsed; as for the *3.0111*-μm line above	9
3.6140628	$5d\ ^2D_{3/2} \rightarrow 6p\ ^2P^o_{3/2}$	Pulsed; as for the *3.011*-μm line above	9
4.2181082	$8s\ ^2S_{1/2} \rightarrow 7p\ ^2P^o_{3/2}$	Pulsed; as for the 0.4555-μm line above	3
7.1853791	$8p\ ^2P^o_{1/2} \rightarrow 8s\ ^2S_{1/2}$	CW; as for the *3.205*-μm line above	13-15,16

[a] Wavelength and spectral assignments are taken from data in Johansson, L., *J. Opt. Soc. Am.*, 52, 441—447, 1962.

[b] Mean value for hyperfine structure components.

Table 1.1.6
COPPER (FIGURE 1.4)

Wavelength (μm)	Transition assignment	Comments	Ref.
0.510554[a,b]	$4p\ ^2P^o_{3/2} \rightarrow 4s^2\ ^2D_{5/2}$	Pulsed; short rise-time high-voltage single, double-, or multiple-pulse excitation of various Cu compounds or of Cu vapor at high temperature; see text for further details	17-98
0.570024[a]	$4p\ ^2P^o_{3/2} \rightarrow 4s^2\ ^2D_{3/2}$	Pulsed; short rise-time high-voltage pulsed excitation of Cu iodide at moderately high temperature (600°C), vapor pressure 1—10 torr; D = 9mm	29
0.578213[a,b]	$4p\ ^2P^o_{1/2} \rightarrow 4s^2\ ^2D_{3/2}$	Pulsed; as for the 0.5106-μm line above	17-82,84-92,94-98

Table 1.1.6 (continued)
COPPER (FIGURE 1.4)

Wavelength (μm)	Transition assignment	Comments	Ref.
1.8199686	4f $^2F^o_{5/2}$ → 4d $^2D_{3/2}$	Pulsed; lases in the afterglow of a slotted Cu hollow-cathode discharge, He or Ne buffer at 8—20 torr used[c]	99
1.8234057	4f $^2F^o_{7/2}$ → 4d $^2D_{5/2}$	Pulsed; lases in the afterglow of a slotted copper hollow-cathode discharge, He or Ne buffer at 8—20 torr used[c]	99

[a] Measured wavelengths from Meggers, W. F., Corliss, C. H., and Scribner, B. F., *Natl. Bur. Stand. (U.S.) Monogr.*, 145(1), 1975.

[b] Very strong self-terminating lines; gain can be greater than 42 dBm $^{-1}$.

[c] Excitation believed to involve recombination of electrons with metal ions which are formed by election impact and change-transfer during the current pulse.

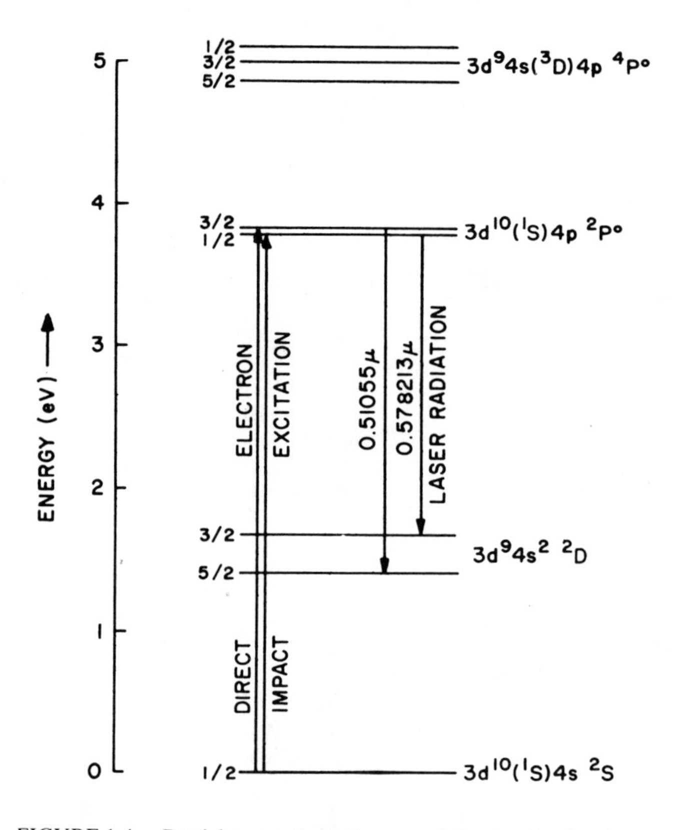

FIGURE 1.4. Partial energy level diagram of Cu showing the strong self-terminating visible laser transitions.[17]

Table 1.1.7
SILVER

Wavelength (μm)	Transition assignment	Comments	Ref.
1.8380629	$4f\ ^2F^o_{5/2,7/2} \rightarrow 5d\ ^2D_{5/2}$	Pulsed; lases in the afterglow of a discharge in a slotted Ag hollow-cathode; He or Ne buffer at 8—20 torr used; excitation by ion-electron recombination appears likely; also in a segmented transversely excited device incorporating recombining silver plasmas in a few torr of He	99,146
1.9371923	$5s^2\ ^2D_{3/2} \rightarrow 5p\ ^2P^o_{1/2}$	Pulsed; lases both during current pulse and in the afterglow of a discharge in a slotted Ag hollow-cathode; He or Ne buffer at 8—20 torr used.	99

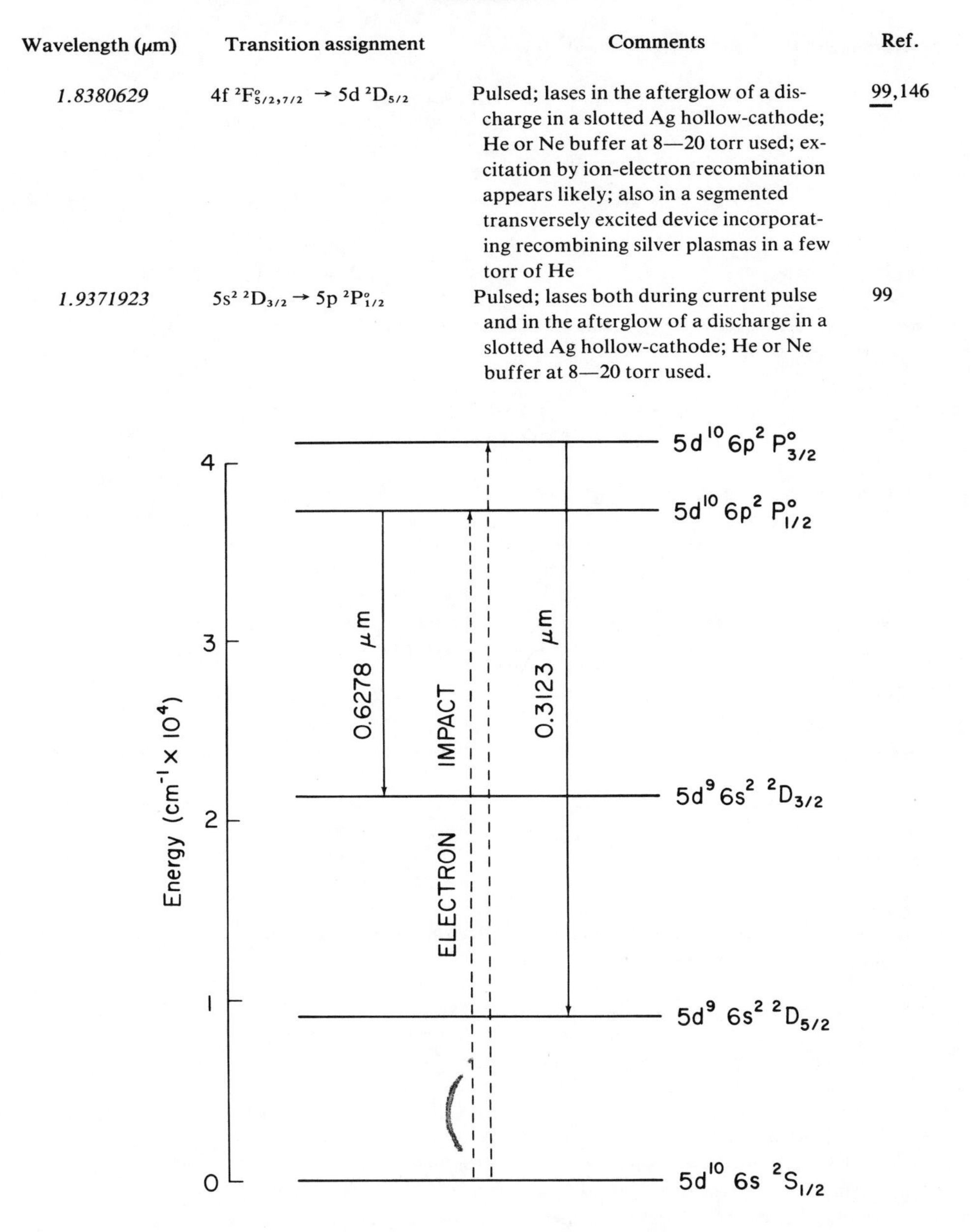

FIGURE 1.5. Partial energy level diagram of Au showing the two self-terminating laser transitions.

Table 1.1.8
GOLD[a] (FIGURE 1.5)

Wavelength (μm)	Transition assignment	Comments	Ref.
0.3122784	$6p\ ^2P^o_{3/2} \rightarrow 6s^2\ ^2D_{5/2}$	Pulsed; Au-coated tube filled with 10 torr of Ne self-heated by repetitive high voltage short-pulse excitation; D = 16 mm	100,102,103
0.6278170	$6p\ ^2P^o_{1/2} \rightarrow 6s\ ^2D_{3/2}$	Pulsed; as for above line, however, optimum buffer pressure 25 torr of Ne or 30 torr of He; also operates with Ar or Xe buffer; D = 16 mm	17,21,22,100-103

[a] Wavelengths and spectral assignments taken from Ehrhardt, J. C. and Davis, S. P., *J. Opt. Soc. Am.*, 61, 1342—1349, 1971.

Table 1.2.1[a]
MAGNESIUM[b] (FIGURE 1.6)

Wavelength (μm)	Transition assignment	Comments	Ref.
1.502499	$4p\ ^3P^o_2 \rightarrow 4s\ ^3S_1$	Pulsed; in a segmented, transversely excited device incorporating recombining Mg plasmas in a few torr of He	146
3.6789254 or *3.6789565*	$5d\ ^3D_1 \rightarrow 5p\ ^3P^o_1$ or[c] $5d\ ^3D_2 \rightarrow 5p\ ^3P^o_1$	CW; in Mg vapor above 450°C in He, Ne, or Ar; D = 10 mm	104
3.6825364[d]	$5d\ ^3D_{3,2,1} \rightarrow 5p\ ^3P^o_2$	CW; in Mg vapor above 450°C in He, Ne, or Ar; D = 10 mm	104
3.86573	?[e]	CW; in Mg vapor above 450°C in He, Ne, or Ar; D = 10 mm	104
4.2013276	$5p\ ^3P^o_2 \rightarrow 5s\ ^3S_1$	CW; in Mg vapor above 450°C in He, Ne, or Ar; D = 10 mm	104
4.3638859[d]	$5d\ ^3D_{3,2,1} \rightarrow 4f\ ^3F^o$	CW; in Mg vapor above 450°C in He, Ne, or Ar; D = 10 mm	104

Note: Lines at 0.9218, 0.9244, 1.0952, and 1.0915 listed as neutral Mg lines in Reference 581 are singly ionized Mg laser lines.

[a] No neutral Be laser transitions have been observed to date, three Be laser transitions listed by Beck et al.[581] as neutral transitions are in fact singly ionized Be lines.

[b] Wavelengths and spectral assignments taken from Risberg, G., *Ark. Fys.*, 28, 381—395, 1965.

[c] The wavelength resolution in Reference 104 was not great enough to determine whether one or both these fine-structure components were oscillating.

[d] Mean calculated wavelength of fine-structure components.

[e] Three possible assignments, slightly outside the stated error limits of the measured wavelength reported in Reference 104, are

$7p\ ^3P^o_2 \rightarrow 6s\ ^3S_1$ at k*3.8670358*μm
$7p\ ^3P^o_1 \rightarrow 6s\ ^3S_1$ at k*3.8681427*μm
$7p\ ^3P^o_0 \rightarrow 6s\ ^3S_1$ at k*3.868638*μm

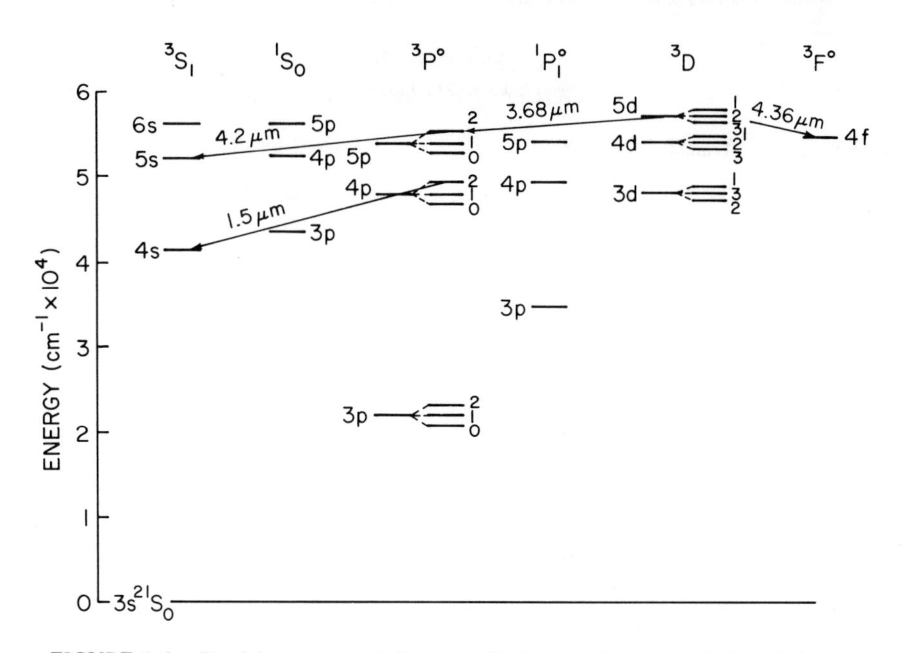

FIGURE 1.6. Partial energy level diagram of Mg-I showing the majority of the reported laser transitions.

Table 1.2.2
CALCIUM[a]

Wavelength (μm)	Transition assignment	Comments	Ref.
0.5349472	$4p'\ ^1F^o_3 \rightarrow 3d\ ^1D_2$[b]	Pulsed; Ca vapor with 0—1000 torr of noble gas in a heat-pipe oven at about 1200 K (Ca density about 5×10^{18} atoms cm^{-3}), excited with a KrF laser (249 nm)	105
0.5857452	$4p^2\ ^1D_2 \rightarrow 4p\ ^1P^o_1$[b]	Pulsed; Ca vapor with 0—1000 torr of noble gas in a heat-pipe oven at about 1200 K (Ca density about 5×10^{18} atoms cm^{-3}, excited with a KrF laser (249 nm)	105
0.644981[c]	$4p'\ ^1D^o_2 \rightarrow 3d\ ^3D_1$	Pulsed; in a hollow-cathode discharge with a DC trickle ionizing discharge; optimum buffer gas pressure ≈25 m torr of Xe	106
5.547327	$4p\ ^1P^o_1 \rightarrow 3d^1D_2$[d]	Short rise-time high-voltage pulsed excitation of Ca vapor at 460°C with 3 torr of He in small-bore tubes; also CW in a 7:1:1 He-Ne-H_2 mixture at 1 torr with Ca vapor at temperatures from 590—650°C; D = 4 mm	106, 107, 108, 109

[a] Unless otherwise indicated, wavelengths and spectral assignments are taken from Risberg, G., *Ark. Fys.*, 37, 231—249, 1968.

[b] Excitation of these transitions is believed to involve the absorption of a 249-nm photon by a Ca quasimolecule which gives some allowed character to the otherwise nonallowed $4s^2\ ^1S_0 \rightarrow 4s\ 6p\ ^1P^o_1$ resonance absorption at 239.856 nm.

[c] Measured wavelength from Meggers, W. F., Corliss, C. H., and Scribner, B. F., *Natl. Bur. Stand. (U.S.) Monogr.*, 145(1), 1975.

[d] When excited with short rise-time high-voltage pulses, this line can have a gain in excess of 300 dBm^{-1}.

Table 1.2.3
STRONTIUM

Wavelength (μm)	Transition assignment	Comments	Ref.
0.638075[a]	$5p'\ ^1D_2^o \rightarrow 4d\ ^3D_1$	Pulsed; in a hollow-cathode discharge operated with a preionizing DC current, 18—85 mtorr of Xe buffer	106
3.0118377	$4d\ ^3D_2 \rightarrow 5p\ ^3P_2^o$	Pulsed; in a self-heated discharge tube containing pieces of Sr, optimum with 80 torr of He; D = 7 or 10 mm	110,111
3.0670208	$4d\ ^3D_1 \rightarrow 5p\ ^3P_2^o$	Pulsed; in a self-heated discharge tube containing pieces of Sr, optimum with 80 torr of He; D = 7 or 10 mm	110,111
6.4566866	$5p\ ^1P_1^o \rightarrow 4d\ ^1D_2$	Pulsed; in a self-heated discharge tube containing pieces of Sr, optimum with 80 torr of He; D = 7 or 10 mm;[b] also CW in $\simeq 10^{-2}$ torr of Ca vapor in a tube at ≃600°C with 0.1—5 torr of H	107,109

[a] Measured wavelength from Meggers, W. F., Corliss, C. H., and Scribner, B. F., *Natl. Bur. Stand. (U.S.) Monogr.*, 145(1), 1975.

[b] The gain of this line when excited by short rise-time high-voltage pulses can be as great as 300 dBm^{-1}.

Table 1.2.4
BARIUM[a] (FIGURE 1.7)

Wavelength (μm)	Transition assignment		Comments	Ref.
0.7120329	$6p'\ ^1D_2^o \rightarrow 5d\ ^3D_1$		Pulsed; in a hollow-cathode discharge with a sustainer DC discharge, optimum with 60 mtorr of Xe or 800—1000 torr of He; also in a discharge tube at 710—900°C with 0.4 torr He, 0.1 to Ne, or 0.04—0.1 torr Xe; D = 2.8 cm	106,112
1.130304[b,c]	$6p\ ^1P_1^o \rightarrow 5d\ ^3D_2$		Pulsed; in Ba vapor in a tube at 500—850°C with He, Ne, Ar, or H at 1—3 torr D = 5—10 mm	113-115
1.50004[b,d]	$6p\ ^1P_1^o \rightarrow 5d\ ^1D_2$		Pulsed. As for the 1.13 μm line; also optically pumped[e]	113-118, 632
1.82041	$6p'\ ^1P_1^o \rightarrow 5d^2\ ^1D_2$		Pulsed; in a self-heated high repetition frequency (5—8kHz) discharge in Ba vapor with 10—15 torr of Ne; D = 4 mm	114
1.9022415[b]	$6d\ ^1D_2 \rightarrow 6p'\ ^3D_3^o$		Pulsed; as for the 1.13-μm line	113
2.1573497[b]	$6p'\ ^1P_1^o \rightarrow 5d^2\ ^3P_2$		Pulsed; as for the 1.13-μm line	113,114
2.32553	$6p\ ^3P_2^o \rightarrow 5d\ ^3D_2$		Pulsed; as for the 1.13-μm line	113,114
2.4764593[b]	$7s\ ^1S_0 \rightarrow 6p'\ ^3D_1^o$		Pulsed; as for the 1.13-μm line	113
2.55157	$6p\ ^3P_2^o \rightarrow 5d\ ^3D_3$		Pulsed; as for the 1.13-μm line	113,114
2.9230381	$6p\ ^3P_1^o \rightarrow 5d\ ^3D_2$[f]		Pulsed; as for the 1.13-μm line	113,114
3.9589222	$7s\ ^1S_0 \rightarrow 6p'\ ^3P_1^o$		Pulsed; as for the 1.13-μm line	113
4.0079678	$6p'\ ^3P_1^o \rightarrow 5d^2\ ^3P_0$		Pulsed; as for the 1.13-μm line	113
4.3285152	$7p\ ^1P_1^o \rightarrow 6d\ ^1D_2$		Pulsed; as for the 1.82-μm line above	114
4.6699795	$6d\ ^3D_1 \rightarrow 6p'\ ^1P_1^o$[g]		Pulsed; as for the 1.13-μm line above	113,114
4.7169143	$10d\ ^3D_2 \rightarrow 9p\ ^1P_1^o$		Pulsed; as for the 1.13-μm line above	113
4.7184144	$6p\ ^3P_2^o \rightarrow 5d\ ^1D_2$		Pulsed; as for the 1.13-μm line above	113,114
5.0322846	$8p\ ^1P_1^o \rightarrow 8s\ ^3S_1$		Pulsed; as for the 1.13-μm line above	113
5.4798	?	[h]	Pulsed; as for the 1.13-μm line above	113
5.5636	?	[h]	Pulsed; as for the 1.13-μm line above	113
5.8899	?	[h]	Pulsed; as for the 1.13-μm line above	113,114
6.4546	?	[h]	Pulsed; as for the 1.13-μm line above	113

Note: Lines at 2.5924 and 2.9057 listed as neutral Ba lines in Reference 581 are singly ionized Ba transitions.

Table 1.2.4 (continued)
BARIUM[a] (FIGURE 1.7)

[a] Wavelengths and spectral assignments are taken from *Natl. Stand. Ref. Data Ser. Natl. Bur. Stand.*, NSRDS-NBS 34, Vol. 3, 1955, Russel, H. N. and Moore, C. E., *J. Res. Natl. Bur. Stand.*, 55, 299—306, 1955.

[b] Lines which operate in an ASE mode.

[c] A gain of 65 dBm^{-1} has been reported for this line.[113]

[d] A gain of 40 dBm^{-1} has been reported for this line.[113]

[e] Optically pumped in a Ba-Tl-Ar mixture[632] by the pair-absorption process: Ba ($6s^2\ ^1S_0$) + Tl ($6p\ ^2P^o_{1/2}$) + $h\nu$ (386.7 nm) → Ba ($6p\ ^1P^o_1$) + Tl($6p\ ^2P^o_{3/2}$).

[f] Two additional assignments are within the experimental error of the reported laser wavelength.

$6d\ ^1D_2 \rightarrow 6p'\ ^1F^o_3$ at *2.9235303* μm
and
$6d'\ ^3P_1 \rightarrow 4f\,^3F^o_2$ at *2.9236192* μm

The assignment given in the table has been suggested[114] as the one most likely to be correct as it terminates on a metastable level in common with many other Ba-I laser lines.

[g] Assignment suggested by the present author.

[h] No energy level combination corresponding to these wavelengths could be found by searching "Atomic Energy Levels," Vol. 3.

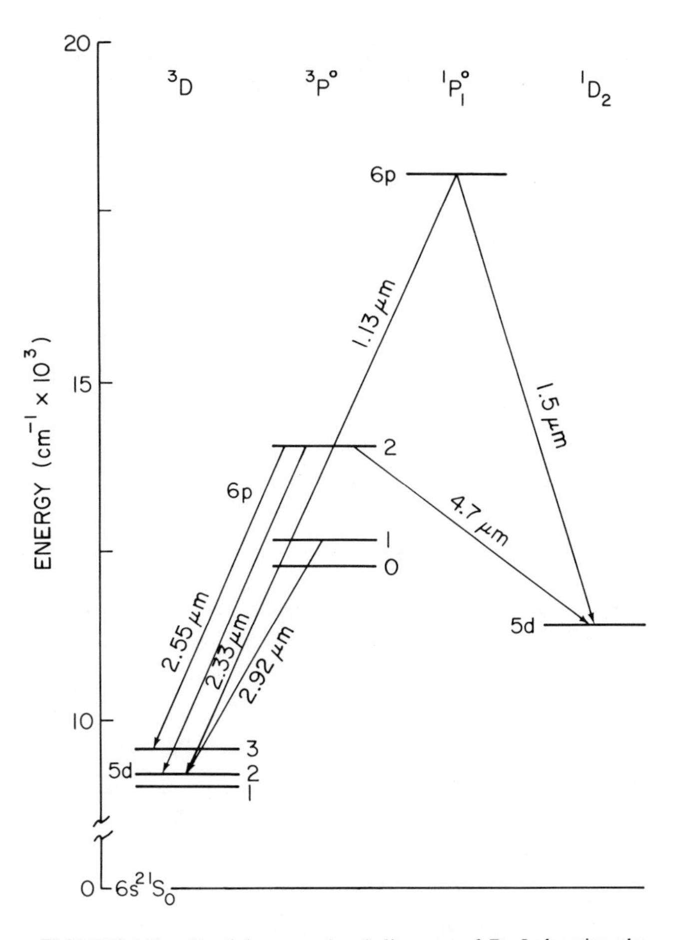

FIGURE 1.7. Partial energy level diagram of Ba-I showing the laser transition between the lowest-lying levels.

1.2B Group IIB
Table 1.2.5
ZINC [a]

Wavelength (μm)	Transition assignment	Comments	Ref.
1.305363	5p $^3P_2^o$ → 5s $^3S_1$	Pulsed; in a segmented transversely excited device incorporating recombining Zn plasmas in a few torr of He	146
1.315059	5p $^3P_1^o$ → 5s $^3S_1$	Pulsed; in a segmented tranversely excited device incorporating recombining Zn plasmas in a few torr of He	146

[a] Wavelength and spectral assignments taken from Johansson, I. and Contreras, R., *Ark. Fys.*, 37, 513—520, 1967.

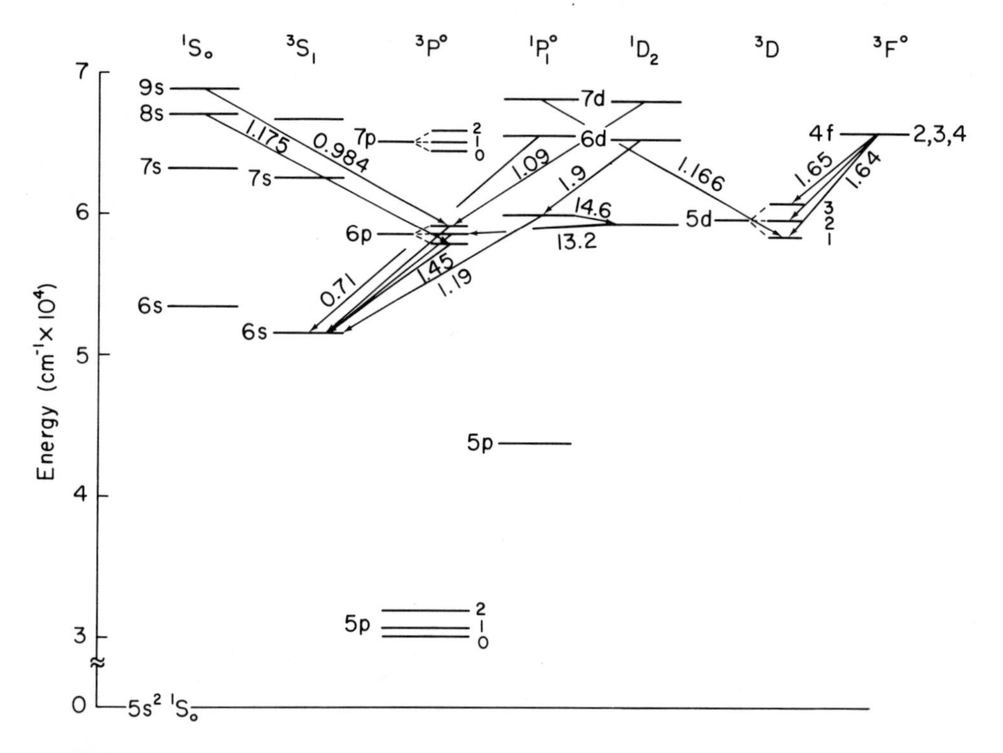

FIGURE 1.8. Partial energy level diagram of Cd-I showing all the classified laser transitions. Wavelengths given in micrometers.

Table 1.2.6
CADMIUM (FIGURE 1.8)

Wavelength (μm)	Transition assignment	Comments	Ref.
0.7133974[a]	7p $^1P_1^o \rightarrow$ 6s $^3S_1$	Pulsed; in 0.001—0.3 torr of Cd with 0.1—20 torr of He or Ne; discharge tube heated in a furnace; D = 15 mm; lases in rising edge of current pulse	119[b]
0.9838743	9s $^1S_0 \rightarrow$ 6p $^3P_2^o$	Pulsed; in 0.001—0.3 torr of Cd with 0.1—20 torr of He or Ne; discharge tube heated in a furnace; D = 15 mm; lases in rising edge of current pulse	119
1.0867911	7d $^1D_2 \rightarrow$ 6p $^3P_2^o$	Pulsed; in 0.001—0.3 torr of Cd with 0.1—20 torr of He or Ne; discharge tube heated in a furnace; D = 15 mm; lases in rising edge of current pulse	119
1.1485	?	Pulsed; in 0.001—0.3 torr of Cd with 0.1—20 torr of He or Ne; discharge tube heated in a furnace; D = 15 mm; lases in rising edge of current pulse	119
1.1554	?	Pulsed; in 0.001—0.3 torr of Cd with 0.1—20 torr of He or Ne; discharge tube heated in a furnace; D = 15 mm; lases in rising edge of current pulse	119
1.1663677[a]	8p $^1P_1^o \rightarrow$ 5d $^3D_1$	Pulsed; in 0.001—0.3 torr of Cd with 0.1—20 torr of He or Ne; discharge tube heated in a furnace; D = 15 mm; lases in rising edge of current pulse	119
1.1745636	8s $^1S_0 \rightarrow$ 6p $^3P_0^o$	Pulsed; 0.001—0.3 torr of Cd with 0.1—20 torr of He or Ne; discharge tube heated in a furnace; D = 15 mm; lases in rising edge of current pulse	119
1.1874246	6p $^1P_1^o \rightarrow$ 6s $^3S_1$	Pulsed; 0.001—0.3 torr of Cd with 0.1—20 torr of He or Ne; discharge tube heated in a furnace; D = 15 mm; lases in rising edge of current pulse	119,<u>120</u>
1.3982714	6p $^3P_2^o \rightarrow$ 6s $^3S_1$	Pulsed; as for the *0.7134*-μm line except that it lases only in the afterglow; also lases in a recombining plasma produced by vaporization of a Cd target with 1.06- or 10.6-μm laser radiation; 5 torr of He buffer used	<u>119</u>,121,146
1.4331602	6p $^3P_1^o \rightarrow$ 6s $^3S_1$	Pulsed; as for the 0.7134-μm line except that it lases only in the afterglow; also lases in a recombining plasma produced by vaporization of a Cd target with 1.06- or 10.6-μm laser radiation; 5 torr of He buffer used	119,<u>120</u>,121, 146

Table 1.2.6 (continued)
CADMIUM (FIGURE 1.8)

Wavelength (μm)	Transition assignment	Comments	Ref.
1.4478302	$6p\ ^3P^o_0 \rightarrow 6s\ ^3S_1$	Pulsed; lases in a recombining plasma produced by vaporization of a Cd target with 1.06- or 10.6-μm laser radiation; 5 torr of He buffer used	121,146,701
1.6404449	$4f\ ^3F^o_2 \rightarrow 5d\ ^3D_1$	Pulsed; as for the *1.398*-μm line above	119,121,146
1.6437081	$4f\ ^3F^o_{2,3} \rightarrow 5d\ ^3D_2$	Pulsed; as for the *0.7134*-μm line except that it lases only in the afterglow	119
1.6486189	$4f\ ^3F^o_{2,3,4} \rightarrow 5d\ ^3D_3$	Pulsed; as for the *0.7134*-μm line	120,119
1.9123124	$6d\ ^1D_2 \rightarrow 6p\ ^1P^o_1$	Pulsed; as for the *0.7134*-μm line	119
13.188714	$5d\ ^1D_2 \rightarrow 6p\ ^3P^o_1$	Pulsed; by dissociation of 0.04 torr of $Cd(CH_3)_2$ in 1.3 torr of He in a transversely excited double-discharge laser	122
14.58202	$6p\ ^1P^o_1 \rightarrow 5d\ ^1D_2$	Pulsed; by dissociation of 0.04 torr of $Cd(CH_3)_2$ in 1.3 torr of He in a transversely excited double-discharge laser	122

[a] Wavelength and spectral assignment taken from data in Burns, K. and Adams, K. B., *J. Opt. Soc. Am.*, 46, 94—99, 1956.

[b] For those lines observed by Dubrovin et al.,[119] the optimum buffer gas pressures were low in the case of lines lasing in the rising edge of the current pulse and were 6 to 8 torr for the afterglow lines; the optimum Cd pressure was ≃0.1 torr.

Table 1.2.7
MERCURY (FIGURE 1.9)

Wavelength (μm)	Transition assignment	Comments	Ref.
0.365015[a]	6d $^3D_3 \rightarrow$ 6p $^3P^o_2$ [b]	Pulsed; in an ASE mode following excitation of Hg vapor in a cell at 570°C with the 266-nm fourth harmonic of a Nd/YAG laser	123
0.365483[a]	6d $^3D_2 \rightarrow$ 6p $^3P^o_2$ [b]	Pulsed; in a ASE mode following excitation of Hg vapor in a cell at 570°C with the 266-nm fourth harmonic of a Nd/YAG laser	123
0.366288[a]	6d $^3D_1 \rightarrow$ 6p $^3P^o_2$ [b]	Pulsed; in a ASE mode following excitation of Hg vapor in a cell at 570°C with the 266-nm fourth harmonic of a Nd/YAG laser	123
0.366328[a]	6d $^1D_2 \rightarrow$ 6p $^3P^o_2$ [b]	Pulsed; in a ASE mode following excitation of Hg vapor in a cell at 570°C with the 266-nm fourth harmonic of a Nd/YAG laser	123
0.435835[a]	7s $^3S_1 \rightarrow$ 6p $^3P^o_1$ [c]	Pulsed; in a ASE mode following excitation of Hg vapor in a cell at 570°C with the 266-nm fourth harmonic of a Nd/YAG laser	123
0.546074[a]	7s $^3S_1 \rightarrow$ 6p $^3P^o_2$ [c]	CW; optically pumped with a Hg lamp, in 10 to 120 torr of N (optimum 25 torr); D = 3 mm; also as for the 0.365-μm line above	123,124, 125,126
0.576959[a]	6d $^3D_2 \rightarrow$ 6p $^1P^o_1$ [b]	Pulsed; as for the 0.365-μm line above	123
0.579065[a]	6d $^1D_2 \rightarrow$ 6p $^1P^o_1$ [b]	Pulsed; as for the 0.365-μm line above	123
0.8677	?	[e]Pulsed; in 0.001 torr of Hg with 0.8—1.2 torr of He; D = 15 mm	127
1.1179812	7p $^1P^o_1 \rightarrow$ 7s $^3S_1$	Pulsed; in 0.09—0.12 torr of Hg with 0.005—0.05 torr of He; D = 6 mm	127,128
1.1290435	7p $^3P^o_2 \rightarrow$ 7s $^3S_1$[b]	Pulsed; as for the 0.365-μm line above	123
1.2222	?	Pulsed; 0.001 torr of Hg with 0.2 torr of Ar; D = 5 mm	129,130
1.2246	?	Pulsed; in a mixture of Hg and Ar; D = 5 mm	129,130
1.2545	?	Pulsed; 0.001 torr of Hg with 0.8—1.2 torr of He; D = 15 mm	127

Table 1.2.7 (continued)
MERCURY (FIGURE 1.9)

Wavelength (μm)	Transition assignment	Comments	Ref.
1.2760	?	Pulsed; 0.001 torr of Hg with 0.2 torr of Ar; 0 = 5 mm	129,130
1.2981	?	Pulsed; 0.001 torr of Hg with 0.8 torr of Ar or 1.2 torr of He; D = 15 mm	127
1.3574217	7p $^1P_1^o$ → 7s ^1S_o	Pulsed; as for the 0.365-μm line above	123
1.3655	Probably the same transition as the one immediately below	Pulsed; as for the 1.2545-μm line above	127
1.3677207	7p $^3P_1^o$ → 7s $^3S_1$ [b]	Pulsed; in 0.09—0.12 torr of Hg with 0.005—0.05 torr of He; D = 6 mm; also as for the 0.365-μm line above	123,128
1.3954389	7p $^3P_0^o$ → 7s $^3S_1$ [b]	Pulsed; as for the 0.365-μm line above	123
1.529954	6p′ $^3P_2^o$ → 7s $^3S_1$	CW; 0.09—0.12 torr of Hg with 0.1—1.0 torr of He, Ne, Kr, or Ar; D = 6—8 mm	127,128, 131-141
1.6924775	5f $^1F_3^o$ → 6d $^1D_2$	Pulsed; 0.09—0.12 torr of Hg with 0.005—0.05 torr of He; D = 6 mm	127,128
1.6946636	5f $^3F_2^o$ → 6d $^3D_1$	Pulsed; 0.09—0.12 torr of Hg with 0.05—0.05 torr of He; D = 6 mm	127,128
1.7077438	5f $^3F_4^o$ → 6d $^3D_3$	Pulsed; 0.09—0.12 torr of Hg with 0.005—0.05 torr of He; D = 6 mm	127,128
1.7114554	5f $^3F_3^o$ → 6d $^3D_2$	Pulsed; in 0.09—0.12 torr of Hg with 0.005—0.05 torr of He; D = 6 mm	127,128
1.7334185	7d $^1D_2$ → 7p $^1P_1^o$	Pulsed; in 0.09—0.3 torr of Hg with 0.005—0.1 torr of He, Ne, Kr, or air; D = 6 mm	128,134
1.8135329	6p′ $^3F_4^o$ → 6d $^3D_3$	CW; in 0.09—0.3 torr of Hg with 0.005—0.1 torr of He, Ne, Kr, or air; D = 6 mm	127,128, 131-136
3.928361[f]	6d $^3D_3$ → 6p′ $^3P_2^o$ or 5g G → 5f $F^{o g}$	Pulsed; in 0.3 torr of Hg with 0.25 torr of Kr; D = 8 mm	128,134

5.8648	6p′ $^1P^o_1 \rightarrow 7d\ ^3D_2$ [h]	Pulsed; in 0.3 torr of Hg with 0.25 torr of Kr; D = 8 mm	128,133, 134,136
5.9833948	6d $^1D_2 \rightarrow 7p\ ^3P^o_1$	Pulsed; by dissociation of 60 torr of $Hg(CH_3)_2$ with 2.4 torr of He in a transversely excited double-discharge laser	142
6.477439 or *6.4887747*	11p $^3P^o_1 \rightarrow 10s\ ^3S_1$ or 9s $^1S_0 \rightarrow 8p\ ^1P^o_1$ [j]	Pulsed; as for the 3.93-μm line	128,133, 136

Note: A line at 3.34 μm reported in Reference 134 and assigned to Hg-I is a neutral Kr line.[128] See Table 1.9.4.

[a] Measured wavelength from Meggers, W. F., Corliss, C. H., and Scribner, B. F., *Natl. Bur. Stand. (U.S.) Monogr.*, 145 (1), 1975.

[b] Excitation of the 6d ^3D, 7p $^3P^o$, 8s ^3S, 6d $^1D_2$, and 7p $^1P^o_1$ levels following optical pumping with 266-nm radiation involves dissociation of electronically excited Hg_2 molecules produced by the absorption of two pump photons.[123]

[c] Excitation of the 7s $^3S_1$ level when pulsed optical pumping with 266-nm radiation is used probably involves a collisional reaction between two excited Hg dimers.[123]

$$Hg^*_2 + Hg^*_2 \rightarrow Hg\,(7s\ ^3S_1) + 3Hg$$

[d] For further details of the CW optically pumped mercury laser see the text.

[e] May be a Hg-II line.

[f] Measured wavelength from Plyler, E. K., Blaine, L. R., and Tidwell, E. D., *J. Res. Natl. Bur. Stand.*, 55, 279—284, 1955.

[g] Assignment as given in Reference 128.

[h] The position of the 6p′ $^1P^o_1$ is not very accurately known.

[j] Several additional assignments are possible for this transition; these other possibilities involve transitions between higher lying states than the ones listed here.

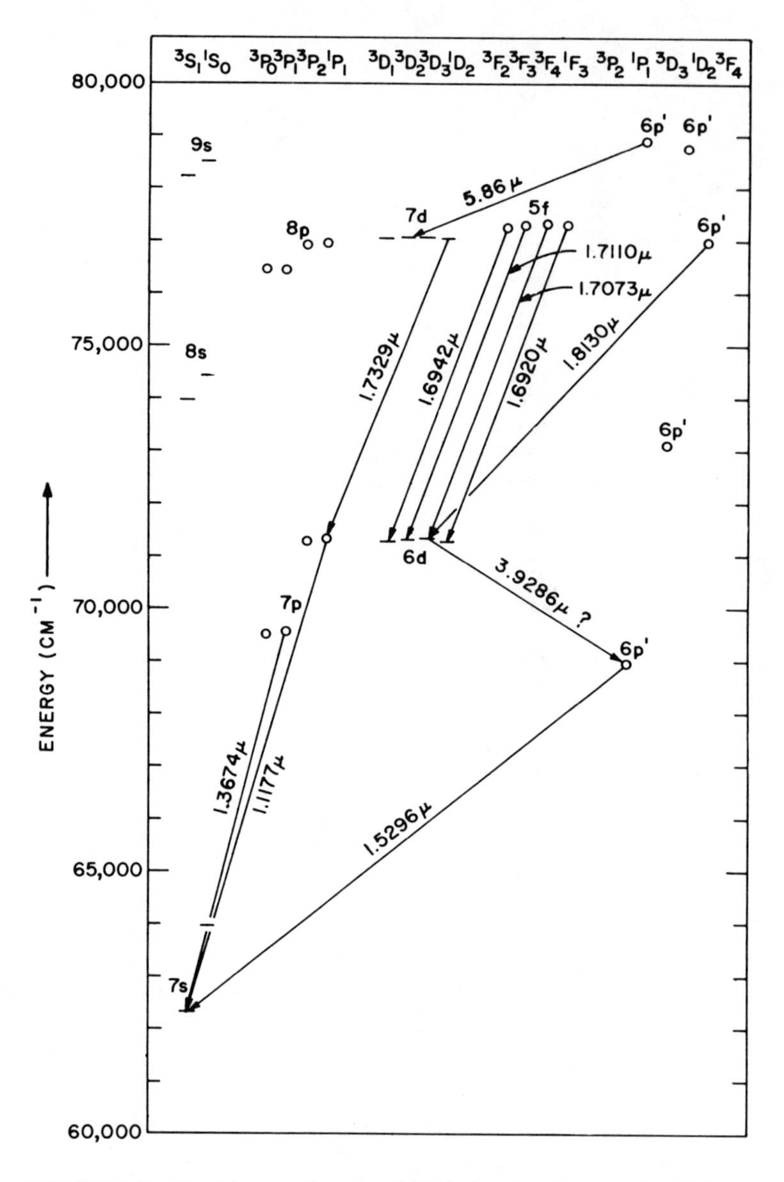

FIGURE 1.9. Partial term diagram of Hg-I showing the majority of the reported laser transitions. Circles represent odd levels, horizontal lines represent even levels. (From Bockasten, K., Garavaglia, M., Lengyel, B. A., and Lundholm, T., *J. Opt. Soc. Am.*, 55, 1051, 1965. With permission.)

1.3A GROUP IIIA

Table 1.3.1 BORON

Wavelength (μm)	Transition assignment	Comments	Ref.
3.601	? [a]	Pulsed; in discharges in mixtures containing about 10 torr of He with about 0.025 torr of B-containing compounds such as B_2H_6, $H_3B\,CO$, B_5H_9, or BBr_3; lases in afterglow	143

[a] A search of spectroscopic data on neutral revealed no likely assignment for this transition. Its positive identification as a neutral atomic B transition must, therefore, remain uncertain.

Table 1.3.2
GALLIUM (FIGURE 1.10)

Wavelength (μm)	Transition assignment	Comments	Ref.
0.4032987[a]	$5s\ ^2S_{1/2} \rightarrow 4p\ ^2P^o_{1/2}$ [b]	Pulsed; by dissociation of GaI_3 with an ArF laser (193 nm); operates with 5—20 torr of GaI_3 with 300 torr of Ar at 160—210°C	144
0.4172042[a]	$5s\ ^2S_{1/2} \rightarrow 4p\ ^2P^o_{3/2}$ [b]	Pulsed; by dissociation of GaI_3 with an ArF laser (193 nm); operates with 5—20 torr of GaI_3 with 300 torr of Ar at 160—210°C	144
1.7367231	$4p^2\ ^4P_{5/2} \rightarrow 5p\ ^2P^o_{3/2}$	Pulsed; by dissociation of 70 torr of $Ga(CH_3)_3$ with 2.8 torr of He in a transversely excited double-discharge laser	142
5.754965[a]	$4d\ ^2D_{3/2} \rightarrow 5p\ ^2P^o_{1/2}$	Pulsed; by dissociation of 70 torr of $Ga(CH_3)_3$ with 2.8 torr of He in a transversely excited double-discharge laser	142
6.1477551[a]	$4d\ ^2D_{3/2} \rightarrow 5p\ ^2P^o_{3/2}$	Pulsed; by dissociation of 70 torr of $Ga(CH_3)_3$ with 22.8 torr of He in a transversely excited double-discharge laser	142

[a] Wavelengths and spectral assignments taken from data in Johansson, I. and Litzen, U., *Ark. Fys.*, 34, 573—587, 1966.

[b] Resonance transition. At the operating temperature some dissociation of GaI_3 into GaI occurs and the latter species may be the predominant one directly photodissociated to yield Ga 5s $^2S_{1/2}$ atoms.

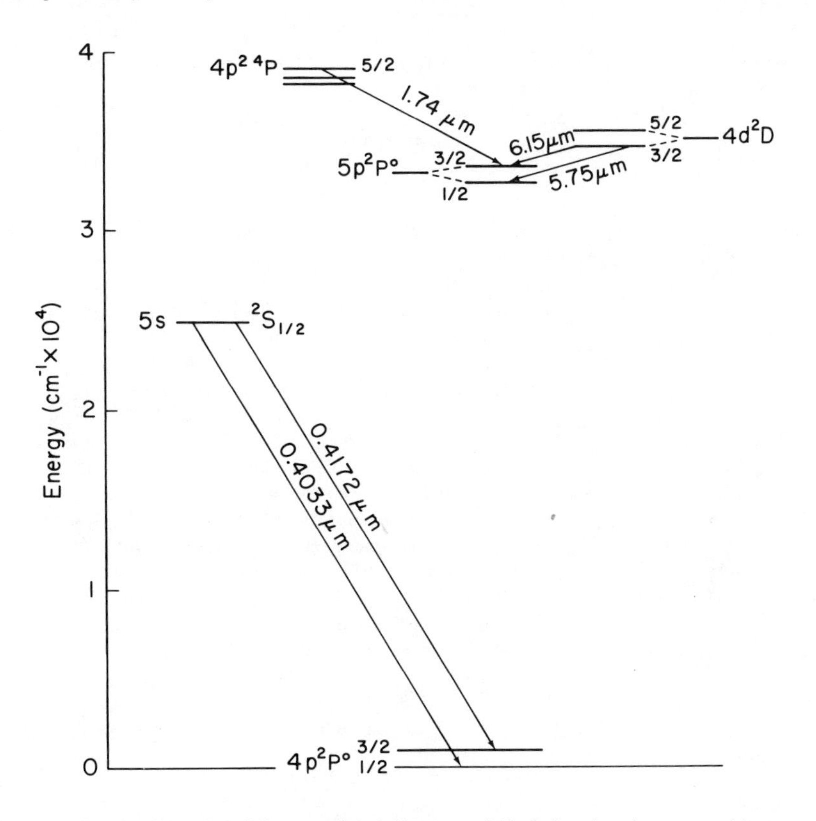

FIGURE 1.10. Partial energy level diagram of Ga-I showing the reported laser transitions. Wavelengths given in microns.

Table 1.3.3
INDIUM[a]

Wavelength (μm)	Transition assignment	Comments	Ref.
0.4101745	6s $^2S_{1/2}$ → 5p $^2P^o_{1/2}$	Pulsed; in an ASE mode following dissociation of InI in a cell at temperatures from 200—600°C with an ArF laser (193 nm); optimum operating temperature 330°C	145
0.4511299	6s $^2S_{1/2}$ → 5p $^2P^o_{3/2}$	Pulsed; as for the line above, but is by far the stronger line of the two	145
1.342996	6p $^2P^o_{1/2}$ → 6s $^2S_{1/2}$	Pulsed; in a segmented transversely excited device incorporating recombining In plasmas in a few torr of He	146
1.431625	6d $^2D_{5/2}$ → 6p $^2P^o_{3/2}$	Pulsed; in a segmented transversely excited device incorporating recombining In plasmas in a few torr of He	146
1.441920	6d $^2D_{3/2}$ → 6p $^2P^o_{3/2}$	Pulsed; in a segmented transversely excited device incorporating recombining In plasmas in a few torr of He	146(E)
1.8736732	5p^2 $^4P_{5/2}$ → 6p $^2P^o_{3/2}$	Pulsed; by dissociation of 70 torr of $In(CH_3)_3$ with 2.8 torr of He in a transversely excited double-discharge laser	142
2.3785794	5p^2 $^4P_{3/2}$ → 6p $^2P^o_{1/2}$	Pulsed; by dissociation of 50 torr of $In(CH_3)_3$ with 2.8 torr of He in a transversely excited double-discharge laser	142

[a] Wavelengths and spectral assignments taken from data in Johansson, I. and Litzen, U., *Ark. Fys.*, 34, 573—587, 1966.

Table 1.3.4
THALLIUM

Wavelength (μm)	Transition assignment	Comments	Ref.
0.377572[a]	7s $^2S_{1/2}$ → 6p $^2P^o_{1/2}$	Pulsed; by dissociation of 0.001—0.5 torr of TlI vapor with an ArF laser (193 nm)	147
0.535065[b]	7s $^2S_{1/2}$ → 6p $^2P^o_{3/2}$	Pulsed; short rise-time high-voltage excitation of more than 0.01 torr of Tl with several torr of Ne or He; D = 1.3,2.0, or 3 mm, or of TlI at 370—440°C with added He, Ne, Ar, or Xe; D = 1.3 mm; Also, in a Tl-Hg or Tl-Cd-Ar mixture excited with a N laser;[c] D = 12 mm	148-152
3.8135916	8p $^2P^o_{1/2}$ → 8s $^2S_{1/2}$	Pulsed; by dissociation of 120 torr of $Tl(CH_3)_3$ with 6 torr of He in a transversely excited, double-discharge laser	142
5.1072522	6d $^2D_{3/2}$ → 7p $^2P^o_{1/2}$	Pulsed; as for line immediately above, but with 60 torr of $Tl(CH_3)_3$ and 2.4 torr of He	142
10.451505	6d $^2D_{3/2}$ → 7p $^2P^o_{3/2}$	Pulsed; as for line immediately above, but with 60 torr of $Tl(CH_3)_3$ and 2.4 torr of He	142

Note: Lines at 0.5152, 0.5949, and 0.6950 μm listed as neutral Tl lines in Reference 581 are ionized Tl laser lines.

[a] Resonance line, measured wavelength from Meggers, W. F., Corliss, C. H., and Scribner, B. F., *Natl. Bur. Stand. (U.S.) Monogr.*, 145(1), 1975.

[b] Measured wavelength and assignment from Seguie, J., *C. R. Acad. Sci. Ser. B*, 263B, 147-150, 1966.

[c] Excitation follows dissociation of an exciplex such as (Tl-Hg)′; optimum operating conditions 0.01 torr Tl(~600°C) with 400 torr Hg(~325°C).

1.4A GROUP IVA

Table 1.4.1
CARBON[a] (FIGURE 1.11)

Wavelength (μm)	Transition assignment	Comments	Ref.
0.8335149	$3p\ ^1S_0 \rightarrow 3s\ ^1P_1^o$	Pulsed; in a mixture of CO_2 and Ne at 4 torr; D = 15 mm	153
0.9405729	$3p\ ^1D_2 \rightarrow 3s\ ^1P_1^o$ [b]	Pulsed; as for the 0.8335-μm line and also in a mixture of 0.05 torr of CO with 2—16 torr of He, optimum 5 torr of He; D = 7 mm	146,153-155
1.0683082	$3p\ ^3D_2 \rightarrow 3s\ ^3P_1^o$ [b]	Pulsed; as for the 0.9406-μm line and also in a mixture of CO_2 and He; D = 16 mm	154,155,156
1.0685345	$3p\ ^3D_1 \rightarrow 3s\ ^3P_0$ [b]	Pulsed; in a mixture of 0.05 torr of CO with 2—16 torr of He, optimum 5 torr; D = 7 mm; also in CO_2 with He; D = 16 mm	153,156
1.0691250	$3p\ ^3D_3 \rightarrow 3s\ ^3P_2^o$ [b]	CW; in 0.01 torr of CO or CO_2 with 2 torr of He; D = 5 mm	153,154,156,157, 158,159
1.0707333	$3p\ ^3D_1 \rightarrow 3s\ ^3P_1^o$	Pulsed; in a mixture of CO_2 and He; D = 16 mm	156
1.454250	$3p\ ^1P_1 \rightarrow 3s\ ^1P_1^o$ [b]	CW; in 0.01 torr of CO or CO_2 with 2 torr of He; D = 5 mm; also pulsed in OCS and several other organic gases; also nuclear pumped by the reaction $^{10}B(n,\alpha)^7Li$ in He-CO and He-CO_2 mixtures	146,153,154,156, 157-161,345,347
2.0655993	$5d\ ^1D_2^o \rightarrow 4p\ ^3P_2$	CW; in 0.02 torr of CO with 1 torr of He; D = 10 mm	158
3.407422	$4d\ ^1D_2^o \rightarrow 4p\ ^1P_1$	CW; in 0.02 torr of CO with 1 torr of He; D = 10 mm	158
3.5117661	$6d\ ^3P_2^o \rightarrow 5p\ ^3D_3$	CW; in 0.02 torr of CO with 1 torr of He; D = 10 mm	158
5.5983205	$4p\ ^3S_1 \rightarrow 3d\ ^3P_1^o$	CW; in 0.02 torr of CO with 1 torr of He; D = 10 mm	14,158

[a] Wavelengths and spectral assignment taken from data in Moore, C. E., *Natl. Stand. Ref. Dat. Ser. U.S. Natl. Bur. Stand.*, NSRDS-NBS3, Sect. 3.

[b] The excitation mechanism for these transitions observed in He-CO and Ne-CO discharges was originally thought to be due to dissociative excitation transfer involving He or Ne metastables,[154] e.g., CO + Ne* ($1s_5$ or $1s_3$) → C′ + O + Ne. However, recent work has indicated that, in fact, their excitation mechanism involves collisional-radiative ion-electron recombination,[155] namely, $C^+ + 2e \rightarrow C' + e$.

Table 1.4.2
SILICON[a]

Wavelength (μm)	Transition assignment	Comments	Ref.
1.1984187	$4p\ ^3D_2 \rightarrow 4s\ ^3P_1^o$	CW; 0.03 torr of $SiCl_4$ with 0.5 torr of Ne; D = 6 mm	162
1.2031507	$4p\ ^3D_3 \rightarrow 4s\ ^3P_2^o$	CW; in 0.04 torr of $SiCl_4$ with 0.5 torr of Ne; D = 6 mm	162
1.588441	$5s\ ^3P_1^o \rightarrow 4p\ ^3D_1$	CW; in 0.03—0.05 torr of $SiCl_4$ with 1—5 torr of Ne; D = 6 mm	162,345

[a] Measured wavelengths and spectral assignments taken from Moore, C. E., *Natl. Stand. Ref. Data Ser. Natl. Bur. Stand.*, NSRDS-NBS 3, 1975, Sect. 2.

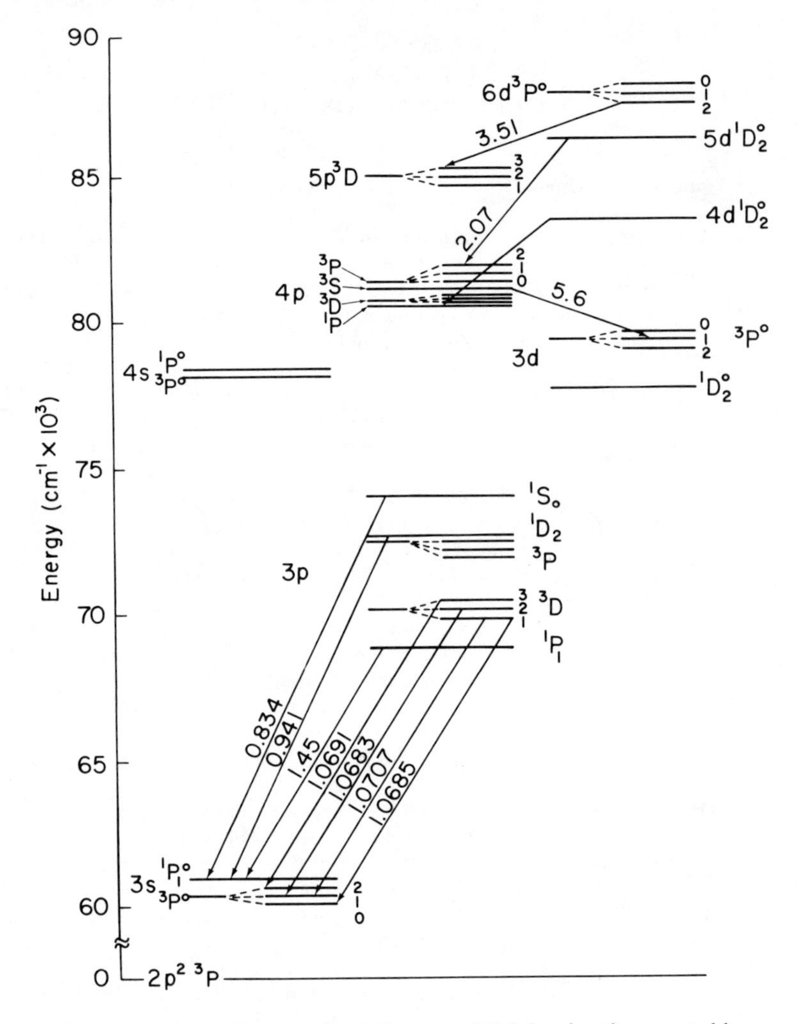

FIGURE 1.11. Partial energy level diagram of C-I showing the reported laser transitions. Wavelengths given in microns.

Table 1.4.3
GERMANIUM[a]

Wavelength (μm)	Transition assignment	Comments	Ref.
1.9814602	6d $^3P^\circ_1$ → 6p $^3P_0$	Pulsed; by dissociation of 40 torr of $GaCl_4$ with 0.4 torr of He in a transversely excited double-discharge laser	142
2.020602	6f $^3G_3$ → 5d $^3P^\circ_2$ [b]	Pulsed; by dissociation of 40 torr of $GaCl_4$ with 0.4 torr of He in a transversely excited double-discharge laser	142

[a] Calculated wavelengths and spectral assignments from Andrew, K. I. and Meissner, K. W., *J. Opt. Soc. Am.*, 49, 146—161, 1959.

[b] The lower level of this transition is incorrectly designated $^3P^\circ_1$ in Atomic Energy Levels, Vol. 2.

Table 1.4.4
TIN

Wavelength (μm)	Transition assignment	Comments	Ref.
0.657903[a]	10d $^3D^o_2 \rightarrow$ 6p $^3P_1$? [b]	Pulsed; in $SnCl_4$ vapor at room temperature; D = 5.6 mm; also in Sn vapor at 0.001—0.1 torr; D = 7 mm	163(E)-165
1.0612556	$5p^3$ $^3D^o_1 \rightarrow$ 6p $^3P_2$	Pulsed; in a transversely excited pin laser with 0.05—3 torr of $SnCl_4$ with 85—250 torr of He; lases in afterglow	166
1.3612294	6p $^1P_1 \rightarrow$ 6s $^1P^o_1$	Pulsed; in a segmented, transversely excited device incorporating recombining tin plasmas in a few torr of He	146
4.6157396	5d $^3D^o_1 \rightarrow$ 6p $^3P_1$	Pulsed; by dissociation of 50 torr of $SnCl_4$ with 2 torr of He in a transversely excited double-discharge laser	122

[a] Measured wavelength.
[b] This assignment is by no means certain; this could be an ionized tin transition; it may be the same line as one at 0.657926 μm reported in *M.I.T. Wavelength Tables.*[577]

Table 1.4.5
LEAD[a] (FIGURE 1.12)

Wavelength (μm)	Transition assignment	Comments	Ref.
0.36395677	$7s(3/2,1/2)^o_1 \rightarrow 6p^2(3/2,3/2)_2$ (7s $^3P^o_1 \rightarrow 6p^2$ $^3P_1$)	Transient laser line requiring fast rise-time high-voltage pulsed excitation; in vapor pressure[b] of ^{208}Pb at a temperature of 800—900°C with a He, Ne, or Ar buffer; D = 2 mm	167
0.40578067	$7s(1/2,1/2)^o_1 \rightarrow 6p^2(3/2,1/2)_2$ (7s $^3P^o_1 \rightarrow 6p^2$ $^3P_2$)	Pulsed; transient laser line requiring fast rise-time high-voltage pulsed excitation; in vapor pressure [b] of ^{208}Pb at a temperature of 800—900°C with a He, Ne, or Ar buffer; D = 2 mm	167,170
0.40621360	6d½ $[3/2]^o_1 \rightarrow 6p^2(3/2,3/2)_2$ (6d $^3D^o{}_1 \rightarrow 6p^2$ $^1D_2$)	Pulsed; transient laser line requiring fast rise-time high-voltage pulsed excitation; in vapor pressure[b] of ^{208}Pb at a temperature of 800—900°C with a He, Ne, or Ar buffer; D = 2 mm	167,170
0.72289653[c]	$7s(1/2,1/2)^o_1 \rightarrow 6p^2(3/2,3/2)_2$ (7s $^3P^o_1 \rightarrow 6p^2$ $^1D_2$)	Transient laser line requiring fast rise-time high-voltage pulsed excitation; in 0.2—2.0 torr vapor pressure of Pb[b] with 3 torr of He; D = 10 mm	167-169,170, 171-173
1.2561370	$7p(1/2,1/2)_1 \rightarrow 7s(1/2,1/2)^o_o$ (7p $^3P_1 \rightarrow$ 7s $^3P^o_o$)	Pulsed; in Pb vapor in a heated tube at 1400°C; D = 5—10 mm	174
1.3103722 or *1.3152769*	$7p(1/2,1/2)_1 \rightarrow 7s(1/2,1/2)^o_1$ (7p $^3P_1 \rightarrow$ 7s $^3P^o_1$) or 7d½ $[5/2]^o_3 \rightarrow 7p(1/2,3/2)_2$ (7d $^3F^o_3 \rightarrow$ 7p $^3D_2$)	Pulsed; in a segmented, transversely excited device incorporating recombining Pb plasmas in a few torr of He	146
1.5335134	$7s(3/2,1/2)_1 \rightarrow 7p(1/2,1/2)_1$ (7s $^1P^o_1 \rightarrow$ 7p $^3P_1$) [d]	Pulsed; in a segmeted, transversely excited device incorporating recombining Pb plasmas in a few torr of He	146(E)
3.1748096	6d½ $[3/2]^o_1 \rightarrow 7p(1/2,1/2)_1$ (6d $^3D^o_1 \rightarrow$ 7p $^3P_1$)	Pulsed; by dissociation of 0.5 torr of $Pb(CH_3)_4$ with 15 torr of He in a transversely excited double-discharge laser	142

Table 1.4.5 (continued)
LEAD[a] (FIGURE 1.12)

Wavelength (μm)	Transition assignment	Comments	Ref.
7.1764192	$6d\frac{1}{2}[3/2]^{o}_{1} \rightarrow 7p(1/2,3/2)_1$ (6d $^3D^o_1$ → 7p $^3D_1$)	Pulsed; by dissociation of 0.5 torr of $Pb(CH_3)_4$ with 15 torr of He in a transversely excited double-discharge laser	142
7.9423392	$6d\frac{1}{2}[3/2]^{o}_{1} \rightarrow 7p(1/2,3/2)_2$ (6d $^3D^o_1$ → 7p $^3D_2$)	Pulsed; by dissociation of 0.06 torr of $Pb(CH_3)_4$ with 0.6 torr of He in a transversely excited double-discharge laser	142

[a] Wavelengths and spectral assignments are taken from data in Wood, D. R. and Andrew, K. C., *J. Opt. Soc. Am.*, 58, 818—829, 1968. Level designations according to "Atomic Energy Levels" are in parentheses.

[b] The vapor pressure of lead can be deduced from Honig, R. E., *R.C.A. Rev.*, 18, 195—204, 1957.

[c] A single-pass gain of 600 dB m^{-1} has been reported for this line.[169]

[d] This appears the most likely assignment for this line, however, there are several other possible ones, namely,

$5f\frac{1}{2}[7/2]_4 \rightarrow 6d\frac{1}{2}[5/2]^o_3$(5f $^3F_4$ → 6d $^3F^o_3$) at 1.53148 μm
$5f\frac{1}{2}[7/2]_3 \rightarrow 6d\frac{1}{2}[5/2]^o_3$(5f $^3F_3$ → 6d $^3F^o_3$) at 1.53276 μm
$5f\frac{1}{2}[5/2]_2 \rightarrow 6d\frac{1}{2}[5/2]^o_3$(5f $^3F_2$ → 6d $^3F^o_3$) at 1.53310 μm

The laser wavelength reported in Reference 146 is 1.532 μm.

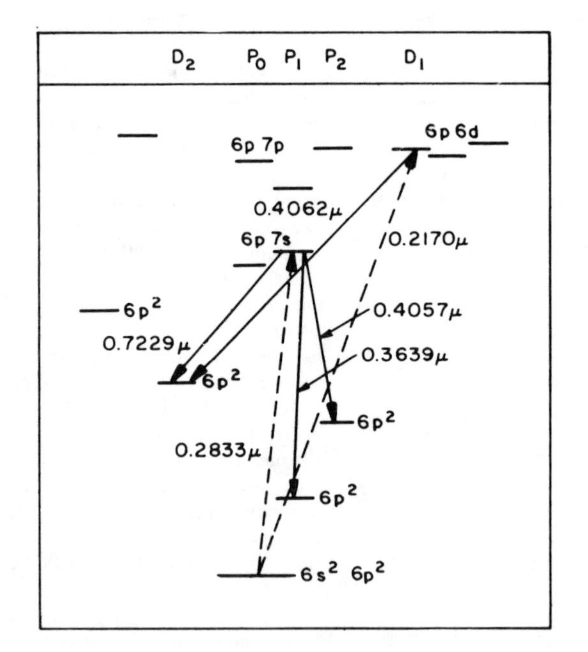

FIGURE 1.12. Partial term diagram of Pb-I showing the self-terminating laser transitions.

1.5A GROUP VA
Table 1.5.1
NITROGEN[a] (FIGURE 1.13)

Wavelength (μm)	Transition assignment	Comments	Ref.
0.4120[b]	?	Pulsed; in a Hg-N mixture at 0.001—0.02 torr; D = 3 mm	175
0.4321334	$5p\ ^4P^o_{3/2} \rightarrow 2p^4\ ^4P_{3/2}$ [c]	Pulsed; in a theta pinch discharge in 1 torr of N; D = 25 mm	176
0.4328395	$5p\ ^4D^o_{5/2} \rightarrow 2p^4\ ^4P_{5/2}$ [c]	Pulsed; in a theta pinch discharge in 1 torr of N; D = 25 mm	176
0.4525[b]	?	Pulsed; in a Hg-N mixture at 0.001—0.02 torr; D = 3 mm	175
0.4750295[b]	$4p\ ^2D^o_{3/2} \rightarrow 3s\ ^2P_{1/2}$? [c]	Pulsed; in a Hg-N mixture at 0.001—0.02 torr; D = 3 mm	175
0.5440[b]	?	Pulsed; in a Hg-N mixture at 0.001—0.02 torr; D = 3 mm	175
0.550042[b]	$8s\ ^2P_{1/2} \rightarrow 3p\ ^2D^o_{3/2}$? [c]	Pulsed; in a Hg-N mixture at 0.001—0.02 torr; D = 3 mm	175
0.5540307[b]	$5d\ ^4P_{3/2} \rightarrow 3p\ ^4D^o_{3/2}$? [c]	Pulsed; in a Hg-N mixture at 0.001—0.02 torr; D = 3 mm	175
0.8594005	$3p\ ^2P^o_{1/2} \rightarrow 3s\ ^2P_{1/2}$	Pulsed; in a mixture of N and He at about 4 torr; D = 15 mm	<u>153</u>,177
0.8629238	$3p\ ^2P^o_{3/2} \rightarrow 3s\ ^2P_{3/2}$ [d]	Pulsed; 0.2—0.7 torr of N with 3 torr of He or 0.15 torr of N only; D = 15 mm; also quasi-CW when nuclear pumped using 75—375 torr of a Ne-N_2 mixture with <0.001 torr of N_2; D = 2.5 cm	153-155,177, <u>178</u>,179
0.9045878	$3p'\ ^2F^o_{7/2} \rightarrow 3s'\ ^2D_{5/2}$	Pulsed; in 0.3 torr of N with 12 torr of He; D = 11 mm	177
0.9187449 or 0.918784	$3p'\ ^2D^o_{5/2} \rightarrow 3s'\ ^2D_{5/2}$ [e] $3p'\ ^2D^o_{5/2} \rightarrow 3s'\ ^2D_{3/2}$	Pulsed; in a mixture of N and He at 4 torr; D = 15 mm	153
0.9386805	$3p\ ^2D^o_{3/2} \rightarrow 3s\ ^2P_{1/2}$ [d]	CW; in 0.02—0.2 torr of N or nitrous oxide with 0.01—0.1 torr of O, H, He, or Ne; D = 3 mm; or in a mixture of N and He at 5 torr; D = 15 mm	153,154,<u>178</u>, 180,181
0.9392789	$3p\ ^2D^o_{5/2} \rightarrow 3s\ ^2P_{3/2}$ [d]	CW; as for the line immediately above, also quasi-CW in a nuclear pumped 75—175 torr Ne-N_2 mixture with <0.01 torr of N_2; D = 2.5 cm	153-155, <u>178</u>, 179,181
1.0563328[f]	$3d\ ^4D_{3/2} \rightarrow 3p\ ^4P^o_{5/2}$	Pulsed; in 0.2 torr of N with 100 torr of He in a transversely excited pin laser	182(E)
1.0623177[f]	$3d\ ^4P_{1/2} \rightarrow 3p\ ^4P^o_{1/2}$	Pulsed; in 0.2 torr of N with 100 torr of He in a transversely excited pin laser	182(E)
1.0623177[f] or 1.0643981	$3d\ ^4P_{1/2} \rightarrow 3p\ ^4P^o_{1/2}$ or $3d\ ^4P_{3/2} \rightarrow 3d\ ^4P^o_{1/2}$	Pulsed; in 0.2 torr of N with 100 torr of He in a transversely excited pin laser	182(E)
1.342961	$3p\ ^2S^o_{1/2} \rightarrow 3s\ ^2P_{1/2}$	Pulsed; in 0.7 torr of N or in 0.15 of N with 3 torr of He; D = 15 mm (or 75 mm?)	178
1.358133	$3p\ ^2S^o_{1/2} \rightarrow 3s\ ^2P_{3/2}$ [d]	CW; in 0.03 torr of nitric or nitrous oxide with 2 torr of He or 1 torr of Ne; D = 5 mm	154,158,159, <u>178</u>
1.45423	? [g]	Pulsed; in 0.2—0.7 torr of N or 0.15 torr of N with 3 torr of He; D = 15 mm (or 75 mm?)	178
1.4553011	$4s\ ^4P_{5/2} \rightarrow 3p\ ^2D^o_{5/2}$	CW; in 0.03 torr of nitric or nitrous oxide with 2 torr of He or 1 torr of Ne	158,159

1.5A GROUP VA

Table 1.5.1 (continued)

NITROGEN[a] (FIGURE 1.13)

Wavelength (μm)	Transition assignment	Comments	Ref.
3.7942	? [h]	Pulsed; in 0.2—0.7 torr of N with 3 torr of He; D = 15 mm (or 75 mm?)	178
3.8154	? [j]	Pulsed; in 0.15 torr of N with 3 torr of He or in 0.2—0.7 torr of N only; D = 15 mm (or 75 mm?)	178

[a] Wavelengths and spectral assignments taken from data in Moore, C. E., *Natl. Stand. Ref. Data Ser. Natl. Bur. Stand.*, NSRDS-NBS 3, 1975, Sect. 5.

[b] There is doubt about the accuracy of the determination of the wavelength of this line and the assignment here to neutral N is probably dubious.

[c] Tentative assignment made by the present author under the assumption that the reported wavelength is accurate.

[d] The excitation mechanism for this transition in a Ne-N or He-N discharge was originally thought to be due to dissociative excitation transfer involving He or Ne metastables,[154] e.g., N_2 + Ne* ($1s_5$ or $1s_3$) → N′ + N + Ne. However, recent work has indicated that in fact the excitation involves collisional-radiative ion-electron recombination,[155] namely, N^+ + 2e → N′ + e.

[e] Alternative assignment suggested by present author.

[f] This line, first observed and assigned by Sutton,[182] is reported by him at a measured wavelength which agrees very well with the calculated wavelength of his assignment based on energy level values for NI reported in Atomic Energy Levels, Vol. 1. However, in view of the revision of the NI energy level scheme reported by Moore (see *a.* above), Sutton's assignment is incorrect. The assignment given here is suggested by the present author on the basis of the reported wavelength; unambiguous assignment would require accurate remeasurement of the laser wavelength.

[g] A possible assignment for this line suggested by the present author, is 4s $^4P_{1/2}$ → 3p at 1.454855 μm.

[h] A possible assignment for this line, suggested by the present author, is 4p $^4D^o_{3/2}$ → 4s $^2P_{1/2}$ at *3.7972035* μm.

[j] A possible assignment for this line, suggested by the present author, is 4p $^4S^o_{3/2}$ → 3d $^4P_{5/2}$ at *3.816051* μm.

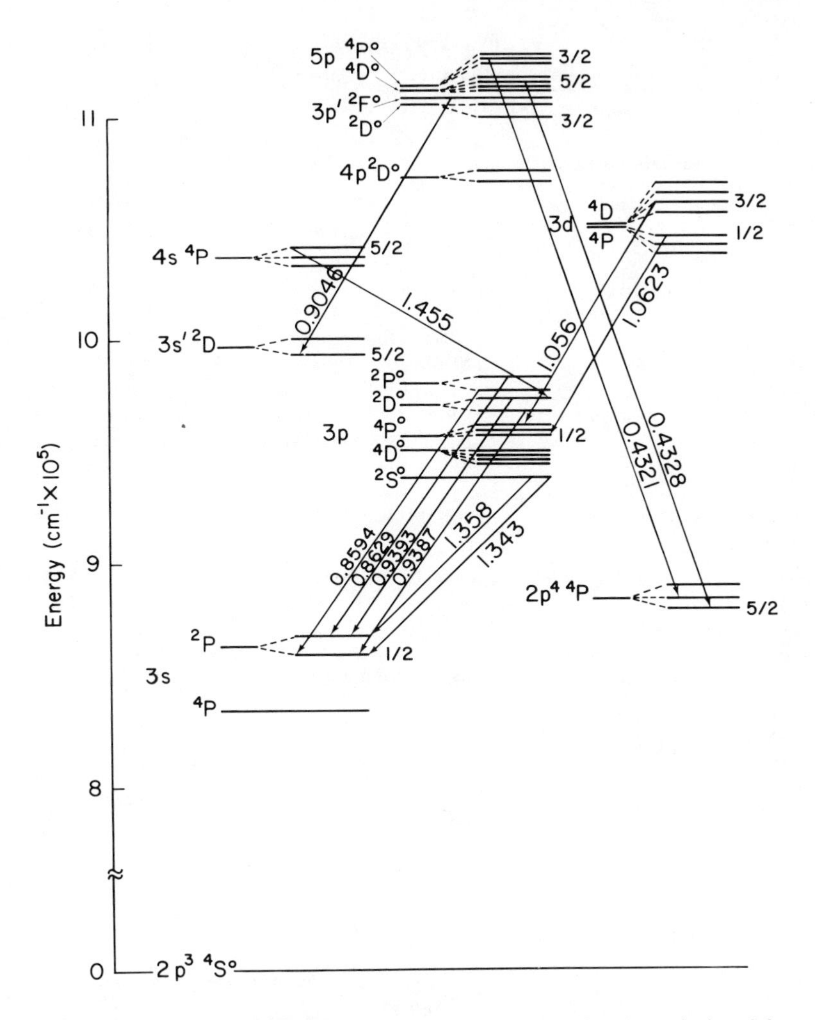

FIGURE 1.13. Partial energy level diagram of N-I showing the majority of the reported laser transitions. Wavelengths given in microns.

Table 1.5.2
PHOSPHORUS[a] (FIGURE 1.14)

Wavelength (μm)	Transition assignment	Comments	Ref.
0.667193	?[b]	Pulsed; in 0.04 torr of P; D = 3mm	183
1.008422	$4p\ ^2P^o_{3/2} \rightarrow 4s\ ^2P_{3/2}$	Pulsed; in P vapor at 0.02—0.1 torr with 0.2—3.5 torr of He; D = 9 mm	184
1.1163455	$4p\ ^4S^o_{3/2} \rightarrow 4s\ ^2P_{1/2}$	Pulsed; in P vapor at 0.02—0.1 torr with 0.2—3.5 torr of He or Ne, lases in afterglow; D = 9 mm	184
1.1186470	$4p\ ^2D^o_{5/2} \rightarrow 4s\ ^2P_{3/2}$	Pulsed; in P vapor at 0.02—0.1 torr with 0.2—3.5 torr of He or Ne, lases in afterglow; D = 9 mm	184
1.1547277	$4p\ ^4S^o_{3/2} \rightarrow 4s\ ^2P_{3/2}$	Pulsed; in P vapor at 0.02—0.1 torr with 0.2—3.5 torr of He or Ne, lases in afterglow; D = 9 mm	184
1.1787698	$4p\ ^4P^o_{3/2} \rightarrow 4s\ ^2P_{1/2}$	Pulsed; in P vapor at 0.02—0.1 torr with 0.2—3.5 torr of He or Ne, lases in afterglow; D = 9 mm	184

Table 1.5.2 (continued)
PHOSPHORUS[a] (FIGURE 1.14)

Wavelength (μm)	Transition assignment	Comments	Ref.
1.5716351	4p $^2S^o_{1/2} \rightarrow$ 4s $^2P_{1/2}$	Pulsed; in P vapor at 0.02—0.1 torr with 0.2—3.5 torr of He or Ne, lases in afterglow; D = 9 mm	184
1.648791	4p $^2S^o_{1/2} \rightarrow$ 4s $^2P_{3/2}$	Pulsed; in P vapor at 0.02—0.1 torr with 0.2—3.5 torr of He or Ne, lases in afterglow; D = 9 mm	184
1.8943842	5s $^2P_{3/2} \rightarrow$ 4p $^2P^o_{1/2}$	Pulsed; P vapor at 0.02—0.1 torr with 0.2—3.5 torr of He or Ne, lases in afterglow; D = 9 mm	184
2.0962339	3d $^2D_{3/2} \rightarrow$ 4p $^2P^o_{1/2}$[c]	Pulsed; in P vapor at 0.02—0.1 torr with 0.2—3.5 torr of He or Ne, lases in afterglow; D = 9 mm	184

Note: A line at 0.784563 μm listed in Reference 581 as a neutral transition is a singly ionized P transition.[213]

[a] Wavelengths and spectral assignments taken from data in Martin, W. C., *J. Opt. Soc. Am.*, 49, 1071—1085, 1959.

[b] Probably an ionized P line.

[c] Assigned by the present author, the assignment given in Reference 184, namely, 5s $^2P_{1/2} \rightarrow$ 4p $^2P^o_{3/2}$ at *2.0596346* μm, is substantially outside the error of the measured wavelength.[184]

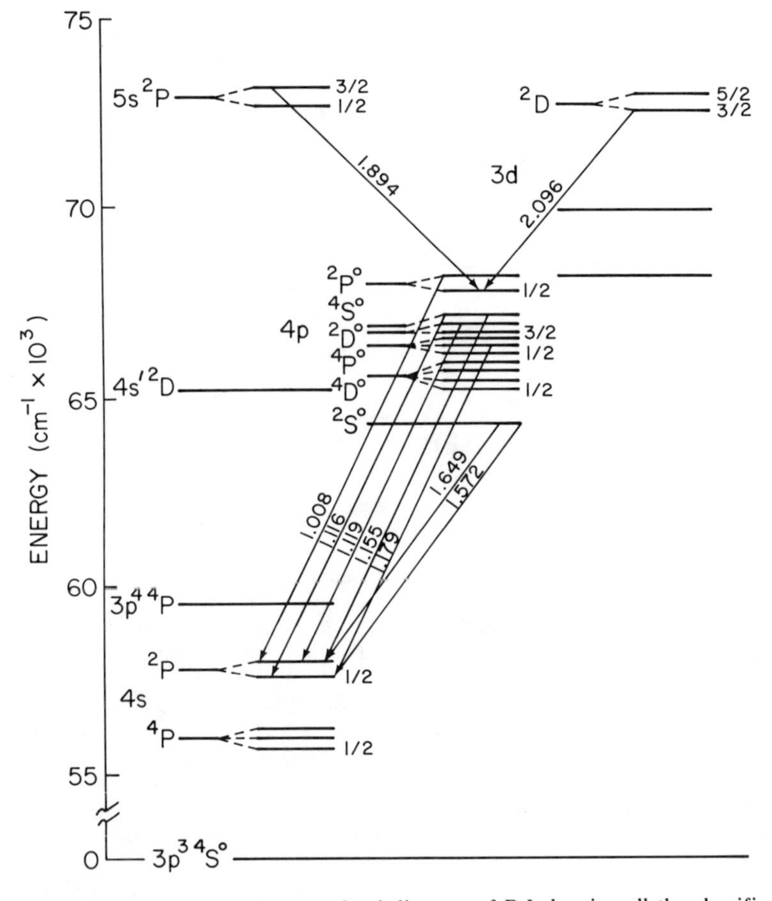

FIGURE 1.14. Partial energy level diagram of P-I showing all the classified laser transitions. Wavelengths given in microns.

Table 1.5.3
ARSENIC (FIGURE 1.15)

Wavelength (μm)	Transition assignment	Comments	Ref.
1.0455985	$5p\ ^2D^o_{5/2} \rightarrow 5s\ ^2P_{3/2}$	Pulsed; double-pulse excitation of Ar vapor at a pressure about 1 mtorr with 8 torr of He or Ne; D = 8 mm	185
1.0617063	$5p\ ^2D^o_{3/2} \rightarrow 5s\ ^2P_{1/2}$	Pulsed; as for the *1.046*-μm line, but with 4 torr of added He or Ne	185
1.1247708	$5p\ ^4P^o_{1/2} \rightarrow 5s\ ^2P_{1/2}$	Pulsed; as for the *1.056*-μm line, but with 4 torr of added Ne	185
1.1522595 or *1.1524056*	$6p\ ^2P^o_{1/2} \rightarrow 4d\ ^2P_{1/2}$ [a] or $5p\ ^4P^o_{5/2} \rightarrow 5s\ ^2P_{3/2}$	Pulsed; as for the *1.046*-μm line, but with 6 torr of added He or 4 torr of Ne. also following dissociation of 0.04 torr of $AsCl_3$ with 1.6 torr of He in a double-discharge transversely excited laser	142,185
1.2945989	$5p\ ^4D^o_{3/2} \rightarrow 5s\ ^2P_{1/2}$	Pulsed; as for the *1.12*-μm line	185
1.4124892	$5p\ ^4D^o_{5/2} \rightarrow 5s\ ^2P_{3/2}$ [b]	Pulsed; as for the *1.046*-μm line and also with 1.4 torr of added Ar	185
1.4258622 or *1.4259232*	$77121_{3/2,5/2} \rightarrow 5p'\ ^2D^o_{3/2}$ or $6p\ ^4D^o_{3/2} \rightarrow 4d\ ^4P_{3/2}$	Pulsed; by dissociation of 0.04 torr of $AsCl_3$ with 1.6 torr of He in a double-discharge, transversely excited laser	142
1.4629079	$5p\ ^2P^o_{1/2} \rightarrow 4p^4\ ^4P_{1/2}$	Pulsed; as for the *1.046*-μm line, but with 8 torr of added He	185
1.8053474	$6p\ ^4P^o_{1/2} \rightarrow 4d\ ^2D_{3/2}$	Pulsed; as for the *1.426*-μm line	142
1.8068806	$5p\ ^4P^o_{3/2} \rightarrow 4p^4\ ^4P_{3/2}$ [c]	Pulsed; as for the 1.046-μm line, but with 8 torr of added He or 4 torr of Ne	185
1.9754647	$75578.7_{3/2,5/2} \rightarrow 6p\ ^4D^o_{3/2}$	Pulsed; as for the *1.426*-μm line	142
2.0282741	$6p\ ^2P^o_{3/2} \rightarrow 6s\ ^2P_{1/2}$ [d]	Pulsed; as for the *1.426*-μm line	142
2.4466627	$4d\ ^2P_{3/2} \rightarrow 5p\ ^4P^o_{3/2}$	Pulsed; as for the *1.426*-μm line	142
2.9813368	$5p\ ^2D^o_{5/2} \rightarrow 5s'\ ^2D_{5/2}$	Pulsed; as for the *1.426*-μm line	142
5.2879276	$4d\ ^4P_{5/2} \rightarrow 5p\ ^4D^o_{3/2}$	Pulsed; as for the *1.426*-μm line	142

[a] Reference 185 lists only the second of these two assignments, however, the accuracy of the wavelength reported there is not sufficiently great to rule out the other possible assignment.

[b] Wavelength reported in Reference 185 is 1.42 ± 0.01 μm, so this may be the transition immediately below whose laser wavelength was accurately measured.[142]

[c] Wavelength reported in Reference 185 is 1.80 ± 0.01, so this may be the transition immediately above whose laser wavelength was accurately measured.[142]

[d] Several additional assignments are possible for this line; these other possibilities involve transitions between higher-lying energy levels than the ones listed.

Table 1.5.4
ANTIMONY

Wavelength (μm)	Transition assignment	Comments	Ref.
12.036591	$53443.3_{5/2} \rightarrow 52612.5^o_{3/2}$	Pulsed; by dissociation of 50 torr of Sb $(CH_3)_3$ with 1.5 torr of He in a transversely excited double-discharge laser (First pulse originally intended for preionization, but in fact serves to dissociate metal complex. Second pulse excites metal atoms.)	142

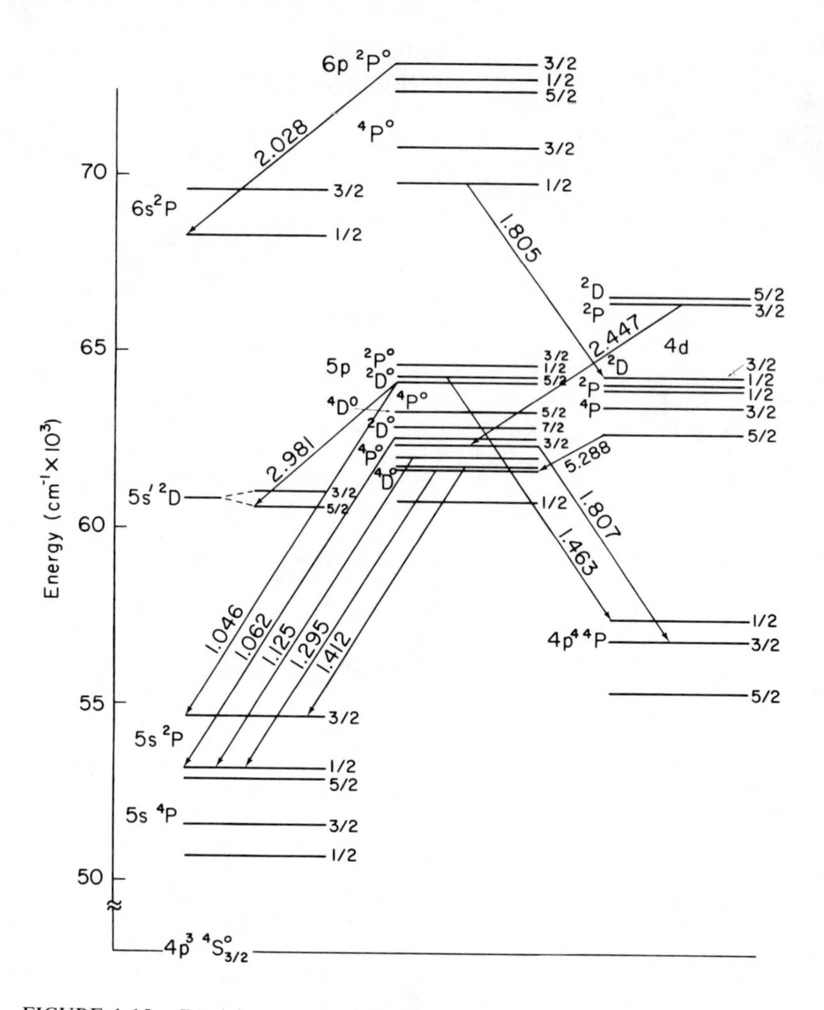

FIGURE 1.15. Partial energy level diagram of As-I showing the majority of the reported laser transitions. Wavelengths given in microns.

Table 1.5.5
BISMUTH

Wavelength (μm)	Transition assignment	Comments	Ref.
0.472252	$7s\ ^4P_{1/2} \rightarrow 6p^3\ ^2D^o_{3/2}$	Pulsed; in an ASE mode in Bi vapor in a self-heated discharge tube with added He, Ne, or Ar; optimum with 32 torr of Ne; D = 8 or 40 mm; also by optically pumping Bi atoms at densities between 10^{16} and 10^{17} cm^{-3} with a XeCl laser ($\lambda \simeq 308$ nm)[b]	186,187,[b] 696
5.3297517	$6d\ ^2D_{5/2} \rightarrow 7p(?)^o_{3/2}$	Pulsed; by dissociation of 60 torr of $Bi(CH_3)_3$ with 2.4 torr of He in a transversely excited, double-discharge laser	142

[a] Mean value of hyperfine component wavelengths listed by Meggers, W. F., Corliss, C. H., and Scribner, B. F., *Natl. Bur. Stand. (U.S.) Monogr.*, 145(1), 1975.

[b] This is a transient laser line; optical pumping results following absorption of XeCl laser photons on the bismuth resonance transition at 306.8 nm.

[c] A line at 0.475 μm listed in Reference 187 is not included as a neutral Bi laser line. This line was generated by stimulated Raman scattering of the 308-nm pump radiation.

Table 1.5.6
VANADIUM[a]

Wavelength (μm)	Transition assignment	Comments	Ref.
2.0200522	f $^6F_{9/2} \rightarrow$ v $^4G^o_{7/2}$ [b]	Pulsed; by dissociation of 90 torr of VCl_4 with 2.3 torr of He in a transversely excited double-discharge laser	122
2.4480996	z $^4P^o_{3/2} \rightarrow$ b $^4D_{1/2}$ [b]	Pulsed; by dissociation of 90 torr of VCl_4 with 2.3 torr of He in a transversely excited double-discharge laser.	122

[a] Calculated wavelengths and spectral assignments from data in Davis, D. S. and Andrew, K. L., *J. Opt. Soc. Am.*, 68, 206—235, 1978; Davis, D. S., Andrew, K. L., and Verges, L., *J. Opt. Soc. Am.*, 28, 235—242, 1978.

[b] Several different assignments are possible for this line; these other possibilities involve transtions between higher-lying levels than the ones listed here.

Table 1.6.1
OXYGEN (FIGURE 1.16)

Wavelength (μm)	Transition assignment	Comments	Ref.
0.55788939	$2p^4\ ^1S_0 \rightarrow 2p^4\ ^1D_2$ [a]	Pulsed; 5—15 torr of O with several atmospheres of Ar, Kr, or Xe or 0.1—10 torr of N_2O with several atmospheres of Ar; excited in each case with a high-energy electron beam	188-192
0.844628 0.844638 0.844672 0.844680	$3p\ ^3P_{0,2,1} \rightarrow 3s\ ^3S^o_1$ [b]	CW; in approximately 0.01—0.04 torr of O with 0.35 torr of Ne, 1.4 torr of Ar, CO or CO_2, or He; D = 7 mm; as an impurity in Br with He or Ne, also in NO and in pure O at 0.1—2 torr; D = 10 mm;[198] also in a transversely excited pin laser in 2 torr of O with 80 torr of He[204]	153,159,178, 193-209
0.88228702	$3p'\ ^1F_3 \rightarrow 3s'\ ^1D^o_2$ [c]	Pulsed; in CO_2 and Ne at 4 torr; D = 15 mm	153
2.6513946[d]	$4d\ ^5D^o \rightarrow 4p\ ^5P$ [e]	Pulsed; in a transversely excited pin laser in 2 torr of O with 80 torr of He	204
2.8944397	$4p\ ^3P_{2,1,0} \rightarrow 4s\ ^3S^o_1$	CW; 0.08 torr of O with 0.5—1.0 torr of He or Ne (probably D = 5 or 7 mm)	203,205
4.5615027	$4p\ ^3P_{2,1,0} \rightarrow 3d\ ^3D^o_{3,2,1}$	CW; as for the line immediately above and also in 0.03—0.3 torr of O with several millitorr of water vapor; D = 15 mm	203,205,206
5.9830082	$7d\ ^3D^o_{3,2,1} \rightarrow 6p\ ^3P_{2,1,0}$	CW; 0.08 torr of O with 0.5—1.0 torr of He or Ne (probably D = 5 or 7 mm)	203,205
6.8175155	$3s'\ ^3D^o_2 \rightarrow 4p\ ^3P_{(2),1}$	Pulsed; in pure O; D = 15 mm (or 75 mm?)	178
6.8598868	$5p\ ^3P_{2,1,0} \rightarrow 5s\ ^3S^o_1$	CW; 0.08 torr of O with 0.5—1.0 torr of He or Ne (probably D = 5 or 7 mm).	203,205
6.8745531	$3s'\ ^3D^o_3 \rightarrow 4p\ ^3P_2$	Pulsed; in pure O; D = 15 mm (or 75 mm?)	178
10.40312	$5p\ ^3P_{2,1,0} \rightarrow 4d\ ^3D^o_{3,2,1}$	CW; 0.08 torr of O with 0.5—1.0 torr of He or Ne (probably D = 5 or 7 mm)	203,205

[a] This is the electric-dipole-forbidden, electric-quadrupole-allowed, auroral line of atomic O which becomes weakly electric-dipole-allowed by virtue of a collision complex such as Ar-O($^1S_0$). This complex may be weakly bound, as in the case of Xe-O($^1S_0$), in which case this laser transition might equally well be referred to as a molecular laser transition.

[b] This is the O-I triplet, as shown in Reference 196. The quartet oscillation is due to the large Doppler width of the line (caused by excitation and dissociation of O molecules) and radiation trapping at the line center causing gain to occur only in the wings of the line.

[c] Tunitskii and Cherkasov[153] have some reservations about the definite assignment of this laser line they observed to O I.

[d] Mean value of wavelength of fine structure components.

[e] Laser oscillation could have been occurring on several components of this transition, but were not capable of being separately resolved.[204]

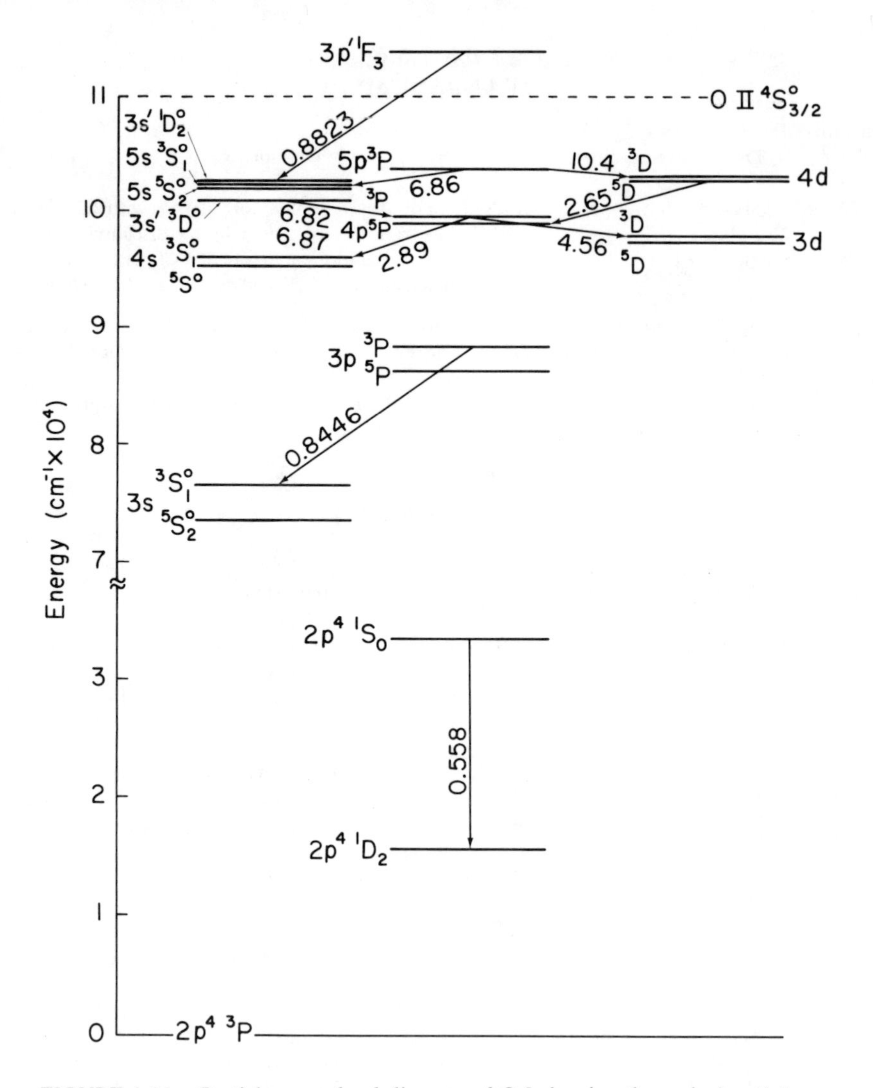

FIGURE 1.16. Partial energy level diagram of O-I showing the majority of the reported laser transitions. Wavelengths given in microns.

Table 1.6.2
SULFUR (FIGURE 1.17)

Wavelength (μm)	Transition assignment	Comments	Ref.
0.7726542	$3p^4\ ^1S_0 \rightarrow 3p^4\ ^1D_2$ [a]	Pulsed; by photodissociation of OCS with a Kr^*_2 laser (146 nm); optimum mix was 1.5 torr OCS, 25 torr SF_6 and 25 torr N_2	207
1.0455451[b]	$4p\ ^3P_2 \rightarrow 4s\ ^3S^o_1$	CW; 0.03 torr of SF_6 or 0.03 torr of SF_6 with 2 torr of He; also in H_2S with He, Ne, or air; D = 5 mm	158,159,208
1.0635993[b]	$4p'\ ^1F_3 \rightarrow 4s'\ ^1D^o_2$	CW; 0.03 torr of SF_6 or 0.03 torr of SF_6 with 2 torr of He; also in H_2S with He, Ne, or air; D = 5 mm	158,159
1.4019620[b]	$3s\ 3p^5\ ^3P^o_2 \rightarrow 4p\ ^3P_2$	Pulsed; in 0.2—0.8 torr of SF_6, CS_2, SO_2, H_2S, or OCS with 1—10 torr of He; optimum with 0.4 torr of SF_6 and 5 torr of He; D = 2.5 cm.	209

Table 1.6.2 (continued)
SULFUR (FIGURE 1.17)

Wavelength (μm)	Transition assignment	Comments	Ref.
1.5422255[b]	4f $^3F_{4,(3)}$ → 3d $^3D^o_3$ [c]	Pulsed; in 5 torr of He containing <0.1 torr of SF_6	210
1.6542665[b]	5p $^5P_3$ → 3d $^5D^o_4$	Pulsed; in 5 torr of He containing <0.1 torr of SF_6	210
2.2801247	4s′ $^3D^o_3$ → 4p $^5P_2$.	Pulsed; in a low-pressure discharge in SO_2 at about 0.1 torr with or without added He	211,345
2.436331[b]	5p $^3P_1$ → 3d $^3D^o_2$	Pulsed; as for the 1.54-μm line	210
3.389503	4s′ $^3D^o_3$ → 4p $^3P_2$ [d]	Pulsed; as for the *2.28*-μm line	211(E)

Note: Lines listed in previous compilations[578,579] at 0.516032 and 0.521962 μm are ionized sulfur lines.[213]

[a] This transition is electric-dipole-forbidden, electric-quadrupole-allowed.

[b] Measured wavelengths and spectral assignments are from Jacobsson, L. R., *Ark. Fys.*, 34, 19—31, 1966.

[c] The 4f $^3F_4$ and 4f $^3F_3$ levels are separated by only 0.006 cm^{-1}, so these two possible assignments lie within a Doppler width of each other.

[d] Incorrectly assigned in Reference 211, assignment here made by present author.

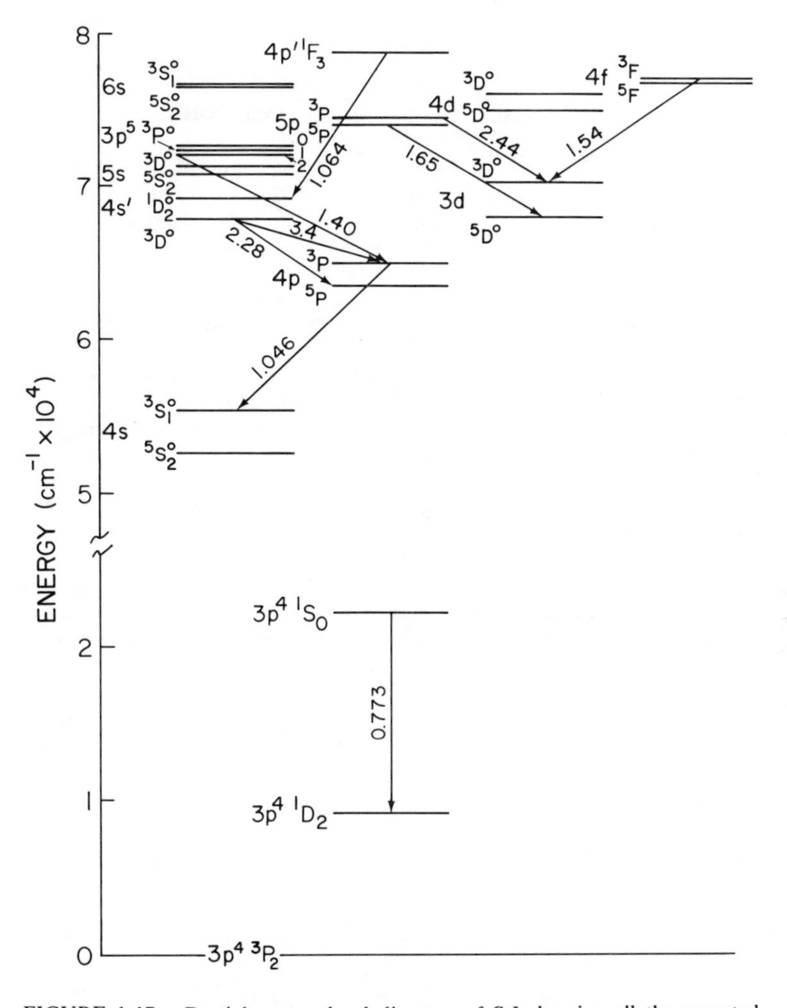

FIGURE 1.17. Partial energy level diagram of S-I showing all the reported laser transitions. Wavelengths given in microns.

Table 1.6.3
SELENIUM

Wavelength (μm)	Transition assignment	Comments	Ref.
0.48884122	$4p^4\ ^1S_0 \rightarrow 4p^4\ ^3P_1$ [a]	Pulsed; by photodissociation of carbonyl selenide with a Xe^*_2 laser (172 nm); operating mixture 1 torr of OCSe with 50 torr of CO[b]	214,215
0.77700379	$4p^4\ ^1S_0 \rightarrow 4p^4\ ^1D_2$	Pulsed; by photodissociation of carbonyl selenide with a Xe^*_2 laser (172 nm); operating mixture 1 torr of OCSe with 50 torr of CO[b]	214,215
6.3687777	$5s'\ ^3D^o_3 \rightarrow 5p\ ^3P_2$	Pulsed; by dissociation of 80 torr of $Se(CH_3)_2$ with 2.0 torr of He in a transversely excited double-discharge laser	122

[a] This transition is electric-dipole-forbidden; because $\Delta S \neq 0$, it becomes electric-quadrupole-allowed through deviations from true LS coupling.
[b] CO buffer used to quench any population in the $Se(^3P_1)$ or $(^1D_2)$ lower laser levels produced either by direct photolysis or by quenching of higher-lying levels.
[c] This transition is electric-dipole-forbidden, electric-quadrupole-allowed.

Table 1.6.4
TELLURIUM[a]

Wavelength (μm)	Transition assignment	Comments	Ref.
0.5454	?	Pulsed; TeI at a temperature of 125—250°C with 0.1—0.25 torr of Ne; D = 6 mm	381
0.5640	?	Pulsed; 0.001—0.002 torr of TeI with 0.2 torr of Ne; D = 3 mm	381
0.63497	?	CW; in a TeI-Ne mixture	380
3.1653653	$5d\ ^3D^o_3 \rightarrow 6p\ ^3P_2$ [c]	Pulsed; by dissociation of 60 torr of $Te(CH_3)_2$ with 1.6 torr of He in a transversely excited double-discharge laser	122(E)
6.7631543	$5d\ ^5D^o_3 \rightarrow 6p\ ^5P_2$ [c]	Pulsed; as for above line, but with 40 torr of $Te(CH_3)_2$ and 1 torr of He	122(E)
7.8071691	$5d\ ^5D^o_4 \rightarrow 6p\ ^5P_3$ [c]	Pulsed; as for above line, but with 40 torr of $Te(CH_3)_2$ and 1 torr of He	122(E)

[a] Spectral assignments and calculated wavelengths are from data in Morillon, C. and Verges, J., *Phys. Scripta*, 12, 129—144, 1975.
[b] Possibly an ionized TeI line.
[c] Although Chou and Cool[122] assigned this transition to TeI, they inadvertently listed a spectral assignment of TeII fairly close to the measured wavelength. The transition listed here is given by the present author as the likely correct assignment; its wavelength agrees very well with the measured wavelength given in Reference 122 and involves low-lying levels.

Table 1.6.5
THULIUM[a] (FIGURE 1.18)

Wavelength (μm)	Transition assignment	Comments	Ref.
1.3058983[b]	$(4,5/2)'_{13/2} \rightarrow (6,1/2)^{o}_{11/2}$ [c]	Pulsed; in Tm vapor above 800°C (Tm vapor pressure ≃0.1 torr) with 2.5 torr of He, 1.5 torr of Ne, or 0.8 torr of Ar	216
1.3104227	$(6,1/2)^{o}_{13/2} \rightarrow (6,3/2)_{15/2}$	Pulsed; in Tm vapor above 800°C (Tm vapor pressure ≃0.1 torr) with 2.5 torr of He, 1.5 torr of Ne, or 0.8 torr or Ar	216
1.3383700	$^{3}[11/2]^{o}_{13/2} \rightarrow (6,3/2)_{15/2}$	Pulsed; in Tm vapor above 800°C (Tm vapor pressure ≃0.1 torr) with 2.5 torr of He, 1.5 torr of Ne, or 0.8 torr of Ar	216
1.4343722	$^{3}[11/2]^{o}_{11/2} \rightarrow (6,3/2)_{11/2}$	Pulsed; in Tm vapor above 800°C (Tm vapor pressure ≃0.1 torr) with 2.5 torr of He, 1.5 torr of Ne, or 0.8 torr of Ar	216
1.4489080	$(6,3/2)^{o}_{11/2} \rightarrow (6,5/2)_{13/2}$	Pulsed; in Tm vapor above 800°C (Tm vapor prassure ≃0.1 torr) with 2.5 torr of He, 1.5 torr of Ne, or 0.8 torr of Ar	216
1.4998810[b] or *1.5036834*	$^{1}[11/2]^{o}_{11/2} \rightarrow (6,5/2)_{11/2}$ $(2,3/2)_{5/2} \rightarrow {}^{3}[5/2]^{o}_{5/2}$ [c]	Pulsed; in Tm vapor above 800°C (Tm vapor pressure ≃0.01 torr) with 2.5 torr of He, 1.5 torr of Ne, or 0.8 torr of Ar	216
1.6383650[b]	$^{3}[9/2]^{o}_{11/2} \rightarrow (6,5/2)_{11/2}$	Pulsed; in Tm vapor above 800°C (Tm vapor pressure ≃0.1 torr) with 2.5 torr of He, 1.5 torr of Ne, or 0.08 torr of Ar	216
1.6758663[b]	$^{3}[9/2]^{o}_{11/2} \rightarrow (7/2,2)_{11/2}$	Pulsed; in Tm vapor above 800°C (Tm vapor pressure ≃0.1 torr) with 2.5 torr of He, 1.5 torr of Ne, or 0.8 torr of Ar	216
1.7323684	$^{1}[11/2]^{o}_{11/2} \rightarrow (7/2,2)_{9/2}$	Pulsed; in Tm vapor above 800°C (Tm vapor pressure ≃0.1 torr) with 2.5 torr of He, 1.5 torr of Ne, or 0.8 torr of Ar	217
1.9589851	$^{3}[11/2]^{o}_{1/2} \rightarrow (6,3/2)_{13/2}$	Pulsed; in Tm vapor above 800°C (Tm vapor pressure ≃0.1 torr) with 2.5 torr of He, 1.5 torr of Ne, or 0.8 torr of Ar	216
1.9722834[b]	$(4,3/2)'_{9/2} \rightarrow {}^{3}[9/2]^{o}_{7/2}$	Pulsed; in Tm vapor above 800°C (Tm vapor pressure ≃0.1 torr with 2.5 torr of He, 1.5 torr of Ne, or 0.8 torr of Ar	216
1.9947227	$(6,1/2)^{o}_{11/2} \rightarrow (6,3/2)_{13/2}$	Pulsed; in Tm vapor above 800°C (Tm vapor pressure ≃0.1 torr) with 2.5 torr of He, 1.5 torr of Ne, or 0.8 torr of Ar	
2.1059135[b] or *2.1129476*	$(5,5/2)_{11/2} \rightarrow {}^{3}[11/2]^{o}_{13/2}$ or $(4,1/2)^{o}_{9/2} \rightarrow (4,5/2)_{9/2}$	Pulsed; in Tm vapor above 800°C (Tm vapor pressure ≃0.1 torr) with 2.5 torr of He, 1.5 torr of Ne, or 0.8 torr of Ar	216
2.3851957[b]	$^{3}[7/2]^{o}_{9/2} \rightarrow (7/2,2)_{9/2}$	Pulsed; in Tm vapor above 800°C (Tm vapor pressure≃0.1 torr) with 2.5 torr of He, 1.5 torr of Ne, or 0.8 torr of Ar	216

Note: Where a level designation is given with a prime, it is to indicate that this is the second-lowest energy level with this designation and has a different core configuration.

[a] Calculated wavelengths and spectral assignments taken from data in Sugar, J., Meggers, W. F., and Camus, P., *J. Res. Natl. Bur. Stand.*, 77A, 1-43, 1973.

[b] Difficult laser line to excite; wavelength not measured very accurately in Reference 216; spectral assignments seems likely to be correct, however, as it involves low-lying levels, and calculated wavelength is in good agreement with experimental values.

[c] Assignment suggested by the present author.

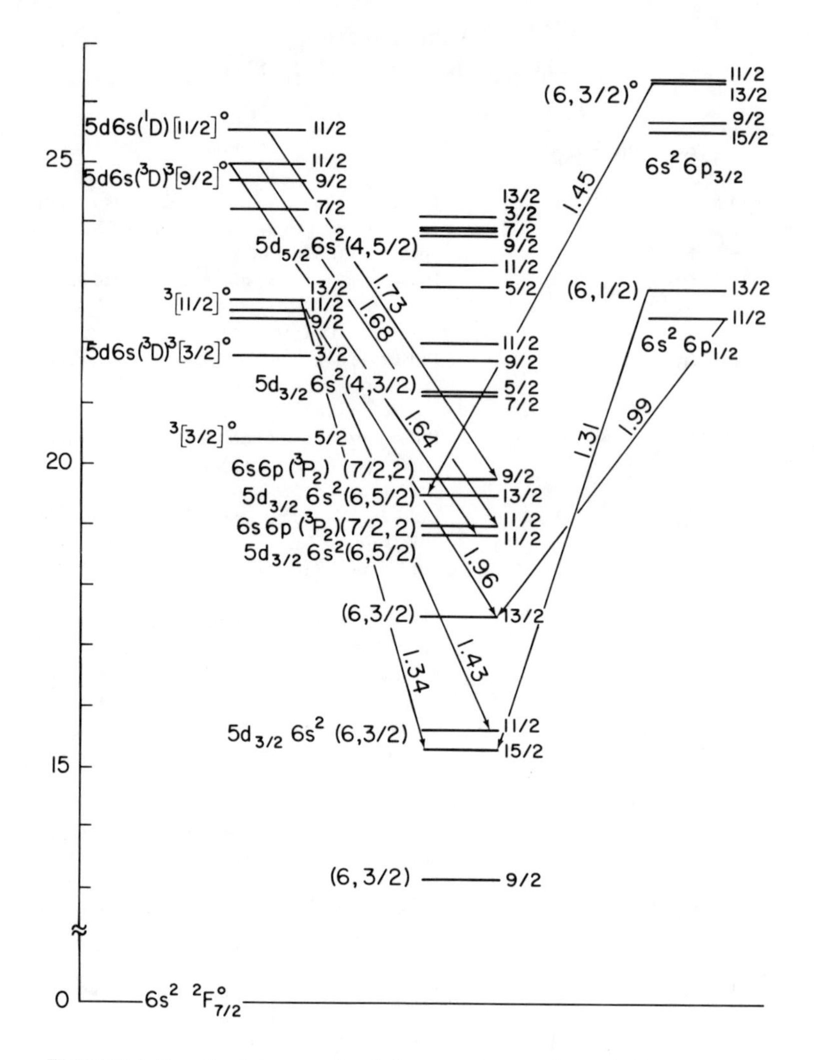

FIGURE 1.18. Partial energy level diagram of neutral atomic Tm showing the majority of the reported laser transitions. Wavelengths given in microns.

1.7A Group VII
Table 1.7.1
FLUORINE[a] (FIGURE 1.19)

Wavelength (μm)	Transition assignment	Comments	Ref.
0.6239651	3p $^4S^o_{3/2}$ → 3s $^4P_{5/2}$	Pulsed; in a 100:1 He-NF_3 or He-F_2 mixture at pressures from 0.25—5 atm excited in a TEA laser	218,219
0.6348508	3p $^4S^o_{3/2}$ → 3s $^4P_{3/2}$	Pulsed; in a vacuum UV photopreionized TEA laser with a 50:1 He-F_2 mixture at 160—1600 torr.	218,219,220
0.6413651	3p $^4S^o_{3/2}$ → 3s $^4P_{1/2}$	Pulsed; as for the 0.624-μm line	218,219
0.6966349	3p $^2P^o_{1/2}$ → 3s $^2P_{3/2}$	Pulsed; in a 100:1 He-NF_3 mixture at optimum pressures from 100—150 torr in a transversely excited pin laser	220,221,222,379

1.7A Group VII

Table 1.7.1 (continued)
FLUORINE[a] (FIGURE 1.19)

Wavelength (μm)	Transition assignment	Comments	Ref.
0.7037469	3p $^2P^o_{3/2}$ → 3s $^2P_{3/2}$	Pulsed; in flowing CF_4, SF_6, or C_2F_6 at 0.03—0.1 torr with 2—10 torr He; D = 25 mm; also in 0.05 torr of HF with 0.3 torr of He; He essential with HF; also in a vacuum-UV photopreionized TEA laser with a 50:1 He-F_2 mixture at 160—1600 torr	160,219,221,222, 223,224,225, 379
0.7127890	3p $^2P^o_{1/2}$ → 3s $^2P_{1/2}$	Pulsed; as for the 0.704-μm line	160,219,221,223, 225,379
0.7202360	3p $^2P^o_{3/2}$ → 3s $^2P_{1/2}$	Pulsed; as for the 0.704-μm line	220-223,379
0.7309033	3p′ $^2F_{7/2}$ → 3s′ $^2D_{5/2}$	Pulsed; in a 100:1 He-NF_3 or He-F_2 mixture at high pressure, with optimum ~1.1 atom, excited in a TEA laser	219
0.7311019	3p $^2S^o_{1/2}$ → 3s $^2P_{3/2}$	Pulsed; in a flowing He or He/Ar mixture at 5 torr containing a small amount of Fl	220-222,225,379
0.7398688	3p $^4P^o_{5/2}$ → 3s $^4P_{5/2}$	Pulsed; in a vacuum-UV photopreionized TEA laser with 50:1 He-F_2-mixture at 160—1600 torr.	218,220
0.7425645	3p $^4P^o_{1/2}$ → 3s $^4P_{3/2}$	Pulsed; in a double-discharge TEA laser with a 0.1% F in He mixture at 0.5—2 atm	226
0.7482723	3p $^4P^o_{3/2}$ → 3s $^4P_{3/2}$	Pulsed; in a double-discharge TEA laser with a 0.1% F in He mixture at 2.5 atm	226
0.7489155	3p $^2S^o_{1/2}$ → 3s $^2P_{1/2}$	Pulsed; in a 100:1 He-F_2 mixture at 10—50 torr; D = 6 mm	220,222,379
0.7514919	3p $^4P^o_{1/2}$ → 3s $^4P_{1/2}$	Pulsed; in a double-discharge TEA laser with a 0.1% F in He mixture at 1—3 atm	226
0.7552235	3p $^4P^o_{5/2}$ → 3s $^4P_{3/2}$	Pulsed; as for the 0.7399-μm line	218—220
0.7754696	3p $^2D^o_{5/2}$ → 3s $^2P_{3/2}$	Pulsed; in a 100:1 He-F_2 mixture at 1—5 torr; D = 6 mm; also in a vacuum UV photo-preionized TEA laser with a 50:1 He-F_2 mixture at 160—1600 torr	220,222,379
0.7800212	3p $^2D^o_{3/2}$ → 3s $^2P_{1/2}$	Pulsed; in 0.05 of HF with 0.3 torr of He; D = 10 cm; also in a 100:1 He-F_2 mixture at 1—5 torr; D = 6 cm, or excited in a TEA laser	220,222,224,379
7.435[b]		Pulsed; in a mixture of UF_6 and He or WF_6 and He in a transversely excited double-discharge pin laser	122

Note: Reference 224 reports 20 unidentified lines, believed to be from atomic F, at wavelengths between 1.5900 and 9.3462 μm, but lists no actual wavelengths. Beck et al.[581] incorrectly list several F lines twice at slightly differently wavelengths. For further details of the F laser, see the text.

[a] Measured wavelengths and spectral assignments are taken from: Liden, K., The arc spectrum of fluorine, *Ark. Fys.*, 229-267, 1949.

[b] Measured wavelength from Reference 122, ± 0.004 μm.

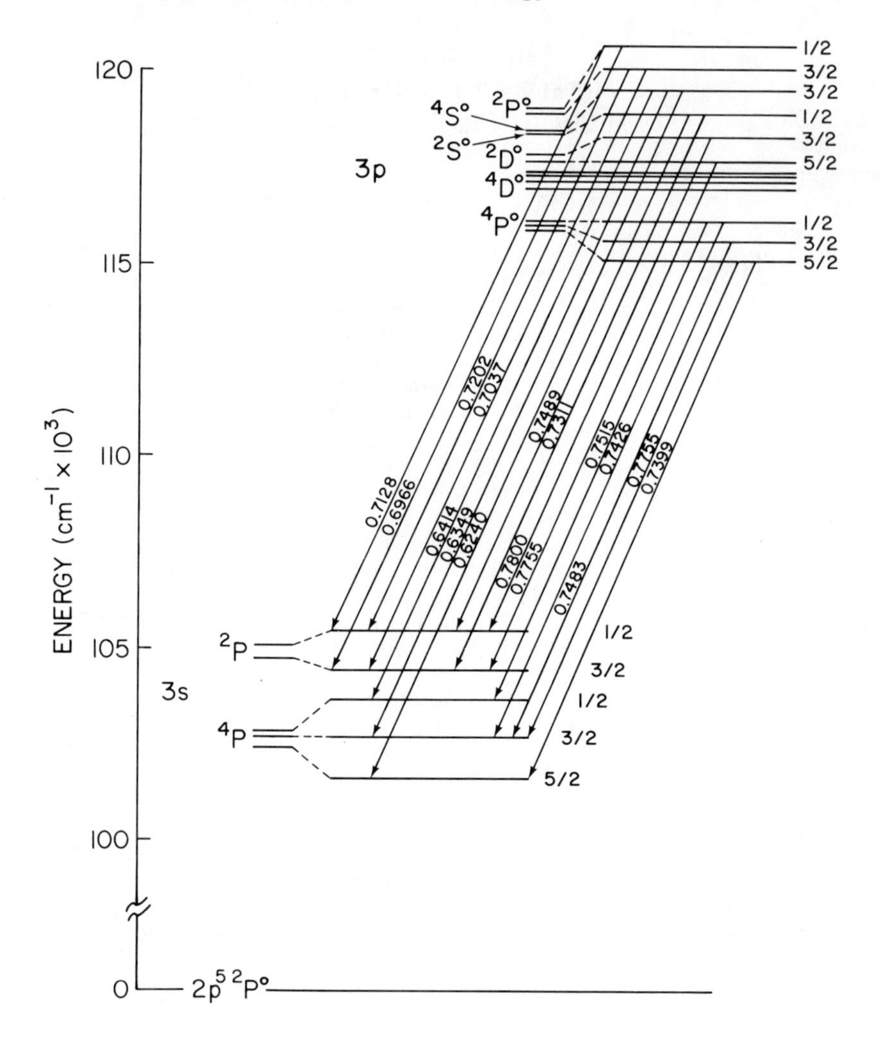

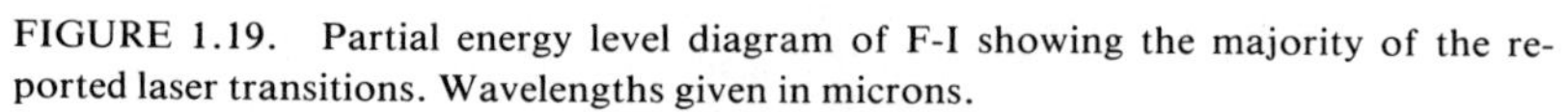

FIGURE 1.19. Partial energy level diagram of F-I showing the majority of the reported laser transitions. Wavelengths given in microns.

Table 1.7.2
CHLORINE[a] (FIGURE 1.20)

Wavelength (μm)	Transition assignment	Comments	Ref.
0.9452098	4p $^2P^o_{3/2}$ → 4s $^2P_{1/2}$	CW; in Cl at 0.01—0.08 torr with 0.3—3 torr of He or Ne; D = 6 mm; also in Freon at 0.001 torr with 0.8 torr of Ne; D = 7 mm	228,230
1.38633	5s $2[2]_{5/2}$ → 4p $^4D^o_{5/2}$	Pulsed; in 0.3 torr of HCl with 0.1 torr of He or Ne; D = 14 mm	230
1.38931	5s $2[2]_{3/2}$ → 4p $^4D^o_{3/2}$	Pulsed; in Cl at 0.01—0.08 torr with 0.3—3 torr of He or Ne; D = 6 mm; also in Freon at 0.001 torr of Ne; D = 7 mm; pulsed; in 0.3 torr of HCl with 0.1 torr of He or Ne; D = 14 mm	230
1.58697	3d $^4F_{9/2}$ → 4p $^4D^o_{7/2}$	CW; in a mixture of Freon (CCl_2F_2) and He at 3.3 torr	160,231,232
1.97553	3d $^4D_{7/2}$ → 4p $^4P^o_{5/2}$	CW; in 0.1 torr of Cl or in HCl or 0.3 torr of silicon tetrachloride with 0.1 torr of He or Ne	162,230,231,233, 234,235

Table 1.7.2 (continued)
CHLORINE[a] (FIGURE 1.20)

Wavelength (μm)	Transition assignment	Comments	Ref.
2.01994	3d $^4D_{5/2} \rightarrow$ 4p $^4P_{3/2}$	CW; in 0.1 torr of Cl or in HCl or 0.3 torr of silicon tetrachloride with 0.1 torr of He or Ne	233,162,230,234, 231
2.44700	3d $^4D_{7/2} \rightarrow$ 4p $^4D^o_{7/2}$	Pulsed; in 0.3 torr of HCl with 0.1 torr of He or Ne; D = 14 mm, or CW in 0.09 torr of Cl with 1.5—7.2 torr of He; D = 25 mm	160,230,235
3.066080	5p $^4D^o_{5/2} \rightarrow$ 5s′ $[1]_{3/2}$	CW; in 0.09 torr of Cl with 2.1 torr of Ar; D = 25 mm[b]	235
3.543052	5p $^4P^o_{5/2} \rightarrow$ 3d $^4P_{5/2}$	Pulsed; in 0.6 torr of Cl with 17—30 torr of He; D = 25 mm	160
3.796602	($^1D_2$) 4p $^2P^o_{3/2} \rightarrow$ 5s $[2]_{3/2}$	Pulsed; in 0.6 torr of Cl with 17—30 torr of He; D = 25 mm	160

Note: Lines at 1.589, 2.499, 2.535, 2.602, 2.784, and 3.801 μm observed to oscillate CW in discharge through He-Freon (CCl_2F_2) mixtures [231] are also possibly Cl-I lines.

[a] Measured wavelengths and spectral assignments are taken from Radziemski, L. J., Jr. and Kaufman, V., *J. Opt. Soc. Am.*, 59, 429—443, 1969; and Humphreys, C. S. and Paul, E., Jr., *J. Opt. Soc. Am.*, 62, 432—439, 1972.

[b] The upper level of this laser transition appears to be excited selectively by excitation transfer from the 4s state of Ar-I.[235]

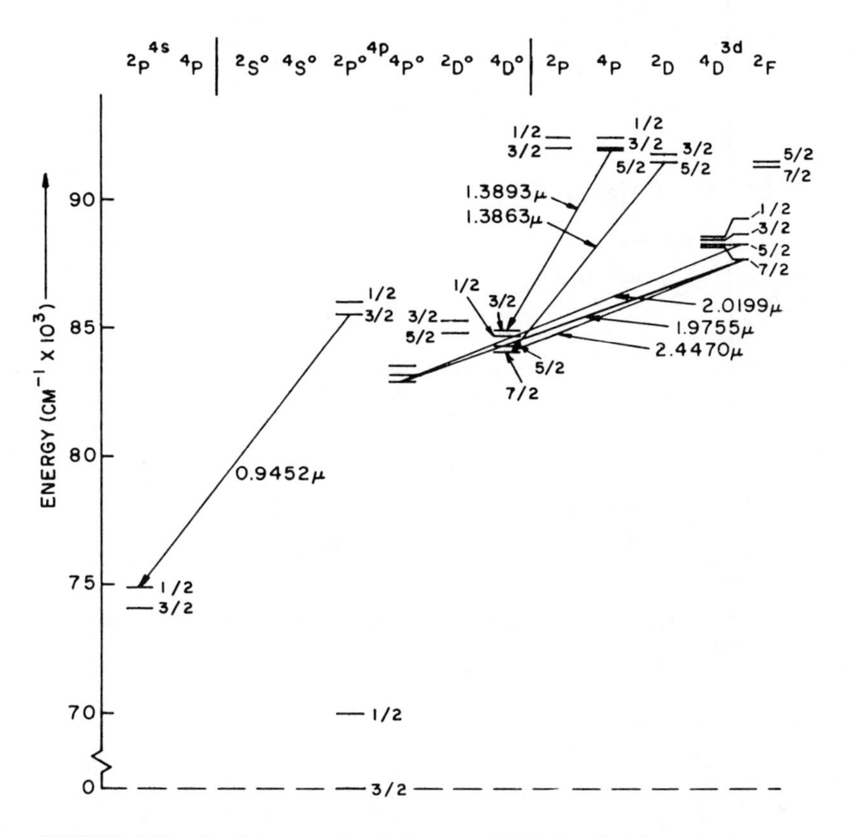

FIGURE 1.20. Partial energy level diagram of Cl-I showing the majority of the reported laser transitions.

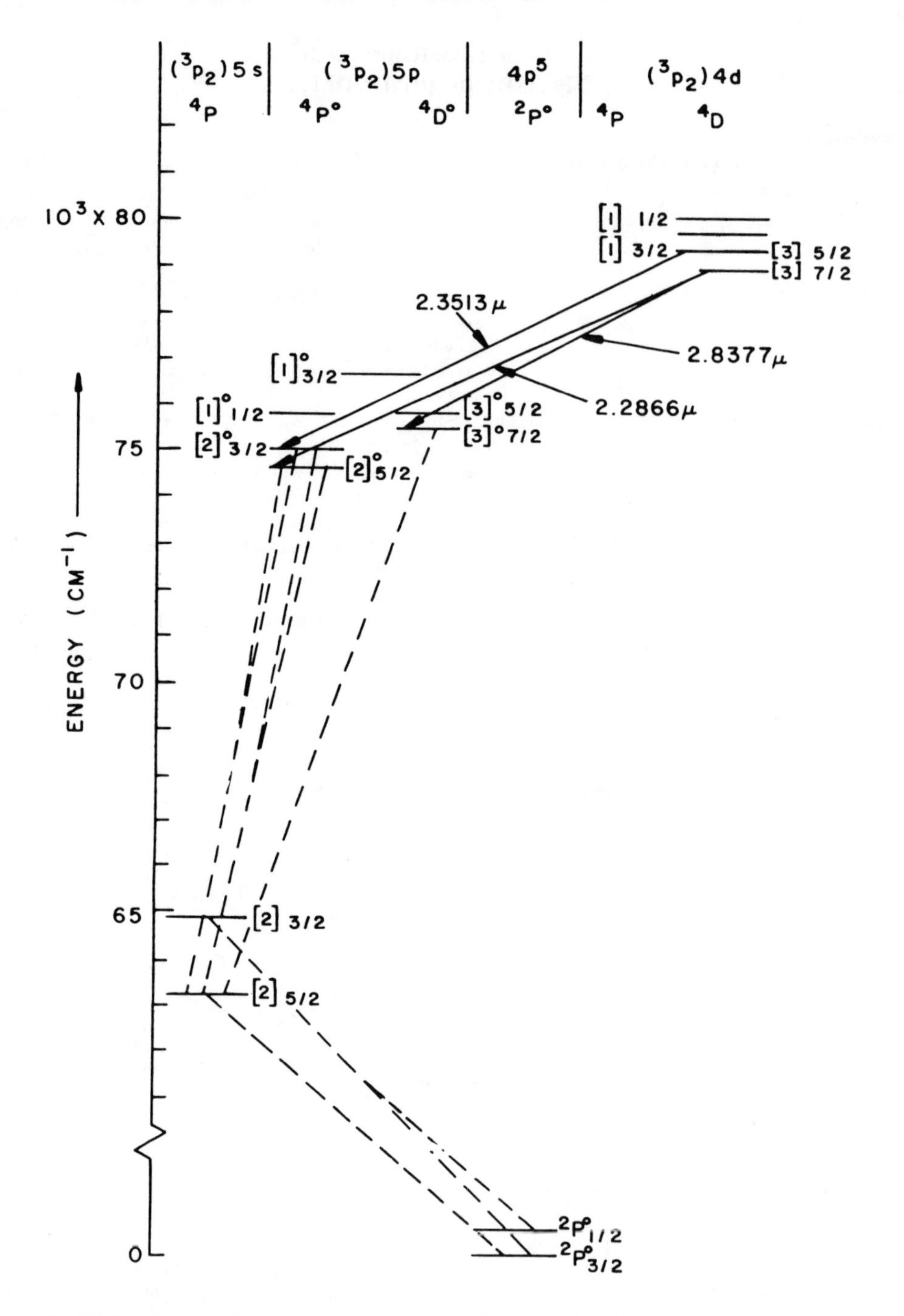

FIGURE 1.21. Partial energy level diagram of Br-I showing the majority of the reported laser transitions.

Table 1.7.3
BROMINE[a] (FIGURE 1.21)

Wavelength (μm)	Transition assignment	Comments	Ref.
1.973362	$4d\ ^4F_{9/2} \rightarrow 5p\ ^4D^o_{7/2}$	CW; in a 14:1 $CBrF_3$-He mixture at 2.8 torr; also pulsed in a 1:100 Br-He mixture at 51 torr; D = 10 mm	231,345
2.286565	$4d\ ^4D_{7/2} \rightarrow 5p\ ^4P^o_{5/2}$	CW; in 0.3 torr of hydrogen bromide; D = 12 mm	236
2.351215	$4d\ ^4D_{5/2} \rightarrow 5p\ ^4P^o_{3/2}$	CW; in 0.3 torr of hydrogen bromide; D = 12 mm	236
2.7135274[b]	$4p^5\ ^2P^o_{1/2} \rightarrow 4p^5\ ^2P^o_{3/2}$[c]	Pulsed; by flash photolysis of IBr at 0.5—5 torr; D = 8 mm; also by flash photolysis of CF_3Br, optimum pressure about 40 torr; D = 7 mm, and as a result of the chemical reaction: $I(5p^5\ ^2P^o_{1/2}) + Br_2 \rightarrow IBr + Br(4p^5\ ^2P^o_{1/2})$	376—378
2.837716	$4d^4D_{7/2} \rightarrow 5p\ ^4D^o_{7/2}$	CW; as for the 2.287-μm line	236

Note: Lines near 0.8446 μm originally thought to be Br lines[159] are in fact O lines[196] (see Table 1.6.1).

[a] Unless otherwise indicated, measured wavelength and spectral assignments are taken from Humphreys, C. J. and Paul, E., Jr., *J. Opt. Soc. Am.*, 62, 432—439, 1972.
[b] Calculated vacuum wavelength from Tech, J. L., *J. Res. Natl. Bur. Stand.*, 67A, 505—554, 1963.
[c] Magnetic dipole transition.

Table 1.7.4
IODINE[a] (FIGURE 1.22)

Wavelength (μm)	Transition assignment	Comments	Ref.
0.98[b]	?	Pulsed; in 0.1 torr of I with a few torr of He; D = 5 mm	237
1.01[b]	?	Pulsed; in 0.1 torr of I with a few torr of He; D = 5 mm	237
1.03[b]	?	Pulsed; in 0.1 torr of I with a few torr of He; D = 5 mm	237
1.06[b]	?	Pulsed; in 0.1 torr of I with a few torr of He; D = 5 mm	237
1.3152443	$5p^5\ ^2P^o_{1/2} \rightarrow 5p^5\ ^2P_{3/2}$[c]	Pulsed; by flash photolysis of tens of torr of various I-containing organic compounds, such as CF_3I, CH_3I, C_3F_7I, with or without a noble gas buffer; D is not critical; also in 0.18 torr of CF_3I with 70 torr of N in a vacuum-UV photopreionized TEA laser;[295] also operates CW as a result of a chemical reaction[332,335,336] and with optical pumping[325]	238—340
1.4545941	$(^3P_0)6d[2]_{3/2} \rightarrow (^3P_2)7p[1]^o_{3/2}$	Pulsed; in 0.3 torr of HI; D = 14 mm	230
1.5533401	$(^3P_2)7s[2]_{3/2} \rightarrow (^3P_2)6p[1]^o_{1/2}$[d]	CW; in 0.05 torr of CH_2I_2 with or without added Ar; D ≃ 12 mm	341
2.598577[e]	$5d'[2]_{3/2} \rightarrow 6p''[1]^o_{3/2}$	CW; in 0.3 torr of HI with 0.3 torr of Ne; D = 14 mm	230
2.757298[e]	$5d[0]_{1/2} \rightarrow 6p[2]^o_{3/2}$	CW; in HI; D = 12 mm or in 0.5 torr of I with 10 torr of He; D = 12.7 mm	342,343

Table 1.7.4 (continued) IODINE[a] (FIGURE 1.22)

Wavelength (μm)	Transition assignment	Comments	Ref.
3.036119[e]	$5d[2]_{3/2} \rightarrow 6p[1]^o_{3/2}$	CW; in I vapor, CH_3I, CF_3I, or HI, with added He, Ar or Xe; optimum 0.5 torr of I with 5 torr of He; D = 2—8 cm	343,344
3.236285[e]	$5d[2]_{5/2} \rightarrow 6p[1]^o_{3/2}$	CW; in HI or I vapor with or without added He; D = 13 mm	131,234, 342,343
3.429573[e,f]	$5d[4]_{7/2} \rightarrow 6p[3]^o_{5/2}$	CW; in I, CH_3I, CF_3I, or HI, with added Helium, Ar, or Xe; optimum 0.4 torr of I with 5 torr of He; D = 2—8 cm	131,234,342 —344
4.3321362	$(^3P_2)5d[1]_{1/2} \rightarrow (^3P_2)6p[2]^o_{3/2}$	CW; in 0.05 torr of CH_2I_2 with added Ar; Ar was essential to obtain laser action; D ≃ 12 mm	341
4.8584726	$(^3P_1)5d[1]_{3/2} \rightarrow (^3P_1)6p[2]^o_{5/2}$	CW; in I vapor, CH_3I, CF_3I, or HI, with added He, Ar, or Xe; optimum, 0.4 torr of I with 5 torr of He; D = 2—8 cm	343,344
4.8629144[f]	$(^3P_2)5d[4]_{3/2} \rightarrow (^3P_2)6p[3]^o_{7/2}$	CW; in I vapor, CH_3I, CF_3, I, or HI, with added He, Ar, or Xe; optimum 0.4 torr of I with 5 torr of He; D = 2—8 cm	343,344
5.498705	$(^3P_0)5d[2]_{5/2} \rightarrow (^3P_0)6p[1]^o_{3/2}$	CW; in I vapor; CH_3I, CF_3, I, or HI, with added He, Ar, or Xe; optimum 0.4 torr of I with 5 torr of He; D = 2—8 cm	343,344
6.721966	$(^1D_2)6s[2]_{3/2} \rightarrow (^3P_2)6p[1]^o_{3/2}$	CW; in I vapor, CH_3I, CF_3, I, or HI, with added He, Ar, or Xe; optimum 0.4 torr of I with 5 torr of He; D = 2—8 cm	343,344
6.9035028	$(^3P_1)5d[3]_{7/2} \rightarrow (^3P_1)6p[2]^o_{5/2}$	CW; in I vapor, CH_3I, CF_3, I, or HI, with added He, Ar, or Xe; optimum 0.4 torr of I with 5 torr of He; D = 2—8 mm	343,344
9.0195724	$(^3P_2)5d[3]_{7/2} \rightarrow (^3P_2)6p[2]^o_{5/2}$	CW; in I vapor, CH_3I, CF_3, I, or HI, with added He, Ar, or Xe; optimum 0.4 torr of I with 5 torr of He; D = 2—8 mm	343,344

[a] Unless otherwise indicated, calculated wavelengths and spectral assignments are taken from data in Minnhagen, L., *Ark. Fys.*, 21, 415—478, 1962.
[b] This line may be an ionized I transition.
[c] Magnetic dipole transition. For further discussion of laser systems based on this transition, see the text.
[d] This assignment is quite likely to be correct, although Reference 341 lists others. For example, a possible alternative assignment is $8p[2]^o_{5/2} \rightarrow 7s[2]_{3/2}$ at *1.5533932* μm.
[e] Measured wavelength and spectral assignment from Humphreys, C. J. and Paul, E., Jr., *J. Opt. Soc. Am.*, 62, 432—439, 1972.
[f] Strongest transitions in CW gas discharge excitation.

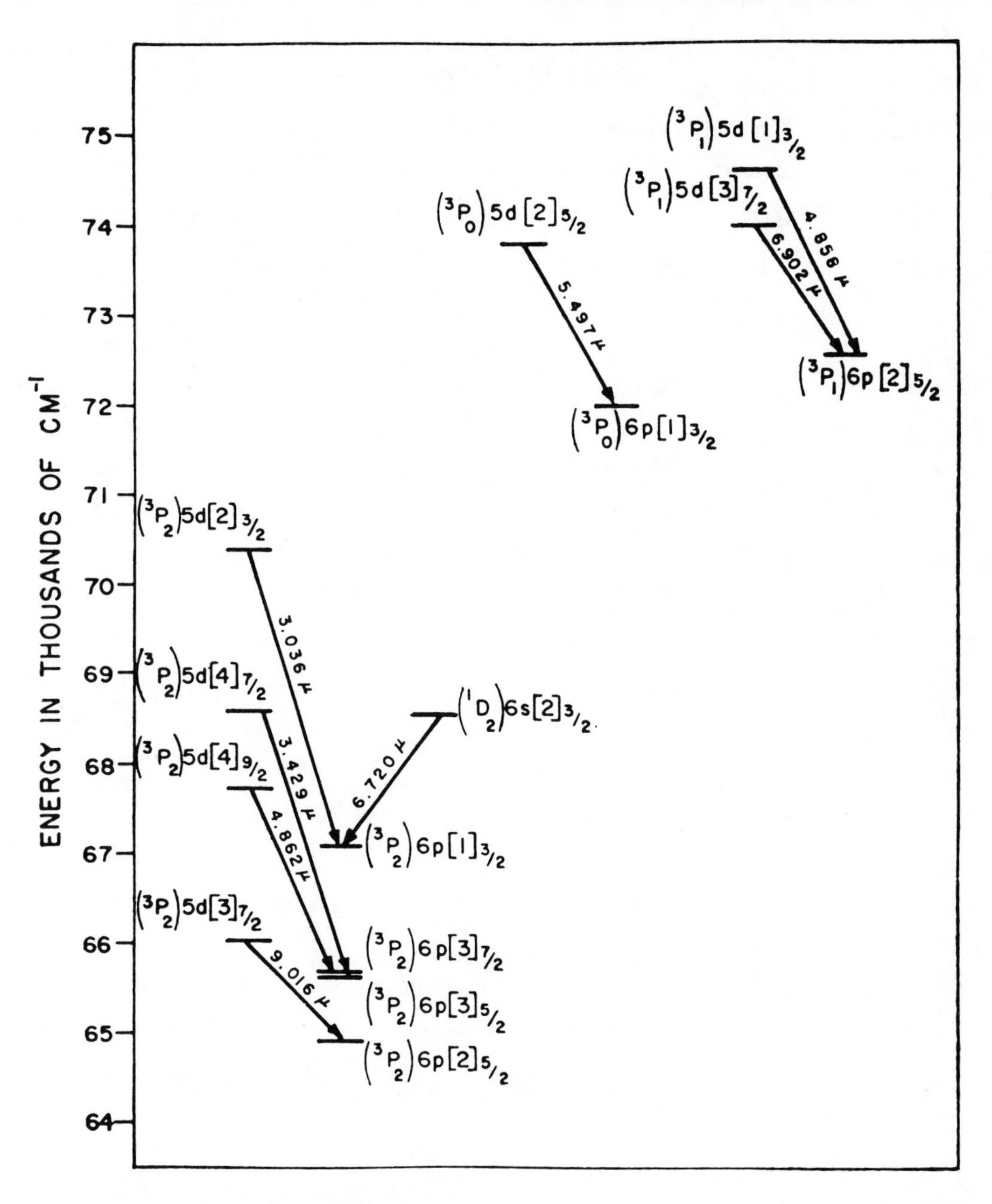

FIGURE 1.22. Partial energy level diagram of neutral atomic I showing the laser lines reported in CW gas discharge-excited operation. (From Kim, H., Paananen, R., and Hanst, P., *IEEE J. Quant. Electron.*, QE-4, 385, 1968. With permission.)

Table 1.7.5
YTTERBIUM[a] (FIGURE 1.23)

Wavelength (μm)	Transition assignment	Comments	Ref.
1.0324559	5d 6s $^1D_2 \rightarrow$ 6s 6p $^3P^o_1$	Pulsed; in Yb vapor above 500°C with 2.5 torr of He, 1.5 torr Ne, or 0.8 torr of Ar	217
1.255136	5d 6s $^1D_2 \rightarrow$ 6s 6p $^3P^o_2$	Pulsed; in Yb vapor above 500°C with 2.5 torr of He, 1.5 torr Ne, or 0.8 torr of Ar	217
1.4283698	$6s^2$ 6p $^1D_2 \rightarrow$ 5d $6s^2$ $^1D^o_2$	Pulsed; in Yb vapor above 500°C with 2.5 torr of He, 1.5 torr Ne, or 0.8 torr of Ar	217
1.4793059	5d 6s $^3D_2 \rightarrow$ 6s 6p $^3P^o_1$	Pulsed; in 0.001—1 torr of Yb vapor with from 0.01—760 torr of noble gas buffer — usually He; D = 7 mm	346
1.7459155	5d $6s^2$ $^3D^o_1 \rightarrow$ 6s 7s $^3S_1$	Pulsed; as for the *1.03*-μm line	217
1.798400	5d 6s $^3D_3 \rightarrow$ 6s 6p $^3P^o_2$	Pulsed; as for the *1.479*-μm line	346
1.9835173	5d 6s $^3D_2 \rightarrow$ 6s 6p $^3P^o_2$	Pulsed; as for the *1.03*-μm line	216,217
2.0041827	$6s^2$ 6p $^1D_2 \rightarrow$ 5d $6s^2$ $^1F^o_3$	Pulsed; as for the *1.03*-μm line	216,217
2.1186997	5d $6s^2$ $^1P^o_1 \rightarrow$ 6s 7s $^3S_1$	Pulsed; as for the *1.03*-μm line	216,217
4.8021974	5d 6s $^3D_3 \rightarrow$ 5d $6s^2$ $^3P^o_2$	Pulsed; as for the *1.03*-μm line	217

Note: A line at 2.7087 μm reported in Reference 160 is in fact an YbII line at 2.4377 μm.[159]

[a] Measured and calculated vacuum wavelengths and spectral assignments are from Wyart, J. F. and Camus, P., *Phys. Scripta*, 20, 43—59, 1979.

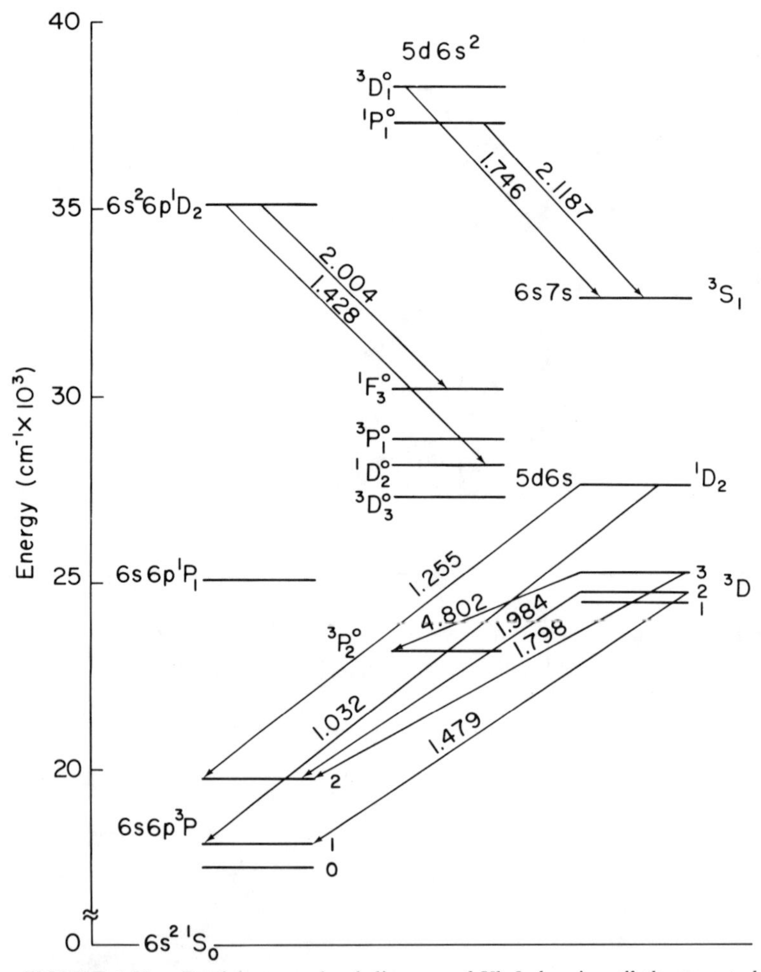

FIGURE 1.23. Partial energy-level diagram of Yb-I showing all the reported laser transitions. Wavelengths given in microns.

1.7B GROUP VII B

Table 1.7.6
MANGANESE[a] (FIGURE 1.24)

Wavelength (μm)	Transition assignment	Comments	Ref.
0.5341065[b]	$y\,^6P^o_{7/2} \rightarrow a\,^6D_{9/2}$	Pulsed; short rise-time high-voltage pulsed excitation of 0.1—2.0 torr of Mn (at a temperature of 1100—1300°C) with 1—2 torr of He or Ne; D = 10 mm; also by single- or double-pulse excitation of $MnCl_2$ at 700—800°C with up to 120 torr of Ne, 80 torr of He, or 20 torr of Xe; optimum is about 15 torr of Ne; D = 16 mm	17,348—355
0.5420368	$y\,^6P^o_{5/2} \rightarrow a\,^6D_{7/2}$	Pulsed; short rise-time, high-voltage pulsed excitation of 0.1—2.0 torr of Mn (at a temperature of 1100—1300°C) with 1—2 torr of He or Ne; D = 10 mm; also by single- or double-pulse excitation of $MnCl_2$ at 700—800°C with up to 120 torr of Ne, 80 torr of He, or 20 torr of Xe; optimum is about 15 torr of Ne; D = 16 mm	17,348,349, 352—354
0.5470640	$y\,^6P^o_{5/2} \rightarrow a\,^6D_{5/2}$	Pulsed; short rise-time high-voltage pulsed excitation of 0.1—2.0 torr of Mn (at a temperature of 1100—1300°C) with 1—2 torr of He or Ne; D = 10 mm	348,349,354
0.5481345	$y\,^6P^o_{3/2} \rightarrow a\,^6D_{5/2}$	Pulsed; short rise-time high-voltage pulsed excitation of 0.1—2.0 torr of Mn (at a temperature of 1100—1300°C) with 1—2 torr of He or Ne; D = 10 mm	17,349
0.5516777	$y\,^6P^o_{3/2} \rightarrow a\,^6D_{3/2}$	Pulsed; short rise-time high-voltage pulsed excitation of 0.1—2.0 torr of Mn (at a temperature of 1100—1300°C) with 1—2 torr of He or Ne; D = 10 mm	17,348,349, 354
0.5537749	$y\,^6P^o_{3/2} \rightarrow a\,^6D_{1/2}$	Pulsed; short rise-time high-voltage pulsed excitation of 0.1—2.0 torr of Mn (at a temperature of 1100—1300°C with 1—2 torr of He or Ne; D = 10 mm	17,348,349,3 54
1.28997	$z\,^6P^o_{7/2} \rightarrow a\,^6D_{9/2}$	Pulsed; as for the 0.534-μm line	17,348,349, 352—354
1.32941	$z\,^6P^o_{7/2} \rightarrow a\,^6D_{7/2}$	Pulsed; as for the 0.534-μm line	17,348,349, 352—354
1.33179	$z\,^6P^o_{5/2} \rightarrow a\,^6D_{7/2}$	Pulsed; as for the 0.534-μm line	17,348,349, 352—354
1.36257	$z\,^6P^o_{5/2} \rightarrow a\,^6D_{1/2}$	Pulsed; as for the 0.534-μm line	17,348,349, 352—354
1.38638	$z\,^6P^o_{3/2} \rightarrow a\,^6D_{3/2}$	Pulsed; as for the 0.534-μm line	17,348,349, 352—354
1.39970	$z\,^6P^o_{3/2} \rightarrow a\,^6D_{1/2}$	Pulsed; as for the 0.534-μm line	17,348,349, 352,353

[a] Measured wavelengths and spectral assignments are from Catalan, M. A., Meggers, W. F., and Garcia-Riquelme, O., *J. Res. Natl. Bur. Stand.*, 68A, 9—59, 1964.

[b] Oscillation has been reported on six hyperfine components of this line.[349]

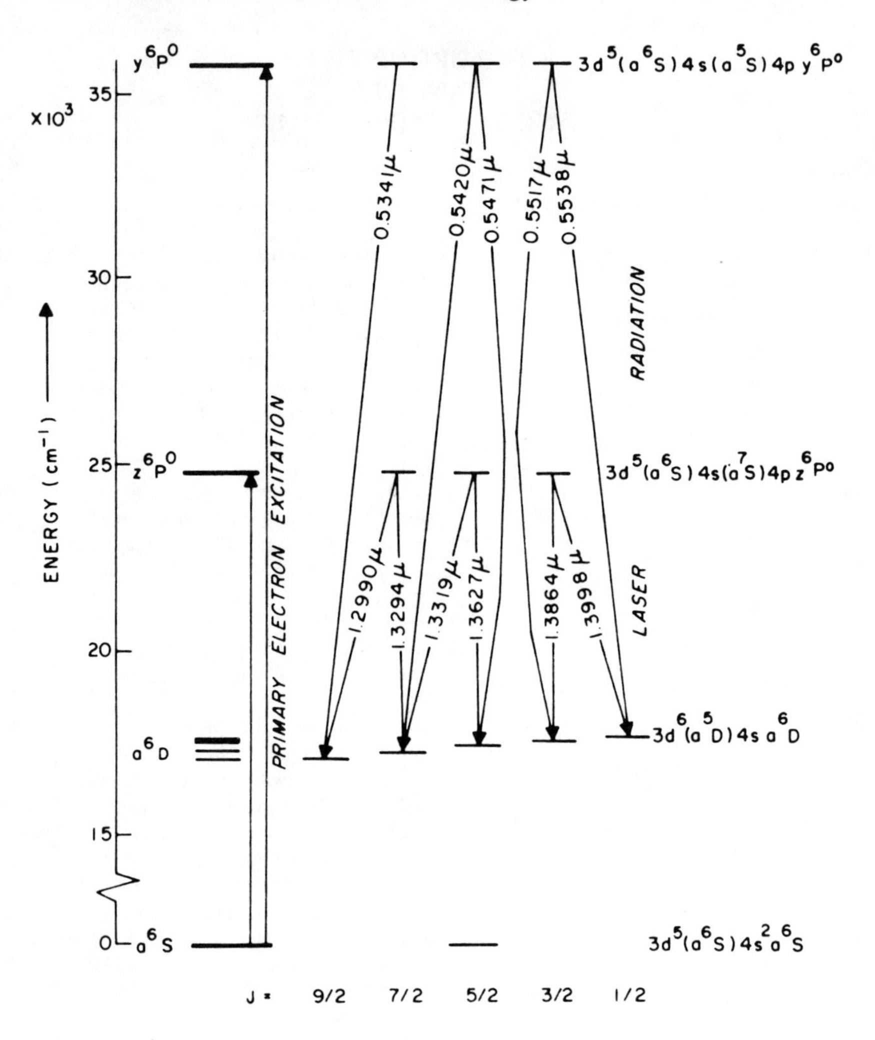

FIGURE 1.24. Partial energy-level diagram of Mn-I showing the self-terminating laser transitions and the primary route for direct electron-impact excitation.

1.8 GROUP XIII
Table 1.8.1
IRON

Wavelength (μm)	Transition assignment	Comments	Ref.
0.299951[a]	$x\,^5F^o_5 \rightarrow a\,^5F_5$	Pulsed; Fe atoms at a density of $\simeq 10^{14}$ atoms cm^{-3}; produced by flash photolysis of $Fe(CO)_5$ in an Ar buffer or by a discharge through 0.1 torr of $Fe(CO)_5$ with 50 torr of Ne, optically pumped with a KrF laser (248 nm)	356
0.301618[a]	$x\,^5F^o_1 \rightarrow a\,^5F_2$	Pulsed; Fe atoms at a density of $\simeq 10^{14}$ atoms cm^{-3}; produced by flash photolysis of $Fe(CO)_5$ in an Ar buffer or by a discharge through 0.1 torr of $Fe(CO)_5$ with 50 torr of Ne, optically pumped with a KrF laser (248 nm)	356

1.8 GROUP XIII

Table 1.8.1 (continued)
IRON

Wavelength (μm)	Transition assignment	Comments	Ref.
0.303164[a]	x $^5F_1^o$ → a $^5F_1$	Pulsed; Fe atoms at a density of ≃10^{14} atoms cm^{-3}; produced by flash photolysis of $Fe(CO)_5$ in an Ar buffer or by a discharge through 0.1 torr of $Fe(CO)_5$ with 50 torr of Ne, optically pumped with a KrF laser (248 cm)	356
0.304043[a]	x $^5F_5^o$ → a $^5F_4$	Pulsed; Fe atoms at a density of ≃10^{14} atoms cm^{-3}; produced by flash photolysis of $Fe(Co)_5$ in an Ar buffer or by a discharge through 0.1 torr of $Fe(CO)_2$ with 50 torr of Ne, optically pumped with a KrF laser (248 nm)	356
0.452862[a]	x $^5D_4^o$ → a $^5P_3$	Pulsed; short rise-time high-voltage pulsed excitation of Fe vapor in a tube at 1680°C with 1.5—3.5 torr of Ne; D = 16 mm	357
6.8487052[b]	w $^5P_3^o$ → e $^5D_4$	Pulsed; by dissociation of 0.08 torr of $Fe(CO)_5$ with 2 torr of He in a transversely excited double-discharge laser	122
8.4927853[c]	e $^5D_4$ → w $^5D_4^o$	Pulsed; by dissociation of 0.08 torr of $Fe(CO)_5$ with 2 torr of He in a transversely excited double-discharge laser	122

[a] Measured wavelength from Meggers, W. F., Corliss, C. H., and Scribner, B. F., *Natl. Bur. Stand. (U.S.) Monogr.*, 145(1), 1975.

[b] Alternative two-electron transition assignment is z $^3I_5^o$ → e $^5D_4$ at *6.8544324* μm. Measured wavelength of laser transition in Reference 122 was 6.847 μm.

[c] Several additional assignments are possible for this line; these other possibilities involve transitions between higher-lying levels than the ones listed here. Measured wavelength in Reference 122 was 8.490 μm.

Table 1.8.2
NICKEL

Wavelength (μm)	Transition assignment	Comments	Ref.
1.396710[a]	?	Pulsed; in Ni vapor produced by sputtering in a slotted hollow-cathode discharge[b]	358
1.4553719	h $^3F_2$ → w $^3P_1^o$	Pulsed; by dissociation of 80 torr of $Ni(CO)_4$ with 2 torr of He in a transversely excited double-discharge laser	122

[a] Measured laser wavelength in Reference 358 was 1.3968 μm. This is probably the unidentified transition at the wavelength listed above reported by Fisher, R. A., Knopf, W. C., Jr., and Kinney, F. E., *Astrophys. J.*, 130, 683—687, 1959. However, two nearby assigned transitions are e $^3D_2$ → y $^1F_3^o$ at *1.3984294* μm and g $^3D_2$ → w $^3D_1^o$ at *1.3987469* μm.

Table 1.8.3
SAMARIUM

Wavelength (μm)	Transition assignment	Comments	Ref.
1.9124005	$5d\,6s^2\ {}^7H^o_2 \rightarrow 5d({}^8D)6s\ {}^9D_3$ [a]	Pulsed; in Sm vapor at about 0.1 torr with He, Ne, or Ar	217
2.0482	?	Pulsed; in Sm vapor at about 0.1 torr with He, Ne, or Ar	217
2.7006079	$6s({}^8F)7s{}^9F_7 \rightarrow 5d\,6p({}^3F^o)?\ {}^9I_6?$	Pulsed; in Sm vapor at about 0.1 torr with He, Ne, or Ar	217
2.9663	?	Pulsed; in Sm vapor about 0.1 torr with He, Ne, or Ar	217
3.4654	?	Pulsed; in Sm vapor about 0.1 torr with He, Ne, or Ar	217
3.5361	? [c]	Pulsed; in Sm vapor about 0.1 torr with He, Ne, or Ar	217
4.1368	?	Pulsed; in Sm vapor about 0.1 torr with He, Ne, or Ar	217
4.8656	?	Pulsed; in Sm vapor about 0.1 torr with He, Ne, or Ar	217

[a] Assignment suggested by the present author; involves low-lying levels and calculated wavelength is within reported error of measured wavelength in Reference 159.

[b] Closest calculated line is $5d\,6s^2\ {}^5F^o_4 \rightarrow 5d({}^6P)6s\ {}^7F_3$ at *2.049928* μm.

[c] Closest calculated line is $31893.78^o_3 \rightarrow 29006.02_2$ at *3.536368* μm.

Table 1.8.4
EUROPIUM (FIGURE 1.25)

Wavelength (μm)	Transition assignment	Comments	Ref.
0.545294[a]	$z\ {}^{10}D_{9/2} \rightarrow a\ {}^{10}D^o_{7/2}$	Pulsed; operates in a true superfluorescent mode	361
0.557714[a]	$z\ {}^{10}D_{9/2} \rightarrow a\ {}^{10}D^o_{11/2}$	Pulsed; operates in a true superfluorescent mode	361
0.605736[a]	$z\ {}^8F_{9/2} \rightarrow a\ {}^8D^o_{9/2}$	Pulsed; operates in a true superfluorescent mode following two-photon excitation of 7.5 torr of Eu vapor	360
1.926[b]	$5d^2\ {}^8G^o_{11/2} \rightarrow z\ {}^8F_{9/2}$	Pulsed; operates in a true superfluorescent mode following two-photon excitation of 7.5 torr of Eu vapor	360
1.961[b]	$5d^2\ {}^8G^o_{9/2} \rightarrow z\ {}^8F_{9/2}$	Pulsed; operates in a true superfluorescent mode following two-photon excitation of 7.5 torr of Eu vapor	360
1.7600985	$y\ {}^8P_{9/2} \rightarrow a\ {}^8D^o_{11/2}$	Pulsed; in about 0.1 torr of Eu vapor with He, Ne, or Ar; also in a self-heated repetitively pulsed discharge with Eu and 15—25 torr of He; D = 1.1 or 2 cm	217, 359
2.5818111	$b\ {}^8D^o_{5/2} \rightarrow z\ {}^8P_{5/2}$	Pulsed; in about 0.1 torr of Eu vapor with He, Ne, or Ar	217
2.7181668	$b\ {}^8D^o_{9/2} \rightarrow z\ {}^8P_{7/2}$	Pulsed; in about 0.1 torr of Eu vapor with He, Ne, or Ar	217
4.3213904	$y\ {}^8P_{9/2} \rightarrow b\ {}^8D^o_{11/2}$	Pulsed; in about 0.1 torr of Eu vapor with He, Ne or Ar	217
4.6948356	$y\ {}^8P_{9/2} \rightarrow b\ {}^8P_{9/2}$	Pulsed; in about 0.1 torr of Eu vapor with He, Ne, or Ar	217
5.0660871	$y\ {}^8P_{7/2} \rightarrow b\ {}^8D^o_{9/2}$	Pulsed; in about 0.1 torr of Eu vapor with He, Ne, or Ar	217

Table 1.8.4 (continued)
EUROPIUM (FIGURE 1.25)

Wavelength (μm)	Transition assignment	Comments	Ref.
5.2825643	$y\ ^8P_{7/2} \rightarrow b\ ^8D^o_{7/2}$	Pulsed; in about 0.1 torr of Eu vapor with He, Ne, or Ar	217
5.4306800	$y\ ^8P_{7/2} \rightarrow b\ ^8D^o_{5/2}$	Pulsed; in about 0.1 torr of Eu vapor with He, Ne, or Ar	217
5.7722389	$y\ ^8P_{5/2} \rightarrow b\ ^8D^o_{7/2}$	Pulsed; in about 0.1 torr of Eu vapor with He, Ne, or Ar	217
5.9495478	$y\ ^8P_{5/2} \rightarrow b\ ^8D^o_{5/2}$	Pulsed; in about 0.1 torr of Eu vapor with He, Ne, or Ar	217
6.0592473	$y\ ^8P_{5/2} \rightarrow b\ ^8D^o_{3/2}$	Pulsed; in about 0.1 torr of Eu vapor with He, Ne, or Ar	217

[a] Measured wavelengths from Meggers, W. F., Corliss, C. H., and Scribner, B. F., *Natl. Bur. Stand. (U.S.) Monogr.*, 145(1), 1975.

[b] Measured wavelength from Reference 360.

1.9 GROUP O NOBLE GASES

Table 1.9.1
HELIUM[a] (FIGURE 1.26)

Wavelength (μm)	Transition assignment	Comments	Ref.
0.7067124	$3s\ ^3S_1 \rightarrow 2p\ ^3P^o_2$	Pulsed; in a 8:7 He/H_2 mix at 12.5 torr in a hollow-cathode discharge	362,[b] 363
0.7067162	$3s\ ^3S_1 \rightarrow 2p\ ^3P^o_1$	Pulsed; in a 8:7 He/H_2 mix at 12.5 torr in a hollow-cathode discharge	362,[b] 363
1.868596	$4f\ ^3F^o \rightarrow 3d\ ^3D$	CW; in 0.4 torr of He; D = 6 mm	364,365
1.954313	$4p\ ^3P^o \rightarrow 3d\ ^3D$	CW; in 0.4 torr of He; D = 6 mm	364,365, 366,367
2.058130[c]	$2p\ ^1P^o_1 \rightarrow 2s\ ^1S_0$	Short rise-time, high-voltage pulsed excitation of 2.7 torr of He; D = 1.3 mm; operates in an ASE mode; also following pulsed excitation of 65 torr of He/NH_3 mixture containing 18% NH_3 in a transversely excited pin laser	148,368
2.060755	$7d\ ^3D \rightarrow 4p\ ^3P^o$	CW; in 8.0 torr of He with a trace of Ar or N;[d] D = 7 mm	14,369,370
4.60535[e]	$5s\ ^1S_0 \rightarrow 4p\ ^1P^o_1$	CW; in 0.2—0.4 torr of He; D = 15 mm	371,372
4.60567(^3He)[e]	$5s\ ^1S_0 \rightarrow 4p\ ^1P^o_1$	CW; in 0.2—0.4 torr of He; D = 15 mm	372
8.529294	$6s\ ^1S_0 \rightarrow 5p\ ^1P^o_1$	CW; in 0.2—0.4 torr of He; D = 15 mm	372
95.77994	$3p\ ^1P^0_1 \rightarrow 3d\ ^1D_2$	Pulsed; in 0.5 torr of He; D = 75 mm or CW in 0.1 torr of He; D = 60 mm	373—375
216.17882	$4p\ ^1P^1_1 \rightarrow 4d\ ^1D_2$	CW; in 0.1 torr of He; D = 60 mm	374,375

[a] Unless otherwise stated, indicated wavelengths and spectral assignments are taken from data in Martin, W. C., *J. Res. Natl. Bur. Stand.*, 64, 19—28, 1960.

[b] It is not clear whether one or both of the lines near *0.7067* μm were observed in Reference 362. The excitation mechanism proposed for these transition involves a two-body recombination process:[362] $H^-(1s^{2\,1}S_0) + He^+(1s\ ^2S_{1/2}) \rightarrow H(1s\ ^2S_{1/2}) + He(1s\ 3s\ ^3S_1)$.

[c] This is a self-terminating laser transition.

[d] The trace of argon or nitrogen is to depopulate the metastable He 2s $^3S_1$ level from which the $4p^3P$ lower laser level is otherwise excited by electron impact.

[e] Measured wavelength from Reference 372.

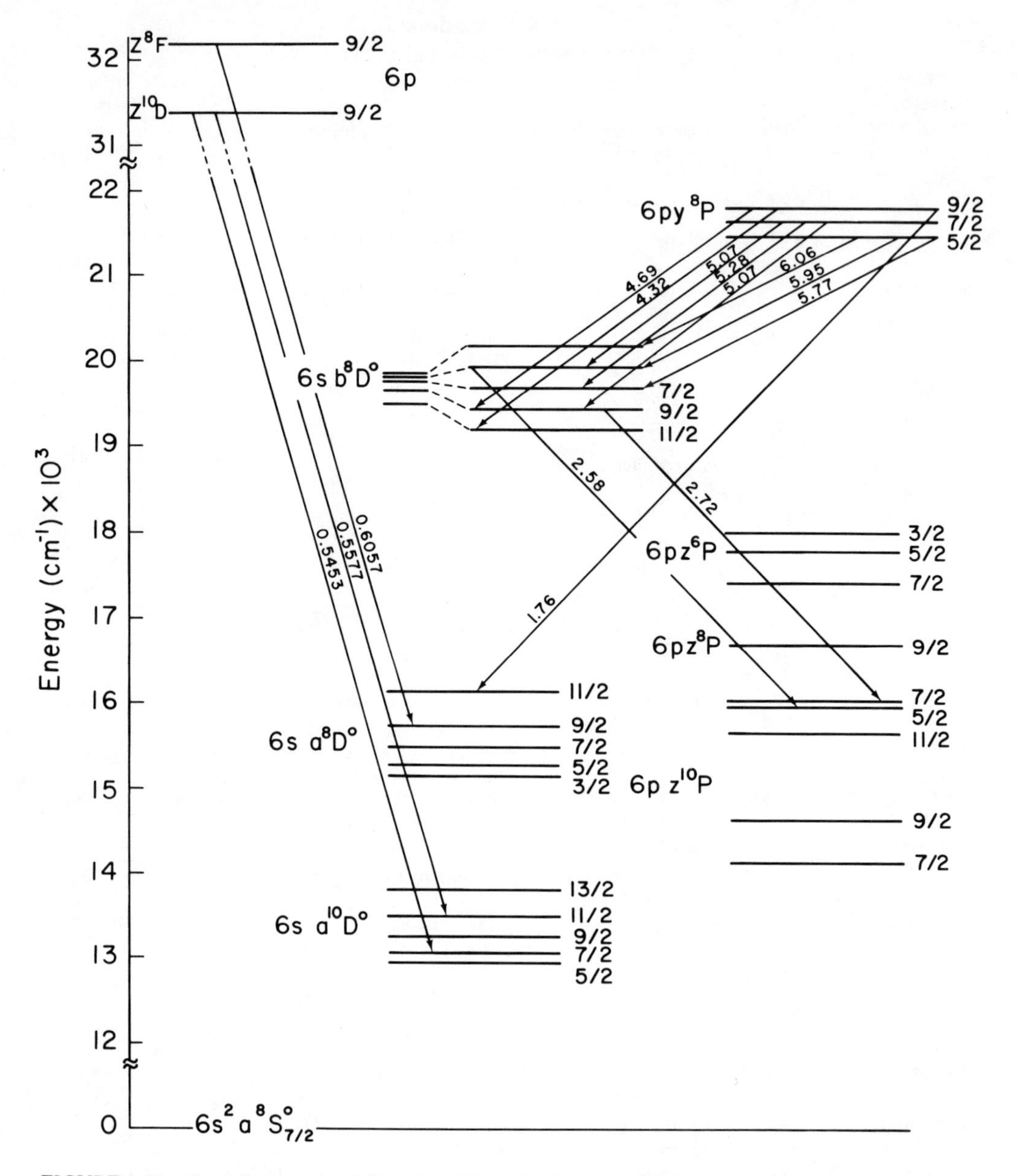

FIGURE 1.25. Partial energy-level diagram of Eu-I showing the majority of the reported laser transitions. Wavelengths given in microns.

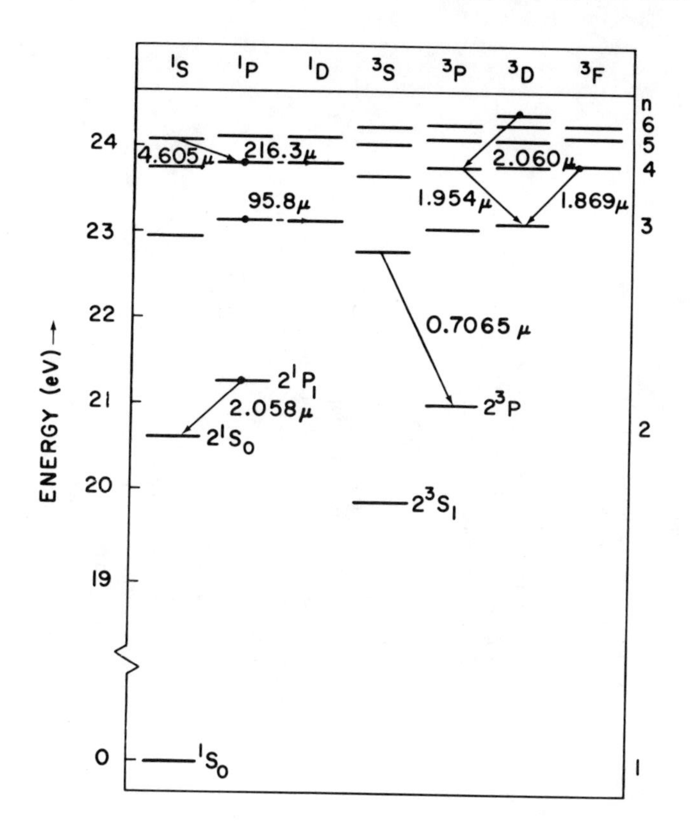

FIGURE 1.26. Partial energy-level diagram of He-I showing the reported laser transitions.

Table 1.9.2
NEON[a] (FIGURE 1.27)

Wavelength (μm)	Transition assignment: Racah	Transition assignment: Paschen	Comments	Ref.
0.54005616[b]	$3p'[1/2]_0 \rightarrow 3s[3/2]_1^o$	$2p_1 \rightarrow 1s_4$	Transient line requiring short rise-time high voltage, high current pulses. In 3 torr of neon; D = 5 mm in a longitudinal discharge system. In a transverse discharge system,[384] 30-35 torr of neon with interelectrode separations of 2.5-10 cm.	17,151,382, 386,397
0.5435161	$5s'[1/2]_1^o \rightarrow 3p[1/2]_1$	$3s_2 \rightarrow 2p_{10}$	CW. In a 7:1 ^3He-^{20}Ne mixture at 1 torr; D = 4 mm	387
0.58524878[b]	$3p'[1/2]_0 \rightarrow 3s'[1/2]_1^o$	$2p_1 \rightarrow 1s_2$	Transient line observed in a pulsed He-Ne mixture requiring trace of argon to destroy Ne-1s metastables. Total pressures 1-200 mtorr; D = 3 mm.	14,388
0.59409633	$5s'[1/2]_1^o \rightarrow 3p[5/2]_2$	$3s_2 \rightarrow 2p_8$	CW. In a 5:1 He-Ne mixture at 3.6 torr-mm. Need to suppress high gain. 3.39 μm($3s_2 \rightarrow 3p_4$) line and strong 0.6328 μm($3s_2 \rightarrow 2p_4$) line. The Ne $3s_2$ level is selectively excited mainly by He 2^1S_0 metastables in an endothermic excitation transfer reaction.	14,389,390[c]
0.59448342[b]	$3p'[3/2]_2 \rightarrow 3s[3/2]_2^o$	$2p_4 \rightarrow 1s_5$	Transient line requiring short rise-time, high voltage, high current pulses. In about 0.3 torr of neon, D = 5 mm and pulse current of 120 A.	382,391,392, 393

0.604[illegible]1348[b]	$5s'[1/2]^\circ_1 \rightarrow 3p[3/2]_1$	$3s_2 \rightarrow 2p_7$	CW. In a 5:1 He-Ne mixture at 3.6 torr-mm. Need to suppress high gain 3.39 μm line and the strong 0.6328 μm line.	14,389
0.61197087	$5s'[1/2]^\circ_1 \rightarrow 3p[3/2]_2$	$3s_2 \rightarrow 2p_6$	CW. In a 5:1 He-Ne mixture at 3.6 torr-mm; need to suppress high gain - 3.39-μm line and the strong 0.6328-μm line	14,389,390, 394[c]
0.61430623[b]	$3p[3/2]_2 \rightarrow 3s[3/2]^\circ_2$	$2p_6 \rightarrow 1s_5$	Transient line requiring short rise-time, high-voltage, high-current pulses; in about 0.3 torr of Ne; D = 5 mm and pulse current of 120 A	382,385,386, 391—398
0.62664950[b]	$3p'[3/2]_1 \rightarrow 3s'[1/2]^\circ_0$	$2p_5 \rightarrow 1s_3$	Pulsed; in an ASE mode; D = 1.3 mm; optimum Ne pressure ≃2 torr	399
0.62937447[b]	$5s'[1/2]^\circ_1 \rightarrow 3p'[3/2]_1$	$3s_2 \rightarrow 2p_5$	CW; in a 5:1 He-Ne mixture at 3.6 torr-mm., need to suppress 3.39 and 0.6328-μm lines	389,394
0.63281646[b]	$5s'[1/2]^\circ_1 \rightarrow 3p'[3/2]_2$	$3s_2 \rightarrow 2p_4$	CW; strongest of the $3s_2 \rightarrow 2p$ lines; in a 5:1 He-Ne mixture at 3.6 torr-mm; the He-Ne mixture ratio and pressure depend on the bore of the discharge tube;[410] the Ne-$3s_3$ level is selectively excited mainly by He 2^1S_0 metastables in an endothermic excitation transfer reaction, as well by direct electron impact;[402,405] also CW nuclear-pumped by the reaction $^3He(n,p)\,^3H$; threshold flux 2×10^{11} n cm^{-2} s^{-1}	400,14,389, 394—452, 631—649

Table 1.9.2 (continued)
NEON[a] (FIGURE 1.27)

Wavelength (μm)	Transition assignment: Racah	Transition assignment: Paschen	Comments	Ref.
0.63518618[b]	$5s'[1/2]^o_1 \rightarrow 3p[1/2]_0$	$3s_2 \rightarrow 2p_3$	CW; as for the 0.6328-μm line; need to suppress the 3.39 and the 0.6328-μm lines	389
0.64028455	$5s'[1/2]^o_1 \rightarrow 3p'[1/2]_1$	$3s_2 \rightarrow 2p_2$	CW; as for the 0.6328-μm line; need to suppress the 0.6328-μm line by the use of a prism or an unstable optical cavity and a discharge current more than optimal for the 0.6328-μm line	394,389,421, 422,453,454
0.73068569	$5s'[1/2]^o_1 \rightarrow 3p[1/2]_0$	$3s_2 \rightarrow 2p_1$	CW; as for 0.6328-μm line; need to suppress the 3.39 and 0.6328-μm lines	389
0.84633569[b]	$3d[3/2]^o_1 \rightarrow 3p[5/2]_2$	$3d_2 \rightarrow 2p_8$	In an ASE mode following fast rise-time high-voltage pulsed excitation of 0.2 torr of Ne; D = 6 mm	456
0.86376895	$3d'[3/2]^o_1 \rightarrow 3p'[3/2]_2$	$3s'_1 \rightarrow 2p_4$	As for 0.8463-μm line, except Ne pressure was 3 torr	456
0.86819216[b]	$3d[3/2]^o_1 \rightarrow 3p[3/2]_1$	$3d_2 \rightarrow 2p_7$	Pulsed; in an ASE mode; D = 1.3, 5, 6, or 6.5 mm; optimum pressure 2—4 torr of Ne, depending on tube diameter	399
0.87740648	$3d'[3/2]^o_1 \rightarrow 3p'[1/2]_1$	$3s'_1 \rightarrow 2p_2$	In an ASE mode following fast rise-time high-voltage pulsed excitation of 3 torr of Ne; D = 6 mm	456
0.88653057[b]	$4s'[1/2]^o_1 \rightarrow 3p[1/2]_1$	$2s_2 \rightarrow 2p_{10}$	CW; observed in a very long discharge tube	423,456,457
0.89910237	$4s'[1/2]^o_0 \rightarrow 3p[1/2]_1$	$2s_3 \rightarrow 2p_{10}$	CW; observed in a very long discharge tube	423,456

0.94892838	$4s[3/2]^{\circ}_{1} \rightarrow 3p[1/2]_{1}$	$2s_4 \rightarrow 2p_{10}$	Pulsed; observed in a hollow-cathode discharge in a Ne-H_2 mixture; need to suppress oscillation on other 2s → 2p transitions	458
0.96680709	$4s[3/2]^{\circ}_{2} \rightarrow 3p[1/2]_{1}$	$2s_5 \rightarrow 2p_{10}$	Pulsed; observed in a hollow-cathode discharge in a Ne-H_2 mixture; need to suppress oscillation on other 2s → 2p transitions	457,458
1.0298238	$4s'[1/2]^{\circ}_{1} \rightarrow 3p[5/2]_{2}$	$2s_2 \rightarrow 2p_8$	CW; observed in a very long discharge tube	423,456,457
1.0623574	$4s'[1/2]^{\circ}_{1} \rightarrow 3p[3/2]_{1}$	$2s_2 \rightarrow 2p_7$	CW; in a He-Ne mixture, 0.15 torr partial pressure of Ne and 2.8 torr total pressure, D probably 5—10 mm; the gain on this transition is only a few tenths of a percent per meter even with oscillation suppressed on competing transitions	14,423,456, 460—462
1.0801000	$4s'[1/2]^{\circ}_{0} \rightarrow 3p[3/2]_{1}$	$2s_3 \rightarrow 2p_7$	CW; in a 10:1 He-Ne mixture at a pD of 7—14 torr-mm.	195,407,456, 459,463
1.0847447	$4s'[1/2]^{\circ}_{1} \rightarrow 3p[3/2]_{2}$	$2s_2 \rightarrow 2p_6$	CW; as for *1.0801*-μm line above.	195,407,456, 459,462,463
1.1146071	$4s[3/2]^{\circ}_{1} \rightarrow 3p[5/2]_{2}$	$2s_4 \rightarrow 2p_8$	CW; as for *1.0801*-μm line; also observed in hollow-cathode discharge in pure Ne and in a mixture of Ne and H; H destroys the Ne-1s metastables	195,407,456, 458,459, 463—467
1.1180588	$4s[3/2]^{\circ}_{2} \rightarrow 3p[5/2]_{3}$	$2s_5 \rightarrow 2p_9$	CW; as for the *1.1146*-μm line	14,401,456—459,461,464, 466,468
1.1393552	$4s[3/2]^{\circ}_{2} \rightarrow 3p[5/2]_{2}$	$2s_5 \rightarrow 2p_8$	CW; as for the *1.0801*-μm line	195,401,456, 458,459,463
1.1412258	$4s'[1/2]^{\circ}_{1} \rightarrow 3p'[3/2]_{1}$	$2s_2 \rightarrow 2p_5$	CW; as for the *1.1146*-μm line	195,401,456, 458,459

Table 1.9.2 (continued)
NEON[a] (FIGURE 1.27)

Wavelength (μm)	Transition assignment: Racah	Transition assignment: Paschen	Comments	Ref.
1.1525900	$4s'[1/2]^{\circ}_{1} \rightarrow 3p'[3/2]_{2}$	$2s_2 \rightarrow 2p_4$	CW; strongest of the 2s → 2p lines; in a 10:1 He-Ne mixture at about 11 torr-mm; the Ne-2s levels are selectively excited mainly by He 2^3S_1 metastables in an exothermic excitation transfer reaction, as well as by direct electron impact;[405,407] observed also in hollow-cathode discharges in pure Ne and mixtures of Ne and H and Ne and O	14,195, 401,405,407, 457,459, 464—467,468 469—477
1.1528174	$4s[3/2]^{\circ}_{1} \rightarrow 3p[3/2]_{1}$	$2s_4 \rightarrow 2p_7$	CW; in a glow discharge in pure Ne, optimum pD is less than 0.5 torr-mm, He suppresses oscillation; observed also in hollow-cathode discharges in pure Ne and a Ne-H mixture	14,367,456, 458,459,464, 466,472,475
1.1604712	$4s'[1/2]^{\circ}_{1} \rightarrow 3p[1/2]_{0}$	$2s_2 \rightarrow 2p_3$	CW; in a He-Ne mixture	195,401,456, 459,461,469
1.1617260	$4s[1/2]^{\circ}_{0} \rightarrow 3p'[3/2]_{1}$	$2s_3 \rightarrow 2p_5$	CW; in a He-Ne mixture	18,195,401, 456,459,465, 469,474
1.1770013	$4s'[1/2]^{\circ}_{1} \rightarrow 3p'[1/2]_{1}$	$2s_2 \rightarrow 2p_2$	CW; as for the *1.0801*-μm line; observed also in hollow-cathode discharges in pure Ne and in mixtures of Ne and H and Ne and O	463,401, 456—459, 464—467, 478
1.1792270	$4s[3/2]^{\circ}_{1} \rightarrow 3p[3/2]_{2}$	$2s_4 \rightarrow 2p_6$	CW; observed in a very long discharge tube, also in a hollow-cathode discharge in a mixture of Ne and H	423,456—459

1.1988192	$4s'[1/2]^{\circ}_{0} \rightarrow 3p'[1/2]_{1}$	$2s_3 \rightarrow 2p_2$	CW; in a He-Ne mixture	195,401,456, 459,465,468
1.2068179	$4s[3/2]^{\circ}_{0} \rightarrow 3p[3/2]_{2}$	$2s_5 \rightarrow 2p_6$	CW; in a He-Ne mixture in glow and hollow-cathode discharges	195,456,458, 459,461,464, 465,474,478
1.2462797	$4s[3/2]^{\circ}_{1} \rightarrow 3p'[3/2]_{1}$	$2s_4 \rightarrow 2p_5$	CW; observed in a very long laser, also in a hollow-cathode discharge in a mixture of Ne and H	423,456,458, 459
1.2588072 or *1.2588088*	$5f[9/2]_{5} \rightarrow 3d[7/2]^{\circ}_{4}$ or $5f[9/2]_{4} \rightarrow 3d[7/2]^{\circ}_{4}$	$5V_5 \rightarrow 3d'_4$ or $5V_4 \rightarrow 3d'_4$	Pulsed, via optical pumping by a He lamp; 0.2 torr of Ne, 3—4 torr of He, D = 5.2 mm	480
1.2598449	$4s[3/2]^{\circ}_{1} \rightarrow 3p'[3/2]_{2}$	$2s_4 \rightarrow 2p_4$	Pulsed, in a hollow-cathode discharge in a mixture of Ne and H	458,459
1.2692672	$4s[3/2]^{\circ}_{1} \rightarrow 3p[1/2]_{0}$	$2s_4 \rightarrow 2p_3$	CW; as for the *1.246*-μm line	366,373,456, 458,466,470, 480
1.2773017	$4s[3/2]^{\circ}_{2} \rightarrow 3p'[3/2]_{1}$	$2s_5 \rightarrow 2p_5$	CW; as for the *1.246*-μm line	456,458
1.2890684	$4s[3/2]^{\circ}_{1} \rightarrow 3p'[1/2]^{\circ}_{1}$	$2s_4 \rightarrow 2p_2$	CW; observed in a very long discharge in a He-Ne mixture	423
1.2915545	$4s[3/2]^{\circ}_{2} \rightarrow 3p'[3/2]_{2}$	$2s_5 \rightarrow 2p_4$	CW; in 0.7-torr He with 0.07-torr Ne; D = 9 mm; also in a hollow-cathode discharge in a mixture of Ne and H	373,423,456, 458,480
1.3222855?[n]	$4s[3/2]^{\circ}_{2} \rightarrow 3p'[1/2]_{1}$	$2s_5 \rightarrow 2p_2$	Pulsed, in a hollow-cathode discharge in a mixture of Ne and H	458
1.4276 ± 5.10^{-4}	Unidentified		CW	481
1.4304 ± 5.10^{-4}	Unidentified		CW	481
1.4321 ± 5.10^{-4}	Unidentified		CW	481
1.4330 ± 5.10^{-4}	Unidentified		CW	481
1.4346 ± 5.10^{-4}	Unidentified		CW	481
1.4368 ± 5.10^{-4}	Unidentified		CW	481
1.4848636	$8p'[1/2]_{1} \rightarrow 5s[3/2]^{\circ}_{2}$	$7p_2 \rightarrow 3s_5$	CW; in a 10:1 He to Ne mixture at 10 torr; D = 5 mm	482
1.4873294	$7p[1/2]_{1} \rightarrow 4p'[1/2]_{0}$?[p]	$6p_{10} \rightarrow 3p_1$?	CW; in a 10:1 He to Ne mixture at 10 torr; D = 5 mm	482
1.4876248	$5p'[1/2]_{1} \rightarrow 3d[3/2]^{\circ}_{1}$	$4p_2 \rightarrow 3d_2$	CW; in a 10:1 He to Ne mixture at 10 torr; D = 5 mm	482

Table 1.9.2 (continued)
NEON[a] (FIGURE 1.27)

Wavelength (μm)	Transition assignment: Racah	Transition assignment: Paschen	Comments	Ref.
1.4892012	$6s'[1/2]^o_1 \rightarrow 4p[3/2]_1$	$4s_2 \rightarrow 3p_7$	CW; in a 10:1 He to Ne mixture at 10 torr; D = 5 mm	482
1.4903576	$6p[1/2]_1 \rightarrow 4p[3/2]_2$?[p]	$5p_{10} \rightarrow 3p_6$?	CW; in a 10:1 He to Ne mixture at 10 torr; D = 5 mm	482
1.4940304	$6s'[1/2]^o_1 \rightarrow 4p[3/2]_1$	$4s_3 \rightarrow 3p_7$	CW; in a 10:1 He to Ne mixture a 10 torr; D = 5 mm	482
1.5234875	$4s'[1/2]^o_1 \rightarrow 3p'[1/2]_0$	$2s_2 \rightarrow 2p_1$	CW; as for the *1.0801*-μm line	195,401,456, 459,463,478, 483
1.6407031	$6s[3/2]^o_2 \rightarrow 4p[5/2]_3$	$4s_5 \rightarrow 3p_9$	Pulsed; via optical pumping by a He lamp; 0.2 torr of Ne, 3—4 torr of He, D = 5.2 mm	479
1.7166616	$4s[3/2]^o_1 \rightarrow 3p'[1/2]_0$	$2s_4 \rightarrow 2p_1$	CW; as for the *1.289*-μm line	423,456
1.8215302	$4p'[1/2]_1 \rightarrow 4s[3/2]^o_1$	$3p_1 \rightarrow 2s_4$	CW; as for the *1.289*-μm line	456
1.8258313 or *1.8258357*[d]	$4f[5/2]_2 \rightarrow 3d[7/2]^o_3$ or $4f[5/2]_3 \rightarrow 3d[7/2]^o_3$	$4Y_2 \rightarrow 3d_4$ or $4Y_3 \rightarrow 3d_4$	CW	14
1.827659[e]	$4f[9/2]_{4,5} \rightarrow 3d[7/2]^o_4$	$4V \rightarrow 3d'_4$	CW; in a 10:1 or 100:1 He-Ne mixture at a pD of about 8 torr-mm	479,484-486
1.828258[e]	$4f[9/2]_4 \rightarrow 3d[7/2]^o_3$	$4V \rightarrow 3d_4$	CW; in a 10:1 He-Ne mixture at a pD of about 8 torr-mm	479,484-486
1.830400[e]	$4f[5/2]_{2,3} \rightarrow 3d[3/2]^o_2$	$4Y \rightarrow 3d_3$	CW; as for the 1.8283-μm line	479,484,486
1.840316[e]	$4f[5/2]_2 \rightarrow 3d[3/2]^o_1$	$4Y \rightarrow 3d_2$	CW; in a 100:1 He-Ne mixture at a pD of about 8 torr-mm	479,484,478, 486
1.859112[e]	$4f[7/2]_3 \rightarrow 3d[5/2]^o_2$	$4Z \rightarrow 3d''_1$	CW; as for the 1.8277-μm line	479,484,486

1.859730[e]	$4f[7/2]_{3,4} \rightarrow 3d[5/2]^o_3$	$4Z \rightarrow 3d'_1$	CW; as for the 1.8277-μm line	483,484,486
1.95740[f]	$4p'[3/2]_2 \rightarrow 4s'[3/2]^o_2$	$3p_4 \rightarrow 2s_5$	CW; in a He-Ne mixture; oscillation is due to cascading from the high-gain 3.39-μm ($3s_2 \rightarrow 3p_4$) line; also in pure Ne in a 10-m long discharge tube at 0.01—0.05 torr; D = 10 mm	14,456,470, 480,487
1.958248	$4p'[1/2]_1 \rightarrow 4s[3/2]^o_2$	$3p_2 \rightarrow 2s_5$	CW; in a He-Ne mixture; oscillation is due to cascade transitions from the well populated Ne-3s levels	14,373,456, 488
2.0355792	$4p'[3/2]_2 \rightarrow 4s[3/2]^o_1$	$3p_4 \rightarrow 2s_4$	CW; as for the 1.9574-μm line; also in pure Ne at 0.01—0.05 torr; D = 10 mm	14,366,373, 456,466,470, 478,487,488, 489,490
2.0359432	$4p'[1/2]_1 \rightarrow 4s[3/2]^o_1$	$3p_2 \rightarrow 2s_4$	CW; as for the *1.958*-μm line	14,456,466, 470,490
2.1023345	$4d'[5/2]^o_2 \rightarrow 4p[3/2]_2$	$4s'''' \rightarrow 3p_6$ (?)	CW; observed in a very long discharge tube	14,374,456
2.10409[f]	$4p'[1/2]^o_0 \rightarrow 4s'[1/2]^o_1$	$3p_1 \rightarrow 2s_2$	CW; as for the *1.958*-μm line, but also oscillates in 250 mtorr of pure Ne in an 8-m long, 10-mm bore discharge tube[489]	14,373,456, 466,470,480, 489
2.17074[f]	$4p[1/2]_0 \rightarrow 4s[3/2]^o_1$	$3p_3 \rightarrow 2s_4$	CW; in pure Ne at 0.05 torr; He suppresses oscillation by selectively populating the $2s_4$ lower laser level	14,373,456, 466,480
2.3266649	$4p[5/2]_2 \rightarrow 4s[3/2]^o_2$	$3p_8 \rightarrow 2s_5$	CW; observed in a very long discharge in pure Ne or a He-Ne mixture	456
2.3957953[g]	$4p'[3/2]_2 \rightarrow 4s[1/2]^o_1$	$3p_4 \rightarrow 2s_2$	CW; in a He-Ne mixture, also in pure Ne; oscillation due to cascading through the high-gain $3s_2 \rightarrow 3p_4$ transition at 3.39 μm	373,456,470, 483,487, 491-494
2.3962995	$4p'[1/2]_1 \rightarrow 4s[1/2]^o_1$	$3p_2 \rightarrow 2s_2$	CW; in a 5:1 He-Ne mixture, total pressure 0.6 torr; D = 8 mm	14,470,478, 493,494

Table 1.9.2 (continued)
NEON[a] (FIGURE 1.27)

Wavelength (μm)	Transition assignment: Racah	Transition assignment: Paschen	Comments	Ref.
2.9676035	$4d[3/2]^o_1 \rightarrow 4p'[3/2]_1$	$4d_2 \rightarrow 3p_5$	CW; in a He-Ne mixture over ranges 0.01—0.2 torr of Ne, 0.00—1.0 torr of He; 0 = 10 mm	490,497
2.9812503	$4d[3/2]^o_2 \rightarrow 4p'[3/2]_1$	$4d_3 \rightarrow 3p_5$	CW; in a He-Ne mixture over ranges 0.01—0.2 torr of Ne, 0.00—1.0 torr of He; D = 10 mm	490,497
3.0267787 or *3.0275836*	$4d[3/2]^o_2 \rightarrow 4p'[1/2]_1$ or $4d[3/2]^o_2 \rightarrow 4p'[3/2]_2$	$4d_3 \rightarrow 3p_2$ or $4d_3 \rightarrow 3p_4$	CW; in a He-Ne mixture over ranges 0.01—0.2 torr of Ne, 0.00—1.0 torr of He; D = 10mm	14,490,497
3.0720016	$5s'[1/2]^o_1 \rightarrow 4p[1/2]_0$	$3s_2 \rightarrow 3p_3$	CW; in a 12:1 He-Ne mixture at a total pressure of 0.65 torr; D = 10 mm; requires wavelength selection to suppress ASE mode operation of the 3.39-μm line	498
3.3182141	$5s[3/2]^o_1 \rightarrow 4p[5/2]_2$	$3s_4 \rightarrow 3p_8$	CW; in a He-Ne mixture over ranges 0.01—0.2 torr of Ne, 0.00—1.0 torr of He; D = 10 mm	470,486,490, 496,497
3.3341754[j]	$5s'[1/2]^o_1 \rightarrow 4p'[3/2]_1$	$3s_2 \rightarrow 3p_5$	CW; in a He-Ne mixture over ranges 0.01—0.2 torr of Ne, 0.00—1.0 torr of He; D = 10 mm	486,490,496, 497
3.3361448[j]	$5s[3/2]^o_2 \rightarrow 4p[5/2]_3$	$3s_5 \rightarrow 3p_9$	CW; in a He-Ne mixture over ranges 0.01—0.2 torr of Ne, 0.00—1.0 torr of He; D = 10 mm	486,490,496, 497

3.3510469 or *3.3520466*	$6d[1/2]^{\circ}_{1} \rightarrow 5p[1/2]_{0}$ or $4p[1/2]_{1} \rightarrow 4s'[1/2]^{\circ}_{1}$ [k]	$6d_5 \rightarrow 4p_3$ or $3p_{10} \rightarrow 2s_2$	CW; in He-Ne mixtures in ratios from 10:1 to 5:1 at total pressures between 0.3—0.5 torr; D = 15 mm	500
3.3813942	$7s'[1/2]^{\circ}_{0} \rightarrow 5p'[3/2]_{1}$	$5s_3 \rightarrow 4p_5$	CW; in a He-Ne mixture over ranges 0.01—0.2 torr of Ne; 0.00—1.0 torr of He; D = 10 mm	490,497
3.3849653	$7s'[1/2]^{\circ}_{0} \rightarrow 5p'[1/2]_{1}$	$5s_3 \rightarrow 4p_2$	CW; in a He-Ne mixture over ranges 0.01—0.2 torr of Ne; 0.00—1.0 torr of He; D = 10 mm	490,497
3.3912244	$5s'[1/2]^{\circ}_{1} \rightarrow 4p'[1/2]_{1}$	$3s_2 \rightarrow 3p_2$	CW; in a 5:1 He-Ne mixture at pD of 3.6 torr-mm; the Ne-$3s_2$ level is selectively excited mainly by excitation transfer from He 2^1S_0 metastables in an endothermic reaction, as well as by direct electron impact from the ground state[405,407]	486,490,493, 496,504-506
3.3922348	$5s'[1/2]^{\circ}_{1} \rightarrow 4p'[3/2]_{2}$	$3s_2 \rightarrow 3p_4$	CW; in a 5:1 He-Ne mixture at pD of 3.6 torr-mm; the Ne-$3s_2$ level is selectively excited from He 2^1S_0 metastables in an endothermic reaction, as well as by direct electron impact from the ground state;[405,407] also in 3 torr of pure Ne with high-voltage fast-pulse excitation; D = 6 mm; this line exhibits very gain	195,370,373, 393,394,401, 403,413,417, 455,483,490, 493,499, 501-505,507
3.4480843	$5s[3/2]^{\circ}_{1} \rightarrow 4p[3/2]_{1}$	$3s_4 \rightarrow 3p_7$	CW; in a He-Ne mixture over ranges 0.01—0.2 torr of Ne, 0.00—1.0 torr of He; D = 10 mm; also in an ASE mode following high-voltage fast-pulse excitation of 3 torr of pure Ne; D = 6 mm	18,455,486, 490,496,497

Table 1.9.2 (continued)
NEON[a] (FIGURE 1.27)

Wavelength (μm)	Transition assignment: Racah	Transition assignment: Paschen	Comments	Ref.
3.4789495	$5s[3/2]^{o}_{1} \rightarrow 4p[3/2]_{2}$	$3s_4 \rightarrow 3p_6$	CW; in He-Ne mixtures in ratios from 10:1 to 5:1 at pressures from 0.3—0.5 torr; D = 15 mm; also in an ASE mode following high-voltage fast-pulse excitation of 3 torr of pure Ne; D = 6 mm	455,500
3.5844556	$5s[3/2]^{o}_{2} \rightarrow 4p[3/2]_{2}$	$3s_5 \rightarrow 3p_6$	CW; in a He-Ne mixture over ranges 0.01—0.2 torr of Ne, 0.00—1.0 torr of He; D = 10 mm	14,486,490, 496
3.6164638	$9p[3/2]_{1,2} \rightarrow 5d[5/2]^{o}_{2}$	$8p_{7,6} \rightarrow 5d''_1$	CW; in He-Ne mixtures in ratios from 10:1 to 5:1 at pressures from 0.3—0.5 torr; D = 15 mm	500
3.7746325	$4p'[1/2]_{0} \rightarrow 3d[3/2]^{o}_{1}$	$3p_1 \rightarrow 3d_2$	CW; in a He-Ne mixture over ranges 0.01—0.2 torr of Ne, 0.00—1.0 torr of He; D = 10 mm	486,490
3.980630[f]	$5s[3/2]^{o}_{1} \rightarrow 4p[1/2]_{0}$	$3s_4 \rightarrow 3p_3$	CW; in a He-Ne mixture over ranges 0.01—0.2 torr of Ne, 0.00—1.0 torr of He; D = 10 mm	490,497
4.2182950	$5s'[1/2]^{o}_{1} \rightarrow 4p'[1/2]_{0}$	$3s_2 \rightarrow 3p_1$	CW; as for the *3.072*-μm line	498
5.103	?		CW; in 0.3 torr of pure Ne; D = 15 mm	500
5.1711388	$5d'[5/2]^{o}_{2} \rightarrow 5p'[3/2]_{1}$	$5s''''_1 \rightarrow 4p_5$	CW; in pure Ne at 0.3 torr or from 0.5—0.6 torr; D = 15 mm	500

5.3258685 or *5.3264331*	$5d[5/2]^{\circ}_{2} \rightarrow 5p[3/2]_{2}$[k] or $5d[3/2]^{\circ}_{1} \rightarrow 5p[3/2]_{1}$	$5d''_{1} \rightarrow 4p_{6}$ or $5d_{2} \rightarrow 4p_{7}$	CW; in pure Ne at 0.3 torr or from 0.5—0.6, torr or in a He-Ne mixture in ratios from 10:1 to 5:1 at total pressures from 0.3—0.5 torr; D = 15 mm	500
5.4048094	$4p'[1/2]_{0} \rightarrow 3d'[3/2]^{\circ}_{1}$	$3p_{1} \rightarrow 3s'_{1}$	CW; in a He-Ne mixture over ranges 0.01—0.2 torr of Ne, 0.00—1.0 torr of He; D = 10 mm. also in an ASE mode following high-voltage fast-pulse excitation of 3 torr of pure Ne; D = 6 mm	455(E),486, 490,496,508
5.515	?		CW; in pure Ne at 0.3 torr or from 0.5—0.6 torr or in a He-Ne mixture in ratios from 10:1 to 5:1 at total pressures from 0.3—0.5 torr; D = 15 mm	500
5.6667372	$4p[1/2]_{0} \rightarrow 3d[3/2]^{\circ}_{1}$	$3p_{3} \rightarrow 3d_{2}$	CW; in a He-Ne mixture over ranges 0.01—0.2 torr of Ne, 0.00—1.0 torr of He; D = 10 mm	490,496,497
5.7067951	$5p'[1/2]_{1} \rightarrow 5s'[1/2]^{\circ}_{0}$	$4p_{2} \rightarrow 3s_{3}$	CW; in a He-Ne mixture having He to Ne ratios from 10:1 to 5:1 at 0.3 torr; D = 15 mm	500
5.7773913	$5p[5/2]_{3} \rightarrow 5s[3/2]^{\circ}_{2}$	$4p_{9} \rightarrow 3s_{5}$	CW; in pure Ne at pressures of 0.3 or from 0.5—0.6 torr or in a He-Ne mixture in ratios from 10:1 to 5:1 at total pressures from 0.3—0.5 torr; D = 15 mm	500
5.8858082	$5p'[3/2]_{1} \rightarrow 5s'[1/2]^{\circ}_{1}$	$4p_{5} \rightarrow 3s_{2}$	CW; in a He-Ne mixture in ratios from 10:1 to 5:1 at pressures from 0.3—0.5 torr; D = 15 mm	500
5.9578742	$5p[5/2]_{2} \rightarrow 5s[3/2]^{\circ}_{1}$	$4p_{8} \rightarrow 3s_{4}$	CW; in a He-Ne mixture in ratios from 10:1 to 5:1 at pressures from 0.3—0.5 torr; D = 15 mm	500
6.7788016	$6s[3/2]^{\circ}_{2} \rightarrow 5p[1/2]_{1}$	$4s_{5} \rightarrow 4p_{10}$	CW; in a He-Ne mixture in ratios from 10:1 to 5:1 at pressures from 0.3—0.5 torr: D = 15 mm	500

Table 1.9.2 (continued)
NEON[a] (FIGURE 1.27)

Wavelength (μm)	Transition assignment: Racah	Transition assignment: Paschen	Comments	Ref.
6.8884755	$7d'[3/2]^o_1 \rightarrow 7p[1/2]_1$	$7s'_1 \rightarrow 6p_{10}$	CW; in a He-Ne mixture in ratios from 10:1 to 5:1 at pressures from 0.3—0.5 torr; D = 15 mm	500
6.9876797	$4p[3/2]_2 \rightarrow 3d[3/2]^o_2$	$3p_6 \rightarrow 3d_3$	CW; in pure Ne at 0.3 or from 0.5—0.6 torr or in a He-Ne mixture in ratios from 10:1 to 5:1 at pressures from 0.3—0.5 torr; D = 15 mm	500
7.098	?		CW; in pure Ne at 0.3 or from 0.5—0.6 torr or in a He-Ne mixture in ratios from 10:1 to 5:1 at 0.5 torr; D = 15 mm	500
7.3228367	$6s[3/2]^o_2 \rightarrow 5p[5/2]_3$	$4s_5 \rightarrow 4p_9$	CW; as for *5.667*-μm line	496,497,490
7.405131	$6s'[1/2]^o_0 \rightarrow 5p'[3/2]_1?\ell$	$4s_3 \rightarrow 4p_5$	CW; in a He-Ne mixture in ratios from 10:1 to 5:1 at 0.5 torr; D = 15 mm	500
7.4222794 or *7.4235357*	$6s'[1/2]^o_0 \rightarrow 5p'[1/2]_1$ or $5p'[1/2]_1 \rightarrow 4d[3/2]_2$	$4s_3 \rightarrow 4p_2$ or $4p_2 \rightarrow 4d_3$	CW; as for *5.667*-μm line	490,497
7.4699904	$4p[3/2]_2 \rightarrow 3d[5/2]^o_2$	$3p_6 \rightarrow 3d''_1$	CW; as for *5.667*-μm line	497
7.4799887	$4p[3/2]_2 \rightarrow 3d[5/2]^o_3$	$3p_6 \rightarrow 3d'_1$	CW; in a He-Ne mixture with 0.15 torr of Ne and 0.3 torr of He; D = 15 mm	486,490,496
7.4995237	$6s[3/2]^o_2 \rightarrow 5p[5/2]_2$	$4s_5 \rightarrow 4p_8$	CW; in a He-Ne mixture with 0.15 torr of Ne and 0.3 torr of He	14,486,497
7.5313000	$6s[3/2]^o_1 \rightarrow 5p[3/2]_1$	$4s_4 \rightarrow 4p_7$	CW; in a He-Ne mixture at 0.15 torr of Ne, 0.3 torr of He; D = 15 mm	14,496
7.5709798	$6d[1/2]^o_1 \rightarrow 5f[3/2]_2$	$6d_5 \rightarrow 5X_2$	CW; in a He-Ne mixture in ratios from 10:1 to 5:1 at pressures from 0.3—0.5 torr; D = 15 mm	500

7.5885028	$6d[1/2]^{\circ}_{0} \rightarrow 5f[3/2]_{1}$	$6d_6 \rightarrow 5X_1$	CW; in pure Ne at 0.3 or from 0.5—0.6 torr or in a He-Ne mixture in ratios from 10:1 to 5:1 at pressures from 0.3—0.5 torr; D = 15 mm	500
7.6163805	$4p[3/2]_1 \rightarrow 3d[5/2]^{\circ}_{2}$	$3p_7 \rightarrow 3d''_1$	CW; in a He-Ne mixture over ranges 0.01—0.2 torr of Ne, 0.00—1.0 to rr of He; D = 10 mm	486,490,496, 497
7.6458386	$5p'[3/2]_1 \rightarrow 4d[5/2]^{\circ}_{2}$	$4p_5 \rightarrow 4d_1$	CW; in a He-Ne mixture over ranges 0.01—0.2 torr of Ne, 0.00—1.0 torr of He; D = 10 mm	497
7.6511521	$4p[5/2]_2 \rightarrow 3d[7/2]^{\circ}_{3}$	$3p_8 \rightarrow 3d_4$	CW; in a He-Ne mixture over ranges 0.01—0.2 torr of Ne, 0.00—1.0 torr of He; D = 10 mm	486,490,496
7.6926284	$4p'[3/2]_2 \rightarrow 3d'[5/2]^{\circ}_{2}$	$3p_4 \rightarrow 3s''''_1$	CW; in a He-Ne mixture over ranges 0.01—0.2 torr of Ne, 0.00—1.0 torr of He; D = 10 mm	497
7.7016367	$4p'[3/2]_2 \rightarrow 3d'[5/2]^{\circ}_{3}$	$3p_4 \rightarrow 3s'''_1$	CW; in a He-Ne mixture over ranges 0.01—0.2 torr of Ne, 0.00—1.0 torr of He; D = 10 mm	486,490,496
7.7408259	$4p[5/2]_2 \rightarrow 3d[3/2]^{\circ}_{2}$	$3p_8 \rightarrow 3d_3$	CW; in a He-Ne mixture over ranges 0.01—0.2 torr of Ne, 0.00—1.0 torr of He; D = 10 mm	497
7.7655499	$4p'[1/2]_1 \rightarrow 3d'[3/2]^{\circ}_{2}$	$3p_2 \rightarrow 3s''_1$	CW; in a He-Ne mixture over ranges 0.01—0.2 torr of Ne, 0.00—1.0 torr of He; D = 10 mm	486,496
7.7815561	$6s[3/2]^{\circ}_{2} \rightarrow 5p[3/2]_1$	$4s_5 \rightarrow 4p_7$	CW; in a He-Ne mixture over ranges 0.01—0.2 torr of Ne, 0.00—1.0 torr of He; D = 10 mm	490,497

Table 1.9.2 (continued)
NEON[a] (FIGURE 1.27)

Wavelength (μm)	Transition assignment: Racah	Transition assignment: Paschen	Comments	Ref.
7.809	?	—	CW; in pure Ne at 0.3 or from 0.5—0.6 torr or in a He-Ne mixture in ratios from 10:1 to 5:1 at pressures from 0.3—0.5 torr; D = 15 mm	500
7.8369230	$6s[3/2]^o_2 \rightarrow 5p[3/2]_2$	$4s_5 \rightarrow 4p_6$	CW; as for the *7.782*-μm line	486,490,496, 497
7.8716418	$8p[3/2]_2 \rightarrow 7s[3/2]^o_2$	$7p_6 \rightarrow 5s_5$	CW; in pure Ne at 0.3 or from 0.5—0.6 torr or in a He-Ne mixture in ratios from 10:1 to 5:1 at pressures from 0.3—0.5 torr; D = 15 mm	500
7.9429190	$8p[5/2]_2 \rightarrow 7s[3/2]^o_2$	$7p_8 \rightarrow 5s_5$	CW; in He-Ne mixture in ratios from 10:1 to 5:1 at pressures from 0.3—0.5 torr; D = 15 mm	500
7.9846694	$8s'[1/2]^o_1 \rightarrow 7p[1/2]_1$	$6s_2 \rightarrow 6p_{10}$	CW; in He-Ne mixtures in ratios from 10:1 to 5:1 at pressures from 0.3—0.5 torr D = 15 mm	500
8.0088892	$4p'[3/2]_1 \rightarrow 3d'[5/2]^o_2$	$3p_5 \rightarrow 3s''''_1$	CW; as for the *7.837*-μm line	486,490,496, 497
8.0621949	$4p[5/2]_3 \rightarrow 3d[7/2]^o_4$	$3p_9 \rightarrow 3d'_4$	CW; as for the *7.837*-μm line	486,490,496, 497
8.116	?		CW; as for the *7.809*-μm line	500
8.1736588	$4p[5/2]_3 \rightarrow 3d[3/2]_2$	$3p_9 \rightarrow 3d_3$	CW; as for the *7.872*-μm line	500
8.3371447 or *8.3496011*	$4p[5/2]_2 \rightarrow 3d[5/2]^o_2$ [k] or $4p[5/2]_2 \rightarrow 3d[5/2]^o_3$	$3p_8 \rightarrow 3d''_1$ or $3p_8 \rightarrow 3d'_1$	CW; in a He-Ne mixture over ranges 0.01—0.2 torr of Ne, 0.00—1.0 torr of He; D = 10 mm	490,497

8.8414054 or *8.8554154*	$4p[5/2]_3 \rightarrow 3d[5/2]^\circ_2$ or $4p[5/2]_3 \rightarrow 3d[5/2]^\circ_3$	$3p_9 \rightarrow 3d''_1$ or $3p_9 \rightarrow 3d'_1$	CW; in a He-Ne mixture over ranges 0.01—0.2 torr of Ne, 0.00—1.0 torr of He; D = 10 mm CW; in a He-Ne mixture over ranges 0.01—0.2 torr of Ne, 0.00—1.0 torr of He; D = 10 mm	490,497
9.0894630	$6s[3/2]^\circ_1 \rightarrow 5p[1/2]_0$	$4s_4 \rightarrow 4p_3$	CW; in a He-Ne mixture over ranges 0.01—0.2 torr of Ne, 0.00—1.0 torr of He; D = 10 mm	486,490,496, 497
10.063422	$4p[1/2]_1 \rightarrow 3d[1/2]^\circ_1$	$3p_{10} \rightarrow 3d_5$	CW; in a He-Ne mixture over ranges 0.01—0.2 torr of Ne, 0.00—1.0 torr of He; D = 10 mm	490,497
10.981641	$4p[1/2]_1 \rightarrow 3d[3/2]^\circ_2$	$3p_{10} \rightarrow 3d_3$	CW; in a He-Ne mixture over ranges 0.01—0.2 torr of Ne, 0.00—torr of He; D = 10 mm	486,490,496, 497
11.860553	$5p[1/2]_1 \rightarrow 5s'[1/2]^\circ_0$	$4p_{10} \rightarrow 3s_3$	CW; in a He-Ne mixture over ranges 0.01—0.2 torr of Ne, 0.00—1.0 torr of He, D = 10 mm	490
11.902069	$5p[1/2]_0 \rightarrow 4d[3/2]^\circ_1$	$4p_3 \rightarrow 4d_2$	CW; in a He-Ne mixture over ranges 0.01—0.2 torr of Ne, 0.00—1.0 torr of He; D = 10 mm	497
12.835306	$5p'[1/2]_0 \rightarrow 4d'[3/2]^\circ_1$	$4p_1 \rightarrow 4s'_1$	CW; in a He-Ne mixture over ranges 0.01—0.2 torr of Ne, 0.00—1.0 torr of He; D = 10 mm	490,497
13.758432^m or *13.758318*	$7s'[1/2]^\circ_1 \rightarrow 6p'[3/2]_2$ or $4d'[5/2]^\circ_3 \rightarrow 4f[5/2]_3$	$5s_2 \rightarrow 5p_4$ or $4s'''_1 \rightarrow 4Y$	CW; in a He-Ne mixture over ranges 0.01—0.2 torr of Ne, 0.00— torr of He; D = 10 mm CW; in a He-Ne mixture over ranges 0.01—0.2 torr of Ne, 0.00—1.0 torr of He; D = 10 mm	486,490,497

Table 1.9.2 (continued)
NEON[a] (FIGURE 1.27)

Wavelength (μm)	Transition assignment: Racah	Transition assignment: Paschen	Comments	Ref.
14.930	?		CW; in a He-Ne mixture over ranges 0.01—0.2 torr of Ne, 0.00—1.0 torr of He; D = 10 mm	490
16.638076	$5p[3/2]_2 \rightarrow 4d[5/2]^\circ_2$	$4p_6 \rightarrow 4d''_1$	CW; in a He-Ne mixture over ranges 0.01—0.2 torr of Ne, 0.00—1.0 torr of He; D = 10 mm	490,497
16.667472	$5p[3/2]_2 \rightarrow 4d[5/2]^\circ_3$	$4p_6 \rightarrow 4d'_1$	CW; in a He-Ne mixture over ranges 0.01—0.2 torr of Ne, 0.00—1.0 torr of He; D = 10 mm	490
16.893261	$5p[3/2]_1 \rightarrow 4d[5/2]^\circ_2$	$4p_7 \rightarrow 4d''_1$	CW; in a He-Ne mixture over ranges 0.01—0.2 torr of Ne, 0.00—1.0 torr of He; D = 10 mm	486,490
16.946194	$5p[5/2]_2 \rightarrow 4d[7/2]^\circ_3$	$4p_3 \rightarrow 4d_4$	CW; in a He-Ne mixture over ranges 0.01—0.2 torr of Ne, 0.00—1.0 torr of He; D = 10 mm	486,490,497
17.157279	$5p'[3/2]_2 \rightarrow 4d'[5/2]^\circ_3$	$4p_4 \rightarrow 4s'''_1$	CW; in a He-Ne mixture over ranges 0.01—0.2 torr of Ne, 0.00—1.0 torr of He; D = 10 mm	486,490
17.188155	$5p'[3/2]_2 \rightarrow 4d'[3/2]^\circ_2$	$4p_4 \rightarrow 4s''_1$	CW; in a He-Ne mixture over ranges 0.01—0.2 torr of Ne, 0.00—1.0 torr of He; D = 10 mm	490

17.803541	$5p'[1/2]_1 \rightarrow 4d'[3/2]^o_2$	$4p_2 \rightarrow 4s''_1$	CW: IN A He-Ne mixture over ranges 0.01—0.2 torr of Ne, 0.00—1.0 torr of He; D = 10 mm	486,490
17.839876	$5p'[3/2]_1 \rightarrow 4d'[5/2]^o_2$	$4p_5 \rightarrow 4s''''_1$	CW; in a He-Ne mixture over ranges 0.01—0.2 torr of Ne, 0.00—1.0 torr of He; D = 10 mm	486,490
17.888255	$5p[5/2]_3 \rightarrow 4d[7/2]^o_4$	$4p_9 \rightarrow 4d'_4$	CW; in a He-Ne mixture over ranges 0.01—0.2 torr of Ne, 0.00—1.0 torr of He; D = 10 mm	486,490,497
18.395067	$5p[5/2]_2 \rightarrow 4d[5/2]^o_2$	$4p_8 \rightarrow 4d''_1$	CW; in a He-Ne mixture over ranges 0.01—0.2 torr of Ne, 0.00—1.0 torr of He; D = 10 mm	486,490,497
20.478165	$6p[1/2]_0 \rightarrow 5d[1/2]^o_1$	$5p_3 \rightarrow 5d_5$	CW; in a He-Ne mixture over ranges 0.01—0.2 torr of Ne, 0.00—1.0 torr of He; D = 10 mm	486,490,497
21.750951	$6p[1/2]_0 \rightarrow 5d[3/2]^o_1$	$5p_3 \rightarrow 5d_2$	CW; in a He-Ne mixture over ranges 0.01—0.2 torr of Ne, 0.00—1.0 torr of He; D = 10 mm	486,490
22.836211	$5p[1/2]_1 \rightarrow 4d[3/2]^o_2$	$4p_{10} \rightarrow 4d_3$	CW; in a He-Ne mixture over ranges 0.01—0.2 torr of Ne, 0.00—1.0 torr of He; D = 10 mm	486,490
25.421617	$6p'[1/2]_0 \rightarrow 5d'[3/2]^o_1$	$5p_1 \rightarrow 5s'_1$	CW; in a He-Ne mixture over ranges 0.01—0.2 torr of Ne, 0.00—1.0 torr of He; D = 10 mm	486,490
28.052222	$6p[3/2]_1 \rightarrow 5d[1/2]^o_1$	$5p_7 \rightarrow 6d_6$	CW; in a He-Ne mixture over ranges 0.01—0.2 torr of Ne, 0.00—1.0 torr of He; D = 10 mm	486,490

Table 1.9.2 (continued)
NEON[a] (FIGURE 1.27)

Wavelength (μm)	Transition assignment: Racah	Transition assignment: Paschen	Comments	Ref.
31.552211	$6p[3/2]_2 \rightarrow 5d[5/2]^{\circ}_3$	$5p_6 \rightarrow 5d'_1$	CW; in a He-Ne mixture over ranges 0.01—0.2 torr of Ne, 0.00—1.0 torr of He; D = 10 mm	490
31.930315	$6p[3/2]_1 \rightarrow 5d[5/2]^{\circ}_2$	$5p_7 \rightarrow 5d''_1$	CW; in pure Ne at 0.05 torr, D = 21 mm	509,510
32.015777	$6p[5/2]_2 \rightarrow 5d[7/2]^{\circ}_3$	$5p_8 \rightarrow 5d_4$	CW; in a He-Ne mixture over ranges 0.01—0.2 torr of Ne, 0.00—1.0 torr of He; D = 10 mm	486
32.518950	$6p'[3/2]_2 \rightarrow 5d'[5/2]^{\circ}_3$	$5p_4 \rightarrow 5s'''_1$	CW; in a He-Ne mixture over ranges 0.01—0.2 torr of Ne, 0.00—1.0 torr of He; D = 10 mm	490
33.815657 or *33.834764*	$6p'[3/2]_1 \rightarrow 5d'[5/2]^{\circ}_2$ or $6p[5/2]_3 \rightarrow 5d[7/2]^{\circ}_4$	$5p_5 \rightarrow 5s''''_1$ or $5p_9 \rightarrow 5d'_4$	CW; in a He-Ne mixture over ranges 0.01—0.2 torr of Ne, 0.00—1.0 torr of He; D = 10 mm	490
34.550549	$6p'[1/2]_1 \rightarrow 5d'[3/2]^{\circ}_2$	$5p_2 \rightarrow 5s''_1$	CW; in a He-Ne mixture over ranges 0.01—0.2 torr of Ne, 0.00—1.0 torr He; D = 10 mm	490
34.678633	$6p[5/2]_2 \rightarrow 5d[5/2]^{\circ}_2$	$5p_8 \rightarrow 5d''_1$	CW; in pure Ne at 0.05 torr, D = 21 mm	509,510
35.608858	$7p[1/2]_0 \rightarrow 6d[3/2]^{\circ}_1$	$6p_3 \rightarrow 6d_2$	CW; in pure Ne at 0.05 torr, D = 21 mm	18,509,510
37.230081	$7p'[1/2]_0 \rightarrow 6d'[3/2]^{\circ}_1$	$6p_1 \rightarrow 6s'_1$	CW; in pure Ne at 0.05 torr, D = 21 mm	509,510

41.738317	$6p[1/2]_1 \rightarrow 5d[3/2]^o_2$	$5p_{10} \rightarrow 5d_3$	CW; in pure Ne at 0.05 torr, D = 21 mm	509,510
50.700169	$7p[3/2]_2 \rightarrow 6d[3/2]^o_2$	$6p_6 \rightarrow 6d_3$	CW; in pure Ne at 0.05 torr, D = 21 mm	509,511
52.429587	$7p'[1/2]_1 \rightarrow 6d'[3/2]^o_2$	$6p_2 \rightarrow 6s''_1$	CW; in pure Ne at 0.02 torr, D = 47 mm	511
53.478223	$7p[3/2]_2 \rightarrow 6d[5/2]^o_3$	$6p_6 \rightarrow 6d'_1$	CW; as for the *34.679*-μm line	509,510
54.041785	$7p[3/2]_1 \rightarrow 6d[5/2]^o_2$	$6p_7 \rightarrow 6d''_1$	CW; in He-Ne mixture, 0.05 torr of Ne with 0.1 torr of He; D = 21 mm	509,510
54.106991	$7p[5/2]_2 \rightarrow 6d[7/2]^o_3$	$6p_8 \rightarrow 6d_4$	CW; in He-Ne mixture, 0.05 torr of Ne with 0.1 torr of He; D = 21 mm	509,510
55.542287	$7p'[3/2]_1 \rightarrow 6d'[5/2]^o_2$	$6p_5 \rightarrow 6s''''_1$	CW; as for the *52.43*-μm line	511
57.364118	$7p[5/2]_3 \rightarrow 6d[7/2]^o_4$	$6p_9 \rightarrow 6d'_4$	CW; in He-Ne mixture, 0.03 torr of Ne with 0.07 torr of He; D = 21 mm	509,510
68.325612	$7p[1/2]_1 \rightarrow 6d[3/2]^o_2$	$6p_{10} \rightarrow 6d_3$	CW; in pure Ne at 0.035 torr, D = 34 mm	14,512
72.118851	$8p'[1/2]_0 \rightarrow 7d'[3/2]^o_1$	$7p_1 \rightarrow 7s'_1$	CW; as for the *52.43*-μm line	511
85.062223	$8p[3/2]_2 \rightarrow 7d[5/2]^o_3$	$7p_6 \rightarrow 7d'_1$	CW; as for the *68.33*-μm line	14,512
86.977698	$8p'[3/2]_2 \rightarrow 7d'[5/2]^o_2$	$7p_4 \rightarrow 7s''''_1$	CW; as for the 68.33-μm line	511
88.487744	$8p[3/2]_1 \rightarrow 7d[5/2]^o_2$	$7p_7 \rightarrow 7d''_1$	CW; as for the 68.33-μm line	14,511
89.872291	$8p[5/2]_3 \rightarrow 7d[7/2]^o_3$	$7p_9 \rightarrow 7d_4$	CW; as for the 68.33-μm line	511
93.02	?		CW; in pure Ne at 0.01 torr; D = 47 mm	511
106.07828	$10p[1/2]_0 \rightarrow 9d[3/2]^o_1$	$9p_3 \rightarrow 9d_2$	CW; in pure Ne at 0.01 torr; D = 47 mm	14,511
124.55626 or *124.79564*	$9p[3/2]_1 \rightarrow 8d[5/2]^o_2$ or $9p[3/2]_2 \rightarrow 8d[5/2]^o_3$	$8p_7 \rightarrow 8d''_1$ or $8p_6 \rightarrow 8d'_1$	CW; in pure Ne at 0.01 torr; D = 47 mm	511
126.1	?		CW; in pure Ne at 0.01 torr; D = 47 mm	511

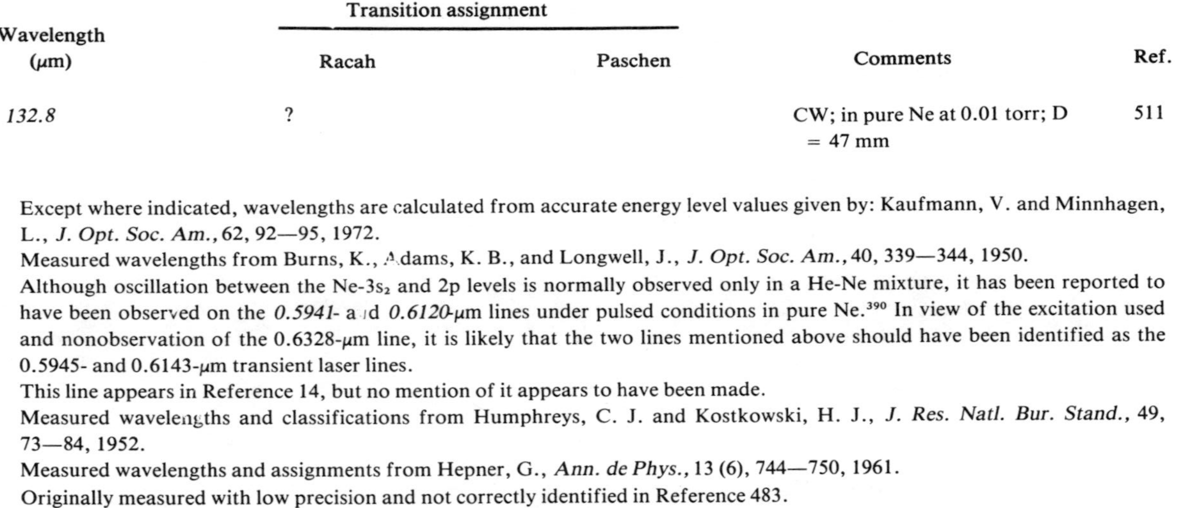

Table 1.9.2 (continued)
NEON[a] (FIGURE 1.27)

Wavelength (μm)	Transition assignment: Racah	Transition assignment: Paschen	Comments	Ref.
132.8	?		CW; in pure Ne at 0.01 torr; D = 47 mm	511

[a] Except where indicated, wavelengths are calculated from accurate energy level values given by: Kaufmann, V. and Minnhagen, L., *J. Opt. Soc. Am.*, 62, 92—95, 1972.

[b] Measured wavelengths from Burns, K., Adams, K. B., and Longwell, J., *J. Opt. Soc. Am.*, 40, 339—344, 1950.

[c] Although oscillation between the Ne-$3s_2$ and 2p levels is normally observed only in a He-Ne mixture, it has been reported to have been observed on the *0.5941*- and *0.6120*-μm lines under pulsed conditions in pure Ne.[390] In view of the excitation used and nonobservation of the 0.6328-μm line, it is likely that the two lines mentioned above should have been identified as the 0.5945- and 0.6143-μm transient laser lines.

[d] This line appears in Reference 14, but no mention of it appears to have been made.

[e] Measured wavelengths and classifications from Humphreys, C. J. and Kostkowski, H. J., *J. Res. Natl. Bur. Stand.*, 49, 73—84, 1952.

[f] Measured wavelengths and assignments from Hepner, G., *Ann. de Phys.*, 13 (6), 744—750, 1961.

[g] Originally measured with low precision and not correctly identified in Reference 483.

[h] Originally assigned to the transition $4d'[3/2]^{\circ}_{1} \rightarrow 4p'[3/2]_2$ ($4s'_1 \rightarrow 3p_4$) in Reference 497; however, this assignment has a wavelength of *2.4395241* μm. Either the reported wavelength or assignment given in Ref. 497 must be incorrect. The assignment given here is suggested by the present author on the assumption that the reported wavelength,[497] *2.864* μm, is correct.

[j] Both these lines were observed in Reference 496. However, it is uncertain whether one or both were observed in oscillation in References 486, 490, and 497.

[k] Preferred assignment of the two alternates.

[l] This transition was not observed in References 490 and 497 as was incorrectly reported previously.[578] However, it is possible that this is the transition observed at *7.400* μm in Reference 500 and previously unidentified.

[m] This is the most likely transition to give the 13.76-μm line;[490] selective excitation of the Ne-$7s'[1/2]^{\circ}_{1}$ state occurs via excitation transfer from He-2p $^1P^{\circ}_1$ atoms.[406]

[n] This transition is reported in Reference 458 where its wavelength is given as 1.3912 μm. This wavelength does not correspond to the transition assigned, nor to any Ne 2p → 2s transition. Either the wavelength reported is incorrect, or the line was not observed at all, or some other unidentified transition of Ne or an impurity was observed.

[p] Assignment given in Reference 482, seems unlikely to be correct, this is a no-parity-change electric-dipole forbidden transition.

Comments: Several previous compilations of neutral gas laser lines[578-581] contain incorrect entries corresponding to data taken from the work of Faust et al.[486,490,497] In some early work of these latter authors, several lines were incorrectly or ambiguously assigned and in many cases more than one measured wavelength for the same transition was reported. These ambiguities were clearly indicated in Reference 490 and partially eliminated in Reference 496.

Excitation transfer is responsible for the selective excitation of the Ne $5s_2$, $3s_2$, and $2s_{2\text{-}4}$ levels via excited He atoms in the He $^1P_1$,He* $^1S_0$, and He* 2^3S_1 states, respectively. The 1/D gain relationship (under optimum discharge conditions exhibited by laser transitions from the Ne-$3s_2$ and $2s_2$ levels in the He-Ne laser) is due to the populations of these levels following the concentration of He 2^1S_0 and 2^3S_1 metastables which follows a 1/D relationship. The 1/D gain relationship is not due to Ne-1s metastables, as stated throughout the laser review and book literature. This can be deduced from an analysis of the results of White and Gordon,[403] (see Labuda).[596] The same 1/D gain relationship is also shown by the 0.4416- and 0.3250-μm laser lines in the He-Cd ion laser. The upper levels of these lines are also selectively excited (in a Penning reaction) by helium metastable (He* 2^3S_1) atoms.[597,598]

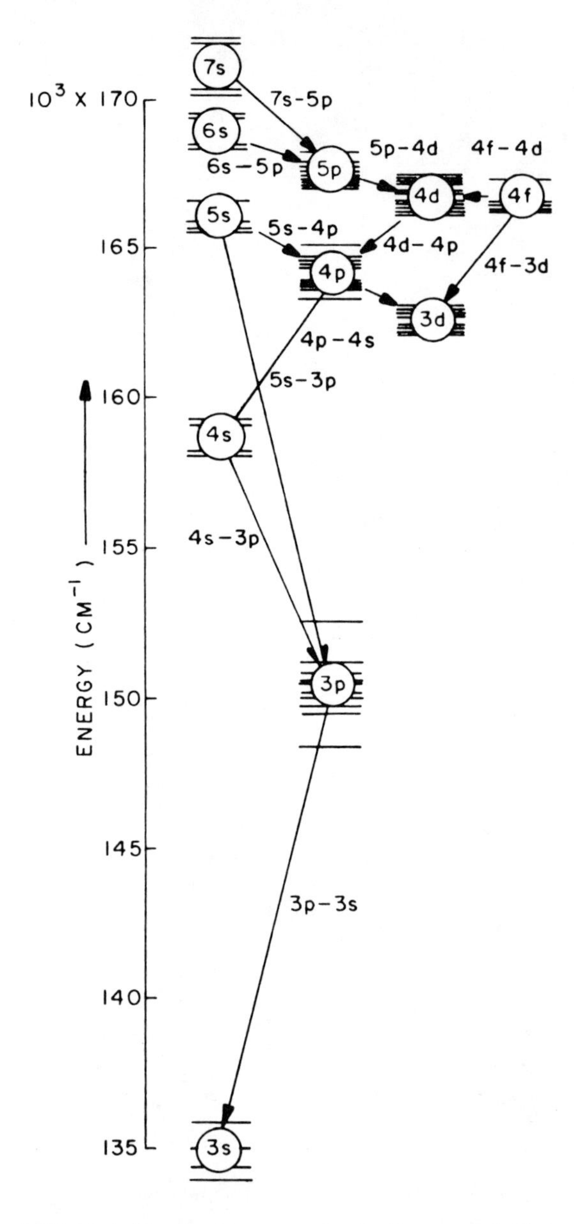

FIGURE 1.27. Partial energy-level diagram of Ne-I showing laser transition groups.

Table 1.9.3
ARGON[a] (FIGURE 1.28)

Wavelength (μm)	Transition assignment: Racah	Transition assignment: Paschen	Comments	Ref.
0.70691661	$4p'\,[3/2]_2 \rightarrow 4s[3/2]^o_2$	$2p_3 \rightarrow 1s_5$	Transient line requiring short rise-time, high-voltage, high-current pulses	392
0.75059341	$4p'\,[1/2]_0 \rightarrow 4s'\,[1/2]^o_1$	$2p_1 \rightarrow 1s_2$	Transient line as above, observed in an Ar-Ne (He) discharge; favors very low Ar pressure	388
0.912297	$4p\,[1/2]_1 \rightarrow 4s\,[3/2]^o_2$	$2p_{10} \rightarrow 1s_5$	Pulsed; in a high-pressure (>1 atm) He-Ar mixture excited in a vacuum UV photo-preionized TEA laser	218,513,514,647
0.965779	$4p\,[1/2]_1 \rightarrow 4s\,[3/2]^o_1$	$2p_{10} \rightarrow 1s_4$	Pulsed; in a high-pressure (>1 atm) He-Ar mixture excited in a vacuum UV photo-preionized TEA laser	513,514,518,647
1.04701	$4p\,[1/2]_1 \rightarrow 4s'\,[1/2]^o_0$	$2p_{10} \rightarrow 1s_3$	Pulsed; in a high-pressure (>1 atm) He-Ar mixture excited in a vacuum UV photo-preionized TEA laser	513
1.14881	$4p\,[1/2]_1 \rightarrow 4s'\,[1/2]^o_1$	$2p_{10} \rightarrow 1s_2$	Pulsed; in a high-pressure (>1 atm) He-Ar mixture excited in a vacuum UV photo-preionized TEA laser	513
1.2115637?[b]	$3d\,[5/2]_3 \rightarrow 4p\,[5/2]_3$	$3d_1' \rightarrow 2p_9$ (?)	Pulsed; in a mixture of Ar and He above 200 torr; D = 11 mm	461
1.21397378	$3d'\,[3/2]^o_1 \rightarrow 4p'\,[3/2]_1$	$3s'_1 \rightarrow 2p_4$	Pulsed; observed in an ASE mode in argon at 0.03 torr; D = 3 mm; also in a pulsed hollow-cathode laser with a 1:75 Ar-He mixture, optimum pressure 30 torr; also as for 0.9123-μm line	513
1.24028269	$3d\,[3/2]^o_1 \rightarrow 4p\,[3/2]_1$	$3d_2 \rightarrow 2p_7$	Pulsed; observed in an ASE mode in argon at 0.03 torr; D = 3 mm; also as for 0.9123-μm line	513,515
1.27022810	$3d'\,[3/2]_1 \rightarrow 4p'\,[1/2]_1$	$3s_1' \rightarrow 2p_2$	Pulsed; as for the 1.214-μm line and also in pure Ar or Ar-He mixtures in a transversely excited pin laser	99,368,513,515-518, 647

Table 1.9.3 (continued)
ARGON[a] (FIGURE 1.28)

Wavelength (μm)	Transition assignment: Racah	Transition assignment: Paschen	Comments	Ref.
1.28027391?[c]	$3d\ [5/2]^{o}_{2} \rightarrow 4p\ [5/2]_2$	$3d_1'' \rightarrow 2p_8$	Pulsed; as for the 1.2116-μm line	461
1.3476544	$7d\ [3/2]^{o}_{2} \rightarrow 5p\ [3/2]_2$	$7d_3 \rightarrow 3p_6$	CW; in Ar at 0.25 torr; D = 2.2 mm	519
1.40936399	$3d\ [3/2]^{o}_{1} \rightarrow 4p\ [1/2]_0$	$3d_2 \rightarrow 2p_5$	Pulsed; in an ASE mode in Ar at 0.04 torr; D = 7 mm	467
1.504605	$3d'\ [3/2]^{o}_{1} \rightarrow 4p'\ [1/2]_0$	$3s_1' \rightarrow 2p_1$	Pulsed; in a Cu hollow-cathode laser with a 1:75 Ar-He mixture; optimum pressure 30 torr	99
1.6180021	$5s\ [3/2]^{o}_{2} \rightarrow 4p'\ [3/2]_2$	$2s_5 \rightarrow 2p_3$	CW; in Ar at 0.05 torr; 0 = 7 mm	195, 369
1.65199	$3d\ [3/2]^{o}_{2} \rightarrow 4p\ [3/2]_1$	$3d_3 \rightarrow 2p_7$	Pulsed; in 5 torr Ar with 3 torr SF_6 in a transversely excited pin laser	518
1.6940584	$3d\ [3/2]^{o}_{2} \rightarrow 4p\ [3/2]_2$	$3d_3 \rightarrow 2p_6$	CW; in 2.5 torr of Ar; D = 7 mm; also pulsed in pure Ar or Ar-SF_6 mixture at low pressures, up to 5 torr; in transversely exicted lasers and in a hollow-cathode laser with a 1:75 Ar-He mixture, optimum pressure 30 torr	99,195,369,467, 513,518
1.7919615[d]	$3d\ [1/2]^{o}_{1} \rightarrow 4p\ [3/2]_2$	$3d_5 \rightarrow 2p_6$	CW; in 0.035 torr of Ar; D = 7 mm; also in low pressure Ar or Ar-SF_6 mixtures in transversely excited lasers or in high-pressure (> 1atm) He-Ar mixtures excited in a vacuum-UV photopreionized TEA laser; also nuclear pumped by the reaction $^3He\ (n,p)^3H$ in a 9:1 ^3He-Ar mixture at pressures from 200—700 torr; D = 2 cm	14,99,195,369,513, 516-518,520(E)-524,647
2.0616228	$3d\ [3/2]^{o}_{2} \rightarrow 4p'\ [3/2]_2$	$3d_3 \rightarrow 2p_3$	CW; in 0.035 torr of Ar; D = 7 mm; also pulsed, in a Cu hollow-cathode laser with a 1:75 Ar-He mixture; optimum pressure 30 torr	99,195,369,520-522

2.0986110	3d $[1/2]^o_1 \rightarrow$ 4p $[1/2]_0$	$3d_5 \rightarrow 2p_5$	CW; in 0.012 torr of Ar; also pulsed in a Cu hollow-cathode laser with a 1:75 Ar-He mixture; optimum pressure 30 torr	14,99,367,491
2.1332885	3d $[1/2]^o_1 \rightarrow$ 4p'$[3/2]_1$	$3d_5 \rightarrow 2p_4$	CW; in 0.01—0.05 torr of Ar; D = 10 mm	497
2.1534205	3d $[3/2]^o_2 \rightarrow$ 4p' $[1/2]_1$	$3d_3 \rightarrow 2p_2$	CW; in 0.018 torr of Ar; also pulsed in 5.0 torr of Ar plus 3 torr of SF_6 in a transversely excited pin laser	367,491,518
2.2077181[e]	3d $[1/2]^o_1 \rightarrow$ 4p' $[3/2]_2$	$3d_5 \rightarrow 2p_3$	CW; in 0.01—0.05 torr of Ar; D = 10 mm; also enhanced by the addition of Cl to an Ar-He mixture; also pulsed in a Cu hollow-cathode laser with a 1:75 Ar-He mixture; optimum pressure 30 torr	99,235,490,497
2.3133204	3d $[1/2]^o_1 \rightarrow$ 4p'$[1/2]_1$	$3d_5 \rightarrow 2p_2$	CW; in 0.01—0.05 torr of Ar; D = 10 mm; also pulsed in 5.0 torr of Ar plus 3 torr of SF_6 in a transversely excited pin laser	467,490,497,518
2.3966520	3d $[1/2]^o_0 \rightarrow$ 4p' $[1/2]_1$	$3d_6 \rightarrow 2p_2$	CW; in 0.01—0.05 torr of Ar; D = 10 mm; enhanced by the addition of Cl to an Ar-He mixture; also pulsed in pure Ar at low-pressure (<1 torr) or a high-pressure He-Ar mixture (>1 atm) excited in a vacuum UV photopreionized laser	235,368,490,497, 513,516
2.5014408	6d' $[3/2]^o_2 \rightarrow$ 6p $[1/2]_1$	$6s''_1 \rightarrow 4p^{10}$	CW; in 0.01—0.5 torr of Ar; D = 10 mm	490,497
2.54946	5p $[5/2]_3 \rightarrow$ 3d $[7/2]^o_3$	$3p_9 \rightarrow 3d_4$	CW; in 0.01—0.5 torr of Ar; D = 10 mm	490,497
or	or	or		
2.5512187	5p $[1/2]_0 \rightarrow$ 5s$[3/2]^o_1$	$3p_5 \rightarrow 2s_4$		
2.5634025	6d'$[3/2]^o_2 \rightarrow$ 6p$[5/2]_3$	$6s''_1 \rightarrow 4p_9$	CW; in 0.01—0.5 torr of Ar; D = 10 mm	497
2.5668023	5p'$[1/2]_0 \rightarrow$ 5s'$[1/2]^o_1$	$3p_1 \rightarrow 3s_2$	CW; in 0.01—0.5 torr of Ar; D = 10 mm	490
2.6550282	5p'$[3/2]_1 \rightarrow$ 3d'$[5/2]^o_2$	$3p_4 \rightarrow 3s''''_1$	CW; in 0.02 torr of Ar with 0.2 torr of He; D = 15 mm	525
2.6843026	5p $[3/2]_1 \rightarrow$ 3d$[5/2]^o_2$	$3p_7 \rightarrow 3d_1''$	CW; in 0.01—0.05 torr of Ar; D = 10 mm	490,497
2.7152859	5p $[3/2]_1 \rightarrow$ 5s$[3/2]^o_2$	$3p_7 \rightarrow 2s_5$	Pulsed; in 0.008—0.014 torr of Ar; D = 4 mm	526
2.7363805	5p'$[1/2]_1 \rightarrow$ 3d'$[3/2]^o_2$	$3p_2 \rightarrow 3s_1''$	CW; in 0.01—0.05 torr of Ar; D = 10 mm	490,497
2.8202417	5p'$[3/2]_1 \rightarrow$ 5s'$[1/2]^{o\prime}_0$	$3p_4 \rightarrow 2s_3$	CW; in 0.01—0.05 torr of Ar; D = 10 mm	490,497
or	or	or		
2.8245953	5p $[3/2]_2 \rightarrow$ 5s$[3/2]^o_1$	$3p_6 \rightarrow 2s_4$		
2.8620231	5p'$[3/2]_2 \rightarrow$ 5s'$[1/2]^o_1$	$3p_3 \rightarrow 2s_2$	CW; in 0.09 torr of Ar with 3 torr of He	522,527

Table 1.9.3 (continued)
ARGON[a] (FIGURE 1.28)

Wavelength (μm)	Transition assignment: Racah	Transition assignment: Paschen	Comments	Ref.
2.8782932	$5p\ [5/2]_3 \rightarrow 5s[3/2]^o_2$	$3p_9 \rightarrow 2s_5$	CW; in 0.01—0.05 torr of Ar; D = 10 mm	490,497
or	or	or		
2.8843088	$5p\ [3/2]_2 \rightarrow 3d[5/2]^o_3$	$3p_6 \rightarrow 3d_1'$		
2.9134037	$5p'[3/2]_1 \rightarrow 5s'[1/2]^o_1$	$3p_4 \rightarrow 2s_2$	CW	527
2.9280662	$5p[1/2]_0 \rightarrow 3d[3/2]^o_1$	$3p_5 \rightarrow 3d_2$	CW; in 0.01—0.05 torr of Ar; D = 10 mm	490,497
2.9796792	$5p\ [5/2]_2 \rightarrow 5s[3/2]^o_1$	$3p_8 \rightarrow 2s_4$	CW; in 0.01—0.05 torr of Ar; D = 10 mm	490,497
3.046207	$5p\ [5/2]_2 \rightarrow 3d[5/2]^o_3$	$3p_8 \rightarrow 3d_1'$	CW; in 0.01—0.05 torr of Ar; D = 10 mm	490,497
3.0996226	$5p\ [5/2]_3 \rightarrow 3d[5/2]^o_3$	$3p_9 \rightarrow 3d'_1$	CW; in 0.01—0.05 torr of Ar; D = 10 mm	490,497
3.1333028	$5p\ [1/2]_1 \rightarrow 5s[3/2]^o_2$	$3p_{10} \rightarrow 2s_5$ (?)	CW; in 0.01—0.05 torr of Ar; D = 10 mm	490
3.1345761	$6p'\ [3/2]_2 \rightarrow 4d[5/2]^o_2$	$4p_3 \rightarrow 4d''_1$	CW; in 0.01—0.05 torr of Ar; D = 10 mm	497
3.631236	$6s'\ [1/2]^o_1 \rightarrow 5p'[3/2]_1$	$3s_2 \rightarrow 3p_4$	Pulsed; in 0.008—0.014 torr of Ar; D = 4 mm	526
3.7013512	$6s'\ [1/2]^o_1 \rightarrow 5p'[1/2]_1$	$3s_2 \rightarrow 3p_2$	Pulsed; in 0.008—0.014 torr of Ar; D = 4 mm	526
3.7086023	$4d\ [3/2]^o_1 \rightarrow 5p[3/2]_1$	$4d_2 \rightarrow 3p_7$	Pulsed; in 0.01—0.015 torr of Ar; D = 8 mm	528
3.7143477	$6s'\ [1/2]^o_1 \rightarrow 5p'[3/2]_2$	$3s_2 \rightarrow 3p_3$	Pulsed; in 0.008—0.014 torr or Ar; D = 4 mm	526
4.2044098	$5p\ [3/2]_2 \rightarrow 3d'[3/2]^o_2$	$3p_6 \rightarrow 3s''_1$	CW; in 0.06 torr of Ar; D = 15 mm	525
4.7151680	$5p\ [5/2]_3 \rightarrow 3d'[5/2]^o_3$	$3p_9 \rightarrow 3s'''_1$	CW; in 0.02 torr of Ar with 0.2 torr of He; D = 15 mm	525
4.9160256	$6p'[3/2]_2 \rightarrow 4d'[3/2]^o_2$	$4p_3 \rightarrow 4s''_1$	CW; in 0.01—0.05 torr or Ar; D = 10 mm	490,497
or	or	or		
4.9207077	$5d\ [5/2]^o_2 \rightarrow 4f[7/2]^i_3$	$5d'_1 \rightarrow 4U$		
5.0235584	$6p'[3/2]_1 \rightarrow 4d'[5/2]^o_2$	$4p_4 \rightarrow 4s''''_1$	Pulsed; in 0.008—0.014 torr of Ar; D = 4 mm	526
5.1216467	$6p\ [5/2]_3 \rightarrow 4d[7/2]^{og}_3$	$4p_9 \rightarrow 4d_4$	CW; in 0.01—0.05 torr of Ar; D = 10 mm	490,497
5.3912471	$5p\ [1/2]_1 \rightarrow 3d'[3/2]^o_2$	$3p_{10} \rightarrow 3s_1''$	CW; in 0.06 torr of Ar; D = 15 mm	525

5.4676586	$5d\,[7/2]^\circ_4 \rightarrow 4f[9/2]_5'$	$5d'_4 \rightarrow 4V_5$	CW; in 0.05 torr of Ar; D = 15 mm	486, 490
or	or	or		
5.4677812	$5d[7/2]^\circ_4 \rightarrow 4f[9/2]_4$	$5d'_4 \rightarrow 4V_4$		
5.8037601	$4d\,[3/2]^\circ_2 \rightarrow 5p[3/2]_2$	$4d_3 \rightarrow 3p_6$	CW; in 0.02 torr of Ar with 0.2 torr of He; D = 15 mm; also pulsed in a 9:3 He-Ar mixture at 100 torr in a transversely excited pin laser	368,525,527
5.8477292	$6p\,[1/2]_0 \rightarrow 6s[3/2]^\circ_1$	$4p_5 \rightarrow 3s_4$	CW; in 0.05 torr of Ar; D = 15 mm	486,490
5.8665236	$6p\,[3/2]_2 \rightarrow 4d[5/2]^\circ_3$	$4p_6 \rightarrow 4d_1'$	CW; in 0.01—0.05 torr of Ar; D = 10 mm	497
6.0530155	$4d\,[1/2]^\circ_1 \rightarrow 5p[5/2]_2$	$4d_5 \rightarrow 3p_8$	CW; in 0.01—0.05 torr of Ar; D = 10 mm	490,497
6.7463414	$6p\,[5/2]_3 \rightarrow 6s[3/2]^\circ_2$	$4p_9 \rightarrow 3s_5$	CW; in 0.02 torr of Ar with 0.2 torr of He; D = 15 mm	525
6.9428548	$4d\,[3/2]^\circ_1 \rightarrow 5p'[3/2]_1$	$4d_2 \rightarrow 3p_4$	CW; in 0.01—0.05 torr of Ar; D = 10 mm	490,497
or	or	or		
6.9448712	$6p'[1/2]_1 \rightarrow 6s'[1/2]^\circ_1{}'$	$4p_2 \rightarrow 3s_2$		
7.2168093	$6p\,[1/2]_1 \rightarrow 6s[3/2]^\circ_2$	$4p^{10} \rightarrow 3s_5$	CW; in 0.05 torr of Ar; D = 10 mm	486,490
7.2932	?		Pulsed; in a 7:3 He-Ar mixture at 100 torr in a transversely excited pin laser	368
7.8001903	$4f\,[3/2]_2 \rightarrow 4d[3/2]^\circ_2$	$4X_2 \rightarrow 4d_3$	CW; in 0.05 torr of Ar; D = 10 mm	490
or	or	or		
7.8026187	$4f\,[3/2]_1 \rightarrow 4d[3/2]^\circ_2$	$4X_1 \rightarrow 4d_3$		
7.8065353	$7s'[1/2]^\circ_1 \rightarrow 4p'[1/2]_1$	$4s_2 \rightarrow 2p_2$	CW; in 0.01—0.05 torr of Ar; D = 10 mm	497
12.140522	$4d'[3/2]^\circ_1 \rightarrow 4f[3/2]_1$	$4s' \rightarrow 4X_1$	CW; in 0.01—0.05 torr of Ar; D = 10 mm	486,490,497
or	or	or		
12.146405	$4d'[3/2]^\circ_1 \rightarrow 4f[3/2]_2$	$4s'_1 \rightarrow 4X_2$		
15.037067	$5d'[3/2]^\circ_2 \rightarrow 5f[5/2]_3$	$5s''_1 \rightarrow 5Y_3$	CW; in 0.05 torr of Ar; D = 10 mm	486,490
or	or	or		
15.042133	$5d'[3/2]^\circ_2 \rightarrow 5f[5/2]_2$	$5s''_1 \rightarrow 5Y_2$		
26.943974	$4d'[3/2]^\circ_2 \rightarrow 4f[5/2]_3$	$4s''_1 \rightarrow 4Y_3$	CW; in 0.05 torr of Ar; D = 10 mm	486,490

[a] Wavelengths and spectral assignments are from data in Minnhagen, L., *J. Opt. Soc. Am.*, 63, 1185-1198, 1973 and Norlen, G., *Phys. Scripta*, 8, 249—268, 1973.

[b] Possibly the same as the Ar-I line at 1.21397 μm.

[c] Possibly the same as the Ar-I line at 1.27023 μm.

[d] This is a very strong line in a high-pressure He-Ar TEA laser. The assignment listed, rather than an alternate listed previously in the literature, was confirmed by Dauger and Stafsudd in line competition experiments.[522]

[e] The assignment listed, rather than an alternate listed previously in the literature was confirmed by Dauger and Stafsudd in line competition experiments.[522]

Table 1.9.3 (continued)
ARGON[a] (FIGURE 1.28)

[f] Preferred assignment.

[g] An alternate assignment, 5d $[7/2]^{\circ}_{3} \rightarrow 4f[9/2]_4$ ($5d'_4 \rightarrow 4V_4$) at *5.1203927* listed in previous compilations of neutral argon transitions has been eliminated; it is considerably outside the reported error of the measured wavelength[486] *5.1218 ± 10⁻⁴*.

[h] This line is mentioned in Reference 497, but not in Reference 490. It is probably, in fact, the line listed immediately above in the table.

Comments: Lines listed in previous compilations[578-580] at 0.8780, 1.0935, and 1.8167 μm have been omitted. The first of these is an Ar-II line,[213] the second is almost certainly an Ar-II line,[213] but is listed here under miscellaneous and unidentified possible neutral laser transitions, (see Table 1.10.1), and the third line is a neutral Kr line (Table 1.9.4). Unidentified laser lines observed in a pulsed Ar-Hg discharge at 1.222, 1.246, and 1.276 μm[129] have been listed as unidentified neutral Hg transitions (Table 1.2.7).

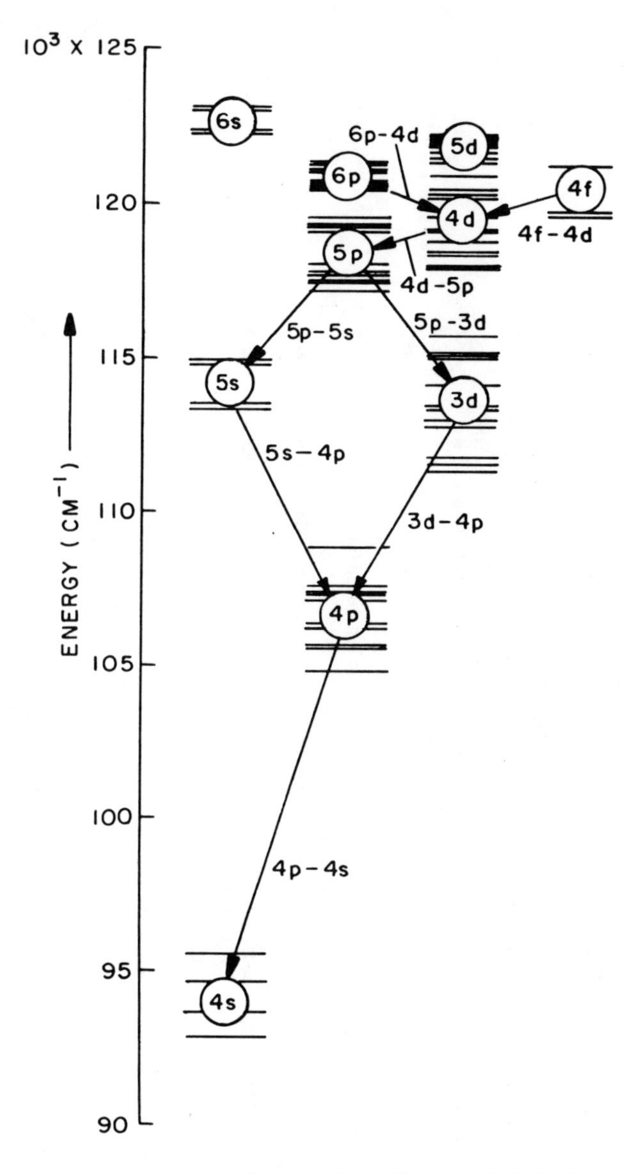

FIGURE 1.28. Partial energy-level diagram of Ar-I showing laser transition groups.

Table 1.9.4
KRYPTON[a] (FIGURE 1.29)

Wavelength (μm)	Transition assignment: Racah	Transition assignment: Paschen	Comments	Ref.
0.760154393	$5p\,[3/2]_2 \rightarrow 5s[3/2]^{\circ}_2$	$2p_6 \rightarrow 1s_5$	Pulsed; in 0.01—0.015 torr of Kr; D = 8 mm	528
0.810436392	$5p\,[5/2]_2 \rightarrow 5s[3/2]^{\circ}_2$	$2p_8 \rightarrow 1s_5$	Transient line requiring short rise-time, high-voltage, high-current pulses; in about 0.1 torr of Kr; D = 3 mm; peak current 1000 A	392,393,467
0.892869155	$5p[1/2]_1 \rightarrow 5s[3/2]^{\circ}_2$	$2p_{10} \rightarrow 1s_5$	Pulsed; in a 7 atm He-Kr mixture excited in a vacuum-UV photopreionized TEA laser	513,514
1.14574813	$6s[3/2]^{\circ}_1 \rightarrow 5p[1/2]_1$	$2s_4 \rightarrow 2p_{10}$	Pulsed; in an ASE mode in 0.04 torr of Kr; D = 7 mm	467
1.31774118	$6s[3/2]^{\circ}_1 \rightarrow 5p[5/2]_2$	$2s_4 \rightarrow 2p_5$	Pulsed; in an ASE mode in 0.06 torr of Kr; D = 7 mm	467
1.36224153	$4d[3/2]^{\circ}_1 \rightarrow 5p[5/2]_2$	$3d_2 \rightarrow 2p_8$	Pulsed; in an ASE mode in 0.03 torr of Kr; D = 7 mm	467
1.44267933	$6s[3/2]^{\circ}_1 \rightarrow 5p[3/2]_1$	$2s_4 \rightarrow 2p_7$	Pulsed; in an ASE mode in 0.08 torr of Kr; D = 7 mm	467
1.47654720	$6s[3/2]^{\circ}_1 \rightarrow 5p[3/2]_2$	$2s_4 \rightarrow 2p_8$	Pulsed; in an ASE mode in 0.2 torr of Kr; D = 7 mm	467
1.49618939	$4d[3/2]^{\circ}_1 \rightarrow 5p[3/2]_1$	$3d_2 \rightarrow 2p_7$	Pulsed; in an ASE mode in 0.01 — 0.015 torr of Kr; D = 2 or 4 mm	529
1.53264796	$4d[3/2]^{\circ}_1 \rightarrow 5p[3/2]_2$	$3d_2 \rightarrow 2p_6$	Pulsed; in an ASE mode in 0.01—0.015 torr of Kr; D = 2 or 4 mm	529
1.68534881	$4d[7/2]^{\circ}_3 \rightarrow 5p[5/2]_3$	$3d_4 \rightarrow 2p_9$	Pulsed; in an ASE mode in 0.07 torr of Kr; D = 7 mm	467
1.68967525	$4d[1/2]^{\circ}_1 \rightarrow 5p[1/2]_1$	$3d_5 \rightarrow 2p_{10}$	Pulsed; in an ASE mode in 0.08 torr of Kr; D = 7 mm	195,369,467
1.69358061	$4d[5/2]^{\circ}_2 \rightarrow 5p[3/2]_1$	$3d''_1 \rightarrow 2p_7$	CW; in 0.05 torr of Kr; D = 7 mm	195,369
1.78427374	$4d[1/2]^{\circ}_0 \rightarrow 5p[1/2]_1$	$3d_6 \rightarrow 2p_{10}$	CW; in 0.07 torr of Kr; D = 7 mm	195,369
1.81673150	$4d[7/2]^{\circ}_4 \rightarrow 5p[5/2]_3$	$3d'_4 \rightarrow 2p_9$	CW; in 0.015 torr of Kr; D = 9 mm	367
1.81850539	$4d'[5/2]^{\circ}_2 \rightarrow 5p'[3/2]_2$	$3s''''_1 \rightarrow 2p_2$	CW; in 0.07 torr of Kr; D = 7 mm	195,369
1.9216572	$8s[3/2]^{\circ}_2 \rightarrow 6p[5/2]_2$	$4s_5 \rightarrow 3p_8$	CW; in 0.035 torr of Kr; D = 7 mm	195,369(E)
2.11654709	$4d[3/2]^{\circ}_2 \rightarrow 5p[3/2]_1$	$3d_3 \rightarrow 2p_7$	CW; in 0.035 torr of Kr; 0 = 7 mm	14,195,369,496

2.19025126	$4d[3/2]^o_2 \rightarrow 5p[3/2]_2$	$3d_3 \rightarrow 2p_6$	CW; in 0.035 torr of Kr; D = 7 mm	14,195,369,496, 530
2.24857754	$6p[3/2]_1 \rightarrow 4d[5/2]^{o\,b}_2$	$3p_7 \rightarrow 3d''_1$	Pulsed; in 0.008—0.014 torr of Kr; D = 4 mm	526
2.42605059	$4d[1/2]^o_1 \rightarrow 5p[3/2]_1$	$3d_5 \rightarrow 2p_7$	CW	367,491
2.52338198	$4d[1/2]^o_1 \rightarrow 5p[3/2]_2$	$3d_5 \rightarrow 2p_6$	CW; in an ASE mode in 1.0 torr of Kr; D = 7 mm; also pulsed in a 93:7 He-Kr mixture at 760 torr in a transversely excited pin laser; also nuclear pumped by the rea ction $^3He(n,p)^3H$	195,368,467,486, 490,496,508, 531
2.6266703	$4d[1/2]^o_0 \rightarrow 5p[3/2]_1$	$3d_6 \rightarrow 2p_7$	CW; in 0.02 torr of Kr; D = 10 mm	490,497
or	or	or		
2.6288137	$7p[3/2]_2 \rightarrow 4d'[5/2]^o_2$	$4p_6 \rightarrow 3s''''_1$		
2.8610550	$6p[5/2]_2 \rightarrow 6s[3/2]^o_2$	$3p_8 \rightarrow 2s_5$	CW; in 0.02 torr of Kr; D = 10 mm	467,486,497
or	or	or	CW; in 0.02 torr of Kr; D = 10 mm	
2.8655717	$6p[5/2]_3 \rightarrow 6s[3/2]^o_2$	$3p_9 \rightarrow 2s_5$		
2.9844656	$6p'[1/2]_1 \rightarrow 5d[5/2]^o_2$	$3p_3 \rightarrow 4d''_1$	CW; in 0.02 torr of Kr; D = 10 mm	490,497
or	or	or		
2.9878091[d]	$6p'[3/2]_1 \rightarrow 6s'[1/2]^o_0$	$3p_4 \rightarrow 2s_3$		
3.0536574	$6p'[3/2]_1 \rightarrow 5d[5/2]^o_2$	$3p_4 \rightarrow 4d''_1$	CW; in 0.02 torr of Kr; D = 10 mm	490,497
3.0663542	$6p[1/2]_1 \rightarrow 6s[3/2]^o_2$	$3p_{10} \rightarrow 2s_5$	CW; in 0.03 torr of Kr; D = 15 mm; also pulsed in a 93:7 He-Kr mixture at 760 torr in a transversely excited pin laser	368,486,490,496
3.1514572	$6p'[1/2]_0 \rightarrow 5d[3/2]^o_1$	$3p_1 \rightarrow 4d_2$	CW; in 0.02 torr of Kr; D = 10 mm	490,497
3.3409635	$4d[1/2]^o_1 \rightarrow 5p[1/2]_0$	$3d_5 \rightarrow 2p_5$	CW; in 0.02 torr of Kr; D = 10 mm	490(E),496,497
3.4679986	$7s[3/2]^o_1 \rightarrow 6p[1/2]_1$	$3s_4 \rightarrow 3p_{10}$	CW; in 0.02 torr of Kr; D = 10 mm	490,497
3.4882957	$6p'[1/2]_1 \rightarrow 7s[3/2]^o_2$	$3p_3 \rightarrow 3s_5$	CW; in 0.02 torr of Kr; D = 10 mm	490,497
or	or	or		
3.4894892	$6p'[1/2]_1 \rightarrow 5d[3/2]^o_1$	$3p_3 \rightarrow 4d_2$		
3.7742128	$7s[3/2]^o_1 \rightarrow 6p[5/2]_2$	$3s_4 \rightarrow 3p_8$	Pulsed; in 0.01—0.015 torr of Kr; D = 8 mm	528
3.9557248	$5d[3/2]^o_1 \rightarrow 6p[5/2]^e_2$	$4d_2 \rightarrow 3p_8$	Pulsed; in 0.01—0.015 torr of Kr; D = 8 mm	528
4.0685162	$7s[3/2]^o_1 \rightarrow 6p[3/2]_1$	$3s_4 \rightarrow 3p_7$	Pulsed; in 0.01—0.015 torr of Kr; D = 8 mm	528
4.1526711	$7s[3/2]^o_1 \rightarrow 6p[3/2]_2$	$3s_4 \rightarrow 3p_6$	Pulsed; in 0.01—0.015 torr of Kr; D = 8mm	528
4.3747938	$5d[3/2]^o_1 \rightarrow 6p[3/2]_2$	$3s_4 \rightarrow 3p_6$	CW; in 0.02 torr of Kr; D = 10 mm	486,490,496,528
4.3766712	$7s[3/2]^o_2 \rightarrow 6p[3/2]_2$	$3s_5 \rightarrow 3p_6$	CW; in 0.02 torr of Kr; D = 10 mm	497

Table 1.9.4 (continued)
KRYPTON[a] (FIGURE 1.29)

Wavelength (μm)	Transition assignment: Racah	Transition assignment: Paschen	Comments	Ref.
4.8773393	$4d[3/2]^{\circ}_{1} \rightarrow 5p'[3/2]_{1}$	$3d_2 \rightarrow 2p_4$	CW; in 0.02 torr of Kr; D = 10 mm	490,497
or	or	or		
4.8831334	$5d[5/2]^{\circ}_{2} \rightarrow 6p[5/2]_{3}$	$4d''_1 \rightarrow 3p_9$		
4.9996952	$4d'[3/2]^{\circ}_{1} \rightarrow 6p[1/2]_{1}$	$3s'_1 \rightarrow 3p_{10}$	CW; in pure Kr at 0.02 torr or a Kr-He mixture with 0.02 torr of Kr and 0.2 torr of He; D = 15 mm	525,529
5.1311509	$6s[3/2]^{\circ}_{1} \rightarrow 5p'[3/2]_{2}$	$2s_4 \rightarrow 2p_2$	CW; in pure Kr at 0.02 torr of a Kr-He mixture with 0.02 torr of Kr and 0.2 torr of He; D = 15 mm	525
5.2999768	$5d[3/2]^{\circ}_{1} \rightarrow 6p[1/2]_{0}$	$4d_2 \rightarrow 3p_5$	CW; in 0.02 torr of Kr; D = 10 mm	486,490,497
or	or	or		
5.3019553	$5d[3/2]^{\circ}_{2} \rightarrow 6p[5/2]_{2}$	$4d_3 \rightarrow 3p_8$		
5.5699805?[f]	$5d[7/2]^{\circ}_{3} \rightarrow 6p[5/2]_{2}$	$4d_4 \rightarrow 3p_8$	CW; in 0.02 torr of Kr; D = 10 mm	14,490,497
5.5862769[d]	$6d[7/2]^{\circ}_{4} \rightarrow 4f[9/2]_{5}$	$5d'_4 \rightarrow 4U$	CW; in 0.03 torr of Kr; D = 15 mm	486,490,496
5.6305126	$6d[3/2]^{\circ}_{2} \rightarrow 4f[5/2]_{3}$	$5d_3 \rightarrow 4T$	CW; in 0.03 torr of Kr; D = 15 mm	486,490,496
7.0580595	$4f[7/2]_{3,4} \rightarrow 5d[7/2]^{\circ}_{4}$	$4W \rightarrow 4d'_4$	CW; in 0.02 torr of Kr; D = 10 mm	490,497
7.3625232	$4f[9/2]_{5} \rightarrow 5d[7/2]^{\circ}_{4}$	$4U_5 \rightarrow 4d'_4$	CW; in pure Kr at 0.02 torr or in a Kr-He mixture with 0.02 torr of Kr and 0.2 torr of He; D = 15 mm	525

[a] Unless otherwise indicated, wavelengths and spectral assignment are taken from data on ^{86}Kr in Kaufman, V. and Humphreys, C. J., *J. Opt. Soc. Am.*, 59, 1614—1628, 1969. Although ^{86}Kr represents only 17.4% of natural isotopic abundance Kr, it is the spectral lines of this isotope that have been selected as wavelength standards by the International Astronomical Union and the International Committee of Weights and Measures.

[b] This assignment is suggested by the present author as being much more likely to be correct than the two alternative assignments involving higher-lying levels listed by Linford.[526] It is also closer to the measured wavelength.

[c] Both these alternative assignments may have been observed as independent laser transitions. Andrade et al.[467] list only the first assignment, while Faust et al.[490] prefer the second.

[d] Calculated wavelength taken from data in Moore, C. E., *Natl. Stand. Ref. Data. Ser. Natl Bur. Stand.*, NSRDS-NBS 35, 2, 1971.

[e] An alternate assignment to the transition $7s[3/2]^{o}_{2} \rightarrow 6p[5/2]_2$ ($3s_5 \rightarrow 3p_6$) at 3.9572660 given in Reference 525 has been eliminated on the basis of the measured wavelength given in Reference 528.

[f] The existence of this laser transition is very doubtful; it is probably the Xe transition at *5.5755* μm; Xe had been used in the same laser tube.

Comments: Kr lines listed at 0.3050,[578-580] 0.7525,[581] and 0.8589 μm[578-580] in previous compilations of neutral laser lines have been omitted; these are singly ionized Kr transitions.[213]

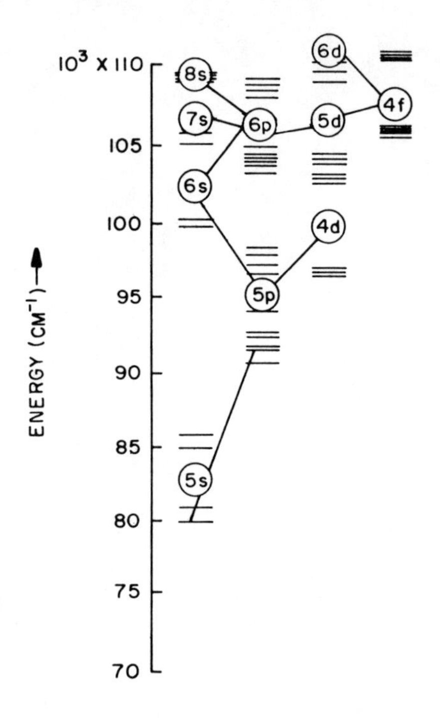

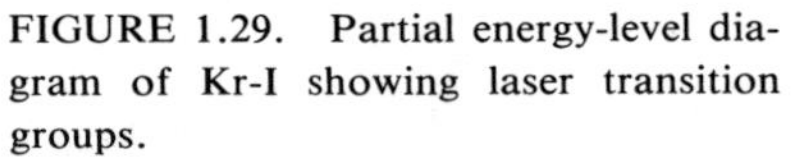
FIGURE 1.29. Partial energy-level diagram of Kr-I showing laser transition groups.

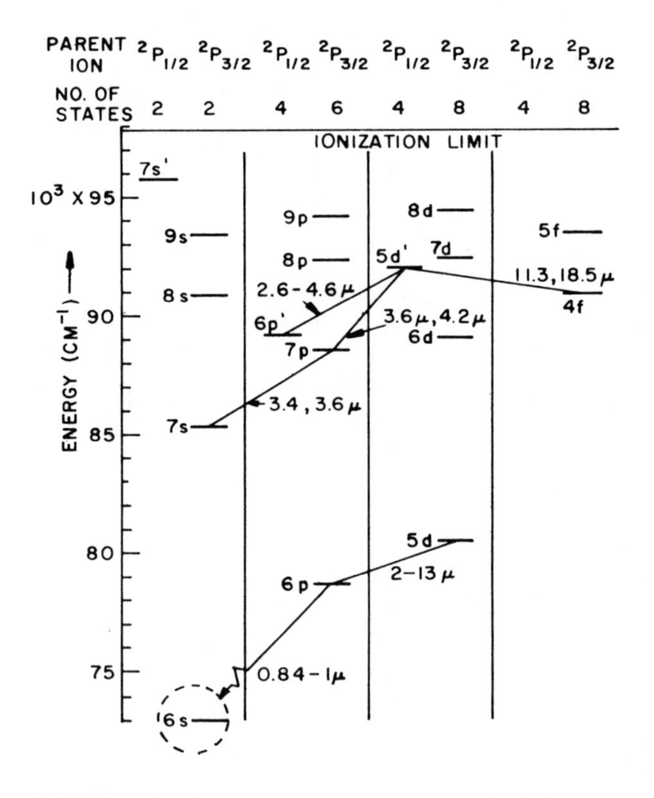

FIGURE 1.30. Partial energy-level diagram of Xe-I showing groups of observed laser transitions.

Table 1.9.5
XENON[a] (FIGURE 1.30)

Wavelength (μm)	Transition assignment: Racah	Transition assignment: Paschen	Comments	Ref.
0.82316376	$6p[3/2]_2 \rightarrow 6s[3/2]^o_2$	$2p_6 \rightarrow 1s_5$	Pulsed; in a 7 atm He-Xe mixture in a vacuum-UV photopreionized TEA laser	218,513,514
0.84091940	$6p[3/2]_1 \rightarrow 6s[3/2]^o_2$	$2p_7 \rightarrow 1s_5$	Transient line which operates in an ASE mode in short rise-time, high-voltage, high-current pulsed discharges; in 0.04 torr of Xe; D = 7 mm	467,532
0.90454514	$6p[5/2]_2 \rightarrow 6s[3/2]^o_2$	$2p_9 \rightarrow 1s_5$	Transient line which operates in an ASE mode as above; in 0.12 torr of Xe; D = 7 mm	393, 467
0.97997039	$6p[1/2]_1 \rightarrow 6s[3/2]^o_2$	$2p_{10} \rightarrow 1s_5$	Pulsed; in 0.2—0.4 torr of Xe; also in a 7 atm He-Xe or a 1 atm Ar-Xe mixture in a vacuum-UV photopreionized TEA laser	393,513,514
1.0634[b,c]	?		Pulsed; in 0.001—0.02 torr of Xe; D = 2.7—4.0 mm	532
1.0950[b]	?		Pulsed in 0.001—0.02 torr of Xe; D = 2.7—4.0 mm	532
1.36570559	$7s[3/2]^o_1 \rightarrow 6p[5/2]_2$	$2s_4 \rightarrow 2p_9$	Pulsed; in an ASE mode in 0.04 torr of Xe; D = 7 mm	467
1.6057666	$7s[3/2]^o_1 \rightarrow 6p[3/2]_2$	$2s_4 \rightarrow 2p_6$	Pulsed; in an ASE mode in 0.1 torr of Xe; D = 7 mm	467
1.7330499	$5d[3/2]^o_1 \rightarrow 6p[5/2]_2$	$3d_2 \rightarrow 2p_9$	Pulsed; in an ASE mode in 0.15 torr of Xe; D = 7 mm; also CW in 0.03—0.1 torr of Xe; D = 4 or 7 mm; also pulsed in a 1 atm He-Xe or Ar-Xe mixture in a vacuum-UV photopreionized TEA laser or in a high-pressure Ar-Xe mixture excited in a E-beam ionizer-sustainer mode laser	467,513,514,533—536,647

Table 1.9.5 (continued)
XENON[a] (FIGURE 1.30)

Wavelength (μm)	Transition assignment: Racah	Transition assignment: Paschen	Comments	Ref.
2.02622395	$5d[3/2]^{o}_{1} \rightarrow 6p[3/2]_{1}$	$3d_2 \rightarrow 2p_7$	CW; operates in an ASE mode even in short discharge tubes; the gain in a 5-mm bore tube is more than 45 dBm^{-1}; in a few mtorr of Xe or in a 100:1 He-Xe mixture at a total pressure of about 10 torr; D = 5 mm. To avoid cataphoretic effects in He-Xe mixture rf-excitation is necessary; clean-up of xenon is a problem;[554] also pulsed in high-pressure (~1 atm) He-Xe, Ar-Xe, or Ne-Xe mixtures in a vacuum-UV photopreionized TEA laser; pulsed in a Cu hollow-cathode discharge; also pulsed, nuclear-pumped by the reaction $^{3}He(n,p)^{3}H$; D = 10 mm	14,99,195,369,467, 513,514,530,531, 533—552,648
2.31933328	$5d[5/2]^{o}_{3} \rightarrow 6p[5/2]_{2}$	$3d'_1 \rightarrow 2p_9$	CW; in 0.01—0.04 torr of Xe, He also added to about 1 torr; D = 7 mm	478,486,508,530
2.48247157	$5d[5/2]^{o}_{3} \rightarrow 6p[5/2]_{3}$	$3d'_1 \rightarrow 2p_8$	CW; in 0.03—0.1 torr of Xe; D = 4 or 7 mm	533
2.5152702	$7d[7/2]^{o}_{4} \rightarrow 7p[5/2]_{3}$	$5d'_4 \rightarrow 3p_9$	Pulsed; in 0.005—0.011 torr of Xe; D = 4 mm	526
2.62690832	$5d[5/2]^{o}_{2} \rightarrow 6p[5/2]_{2}$	$3d''_1 \rightarrow 2p_9$	CW; as for the 2.32-μm line; also pulsed in a high-pressure Ar-Xe mixture containing <5% Xe excited in a E-beam ionizer-sustainer mode	486,508,530,535, 553
2.65108645	$5d[3/2]^{o}_{1} \rightarrow 6p[1/2]_{0}$	$3d_2 \rightarrow 2p_5$	CW; as for the 2.32-μm line; also in high-pressure Ar-Xe or He-Xe mixtures containing <5% Xe excited in E-beam ionizer-sustainer and other types of pulsed transversely excited lasers	467,478,508,530, 533—535,536, 545,547,555,648
2.6608397[d]	$5d'[3/2]^{o}_{1} \rightarrow 6p'[1/2]_{0}$	$3s_1 \rightarrow 2p_1$	CW; as for the 2.32-μm line	486,530
2.6672615	$7d[1/2]^{o}_{1} \rightarrow 6p'[3/2]_{1}$	$3p_{10} \rightarrow 2p_4$	CW; in 0.01—0.06 torr of Xe; D = 15 mm	525

2.8590043	$7p[3/2]_2 \rightarrow 7s[3/2]^o_2$	$3p_6 \rightarrow 2s_5$	Pulsed; in 0.005—0.011 torr of Xe; D = 4 mm	526
3.10692302	$5d[5/2]^o_3 \rightarrow 6p[3/2]_2$	$3d'_1 \rightarrow 2p_6$	CW; as for the 2.32-μm line	478,486,508,530, 534,545,553
3.27392788	$5d[3/2]^o_2 \rightarrow 6p[1/2]_1$	$3d_3 \rightarrow 2p_{10}$	CW; in a 250:1 He-Xe mixture in an rf-discharge at about 0.4 torr; D = 11 mm[478]	367,478,491,530, 553
3.3094055	$8p[5/2]_3 \rightarrow 6d[5/2]^o_2$	$4p_8 \rightarrow 4d'_1$	CW; as for the *2.6673*-μm line above	525
3.36666991	$5d[5/2]^o_2 \rightarrow 6p[3/2]_1$	$3d''_1 \rightarrow 2p_7$	CW; in 0.01—0.04 torr of Xe, He added to 1 torr; D = 7 mm; also weakly in a pulsed high pressure Ar-Xe Mixture with <1% Xe excited in an E-beam ionizer-sustainer laser	478,486,508,530, 534,536,545
3.4023945	$6p'[3/2]_1 \rightarrow 7s[3/2]^o_1$	$2p_4 \rightarrow 2s_4$	CW; in pure Xe at 0.01—0.06 torr or in 0.015 torr of Xe with 0.3 torr of He; D = 15 mm	525
3.4344638[e]	$7p[5/2]_2 \rightarrow 7s[3/2]^o_1$	$3p_9 \rightarrow 2s_4$	CW; in 0.01—0.04 torr of Xe, He added to 1 torr; D = 7 mm; also pulsed in a high repetition rate transversely excited He-Xe mixture containing <1% Xe or in a high-pressure He-Xe mixture excited in an E-beam ionizer-sustainer laser	14,486,490,508, 536,549(E),551
3.50702520[f]	$5d[7/2]^o_3 \rightarrow 6p[5/2]_2$	$3d_4 \rightarrow 2p_9$	CW; operates in an ASE mode even in short discharge tubes, gain as high as 400 dBm^{-1} in a 0.75-mm bore tube;[560] operates in a few tens of mtorr of Xe and with He to about 10 torr with D = 5 mm; clean-up of Xe is a problem;[554] also pulsed in a high-pressure He-Xe mixture containing <1% Xe excited in an E-beam ionizer-sustainer laser or nuclear pumped by the reaction $^3He(n,p)^3H$ in a 10-mm bore tube or in a 200 torr 20:1 He-Xe mixture in a neutron activated ^{235}U lined tube; D = 19 mm	195,478,486,496, 508,530,534,535, 545—547,552—554,556—568
3.6219081	$5d'[3/2]^o_2 \rightarrow 7p[3/2]_2$	$3s''''_1 \rightarrow 3p_6$	CW; in 0.01—0.04 torr of Xe, He added to 1.0 torr; D = 7 mm	486
3.6518315	$7p[1/2]_1 \rightarrow 7s[3/2]^o_2$	$3p_{10} \rightarrow 2s_5$	CW; in 0.02 torr of Xe; D = 7 mm. also pulsed in high-pressure He-Xe mixture with ≤5% Xe excited in E-beam ionizer-sustainer and various other transversely excited lasers; also nuclear pumped by the reaction $^3He(n,p)^3H$ in a 10-mm bore tube.	490,534—536,547, 549,551,552,648

Table 1.9.5 (continued)
XENON[a] (FIGURE 1.30)

Wavelength (μm)	Transition assignment: Racah	Transition assignment: Paschen	Comments	Ref.
3.6798859	$5d[1/2]^o_1 \rightarrow 6p[1/2]_1$	$3d_5 \rightarrow 2p_{10}$	CW; in 0.01—0.04 torr of Xe, He added up to 1.0 torr; D = 7 mm	486,508,553
3.6858866	$5d[5/2]^o_2 \rightarrow 6p[3/2]_2$	$3d'' \rightarrow 2p_6$	CW; in 0.01—0.04 torr of Xe, He added up to 1.0 torr; D = 7 mm	486,508,534,545
3.8696535	$5d'[5/2]^o_3 \rightarrow 6p'[3/2]_2$	$3s''_1 \rightarrow 2p_3$	CW; in 0.01—0.4 torr of Xe, He added up to 1.0 torr; D = 7 mm	486,553
3.8950221	$5d[7/2]^o_3 \rightarrow 6p[5/2]_3$	$3d_4 \rightarrow 2p_8$	CW; in 0.01—0.4 torr of Xe, He added up to 1.0 torr; D = 7 mm	486,508,534
3.9966035	$5d[1/2]^o_0 \rightarrow 6p[1/2]_1$	$3d_6 \rightarrow 2p_{10}$	CW; in 0.01—0.4 torr of Xe, He added up to 1.0 torr; D = 7 mm	486,534,537,545, 546
4.0207278	$7p[1/2]_1 \rightarrow 7s[3/2]^o_1$	$3p_{10} \rightarrow 3s_4$	CW; in 0.01—0.06 torr of Xe or in 0.015 torr of Xe with 0.3 torr of He; D = 15 mm; also pulsed in a high-pressure, high repetition rate TEA laser with a 200:1 He-Xe mixture at 300 torr	525,570
4.1526299	$5d'[5/2]^o_2 \rightarrow 7p[3/2]_1$	$3s''_1 \rightarrow 3p_7$	CW; as for the *3.68*-μm line	486
4.182[g]	?		CW; in 0.01—0.06 torr of Xe or 0.015 torr Xe with 0.3 torr of He; D = 15 mm	525
4.5393674	$5d[3/2]^o_2 \rightarrow 6p[5/2]_2$	$3d_3 \rightarrow 2p_9$	CW; in 0.01—0.04 torr of Xe, He added to give total pressure of 1 torr; D = 7 mm, or in pure Xenon at 0.01—0.06 torr; D = 15 mm	498,525,553
4.5678667 or *4.5706441*	$8p[5/2]_2 \rightarrow 6d[3/2]^o_1$ or $8s[3/2]^o_1 \rightarrow 7p[3/2]_1$	$4p_9 \rightarrow 4d_2$ or $3s_4 \rightarrow 3p_7$	CW; as for the *4.182*-μm line	525
4.6109078	$5d'[3/2]^o_2 \rightarrow 6p'[1/2]_1$	$3s''''_1 \rightarrow 2p_2$	CW; as for the *4.539*-μm line	486
5.0243255	$5d'[5/2]^o_2 \rightarrow 6p'[3/2]_2$	$3s''_1 \rightarrow 2p_3$	CW; in 0.01—0.06 torr of Xe; D = 15 mm	525,526
5.3566973	$5d[1/2]^o_1 \rightarrow 6p[5/2]_2$	$3d_5 \rightarrow 2p_9$	CW; in pure Xe at 0.01—0.06 torr or in 0.015 torr of Xe with 0.3 torr of He; D = 15 mm; lasing is more easily achieved if the *3.507*-μm line is suppressed	498,525

5.4749269	$7d[5/2]^o_3 \rightarrow 4f[5/2]_3$	$5d'_1 \rightarrow 4U$	CW; as for the *4.182*-μm line	525
5.501[g]	?		CW; as for the *4.182*-μm line	525
5.5754726	$5d[7/2]^o_4 \rightarrow 6p[5/2]_3$	$3d'_4 \rightarrow 2p_8$	CW; in 0.01—0.04 torr of Xe, He added up to 1 torr; D = 7 mm	195,486,496,508, 534,545,553,567
5.6034328	$7d[7/2]^o_3 \rightarrow 4f[9/2]_4$	$5d_4 \rightarrow 4Z$	CW; in 0.01—0.06 torr of Xe; D = 15 mm	525
5.692[g]	?	—	CW; as for the *4.182*-μm line	525
5.913[g]	?	—	CW; in 0.015 torr of Xe with 0.3 torr of He; D = 15 mm	525
6.132[g]	?	—	CW; in 0.01—0.06 torr of Xe; D = 15 mm	525
6.3120378 or *6.3154334*	$7d[7/2]^o_4 \rightarrow 4f[9/2]_5$ or $7d[7/2]^o_4 \rightarrow 4f[9/2]_4$	$5d'_4 \rightarrow 4T$ or $5d'_4 \rightarrow 4Z$	CW; in 0.01—0.06 torr of Xe; D = 15 mm	525
7.240[g]	?		CW; in 0.015 torr of Xe with 0.3 torr of He; D = 15 mm	525
7.3168036	$5d[3/2]^o_2 \rightarrow 6p[3/2]_1$	$3d_3 \rightarrow 2p_7$	CW; in 0.01—0.04 torr of Xe, He added up to 1.0 torr; D = 7 mm	486,508,545,553
7.4313142	$6d[3/2]^o_1 \rightarrow 7p[3/2]_2$	$4d_2 \rightarrow 3p_6$	CW; in 0.01—0.06 torr of Xe or 0.015 torr of Xe with 0.3 torr of He; D = 15 mm	525
9.0067086	$5d[3/2]^o_2 \rightarrow 6p[3/2]_2$	$3d_3 \rightarrow 2p_6$	CW; as for the *7.3168*-μm line above	14,486,508,534, 545,553
9.7031910	$5d[1/2]^o_1 \rightarrow 6p[3/2]_1$	$3d_5 \rightarrow 2p_7$	CW; as for the *7.3168*-μm line above	486,508
11.298683	$5d'[5/2]^o_3 \rightarrow 4f[9/2]_4$[g]	$3s'''_1 \rightarrow 4Z$	CW; as for the *7.3168*-μm line above	486,490
12.266358	$5d[1/2]^o_0 \rightarrow 6p[3/2]_1$	$3d_6 \rightarrow 2p_7$	CW; as for the *7.3168*-μm line above	486,508
12.917304	$5d[1/2]^o_1 \rightarrow 6p[3/2]_2$	$3d_5 \rightarrow 2p_6$	CW; as for the *7.3168*-μm line above	486,508
18.505324	$5d'[3/2]^o_2 \rightarrow 4f[5/2]_3$	$3s''''_1 \rightarrow 4U$	CW; as for the *7.3168*-μm line above	486,490
75.561687	$6p[1/2]_0 \rightarrow 5d[1/2]^o_1$	$2p_5 \rightarrow 3d_5$	CW; in a 100:1 He-Xe mixture at 35 mtorr of Xe or in a 3:1 Kr-Xe mixture at 15—20 mtorr of Xe; D = 6 mm	571

Note: Lines at 6.384 and 8.191 μm included by Willett in one compilation of neutral Xe laser transitions,[578] but omitted in a second,[580] are not included here. These lines were apparently only mentioned in a U.S. Government contract report and never reported in the literature.

Table 1.9.5 (continued)
XENON[a] (FIGURE 1.30)

[a] Unless otherwise indicated, measured and calculated wavelengths are taken from data on ^{136}Xe in Humphreys, C. J. and Paul, E., Jr., *J. Opt. Soc. Am.*, 60, 1302—1310, 1970. ^{136}Xe is the heaviest isotope of Xe and the measured wavelengths for this isotope are the most accurately measured. In any case, the isotope shift in Xe is extremely small, e.g., 1500 kHz ($2.10^{-8}\mu m$) between the ^{134}Xe—^{136}Xe transitions at 2.03 μm.[548] Thus, the wavelength values given in the table probably only differ in the eighth decimal place between the various isotopes.

[b] Wavelength measured in Reference 532.

[c] May be an ionized Xe transition.[213]

[d] Calculated wavelength from data in Moore, C. E., *Natl. Stand. Ref. Data. Ser. Natl. Bur. Stand.*, NSRDS-NBS 35, 3, 1971.

[e] This line was observed in Reference 549, but was incorrectly reported there as the transition at 3.507 μm.[553] This latter transition is rarely, or only weakly, observed in high-pressure transversely excited He-Xe lasers.

[f] See d.

[g] Vacuum wavelength measured in Reference 525.

Table 1.10.1
UNIDENTIFIED, POSSIBLE NEUTRAL LASER TRANSITIONS

Wavelength (μm)	Likely species	Occurrence	Excitation if not pulsed	Ref.
0.247739[a]	Xe	In Xe at 0.001—0.1 torr; D = 4 mm; probably an ionized Xe line[213]	—	572
0.274139[a]	Kr	In Kr at 0.001—0.1 torr; D = 4 mm	—	572
0.304970[a]	Kr	In Kr at low pressure	—	388,572
0.3300	Ne	In Ne using fast pulse excitation	—	573
0.332437	Ne	Probably an ionized Ne line	—	574
0.3545	Hg or Ar	In Hg-Ar mixture at low pressure[b]	—	130
0.46453	—	In Ar at low pressure, weak line	—	388,583
0.521937[a]	S	In sulfur dioxide or hexafluoride; probably S-II or S-III	—	212
0.6072	Kr	In Kr; D = 5 m	—	575
0.7065	Ar (or Hg?)	In a Hg-Ar mixture at low pressure; D = 3 mm; is probably the Ar transition at *0.7069* μm	—	130
0.8390	Cd(II?)	In a Cd-He mixture; almost certainly a Cd-II line[213]	—	120
0.8569	Xe	In Xe at 1—20 mtorr; D = 2.7—4 mm	—	532
1.0935	Ar(II?)	In Ar at 1—20 mtorr; D = 2.7—4.0 mm; probably Ar-II	—	532
1.1869	Cd(II?)	In mixtures of Cd and Ne or Cd and He; D = 12 mm; probably Cd-II[213]	—	120
1.3585	—	In 150 torr of He in a TEA laser	—	368
1.589	Cl	In a mixture of Freon (CCl_2F_2) and He at 3.3 torr	CW	231
1.977	Cl	As for the 1.589-μm line, probably the Cl-I line at 1.97553 μm	CW	231
2.021	Cl?	As for the 1.589-μm line	CW	231
2.499	—	As for the 1.589-μm line	CW	231
2.535	—	As for the 1.589-μm line	CW	231
2.602	—	As for the 1.589-μm line	CW	231
2.784	—	As for the 1.589-μm line	CW	231
3.425	—	In a He-Ne mixture in ratios from 10:1 to 5:1 at pressures from 0.3—0.5 torr; D = 15 mm	CW	500
3.801	—	In a mixture of Freon (CCl_2F_2) and He at 3.3 torr	CW	231
3.806	—	In a He-Ne mixture in ratios from 10:1 to 5:1 at a pressure of 0.5 torr; D = 15 mm	CW	500
4.247	—	In a He-Ne mixture in ratios from 10;1 to 5:1 at pressures from 0.3—0.5 torr.	CW	500
10.604	—	In $CBrF_3$ and He mixture at 2.8 torr	CW	231
14.78	—	In ammonia at 0.5—1.0 torr; D = 10 cm	—	576
15.04	—	In ammonia at 0.5—1.0 torr; D = 10 cm		576
15.08	—	In ammonia at 0.5—1.0 torr; D = 10 cm	—	576
15.41	—	In ammonia at 0.5—1.0 torr, D = 10 cm	—	576
15.47	—	In ammonia at 0.5—1.0 torr, D = 10 cm	—	576
18.21	—	In ammonia at 0.5—1.0 torr, D = 10 cm	—	576
21.46	—	In ammonia at 0.5—1.0 torr, D = 10 cm	—	576
22.54	—	In ammonia at 0.5—1.0 torr, D = 10 cm	—	576
22.71	—	In ammonia at 0.5—1.0 torr, D = 10 cm	—	576
23.68	—	In ammonia at 0.5—1.0 torr, D = 10 cm	—	576
23.86	—	In ammonia at 0.5—1.0 torr, D = 10 cm	—	576
24.92	—	In ammonia at 0.5—1.0 torr, D = 10 cm	—	576
25.12	—	In ammonia at 0.5—1.0 torr, D = 10 cm	—	576
26.27	—	In ammonia at 0.5—1.0 torr, D = 10 cm	—	576
30.69	—	In ammonia at 0.5—1.0 torr, D = 10 cm	—	576

Table 1.10.1 (continued)
UNIDENTIFIED, POSSIBLE NEUTRAL LASER TRANSITIONS

Wavelength (μm)	Likely species	Occurrence	Excitation if not pulsed	Ref.
31.47	—	In ammonia at 0.5—1.0 torr, D = 10 cm	—	576
31.92	—	In ammonia at 0.5—1.0 torr, D = 10 cm	—	576
32.13	—	In ammonia at 0.5—1.0 torr, D = 10 cm	—	576

[a] Probably an ion line, measured wavelength from Reference 213.
[b] Probably measured with a large error and may be the Ar-II transition at 0.35112 μm. If the wavelength measurement were accurate, there are two possible assignments: Ar-III 4d $^2F_{5/2} \rightarrow 4p^2D^{\circ}_{3/2}$ at 0.354560 μm or an unclassified Hg transition at 0.354506 μm.[577]

Comments: Many of the lines listed as unidentified possible neutral laser lines by Willett[578-580] have now been eliminated from this category. Several have now been classified and placed in the appropriate element table and many have now been positively identified as ionized transitions (see Reference 213), e.g., lines listed at 0.269182, 0.335004, 0.34836, 0.364546, 0.366920, 0.3760, 0.380327, 0.397293, 0.465040, 0.4764, 0.495410, and 0.595573 μm.[578] Several other lines which remain to be positively identified, but where there appears little doubt that these are lines of a specific element have been placed in the appropriate element table. Lines at 0.4120, 0.4545, 0.4750, 0.5440, 0.5500, 0.5540, 0.5679, and 0.5690 observed in Hg-N_2 mixtures and in CO[175] have been listed as possible O-II lines by Willett.[579,580] These transitions were observed under conditions where characteristic lines would be expected to occur; it seems likely that these lines were inaccurately measured and in fact correspond to neutral or singly ionized O, N, or C transitions of similar wavelength listed here and in Reference 213.

METHODS FOR EXCITATION OF LASER ACTION IN NEUTRAL ATOMIC SPECIES

Several different methods exist for exciting laser action in neutral species. Many lines have been observed only by the use of a single pumping scheme, although others will oscillate when pumped in several diverse ways. For example, the 1.315-μm transition in atomic I has been optically pumped by photodissociation of numerous I-containing organic compounds,[238-294,296-331] excited in a pulsed gas discharge,[295] and pumped by a chemical reaction.[332,335,336] The excitation methods that have been reported for neutral atomic gas lasers in general are

1. Weakly ionized DC and RF-excited discharges
2. Pulsed discharges
3. Short rise-time pulsed discharges
4. Transversely excited discharges — usually in high-pressure gases
5. Pulsed electron-beam excitation
6. Excitation in recombining plasmas
7. Direct optical pumping
8. Photodissociative optical pumping
9. Excitation as a result of chemical reactions
10. Nuclear pumping

Weakly Ionized DC and RF-Excited Discharges

The most commonly used CW laser medium for neutral species is the weakly ionized plasma of the positive column of the glow discharge. The current densities involved in such discharges in which CW-oscillation has been reported are typically 100 to 200 mA/cm². The properties of the plasma of the positive column are determined by the electric field existing along the column. In a steady, unstriated, uniform positive column, the longitudinal electric field has such a value that the number of electrons and ions produced is equal to the diffusion loss of charged particles to the walls of the discharge tube. The electron temperature in the plasma adjusts itself to that value required to maintain the positive ion concentration against the flow of positive ions to the walls where they recombine with electrons. A theoretical treatment of the positive column in gases[584,612] and gas mixtures,[613] where volume-ionization and electron-metastable collisions are not significant, leads to the important result that the average electron temperature is determined primarily by the pressure and tube-diameter product (pD). A low pD results in a high electron temperature, and a high pD results in a low electron temperature (see Reference 441). To reproduce given discharge conditions in a discharge in a single gas, all that is necessary is that the pD be maintained constant (at constant electron concentration). In the weakly ionized plasma of the positive column, the electron concentration is directly proportional to the current density. (Note that the pressure, p, which enters into these considerations is the actual pressure in the operating region of the discharge. Thermal drive-out will cause this pressure to be less than the filling pressure of the laser by an amount which will depend on the elevation of the temperature of the gas in the discharge region above ambient.[585])

Hollow-cathode discharges have been used successfully for exciting CW laser action in Ne.[134,437,440,459,464] Hollow-cathode discharges can be operated at low sustaining voltages and at high pressures; they yield quite different plasma conditions from those in the positive column. The plasma in such discharges constitutes an extended negative glow region and contains larger numbers of both low- and high-energy electrons than the positive column. In particular, hollow-cathode discharges in the noble gases contain large concentrations of thermal energy positive ions which make them especially useful for exciting metal-vapor ion laser transitions in elements such as Cd, Zn, Se, Te, I, and As,[213] Ag, Cu,[611] and Au.[593] Figure 1.31 shows the experimentally determined electron energy distributions in a positive column and in a hollow-cathode discharge in He under identical conditions of current and pressure.[594] Further details of the use of hollow-cathode discharges in laser systems are given in Reference 580.

Because of the basic importance of the pD product in longitudinal laser discharges in pure gases or gas mixtures, Tables 1.1.1 to 1.10.1 include values of the optimum pD for giving population inversion on a large number of transitions.

Pulsed Discharges

The use of pulsed discharges, where current densities of tens of thousands of A cm^{-2} can readily be obtained and extensive ionization results, allows gas discharge plasma conditions to be obtained which could not be sustained on a DC basis. Laser action that results in pulsed discharges can be broadly divided into laser action that occurs during the current pulse, and laser action that occurs during the afterglow.

The electron temperature during the current pulse is higher than in the afterglow, and direct excitation of upper laser levels by electron impact on ground state atoms (or metastables) is favored. However, in pulsed afterglow discharges, first used by Boot et al.[474] to obtain laser oscillation in neutral atomic species, electrons rapidly thermalize, and recombination and dissociative-recombination processes, involving long-lifetime species of atomic and molecular ions and excited atoms, dominate. In both pure gases and mixtures of gases, these processes lead to transient laser-oscillation with high-gain and high-output power, but at low equivalent CW power.

Conditions in a pulsed discharge or in the afterglow are not readily specified by a simple pD relationship. They are determined by the voltage at breakdown of the gas, the gas, gas pressure, the energy dissipated in the discharge, the external-circuit resistance, and the shape of the leading edge and trailing edge of the current pulse.

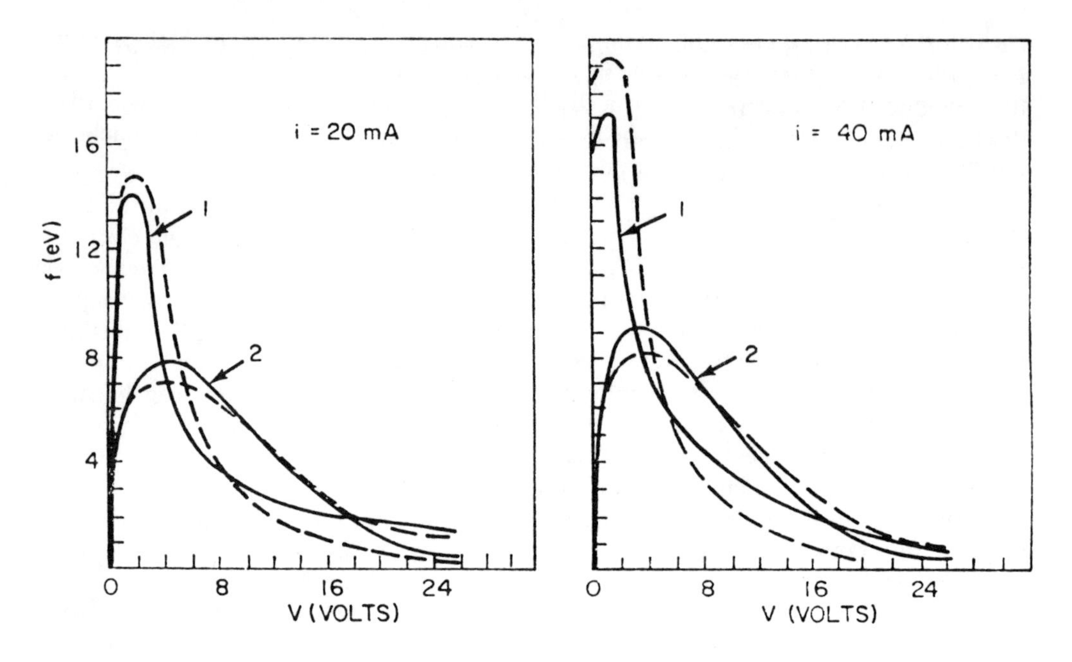

FIGURE 1.31. Comparison of electron energy distributions in a cylindrical hollow-cathode discharge and in a positive column in He at two discharge currents. The dashed curves correspond to Maxwellian distribution functions. The hollow cathode and positive column diameters were both 2 cm, and the operating pressure was 0.9 torr. (From Borodin, V. S. and Kagan, Yu. M., *Sov. Phys. Tech. Phys.*, 11, 131, 1966. With permission.)

Pulsed hollow-cathode discharges are also effective for exciting several neutral gas lasers. Ar, Xe, Cu, Ag,[99] Ca, Sr, and Ba,[106] and Ni[358] lasers have been excited in this way, the metal vapor needed for laser action frequently being generated by sputtering from the cathode.

Short Rise-Time Pulsed Discharges

Several important laser transitions occur between a neutral resonance level (a level directly connected optically to the ground state) and an intermediate lower lying level which is metastable or quasimetastable. The electron-impact excitation of the resonance level generally has a very large cross section for electrons above threshold, and preferential excitation of the upper rather than the lower laser level is readily obtained. However, because of the adverse lifetime ratio of the laser levels, any population inversion will be destroyed by spontaneous or stimulated emission on a timescale on the order of the lifetime of the upper laser level.

Short rise-time pulsed discharges are used almost exclusively for exciting such self-terminating transient gas lasers. In these discharges, the pulse rise-time must be comparable to the radiative life-time of the upper laser level to produce population inversion.[17] The current densities at the peak of the pulse are of the order of hundreds to thousands of amperes per square centimeter. Excitation conditions in the discharge are dependent on the energy dissipated per unit time, the gas pressure, the breakdown voltage of the gas, the anode-cathode geometry, and the shape of the electrodes. Because of the finite breakdown time of a longitudinal column of gas, multielectrode or transversely excited discharge-structures[20,44,48,384] have a definite advantage over longitudinally-excited, two-electrode discharge-structures as a means of short rise-time pulsed excitation.

Transversely Excited Discharges

Apart from their separate utility in short-rise time excitation, the use of transversely excited discharges allows the attainment of electron temperatures in high-pressure gases which would be unattainable in practice with longitudinal excitation. The electron temperature achieved in such a discharge is a roughly monotonically increasing function of E/p, where E is the field strength in the discharge. Attempts to achieve comparable E/p values at high pressures in longitudinal discharges would invariably lead to a glow-to-arc transition, increased ionization, and a likely fall in electron temperature. Transversely excited discharges fall into two broad categories: pin discharges and volume-excited structures. Pin discharges, originally developed by Beaulieu for CO_2 lasers,[586] involve numerous separate discharges transverse to the laser axis between individual pin cathodes and a common bar anode or separate pin anodes. The system is constructed and operated so as to produce individual glow, rather than arc, discharges between cathodes and anode(s). Volume-excited structures involve continuous or segmented electrode structures designed in such a way that uniform transverse excitation of a high-pressure gas can be accomplished. This involves inhibition of arc-formation — the main problem with attempting to achieve uniform high pressure discharges over large volumes. The first successful approach to the construction of such volume-excited structures was taken by Dumachin et al.,[587,588] who used numerous separate trigger electrodes near the segmented cathode to produce uniform preionization of a layer of gas near the cathode, and encourage uniform breakdown between cathode and anode. Numerous "double-discharge" transversely excited structures of various designs based on this idea have appeared in the literature. An alternative approach to uniform excitation of a large volume of high-pressure gas is to use an electrode structure incorporating two electrodes formed in a Rogowski profile,[589] or one Rogowski and one planar electrode. This approach, first developed by Lamberton and Pearson[590,591] inhibits arc formation by ensuring high electric field uniformity between anode and cathode. Even so, such profiled electrode structures are usually operated in a preionized mode, either in a double-discharge structure or by vacuum-UV photopreionization.[592] In the latter technique, an auxiliary array of spark discharges placed to the side of the main discharge volume, or behind a mesh main electrode, injects vacuum UV photons into the main discharge volume before the main discharge fires.

Pulsed Electron-Beam Excitation

Pulsed electron-beam excitation, first used by Hodgson and Dreyfus[595] to obtain laser oscillation in the Lyman and Werner bands of molecular hydrogen, has been used successfully to excite laser action in Xe in a high-pressure Ar-Xe mixture[535] and O in high pressure O-noble gas or O-N_2O mixtures.[188-192] This means of excitation can generally be regarded as an alternative, yet in some ways similar, means of excitation to transversely excited discharge excitation for large volumes of high-pressure gas. However, it does in principle allow the deposition of large numbers of electrons at controlled average electron temperature without any constraints imposed by gas-discharge breakdown characteristics and tendencies towards arc formation. When used in a pulser-sustainer mode,[535] the bulk of the excitation energy is provided by a conventional pulsed discharge from a storage capacitor after initial ionization by the electron-beam source. Sustainer voltage does not need to be as high as would be needed to initiate gas breakdown in a conventional, transversely excited discharge structure.

Excitation in Recombining Plasmas

Although, as will be discussed later, several laser transitions are excited by ion-electron recombination in gas-discharge plasmas, Silfvast et al. have reported laser oscillation in a recombining laser-produced Cd plasma[121,700] produced by focusing a Nd or CO_2 laser onto a Cd target. Laser action has also been observed in expanding plasmas of Ar, Kr[700] and Xe[700,701] produced by focusing a CO_2 laser into the high-pressure gas.

Silfvast et al.[146] have also reported laser action in several neutral metal vapors following vaporization of electrode material in a transversely excited, segmented structure, and subsequent recombination of the generated plasma. In so far as this laser is electrically excited and subject to the constraints imposed by gas breakdown and E/p ratio, it is akin to pulsed afterglow lasers mentioned previously.

Direct Optical Pumping

The earliest example of a direct optically pumped neutral gas laser was the Cs laser of Rabinowitz et al.,[15,16] where a fortuitous coincidence between the He line at 0.3888 μm and the Cs resonance absorption at 0.388851 μm was used to good effect to preferentially populate the cesium 8p $^2P^o_{1/2}$ level. A more practically useful optically pumped laser is the Hg laser, where inversion follows sequential absorption of two-pump photons from a mercury lamp.[124-126] However, the use of incoherent pump sources to generate inversion in this way is not of widespread usefulness; absorption lines in atomic gases are narrow, and highly efficient coincidental line sources are rarely available. The development of high-energy, tunable, short-wavelength lasers such as the noble gas halide exciplexes, has provided a much more practical optical pumping source. Although these sources have so far proved most useful for photodissociative pumping, they have been used successfully to generate population inversion following resonant absorption in Ca, Bi,[187] and two-photon excitation of Eu.[360]

Photodissociative Optical Pumping

The broad electronic absorption bands of molecules allow large amounts of optical power to be coupled into a gas from either an intense, incoherent continuum source or a laser. If the electronic state thereby excited dissociates to yield an excited atomic product, the achievement of a population inversion between states of this atom is possible. Photodissociative optical pumping with intense incoherent sources has proved successful with several organic I-[238-294,296-331] and Br-[376-378] containing organic molecules to yield excited I and Br atoms. Optical pumping with one or the other of ruby, dye, neodymium, noble gas excimer, or noble gas-halide exciplex lasers has produced laser oscillation in sodium[3,4] (following photodissociation of NaI or NBr), K[3,9] (KI, KBr, K_2), Rb[3,9,12] (RbI, RbBr, Rb_2), Cs (CsI, CsBr, Cs_2), Ba (BaTl),[632] mercury[123] (Hg_2), Ga[144] (GaI_3 or GaI), In[145] (InI), Tl (TlI), sulfur[267] (OCS), and Se[214,215] (COSe).

Large volumes of gas can, in principle, be excited in this way. The optimum operating pressure of the laser will, in general, eventually need to be reduced as its excited volume is increased, to allow penetration of the pump radiation, unless the latter is so intense as to lead to bleaching effects.

Excitation as a Result of Chemical Reactions

Chemical pumping has proved of most widespread usefulness in generating laser oscillation in chemically formed molecular species.[600] There has, however, been an intensive effort during the last decade to operate a chemically pumped electronic-transition laser. This has fairly recently been successfully accomplished in I[332,335,336] and Br;[378] further details will be given in a later section.

Nuclear Pumping

The original idea of using nuclear energy for generating laser action has been around since the early days of the laser. Much of the early work on the subject is discussed by Thom and Schneider,[601] who made a serious study of the feasibility of direct nuclear pumped (DNP) lasers.[601,602] Following a series of experiments which reported enhancement of laser output by nuclear reactions or beams of heavy particles,[603-608] laser action was successfully produced in CO,[609] and He-Xe[610] as a direct result of fission products interacting with the gas. DNP laser action in neutral atomic species has subsequently been extended to N in Ne-N_2 mixtures,[179] Ar in ^{3}He-Ar mixtures,[523] Kr in ^{3}He-Kr mixtures,[531] and C in He-CO and He-CO_2 mixtures.[347] Although DNP lasers have not, to date, produced substantial power outputs, they possess the potential for very high specific energy output (joules per unit volume) and promise to be useful in extra-terrestrial applications where a laser pumped with its own small nuclear reactor and which needs no additional power source is attractive. The potential for high specific energy output stems from the ability of high-energy fission fragments to excite high-pressure gases uniformly, and by secondary ionization yield electron-energy and excited-atom distributions which may be very different from those achievable by gas-discharge excitation. Specific details of the various collisional processes which occur in the excitation of DNP lasers and of the various schemes which have been considered for excitation of large volume gaseous laser media will be given later.

SPECIFIC EXCITATION PROCESSES IN NEUTRAL GAS LASERS

Excitation Transfer (Atom-Atom)

This so-called "collision of the second kind" involves the transfer of potential energy from one excited atom to another in a collision. The reaction can be represented as

$$A' + B \rightleftarrows A + B' \pm \Delta E_\infty \quad (1)$$

where ΔE_∞ is the difference in potential energy at infinite separation between the excited species A′ and B′, and A and B can be (and usually are) ground state atoms. When A′ is a metastable species (A*) and B′ can decay radiatively, the reaction proceeds in the forward direction. It has been found experimentally that the probability of excitation transfer occurring in a collision appears to be dependent on at least four factors:

1. The magnitude of ΔE_∞
2. Spin conservation (the Wigner spin rule)
3. The size of the colliding atoms
4. The relative velocity of the colliding atoms A′ and B (determined by the gas temperature (T_g) of the discharge and the mass of the colliding atoms)

Theory predicts that with $\Delta E_\infty \rightarrow 0$ (a resonant collision), cross sections two orders of magnitude larger than gas kinetic occur. With $\Delta E_\infty > 0.1$ eV, the cross section for excitation transfer becomes negligible,[614] as illustrated in Figure 1.32.

Total spin conservation ($\Delta S = 0$) is favored in a collision involving excitation transfer. Given the reactions

$$A'(\uparrow\uparrow) + B(\uparrow\downarrow) \nearrow A(\uparrow\downarrow) + B'(\uparrow\uparrow) \pm \Delta E_\infty \quad (2a)$$

$$A'(\uparrow\uparrow) + B(\uparrow\downarrow) \searrow A(\uparrow\downarrow) + B'(\uparrow\downarrow) \pm \Delta E_\infty \quad (2b)$$

2a is the more likely reaction channel.

Excitation transfer, which occurs in the He-Ne laser between at least three excited states of He and Ne, provides most of the information on the importance of the four factors (1) to (4). The three energy-coincidences between excited states of He and Ne, which lead to three s-p groups of laser transitions, are shown in Figure 1.33. The He states involved are the $2p^1P^\circ_1$, $2s^1S_0$ and $2s^3S_1$. The $2s^1S_0$ and $2s^3S_1$ states are metastable (He* 2^1S_0 and He* 2^3S_1) and lead to strong selective excitation of the Ne-5s and Ne-4s levels (3s and 2s in Paschen notation).

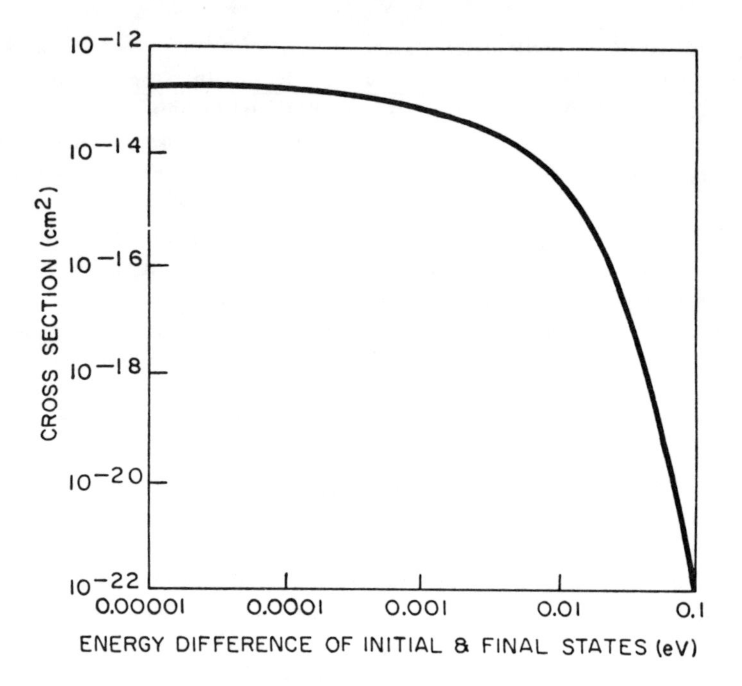

FIGURE 1.32. Theoretical variation of the cross section for excitation transfer as a function of the energy discrepancy between the initial and final states of the colliding atoms. (From Mott, N. F. and Massey, H. S. W., *The Theory of Atomic Collisions*, 3rd ed., Oxford University Press, 1965. With permission.)

Detailed energy-coincidences between the He* 2^1S_0 and He* 2^3S_1 states and Ne are given in Figures 1.34 and 1.35. Figure 1.34 illustrates the energy-level-coincidence for the excitation-transfer reaction from He 2^1S_0 metastables to the Ne-3s states. Here preferential selective excitation of the Ne-$3s_2$ level occurs in the endothermic reaction

$$\text{He}^*\ 2^1S_0\ (\uparrow\downarrow) + \text{Ne}\ ^1S_0\ (\uparrow\downarrow) \rightarrow \text{He}\ ^1S_0\ (\uparrow\downarrow) + \text{Ne}\ 3s_2\ (\uparrow\downarrow) - \Delta E_\infty(386\ \text{cm}^{-1}) \tag{3}$$

in which total spin is conserved, and a ΔE_∞ of 386 cm^{-1} (about 2 kT) has to be provided in kinetic energy supplied by the gas discharge.

Figure 1.35 illustrates the similar coincidence between He 2^3S_1 metastables and the 2s-states of Ne. Unlike excitation transfer to the 3s-levels, selective excitation is more evenly distributed to all the four Ne-2s levels.[405] On the basis of energy-level coincidences, in

$$\text{He}^*\ 2^3S_1\ (\uparrow\uparrow) + \text{Ne}\ ^1S_0(\uparrow\downarrow) \rightarrow \text{He}\ ^1S_0(\uparrow\downarrow) + \text{Ne}\ 2s_{2,3,4,5} + \Delta E_\infty(313\ \text{to}\ 1247\ \text{cm}^{-1}) \tag{4}$$

since the reaction is exothermic (with ΔE_∞ positive), kinetic energy does not have to be supplied by the gas for the reaction to proceed. Although ΔE_∞ of 1247 cm^{-1} for the coincidence with the Ne-$2s_5$ level is larger than the ΔE_∞ of 800 cm^{-1} (0.1 eV) — when excitation transfer is predicted to be negligible — selective excitation to this state from He 2^3S_1 metastables does occur. In Equation 4 total spin is conserved in excitation transfer to three of the Ne states ($2s_3$, $2s_4$, and $2s_5$).

Although total spin is not conserved in excitation transfer to the $2s_2$ state, it has been shown that the $2s_2$ state in Ne is intermediate between a ^1P state and a 3/4 j-j coupling state.[615] This reduces the effectiveness of the Wigner spin rule and accounts for the relatively high excitation transfer cross section for this state.

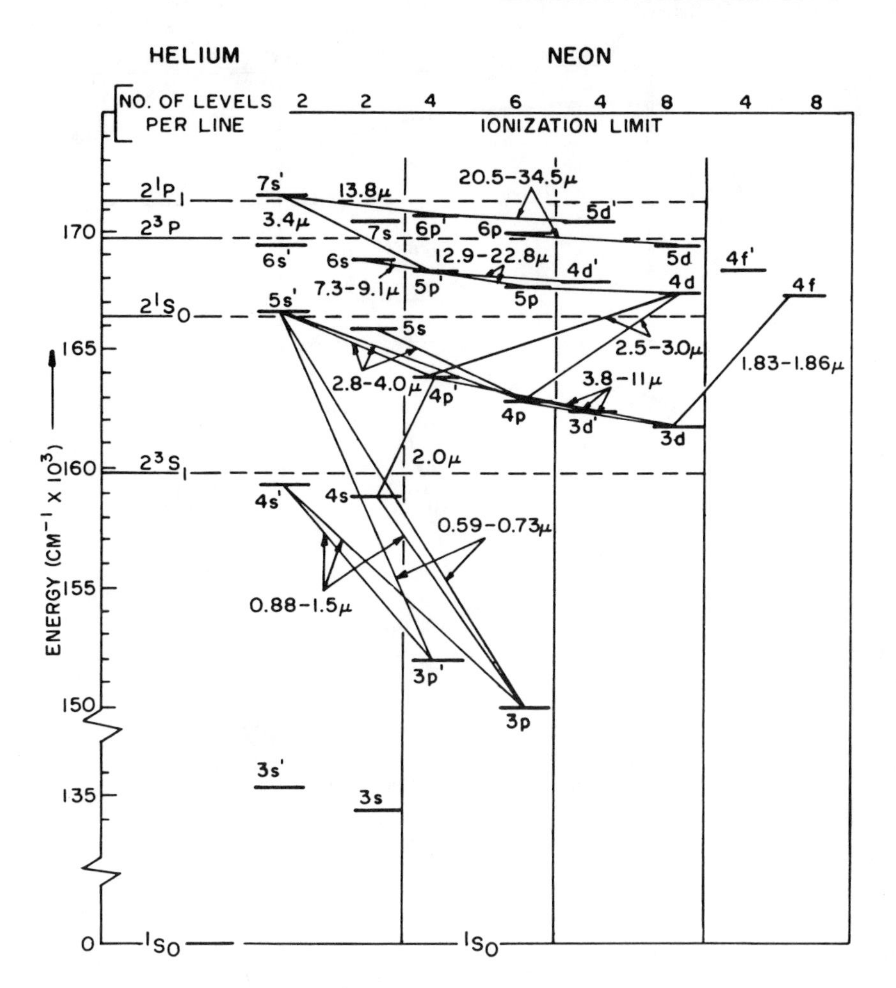

FIGURE 1.33. Energy-level diagram for Ne showing groups of laser transitions observed in a He-Ne mixture and energy coincidences with He(2^3S_1, 2^1S_0, $2^3P_{2,1,0}$, and 2^1P_1) levels. (This figure is a modification of a figure in Reference 486.)

Jones and Robertson[407] have shown experimentally that the cross section (σ) for excitation transfer between He* 2^1S_0 and He* 2^3S_1 atoms and ground-state Ne atoms follows the relationship

$$\sigma = \sigma_0 \exp(-E_a/kT) \tag{5}$$

Plots of the experimental results which lead to this relationship are given in Figures 1.36 and 1.37. In Equation 5, E_a has the form of an activation energy (or dissociation energy) in which $E_a \neq \Delta E_\infty$. This observed relationship indicates that the interaction potentials between the He* 2^3S_1 and Ne$^1S_0$ states might have the form represented in Figures 1.38A and 1.38B, respectively. The larger activation energy for the triplet state reaction has been confirmed by Chen et al.,[616] although their data suggest that E_A in this case is >0.065 eV. There may also be an actual crossing of potential energy curves in the case of the singlet reaction.[617]

Dissociative Excitation Transfer (Atom-Molecule)

This takes the form

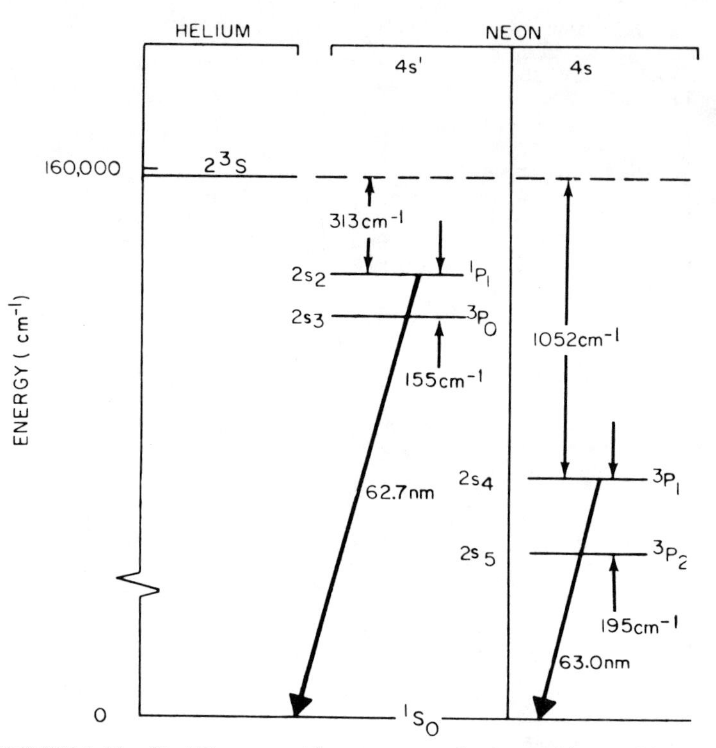

FIGURE 1.34. He 2^1S_0 metastable state energy-level coincidence with Ne 3S-levels.

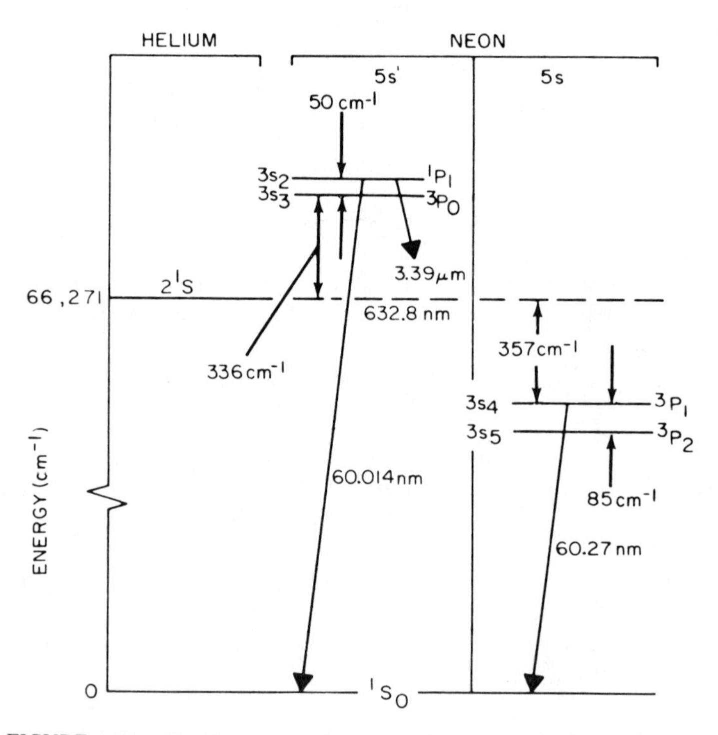

FIGURE 1.35. He 2^3S_1 metastable state energy-level coincidence with Ne 2s-levels.

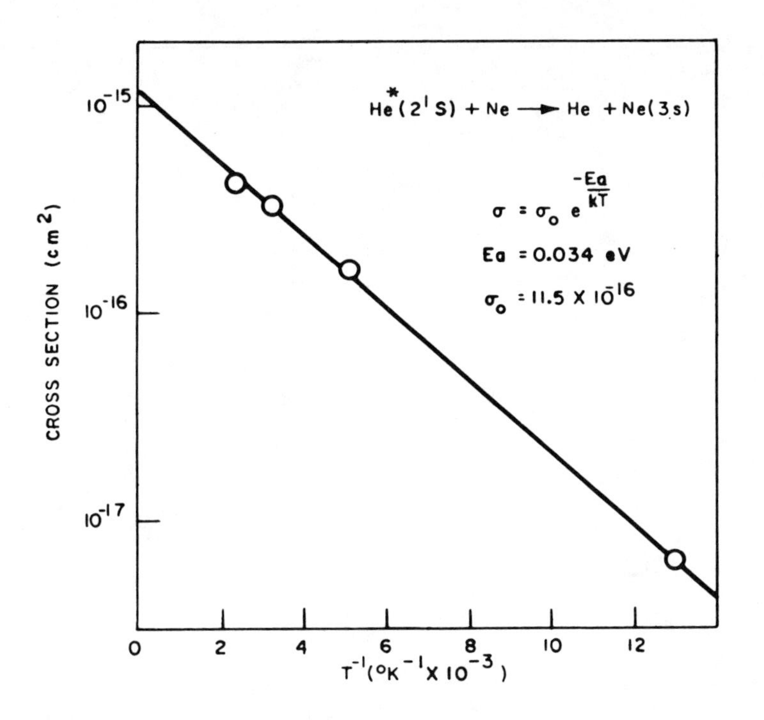

FIGURE 1.36. Variation of cross section with gas temperature for excitation transfer from He 2^1S_0 metastables to the Ne-$3s_2$ level.

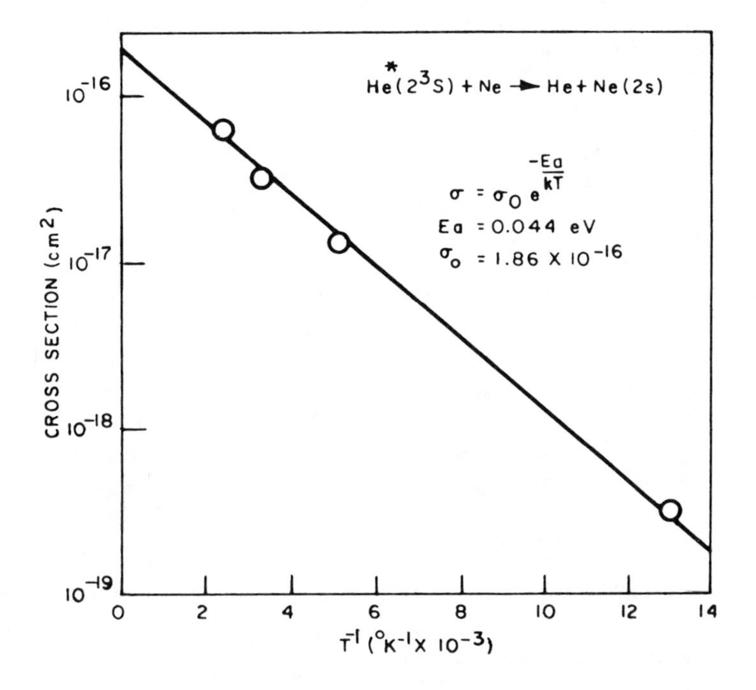

FIGURE 1.37. Variation of cross section with gas temperature for excitation transfer from He 2^3S_1 metastables to the Ne-2s levels.

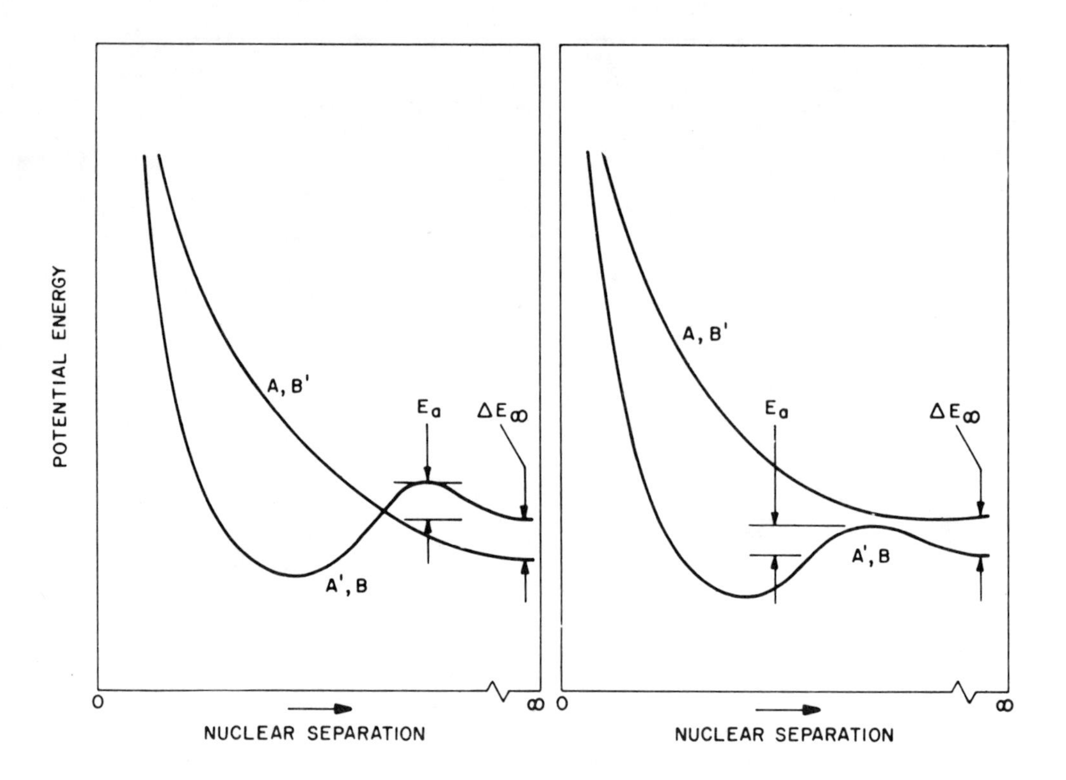

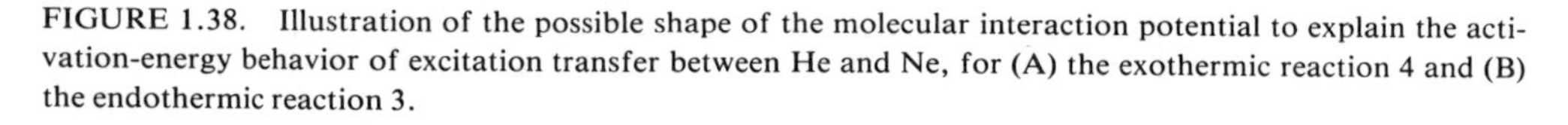

FIGURE 1.38. Illustration of the possible shape of the molecular interaction potential to explain the activation-energy behavior of excitation transfer between He and Ne, for (A) the exothermic reaction 4 and (B) the endothermic reaction 3.

$$M' + AB \rightarrow M + A' + B + \text{kinetic energy} \quad (6a)$$

$$M' + AB \rightarrow M + A + B' + \text{kinetic energy} \quad (6b)$$

where M′ is an excited state (usually metastable M*). In these reactions, it is assumed that relatively long-range interactions between M′ and the molecule AB cause vertical transitions between the ground state and repulsive states $(AB)^+$ or (AB′) (see Figure 1.39). Since several repulsive states can exist in the vicinity of M′ and since more than two atoms are involved in the collision, there is no strong requirement for close energy-coincidence as in atom-atom excitation transfer. Any excess potential energy appears as kinetic energy of the dissociated atoms A′ and B′.

In the Ne-O_2 laser, dissociative excitation transfer occurs via the reaction

$$Ne^* + O_2 \rightarrow Ne\ ^1S_0 + O + O'\ (3p\ ^3P) + K.E. \quad (7)$$

leading to direct selective excitation of the 3p ^3P triplet state of atomic O and population inversion occurring between the O-I 3p ^3P and 3s $^3S_1^o$ states (see Figure 1.40).

In the Ar-O_2 laser, dissociative excitation transfer occurs via reactions

$$Ar^* + O_2 \rightarrow Ar\ ^1S_0 + O + O'\ (2^1S_0) + K.E. \quad (8a)$$

$$Ar^* + O_2 \rightarrow Ar\ ^1S_0 + O + O'\ (2^1D_2) + K.E. \quad (8b)$$

Subsequently, inelastic electron collisions with atoms in the O-I 2^1S_0 and 2^1D_2 states leads to population inversion occurring between the O-I 3p ^3P and 3s $^3S_0$ levels.[14,193]

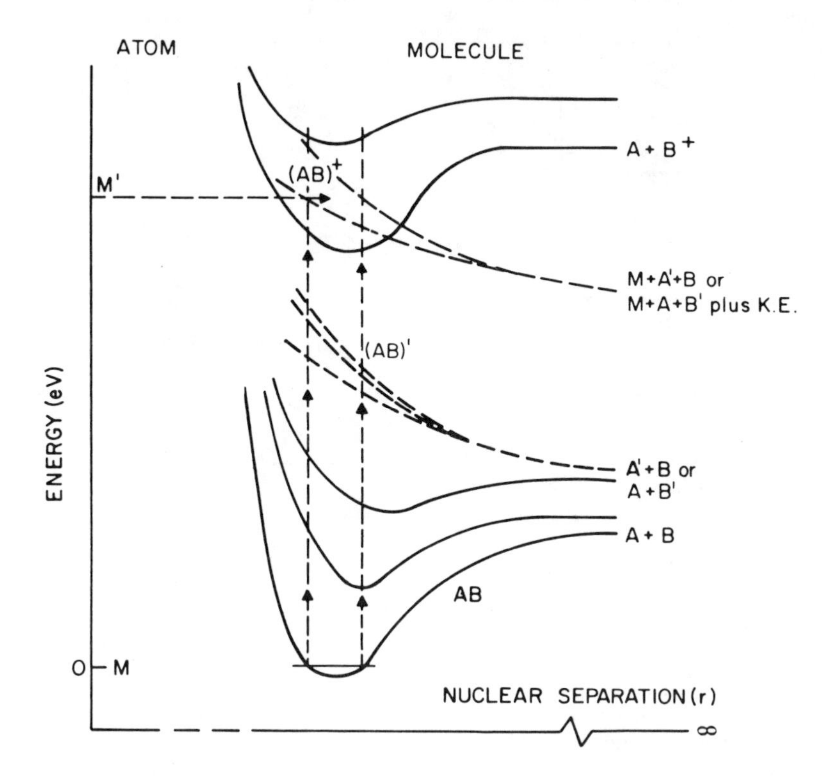

FIGURE 1.39. Molecular potentials illustrating the dissociation of an atomic molecule into repulsive states in dissociative excitation transfer. The dashed-curves represent a number of repulsive molecular potentials which have common dissociation limits. M′ indicates a metastable state.

The O atom laser transition at 0.8446 μm which thereby results is of some interest. Oscillation on four closely spaced wavelengths which lie in the wavelength interval occupied by the O-I(3p $^3P_{0,2,1} \rightarrow$ 3s $^3S_1^0$) triplet was reported[159] in a mixture of Br and He or Ne, with O as an impurity. The coincidental presence of a Br-I emission line at 0.8844655 μm added to the confusion. However, it was subsequently determined that the four observed lines were indeed O lines, displaced from the centers of the O triplet spontaneous emission peaks, as shown in Figure 1.41. Oscillation is prevented at the line centers by strong radiation trapping, but is made possible off-center by the extensive linewidth of each component under which gain is possible. The large linewidth of the oxygen triplet (more than five times that expected for the gas temperature in the discharge) is due to the large amounts of kinetic energy carried by excited 3p ^3P atoms following dissociative excitation-transfer.[197,203]

If a strongly allowed optical transition connects the ground state with an excited state, it follows from the Born approximation,[614] where electron exchange can be neglected, that selective excitation of that excited state can occur in inelastic electron-impact collisions. In the noble gases, the main selective excitation from the ground state occurs through the reactions,

$$(np)^6 + e^- + \text{K.E.} \nearrow (np)^5\ ms + e^- \quad (9a)$$

$$(np)^6 + e^- + \text{K.E.} \searrow (np)^5\ md + e^- \quad (9b)$$

where m = n + 1, n + 2, etc., and n = 2 for Ne, 3 for Ar, 4 for Kr, and 5 for Xe. Under pD-discharge conditions sufficient to produce radiation trapping on the reso-

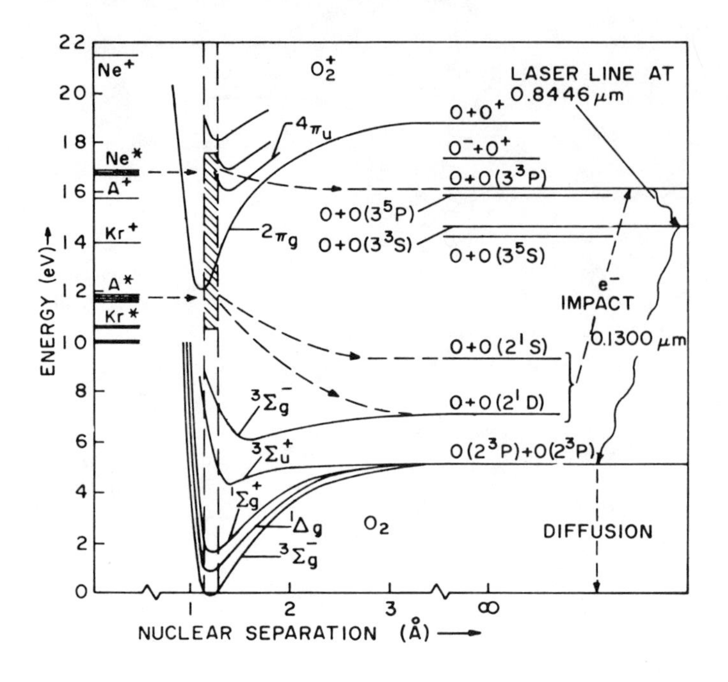

FIGURE 1.40. Partial energy-levels relevant to the Ne-O_2 and Ar-O_2 laser operating at 0.8446 μm. The dominant excitation paths are indicated by the arrows.[193]

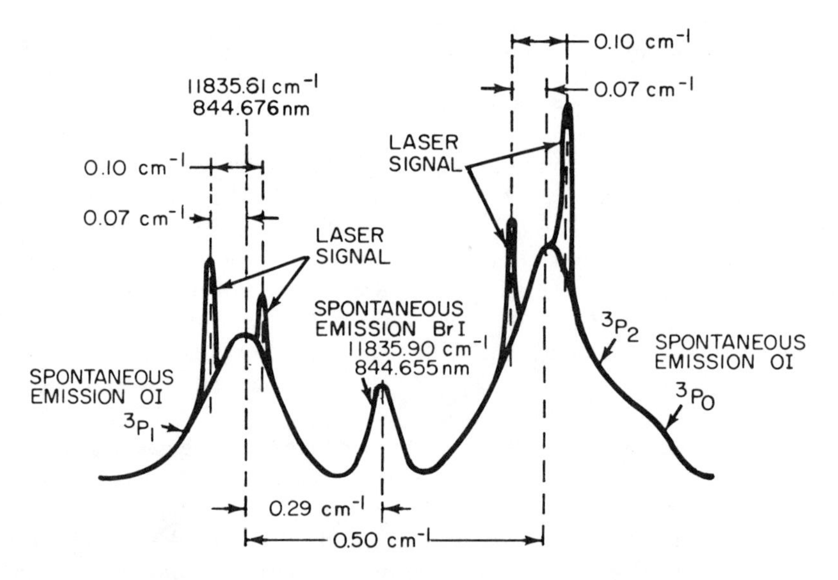

FIGURE 1.41. Oscillation on four lines in the O triplet at 0.8446 μm in Ar-Br_2 and Ar-O_2 lasers shown relative to spontaneous emission lines of Br-I and O-I. (From Tunitskii, L. N. and Cherkasov, E. M., *J. Opt. Soc. Am.*, 56, 1783, 1966. With permission.)

nance state, and low enough to prevent excitation transfer to other electron configurations in collisions, radiative decay can only occur through the transitions labeled LASER in Figure 1.42.[14] This type of selective excitation in a system, where radiative decay of the lower laser level is allowed and life times are favorable, can lead to the realization of continuous oscillation.

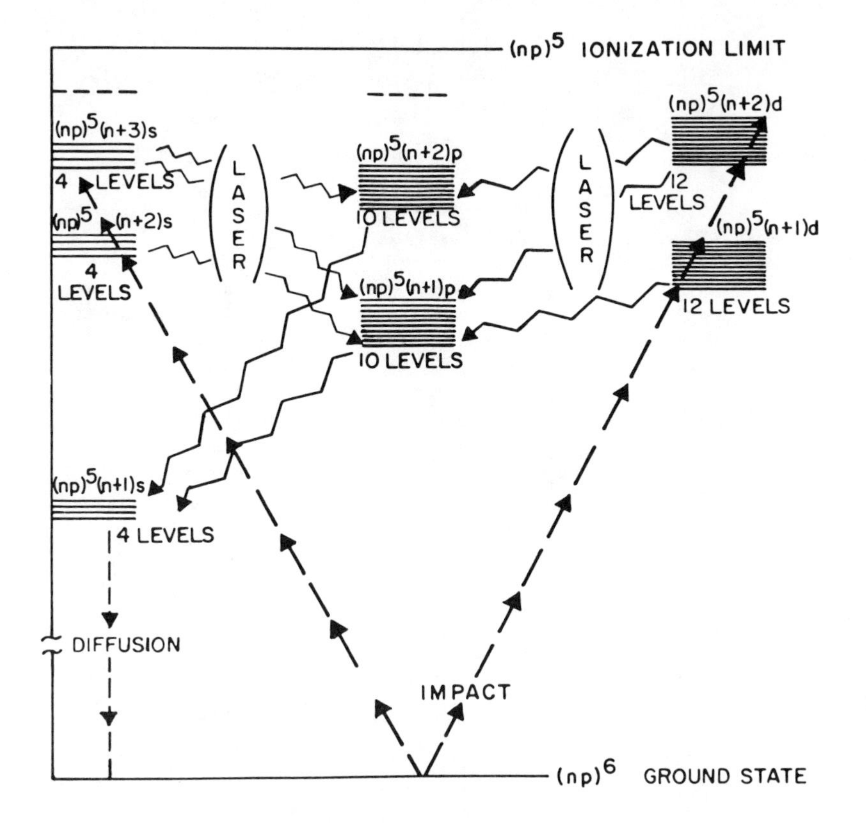

FIGURE 1.42. Schematic indication of electron configurations relevant to pure Ne, Ar, Kr, and Xe lasers. The dominant excitation paths are indicated by the arrows.[14]

A number of gas laser systems selectively excited by electron impact are intrinsically self-terminating or transient in nature. These transitions occur in the elements Cu, Au, Ca, Sr, Ba, Tl, Pb, Bi, Mn, and Eu, as well as the noble gases He, Ne, Ar, Kr, and Xe. They are listed in Table 1.11. Figure 1.43 shows a typical schematic energy level diagram for a self-terminating laser. The large electric-dipole-moment matrix element connecting Level 1 and the ground state — the resonance transition — allows very efficient electron impact excitation of Level 1. Level 2 is not connected optically to the ground state (or is only weakly connected), so its electron impact excitation is inefficient except for electrons close to its threshold energy.[618] If the ground state density is high enough, the effective lifetime of Level 1 will be controlled by its spontaneous emission to Level 2 because of trapping of the resonance transition.[619,620] Inversion of the system can therefore only be maintained for times on the order of the effective radiative lifetime of Level 2. The occurrence of laser action will terminate the inversion even more rapidly. To obtain laser action, pulsed excitation which is faster than the effective radiative lifetime of Level 2 is generally necessary. Following one excitation pulse, a second cannot be applied until Level 2 has been quenched by collisions or diffusion to the walls, or by physical removal of atoms in Level 2 in a fast flow scheme. CW laser action has in fact been obtained on the optically pumped self-terminating iodine laser transition at 1.315 μm in this way.[325]

The energy levels relevant to the very high-gain 0.7229-μm lead-vapor laser are shown in Figure 1.44. This laser, and others like it, particularly the Cu-vapor laser, are attractive because of their high quantum efficiency and potentially high practical efficiency. The quantum efficiency of the 0.7229-μm laser in Figure 1.44 is ≃ 39%. These systems are frequently called cyclic lasers because of the upper level excitation/

Table 1.11
SELF-TERMINATING LASER TRANSITIONS

Element	Wavelength (μm)	Transition assignment	Ref.
Cu	0.5106	$4p\ ^2P^o_{3/2} \rightarrow 4s^2\ ^2D_{5/2}$	17-98
Cu	0.5700	$4p\ ^2P^o_{3/2} \rightarrow 4s^2\ ^2D_{3/2}$	29
Cu	0.5782	$4p\ ^2P^o_{1/2} \rightarrow 4s^2\ ^2D_{3/2}$	17-82,84-92,94-98
Au	0.3123	$6p\ ^2P^o_{3/2} \rightarrow 6s^2\ ^2D_{5/2}$	100,102,103
Au	0.6278	$6p\ ^2P^o_{1/2} \rightarrow 6s^2\ ^2D_{3/2}$	17,21,22,100-103
Ca	0.5349	$4p'\ ^1F^o_3 \rightarrow 3d\ ^1D_2$	105
Ca	*5.547*	$4p\ ^1P^o_1 \rightarrow 3d\ ^1D_2$	107,106,108
Sr	*6.457*	$5p\ ^1P^o_1 \rightarrow 4d\ ^1D_2$	107,109
Ba	1.130	$6p\ ^1P^o_1 \rightarrow 5d\ ^3D_2$	113-115
Ba	1.500	$6p\ ^1P^o_1 \rightarrow 5d\ ^1D_2$	113-118
Ba	*2.923*	$6p\ ^3P^o_1 \rightarrow 5d\ ^3D_2$	113,114
Tl	0.5351	$7s\ ^2S_{1/2} \rightarrow 6p\ ^2P^o_{3/2}$	148-152
Pb	0.3640	$7s(3/2, 1/2)^o_1 \rightarrow 6p^2\ (3/2, 3/2)_2$	167
Pb	0.4058	$7s(1/2, 1/2)^o_1 \rightarrow 6p^2(3/2,1/2)_2$	167,170
Pb	0.7229	$7s(1/2,1/2)^o_1 \rightarrow 6p^2(3/2,3/2)_2$	17,167,168,169, 171-173
Bi	0.4724	$7s\ ^4P_{1/2} \rightarrow 6p^3\ ^2D^o_{3/2}$	186,187
Mn	0.5341	$y\ ^6P^o_{7/2} \rightarrow a\ ^6D_{9/2}$	17,348-355
Mn	0.5420	$y\ ^6P^o_{5/2} \rightarrow a\ ^6D_{7/2}$	17,348,349,352-354
Mn	0.5471	$y\ ^6P^o_{5/2} \rightarrow a\ ^6D_{5/2}$	348,349,354
Mn	0.5481	$y\ ^6P^o_{3/2} \rightarrow a\ ^6D_{5/2}$	17,349
Mn	0.5517	$y\ ^6P^o_{3/2} \rightarrow a\ ^6D_{3/2}$	17,348,349,354
Mn	0.5538	$y\ ^6P^o_{3/2} \rightarrow a\ ^6D_{1/2}$	17,348,349,354
Mn	1.290	$z\ ^6P^o_{7/2} \rightarrow a\ ^6D_{9/2}$	17,348,349,352-354
Mn	1.329	$z\ ^6P^o_{7/2} \rightarrow a\ ^6D_{7/2}$	17,348,349,352-354
Mn	1.332	$z\ ^6P^o_{5/2} \rightarrow a\ ^6D_{7/2}$	17, 348,349,352-354
Mn	1.363	$z\ ^6P^o_{5/2} \rightarrow a\ ^6D_{1/2}$	17,348,349,352-354
Mn	1.386	$z\ ^6P^o_{3/2} \rightarrow a\ ^6D_{3/2}$	17,348,349,352-354
Mn	1.400	$z\ ^6P^o_{3/2} \rightarrow a\ ^6D_{1/2}$	17,348,349,352-354
Eu	*1.760*	$y\ ^8P_{9/2} \rightarrow a\ ^8D^o{}_{11/2}$	217,359
He	2.058	$2p\ ^1P^o_1 \rightarrow 2s\ ^1S_0$	148,368
Ne	0.5401	$2p_1 \rightarrow 1s_4$	17,151,382-386, 388,397
Ne	0.5852	$2p_1 \rightarrow 1s_2$	14,383
Ne	0.5945	$2p_4 \rightarrow 1s_5$	382,391-393
Ne	0.6143	$2p_6 \rightarrow 1s_5$	382,385,386,391-398
Ar	*0.7069*	$2p_3 \rightarrow 1s_5$	392
Ar	*0.7506*	$2p_1 \rightarrow 1s_2$	388
Ar	0.9123	$2p_{10} \rightarrow 1s_5$	218,513,514
Ar	0.9658	$2p_{10} \rightarrow 1s_4$	218,513,514
Ar	1.047	$2p_{10} \rightarrow 1s_3$	513
Ar	1.149	$2p_{10} \rightarrow 1s_2$	513
Kr	0.7602	$2p_6 \rightarrow 1s_5$	528
Kr	0.8104	$2p_8 \rightarrow 1s_5$	392,393,467
Kr	0.8929	$2p_{10} \rightarrow 1s_5$	513,514
Xe	0.8232	$2p_6 \rightarrow 1s_5$	218,513,514
Xe	0.8409	$2p_7 \rightarrow 1s_5$	467,532
Xe	0.9045	$2p_9 \rightarrow 1s_5$	393,467
Xe	0.9800	$2p_{10} \rightarrow 1s_5$	393,513,514

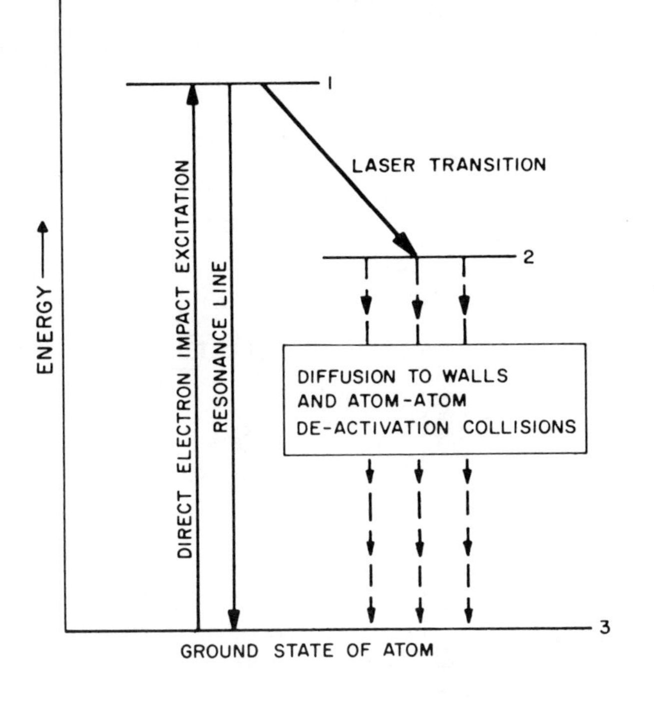

FIGURE 1.43. Generalized partial energy-level diagram for a self-terminating laser.

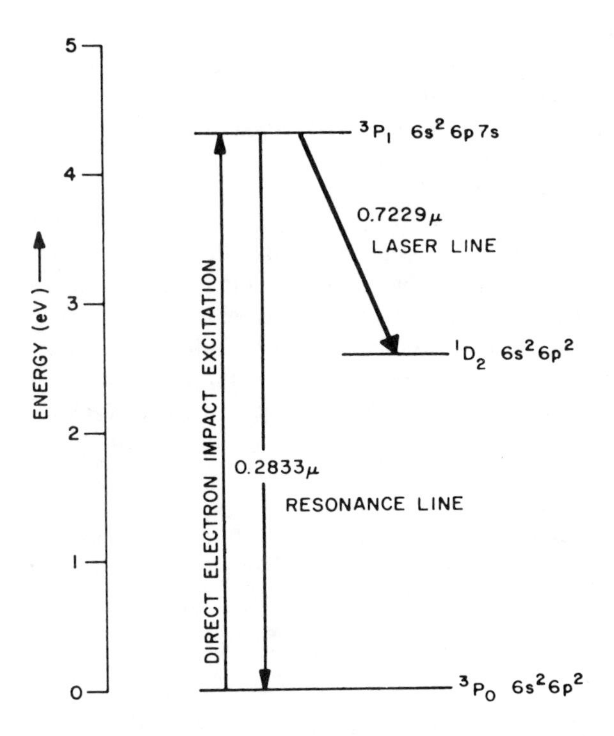

FIGURE 1.44. Energy levels relevant to the 0.7229 μm self-terminating Pb laser showing the dominant election-impact excitation path and the 0.2833-μm resonance line.

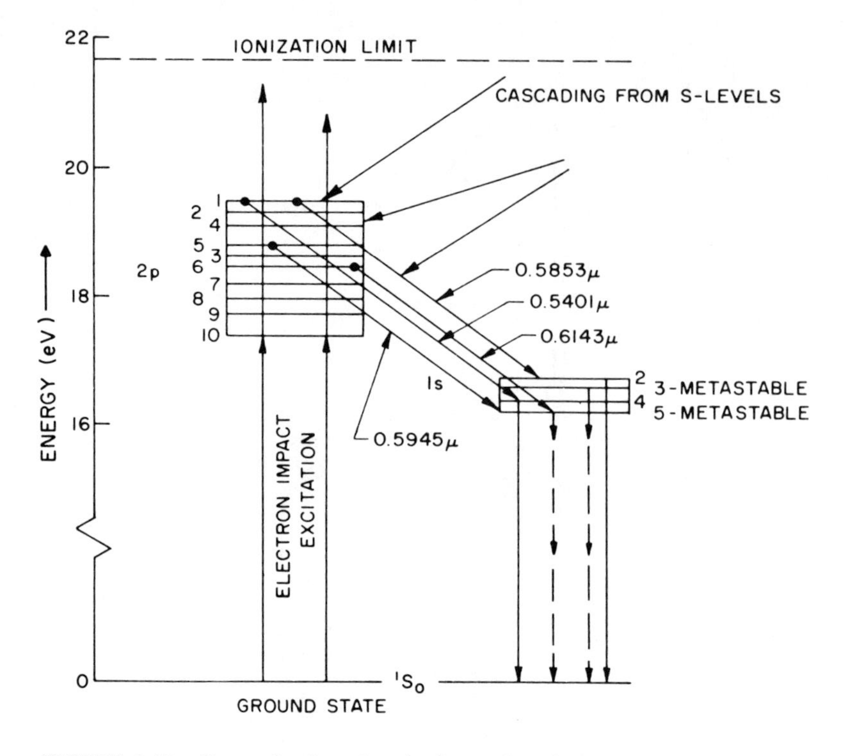

FIGURE 1.45. Energy levels and excitation paths relevant to self-terminating laser transitions in Ne.

lower level quenching cycle which operates within them. In the noble gases, in which transient oscillation occurs on p-s transitions, selective excitation of the p levels occurs by radiative cascading from higher levels as well as by direct electron impact from the ground state. Transient laser transitions observed in neon are shown in Figure 1.45.

Radiative Cascade-Pumping

Cascading through strong laser transitions can be used to provide selective excitation of levels having the same parity as the ground state, which do not favor excitation by direct electron impact. In the He-Ne laser, cascade pumping occurs via strong s-p transitions from the well-populated Ne-3s$_2$ level to give selective excitation of the Ne-3p levels, and oscillation occurs in the near-IR on a number of 3p-2s transitions at about 2 μm. As illustrated in Figure 1.46, extensive pumping of the 3p-levels is followed by cascade-pumping of the Ne-2s levels which gives population inversion on a number of 2s-2p transitions and oscillation in the near-IR around 1.2 μm.

Recombination Pumping

Positive-negative charge recombination takes the general form

$$A^+ + B^- \rightarrow A' + B + \text{kinetic energy} \quad (10a)$$

$$A^+ + B^- \rightarrow A + B' + \text{kinetic energy} \quad (10b)$$

The excess energy appears as kinetic energy of either of the neutral particles A′ or B′ of about 1 eV. Given a particular A^+ and B^-, there is a strong preference for A′ and B or A and B′.[614]

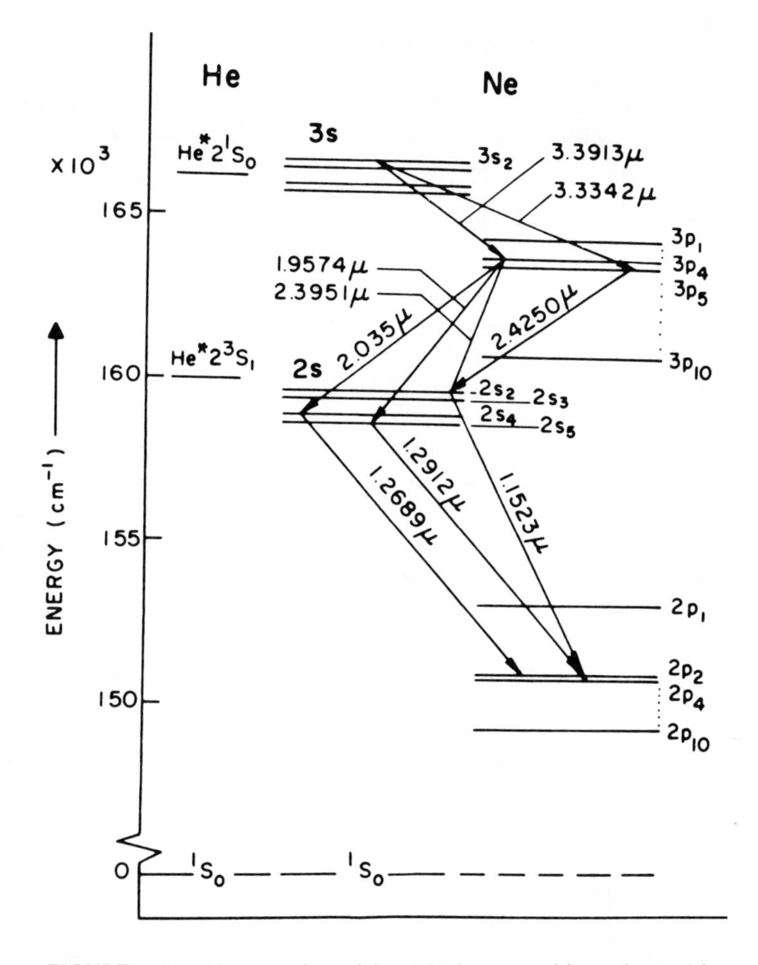

FIGURE 1.46. Ne 3p → 2s and 2s → 2p laser transitions observed in cascade following selective excitation of the $3p_4$ and $3p_5$ levels by the strong 3.3913- and 3.3342-μm Ne-I lines.

In the pulsed Na-H laser selective excitation occurs via the reaction

$$Na^+ + H^- \rightarrow Na'\ (4s) + H \tag{11}$$

with H^- supplied in the electron-molecule dissociation-reaction

$$H_2 + e^- \rightarrow H^- + H \tag{12a}$$

$$H_2 + e^- \rightarrow H^- + H' \tag{12b}$$

$$H_2 + e^- \rightarrow H^- + H^+ + e^- \tag{12c}$$

In a gas discharge above a pressure of a few torr, recombination of positive and negative ions is believed to be via three-body processes.[621] A number of lasers operate in the afterglow of high-pressure discharges, where electron-ion recombination is the selective excitation mechanism.

Selective excitation in the pure O laser operating in the afterglow of a pulsed discharge[201] occurs in the electron-ion recombination reactions

$$e^- + O_2^+ \rightarrow O'\ (3p) + O + \text{kinetic energy} \tag{13}$$

$$e^- + O_3^+ \rightarrow O'\ (3p) + O_2 + \text{kinetic energy} \quad (14)$$

Electron-ion recombination is probably responsible for selective excitation and laser oscillation in the afterglow of pulsed, high-pressure discharges in mixtures of noble gases. In two specific cases, the reactions

$$(\text{He Ne})^+ + e^- \rightarrow \text{He}' + \text{Ne}'\ (2s) \quad (15)$$

and

$$(\text{He Ar})^+ + e^- \rightarrow \text{He}' + \text{Ar}'\ (3d', 3d'') \quad (16)$$

are responsible for selective excitation and oscillation in pulsed He-Ne and He-Ar lasers.[461] Metastable He 2^3S_1 atoms are essential for the formation of the molecular ions $(\text{He Ne})^+$ and $(\text{He Ar})^+$. Electron-ion recombination also appears to be the dominant excitation mechanism of laser action in Ar, Kr, and Xe plasmas produced by firing CO_2 laser radiation into high-pressure gas.[699,700] The expanding laser-produced plasma rapidly cools and a high recombination rate thereby results followed by direct and radiative and collisionally induced cascade population of the upper laser levels.

Electron-ion recombination also appears to be the dominant excitation mechanism of several laser transitions in silver,[99,146] Mg,[146] Zn,[146] Cd,[121,146,701] In,[146] Sn,[146] and Pb.[146] In these lasers, a plasma generated from the vaporized metal either with a focused Nd laser[121] or in a transversely electrically excited device[146] is cooled and prevented from expanding too rapidly by a helium background gas. The rate controlling step in the laser-excited cadmium laser, at least, appears to be the recombination reaction itself and these lasers need not be self-terminating, but could, in principle, be made to operate CW.

Direct Optical Pumping

Because of the narrow linewidth of absorption lines in atomic gases, it is not generally possible for significant population inversions to be generated with incoherent line or continuum pump sources. However, if a coincidence between a strong line pump source and a resonance transition can be found, where the pump line is within a Doppler width of the resonance transition, selective excitation of the resonance state is possible. Laser oscillation was first observed using this scheme in Cs, where a fortuitous overlap between the ($3p\ ^3P_2 \rightarrow 3s\ ^3S_1$) He line at 0.3888 μm and the Doppler-broadened Cs resonance line ($6s\ ^2S_{1/2} \rightarrow 8p\ ^2P^o_{1/2}$) at 0.388851 μm, as shown in Figure 1.47, allowed selective excitation of the Cs $8p\ ^2P^0_{1/2}$ level and subsequent laser oscillation at 3.2 and 7.2 μm. Direct optical pumping from well populated excited states of neon followed by laser action has also been reported.[479]

A more significant laser system which uses incoherent direct optical pumping is the mercury laser developed by Djeu and Burnham.[124] The pumping scheme for this laser is shown in Figure 1.48. Sequential pumping of Hg atoms by two photons of wavelengths 253.7 and 404.7 nm, derived from a separate Hg lamp, selectively excites the upper laser level. N, CO, or water vapor added to the laser gas collisionally depopulates the lower laser level and effectively channels excitation into the intermediate state in the optical pumping process. This laser appears to be an attractive source of green radiation; a thorough investigation of its kinetics and potentialities has been given by Holmes and Siegman.[126] The principle limitation to efficient operation is the inability of the pumping lamp to generate pump photons at 404.7 nm efficiently.

Pulsed optically pumped neutral gas lasers which utilize direct optical excitation of a resonance state with noble gas halide exciplex lasers have been reported in Ca,[103] Bi,[187] and Fe.[356] In Bi the excitation scheme is

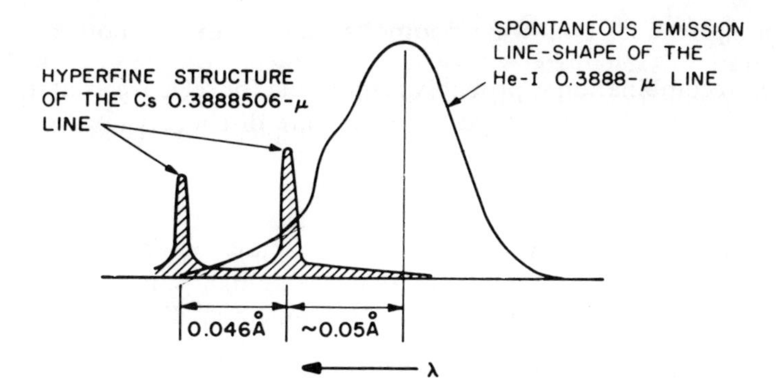

FIGURE 1.47. Coincidence of the line shapes of the Cs resonance line at 0.488851 μm. and the He-I line ($3p^3P \rightarrow 3s^3S_1$) at 0.3888 μm.[15]

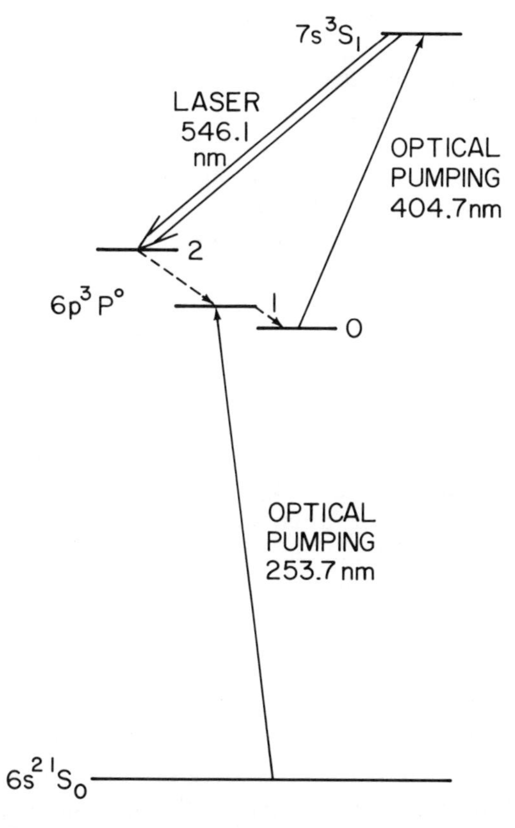

FIGURE 1.48. Energy levels of Hg-I relevant to the optically pumped 546.1 nm laser.

$$\mathrm{Bi}(6p^3\ {}^4S^\circ_{3/2}) + h\nu\ (\mathrm{XeC\ell\ laser,\ 306.772\ nm}) \rightarrow \mathrm{Bi}'\ (7s\ {}^4P_{1/2}) \quad (17)$$

The development of high-energy, quasitunable UV lasers make laser-pumped lasers of this type attractive for generating high power outputs at characteristic atomic emission frequencies.

Photodissociative Optical Pumping

The electronic absorption bands of molecules which lead to dissociation are broad, and allow a large amount of energy to be coupled into a gas even with a broad-band incoherent pump source. Following photodissociation, one or both of the resultant fragments will be in excited states; generally for dissociation though the lowest-lying dissociative or predissociative potential surface, one fragment will be in the ground and the other in an excited electronic state. For example, in the schematic reaction

$$AB + h\nu \rightarrow A' + B + \text{kinetic energy} \quad (18)$$

The classic example of this type of laser is the atomic iodine laser based on photodissociation of various iodine containing organic compounds, particularly the perfluoroalkyliodides. For example, in the case of trifluoromethyliodide,

$$CF_3I + h\nu\ (\lambda \sim 270\ \text{nm}) \rightarrow I^*\ (^2P^o_{1/2}) + CF_3 \quad (19)$$

The yield of ground-state I atoms ($^2P^o_{3/2}$) from this photodissociation is small,[622-624] so an automatic population inversion is generated on the magnetic dipole transition

$$I^*\ (^2P^o_{1/2}) \rightarrow I\ (^2P^o_{3/2}) + h\nu\ (\lambda = 1.315\ \mu m) \quad (20)$$

Variants from the classic photodissociative process exist in several lasers; the lower state of the molecule being dissociated may have a deep potential minimum, as in the case of the alkali metal halides,[3,4] InI[145] or GaI_3 (or GaI)[144]; have only a fairly shallow potential minimum, as in the case of the alkali-metal dimers[1] or TlHg[152]; or be repulsive (unbound), as in the case of the collision induced absorption of atomic Ca in Ca/noble gas mixtures.[105] The excitation scheme in this laser involves absorption of a 249-nm photon from an unbound $^1\Sigma^+$ state of CaM, where M is a noble gas atom, to various higher lying repulsive states of CaM which yield excited Ca atoms. A typical arrangement of potential surfaces and atomic energy levels which result from dissociation is shown in Figure 1.49. These curves are in fact appropriate to the TlHg system, but typify the general scheme for photodissociation from a lower state with a potential minimum. The ground potential surface $(AB)_1$ correlates with the A_1 state of atom A, while the potential surface $(AB)_3$, which dissociates following absorption of a photon and collisions, correlates with the A_3 state of A.

In the high-gain optically pumped atomic mercury photodissociation laser,[123] the production of excited atoms appears in most cases to involve sequential absorption of two-pump photons.

$$Hg_2 + h\nu\ (266\ \text{nm}) \rightarrow Hg_2'\ (^3I_u \text{ or } {}^3O_u^-) \quad (21)$$

$$Hg_2' + h\nu\ (266\ \text{nm}) \rightarrow 2Hg'\ (6d\ {}^3D,\ 7p\ {}^3P^o_1, 8s\ {}^3S_1,\ 6d\ {}^1D_2 \text{ or } 7p\ {}^1P^o_1) \quad (22)$$

The availability of high-energy pulsed UV lasers, notably the noble gas-halides, makes photodissociatively pumped laser action in many additional atomic species potentially possible in the future.

Excitation as a Result of Chemical Reactions

Only one atomic laser transition has been made to operate solely as a result of chemical reactions. This laser is based on the nearly resonant excitation transfer between molecular O and atomic I.[625]

$$O_2\ (^1\Delta) + I(^2P^o_{3/2}) \rightleftarrows O_2\ (^3\Sigma) + I^*\ (^2P^o_{1/2}) \quad (23)$$

This reaction yields a population inversion provided more than 16% of the oxygen is in the ($^1\Delta$) state.[626] In successful I lasers based on the above reaction,[332,335,336] the O_2

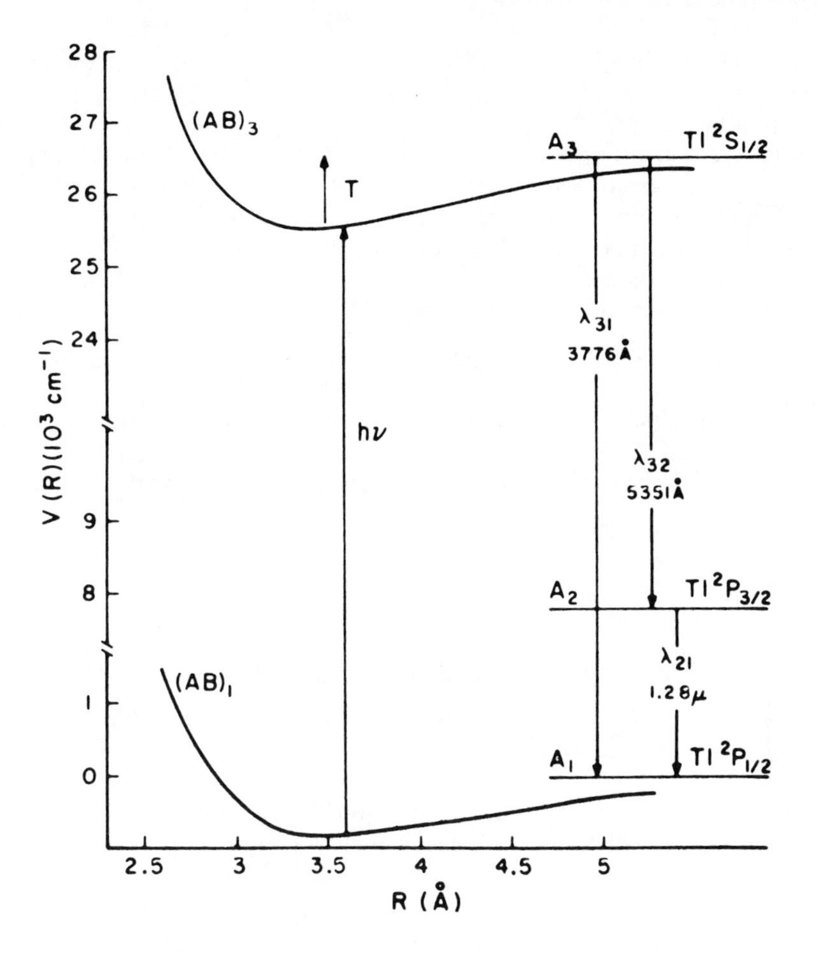

FIGURE 1.49. Energy level diagram showing molecular potential curves and corresponding atomic levels of a photodissociation laser system, as exemplified by the Tl-Hg system. (From Chilukuri, S., *Appl. Phys. Lett.*, 34, 284, 1979. With permission.)

($^1\Delta$) is generated by the reaction of chlorine and H_2O_2 in an alkaline environment, namely

$$2H_2O_2 \xrightarrow{OCl} 2H_2O + O_2\ (^1\Delta) \quad (24)$$

The resultant reaction scheme being

$$O_2(^1\Delta) + O_2(^1\Delta) \rightarrow O_2(^1\Sigma) + O_2(^3\Sigma) \quad (25)$$

$$O_2(^1\Sigma) + I_2 \rightarrow O_2(^3\Sigma) + 2I\ (^2P^\circ_{3/2}) \quad (26)$$

$$O_2(^1\Delta) + I(^2P^\circ_{3/2}) \rightleftarrows O_2(^3\Sigma) + I^*(^2P^\circ_{1/2}) \quad (27)$$

$$I^*(^2P^\circ_{1/2}) + O_2(^1\Delta) \rightarrow O_2(^1\Sigma) + I(^2P^\circ_{3/2}) \quad (28)$$

An atomic gas laser which involves partial chemical pumping has been reported by Spencer and Wittig.[387] Photolytically produced excited I atoms react with molecular bromine

$$I^*(^2P^\circ_{1/2}) + Br_2 \rightarrow IBr + Br^*(^2P^\circ_{1/2}) \qquad (29)$$

Nuclear Pumping

In a direct nuclear-pumped laser, energetic fission fragments which are either injected into a gas or are produced directly within it by collisions with other energetic particles, excite and ionize gas atoms leading to a population inversion. Two early systems of this type used laser tubes lined wih fissionable material which was bombarded with neutrons from a reactor.[610,179] In the case of a tube lined with ^{235}U, the pertinent reactions which lead to inversion are[601]

$$^{235}U + n \rightarrow ff + FF + \sim 180\ \text{MeV} \qquad (30)$$

where ff is a light fission fragment with energy ~65 MeV and FF is a heavy fission fragment with energy ~97 MeV.

The fission fragments then create ions or metastables, e.g., in the case of helium:

$$ff + He \rightarrow He^+ + e + ff \qquad (31)$$

$$ff + He \rightarrow He^* + ff \qquad (32)$$

Under typical laser operating conditions, most of the He ions form molecular ions in the reaction:[627]

$$He^+ + 2He \rightarrow He_2^+ + He \qquad (33)$$

Excitation of the laser gas, e.g., Xe, in the nuclear-created plasma involves a large number of reactions,[628] principally processes such as Penning ionization,

$$He^* + Xe \rightarrow Xe^{+(\prime)} + He \qquad (34)$$

charge transfer,

$$He_2^+ + Xe \rightarrow He_2 + Xe^+ \qquad (35)$$

and three-body recombination

$$Xe^+ + e + (\text{third body}) \rightarrow Xe' \qquad (36)$$

The latter reaction may excite the upper laser level directly or this may involve a subsequent cascade process. Direct electron impact in the nuclear created plasma also contributes to the excitation of the upper laser level.

In the case of a laser tube lined with ^{10}B, the initial nuclear pumping reaction is

$$^{10}B + n \rightarrow \alpha + {}^7Li + 2.3\ \text{MeV}\ (\alpha = 1.46\ \text{MeV};\ {}^7Li = 0.84\ \text{MeV}) \qquad (37)$$

The energetic fission fragments deposit their energy in the laser gas.

Xe,[552] Kr,[531] Ar,[523] and Ne[631] lasers, all but the latter in a pulsed mode, have been successfully pumped by the volume nuclear reaction:

$$^3He + n \rightarrow {}^3H + p + 0.76\ \text{MeV}\ (p = 0.57\ \text{MeV},\ {}^3H(\text{triton}) = 0.17\ \text{MeV}) \qquad (38)$$

Figure 1.50 shows the major processes which result from this reaction in a ^{3}He-Ar mixture. Neutron flux levels needed to reach the threshold of oscillation vary from 2 $\times 10^{11}$ n cm^{-2} sec^{-1} for CW laser action in ^{3}He-Ne mixtures[631] to 1.1×10^{17}n cm^{-2} sec^{-1} in the case of ^{3}He-Kr mixtures.[531] Lasers of the volumetrically excited kind have been

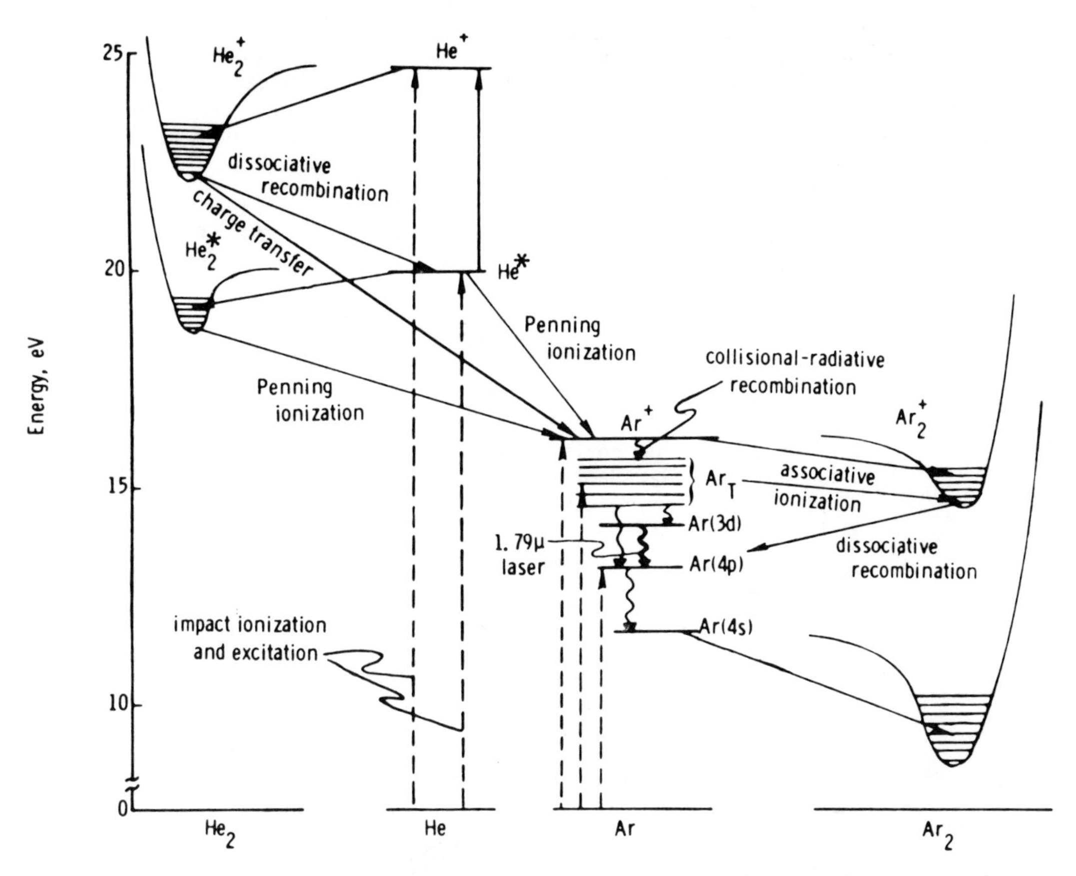

FIGURE 1.50. Energy-level diagram which shows the major energy pathways of the ^{3}He-Ar nuclear-pumped laser system. (From Wilson, J. W., DeYoung, R. J., and Harries, W. L., *J. Appl. Phys.*, 50, 1226, 1979. With permission.)

considered in some detail by Wilson et al.[524,629,630] The ^{3}He reaction has several advantages over the ^{235}U or ^{10}B reactions for gas laser pumping. With coating sources external to the laser gas, much of the kinetic energy of the fission fragments is lost in escaping the coating material. Also, the energetic particles, which are initially created in the wall, travel inward towards the laser tube axis. This effect may cause inhomogeneous excitation of the laser medium. However, one drawback of the ^{3}He reaction is that the fission fragments are not very energetic and must make many collisions to loose their energy. Thus, there is a trade-off between laser size and operating pressure. High pressure is desirable to present escape of fission fragments from the laser volume before they have deposited all their energy; on the other hand, input neutrons will not penetrate a high-pressure laser medium uniformly if it is too large. Even so, DeYoung and Winters[629] estimate that a 1% efficient laser of this kind could have a specific CW power output of 82 kW/ℓ^{-1}. An alternative approach to creating a large volume high-power DNP laser has been suggested by Wilson and DeYoung. By seeding the laser medium with gaseous $^{235}UF_6$, very energetic fission fragments can be created homogeneously by Reaction 30. These energetic, highly charged fission fragments rapidly deposit their energy in the laser medium.

CHARACTERISTICS OF IMPORTANT NEUTRAL GAS LASERS

The 0.6328-μm Helium-Neon Laser

The output power of the 0.6328-μm He-Ne laser is a function of a number of dis-

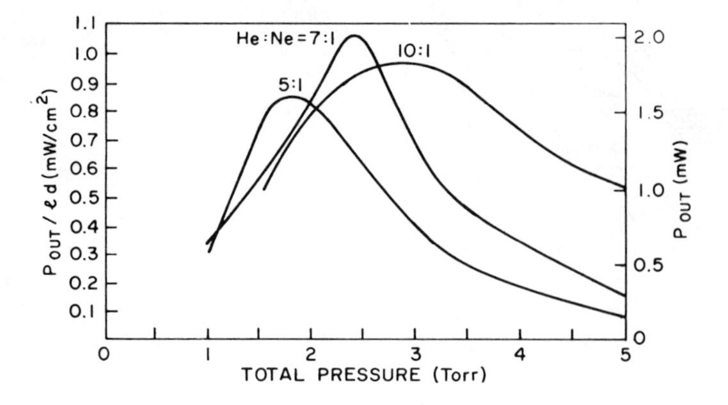

FIGURE 1.51. Output power and normalized output power vs. total pressure for a discharge tube diameter (d) of 1.5 mm and a plasma length (ℓ) of 12.5 cm. (From Field, R. L., Jr., *Rev. Sci. Instrum.,* 38, 1720, 1967. With permission.)

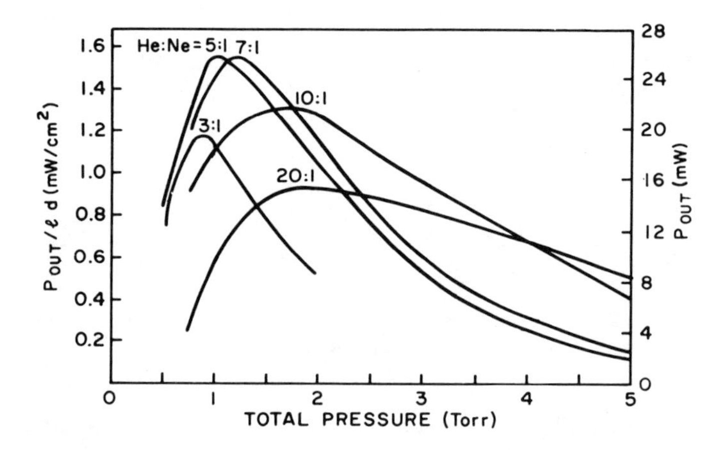

FIGURE 1.52. Output power and normalized output power vs. total pressure for a discharge tube diameter (d) of 3.0 mm and a plasma length (ℓ) of 55 cm. (From Field, R. L., Jr., *Rev. Sci. Instrum.,* 38, 1720, 1967. With permission.)

charge parameters. These include gas pressure, tube diameter (bore), mixture ratio, current (electron density), and gas temperature. For maximum output power at 0.6328 μm, the (very high-gain) $3s_2 \rightarrow 3p_4$ transition at 3.39 μm must be suppressed by the use of wavelength selection, selective cavity-mirror reflectivities, or inhomogenous longitudinal magnetic fields.

Figures 1.51, 1.52, and 1.53 show the effect of total pressure, mixture ratio, and tube diameter on the output power realizable at 0.6328 μm in a longitudinally excited positive column with oscillation eliminated on the competing 3.39-μm line.[410] Cavity and mirror data applicable to the results given in the figures for a spherical-mirror and high-reflectivity prism-cavity configuration are given in Table 1.12.

The relationship between output power, gain, cavity loss, optimum mirror transmission, and cavity configuration can be determined from "The output power of a 6328-Å He-Ne gas laser", and "On the optimum geometry of a 6328-Å laser oscillator", by Smith,[414,415] and "Dependence of He-Ne laser output power on discharge current, gas pressure and tube radius", by Sakurai.[634] The parametric behavior of hollow-cath-

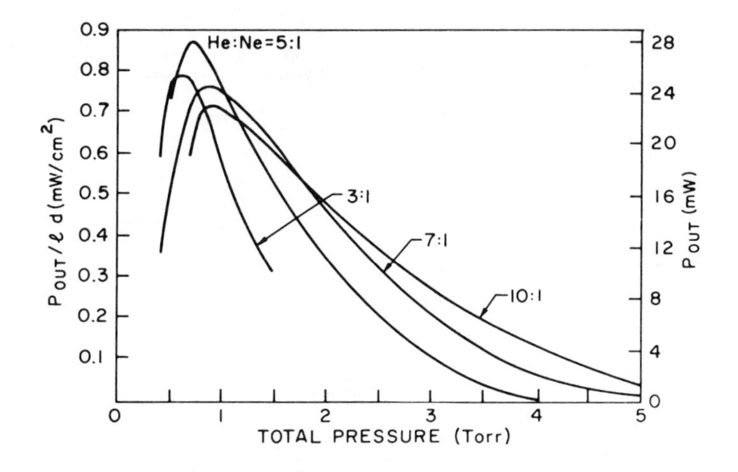

FIGURE 1.53. Output power and normalized output power vs. total pressure for a discharge tube diameter (d) of 5.0 mm and plasma length (ℓ) of 65 cm. (From Field, R. L., Jr., *Rev. Sci. Instrum.*, 38, 1720, 1967. With permission.)

Table 1.12
0.6328-μm CAVITY AND MIRROR DATA

Tube I.D. (d) (mm)	Mirror radius (m)	Mirror transmission (%)	Cavity length (ℓ) (cm)
1.5	0.5	1.1	22
3.0	2	1.0	70
5.0	2	1.0	80
8.0	10	1.1	215

ode excited,[437] waveguide,[452] and flat-plate 0.6328-μm He-Ne lasers[449] is in each case somewhat different from conventional longitudinally excited structures; for further details the appropriate references should be consulted. He-Ne lasers of these types are not presently in widespread use, however, flat-plate structures are under extensive development[449-451] because of their attractive features for large-scale manufacturing.

The results given in Figures 1.51, 1.52, and 1.53 are summarized in Figure 1.54, in which the total pressure and tube diameter product is plotted vs. capillary (tube) diameter for various He-Ne pressure ratios at each diameter. Under optimum conditions pD = 3.6 torr-mm,[402,415] in close agreement with that stated by Smith[414] to be optimum for output at 0.6328 μm. Figure 1.55 shows the optimum He to Ne pressure ratio vs. tube diameter. It should be noted that Figures 1.51 to 1.55 are related to He-Ne mixtures in which the helium is ^{4}He. A 25% increase in power can be achieved by substituting isotopic ^{3}He for ^{4}He.

Values of optimum current (for oscillation at 0.6328 μm) vs. discharge tube diameter are plotted in Figure 1.56. This figure is a modification of that given by Field.[410] Curve B exhibits the constancy of current density under optimum discharge conditions found to apply to the 0.6328-μm He-Ne laser when oscillation at 3.39 μm is suppressed. Curve A (an addition to the figure given by Field) is a plot of the relationship

$$I_{optimum}\ (\text{mA}) = 3.5 + 1.5\ D^2 \tag{39}$$

between optimum current for oscillation at 0.6328 μm and tube diameter, where D is in mm, when oscillation at 3.39 μm is not suppressed.[633]

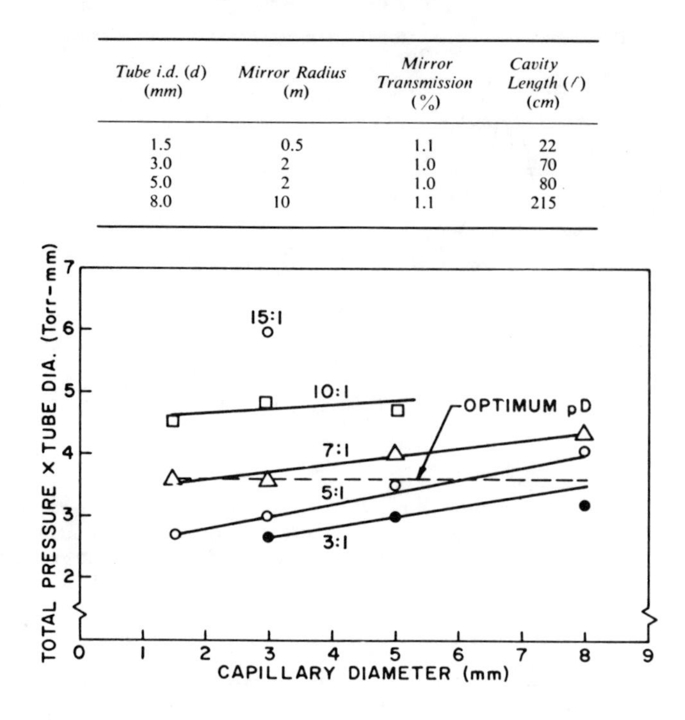

Tube i.d. (*d*) (*mm*)	*Mirror Radius* (*m*)	*Mirror Transmission* (%)	*Cavity Length* (*l*) (*cm*)
1.5	0.5	1.1	22
3.0	2	1.0	70
5.0	2	1.0	80
8.0	10	1.1	215

FIGURE 1.54. Optimum total pressure × tube diameter (torr-mm) vs. capillary diameter (mm) for various He to Ne mixture-ratios. The dashed line at a pD of 3.6 torr-mm represents experimentally observed optimum conditions to give maximum output power at 0.6328 μm. (From Field, R. L., Jr., *Rev. Sci. Instrum.*, 38, 1720, 1967. With permission.)

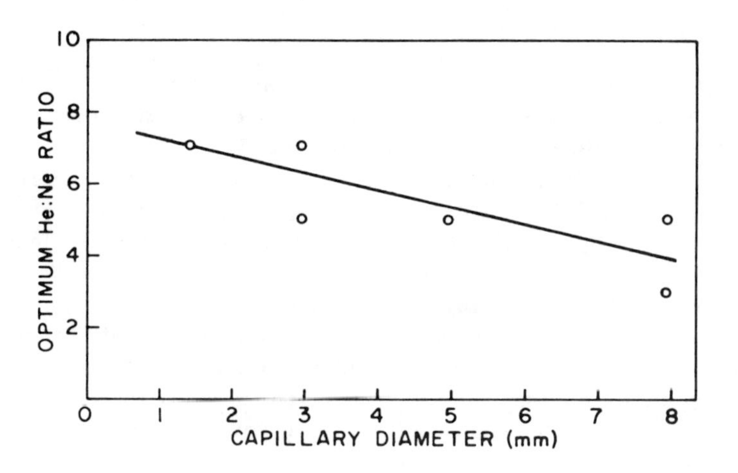

FIGURE 1.55. Optimum He-Ne ratio as a function of capillary diameter. (From Field, R. L., Jr., *Rev. Sci. Instrum.*, 38, 1720, 1967. With permission.)

Under optimum discharge conditions and where the pD is maintained constant, the saturated gain (G_s) exhibits a 1/D dependence. This is due to the way in which the population of the upper laser level follows the concentration of He 2^1S_0 metastables in the discharge, the concentration of which is proportional to the pressure (p), or inversely proportional to D.[402,596]

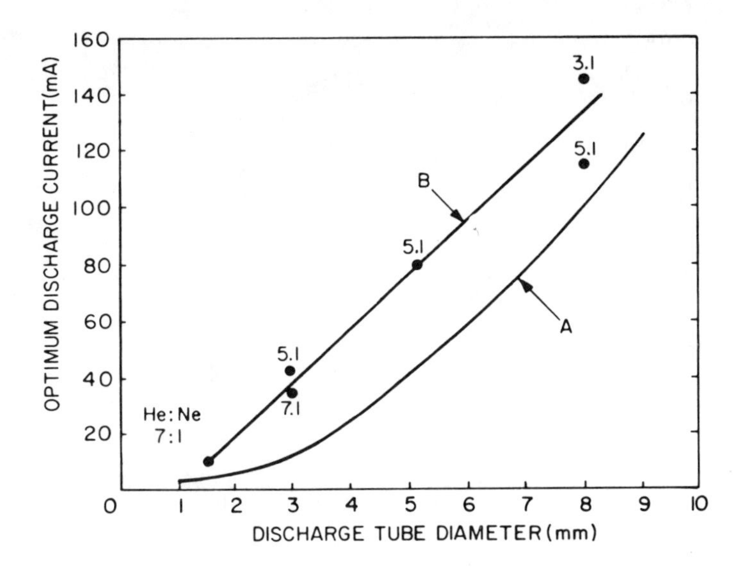

FIGURE 1.56. Optimum discharge current in a He-Ne laser as a function of capillary diameter. Curve B is the observed linear relationship between optimum discharge and capillary diameter when oscillation at 3.39 μm is suppressed.[410] Curve A is a curve of the same relationship, when oscillation at 3.39 μm is not suppressed (see Equation 39).[633]

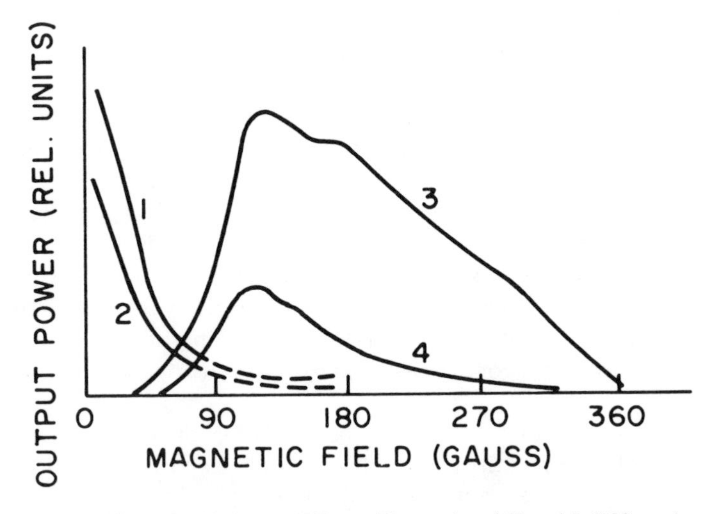

FIGURE 1.57. Competition of emission at 3.39 μm (Curves 1 and 2) and 0.6328 μm (curves 3 and 4) with the plasma in a longitudinal magnetic field. Curves 1 and 3 and 2 and 4 were recorded simultaneously. The rf-drive power for 1 and 3 was greater than for 2 and 4. Excitation conditions: rf-excitation; He to Ne ratio, 7:1; total pressure, 0.8 torr; tube diameter, 4 mm. (From Alekseeva, A. N. and Gordeev, D. V., *Opt. Spectrosc.*, 23, 520, 1967. With permission.)

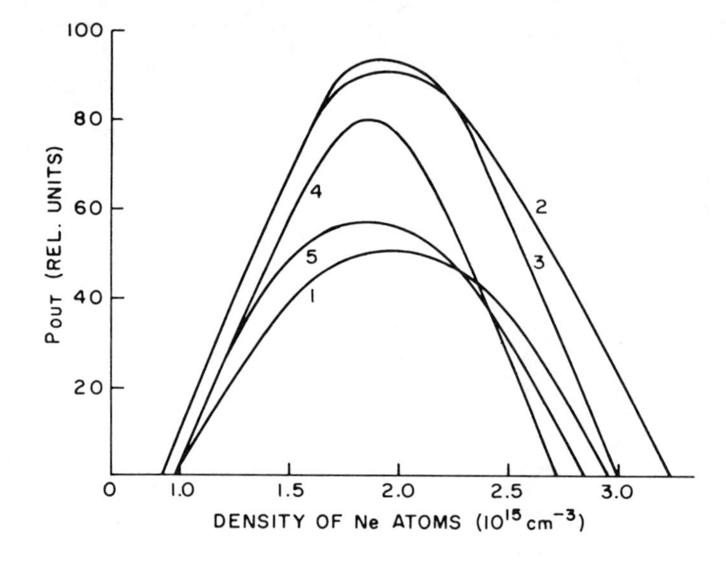

FIGURE 1.58. Output power of a 0.6328 μm He-Ne laser as a function of atom density in He to Ne mixture for various wall temperatures of the discharge tube. Curve 1, 293 K; 2, 425 K ; 3, 483 K ; and 5, 780 K. Excitation conditions: DC-discharge current 70 mA; He to Ne ratio 8:1; D = 8 mm. (From Gonchukov, G. A., Ermakov, G. A., Mikhenko, G. A., and Protsenko, E. D., *Opt. Spectrosc.*, 20, 601, 1966. With permission.)

In a 5:1 He-Ne mixture without suppression of the 3.39-μm line, with an optimum discharge current given by Equation 39,

$$G_s = 3.0 \times 10^{-2}\ \ell/D \text{ percent} \quad (40)$$

where ℓ is the plasma length in cm and D is the tube diameter in cm.[495] The single-pass gain (G_0) obtainable on the Ne $3s_2 \rightarrow 2p_4$ transition at 0.6328 μm has a more complicated dependence on discharge conditions than Equation 40.[416]

For He-Ne lasers designed specifically for operation in a TEM_{oo} mode, the optimum operating conditions vary somewhat from those for multimode operation. Because the unsaturated gain of a He-Ne mixture on the tube axis peaks at a lower pressure and current and for a higher He to Ne ratio than on the periphery of the positive column,[445] the optimum operating conditions for TEM_{oo} mode operation are correspondingly adjusted. The exact optimum conditions for TEM_{oo} mode operation depend on resonator geometry.[435]

The effect of the competing Ne $3s_2 \rightarrow 3p_4$ transition at 3.39 μm on oscillation at 0.6328 μm is reduced by Zeeman splitting the 3.39-μm line by means of a longitudinal magnetic field over the active length of the discharge.[413,417,429] The variation of output power at 0.6328 and 3.39 μm vs. magnetic field is given in Figure 1.57. An approximate 50% increase in power at 0.6328 μm over that given in Figures 1.51 to 1.53 can be achieved, if a weak inhomogenous magnetic field is used in conjunction with prism wavelength selection.[410]

The output power at 0.6328 μm is also affected by the gas temperature in the active region of the plasma.[412,418] This is shown in Figure 1.58. The variation of output power with discharge tube wall temperature reflects the effect of the gas temperature on numerous collisional and diffusive effects in the He-Ne laser.

The Near-IR Xe, Xe-He, and Xe-Ar Lasers

A number of 5d → 6p transitions in neutral Xe exhibit high gain when operated in

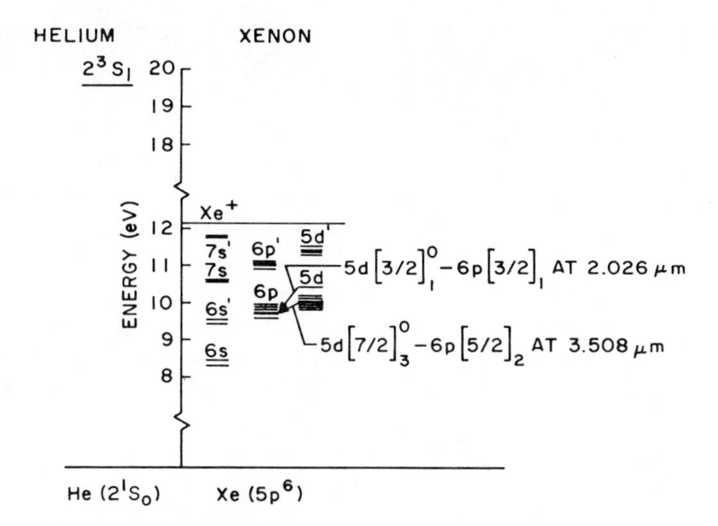

FIGURE 1.59. Partial energy-level diagram of Xe and of He, showing disposition of the two very high-gain 5d-6p laser transitions at 3.5 and 2.026 μm and the He 2^3S_1 metastable state.

a CW mode at low Xe pressures. Two of these transitions, the $5d[7/2]^o_3 \rightarrow 6p[3/2]_2$ transition at 3.507 μm and the $5d[3/2]^o_1 \rightarrow 6p[3/2]_1$ transition at 2.026 μm, exhibit extremely high gain. High power and efficient pulsed laser oscillation has been obtained on the $5d[3/2]^o_1 \rightarrow 6p[5/2]_2$ at 1.73 μm in transversely excited high-pressure Xe - Ar mixtures dilute in Xe. Transversely excited laser action with high peak and average power has also been obtained at 2.026 μm and on 5d $\rightarrow$ 6p transitions at 2.65, 3.43, and 3.65 μm in high pulse repetition frequency He-Xe lasers operating with a transverse gas flow. These lasers are of technological importance because they provide coherent output in atmospheric windows. Unfortunately, when operated in a CW mode, their power outputs saturate at low values.

The atomic processes that lead to high gain in pure Xe and Xe-noble gas mixtures appear to involve electron-impact excitation from the atom ground state and low-lying metastables, together with contributions from recombination/cascade pumping from higher-lying levels and excitation transfer from metastable species. This latter process has been shown to occur in a Xe-Kr laser where Kr 5s metastables transfer their energy to Xe 5d levels.[635,636] The Kr $5s[3/2]^o_2$ and Xe $5d[1/2]^o_1$ levels are within 15 cm^{-1} of each other. However, as shown in Figure 1.59, direct excitation transfer from the lowest He metastable (2s $^3S_1$) cannot occur. It is possible that He metastables are involved in the excitation of Xe 5d levels by some indirect process; they are certainly involved in Penning ionization

$$He^*(2s\,^3S_1) + Xe \rightarrow He + Xe^+ + e \qquad (41)$$

which increases the electron density in a Xe-He discharge above its value for pure Xe without too large a concomitant reduction in electron temperature. For example, an electron temperature of 9 eV (104,000 K) in pure Xe at 0.05 torr is only reduced to about 8 eV by adding 4 torr of He in an 11-mm tube, while the electron density increases by a factor of three.[540]

Population inversion occurs readily between the Xe 5d and 6p levels because of favorable lifetime ratios for these levels. The lifetimes of the upper levels vary from between about 5 and 170 times longer than the lower levels,[637,638] so that given equal excitation of both states, population inversion is easily achieved in a discharge.

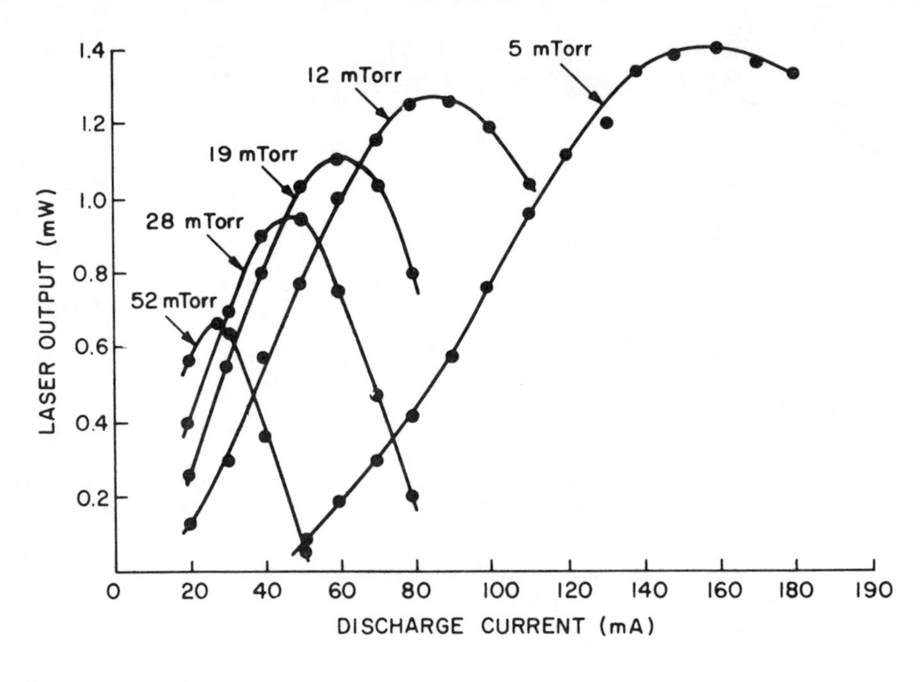

FIGURE 1.60. 3.5-μm output power vs. discharge current and gas pressure of a pure-Xe laser; D = 6 mm.[554]

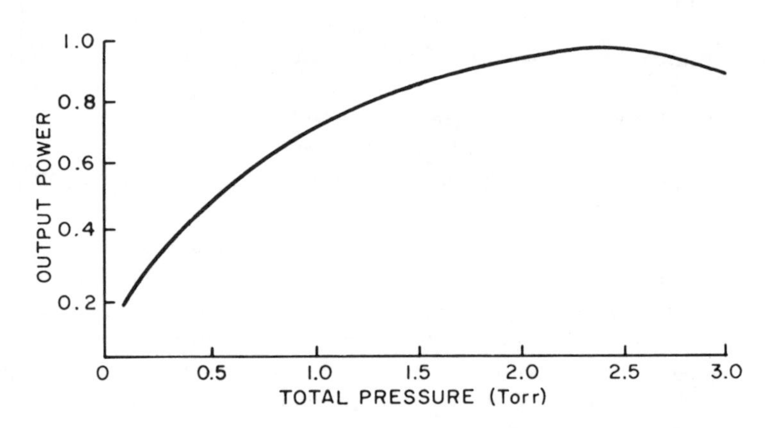

FIGURE 1.61. 3.5-μm laser output power vs. total pressure of a 200:1, He to Xe gas mixture. The output power is related to the maximum power at 2.5 torr. Excitation conditions: 30 MHz rf-excitation at approximately 100 W, D = 12 mm. (From Kuznetsov, A. A. and Mash, D. I., *Radio Eng. Electron. Phys.*, 10, 319, 1965. With permission.)

Optical gains as high as 10^{40} m^{-1} in a 3/4-mm bore or 10^{6} m^{-1} in a 7-mm bore tube have been observed at 3.507 μm for DC discharges in Xe-He mixtures.[560,556] In pure Xe, gains as high as 10^{5} m^{-1} in a 7-mm bore tube have been observed for this line.[557] For the 2.026-μm line, gains in excess of 200 m^{-1} have been reported for an RF-excited Xe-He discharge in a 7 mm-bore tube. In the case of this line, He apears to be essential in producing high gain.[578]

The dependence of output power on discharge current and pressure for a DC-excited laser in pure Xe operating at 3.5 μm is presented in Figure 1.60. A curve of output power at 3.5 μm vs. total (He-Xe) pressure using rf-excitation is given in Figure 1.61. Strong cataphoretic effects in a DC-excited Xe-He laser make the use of rf-excitation

Table 1.13
XENON-LASER DISCHARGE CONDITIONS

Line (μm)	pD (torr-mm)	Mixture ratio, He to Xe	Excitation	Ref.
3.507	<0.03	Pure Xe	DC	554, 557
	<30	200:1	r.f.	545
	<13	60:1	r.f	557
2.026	~15	100:1	r.f	537
	~20	500:1	r.f	644
	~35	25:1	r.f	508

desirable. Rapid gas clean-up in discharges in both pure Xe and He-Xe mixtures make it essential to use some form of pressure control.[554]

Representative discharge conditions for the two high-gain 3.5- and 2.026-μm Xe lines are tabulated in Table 1.13.

In the 3.5-μm pure Xe laser, there is no optimum gas pressure in the range 10 to 150 mtorr. The optical gain increases monotonically as the pressure decreases. The observed dependence between Xenon pressure and tube diameter is of the form

$$pD^n = \text{a constant (C)} \tag{42}$$

where $n \cong 3.2 \pm 0.2$, and C depends on the gain.[559]

For large-diameter tubes (I.D. $\geqslant$7 mm), the gain is proportional to 1/D. For I.D. $\leqslant$3 mm, the gain varies as $(D)^{-n}$, where $n > 1$ for pressures of Xe, $p > 50$ mtorr, and $n < 1$ for $p < 50$ mtorr.

In the He-Xe 2.026-μm laser, for tube-diameters between 3 and 5 mm, the gain is approximately proportional to 1/D.[537]

The low saturation output power of the Xe laser operating in a CW mode can be largely explained in terms of the electron-impact excitation of the 5d levels. Bennett[195] has shown on theoretical grounds that the cross section for the process

$$Xe(5p^6\ {}^1S_0) + e \rightarrow Xe(5d) + e \tag{43}$$

should be an order of magnitude larger than the cross section for the process

$$Xe(5p^6\ {}^1S_0) + e \rightarrow Xe(6s \text{ or } 6s') + e \tag{44}$$

for electrons above threshold, while the process

$$Xe(5p^6\ {}^1S_0) + e \rightarrow Xe(6p) + e \tag{45}$$

should have a much smaller cross section because there is no parity change involved.

The primary excitation reaction (see Equation 43) requires high-energy electrons; unfortunately, electrons with energy more than 1 to 2 eV above threshold are more likely instead to produce ionization. The preference of the laser for operation at low pressures is accounted for by the increasing electron temperature that occurs in this case. At these low pressures, as for example in Figure 1.60, the Xe density is not very great, so output powers are low. Increasing current through the discharge increases the population of 6s and 6s′ levels which are either metastable or quasimetastable because of radiation trapping at pressures above about 15 mtorr. Excitation of the 6p lower laser levels from these 6s and 6s′ levels will reduce the available inversion. In addition, the 6s and 6s′ levels are easily ionized, which may cause a reduction in electron temperature with increasing current.

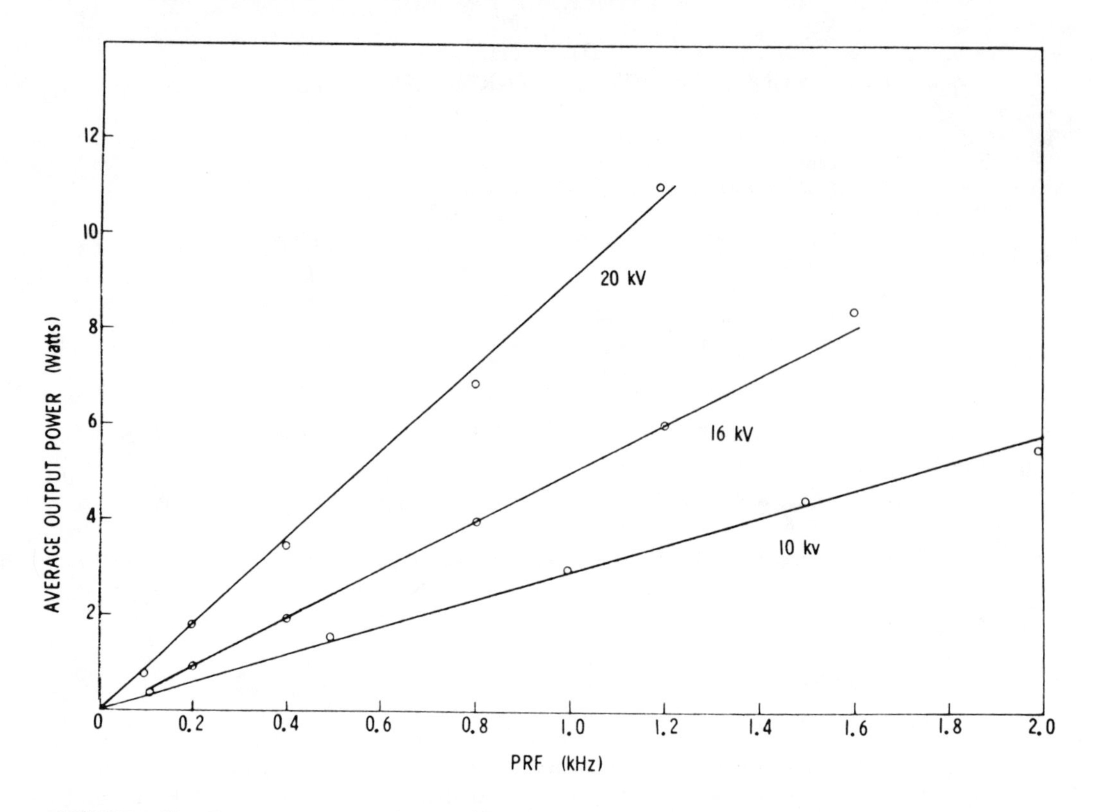

FIGURE 1.62. Output power vs. pulse repetition frequency at different applied voltages for a transversely excited pulsed Xe laser with transverse gas flow. (From Fahlen, T. S. and Targ, R., *IEEE J. Quant. Electron.*, QE-9, 609, 1973. With permission.)

High-voltage transversely excited pulsed Xe lasers, operating at pressures 100 or more times the CW values, overcome these problems for two main reasons. During the breakdown period of the discharge, the electron temperature is high and efficient excitation of 5d levels occurs; higher operating pressures mean greater specific energy outputs. Additionally, if such a laser is operated at a high pulse repetition frequency in a gas transport mode, undesirable metastables are swept from the active volume between pulses and substantial average powers are obtained, as shown in Figure 1.62.

The Red Fluorine Laser

Several strong laser lines in the vicinity of 700 nm can be obtained from pulsed, longitudinally excited or transversely excited lasers operating on He-NF_3 and He-F_2 mixtures dilute in the fluorine containing material. Output powers up to 250 kW in 2- to 3-nsec pulses[219] or 0.5 mJ per pulse in 30-nsec pulses over all lines have been reported to date. The spectral output of these lasers varies in a characteristic way with increasing pressure as shown in Figure 1.63. As the pressure is increased, lasing tends to switch from 3p ($^2P^o$,$^2S^o$) → 3s ^2P transitions to 3p ($^4S^o$, $^4P^o$) → 3s ^4P transitions. This seems to indicate that, as the pressure is increased, collisions of the excited fluorine atoms transfer population from doublet to quartet states. Angular momentum is conserved in these transfers so, for example, a decrease in the intensity of the ^2S(731.1 nm) is matched by a growth in intensity of the ^4S(634.9 nm) line.

High-resolution studies of the laser lines in the doublet manifold indicate strongly that the excitation mechanism for the He-F_2 system involves dissociation of HeF exciplexes,[379] e.g., in the following series of reactions

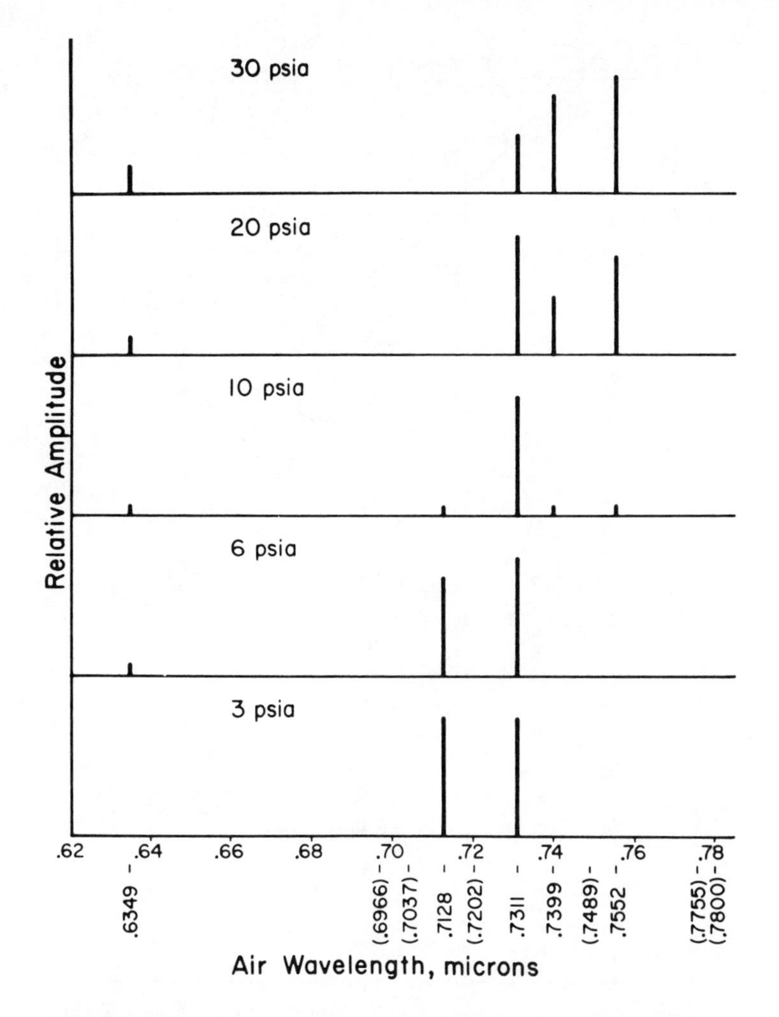

FIGURE 1.63. Schematic comparison of the laser intensity on various neutral fluorine lines at several pressures. Line intensities are roughly indicated by height. (From Loree, T. R. and Sze, R. C., *Opt. Commun.*, 21, 255, 1977. With permission.)

$$F_2 + e \rightarrow 2F + e \quad (46)$$

$$F + He^*(2s\ ^1S \text{ or } 2p\ ^1P^\circ) \rightarrow (HeF)' \quad (47)$$

$$(HeF)' \rightarrow He + F\ (3p(^2S^\circ, ^2P^\circ, ^2D^\circ)) \quad (48)$$

The excitation mechanism for F lasers using He-NF_3[219] or He-HF[224] mixtures was originally believed to involve dissociation of the F containing molecule by He (2s ^{1}S) metastables. However, at least for He-NF_3 mixtures, the reaction

$$He(2s\ ^1S) + NF_3 \rightarrow He + NF_2 + F' \quad (49)$$

appears not to be the primary laser pumping mechanism.[227] Direct electron impact may still be important, for example,

$$NF_3 + e \rightarrow NF_2 + F' + e \quad (50)$$

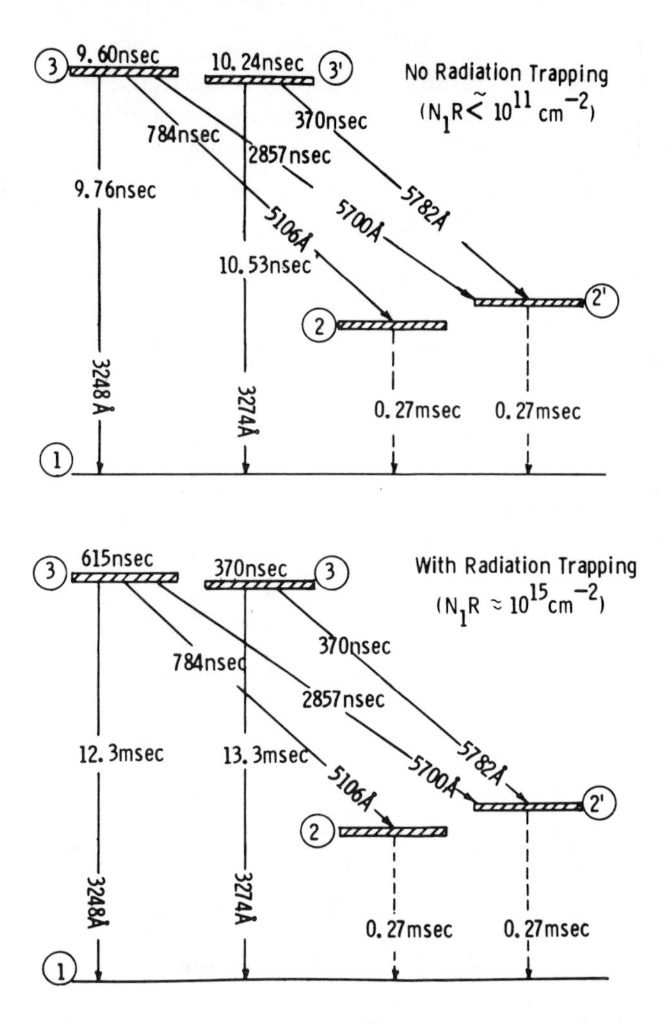

FIGURE 1.64. Effective transition lifetimes for a Cu laser system with and without resonance trapping. (From Weaver, L. A., Liu, C. S., and Sucov, E. W., *IEEE J. Quantum Electron.*, QE-10, 140, 1974. With permission.)

The 510.6- and 578.2-nm Copper Vapor Lasers

In the last decade, no neutral gas laser except the 1.315-μm I laser has been the subject of such intensive technological development as the Cu vapor laser, which has intense, very high-gain self-terminating green and yellow laser lines. The excitation of laser action in the copper atoms is relatively straight forward. In a fast-rise-time pulsed discharge, inversion results as shown in Figure 1.4. The 4p $^2P^o$ upper laser levels, which are connected to the ground state by strong resonance transitions, are efficiently excited by direct electron impact, and provided the threshold for oscillation is reached before spontaneous emission branching to the 4s^2 ^2D lower level reduces the inversion, self-terminating laser action results. At sufficiently high Cu atom ground-state densities, the 4p $^2P_{3/2}{}^o$ and 4p $^2P_{1/2}{}^o$ levels become resonance trapped and their effective lifetimes, determined by spontaneous emission to the lower laser levels, are 615 and 370 nsec, respectively.[25] Resonance trapping becomes effective when the copper atom ground state density (N_1)X tube radius (R) factor exceeds 10^{13} cm^{-2}. The laser tube temperature necessary to produce this number density of vaporized copper atoms is about 500°C in a 9-mm bore tube. The graphic difference in effective lifetime of the levels of the Cu laser with and without resonance trapping is shown in Figure 1.64. For N_1R values $\leqslant 10^{12}$ cm^{-2}, resonance trapping is ineffective; to create a population inversion at these densities, the excitation current pulse must fight an upper-level lifetime of only 10 nsec — a technologically difficult requirement.

Although a substantial power output (40 kW) and good efficiency (1.2%) was obtained[19] very shortly after the first Cu laser was reported, interest in this laser waned for several years because of the technological difficulties involved in maintaining the laser tube at the high temperatures, up to 1800°C, needed to generate large copper vapor pressures for high power operation — the vapor pressure of copper is only about 1 torr at 1600°C. However, the report of high gain laser action in pulsed discharges through cuprous iodide (CuI) at temperatures of only about 600°C, where the CuI vapor pressure is 1 to 10 torr, by Weaver et al.,[25] stimulated investigation of a whole class of Cu lasers which would operate at manageable temperatures. Subsequently, laser action has been studied in discharges through cuprous chloride (CuCl),[27,35,42,47,52-56,65,69,70,72-74,80,84,91,95,96] where the optimum operating temperature is ≃400°C; cuprous bromide (CuBr),[38,41,42,47,51,81,82,84,92] optimum temperature ≃500°C; (these optimum temperatures for the Cu halides all apparently correspond to a vapor pressure of about 0.03 torr[645]) cuprous oxide (Cu_2O),[46,64] optimum temperature ≃1350°C; Cu (II) acetylacetonate (CuAcAc, Cu $(C_5H_7O_2)_2$),[66,83,95] optimum temperature 30 to 140°C; copper nitrate ($Cu(NO_3)_2$),[66] optimum temperature ≃190°C; and copper (II) acetate ($Cu(C_2H_3O_2)_2$),[95] optimum temperature 159°C. Typically, these lasers operate with several torr of added Ne or He. During single-pulse excitation of these Cu compounds, electron impact serves both to dissociate the molecule to generate Cu atoms and to excite these atoms to the upper laser level. However, Chen et al.[27] determined that much more efficient laser operation could be obtained by using double-pulse excitation of the Cu-containing lasant vapor. The first pulse dissociates molecules to generate Cu atoms, and, after an appropriate delay, a second pulse excites these atoms to the upper laser level, for example,

$$\mathrm{CuBr + e\ (\sim 2.5\ eV) \rightarrow Cu + Br + e} \tag{51}$$

$$\mathrm{Cu + e\ (\sim 3.8\ eV) \rightarrow Cu'\ (^3P) + e} \tag{52}$$

The double-pulse excitation technique has been widely used, although at high enough single-excitation pulse repetition frequencies it becomes less necessary. In this case, each excitation pulse both dissociates lasant molecules and excites atoms remaining from the previous pulse. It is unclear which Cu-containing lasant gives best laser performance; various authors indicate that CuBr > CuCl > CuI,[56] CuBr > CuI > CuCl > Cu $(NO_3)_2$ > CuAcAc,[66] CuAcAc > CuAc > CuCl,[95] or CuBr > CuI > CuCl.[41] However, CuCl and CuI are the most widely used lasants.

Parallel with the development of Cu lasers using Cu halides and other compounds, there has been renewed interest in lasers operating in copper vapor produced from the heated metal.[28,32-34,40,44] In high-pulse-repetition-frequency devices, self-heating of the discharge tube can be used to maintain the active region at the operating temperature of 1500°C, where the copper vapor pressure reaches about 0.4 torr. Operating frequencies up to 150 kHz have been reported.[57] It appears that, despite the technological problems of having the laser tube operating at high temperature, Cu vapor lasers based on vaporization of Cu metal are somewhat superior to those using more exotic lasants in both power output and efficiency. Average output powers up to 43.5 W have been reported.[58]

Although the Cu vapor laser is self-terminating, under resonance-trapped conditions, the requirements on excitation pulse rise time are not as critical as they are, for

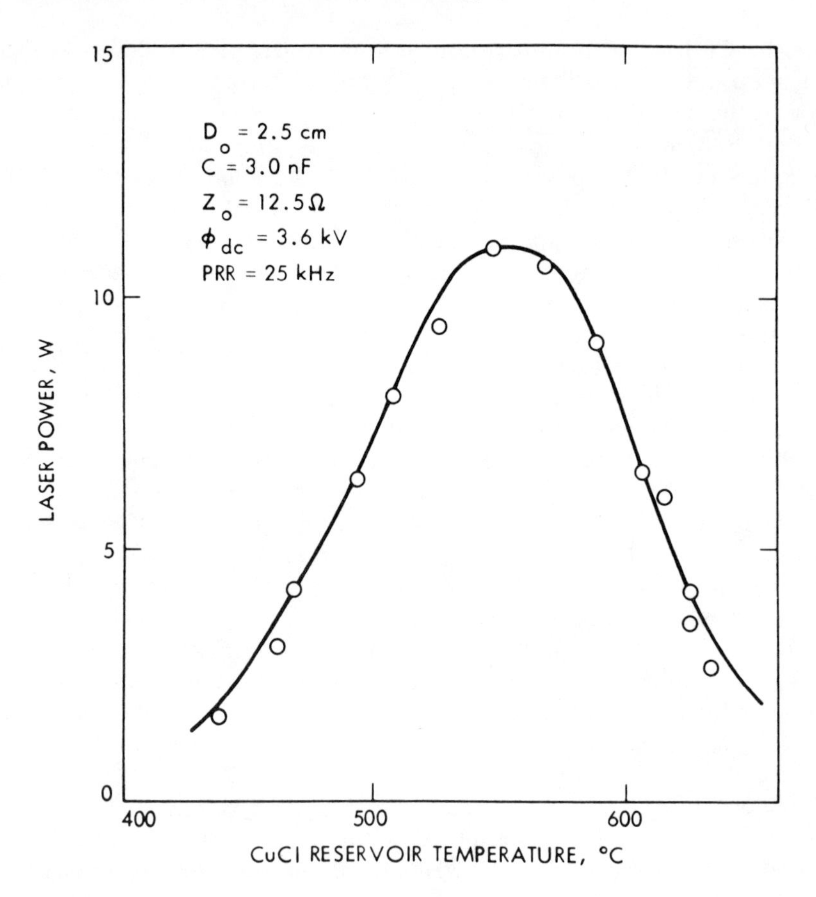

FIGURE 1.65. Dependence of laser power on copper chloride reservoir temperature in a continuously pulsed (25 kHz) CuCl laser. The CuCl vapor pressure in the laser tube is assumed to be proportional to reservoir temperature. Some parameters of the Blumlein discharge circuit are also indicated. (From Nerheim, N. M., Bhanji, A. M., and Russell, G. R., *IEEE J. Quantum Electron.*, QE-14, 686, 1978. With permission.)

example, in the molecular N laser. As a consequence, although some Cu vapor lasers have been transversely excited[44,48,55,76,81] or operated in pulsed hollow-cathode discharges,[30,50] most are operated in longitudinal discharges where the inductance of the discharge channel should be kept low.[70,72]

Irrespective of what lasant is used, the output characteristics of a Cu vapor laser are a function of several parameters, such as lasant pressure, buffer gas pressure, tube diameter, current pulse rise time, and pulse repetition frequency (or pulse separation when operated in the double-pulse excitation mode).

In a Cu laser operating with Cu vapor and a He or Ne buffer, operating temperatures of about 1400°C are necessary to ensure resonance trapping; laser operation has been reported up to temperatures of 1850°C,[34] however, the optimum operating temperature may be anywhere from 1500 to 1700°C, depending on tube size, buffer gas pressure, and excitation geometry. Some typical examples of the parametric behavior of Cu halide lasers are shown in Figures 1.65, 1.66, and 1.67. Figure 1.65 shows the variation of laser power with CuCl temperature (and consequently vapor pressure). Figure 1.66 shows both the variation of total laser power as a function of buffer gas pressure and as a function of pulse separation in transversely excited CuI and CuBr lasers excited in the double-pulse mode. Figure 1.67 shows the relative variation of

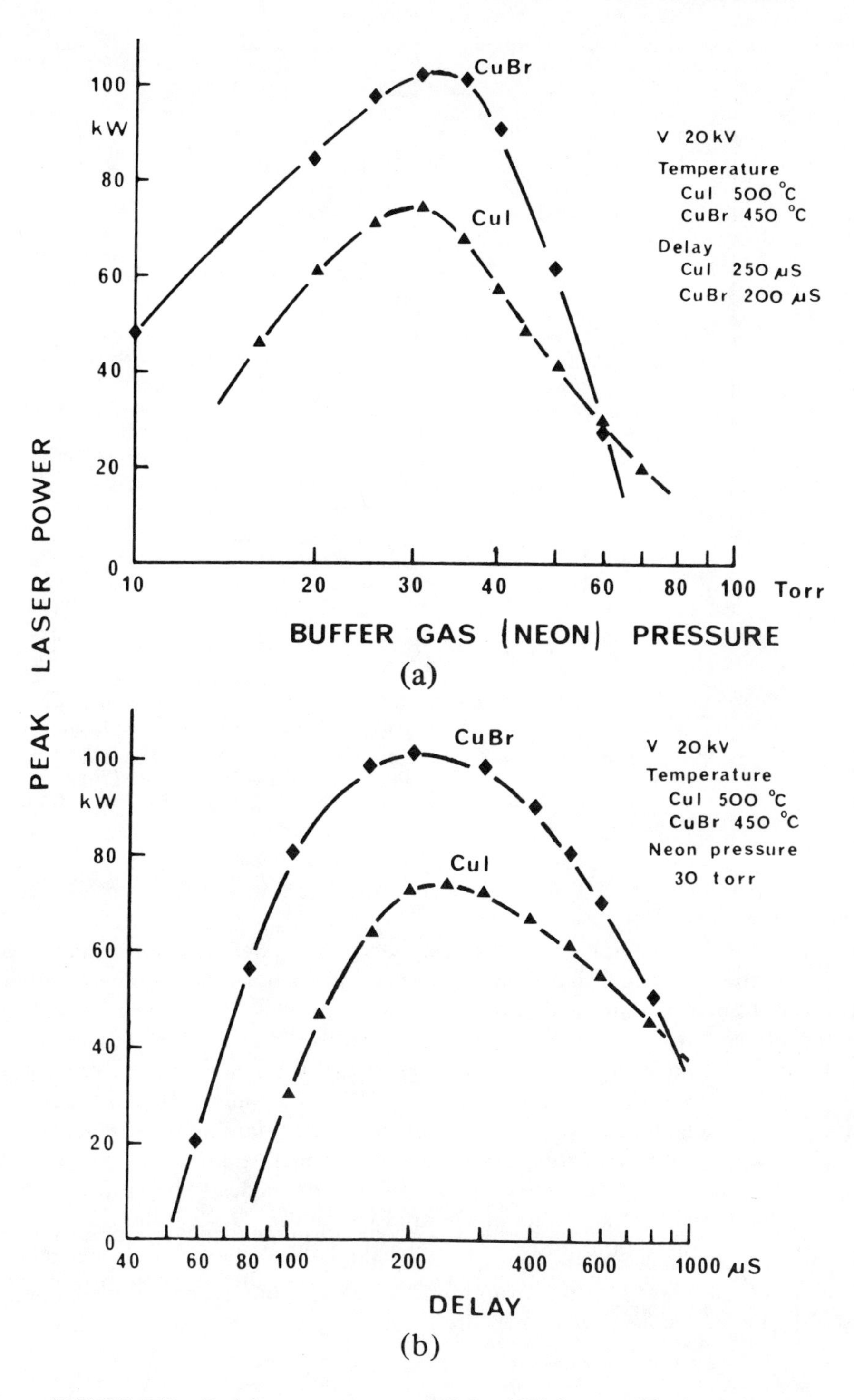

FIGURE 1.66. Peak laser output power (510.6 and 578.2 nm combined) as a function of (a) buffer gas (neon) pressure and (b) delay between "dissociation" and "excitation" pulses in a double-pulsed copper laser using CuI and CuBr as lasants. (From Piper, J. A., *IEEE J. Quantum Electron.*, QE-14, 405, 1978. With permission.)

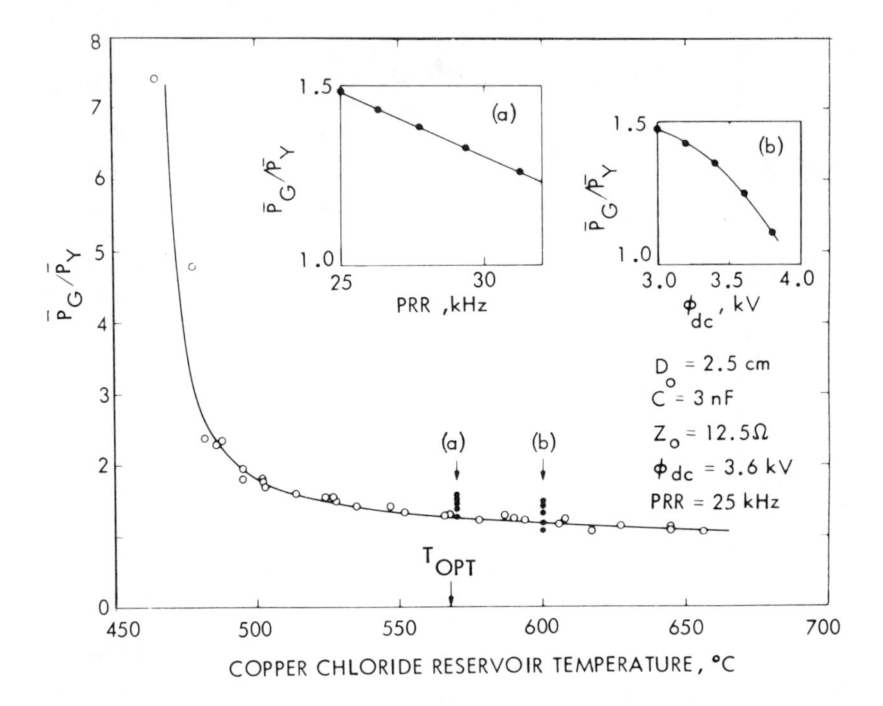

FIGURE 1.67. The ratio of laser power at 510.6 and 578.2 nm as a function of CuCl reservoir temperature in a continuously pulsed (25 kHz) copper chloride laser. The effects of varying the pulse repetition rate (PRR) and supply voltage (ϕ_{DC}) are shown in the insets. Z_0 is the impedance of the Blumlein discharge circuit used. (From Nerheim, N. M., Bhanji, A. M., and Russell, G. R., *IEEE J. Quantum Electron.*, QE-14, 686, 1978. With permission.)

laser power on the green (510.6 nm) and yellow (578.2 nm) laser lines as a function of various parameters.

The parametric behavior of the Cu laser can be very well described qualitatively in terms of a kinetic scheme that involves a series of phenomena. In repetitively pulsed operation of either Cu atoms or Cu-containing molecules, the spacing between pulses must be adjusted to allow the $4s^2\ ^2D$ metastable lower laser levels to be removed. This can occur by homogeneous collisional quenching by other atoms, molecules, or by low-energy electrons or by diffusion to the laser tube walls.[93] In the "dissociation-excitation" double-pulse mode of operation with Cu-containing molecular lasants, the dissociation pulse produces substantial numbers of metastable Cu atoms which must be allowed to decay before the excitation pulse is applied, as shown in Figure 1.68. The $^2D_{3/2}$ level apparently empties more rapidly than the $^2D_{5/2}$, so the yellow laser line will persist to shorter interpulse separations than the green.[41,57,69] If the separation between dissociation and excitation pulses is too long, the Cu atom ground state is removed by recombination,

$$Cu + X + M \rightarrow CuX + M \tag{53}$$

where M is a third body or the discharge tube wall.

The variation of laser power with lasant pressure, although strongly influenced by available number densities for excitation and resonance trapping effects, is also connected with variations in electron temperature and density. There is evidence that the operating pressures for optimum laser operation correspond to conditions giving minimum average electron temperature.[39] Increase of the lasant pressure above optimum

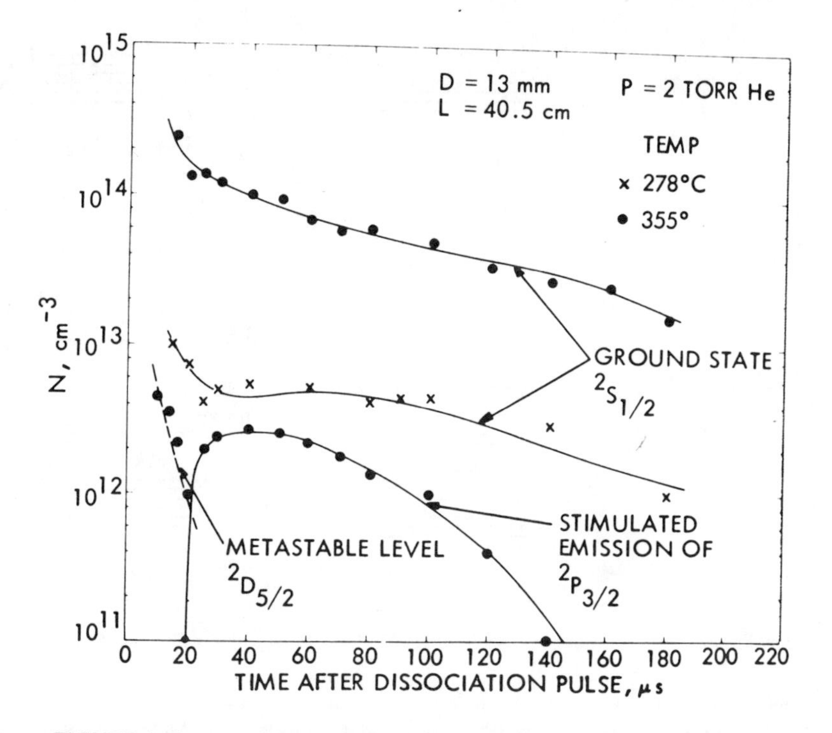

FIGURE 1.68. Measured densities of copper ground state and metastable level as a function of time after the dissociation pulse in a double-pulsed CuCl laser. The variation in laser energy at 510.6 nm is also shown as a function of pulse separation. (From Nerheim, N. M., *J. Appl. Phys.*, 48, 3244, 1977. With permission.)

was found to increase the electron temperature and lead to excessive ionization of Cu atoms, thereby reducing the population of both the copper atom ground state and upper laser levels. The buffer gas in Cu vapor lasers plays several roles: it assists in optimizing the electron temperature and density conditions for optimum laser action, minimizes diffusional loss of lasant from the hot zone of the discharge tube, and collisionally quenches metastable Cu atoms.[33] The buffer gas may also indirectly excite the upper laser levels by Penning ionization and charge exchange reactions, followed by recombination and cascade pumping of the upper laser levels.[63] On the other hand, ionization of copper atoms by noble gas metastables or ions also depletes the atom population.[39]

In single-pulse excitation of Cu atoms in a Cu vapor laser, early in the pulse the electron temperature is too low to excite the upper laser levels efficiently. As the electrons heat to their optimum temperature (5 to 10 eV),[86] laser action commences and then terminates as a result of radiative population of the lower levels plus collisional removal of the upper laser levels both in a downward direction and by ionization.[24,86] As a consequence, the pulse duration emitted by a Cu laser, which is typically a few tens of nanoseconds, although values as large as 185 nsec have been reported,[24] is shorter than the effective lifetime of the upper laser levels.

The 1.315-μm Atomic Iodine Photodissociation Laser

Although the 1.315-μm atomic I laser has been made to operate CW as a result of chemical reactions[332,335,336] and will operate following photopreionized transverse electric discharge excitation of CF_3I in an excess of N, it has been studied predominantly as a photodissociation laser. At the present time, this laser is the neutral gas laser with the highest peak power and energy output, 300-J, 1-nsec pulses have been reported,[302]

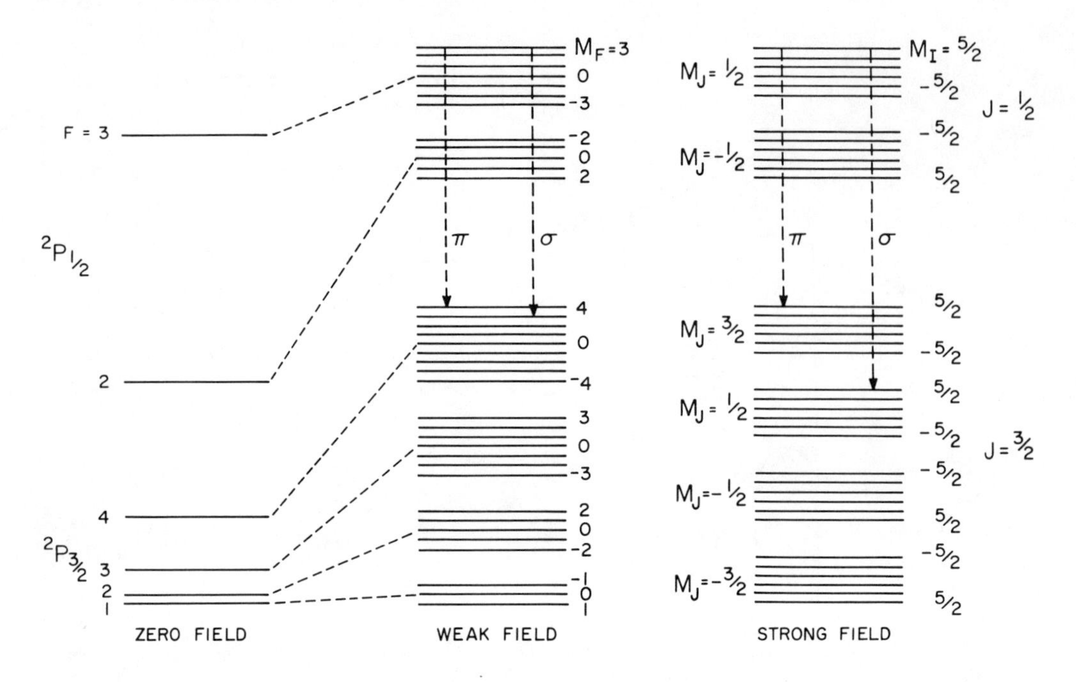

FIGURE 1.69. Schematic diagram of the hyperfine structure of the iodine $^2P^\circ_{1/2} \rightarrow {}^2P^\circ_{3/2}$ transition in zero, weak, and strong external magnetic fields. (From Davis, C. C., Pirkle, R. J., McFarlane, R. A., and Wolga, G. J., *IEEE J. Quantum Electron.*, QE-12, 334, 1976. With permission.)

and 1000-J, 1.4% efficient operation in pulses 70-μsec long,[333] 100-psec pulses can also be generated and amplified.[322] A large number of I-containing organic compounds with the general formula RI dissociate following UV irradiation to yield excited iodine atoms.

$$RI + h\nu \rightarrow R + I^*(^2P^\circ_{1/2}) \qquad (54)$$

Compounds which dissociate in this way and have been used in I lasers include CH_3I, CF_3I;[238] C_2H_5I, n-C_3H_7I, n-C_4H_9I, i-C_4H_9I, C_2F_5I, n-C_3F_7I;[239] i-C_3F_7I;[263] CF_3 CH_2I;[254] CD_3I;[286] $(CF_3)_3$ CI;[312] $(CF_3)_2AsI$, $CF_3(C_2F_5)AsI$, $CF_3(C_3F_7)AsI$, $(C_2F_5)_2AsI$, $(C_3F_7)_2AsI$, $(CF_3)_2PI$, $CF_3(C_2F_5)PI$, $CF_3(C_3F_7)PI$, $(C_2F_5)PI$, $(C_3F_7)_2PI$, $CF_3(CH_3)PI$, $CF_3PI(CN)$, $CF_3(CF_2Cl)(CFH)PI$, F_3PI, OPF_2I, and $(CF_3)_2$ SbI.[272] Apart from some interest in $(CF_3)_2AsI$ by Soviet workers,[311] most I photodissociation lasers presently use i- or n-C_3F_7I, although CF_3I, C_2F_5I, and CD_3I also operate well in these systems.

The laser transition itself occurs on the magnetic dipole transition between the two spin-orbit split components of the I atom ground state

$$I^*(^2P^\circ_{1/2}) \rightarrow I(^2P^\circ_{3/2}) + h\nu \; (\lambda = 1.315 \; \mu m) \qquad (55)$$

The transition is electric dipole forbidden as it involves no parity change.[650] Because of the nuclear spin of the ^{127}I nucleus, there is a magnetic dipole interaction between electrons and nucleus, and the laser transition exhibits significant hyperfine structure, as shown in Figure 1.69. Laser action has certainly been observed on four of the hyperfine components of the transition[318] which in order of intensity are $3 \rightarrow 4 > 2 \rightarrow 2 > 3 \rightarrow 2 > 2 \rightarrow 3$; oscillation may also occur on the remaining two. Because photodissociation lasers are generally excited by high peak-current flashlamps, they may be exposed to potentially strong magnetic fields during operation. Several authors have given closé attention to this phenomenon.[261,274,279,308] However, in high-energy-storage Iodine laser systems, where the stimulated emission cross section is reduced by pres-

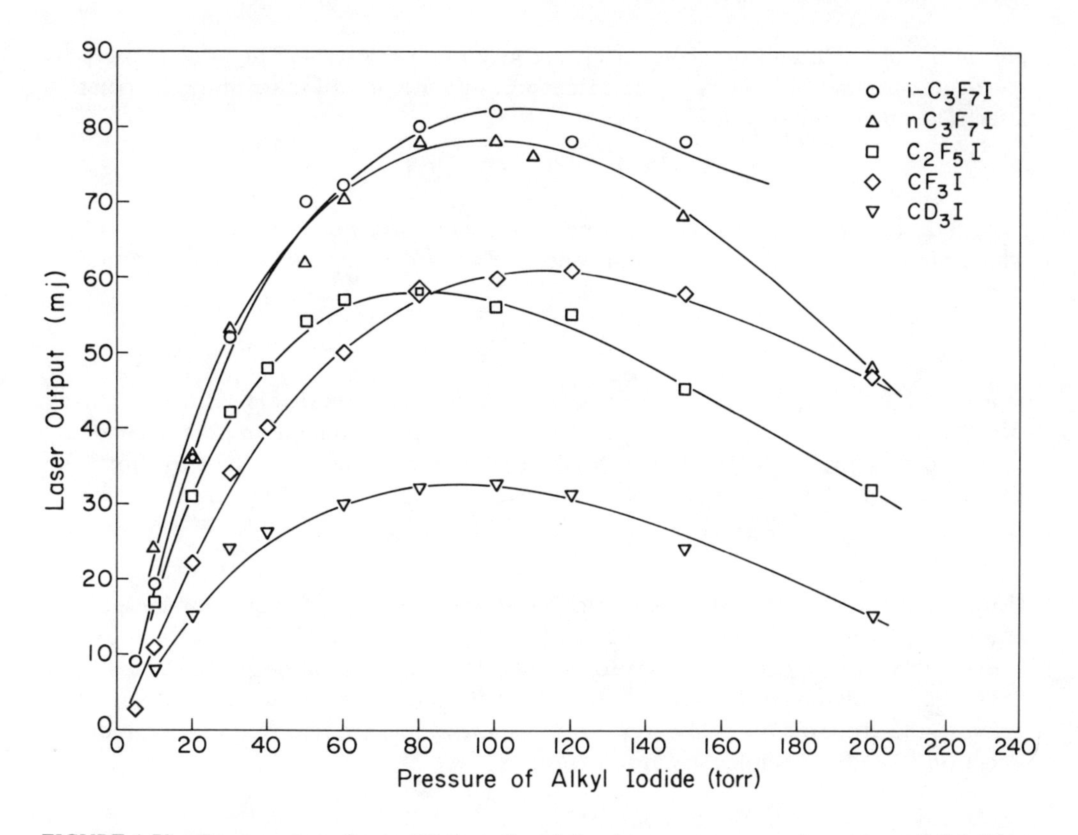

FIGURE 1.70. Pressure dependence of I photodissociation laser energy output for various alkyl iodides, observed using fast-flash excitation. Active region of laser was 15 cm long by 10-mm bore, flash duration was ∼3 μsec with an energy ∼400 J. (From Davis, C. C., Pirkle, R. J., McFarlane, R. A., and Wolga, G. J., *IEEE J. Quantum Electron.*, QE-12, 334, 1976. With permission.)

sure-broadening the laser transition with up to several atmospheres of buffer gas to prevent spurious laser oscillation, the hyperfine and magnetic field induced structure is smeared out.[314] A large number of buffer gases, such as CO_2, N_2, CO, SF_6, CF_3Br, CF_2Cl_2, $(CF_3)_2$ CO, and the noble gases, have been used in this way.[296,314] CO_2 and Ar appear most desirable in this application, each offers a compromise between pressure-broadening capability and relatively small cross section for collisional deactivation of excited I atoms.

The dependence of I photodissociation laser output energy on lasant pressure for the five-most widely used materials is shown in Figure 1.70. The optimum operating pressure falls as the diameter of the laser tube is increased, otherwise optical thickness effects prevent homogeneous photolysis by the external flashlamp(s). The kinetic processes which occur in an iodine laser following photolysis are many, and have been the subject of numerous experimental and theoretical investigations. The reactions of principal importance are, using C_3F_7I as an example,

$$C_3F_7I + h\nu(\lambda \sim 270 \text{ nm}) \rightarrow C_3F_7 + I^* (^2P^\circ_{1/2}) \qquad (56)$$

$$C_3F_7I + h\nu \rightarrow C_3F_7 + I(^2P^\circ_{1/2}) \qquad (57)$$

The C_3F_7 radicals are themselves almost certainly vibrationally and translationally hot.[652-654] The branching ratio between Equations 56 and 57 is of critical importance in determining whether a given I-containing compound will operate well in a laser system. Some experimentally measured fractional yields of I* following broadband photolysis of C_3F_7I and other materials are given in Table 1.14. The table graphically illustrates, for example, why i-C_3H_7I does not operate in an iodine photodissociation laser. The quenching reaction

$$I^* + C_3F_7I \rightarrow I + C_3F_7I \qquad (58)$$

strongly influences the efficiency of the iodide as a laser material. Molecular iodine, which builds up in the laser medium following photolysis and three-body recombination,

$$I + I + M \rightarrow I_2 + M \qquad (59)$$

is a very efficient quencher of excited I atoms, as shown in Table 1.15, and plays a dominant role in the laser kinetics.[245,252,260,263,273,277,309] If repetitive photolysis of the same iodide sample is carried out, the laser output degrades from shot to shot[308] because of the rapid quenching reaction

$$I^* + I_2 \rightarrow I + I_2 \qquad (60)$$

However, recirculation of the lasant, coupled with removal of molecular iodide, will allow long-term repetitively pulsed operation.[315] The most desirable lasants i- and *n*-C_3F_7I have very small cross sections for quenching of excited iodine atoms.

The I photodissociation laser is essentially self-terminating because it involves a transition to the ground state. However, ground state atoms are removed fairly rapidly by Reaction 59 and by chemical recombination

$$C_3F_7 + I(+M) \rightarrow C_3F_7I(+M) \qquad (61)$$

It is very desirable for this reaction to occur if repetitive photolysis of the same sample is desired. The presence of a third body is necessary to stabilize the highly vibrationally excited molecule produced by the recombination of the radical and the iodine atom. Interestingly enough, although the I laser seems to be intrinsically self-terminating, CW optically pumped operation has been achieved by rapidly removing ground state atoms in a longitudinal flow system.[325]

An undesirable reaction is radical recombination, which removes lasant molecules.

$$2C_3F_7(+M) \rightarrow C_6F_{17}(+M) \qquad (62)$$

This reaction is relatively inefficient in the case of i-C_3F_7I. Other chemical products can result if the lasant is heated too much by photolysis,[308] however, in lasers using an excess of buffer gas this problem is largely eliminated.

Neutral Gas Lasers as Frequency and Length Standards

On account of their inherently narrow spontaneous emission linewidths, very adequate output powers and good spatial and temporal coherence properties when compared to other types of lasers, neutral gas lasers are very attractive for applications in metrology. The passive amplitude and frequency stability of well-constructed DC gas-discharge-excited neutral gas lasers can be very good. For example, by very carefully isolating a He-Ne laser structure constructed of low-expansion materials from all extraneous sources of disturbance, Jaseja and co-workers[655,656] were able to obtain a short-term frequency stability at 632.8 nm of 1 part in 10^{13}. However, to accomplish this, the laser had to be thermally and acoustically isolated on a massive shock-proof table in the wine cellar of a house remote from city noise on a still, windless day! The long-term frequency stability of passively stabilized lasers can exceed 1 part in 10^{10},

Table 1.14
FRACTIONAL YIELDS p* OF EXCITED IODINE ATOMS PRODUCED FOLLOWING BROAD-BAND UV PHOTOLYSIS OF VARIOUS ALKYL IODIDES[622,623]

Compound	p*[a]
CH_3I	0.92 ± 0.02
C_2H_5I	0.69 ± 0.05
n-C_3H_7I	0.67 ± 0.04
i-C_3H_7I	<0.10
n-C_4H_9I	0.82 ± 0.04
s-C_4H_9I	<0.10
i-C_4H_9I	0.69 ± 0.04
t-C_4H_9I	<0.10
CD_3I	0.99
CF_3I	0.91 ± 0.03; 0.910 ± 0.005[b]
C_2F_5I	>0.9
n-C_3F_7I	>0.99; 0.978 ± 0.012[b]
i-C_3F_7I	0.90 ± 0.02
$(CF_3)_3I$	0.877 ± 0.013[b]
HI	0.10 ± 0.05

[a] $p^* = [I^*]_0/([I^*]_0 + [I]_0)$ where $[I^*]_0$ and $[I]_0$ are the initial concentrations of excited and ground-state I atoms produced following photolysis.

[b] Measured in Reference 651 using 266-nm laser photolysis.

Table 1.15
RATE CONSTANTS FOR COLLISIONAL DEACTIVATION OF $I^*(^2P^o_{1/2})$ AT 300 K[308]

Quenching species	k (cm^3-molecule^{-1} sec^{-1})
He	$<5 \times 10^{-18}$
Ar	$<2 \times 10^{-18}$
Xe	$<1.6 \times 10^{-18}$
$I(^2P_{3/2})$	$<1.6 \times 10^{-14}$
H_2	8.8×10^{-14}
D_2	2.2×10^{-15}
N_2	1.5×10^{-16}
CO	1.2×10^{-15}
CO_2	1.3×10^{-16}
SF_6	3.1×10^{-15}
H_2O	2.3×10^{-12}
NO	1.1×10^{-11}
O_2	2.5×10^{-11}
i-C_3H_7	3.0×10^{-10}
I_2	3.6×10^{-11}
CH_3I	5.7×10^{-13}
C_2H_5I	6.1×10^{-13}
n-C_3H_7I	6.3×10^{-13}
i-C_3H_7I	6.3×10^{-13}
CD_3I	1.8×10^{-14}
CF_3I	3.5×10^{-16}
C_2F_5I	1.1×10^{-16}
n-C_3F_7I	$\sim 5 \times 10^{-16}$
i-C_3F_7I	$\sim 3 \times 10^{-16}$

Table 1.16
FREQUENCIES AND WAVELENGTHS OF NEUTRAL GAS LASER LINES

Species	Approximate wavelength (μm)	Hyperfine component	ν (MHz)	λ_{vac} (fm)	Ref.
^{20}Ne	0.633	$^{127}I_2$i	473612213.5 ± 0.7[a]	632991399.8 ± 0.9	668
^{20}Ne	0.633	$^{129}I_2$k	473612309.6 ± 0.7[a]	632991271.4 ± 0.9	668
^{20}Ne	0.633	$^{129}I_2$m	473612268.0 ± 0.5[a]	632991327.0 ± 0.7	672
^{22}Ne	0.633	$^{129}I_2$B	473613202.1 ± 0.7[a]	632990078.6 ± 0.9	668
^{20}Ne	1.15	[b]	260103284 ± 30	1152590053 ± 133[a]	687
Xe	2.03	—	147915850 ± 15	2026777103 ± 206[a]	689
^{22}Ne	3.39	[b]	88376181.627 ± 0.05	3392231376 ± 12	691
Xe	3.51	—	8545997 ± 3	3507985824 ± 124[a]	689

[a] Calculated from measured wavelength (or frequency) using c = 299792458 msec^{-1}.
[b] Locked to a methane reference absorption.

but construction with low expansion materials and long-term temperature control of the laser cavity structure to within a millidegree is necessary to accomplish this.[657]

Alternatively, to obtain a stable frequency output, the laser output can be actively locked to the center of the gain profile by the use of the Lamb dip[658] or by Zeeman splitting the gain profile.[659] The laser frequency can also be locked to an external resonance cavity or to a suitable absorption reference. For further details of these and other schemes for stabilizing the frequency of gas lasers, the articles by Wallard,[660] Birnbaum,[661] Laures,[662] and White[663] should be consulted. The last three articles review the early work in the field.

Unfortunately, to serve as a frequency or length standard, a laser must have a reproducible frequency, irrespective of by whom or where in the world it is constructed. In particular, differences in operating pressure, gas composition, including the presence of impurities,[664] can affect laser frequency even in a laser which is adjusted so it oscillates at the peak of the spontaneous emission profile. Intercomparisons of various He-Ne lasers oscillating at the center of the 632.8-nm line indicate that absolute frequency variations up to about 1 part in 10^7 can be expected. Consequently, lasers for use as frequency or length standards are actively stabilized against well-characterized absorption resonances which permit accurate reproducibility of the stabilized laser frequency. The two laser systems most widely used in this way are the 632.8-nm He-Ne line locked to absorption resonances in $^{127}I_2$ or $^{129}I_2$ and the 3.39-μm He-Ne line locked to an absorption resonance in methane.

In the I-stabilized He-Ne laser, ^{3}He-^{20}Ne or ^{3}He-^{22}Ne mixtures are generally used which are locked to any one of several hyperfine absorption lines in the spectrum of $^{127}I_2$ or $^{129}I_2$. In the $^{127}I_2$ case, there are 21 hyperfine absorption features which can be used as a frequency reference, these components are designated a,b,c, . . . s,t,u by Hanes and Dahlstrom.[665] In the $^{129}I_2$ case, there is a characteristic triplet of peaks called A, B, and C. The wavelengths of these various absorption features have been measured against the ^{86}Kr-606-nm primary standard line[666] by several investigators; some representative values are given in Table 1.16. Small differences in wavelength values reported by different laboratories are partially accounted for by varying interpretations of exactly which point on the ^{86}Kr line profile should be used as the reference standard. There has been, and continues to be, substantial interest in I-stabilized He-Ne lasers for use as length standards; current state of the art is a 400-sec absolute frequency stability of about 4 parts in 10^{13} and a reproducibility of a few parts in 10^{11}.[676] For further details of I stabilized He-Ne lasers, References 668 to 675 and references contained therein should be consulted.

The 3.39-μm He-Ne laser transition stabilized with respect to the $F_1^{(2)}$ component of the P(7) line of the ν_3 mode of methane has also been the subject of much investigation as a frequency standard. By the use of saturated-absorption techniques, absolute frequency stability over 10-sec periods of between 1 part in 10^{13} and 1 part in 10^{14},[677-679] and reproducibility, measured by comparison of lasers from different laboratories, of about 2 parts in 10^{11},[680] have been obtained. For further details of methane-stabilized He-Ne lasers, References 677 to 682 and references therein should be consulted.

To conclude, it is appropriate to note that absolute frequency measurements of several IR and visible laser transitions have been made during the last several years. These frequency measurements are derived from a series of harmonic generation and mixing experiments which eventually synthesize a frequency, itself directly related to a cesium atomic clock frequency standard, which can be heterodyned with the laser whose frequency is to be measured. The 3.39-,[683-686] 1.15-μm,[687] and several I-stabilized doubled He-Ne wavelengths near 576 nm[688] have been measured in this way, as have the 2.03- and 3.5-μm lines of the neutral Xe laser.[689] Some representative frequency values are collected together in Table 1.16. The combination of accurate frequency and wavelength measurement of the 3.39-μm methane-stabilized He-Ne line have provided a value for the speed of light 100 times more accurate than had been obtained previously by unrelated methods. The previously accepted value was 299792.5 $^{+}$ 3 km sec^{-1}.[690] However, as a result of recent experiments, the refined value is now 299792458.05 ± 0.75 $msec^{-1}$.[690] This value is a weighted average of values reported in References 685, 686, and 690 to 695. The precision of this value is such that in June of 1979 the Comité Consultatif pour la Définition du Mètre, in Recommendation M2, has proposed a new definition of the metre, based on the second: "The metre is the length equal to the distance travelled in a time interval of 1/299792458 of a second by plane electromagnetic waves in vacuum".

REFERENCES

1. **Dezenberg, G. J. and Willett, C. S.,** New unidentified high-gain oscillation at 486.1 and 434.0 nm in the presence of neon, *IEEE J. Quantum Electron.*, QE-7, 491—493, 1971.
2. **Bockasten, K., Lundholm, T., and Andrade, O.,** Laser lines in atomic and molecular hydrogen, *J. Opt. Soc. Am.*, 56, 1260—1261, 1966.
3. **Ehrlich, D. J. and Osgood, R. M.,** Alkali-metal resonance-line lasers based on photodissociation, *Appl. Phys. Lett.*, 34, 655—658, 1979.
4. **White, J. C.,** Inversion of the Na resonance line by selective photodissociation of NaI, *Appl. Phys. Lett.*, 33, 325—327, 1978.
5. **Tibilov, A. S. and Shukhtin, A. M.,** Laser action with sodium lines, *Opt. Spectrosc.*, 21, 69—70, 1966.
6. **Pogorelyi, P. A. and Tibilov, A. S.,** On the mechanism of laser action in Na-H_2 mixtures, *Opt. Spectrosc.*, 25, 301—305, 1968.
7. **Tibilov, A. S. and Shukhtin, A. M.,** Investigation of generation of radiation in the Na-H_2 mixture, *Opt. Spectrosc.*, 25, 221—224, 1968.
8. **Mishakov, V. G., Tibilov, A. S., and Shukhtin, A. M.,** Laser action in Na-H_2 and K-H_2 mixtures with pulsed injection of metal vapor into a gas-discharge plasma, *Opt. Spectrosc.*, 31, 176—177, 1971.
9. **Sorokin, P. P. and Lankard, J. R.,** Infrared lasers resulting from giant-pulse laser excitation of alkali metal molecules, *J. Chem. Phys.*, 54, 2184—2190, 1971.
10. **Grishkowsky, D. R., Sorokin, P. P., and Lankard, J. R.,** An atomic 16 micron laser, post-deadline Paper 5.4, 9th Int. Quantum Electronics Conf., Amsterdam, June 14 to 18, 1976; *Opt. Commun.*, 18, 205—206, 1977.

11. **Grishkowsky, D. R., Lankard, J. R., and Sorokin, P. P.**, An atomic Rydberg state 16-μ laser, *IEEE J. Quantum Electron.*, QE-13, 392—396, 1977.
12. **Sorokin, P. P. and Lankard, J. R.**, Infrared lasers resulting from photodissociation of Cs_2 and Rb_2, *J. Chem. Phys.*, 51, 2929—2931, 1969.
13. **Jacobs, S., Rabinowitz, P., and Gould, G.**, Coherent light amplification in optically pumped Cs vapor, *Phys. Rev. Lett.*, 7, 415—417, 1961.
14. **Bennett, W. R., Jr.**, Inversion mechanisms in gas lasers, *Appl. Opt.*, Suppl. Chem. Lasers, 3—33, 1965.
15. **Rabinowitz, P. and Jacobs, S.**, The optically pumped cesium laser, in *Quantum Electronics III*, Grivet, P. and Bloembergen, N., Eds., Columbia University Press, New York, 1964, 489—498.
16. **Rabinowitz, P., Jacobs, S., and Gould, G.**, Continuously optically pumped Cs laser, *Appl. Opt.*, 1, 511—516, 1962.
17. **Walter, W. T., Solimene, N., Piltch, M., and Gould, G.**, Efficient pulsed gas discharge lasers, *IEEE J. Quantum Electron.*, QE-2, 474—479, 1966.
18. **Walter, W. T., Solimene, N., Piltch, M., and Gould, G.**, Pulsed-laser action in atomic copper vapor, *Bull. Am. Phys. Soc.*, 11, 113, 1966.
19. **Walter, W. T.**, 40-kW pulsed copper laser, *Bull. Am. Phys. Soc.*, 12, 90, 1967.
20. **Leonard, D. A.**, A theoretical description of the 5106-Å pulsed copper vapor laser, *IEEE J. Quantum Electron.*, QE-3, 380—381, 1967.
21. **Walter, W. T.**, Metal vapor lasers, Paper 126-2, Int. Quantum Electronics Conf., Miami, Florida, May 14 to 17, 1968; abstract, *IEEE J. Quantum Electron.*, QE-4, 355—356, 1968.
22. **Asmus, J. E. and Moncur N. K.**, Pulse broadening in a MHD copper vapor laser, *Appl. Phys. Lett.*, 13, 384—385, 1968.
23. **Isaev, A. A., Kazaryan, M. A., and Petrash, G. G.**, Effective pulsed copper-vapor laser with high average generation power, *JETP Lett.*, 16, 27—29, 1972.
24. **Russell, G. R., Nerheim, M. M., and Pivirotto, T. J.**, Supersonic electrical-discharge copper vapor laser, *Appl. Phys. Lett.*, 21, 656—657, 1972.
25. **Liu, C. S., Sucov, E. W., and Weaver, L. A.**, Copper superradiant emission from pulsed discharges in copper iodide vapor, *Appl. Phys. Lett.*, 23, 92, 1973.
26. **Ferrar, C. M.**, Copper-vapor laser with closed-cycle transverse vapor flow, *IEEE J. Quantum Electron.*, QE-9, 856—857, 1973.
27. **Chen, C. J., Nerheim, N. M., and Russell, G. R.**, Double-discharge copper vapor laser with copper chloride as a lasant, *Appl. Phys. Lett.*, 23, 514—515, 1973.
28. **Isaev, A. A., Kazaryan, M. A., and Petrash, G. G.**, Copper vapor pulsed laser with a repetition frequency of 10kHz, *Opt. Spectrosc.*, 35, 307—308, 1973.
29. **Weaver, L. A., Liu, C. S., and Sucov, E. W.**, Superradiant emission at 5106, 5700 and 5782 Å in pulsed copper iodide discharges, *IEEE J. Quantum Electron.*, QE-10, 140—147, 1974.
30. **Fahlen, T. S.**, Hollow-cathode copper-vapor laser, *J. Appl. Phys.*, 45, 4132—4133, 1974.
31. **Liberman, I., Babcock, R. V., Liu, C. S., George, T. V., and Weaver, L. A.**, High-repetition-rate copper iodide laser *Appl. Phys. Lett.*, 25, 334—335, 1974.
32. **Bokhan, P. A. and Solomonov, V. I.**, Mechanism of laser action in copper vapor, *Sov. J. Quantum Electron.*, 3, 481—483, 1974.
33. **Ferrar, C. M.**, Buffer gas effects in a rapidly pulsed copper vapor laser, *IEEE J. Quantum Electron.*, QE-10, 655—656, 1974.
34. **Anderson, R. S., Springer, L., Bricks, B. G., and Karras, T. W.**, A discharge heated copper vapor laser, *IEEE J. Quantum Electron.*, QE-11, 172—174, 1975.
35. **Chen, C. J. and Russell, G. R.**, High-efficiency multiply pulsed copper vapor laser utilizing copper chloride as a lasant, *Appl. Phys. Lett.*, 26, 504—505, 1975.
36. **Bokhan, P. A., Nikolaev, V. N., and Solomonov, V. I.**, Sealed Copper vapor laser, *Sov. J. Quantum Electron.*, 5, 96—98, 1975.
37. **Smilanski, I., Levin, L. A., and Erez, G.**, A copper laser using CuI vapor, *IEEE J. Quantum Electron.*, QE-11, 919—920, 1975.
38. **Shukhtin, M., Fedotov, G. A., and Mishakov, V. G.**, Lasing with CuI lines using copper bromide vapor, *Opt. Spectrosc.*, 39, 681, 1976.
39. **Sovero, E., Chen, C. J., and Culick, F. E. C.**, Electron temperature measurements in a copper chloride laser utilizing a microwave radiometer, *J. Appl. Phys.*, 47, 4538—4542, 1976.
40. **Shukhtin, A. M., Fedotov, G. A., and Mishakov, V. G.**, Stimulated emission on copper lines during pulsed production of vapor without the use of a heating element, *Opt. Spectrosc.*, 40, 237—238, 1976.
41. **Abrosimov, G. V., Vasil'tsov, V. V., Voloshin, V. N., Korneev, A. V., and Pis'mennyi, V. D.**, Pulsed laser action on self-limiting transitions of copper atoms in copper halide vapor, *Sov. Tech. Phys. Lett.*, 2, 162—163, 1976.

42. **Akirtava, O. S., Dzhikiya, V. L., and Oleinik, Yu. M.,** Laser utilizing CuI transitions in copper halide vapors, *Sov. J. Quantum Electron.,* 5, 1001—1002, 1976.
43. **Subotinov, N. V., Kalchev, S. D., and Telbizov, P. K.,** Copper vapor laser operating at a high pulse repetition frequency, *Sov. J. Quantum Electron.,* 5, 1003—1004, 1976.
44. **Aleksandrov, I. S., Babeiko, Yu. A., Babaev, A. A., Buzhinskii, O. I., Vasil'ev, L. A., Efimov, A. V., Krysanov, S. I., Nikolaev, G. N., Slivitskii, A. A., Sokolov, A. V., Tatarintsev, L. V., and Tereshchenkov, V. S.,** Stimulated emission from a transverse discharge in copper vapor, *Sov. J. Quantum Electron.,* 5, 1132—1133, 1976.
45. **Alaev, M. A., Baranov, A. I., Vereshchagin, N. M., Gnedin, I. N., Zherebtsov, Yu. P., Moskalenko, V. F., and Tsukanov, Yu. M.,** Copper vapor laser with a pulse repetition frequency of 100 kHz, *Sov. J. Quantum Electron.,* 6, 610—611, 1976.
46. **Anderson, R. S., Bricks, B. G., and Karras, T. W.,** Copper oxide as the metal source in a discharge heated copper vapor laser, *Appl. Phys. Lett.,* 29, 187—189, 1976.
47. **Isaev, A. A., Kazaryan, M. A., Lemmerman, G. Yu., Petrash, G. G., and Trofimov, A. N.,** Pulse stimulated emission due to transitions in copper atoms excited by discharges in cuprous bromide and chloride vapors, *Sov. J. Quantum Electron.,* 6, 976—977, 1976.
48. **Babeiko, Yu. A., Vasil'ev, L. A., Orlov, V. K., Sokolov, A. V., and Tatarintsev, L. V.,** Stimulated emission from copper vapor in a radial transverse discharge, *Sov. J. Quantum Electron.,* 6, 1258—1259, 1976.
49. **Fedorov, A. I., Sergeenko, V. P., Tarasenko, V. F., and Sedoi, V. S.,** Copper vapor laser with pulse production of the vapor, *Sov. Phys.,* J., 20, 251—253, 1976.
50. **Smilanski, I., Kerman, A., Levin, L. A., and Erez, G.,** A hollow-cathode copper halide laser, *IEEE J. Quantum Electron.,* QE-13, 24—36, 1977.
51. **Liu, G. S., Feldman, D. W., Pack, J. C., and Weaver, L. A.,** Axial cataphoresis effects in continuously pulsed copper halide lasers, *J. Appl. Phys.,* 48, 194—195, 1977.
52. **Nerheim, N. M.,** A parametric study of the copper chloride laser, *J. Appl. Phys.,* 48, 1186—1190, 1977.
53. **Anderson, R. S., Bricks, B. G., and Karras, T. W.,** Steady multiply pulsed discharge-heated copper-vapor laser with copper halide lasant, *IEEE J. Quantum Electron.,* QE-13, 115—117, 1977.
54. **Vetter, A. A. and Nerheim, N. M.,** Addition of HCl to the double-pulse copper chloride laser, *Appl. Phys. Lett.,* 30, 405—407, 1977.
55. **Abrosimov, G. V. and Vasil'tsov, V. V.,** Stimulated emission due to transitions in copper atoms formed in transverse discharge in copper halide vapors, *Sov. J. Quantum Electron.,* 7, 512—513, 1977.
56. **Gabay, S., Smilanski, I., Levin, L. A., and Erez, G.,** Comparison of CuCl, CuBr, and CuI as lasants for copper-vapor lasers, *IEEE J. Quantum Electron.,* QE-13, 364—366, 1977.
57. **Fahlen, T. S.,** High pulse rate, mode-locked copper vapor laser, *IEEE J. Quantum Electron.,* QE-7, 546—547, 1977.
58. **Isaev, A. A. and Lemmerman, G. Yu.,** Investigation of a copper vapor pulsed laser at elevated powers, *Sov. J. Quantum Electron.,* 7, 799—801, 1977.
59. **Batenin, V. M., Burmakin, V. A., Vokhmin, P. A., Evtyunin, A. I., Klimovskii, I. I., Lesnoi, M. A., and Selezneva, L. A.,** Time dependence of the electron density in a copper vapor laser, *Sov. J. Quantum Electron.,* 7, 891—893, 1977.
60. **Bokhan, P. A., Solomonov, V. I., and Shcheglov, V. B.,** Investigation of the energy characteristics of a copper vapor laser with a longitudinal discharge, *Sov. J. Quantum Electron.,* 7, 1032—1033, 1977.
61. **Klimovski, I. I. and Vokhmin, P. A.,** Connection of the copper vapor laser emission pulse characteristics with plasma parameters, *Proc. 13th Int. Conf. Phenomena in Ionized G Gases,* September, 1977, Berlin, East Germany, East German Physical Society.
62. **Rodin, A. V. and Zemtsov, Yu L.,** Electron energy distribution function for copper vapor, *Proc. 13th Int. Conf. Phenomena in Ionized Gases,* September, 1977, Berlin, East Germany, East German Physical Society,
63. **Elaev, V. F., Kirilov, A. E., Polunin, Yu. P., Soldatov, A. N., and Fedorov, V. F.,** Experimental investigation of the pulse discharge in Cu + Ne mixture in high repetiton rate regime, *Proc. 13th Int. Conf. Phenomena in Ionized Gases,* September, 1977, Berlin, East Germany, East German Physical Society,
64. **Shukhtin, A. M., Mishakov, V. G., and Fedotov, G. A.,** Production of Cu vapor from Cu_2O dust in a pulsed discharge, *Sov. Tech. Phys. Lett.,* 3, 304—305, 1977.
65. **Nerheim, N. M.,** Measurement of copper ground-state and metastable level population densities in copper-chloride laser, *J. Appl. Phys.,* 48, 3244—3250, 1977.
66. **Andrews, A. J., Webb, C. E., Tobin, R. C., and Denning, R. G.,** A copper vapor laser operating at room temperature, *Opt. Commun.,* 22, 272—274, 1977.

67. **Fedorov, A. J., Sergeenko, V. P., and Tarasenko, V. F.,** Apparatus for investigating stimulated emission from explosively formed metal vapors, *Sov. J. Quantum Electron.*, 7, 1166—1167, 1977.
68. **Liu, C. S., Feldman, D. W., Pack, J. L., and Weaver, L. A.,** Kinetic Processes in Continuously Pulsed Copper Halide Lasers, *IEEE J. Quantum Electron.*, QE-13, 744—751, 1977.
69. **Gokay, M. C., Jenkins, R. S., and Cross, L. A.,** Output characteristics of the CuCl double-pulse laser at small pumping pulse delays, *J. Appl. Phys.*, 48, 4395—4396, 1977.
70. **Vetter, A. A.,** Quantitative effect of initial current rise on pumping the double-pulsed copper chloride laser, *IEEE J. Quantum Electron.*, QE-13, 889—891, 1977.
71. **Kneipp, H. and Rentsch, M.,** Discharge-heated copper vapor laser, *Sov. J. Quantum Electron.*, 7, 1454—1455, 1977.
72. **Vetter, A. A. and Nerheim, N. M.,** Effect of dissociation pulse circuit inductance on the CuCl laser, *IEEE J. Quantum Electron.*, QE-14, 73—74, 1978.
73. **Cross, L. A., Jenkins, R. S., and Gokay, M. C.,** The effects of a weak axial magnetic field on the total energy output of the CuCl double-pulse laser, *J. Appl. Phys.*, 49, 453—454, 1978.
74. **Nerheim, N. M., Vetter, A. A., and Russell, G. R.,** Scaling a double-pulsed copper chloride laser to 10 mJ, *J. Appl. Phys.*, 49, 12—15, 1978.
75. **Gordon, E. B., Egorov, V. G., and Pavcenko, V. S.,** Excitation of metal vapor lasers by pulse trains, *Sov. J. Quantum Electron.*, 8, 266—268, 1978.
76. **Bokhan, P. A. and Shcheglov, V. B.,** Investigation of a transversely excited pulsed copper vapor laser, *Sov. J. Quantum Electron.*, 8, 219—222, 1978.
77. **Zemskov, K. I., Kazaryan, M. A., Mokerov, V. G., Petrash, G. G., and Petrova, A. G.,** Coherent properties of a copper vapor laser and dynamic holograms in vanadium dioxide films, *Sov. J. Quantum Electron.*, 8, 245—247, 1978.
78. **Burmakin, V.A., Evtyunin, A.N., Lesnoi, M.A., and Bylkin, V.I.,** Long-life sealed copper vaper laser, *Sov. J. Quantum Electron.*, 8, 574—576, 1978.
79. **Gridnev, A. G., Gorblinova, T. M., Elaev, V. F., Evtushenko, G. S., Ospiova, V., and Soldatov, A. N.,** Spectroscopic investigation of a gas discharge plasma of a Cu + Ne laser, *Sov. J. Quantum Electron.*, 8, 656—658, 1978.
80. **Tenenbaum, J., Smilanski, I., Gabay, S., Erez, G., and Levin, L. A.,** Time dependence of copper-atom concentration in ground and metastable states in a pulsed CuCl laser, *J. Appl. Phys.*, 49, 2662—2665, 1978.
81. **Piper, J. A.,** A transversely excited copper halide laser with large active volume, *IEEE J. Quantum Electron.*, QE-14, 405—407, 1978.
82. **Chen, C. J., Bhanji, A. M., and Russell, G. R.,** Long duration high-efficiency operation of a continuously pulsed copper laser utilizing copper bromide as a lasant, *Appl. Phys. Lett.*, 33, 146—148, 1978.
83. **Gokay, M. C., Soltanoalkotabi, M., and Cross, L. A.,** Copper acetylacetonate as a source in the 5106 — Å neutral copper laser, *J. Appl. Phys.*, 49, 4357—4358, 1978.
84. **Nerheim, N. M., Bhanji, A. M., and Russell, G. R.,** A continuously pulsed copper halide laser with a cable-capacitor Blumlein discharge circuit, *IEEE J. Quantum Electron.*, QE-14, 686—693, 1978.
85. **Babeiko, Yu. A., Vasil'ev, L. A., Sokolov, A. V., Sviridov, A. V., and Tatarintsev, L. V.,** Coaxial copper-vapor laser with a buffer gas at above atmospheric pressure, *Sov. J. Quantum Electron.*, 8, 1153—1154, 1978.
86. **Kushner, M. J. and Culick, F. E. C.,** Extrema of electron density and output pulse energy in a CuCl/Ne discharge and a Cu/CuCl double-pulsed laser, *Appl. Phys. Lett.*, 33, 728—731, 1978.
87. **Bokhan, P. A., Gerasimov, V. A., Solomonov, V. I., and Shcheglov, V.B.,** Stimulated emission mechanism of a copper vapor laser, *Sov. J. Quantum Electron.*, 8, 1220—1227, 1978.
88. **Kazaryan, M. A. and Trofimov, A. N.,** Gas-discharge tubes for metal halide vapor lasers, *Sov. J. Quantum Electron.*, 8, 1390—1391, 1978.
89. **Bokhan, P. A. and Gerasimov, V. A.,** Optimization of the excitation conditions in a copper vapor laser, *Sov. J. Quantum Electron.*, 9, 273—275, 1979.
90. **Smilanski, I., Erez, G., Kerman, A., and Levin, L. A.,** High-power, high-pressure, discharge-heated copper vapor laser, *Opt. Commun.*, 30, 70—74, 1979.
91. **Kushner, M. J. and Culick, F. E. C.,** A continuous discharge improves the performance of the Cu/CuCl double pulse laser, *IEEE J. Quantum Electron.*, QE-15, 835—837, 1979.
92. **Tennenbaum, J., Smilanski, I., Gabay, S., Levin, L. A., and Erez, G.,** Laser power variation and time dependence of populations in a burst-mode CuBr laser, *J. Appl. Phys.*, 50, 57—61, 1979.
93. **Miller, J. L. and Kan, T.,** Metastable decay rates in a Cu-metal-vapor laser, *J. Appl. Phys.*, 50, 3849—3851, 1979.
94. **Kan, T., Ball, D., Schmitt, E., and Hill, J.,** Annular discharge copper vapor laser, *Appl. Phys. Lett.*, 35, 676—677, 1979.

95. **Gokay, M. C. and Cross, L. A.**, Comparison of copper acetylacetonate, copper (II) acetate, and copper chloride as lasants for copper vapor lasers, *IEEE J. Quantum Electron.*, QE-15, 65—66, 1979.
96. **Kazaryan, M. A. and Trofimov, A. N.**, Kinetics of metal salt vapor lasers, *Sov. J. Quantum Electron.*, 9, 148—152, 1979.
97. **Babeiko, Yu. A., Vasil'ev, L. A., Sviridov, A. V., Sokolov, A. V., and Tatarintsev, L. V.**, Efficiency of a copper vapor laser, *Sov. J. Quantum Electron.*, 9, 651—653, 1979.
98. **Hargrove, R. S., Grove, R., and Kan, T.**, Copper vapor laser unstable resonator oscillator and oscillator-amplifier characteristics, *IEEE J. Quantum Electron.*, QE-15, 1228—1233, 1979.
99. **Solanki, R., Latush, E. L., Fairbank, W. M., Jr., and Collins, G. J.**, New infrared laser transitions in copper and silver hollow cathode discharges, *Appl. Phys. Lett.*, 34, 568—570, 1979.
100. **Isaev, A. A., Kazaryan, M. A., and Petrash, G. G.**, *Kratk. Soobshch. Fiz.*, 3, 3, 1972.
101. **Fahlen, T. S.**, Self-heated, multiple-metal-vapor laser, *IEEE J. Quantum Electron.*, QE-12, 200—201, 1976.
102. **Markova, S. V. and Cherezov, V. M.**, Investigation of pulse stimulated emission from gold vapor, *Sov. J. Quantum Electron.*, 7, 339—342, 1977.
103. **Markova, S. V., Petrash, G. G., and Cherezov, V. M.**, Ultraviolet-emitting gold vapor laser, *Sov. J. Quantum Electron.*, 8, 904—906, 1978.
104. **Cahuzac, P.**, New infrared laser lines in Mg vapor, *IEEE J. Quantum Electron.*, QE-8, 500, 1972.
105. **Trainor, D. W. and Mani, S. A.**, Atomic calcium laser: pumped via collision-induced absorption, *Appl. Phys. Lett.*, 33, 648—650, 1978.
106. **Baron, K. U. and Stadler, B.**, Hollow cathode-excited laser transitions in calcium, strontium and barium, Paper R9, presented at the IX Int. Quantum Electronics Conf., Amsterdam, June 14 to 18, 1976.
107. **Deech, J. S. and Sanders, J. H.**, New self-terminating laser transitions in calcium and stronium, *IEEE J. Quantum Electron.*, QE-4, 474, 1968.
108. **Klimkin, V. M. and Kolbycheva, P. D.**, Tunable single-frequency calcium-hydrogen laser emitting at 5.54 μm, *Sov. J. Quantum Electron.*, 7, 1037—1039, 1977.
109. **Klimkin, V. M., Monastyrev, S. S. and Prokop'ev, V. E.**, Selective relaxation of long-lived states of metal atoms in a gas discharge plasma. Stationary generation on $^1P_1^o$-$^1D_2$ transitions of calcium and strontium, *JETP Lett.*, 20, 110—111, 1974.
110. **Platanov, A. V., Soldatov, A. N., and Filonov, A. G.**, Pulsed strontium vapor laser, *Sov. J. Quantum Electron.*, 8, 120—121, 1978.
111. **Bokhan, P. A. and Burlakov, V. D.**, Mechanism of laser action due to $4d^3D_{1,2} \rightarrow 5p^3P_2^o$ transitions in a strontium atom, *Sov. J. Quantum Electron.*, 9, 374—376, 1979.
112. **Baron, K. U. and Stadler, B.**, New visible laser transitions in BaI and BaII, *IEEE J. Quantum Electron.*, QE-11, 852—853, 1975.
113. **Cahuzac, P.**, New infrared laser lines in barium vapor, *Phys. Lett.*, 32a, 150—151, 1970.
114. **Isaev, A. A., Kazaryan, M. A., Markova, S. V., and Petrash, G. G.**, Investigation of pulse infrared stimulated emission from barium vapor, *Sov. J. Quantum Electron.*, 5, 285—287, 1975.
115. **Bricks, B. G., Karras, T. W., and Anderson, R. S.**, An investigation of a discharge-heated barium laser, *J. Appl. Phys.*, 49, 38—40, 1978.
116. **Isaev, A. A., Kazaryan, M. A., and Petrash, G. G.**, Emission of laser pulses due to transitions from a resonance to a metastable level in barium vapor, *Sov. J. Quantum Electron.*, 3, 358—359, 1974.
117. **Cross, L. A. and Gokay, M. C.**, A pulse repetition frequency scaling law for the high repetition rate neutral barium laser, *IEEE J. Quantum Electron.*, QE-14, 648, 1978.
118. **Bokhan, P. A. and Solomonov, V. I.**, Barium vapor laser with a high average output power, *Sov. Tech. Phys. Lett.*, 4, 486—487, 1978.
119. **Dubrovin, A. N., Tibilov, A. S., and Shevtsov, M. K.**, Lasing on Cd, Zn and Mg lines and possible applications, *Opt. Spectrosc.*, 32, 685, 1972.
120. **Tibilov, A. S.**, Generation of radiation in He-Cd and Ne-Cd mixtures, *Opt. Spectrosc.*, 19, 463—464, 1965.
121. **Silfvast, W. T., Szeto, L. H., and Wood, O. R.**, II, Recombination lasers in Nd and CO_2 laser-produced cadmium plasmas, *Opt. Lett.*, 4, 271—273, 1979.
122. **Chou, M. S. and Cool, T. A.**, Laser operation by dissociation of metal complexes. II. New transitions in Cd, Fe, Ni, Se, Sn, Te, V and Zn, *J. Appl. Phys.*, 48, 1551—1555, 1977.
123. **Komine, H. and Byer, R. L.**, Optically pumped atomic mercury photodissociation laser, *J. Appl. Phys.*, 48, 2505—2508, 1977.
124. **Djeu, N. and Burnham, R.**, Optically pumped CW Hg laser at 546.1 nm, *Appl. Phys. Lett.*, 25, 350—351, 1974.
125. **Artusy, M., Holmes, N., and Siegman, A. E.**, D.C.-Excited and sealed-off operation of the optically pumped 546.1 nm Hg laser, *Appl. Phys. Lett.*, 28, 1331—1334, 1976.
126. **Holmes, N. C. and Siegman, A. E.**, The optically pumped mercury vapor laser, *J. Appl. Phys.*, 49, 3155—3170, 1978.

127. **Bloom, A. L., Bell, W. E., and Lopez, F. O.,** Laser spectroscopy of a pulsed mercury-helium discharge, *Phys. Rev.,* 135, A578—A579, 1964.
128. **Bockasten, K., Garavaglia, M., Lengyel, B. A., and Lundholm, T.,** Laser lines in Hg I, *J. Opt. Soc. Am.,* 55, 1051—1053, 1965.
129. **Heard, H. G. and Peterson, J.,** Laser action in mercury rare gas mixtures, *Proc. IEEE,* 52, 414, 1964.
130. **Heard, H. G. and Peterson, J.,** Mercury-rare gas visible-UV laser, *Proc. IEEE,* 52, 1049—1050, 1964.
131. **Rigden, J. D. and White, A. D.,** Optical laser action in iodine and mercury discharges, *Nature (London),* 198, 774, 1963.
132. **Paananen, R. A., Tang, C. L., Horrigan, F. A., and Statz, H.,** Optical laser action in He-Hg rf discharges, *J. Appl. Phys.,* 34, 3148—3149, 1963.
133. **Armand, M. and Martinot-Lagarde, P.,** Effect laser sur la vapeur de mercure dans un melange He-Hg, *C. R. Acad. Sci. Ser. B,* 258, 867—868, 1964.
134. **Doyle, W. M.,** Use of time resolution in identifying laser transitions in mercury rare gas discharge, *J. Appl. Phys.,* 35, 1348—1349, 1964.
135. **Chebotayev, V. P.,** Isotopic structure of the 1.5295 millimicron laser line of mercury, *Opt. Spectrosc.,* 25, 267—268, 1968.
136. **Convert, G., Armand, M., and Martinot-Lagarde, P.,** Effect laser dans des melanges mercure-gas rares, *C. R. Acad. Sci. Ser. B,* 257, 3259—3260, 1964.
137. **Beterov, I. M., Klement'ev, V. M., and Chebotaev, V. P.,** A mercury laser secondary frequency standard in the microwave region, *Radio Eng. Electron. Phys. (USSR),* 14, 1790—1792, 1969.
138. **Bikmukhametov, K. A., Klement'ev, V. M., and Chebotaev, V. P.,** Investigation of the stability of the oscillation frequency of a mercury laser emitting at $\lambda = 1.53\ \mu$, *Sov. J. Quantum Electron.,* 2, 254—256, 1972.
139. **Bikmukhametov, K. A., Klement'ev, V. M., and Chebotaev, V. P.,** Collision broadening of the 1.53 μm line of mercury in an Hg-He, Hg-Ne mixture, *Opt. Spectrosc.,* 34, 616—617, 1973.
140. **Klement'ev, V. M. and Solov'ev, M. V.,** Mercury-vapor laser, *J. Appl. Spectrosc.,* 18, 29—32, 1973.
141. **Bikmukhametov, K. A., Klement'ev, V. M., and Chebotaev, V. P.,** Experimental investigation of the dependences of the collision broadening and shift of the emission line of a mercury laser on He and Ne pressures, *Sov. J. Quantum Electron.,* 5, 278—281, 1975.
142. **Chou, M. S. and Cool, T. A.,** Laser operation by dissociation of metal complexes: new transitions in As, Bi, Ga, Hg, In, Pb, Sb, and Tl, *J. Appl. Phys.,* 47, 1055—1061, 1976.
143. **Stricker, J. and Bauer, S. H.,** An atomic boron laser-pumping by incomplete autoionization or ion-electron recombination, Paper TCl, 4th Conf. Chemical and Molecular Lasers, St. Louis, Mo., October 21 to 23, 1974, abstract, *IEEE J. Quantum Electron.,* QE-11, 701—702, 1975.
144. **Hemmati, H. and Collins, G. J.,** Atomic gallium photodissociation laser, *Appl. Phys. Lett.,* 34, 844—845, 1979.
145. **Burnham, R.,** Atomic indium photodissociation laser at 451 nm, *Appl. Phys. Lett.,* 30, 132—133, 1977.
146. **Silfvast, W. T., Szeto, L. H., and Wood, O. R.,** II, Simple metal vapor recombination lasers using segmented plasma excitation, *Appl. Phys. Lett.,* in press.
147. **Ehrlich, D. J., Maya, J., and Osgood, R. M., Jr.,** Efficient thallium photodissociation laser, *Appl. Phys. Lett.,* 33, 931—933, 1978.
148. **Isaev, A. A., Ischenko, P. I., and Petrash, G. G.,** Super-radiance at transitions terminating at metastable levels of helium and thallium, *JETP Lett.,* 6, 118—121, 1967.
149. **Isaev, A. A. and Petrash, G. G.,** Pulsed superradiance at the green line of thallium in TlI-vapor, *JETP Lett.,* 7, 156—158, 1968.
150. **Isaev, A. A., Kazaryan, M. A., and Petrash, G. G.,** Mechanism of pulsed lasing of the green thallium line in a thallium iodide vapor discharge, *Opt. Spectrosc.,* 31, 180—183, 1971.
151. **Korolev, F. A., Odintsov, A. I., Turkin, N. G., and Yakunin, V. P.,** Spectral structure of pulse superluminescence lines of gases, *Sov. J. Quantum Electron.,* 5, 237—239, 1975.
152. **Chilukuri, S.,** Selective optical excitation and inversions via the excimer channel: superradiance at the thallium green line, *Appl. Phys. Lett.,* 34, 284—286, 1979.
153. **Tunitskii, L. N. and Cherkasov, E. M.,** New oscillation lines in the spectra of Nl and Cl, *Sov. Phys. Tech. Phys.,* 13, 1696—1697, 1969.
154. **Atkinson, J. B. and Sanders, J. H.,** Laser action in C and N following dissociative excitation transfer, *J. Phys. B,* 1, 1171—1179, 1968.
155. **Cooper, G. W. and Verdeyen, J. T.,** Recombination pumped atomic nitrogen and carbon afterglow lasers, *J. Appl. Phys.,* 48, 1170—1175, 1977.
156. **Voitovich, A. P. and Dubovik, M. V.,** Time and power characteristics of pulsed atomic gas lasers in a magnetic field, *J. Appl. Spectrosc.,* 27, 1399—1403, 1978.

157. **Boot, H. A. H. and Clunie, D. M.,** Pulsed gaseous maser, *Nature (London)* 197, 173—174, 1963.
158. **Patel, C. K. N., McFarlane, R. A., and Faust, W. L.,** Further infrared spectroscopy using stimulated emission techniques, in *Quantum Electronics III,* Grivet, P. and Bloembergen, N., Eds., Columbia University Press, New York, 1964, 561—572.
159. **Patel, C. K. N., McFarlane, R. A., and Faust, W. L.,** Optical maser action in C, N, O, S and Br on dissociation of diatomic and polyatomic molecules, *Phys. Rev.,* 133, A1244—A1248, 1964.
160. **English, J. R., III, Gardner, H. C., and Merritt, J. A.,** Pulsed stimulated emission from N, C, Cℓ and F atoms, *IEEE J. Quantum Electron.,* QE-8, 843—844, 1972.
161. **DePoorter, G. C. and Balog, G.,** New infrared laser line in OCS and new method for C atom lasing, *IEEE J. Quantum Electron.,* QE-8, 917—918, 1972.
162. **Shimazu, M. and Suzaki, Y.,** Laser oscillations in silicon tetrachloride vapor, *Jpn. J. Appl. Phys.,* 4, 819, 1965.
163. **Cooper, H. G. and Cheo, P. K.,** Laser transitions in BII, BrII, and Sn, *IEEE J. Quantum Electron.,* QE-2, 785, 1966.
164. **Carr, W. C. and Grow, R. W.,** A new laser line in tin using stannic chloride vapor, *Proc. IEEE,* 55, 1198, 1967.
165. **Zhukov, V. V., Latush, E. L., Mikhalevskii, V. S., and Sem, M. F.,** New laser transitions in the spectrum of tin and population-inversion mechanism, *Sov. J. Quantum Electron.,* 5, 468—469, 1975.
166. **Sutton, D. G., Galvan, L., and Suchard, S. N.,** Two-electron laser transition in Sn(I)?, *IEEE J. Quantum Electron.,* QE-11, 312, 1975.
167. **Isaev, A. A. and Petrash, G. G.,** New generation and superradiance lines of lead vapor, *JETP Lett.,* 10, 119—121, 1969.
168. **Fowles, G.R. and Silfvast, W. T.,** High gain laser transition in lead vapor, *Appl. Phys. Lett.,* 6, 236—237, 1965.
169. **Silfvast, W. T. and Deech, J. S.,** Six db/cm single pass gain at 7229 Å in lead vapor, *Appl. Phys. Lett.,* 11, 97—99,1967.
170. **Anderson, R. S., Bricks, B. G., Karras, T. W., and Springer, L. W.,** Discharge-heated lead vapor laser, *IEEE J. Quantum Electron.,* QE-12, 313—315, 1976.
171. **Kirilov, A. E., Kukharev, V. N., Soldatov, A. N., and Tarasenko, V. F.,** Lead vapor lasers, *Sov. Phys. J.,* 20, 1381—1384, 1977.
172. **Feldman, D. W., Liu, C. S., Pack, J. L., and Weaver, L. A.,** Long-lived lead-vapor lasers, *J. Appl. Phys.,* 49, 3679—3683, 1978.
173. **Kirilov, A. E., Kukharev, V. N., and Soldatov, V. N.,** Investigation of a pulsed λ = 722.9 nm Pb laser with a double-section gas-discharge tube, *Sov. J. Quantum Electron.,* 9, 285—287, 1979.
174. **Piltch, M. and Gould, G.,** High temperature alumina discharge tube for pulsed metal vapor lasers, *Rev. Sci. Instrum.,* 37, 925—927, 1966.
175. **Heard, H. G. and Peterson, J.,** Visible laser transitions in ionized oxygen, nitrogen and carbon monoxide, *Proc. IEEE,* 52, 1258, 1964.
176. **Hitt, J. S. and Haswell, W. T.,** III, Stimulated emission in the theta pinch discharge, *IEEE J. Quantum Electron.,* QE-2, xlii, 1966.
177. **Chou, M. S. and Zawadzkas, G. A.,** Observation of new atomic nitrogen laser transition at 9046 Å, *Opt. Commun.,* 26, 92, 1978.
178. **McFarlane, R. A.,** Stimulated emission spectroscopy of some diatomic molecules, in *Physics of Quantum Electronics,* Kelley, P. L., Lax, B., and Tannenwald, P. E., Eds., McGraw-Hill, New York, 1966, 655—663.
179. **DeYoung, R. J., Wells, W. E., Miley, G. H., and Verdeyen, J. T.,** Direct nuclear pumping of a Ne-N_2 laser, *Appl. Phys. Lett.,* 28, 519—521, 1976.
180. **Janney, G. M.,** New infrared laser oscillations in atomic nitrogen, *IEEE J. Quantum Electron.,* QE-3, 133, 1967.
181. **Janney, G. M.,** Correction to near infrared laser oscillations in atomic nitrogen, *IEEE J. Quantum Electron.,* QE-3, 339, 1967.
182. **Sutton, D. G.,** New laser oscillation in the N atom quartet manifold, *IEEE J. Quantum Electron.,* QE-12, 315—316, 1976.
183. **Cheo, P. K. and Cooper, H. G.,** UV and visible laser oscillations in fluorine, phosphorus and chlorine, *Appl. Phys. Lett.,* 7, 202—204, 1965.
184. **Fowles, G. R., Zuryk, J. A., and Jensen, R. C.,** Infrared laser lines in neutral atomic phosphorus, *IEEE J. Quantum Electron.,* QE-10, 394—395, 1974.
185. **Fowles, G. R., Zuryk, J. A., and Jensen, R. C.,** Infrared laser lines in arsenic vapor, *IEEE J. Quantum Electron.,* QE-10, 849, 1974.
186. **Markova, S. V., Petrash, G. G., and Cherezov, V. M.,** Pulse stimulated emission of the 472.2nm line of the bismuth atom, *Sov. J. Quantum Electron.,* 7, 657, 1977.

187. **Burnham, R.**, Optically pumped bismuth lasers at 472 and 475 nm, Paper X.7, 10th Int. Quantum Electronics Conf., Atlanta, Georgia, May 29 to June 1, 1978.
188. **Murray, J. R., Powell, H. T., and Rhodes, C. K.**, Inversion of the auroral green transition of atomic oxygen by argon excimer transfer to nitrous oxide, post-deadline Paper P. 8, 13th Int. Quantum Electronics Conf., San Francisco, June 10 to 13, 1974.
189. **Powell, H. T., Murray, J. R., and Rhodes, C. K.**, Laser oscillation on the green bands of xenon oxide, Paper MA2, 4th Conf. on Chemical and Molecular Lasers, St. Louis, Mo. October 21 to 23, 1974.
190. **Powell, H. T., Murray, J. R., and Rhodes, C. K.**, Laser oscillation on the green bands of XeO and KrO, *Appl. Phys. Lett.*, 25, 730—732, 1974.
191. **Powell, H. T., Murray, J. R., and Rhodes, C. K.**, Collision-induced auroral line lasers, Paper 7.8, Conf. Laser Engineering and Applications, Washington, D.C., May 28 to 30, 1975.
192. **Hughes, W. M., Olson, N. T., and Hunter, R.**, Experiments on 558-nm argon oxide laser systems, *Appl. Phys. Lett.*, 28, 81—83, 1976.
193. **Bennett, W. R., Jr., Faust, W. L., McFarlane, R. A., and Patel, C. K. N.**, Dissociative excitation transfer and optical maser oscillation in Ne-O_2 and Ar-O_2rf discharges, *Phys. Rev. Lett.*, 8, 470—473, 1962.
194. **Tunitskii, L. N. and Cherkasov, E. M.**, Method for varying the frequency of a gas laser, *Sov. Phys. Tech. Phys.*, 12, 1500—1501, 1968.
195. **Bennett, W. R., Jr.**, Gaseous optical masers, in *Appl. Optics Supplement on Optical Masers*, Heavens, O. S., Ed., 1962, 24—61.
196. **Tunitskii, L. N. and Cherkasov, E. M.**, Interpretation of oscillation lines in Ar-Br_2 laser, *J. Opt. Soc. Am.*, 56, 1783—1784, 1966.
197. **Rautian, S. G. and Rubin, P. L.**, On some features of gas lasers containing mixtures of oxygen and rare gases, *Opt. Spectrosc.*, 18, 180—181, 1965.
198. **Tunitskii, L. N. and Cherkasov, E. M.**, Pulsed mode generation in an argon-oxygen laser, *Opt. Spectrosc.*, 27, 344—346, 1969.
199. **Feld, M. S., Feldman, B. J., and Javan, A.**, Frequency shifts of the fine structure oscillations of the 8446-Å atomic oxygen laser, *Bull. Am. Phys. Soc.*, 12, 15, 1967.
200. **Tunitskii, L. N. and Cherkasov, E. M.**, The mechanism of laser action in oxygen-inert gas mixtures, *Opt. Spectrosc.*, 23, 154—157, 1967.
201. **Tunitskii, L. N. and Cherkasov, E. M.**, Pure oxygen laser, *Sov. Phys. Tech. Phys.*, 13, 993—994, 1969.
202. **Kolpakova, I. V. and Redko, T. P.**, Some remarks on the operation of the neon-oxygen gas laser, *Opt. Spectrosc.*, 23, 351—352, 1967.
203. **Feld, M. S., Feldman, B. J., Javan, A., and Domash, L. H.**, Selective reabsorption leading to multiple oscillations in the 8446Å atomic oxygen laser, *Phys. Rev.*, A7, 257—262, 1973.
204. **Sutton, D. G., Galvan, L., and Suchard, S. N.**, New laser oscillation in the oxygen atom, *IEEE J. Quantum Electron.*, QE-11, 92, 1975.
205. **Flynn, G. W., Feld, M. S., and Feldman, B. J.**, New infrared-laser transition and g-values in atomic oxygen, *Bull. Am. Phys. Soc.*, 12, 15, 1967.
206. **Powell, F. X. and Djeu, N. I.**, CW atomic oxygen laser at 4.56 μ, *IEEE J. Quantum Electron.*, QE-7, 176—177, 1971.
207. **Powell, H. T., Prosnitz, D., and Schleicher, B. R.**, Sulfur $^1S_0$-$^1D_2$ laser by OCS photodissociation, *Appl. Phys. Lett.*, 34, 571—573, 1979.
208. **Martinelli, R. U. and Gerritsen, H. J.**, Laser action in sulphur using hydrogen sulphide, *J. Appl. Phys.*, 37, 444—445, 1966.
209. **Ultee, C. J.**, Infrared laser emission from discharges through gaseous sulfur compounds, *J. Appl. Phys.*, 44, 1406, 1973.
210. **Hocker, L. O.**, New infrared laser transitions in neutral sulfur, *J. Appl. Phys.*, 48, 3127—3128, 1977.
211. **Hubner, G. and Wittig, C.**, Some new infrared laser transitions in atomic oxygen and sulfur, *J. Opt. Soc. Am.*, 61, 415—416, 1971.
212. **Cooper, H. G. and Cheo, P. K.**, Ion laser oscillations in sulphur, in *Physics of Quantum Electronics*, Kelley, P. L., Lax, B., and Tannewald, P. E., Eds., McGraw-Hill, New York, 1966, 690—697.
213. **Davis, C. C. and King, T. A.**, Gaseous ion lasers, in *Advances in Quantum Electronics*, Vol. 3, Academic Press, London, 1977, 169—454.
214. **Powell, H. T. and Ewing, J. J.**, Forbidden transition selenium atom photodissociation lasers, Paper X.2, 10th Int. Quantum Electronics Conf., Atlanta, Ga., May 29 to June 1, 1978.
215. **Powell, H. T. and Ewing, J. J.**, Photodissociation lasers using forbidden transitions of selenium atoms, *Appl. Phys. Lett.*, 33, 165—167, 1978.
216. **Cahuzac, P.**, Emission laser infrarouges dans les vapeurs de thulium et d'ytterbium, *Phys. Lett.*, 27A, 473—474, 1968.
217. **Cahuzac, P.**, Infrared laser emission from rare-earth vapors, *Phys. Lett.*, 31A, 541—542, 1970.

218. **Chapovsky, P. L., Kochubei, S. A., Lisitsyn, V. N., and Razhev, A. M.**, Excimer ArF/XeF lasers providing high-power stimulated radiation in Ar/Xe and F lines, *Appl. Phys.*, 14, 231—233, 1977.
219. **Lisitsyn, V. N. and Razhev, A. M.**, High-power, high-pressure laser based on red fluorine lines, *Sov. Tech. Phys. Lett.*, 3, 350—351, 1977.
220. **Loree, T. R. and Sze, R. C.**, The atomic fluorine laser: spectral pressure dependence, *Opt. Commun.*, 21, 255—257, 1977.
221. **Bigio, I. J. and Begley, R. F.**, High power visible laser action in neutral atomic fluorine, *Appl. Phys. Lett.*, 28, 263—264, 1976.
222. **Hocker, L. O. and Phi, T. B.**, Pressure dependence of the atomic fluorine transition intensities, *Appl. Phys. Lett.*, 29, 493—494, 1976.
223. **Kovacs, M. A. and Ultee, C. J.**, Visible laser action in fluorine I, *Appl. Phys. Lett.*, 17, 39—40, 1970.
224. **Jeffers, W. Q. and Wiswall, C. E.**, Laser action in atomic fluorine based on collisional dissociation of HF, *Appl. Phys. Lett.*, 17, 444—447, 1970.
225. **Florin, A. E. and Jensen, R. J.**, Pulsed laser oscillation at 0.7311μ from F atoms, *IEEE J. Quantum Electron.*, QE-7, 472, 1971.
226. **Sumida, S., Obara, M., and Fujioka, T.**, Novel neutral atomic fluorine laser lines in a high-pressure mixture of F_2 and He, *J. Appl. Phys.*, 50, 3884—3887, 1979.
227. **Lawler, J. E., Parker, J. W., Anderson, L. W., and Fitzsimmons, W. A.**, Experimental investigation of the atomic fluorine laser, *IEEE J. Quantum Electron.*, QE-15, 609—613, 1979.
228. **Paananen, R. A. and Horrigan, F. A.**, Near infra-red lasering in $NeCl_2$ and He-Cl_2, *Proc. IEEE*, 52, 1261—1262, 1964.
229. **Shimazu, M. and Suzaki, Y.**, Laser oscillation in the mixtures of freon and rare gases, *Jpn. J. Appl. Phys.*, 4, 381—382, 1965.
230. **Jarrett, S. M., Nunez, J., and Gould, G.**, Laser oscillation in atomic Cl in HCl and HI gas discharges, *Appl. Phys. Lett.*, 8, 150—151, 1966.
231. **Trusty, G. L., Yin, P. K., and Koozekanani, S. K.**, Observed laser lines in freon-helium mixtures, *IEEE J. Quantum Electron.*, QE-3, 368, 1967.
232. **Pollack, M. A.**, private communication to C. S. Willett.
233. **Paananen, R. A., Tang, C. L., and Horrigan, F. A.**, Laser action in Cl_2 and He-Cl_2, *Appl. Phys. Lett.*, 3, 154—155, 1963.
234. **Bockasten, K.**, On the classification of laser lines in chlorine and iodine, *Appl. Phys. Lett.*, 4, 118—119, 1964.
235. **Dauger, A. B. and Stafsudd, O. M.**, Observation of CW laser action in chlorine, argon and helium gas mixtures, *IEEE J. Quantum Electron.*, QE-6, 572—573, 1970.
236. **Jarrett, S. M., Nunez, J., and Gould, G.**, Infrared laser oscillation in HBr and HI gas discharges, *Appl. Phys. Lett.*, 7, 294—296, 1965.
237. **Jensen, R. C. and Fowles, G. R.**, New laser transitions in iodine-inert-gas mixtures, *Proc. IEEE*, 52, 1350, 1964.
238. **Kasper, J. V. V. and Pimentel, G. C.**, Atomic iodine photodissociation laser, *Appl. Phys. Lett.*, 5, 231—233, 1964.
239. **Kasper, J. V. V., Parker, J. H., and Pimentel, G. C.**, Iodine-atom laser emission in alkyl iodide photolysis, *J. Chem. Phys.*, 43, 1827—1828, 1965.
240. **Pollack, M. A.**, Pressure dependence of the iodine photodissociation laser peak output, *Appl. Phys. Lett.*, 8, 36—38, 1966.
241. **Andreeva, T. L., Dudkin, V. A., Malyshev, V. I., Mikhailov, G. V., Sorokin, V. N., and Novikova, L. A.**, Gas laser excited in the process of photodissociation, *Sov. Phys. JETP*, 22, 969—970, 1966.
242. **DeMaria, A. J. and Ultee, C. J.**, High-energy atomic iodine photodissociation laser, *Appl. Phys. Lett.*, 9, 67—69, 1966.
243. **Gregg, D. W., Kidder, R. E., and Dobler, C. V.**, Zeeman splitting used to increase energy from a Q-switched laser, *Appl. Phys. Lett.*, 13, 297—298, 1968.
244. **Ferrar, C. M.**, Q-switching and mode locking of a CF_3I photolysis laser, *Appl. Phys. Lett.*, 12, 381—383, 1968.
245. **Zalesskii, V. Yu and Venediktov, A. A.**, Mechanism of generation termination at the $5^2P_{1/2}$-$5^2P_{3/2}$ transition, *Sov. Phys. JETP*, 28, 1104—1107, 1969.
246. **O'Brien, D. E. and Bowen, J. R.**, Kinetic model for the iodine photodissociation laser, *J. Appl. Phys.*, 40, 4767—4769, 1969.
247. **Andreeva, T. L., Malyshev, V. I., Maslov, A. I., Sobel'man, I. I., and Sorokin, V. N.**, Possibility of obtaining excited iodine atoms as a result of chemical reactions, *JETP Lett.*, 10, 271—274, 1969.
248. **Zalesskii, V. Yu and Moskalev, E. I.**, Optical probing of a photodissociation laser, *Sov. Phys. JETP*, 30, 1019—1023, 1970.

249. **Belousova, I. M., Danilov, O. B., Sinitsina, I. A., and Spiridonov, V. V.,** Investigation of the optical inhomogeneities of the active medium of a CF_3I photodissociation laser, *Sov. Phys. JETP*, 31, 791—793, 1970.
250. **Velikanov, S. D., Kormer, S. B., Nikolaev, V. D., Sinitsyn, M. V. Solov'ev, Yu A., and Urlin, V. D.,** Lower limit of the luminescence spectral linewidth of the $5^2P_{1/2}$ - $5^2P_{3/2}$ transition in atomic iodine in a photodissociation laser, *Sov. Phys. Dokl.*, 15, 478—480, 1970.
251. **Gensel, P., Hohla, K., and Kompa, K. L.,** Energy storage of CF_3I photodissociation laser, *Appl. Phys. Lett.*, 18, 48—50, 1971.
252. **Belousova, I. M., Danilov, O. B., Kladovikova, N. S., and Yachnev, I. L.,** Quenching of excited atoms in a photodissociation laser, *Sov. Phys. Tech. Phys.*, 15, 1212—1213, 1971.
253. **O'Brien, D. E. and Bowen, J. R.,** Parametric studies of the iodine photodissociation laser, *J. Appl. Phys.*, 42, 1010—1015, 1971.
254. **DeWolf Lanzerotti, M. Y.,** Iodine-atom laser emission in 2-2-2 Trifluoroethyliodide, *IEEE J. Quantum Electron.*, QE-7, 207—208, 1971.
255. **Zalesskii, V. Yu. and Krupenikova, T. I.,** Deactivation of metastable iodine atoms by collision with perfluoroalkyliodide molecules, *Opt. Spectrosc.*, 30, 439—443, 1971.
256. **Andreeva, T. L., Kuznetsova, S. V., Maslov, A. I., Sobel'man, I. I., and Sorokin, V. N.,** Investigation of reactions of excited iodine atoms with the aid of a photodissociation laser, *JETP Lett.*, 13, 449—452, 1971.
257. **Hohla, K.,** Photochemical iodine laser: kinetic foundations for giant pulse operation, IPP Report IV/33, Max-Planck-Institüt für Plasma Physik, Garching bei Munich, West Germany, December 1971.
258. **Belousova, I. M., Kiselev, V. M., and Kurzenkov, V. N.,** Induced emission spectrum of atomic iodine due to the hyperfine structure of the transition $^2P_{1/2}$ - $^2P_{3/2}$ (7603 cm $^{-1}$), *Opt. Spectrosc.*, 33, 112—114, 1972.
259. **Belousova, I. M., Kiselev, V. M., and Kurzenkov, V. N.,** Line width for induced emission due to the $^2P_{1/2}$ - $^2P_{3/2}$ transition of atomic iodine, *Opt. Spectrosc.*, 33, 115—116, 1972.
260. **Zalesskii, V. Yu.,** Kinetics of a CF_3I photodissociation laser, *Sov. Phys. JETP*, 34, 474—480, 1972.
261. **Hwang, W. C. and Kasper, J. V. V.,** Zeeman effects in the hyperfine structure of atomic iodine photodissociation laser emission, *Chem. Phys. Lett.*, 13, 511—514, 1972.
262. **Hohla, K. and Kompa, K. L.,** Energy transfer in a photochemical iodine laser, *Chem. Phys. Lett.*, 14, 445—448, 1972.
263. **Hohla, K. and Kompa, K. L.,** Kinetische Prozesse in einem photochemischen Jodlaser, *Z. Naturforsch.*, 27a, 938—947, 1972.
264. **Filyukov, A. A. and Karpov, Ya.,** A criterion for probable laser quenching, *Sov. Phys. JETP* , 35, 63—65, 1972.
265. **Gavrilina, L. K., Karpov, V. Ya., Leonov, Yu. S., Sautkin, V. A., and Filyukov, A. A.,** Selective pumping effect of a photodissociative laser, *Sov. Phys. JETP*, 35, 258—259, 1972.
266. **Hohla, K. and Kompa, K. L.,** Gigawatt photochemical laser, *Appl. Phys. Lett.*, 22, 77—78, 1973.
267. **Aldridge, F. T.,** High-pressure iodine photodissociation laser, *Appl. Phys. Lett.*, 22, 180—182, 1973.
268. **Alekseev, V. A., Andreeva, T. L., Volkov, V. N., and Yukov, E. A.,** Kinetics of the generation spectrum of a photodissociation laser, *Sov. Phys. JETP*, 36, 238—242, 1973.
269. **Yukov, E. A.,** Elementary processes in the active medium of an iodine photodissociation laser, *Sov. J. Quantum Electron.*, 3, 117—120, 1973.
270. **Gusinow, M. A., Rice, J. K., and Padrick, T. D.,** The apparent late-time gain in a photodissociation iodine laser, *Chem. Phys. Lett.*, 21, 197—199, 1973.
271. **Hohla, K.,** The iodine laser, a high power gas laser, in *Laser Interaction and Related Plasma Phenomena*, Vol. 3A, Schwarz, H. J. and Hora, H., Eds., Plenum Press, New York, 1974.
272. **Birich, G. N., Drozd, G. I., Sorokin, V. N., and Struk, I. I.,** Photodissociation iodine laser using compounds containing group-V atoms, *JETP Lett.*, 19, 27—29, 1974.
273. **Belousova, I. M., Gorshkov, N. G., Danilov, O. B., Zalesskii, V. Yu., and Yachnev, I. L.,** Accumulation of iodine molecules in flash photolysis of CF_3I and n-C_3F_7I vapor, *Sov. Phys. JETP*, 38, 254—257, 1974.
274. **Belousova, I. M., Bobrov, B. D., Kiselev, V. M., Kurzenkov, V. N., and Krepostnov, P. I.,** Photodissociative I^{127} laser in a magnetic field, *Sov. Phys. JETP*, 38, 258—263, 1974.
275. **Basov, N. G., Golubev, L. E., Zuev, V. S., Katulin, V. A., Netemin, V. N., Nosach, V. Yu., Nosach, O. Yu., and Petrov, A. L.,** Iodine laser emitting short pulses of 50 J energy and 5 nsec duration, *Sov. J. Quantum Electron.*, 3, 524, 1974.
276. **Golubev, L. E., Zuev, V. S., Katulin, V. A., Nosach, V. Yu., and Nosach, O. Yu.,** Investigation of optical inhomogeneities which appear in an active medium of a photodissociation laser during coherent emission, *Sov. J. Quantum Electron.*, 3, 464—467, 1974.
277. **Kuznetsova, S. V. and Maslov, A. I.,** Investigation of the reactions of atomic iodine in a photodissociation laser using n-C_3F_7I and i-C_3F_7I molecules, *Sov. J. Quantum Electron.*, 3, 468—471, 1974.

278. **Palmer, R. E. and Gusinow, M. A.,** Late-time gain of the CF_3I iodine photodissociation laser, *J. Appl. Phys.*, 45, 2174—2178, 1974.
279. **Belousova, I. M., Bobrov, B. D., Kiselev, V. M., Kurzenkov, V. N., and Krepostnov, P. I.,** I^{127} atom in a magnetic field, *Opt. Spectrosc.*, 37, 20—24, 1974.
280. **Palmer, R. E. and Gusinow, M. A.,** Gain versus time in the CF_3I iodine photodissociation laser, *IEEE J. Quantum Electron.*, QE-10, 615—616, 1974.
281. **Silfvast, W. T., Szeto, L. H., and Wood, O. R.,** II, C_3F_7I photodissociation laser initiated by a CO_2-laser-produced plasma, *Appl. Phys. Lett.*, 25, 593—595, 1974.
282. **Belousova, I. M., Bobrov, B. D., Kiselev, V. M., and Kurzenkov, V. N.,** Characteristics of the stimulated emission from iodine atoms in pulsed magnetic fields, *Sov. J. Quantum Electron.*, 4, 767—769, 1974.
283. **Hohla, K., Fuss, W., Volk, R., and Witte, K. J.,** Iodine laser oscillator in gain switch mode for ns pulses, *Opt. Commun.*, 13, 114—116, 1975.
284. **Hohla, K., Brederlow, G., Fuss, W., Kompa, K. L., Raeder, J., Volk, R., Witkowski, S., and Witte, K. J.,** 60J 1-nsec iodine laser, *J. Appl. Phys.*, 46, 808—809, 1975.
285. **Zalesskii, V. Yu.,** Analytic estimate of the maximum duration of stimulated emission from a CF_3I photodissociation laser, *Sov. J. Quantum Electron.*, 4, 1009—1014, 1975.
286. **Butcher, R. J., Donovan, R. J., Fotakis, C., Fernie, D., and Rae, A. G. A.,** Photodissociation laser isotope effects, *Chem. Phys. Lett.*, 30, 398—402, 1975.
287. **Baker, H. J. and King, T. A.,** Mode-beating in gain-switch iodine photodissociation laser, *J. Phys. D*, 8, L31—L33, 1975.
288. **Baker, H. J. and King, T. A.,** Iodine photodissociation laser oscillator characteristics, *J. Phys. D.*, 8, 609—619, 1975.
289. **Aldridge, F. T.,** Stimulated emission cross section and inversion lifetime in a three-atmosphere iodine photodissociation laser, *IEEE J. Quantum Electron.*, QE-11, 215—217, 1975.
290. **Antonov, A. V., Basov, N. G., Zuev, V. S., Katulin, V. A., Korol'kov, K. S., Mikhailov, G. V., Netemin, V. N., Nikolaev, F. A., Nosach, V. Yu., Nosach, O. Yu., Petrov, A. L., and Shelobolin, A. V.,** Amplifier with a stored energy over 700 J designed for a short-pulse iodine laser, *Sov. J. Quantum Electron.*, 5, 123, 1975.
291. **Ishii, S., Ahlborn, B., and Curzon, F. L.,** Gain switching and Q spoiling of iodine laser with a shock wave, *Appl. Phys. Lett.*, 27, 118—119, 1975.
292. **Borovich, B. L., Zuev, V. S., Katulin, V. A., Nosach, V. Yu., Nosach, O. Yu., Startsev, A. V., and Stoilov, Yu. Yu.,** Characteristics of iodine laser short-pulse amplifier, *Sov. J. Quantum Electron.*, 5, 695—702, 1975.
293. **Zalesskii, V. Yu., and Polikarpov, S. S.,** Investigation of the conditions governing the stimulated emission threshold of a CF_3I (iodine) laser, *Sov. J. Quantum Electron.*, 5, 826—831, 1975.
294. **Pirkle, R. J., Davis, C. C., and McFarlane, R. A.,** Self-mode-locking of an iodine photodissociation laser, *J. Appl. Phys.*, 46, 4083—4085, 1975.
295. **Pleasance, L. D. and Weaver, L. A.,** Laser emission at 1.32 μm from atomic iodine produced by electrical dissociation of CF_3I, *Appl. Phys. Lett.*, 27, 407—409, 1975.
296. **Beverly, R. E., III,** Pressure-broadened iodine-laser-amplifier kinetics and a comparison of diluent effectiveness, *Opt. Commun.*, 15, 204—208, 1975.
297. **Gusinow, M. A.,** The enhancement of the near UV flashlamp spectra with special emphasis on the iodine photodissociation laser, *Opt. Commun.*, 15, 190—192, 1975.
298. **Skribanowitz, N. and Kopainsky, B.,** Pulse shortening and pulse deformation in a high-power iodine laser amplifier, *Appl. Phys. Lett.*, 27, 490—492, 1975.
299. **Pirkle, R. J., Jr., Davis, C. C., and McFarlane, R. A.,** Comparative performance of CF_3I, CD_3I and CH_3I in an atomic iodine photodissociation laser, *Chem. Phys. Lett.*, 36, 805—807, 1975.
300. **Gal'pern, M. G., Gorbachev, V. A., Katulin, V. A., Lebedev, O. L., Luk'yanets, E. A., Mekhryakova, N. G., Mizin, V. M., Nosach, V. Yu., Petrov, A. L., and Semenoskaya, G. G.,** Bleachable filter for the iodine laser emitting at λ = 1.35 μm, *Sov. J. Quantum Electron.*, 5, 1384—1385, 1975.
301. **Basov, N. G. and Zuev, V. S.,** Short-pulsed iodine laser, *Nuovo Cimento*, 31, 129—151, 1976.
302. **Brederlow, G., Witte, K. J., Fill, E., Hohla, K., and Volk, R.,** The Asterix III pulsed high-power iodine laser, *IEEE J. Quantum Electron.*, QE-12, 152—155, 1976.
303. **Jones, E. D., Palmer, M. A., and Franklin, F. R.,** Subnanosecond high-pressure iodine photodissociation laser oscillator, *Opt. Quantum Electron.*, 8, 231—235, 1976.
304. **Katulin, V. A., Nosach, V. Yu., and Petrov, A. L.,** Iodine laser with active Q switching, *Sov. J. Quantum Electron.*, 6, 205—208, 1976.
305. **Ishii, S. and Ahlborn, B.,** Elimination of compression waves induced by pump light in iodine lasers, *J. Appl. Phys.*, 47, 1076—1078, 1976.
306. **Swingle, J. C., Turner, C. E., Jr., Murray, J. R., George, E. V., and Krupke, W. F.,** Photolytic pumping of the iodine laser by XeBr*, *Appl. Phys. Lett.*, 28, 387—388, 1976.

307. **Brederlow, G., Fill, E., Fuss, W., Hohla, K., Volk, R., and Witte, K. J.**, High-power iodine laser development at the Institut fur Plasmaplysik, Garching, *Sov. J. Quantum Electron.*, 6, 491—495, 1976.
308. **Davis, C. C., Pirkle, R. J., McFarlane, R. A., and Wolga, G. J.**, Output mode spectra, comparative parametric operation, quenching, photolytic reversibility, and short-pulse generation in atomic iodine photodissociation lasers, *IEEE J. Quantum Electron.*, QE-12, 334—352, 1976.
309. **Arkhipova, E. V., Borovich, B. L., and Zapol'skii, A. K.**, Accumulation of excited iodine atoms in iodine photodissociation laser. Analysis of kinetic equations, *Sov. J. Quantum Electron.*, 6, 686—696, 1976.
310. **Nosach, O. Yu. and Orlov, E. P.**, Some features of formation of the angular spectrum of the stimulated radiation emitted from iodine laser, *Sov. J. Quantum Electron.*, 6, 770—777, 1976.
311. **Andreeva, T. L., Birich, G. N., Sorokin, V. N., and Struk, I. I.**, Investigations of photodissociation iodine lasers utilizing molecules with bonds between iodine atoms and group V elements. I. Experimental investigation of $(CF_3)_2$ AsI iodine laser, *Sov. J. Quantum Electron.*, 6, 781—789, 1976.
312. **Zalesskii, V. Yu. and Kokushkin, A. M.**, Tracing chemical changes in the active media of iodine photodissociation laser, *Sov. J. Quantum Electron.*, 6, 813—817, 1976.
313. **Katulin, V. A., Nosach, V. Yu., and Petrov, A. L.**, Nanosecond iodine laser with an output energy of 200J, *Sov. J. Quantum Electron.*, 6, 998—999, 1976.
314. **Fuss, W. and Hohla, K.**, Pressure broadening of the 1.3 μm iodine laser line, *Z. Naturforsch. Teil A*, 31, 569—577, 1976.
315. **Fuss, W. and Hohla, K.**, A closed cycle iodine laser, *Opt. Commun.*, 18, 427—430, 1976.
316. **Fill, E. and Hohla, K.**, A saturable absorber for the iodine laser, *Opt. Commun.*, 18, 431—436, 1976.
317. **Kamrukov, A. S., Kashnikov, G. N., Kozlov, N. P., Malashchenko, V. A., Orlov, V. K., and Protasov, Yu, S.**, Investigation of an iodine laser excited optically by high-current plasmadynamic discharges, *Sov. J. Quantum Electron.*, 6, 1101—1104, 1976.
318. **Mukhtar, E. S., Baker, H. J., and King, T. A.**, Selection of oscillation frequency in the 1.315 μm iodine laser, *Opt. Commun.*, 19, 193—196, 1976.
319. **Alekhin, B. V., Lazhintsev, B. V., Nor-arevyan, V. A., Petrov, N. N., and Sukhanov, L. V.**, Short-pulse photodissociation laser with magnetic-field modulation of gain, *Sov. J. Quantum Electron.*, 6, 1290—1292, 1976.
320. **Baker, H. J. and King, T. A.**, Line broadening and saturation parameters for short-pulse high-pressure iodine photodissociation laser systems, *J. Phys. D*, 9, 2433—2445, 1976.
321. **Padrick, T. D. and Palmer, R. E.**, Use of titanium doped quartz to eliminate carbon deposits in an atomic iodine photodissociation laser, *J. Appl. Phys.*, 47, 5109—5110, 1976.
322. **Fill, E., Hohla, K., Schappert, G. T., and Volk, R.**, 100-ps pulse generation and amplification in the iodine laser, *Appl. Phys. Lett.*, 29, 805—807, 1976.
323. **Olsen, J. N.**, Pulse shaping in the iodine laser, *J. Appl. Phys.*, 47, 5360—5364, 1976.
324. **Belousova, I. M., Bobrov, B. D., Grenishin, A. S., and Kiselev, V. M.**, Magnetic-field control of the duration of pulses emitted from an iodine photodissociation laser, *Sov. J. Quantum Electron.*, 7, 249—255, 1977.
325. **Andreeva, T. L., Birich, G. N., Sobel'man, I. I., Sorokin, V. N., and Struk, I. I.**, Continuously pumped continuous-flow iodine laser, *Sov. J. Quantum Electron.*, 7, 1230—1234, 1977.
326. **Mukhtar, E. S., Baker, H. J., and King, T. A.**, Pressure-induced frequency shifts in the atomic iodine laser, *Opt. Commun.*, 24, 167—169, 1978.
327. **Kiselev, V. M., Bobrov, B. D., Grenishin, A. S., and Kotlikova, T. N.**, Faraday rotation in the active medium of an iodine photodissociation laser oscillator or amplifier, *Sov. J. Quantum Electron.*, 8, 181—184, 1978.
328. **Babkin, V. I., Kuznetsova, S. V., and Maslov, A. I.**, Simple method for determination of stimulated emission cross section of $^2P_{1/2}$ (F = 3) $\rightarrow ^2P_{3/2}$ (F' = 4) transition in atomic iodine, *Sov. J. Quantum Electron.*, 8, 285—289, 1978.
329. **Katulin, V. A., Nosach, V. Yu., and Petrov, A. L.** Q-Switched iodine laser emitting at two frequencies, *Sov. J. Quantum Electron.*, 8, 380—382, 1978.
330. **Saito, H., Uchiyama, T., and Fujioka, T.**, Pulse propagation in the amplifier of a high-power iodine laser, *IEEE J. Quantum Electron.*, QE-14, 302—309, 1978.
331. **Gaidash, V. A., Mochalov, M. R., Shemyakin, V. I., and Shurygin, V. K.**, Modulation of the transmission of an atomic iodine switch, *Sov. J. Quantum Electron.*, 8, 530—531, 1978.
332. **McDermott, W. E., Pchelkin, N. R., Benard, D. G., and Bousek, R. R.**, An electronic transition chemical laser, *Appl. Phys. Lett.*, 32, 469—470, 1978.
333. **Antonov, A. S., Belousova, I. M., Gerasimov, V. A., Danilov, O. B., Zhevlakov, A. P., Sapelkin, N. V., and Yachnev, I. L.**, Flashlamp-excited photodissociation 1000 J laser with 1.4% efficiency, *Sov. Tech. Phys. Lett.*, 4, 459, 1978.

334. **Vinokurov, G. N. and Zalesskii, V. Yu.**, Chemical kinetics and gasdynamics of a Q-switch iodine laser with an optically thick active medium, *Sov. J. Quantum Electron.*, 8, 1191—1197, 1978.
335. **Benard, D. J., McDermott, W. E., Pchelkin, N. R., and Bousek, R. R.**, Efficient operation of a 100-W transverse-flow oxygen-iodine chemical laser, *Appl. Phys. Lett.*, 34, 40—41, 1979.
336. **Richardson, R. J. and Wiswall, C. E.**, Chemically pumped iodine laser, *Appl. Phys. Lett.*, 35, 138—139, 1979.
337. **Witte, K. J., Burkhard, P., and Lüthi, H. R.**, Low pressure mercury lamp pumped atomic iodine laser of high efficiency, *Opt. Commun.*, 28, 202—206, 1979.
338. **Riley, M. E., Padrick, T. D., and Palmer, R. E.**, Multilevel paraxial Maxwell-Bloch equation description of short pulse amplification in the atomic iodine laser, *IEEE J. Quantum Electron.*, QE-15, 178—189, 1979.
339. **Zuev, V. S., Netemin, V. N., and Nosach, O. Yu.**, Wavefront instability of iodine laser radiation and dynamics of optical inhomogeneity evolution in the active medium, *Sov. J. Quantum Electron.*, 9, 522—524, 1979.
340. **Palmer, R. E., Padrick, T. D., and Palmer, M. A.**, Diffraction-limited atomic iodine photodissociation laser, *Opt. Quantum Electron.*, 11, 61—70, 1979.
341. **Djeu, N. and Powell, F. X.**, More infrared laser transitions in atomic iodine, *IEEE J. Quantum Electron.*, QE-7, 537—538, 1971.
342. **Jarrett, S. M., Nunez, N., and Gould, G.**, Infra-red laser oscillation in HBr and HI gas discharges, *Appl. Phys. Lett.*, 7, 294—296, 1965.
343. **Kim, H. H. and Marantz, H.**, A study of the neutral atomic iodine laser, *Appl. Opt.*, 9, 359—368, 1970.
344. **Kim, H., Paananen, R., and Hanst, P.**, Iodine infrared laser, *IEEE J. Quantum Electron.*, QE-4, 385—386, 1968.
345. **Brandelik, J. E. and Smith, G. A.**, Br, C, Cl, S and Si laser action using a pulsed microwave discharge, *IEEE J. Quantum Electron.*, QE-16, 7—10, 1980.
346. **Klimkin, V. M.**, Investigation of an ytterbium vapor laser, *Sov. J. Quantum Electron.*, 5, 326—329, 1975.
347. **Prelas, M. A., Akerman, M. A., Boody, F. P., and Miley, G. H.**, A direct nuclear pumped 1.45 μ atomic carbon laser in mixtures of He-CO and He-CO_2, *Appl. Phys. Lett.*, 31, 428—430, 1977.
348. **Piltch, M., Walter, W. T., Solimene, N., Gould, G., and Bennett, W. R., Jr.**, Pulsed laser transitions in manganese vapor, *Appl. Phys. Lett.*, 7, 309—310, 1965.
349. **Silfvast, W. T. and Fowles, G. R.**, Laser action on several hyperfine transitions in MnI, *J. Opt. Soc. Am.*, 56, 832—833, 1966.
350. **Chen, C. J.**, Manganese laser, *J. Appl. Phys.*, 44, 4246—4247, 1973.
351. **Chen, C. J.**, Manganese laser using manganese chloride as lasant, *Appl. Phys. Lett.*, 24, 499—500, 1974.
352. **Isakov, V. K., Kapugin, M. M., and Potapov, S. E.**, $MnCl_2$-vapor laser (energy characteristics), *Sov. Tech. Phys. Lett.*, 2, 292—293, 1976.
353. **Isakov, V. K., Kalugin, M. M., and Potapov, S. E.**, Output spectrum of a manganese-chloride laser, *Sov. Tech. Phys. Lett.*, 4, 333—334, 1978.
354. **Bokhan, P. A., Burlakov, V. D., Gerasimov, V. A., and Solomonov, V. I.**, Stimulated emission mechanism and energy characteristics of manganese vapor laser, *Sov. J. Quantum Electron.*, 6, 672—675, 1976.
355. **Isaev, A. A., Kazaryan, M. A., Petrash, G. G., and Cherezov, V. M.**, An investigation of pulse manganese vapor laser, *Sov. J. Quantum Electron.*, 6, 978—980, 1976.
356. **Trainor, D. W. and Mani, S. A.**, Pumping iron: a KrF laser pumped atomic iron laser, *J. Chem. Phys.*, 68, 5481—5485, 1978.
357. **Linevsky, M. J. and Karras, T. W.**, An iron-vapor laser, *Appl. Phys. Lett.*, 33, 720—721, 1978.
358. **Solanki, R., Collins, G. J., and Fairbank, W. M., Jr.**, IR laser transitions in a nickel hollow cathode discharge, *IEEE J. Quantum Electron.*, QE-15, 525, 1979.
359. **Bokhan, P. A., Klimkin, V. M., Prokop'ev, V. E., and Solomonov, V. I.**, Investigation of a laser utilizing self-terminating transitions in europium atoms and ions, *Sov. J. Quantum Electron.*, 7, 81—82, 1977.
360. **Cahuzac, P., Sontag, H., and Toschek, P. E.**, Visible superfluorescence from atomic europium, *Opt. Commun.*, 31, 37—41, 1979.
361. **Breichignac, C. and Cahuzac, P.**, to be published.
362. **Pixton, R. M. and Fowles, G. R.**, Visible laser oscillations in helium at 7065Å, *Phys. Lett.*, 29A, 654—655, 1969.
363. **Schuebel, W. K.**, Laser action in AlII and HeI in a slot cathode discharge, *Appl. Phys. Lett.*, 30, 516—519, 1977.

364. **Abrams, R. L. and Wolga, G. J.,** Near infrared laser transitions in pure helium, *IEEE J. Quantum Electron.,* QE-5, 368, 1967.
365. **Abrams, R. L. and Wolga, G. J.,** Direct demonstration of the validity of the Wigner spin rule for helium-helium collisions, *Phys. Rev. Lett.,* 19, 1411—1414, 1967.
366. **Cagnard, R., der Agobian, R., Otto, J. L., and Echard, R.,** L'emission stimulee de quelque transitions infrarouges de l'helium et du neon, *C. R. Acad. Sci. Ser. B,* 257, 1044—1047, 1963.
367. **der Agobian, R., Otto, J. L., Cagnard, R., and Echard, R.,** Emission stimulée de nouvelles transitions infrarouges dans les gaz rares, *J. Phys.,* 25, 887—897, 1964.
368. **Wood, O. R., Burkhardt, E. G., Pollack, M. A., and Bridges, T. J.,** High pressure laser action in 13 gases with transverse excitation, *Appl. Phys. Lett.,* 18, 261—264, 1971.
369. **Patel, C. K. N., Bennett, W. R., Jr., Faust, W. L., and McFarlane, R. A.,** Infrared spectroscopy using stimulated emission techniques, *Phys. Rev. Lett.,* 9, 102—104, 1962.
370. **Bennett, W. R., Jr. and Kindlmann, P. J.,** Collision cross sections and optical maser considerations for helium, *Bull. Am. Phys. Soc.,* 8, 87, 1963.
371. **Brochard, J. and Liberman, S.,** Emission stimulee de nouvelles transitions infrarouge de l'helium et du neon, *C. R. Acad. Sci. Ser. B,* 260, 6827—6829, 1965.
372. **Brochard, J., Lespritt, J. F., and Liberman, L.,** Measurement of the isotopic separation of two infrared laser lines of HeI, *C. R. Acad. Sci. Ser. B,* 600—602, 1970.
373. **Mathias, L. E. S., Crocker, A., and Wills, M. S.,** Pulsed laser emission from helium at 95 μm, *IEEE J. Quantum Electron.,* QE-3, 170, 1967.
374. **Levine, J. S. and Javan, A.,** Far infra-red continuous wave oscillation in pure helium, *Appl. Phys. Lett.,* 14, 348—350, 1969.
375. **Turner, R. and Murphy, R. A.,** The far infrared helium laser, *Infrared Phys.,* 16, 197—200, 1976.
376. **Giuliano, C. R. and Hess, L. D.,** Reversible photodissociative laser system, *J. Appl. Phys.,* 40, 2428—2430, 1969.
377. **Campbell, J. D. and Kasper, J. V. V.,** Hyperfine structure in the CF_3Br photodissociation laser, *Chem. Phys. Lett.,* 10, 436—437, 1971.
378. **Spencer, D. J. and Wittig, C.,** Atomic bromine electronic-transition chemical laser, *Opt. Lett.,* 4, 1—3, 1979.
379. **Hocker, L. O.,** High resolution study of the helium-fluorine laser, *J. Opt. Soc. Am.,* 68, 262—265, 1978.
380. **Bell, W. E., Bloom, A. L., and Goldsborough, J. P.,** New laser transitions in antimony and tellurium, *IEEE J. Quantum Electron.,* QE-2, 154, 1966.
381. **Webb, C. E.,** New pulsed laser transitions in TeII, *IEEE J. Quantum Electron.,* QE-4, 426—427, 1968.
382. **Clunie, D. M., Thorn, R. S. A., and Trezise, K. E.,** Asymmetric visible super-radiant emission from a pulsed neon discharge, *Phys. Lett.,* 14, 28—29, 1965.
383. **Leonard, D. A., Neal, R. A., and Gerry, E. T.,** Observation of a super-radiant self-terminating green laser transition in neon, *Appl. Phys. Lett.,* 7, 175, 1965.
384. **Leonard, D. A.,** The 5401 Å pulsed neon laser, *IEEE J. Quantum Electron.,* QE-3, 133—135, 1967.
385. **Isaev, A. A., Kazaryan, M. A., and Petrash, G. G.,** Shape and duration of superradiance pulses corresponding to neon lines, *Sov. J. Quantum Electron.,* 2, 49—51, 1972.
386. **Magda, I. I., Tkach, Yu V., Lemberg, E. A., Skachek, G. V., Gadetskii, N. P., Sidel'nikova, A. V., Dyatlova, V. V., and Bessarab, Ya. Ya.,** High power nitrogen and neon pulsed gas lasers, *Sov. J. Quantum Electron.,* 3, 260—261, 1973.
387. **Perry, D. L.,** CW laser oscillation at 5433 Å in neon, *IEEE J. Quantum Electron.,* QE-7, 102, 1971.
388. **Bridges, W. B. and Chester, A. N.,** Visible and uv laser oscillation at 118 wavelengths in ionized neon, argon, krypton, xenon, oxygen and other gases, *Appl. Opt.,* 4, 573—580, 1965.
389. **White, A. D. and Rigden, J. D.,** The effect of super-radiance at 3.39 μ on the visible transitions in the He-Ne maser, *Appl. Phys. Lett.,* 2, 211—212, 1963.
390. **Heard, H. G. and Peterson, J.,** Super-radiant yellow and orange transitions in pure neon, *Proc. IEEE,* 52, 1258, 1964.
391. **Rosenberger, D.,** Laser-ubergange and superstrahlung bei 6143 Å in einer gepulsten Neon-entladungen, *Phys. Lett.,* 13, 228—229, 1964.
392. **Ericsson, K. G. and Lidholt, L. R.,** Super-radiant transitions in argon, krypton and xenon, *IEEE J. Quantum Electron.,* QE-3, 94, 1967.
393. **Rosenberger, D.,** Superstrahlung in gepulsten argon-, krypton und xenon-entladungen, *Phys. Lett.,* 14, 32, 1965.
394. **Bloom, A. L.,** Observation of new visible gas laser transition by removal of dominance, *Appl. Phys. Lett.,* 2, 101—102, 1963.
395. **Ericsson, G. and Lidholt, R.,** Generation of short light pulses by superradiance in gases, *Ark. Fys.,* 37, 557—568, 1967.

396. **Abrosimov, G. V.**, Spatial and temporal coherence of the radiation of pulsed neon and thallium gas lasers, *Opt. Spectrosc.*, 31, 54—56, 1971.
397. **Isaev, A. A., Kazaryan, M. A., and Petrash, G. G.**, Shape and duration of superradiance pulses corresponding to neon lines, *Sov. J. Quantum Electron.*, 2, 49—52, 1972.
398. **Odintsov, A. I., Turkin, N. G., and Yakunin, V. P.**, Spatial coherence and angular divergence of pulsed superradiance of neon, *Opt. Spectrosc.*, 38, 244—245, 1975.
399. **Isaev, A. A. and Petrash, G. G.**, Mechanism of pulsed superradiance from 2p-1s transitions in neon, *Sov. Phys. JETP*, 29, 607—614, 1969.
400. **White, A. D. and Rigden, J. D.**, Continuous gas maser operation in the visible, *Proc. IRE*, 50, 1796, 1962.
401. **Rigden, J. D. and White, A. D.**, The interaction of visible and infrared maser transitions in the helium-neon system, *Proc. IEEE*, 51, 943, 1963; *Quantum Electronics III*, Grivet, P. and Bloembergen, N., Eds., Columbia University Press, New York, 1964, 499—505.
402. **White, A. D. and Gordon, E. I.**, Excitation mechanism and current dependence of population inversion in He-Ne lasers, *Appl. Phys. Lett.*, 3, 197—198, 1963.
403. **Gordon, E. I. and White, A. D.**, Similarity laws for the effect of pressure and discharge diameter on gain of He-Ne lasers, *Appl. Phys. Lett.*, 3, 199—201, 1963.
404. **Labuda, E. F. and Gordon, E. I.**, Microwave determination of average electron energy and density in He-Ne discharges, *J. Appl. Phys.*, 35, 1647—1648, 1964.
405. **Young, R. T., J., Willett, C. S., and Maupin, R. T.**, The effect of helium on population inversion in the He-Ne laser, *J. Appl. Phys.*, 41, 2936—2941, 1970.
406. **Young, R. T., Jr., Willett, C. S., and Maupin, R. T.**, An experimental determination of the relative contributions of resonance and electron impact collision to the excitation if Ne atom in He-Ne laser discharges, *Bull. Am. Phys. Soc.*, 13, 206, 1968.
407. **Jones, C. R. and Robertson, W. W.**, Temperature dependence of reaction of the helium metastable atom, *Bull. Am. Phys. Soc.*, 13, 198, 1968. (At the 20th Gaseous Electronics Conference, San Francisco, October 1967, material was presented on the potential barrier behaviour of the reaction $He^*2^1S_0 + Ne^1S_0 \rightarrow He^1S_0 + Ne3s_2 - \Delta E_\infty(386\ cm^{-1})$, as well as the reaction involving $He^*2\ S_1$ metastables stated in the abstract in this reference.)
408. **Korolev, F. A., Odintsov, A. I., and Mitsai, V. N.**, A study of certain characteristics of a He-Ne laser, *Opt. Spectrosc.*, 19, 36—39, 1965.
409. **Suzuki, N.**, Vacuum uv measurements of helium-neon laser discharge, *Jpn. J. Appl. Phys.*, 4, 285—291, 1965.
410. **Field, R. L., Jr.**, Operating characteristics of dc-excited He-Ne gas lasers, *Rev. Sci. Instrum.*, 38, 1720—1722, 1967.
411. **Suzuki, N.**, Spectroscopy of He-Ne laser discharges, *Jpn. J. Appl. Phys.*, 4(Suppl. 1) 642—647, 1965.
412. **Gonchukov, G. A., Ermakov, G. A., Mikhenko, G. A., and Protsenko, E. D.**, Temperature effects in the He-Ne laser, *Opt. Spectrosc.*, 20, 601—602, 1966.
413. **Alekseeva, A. N. and Gordeev, D.V.**, The effect of longitudinal magnetic field on the output of a helium-neon laser, *Opt. Spectrosc.*, 23, 520—524, 1967.
414. **Smith, P. W.**, The output power of a 6328-Å He-Ne gas laser, *IEEE J. Quantum Electron.*, QE-2, 62—68, 1966.
415. **Smith, P. W.**, On the optimum geometry of a 6328 Å laser oscillator, *IEEE J. Quantum Electron.*, QE-2, 77—79, 1966.
416. **Herziger, G., Holzapfel, W., and Seelig, W.**, Verstarkung einer He-Ne gasentladung fur die laser wellenlange, λ = 6328 AE, *Z. Phys.*, 189, 385—400, 1966.
417. **Bell, W. E. and Bloom, A. L.**, Zeeman effect at 3.39 microns in a helium-neon laser, *Appl. Optics*, 3, 413—415, 1964.
418. **Belusova, I. M., Danilov, O. B., Elkina, I. A., and Kiselev, K. M.**, Investigation of the causes of gas temperature effects on the generation of power of a He-Ne laser at 6328 A, *Opt. Spectrosc.*, 16, 44—47, 1969.
419. **Allen, L. and Jones, D. G. C.**, *Principles of Gas Lasers*, Plenum Press, New York, 1967, 73—103.
420. **Bloom, A. L.**, *Gas Lasers*, John Wiley & Sons, New York, 1968, 52—59.
421. **White, A. D.**, Anomalous behaviour of the 6402.84 Å gas laser, *Proc. IEEE*, 52, 721, 1964.
422. **Bloom, A. L. and Hardwick, D. L.**, Operation of He-Ne lasers in the forbidden resonator region, *Phys. Lett.*, 20, 373—375, 1966.
423. **Zitter, R. N.**, 2s-2p and 3p-3s neon transitions in a very long laser, *Bull. Am. Phys. Soc.*, 9, 500, 1964.
424. **Patel, C. K. N.**, Gas lasers, in *Lasers*, Vol. 1, Levine, A. K., Ed., Marcel Dekker, New York, 1968, 39—50.
425. **Yariv, A.**, *Introduction to Optical Electronics*, 2nd ed., Holt, Rinehart & Winston, New York, 1976, 170—172.

426. **Uchida, T.**, Frequency spectra of the He-Ne optical masers with external concave mirrors, *Appl. Opt.*, 4, 129—131, 1965.
427. **Massey, G. A., Oshman, M. K., and Targ, R.**, Generation of single-frequency light using the FM laser, *Appl. Phys. Lett.*, 6, 10—11, 1965.
428. **Steier, W. H.**, Coupling of high peak power pulses from He-Ne lasers, *Proc. IEEE*, 54, 1604—1606, 1966.
429. **Kuznetsov, A. A., Mash, D. I., and Skuratova, N. V.**, Effect of an axial magnetic field on the output power of a neon-helium laser simultaneously generating lines of 3.39 and 0.6328 microns, *Radio Eng. Electron. Phys.*, 12, 140—143, 1967.
430. **Lee, P. H. and Skolnick, M. L.**, Saturated neon absorption inside a 6328 Å laser, *Appl. Phys. Lett.*, 10, 303—305, 1967.
431. **Hochuli, U., Haldemann, P., and Hardwick, D.**, Cold cathodes for He-Ne gas lasers, *IEEE J. Quantum Electron.*, QE-3, 612—614, 1967.
432. **Carlson, F. P.**, On the optimal use of dc and RF excitation in He-Ne lasers, *IEEE J. Quantum Electron.*, QE-4, 98—99, 1968.
433. **Suzuki, T.**, Discharge current noise in He-Ne laser and its suppression, *Jpn. J. Appl. Phys.*, 9, 309—310, 1970.
434. **Ehlers, K. W. and Brown, I. G.**, Regeneration of helium neon lasers, *Rev. Sci. Instrum.*, 41, 1505—1506, 1970.
435. **Leontov, V. G., Ostapchenko, E. P., and Sedov, G. S.**, Optimum lasing conditions of a He-Ne laser operating in the TEM_{00} axial mode, *Opt. Spectrosc.*, 418—419, 1971.
436. **Sakurai, T.**, Discharge current dependence of saturation parameter of a He-Ne gas laser, *Jpn. J. Appl. Phys.*, 11, 1832—1836, 1972.
437. **Wang, S. C. and Siegman, A. E.**, Hollow-cathode transverse discharge He-Ne and $He\text{-}Cd^+$ lasers, *Appl. Phys.*, 2, 143—150, 1973.
438. **Zborovskii, V. A., Molchanov, M. I., Turkin, A. A., and Yaroshenko, N. G.**, Measurement of the natural line width of a travelling-wave He-Ne laser in the 0.63 μm region, *Opt. Spectrosc.*, 34, 704—705, 1973.
439. **Dote, T., Yamaguchi, N., and Nakamura, T.**, Effects of RF electric field on He-Ne laser output, *Phys. Lett.*, 45A, 29—30, 1973.
440. **Belousova, I. M. and Znamemskii, V. B.**, Properties of the lasing mechanism of a helium-neon mixture excited by a discharge in a hollow cathode, *J. Appl. Spectrosc.*, 25, 1109—1114, 1976.
441. **Young, R. T., Jr.**, Calculation of average electron energies in He-Ne discharges, *J. Appl. Phys.*, 36, 2324—2325, 1965.
442. **Crisp, M. D.**, A magnetically polarized He-Ne laser, *Opt. Commun.*, 19, 316—319, 1976.
443. **Ihjima, T., Kuroda, K., and Ogura, I.**, Radial profiles of upper- and lower-laser-level emission in an oscillating He-Ne laser, *J. Appl. Phys.*, 48, 437—439, 1977.
444. **Muller, Ya. N., Geller, V. M., Lisitsyna, L. I., and Grif, G. I.**, Microwave-pumped helium-neon laser, *Sov. J. Quantum Electron.*, 7, 1013—1015, 1977.
445. **Leontov, V. G. and Ostapchenko, E. P.**, Effect of excitation conditions on the radial distribution of population inversion in the active element of a He-Ne laser, *Opt. Spectrosc.*, 43, 321—325, 1977.
446. **Honda, T. and Endo, M.**, International intercomparison of laser power at 633 nm, *IEEE J. Quantum Electron.*, QE-14, 213—214, 1978.
447. **Otieno, A. V.**, Homogeneous saturation of the 6328 Å neon laser transition due to collisions in the weak collision model, *Opt. Commun.*, 26, 207—210, 1978.
448. **Ferguson, J. B. and Morris, R. H.**, Single-mode collapse in 6328 Å He-Ne lasers, *Appl. Opt.*, 17, 2924—2929, 1978.
449. **Chance, D. A., Chastang, J. -C. A., Crawford, V. S., Horstmann, R. E., and Lussow, R. O.**, HeNe parallel plate laser development, *IBM J. Res. Dev.*, 23, 108—118, 1979.
450. **Chance, D. A., Brusic, V., Crawford, V. S., and Macinnes, R. D.**, Cathodes for HeNe lasers, *IBM J. Res. Develop.*, 23, 119—127, 1979.
451. **Ahearn, W. E. and Horstmann, R. E.**, Nondestructive analysis for HeNe lasers, *IBM J. Res. Develop.*, 23, 128—131, 1979.
452. **Schuöcker, D., Reif, W., and Lagger, H.**, Theoretical description of discharge plasma and calculation of maximum output intensity of He-Ne waveguide lasers as a function of discharge tube radius, *IEEE J. Quantum Electron.*, QE-15, 232—239, 1979.
453. **Tobias, I. and Strouse, W. M.**, The anomalous appearance of laser oscillation at 6401 Å, *Appl. Phys. Lett.*, 10, 342—344, 1967.
454. **Schlie, L. A. and Verdeyn, J. T.**, Radial profile of Ne $1s_5$ atoms in a He-Ne discharge and their lens effect on lasing at 6401 Å, *IEEE J. Quantum Electron.*, QE-5, 21—29, 1969.
455. **Sanders, J. H. and Thomson, J. E.**, New high-gain laser transitions in neon, *J. Phys. B*, 6, 2177—2183, 1973.

456. **Zitter, R. N.**, 2s-2p and 3p-2s transitions of neon in a laser ten meters long, *J. Appl. Phys.*, 35, 3070—3071, 1964.
457. **Schearer, L. D.**, Polarization transfer between oriented metastable helium atoms and neon atoms, *Phys. Lett.*, 27a, 544—545, 1968.
458. **Chebotayev, V. P. and Vasilenko, L. S.**, Investigation of a neon-hydrogen laser at large discharge current, *Sov. Phys. JETP*, 21, 515—516, 1965.
459. **Chebotayev, V. P.**, Effect of hydrogen and oxygen on the operation of a neon maser, *Radio Eng. Electron. Phys.*, 10, 316—318, 1965.
460. **McClure, R. M., Pizzo, R., Schiff, M., and Zarowin, C. B.**, Laser oscillation at 1.06 microns in He-Ne, *Proc. IEEE*, 52, 851, 1964.
461. **Shtyrkov, E. I. and Subbes, E. V.**, Characteristics of pulsed laser action in helium-neon and helium-argon mixtures, *Opt. Spectrosc.*, 21, 143—144, 1966.
462. **Itzkan, I. and Pincus, G.**, 1.0621-μ He-Ne laser, *Appl. Opt.*, 5, 349, 1966.
463. **McFarlane, R. A., Patel, C. K. N., Bennett, W. R., Jr., and Faust, W. L.**, New helium-neon optical maser transitions, *Proc. IRE*, 50, 2111—2112, 1962.
464. **Chebotayev, V. P. and Pokasov, V. V.**, Operation of a laser on a mixture of He-Ne with discharge in a hollow cathode, *Radio Eng. Electron. Phys.*, 10, 817—819, 1965.
465. **Petrash, G. G. and Knyazev, I. N.**, Study of pulsed laser generation in neon and mixtures of neon and helium, *Sov. Phys. JETP*, 18, 571—575, 1964.
466. **der Agobian, R., Otto, J. L., Cagnard, R., and Echard, R.**, Cascades de transitions stimulées dans le neon pur, *C. R. Acad. Sci. Ser. B*, 259, 323—326, 1964.
467. **Andrade, O., Gallardo, M., and Bockasten, K.**, High-gain laser lines in noble gases, *Appl. Phys. Lett.*, 11, 99—100, 1967.
468. **Javan, A., Bennett, W. R., Jr., and Herriott, D. R.**, Population inversion and continuous optical maser oscillation in a gas discharge containing a He-Ne mixture, *Phys. Rev. Lett.*, 6, 106—110, 1961.
469. **Herriott, D. R.**, Optical properties of a continuous helium-neon optical maser, *J. Opt. Soc. Am.*, 52, 31—37, 1962.
470. **der Agobian, R., Cagnard, R., Echard, E., and Otto, J. L.**, Nouvelle cascade de transitions stimulee du neon, *C. R. Acad. Sci. Ser. B*, 258, 3661—3663, 1964.
471. **Bennett, W. R., Jr. and Knutson, J. W., Jr.**, Simultaneous laser oscillation on the neon doublet at 1.1523 μ, *Proc. IEEE*, 52, 861—862, 1964.
472. **Patel, C. K. N.**, Optical power output in He-Ne and pure Ne maser, *J. Appl. Phys.*, 33, 3194—3195, 1962.
473. **Chebotayev, V. P.**, Operating condition of an optical maser containing a helium-neon mixture, *Radio Eng. Electron. Phys.*, 10, 314—316, 1965.
474. **Boot, H. A. H., Clunie, D. M., and Thorn, R. S. A.**, Pulsed laser operation in a high-pressure helium-neon mixture, *Nature (London)*, 198, 773—774, 1963.
475. **Smith, J.**, Optical maser action in the negative glow region of a cold cathode glow discharge, *J. Appl. Phys.*, 35, 723—724, 1964.
476. **Cool, T. A.**, A fluid mixing laser, *Appl. Phys. Lett.*, 9, 418—420, 1966.
477. **Mustafin, K. S., Seleznev, V. A., and Shtyrkov, E. I.**, Stimulated emission in the negative glow region of a glow discharge, *Opt. Spectrosc.*, 21, 429—430, 1966.
478. **Kuznetsov, A. A., Mash, D. I., Milinkis, B. M., and Chirina, L. P.**, Operating conditions of an optical quantum generator (laser) in helium-neon and xenon-helium gas mixtures, *Radio Eng. Electron. Phys.*, 9, 1576, 1964.
479. **Lisitsyn, V. N., Fedchenko, A. I., and Chebotayev, V. O.**, Generation due to upper neon transitions in a He-Ne discharge optically pumped by a helium lamp, *Opt. Spectrosc.*, 27, 157—161, 1969.
480. **der Agobian, R., Otto, J. L., Echard, R., and Cagnard, R.**, Emission stimulee de nouvelles transitions infrarouges du neon, *C. R. Acad. Sci. Ser. B*, 257, 3844—3847, 1963.
481. **Blau, E. J., Hochheimer, B. F., Massey, J. T., and Schulz, A. G.**, Identification of lasing energy levels by spectroscopic techniques, *J. Appl. Phys.*, 34, 703, 1963.
482. **Smith, D. S. and Riccius, H. D.**, Observation of new helium-neon laser transitions near 1.49 μm, *IEEE J. Quantum Electron.*, QE-13, 366, 1977.
483. **Gires, F., Mayer, H., and Pailette, M.**, Sur quelques transitions présentant l'effet laser dans le melange helium-neon, *C. R. Acad. Sci. Ser. B*, 256, 3438—3439, 1963.
484. **McFarlane, R. A., Faust, W. L., and Patel, C. K. N.**, Oscillation of f-d transitions in neon in a gas optical maser, *Proc. IEEE*, 51, 468, 1963.
485. **Lisitsyn, V. N. and Chebotayev, V. P.**, The generation of laser action in the 4f-3d transitions of neon by optical pumping of a He-Ne discharge with helium lamp, *Opt. Spectrosc.*, 20, 603—604, 1966.
486. **McFarlane, R. A., Faust, W. L., Patel, C. K. N., and Garrett, C. G. B.**, Gas maser operation at wavelengths out to 28 microns, in *Quantum Electronics III*, Grivet, P. and Bloembergen, N., Eds., Columbia University Press, New York, 1964, 573—586.

487. **Rosenberger, D.**, Oscillation of three 3p-2s transitions in a He-Ne mixtures, *Phys. Lett.*, 9, 29—31, 1964.
488. **Bennett, W. R., Jr. and Pawilkowski, A. T.**, Additional cascade laser transitions in He-Ne mixtures, *Bull. Am. Phys. Soc.*, 9, 500, 1964.
489. **Smiley, V. N.**, New He-Ne and Ne laser lines in the infra-red, *Appl. Phys. Lett.*, 4, 123—124, 1964.
490. **Faust, W. L., McFarlane, R. A., Patel, C. K. N., and Garrett, C. G. B.**, Noble gas optical maser lines at wavelengths between 2 and 35 μ, *Phys. Rev.*, 133, A1476—A1478, 1964.
491. **Otto, J. L., Cagnard, R., Echard, R., and der Agobian, R.**, Emission stimulée de nouvelles transition infrarouges dan les gas rares, *C. R. Acad. Sci. Ser. B*, 258, 2779—2780, 1964.
492. **Grudzinski, R., Pailette, M. R., and Becrelle, J.**, Étude des transitions laser complees dans un melange helium-neon, *C. R. Acad. Sci. Ser. B*, 258, 1452—1454, 1964.
493. **Bergman, K. and Demtroder, W.**, A new cascade laser transition in He-Ne mixture, *Phys. Lett.*, 29a, 94—95, 1969.
494. **Gerritsen, H. J. and Goedertier, P. V.**, A gaseous (He-Ne) cascade laser, *Appl. Phys. Lett.*, 4, 20—21, 1964.
495. **Smiley, V. N.**, A long gas phase optical maser cell, in *Quantum Electronics III*, Grivet, P. and Bloembergen, N., Eds., Columbia University Press, New York, 1964, 587—591.
496. **McMullin, P. G.**, Precise wavelength measurement of infrared optical maser lines, *Appl. Opt.*, 3, 641—642, 1964.
497. **Patel, C. K. N., McFarlane, R. A., and Faust, W. L.**, Further infrared spectroscopy using stimulated emission techniques, in *Quantum Electronics III*, Grivet, P. and Bloembergen, N., Eds., Columbia University Press, New York, 1964, 561—572.
498. **Brunet, H. and Laures, P.**, New infrared gas laser transitions by removal of dominance, *Phys. Lett.*, 12, 106—107, 1964.
499. **Bloom, A. L., Bell, W. E., and Rempel, R. C.**, Laser action at 3.39 μ in a helium-neon mixtures, *Appl. Opt.*, 2, 317—318, 1963.
500. **Brochard, J. and Liberman, S.**, Emission stimulée de nouvelles transitions infrarouge de l'helium et du neon, *C. R. Acad. Sci. Ser. B*, 260, 6827—6829, 1965.
501. **Herceg, J. E. and Miley, G. H.**, A laser utilizing a low-voltage arc discharge in helium-neon, *J. Appl. Phys.*, 39, 2147—2149, 1968.
502. **Konovalov, I. P., Popov, A. I., and Protsenko, E. D.**, Measurement of spectral characteristics of the $5s'[1/2]_1^\circ \rightarrow 4p[3/2]_2$ Ne (3.39 μm) transition, *Opt. Spectrosc.*, 33, 6—10, 1972.
503. **Mazanko, I. P., Ogurok, N. D. D., and Sviridov, M. V.**, Measurement of the saturation parameter of a neon-helium mixture at the 3.39-μm wavelength, *Opt. Spectrosc.*, 35, 327—328, 1973.
504. **Watanabe, S., Chihara, M., and Ogura, I.**, Decay rate measurements of upper laser levels in He-Ne and He-Se lasers, *Jpn. J. Appl. Phys.*, 13, 164—169, 1974.
505. **Balakin, V. A., Konovalov, I. P., Ocheretyanyi, A. I., Popov, A. I., and Protsenko, E. D.**, Switching of the emission wavelength of a helium-neon laser in the 3.39 μ region, *Sov. J. Quantum Electron.*, 5, 230—231, 1975.
506. **Balakin, V. A., Konovalov, I. P., and Protsenko, E. D.**, Measurement of the spectral characteristics of the 3.3912 μ ($3s_2$-$3p_2$) Ne line, *Sov. J. Quantum Electron.*, 5, 581—583, 1975.
507. **Popov, A. I. and Protsenko, E. D.**, Laser gain due to $5s'[1/2]_1^\circ$ - $4p'[3/2]_2$ transition in neon at λ = 3.39 μ, *Sov. J. Quantum Electron.*, 5, 1153—1154, 1976.
508. **Faust, W. L., McFarlane, R. A., Patel, C. K. N., and Garrett, C. G. B.**, Gas maser spectroscopy in the infrared, *Appl. Phys. Lett.*, 4, 85—88, 1962.
509. **Patel, C. K. N., McFarlane, R. A., and Garrett, C. G. B.**, Laser action up to 57.355 μ in gaseous discharges (Ne, He-Ne), *Appl. Phys. Lett.*, 4, 18—19, 1964.
510. **Patel, C. K. N., McFarlane, R. A., and Garrett, C. G. B.**, Optical-maser action up to 57.355 μm in neon, *Bull. Am. Phys. Soc.*, 9, 65, 1964.
511. **Patel, C. K. N., Faust, W. L., McFarlane, R. A., and Garrett, C. G. B.**, CW optical-maser action up to 133 μ (0.133 mm) in neon discharges, *Proc. IEEE*, 52, 713, 1964.
512. **McFarlane, R. A., Faust, W. L., Patel, C. K. N., and Garrett, C. G. B.**, Neon gas maser lines at 68.329 μ and 85.047 μ, *Proc. IEEE*, 52, 318, 1964.
513. **Chapovsky, P. L., Lisitsyn, V. N., and Sorokin, A. R.**, High-pressure gas lasers on ArI, XeI and KrI transitions, *Opt. Commun.*, 26, 33—36, 1976.
514. **Kochubei, S. A., Lisitsyn, V. N., Sorokin, A. R., and Chapovskii, P. L.**, High-pressure tunable atomic gas lasers, *Sov. J. Quantum Electron.*, 7, 1142—1144, 1978.
515. **Bockasten, K., Lundholm, T., and Andrade, O.**, New near infrared laser lines in argon 1, *Phys. Lett.*, 22, 145—146, 1966.
516. **Brisbane, A. D.**, High gain pulsed laser, *Nature (London)*, 214, 75, 1967.
517. **Bockasten, K. and Andrade, O.**, Identification of high gain laser lines in argon, *Nature (London)*, 215, 382, 1967.

518. **Sutton, D. G., Galvan, L., Valenzuela, P. R., and Suchard, S. N.,** Atomic laser action in rare gas-SF_6 mixture, *IEEE J. Quantum Electron.,* QE-11, 54—57, 1975.
519. **Horrigan, F. A., Koozekanani, S. H., and Paananen, R. A.,** Infrared laser action and lifetimes in argon II, *Appl. Phys. Lett.,* 6, 41—43, 1965.
520. **Dauger, A. B. and Stafsudd, O. M.,** Characteristics of the continuous wave neutral argon laser, *Appl. Opt.,* 10, 2690—2697, 1971.
521. **Willett, C. S.,** Comments on: characteristics of the continuous wave neutral argon laser, *Appl. Opt.,* 11, 1429—1431, 1972.
522. **Dauger, A. B. and Stafsudd, O. M.,** Line competition in the neutral argon laser, *IEEE J. Quantum Electron.,* QE-8, 912—913, 1972.
523. **Jalufka, N. W., DeYoung, R. J., Hohl, F., and Williams, M. D.,** Nuclear-pumped ^{3}He-Ar laser excited by the ^{3}He(n,p)^{3}H reaction, *Appl. Phys. Lett.,* 29, 188—190, 1976.
524. **Wilson, J. W., DeYoung, R. J., and Harries, W. L.,** Nuclear-pumped ^{3}He-Ar laser modelling, *J. Appl. Phys.,* 50, 1226—1234, 1979.
525. **Liberman, S.,** Emission stimulée de nouvelles transitions infrarouge de l'argon, du krypton et du xenon, *C. R. Acad. Sci. Ser. B,* 261, 2601—2604, 1965.
526. **Linford, G. J.,** New pulsed and CW laser lines in the heavy noble gases, *IEEE J. Quantum Electron.,* QE-9, 611—612, 1973.
527. **Brochard, J., Cahuzac, P., and Vetter, R.,** Mesure des ecouts isotopiques de six raies laser infrarouges dans l'argon, *C. R., Acad. Sci. Ser. B,* 265, 467—470, 1967.
528. **Linford, G. J.,** High-gain neutral laser lines in pulsed noble-gas discharges, *IEEE J. Quantum Electron.,* QE-8, 477—482, 1972.
529. **Linford, G. J.,** New pulsed laser lines in krypton, *IEEE J. Quantum Electron.,* QE-9, 610—611, 1973.
530. **Walter, W. T. and Jarrett, J. M.,** Strong 3.27 μ oscillation in xenon, *Appl. Opt.,* 3, 789—790, 1964.
531. **DeYoung, R. J., Jalufka, N. W., and Hohl, F.,** Nuclear-pumped lasing of ^{3}He-Xe and ^{3}He-Kr, *Appl. Phys. Lett.,* 30, 19—21, 1977.
532. **Sinclair, D. C.,** Near-infrared oscillations in pulsed noble-gas-ion lasers, *J. Opt. Soc. Am.,* 55, 571, 1965.
533. **Courville, G. E., Walsh, P. J., and Wasko, J. H.,** Laser action in Xe in two distinct current regions of ac and dc discharges, J. *Appl. Phys.,* 35, 2547—2548, 1964.
534. **Clark, P. O.,** Pulsed operation of the neutral xenon laser, *Phys. Lett.,* 17, 190—192, 1965.
535. **Newman, L. A. and DeTemple, T. A.,** High-pressure infrared Ar-Xe laser system: ionizer-sustainer mode of excitation, *Appl. Phys. Lett.,* 27, 678—680, 1975.
536. **Lawton, S. A., Richards, J. B., Newman, L. A., Specht, L., and DeTemple, T. A.,** The high-pressure neutral infrared xenon laser, *J. Appl. Phys.,* 50, 3888—3898, 1979.
537. **Patel, C. K. N., Faust, W. L., and McFarlane, R. A.,** High gain gaseous (Xe-He) optical maser, *Appl. Phys. Lett.,* 1, 84—85, 1962.
538. **Tang, C. L.,** Relative probabilities for the xenon laser transitions, *Proc. IEEE,* 219—220, 1963.
539. **Fork, R. L. and Patel, C. K. N.,** Broadband magnetic field tuning of optical masers, *Appl. Phys. Lett.,* 2, 180—181, 1963.
540. **Aisenberg, S.,** The effect of helium on electron temperature and electron density in rare gas lasers, *Appl. Phys. Lett.,* 2, 187—189, 1963.
541. **Patel, C. K. N.,** Determination of atomic temperature and Doppler broadening in a gaseous discharge with population inversion, *Phys. Rev.,* 131, 1582—1584, 1963.
542. **Patel, C. K. N., McFarlane, R. A., and Faust, W. L.,** High gain medium for gaseous optical masers, in *Quantum Electronics III,* Grivet, P. and Bloembergen, N., Eds., Columbia University Press, New York, 1964,507—514.
543. **Faust, W. L. and McFarlane, R. A.,** Line strengths for noble-gas maser transitions; calculations of gain/inversion at various wavelengths, *J. Appl. Phys.,* 35, 2010—2015, 1964.
544. **Smiley, V. N., Lewis, A. L., and Forbes, D. K.,** Gain and bandwidth narrowing in a regenerative He-Xe laser amplifier, *J. Opt. Soc. Am.,* 55, 1552—1553, 1965.
545. **Kuznetsov, A. A. and Mash, D. I.,** Operating conditions of an optical maser with a helium-xenon mixture in the middle infrared region of the spectrum, *Radio Eng. Electron. Phys.,* 10, 319—320, 1965.
546. **Moskalenko, V. F., Ostapchenko, E. P., and Pugnin, V. I.,** Mechanism of xenon-level population inversion in the positive column of a helium-xenon mixture, *Opt. Spectrosc.,* 23, 94—95, 1967.
547. **Schwarz, S. E., DeTemple, T. A., and Targ, R.,** High-pressure pulsed xenon laser, *Appl. Phys. Lett.,* 17, 305—306, 1970.
548. **Shafer, J. H.,** Optical heterodyne measurement of xenon isotope shifts, *Phys. Rev.,* A3, 752—757, 1971.
549. **Targ, R. and Sasnett, M. W.,** High-repetition-rate xenon laser with transverse excitation, *IEEE J. Quantum Electron.,* QE-8, 166—169, 1972.

550. **Dandawate, V. D., Thomas, G. C., and Zembrod, A.**, Time behavior of a TEA xenon laser, *IEEE J. Quantum Electron.*, QE-8, 918—919, 1972.
551 **Fahlen, T. S. and Targ, R.**, High-average-power xenon laser, *IEEE J. Quantum Electron.*, QE-9, 609, 1973.
552. **Mansfield, C. R., Bird, P. F., Davis, J. F., Wimett, T. F., and Helmick, H. H.**, Direct nuclear pumping of a ^{3}He-Xe laser, *Appl. Phys. Lett.*, 30, 640—641, 1977.
553. **Liberman, L.**, Sur la structure hyperfine de quelques raies laser infrarouges de xenon 129, *C. R. Acad. Sci. Ser. B*, 266, 236—239, 1968.
554. **Armstrong, D. R.**, A method for the control of gas pressure in the xenon laser, *IEEE J. Quantum Electron.*, QE-4, 968—969, 1968.
555. **Culshaw, W. and Kannelaud, J.**, Mode interaction in a Zeeman laser, *Phys. Rev.*, 156, 308—319, 1967.
556. **Paananen, R. A. and Bobroff, D. L.**, Very high gain gaseous (Xe-He) optical maser at 3.5 μ, *Appl. Phys. Lett.*, 2, 99—100, 1963.
557. **Bridges, W. B.**, High optical gain at 3.5 μ in pure xenon, *Appl. Phys. Lett.*, 3, 45—47, 1963.
558. **Markin, E. P. and Nikitin, V. V.**, The 3.5 μ Xe-Ne laser, *Opt. Spectrosc.*, 17, 519, 1964.
559. **Clark, P. O.**, Investigation of the operating characteristics of the 3.5 μ xenon laser, *IEEE J. Quantum Electron.*, QE-1, 109—113, 1965.
560. **Kluver, J. W.**, Laser amplifier noise at 3.5 microns in helium-xenon, *J. Appl. Phys.*, 37, 2987—2999, 1966.
561. **Freiberg, R. J. and Weaver, L. A.**, Effects of lasering upon the electron gas and excited-state populations in xenon discharges, *J. Appl. Phys.*, 38, 250—262, 1967.
562. **Aleksandrov, E. B. and Kulyasov, V. N.**, Determination of the elementary-emitter spectrum latent in an inhomogeneously broadened spectral line, *Sov. Phys. JETP*, 28, 396—400, 1969.
563. **Fork, R. L., Dienes, A., and Kluver, J. W.**, Effects of combined RF and optical fields on a laser medium, *IEEE J. Quantum Electron.*, QE-5, 607—616, 1969.
564. **Wang, S. C., Byer, R. L., and Siegman, A. E.**, Observation of an enhanced Lamb dip with a pure Xe gain cell inside a 3.51 μ He-Xe laser, *Appl. Phys. Lett.*, 17, 120—122, 1970.
565. **Kasuya, T.**, Broad band frequency tuning of a He-Xe laser with a superconducting solenoid, *Appl. Phys.*, 2, 339—343, 1973.
566. **Linford, G. J., Peressini, E. R., Sooy, W. R., and Spaeth, M. L.**, Very long lasers, *Appl. Opt.*, 13, 379—390, 1974.
567. **Aleksandrov, E. B., Kulyasov, V. N., and Kharnang, K.**, Tunable single-frequency xenon laser operating at the two infrared transitions λ = 5.57 μm and 3.51 μm, *Opt. Spectrosc.*, 38, 439—440, 1975.
568. **Wolff, P. A., Abraham, N. B., and Smith, S. R.**, Measurement of radial variation of 3.51 μm gain in xenon discharge tubes, *IEEE J. Quantum Electron.*, QE-13, 400—403, 1977.
569. **Vetter, R.**, Ecarts isotopiques dans la transition laser a λ = 3.99 μm du xenon, Accroissement de l'effet de volume pour les couples (136,134) et (126,124), *Phys. Lett.*, 42A, 231—232, 1972.
570. **Olson, R. A., Bletzinger, P., and Garscadden, A.**, New pulsed Xe-neutral laser line, *IEEE J. Quantum Electron.*, QE-12, 316—317, 1976.
571. **Petrov, Yu. N. and Prokhorov, A. M.**, 75-micron laser, *Sov. Phys. JETP Lett.*, 1, 24—25, 1965.
572. **Cheo, P. K. and Cooper, H. G.**, Ultraviolet ion laser transitions between 2300 Å and 4000 Å, *J. Appl. Phys.*, 36, 1862—1865, 1965.
573. **Dreyfus, R. W. and Hodgson, R. T.**, Electron-beam gas laser excitation, paper presented at the 7th Int. Quantum Electronics Conf., Montreal, 8 to 11 May, 1972, abstract, *IEEE J. Quantum Electron.*, QE-8, 537—538, 1972.
574. **Dana, L., Laures, P., and Rocherolles, R.**, Raies laser ultraviolettes dans le neon, l'argon et le xenon, *C. R.*, 260, 481—484, 1965.
575. **Cottrell, T. H. E., Sinclair, D. C., and Forsyth, J. M.**, *IEEE J. Quantum Electron.*, QE-2, 703, 1966.
576. **Akitt, D. P. and Wittig, C. F.**, Laser emission in ammonia, *J. Appl. Phys.*, 40, 902—903, 1969.
577. **Harrison, G. R.**, *M.I.T. Wavelength Tables*, M.I.T. Press, Cambridge, Mass., 1969.
578. **Willett, C. S.**, Neutral gas lasers, in *Handbook of Lasers*, CRC Press, Cleveland, Ohio, 1971.
579. **Willett, C. S.**, Laser lines in atomic species, in *Progress in Quantum Electronics*, Vol. 1(Part 5), Pergamon Press, Oxford, 1971.
580. **Willett, C. S.**, *Introduction to Gas Lasers: Population Inversion Mechanisms*, Pergamon Press, Oxford, 1974.
581. **Beck, R., Englisch, W., and Gurs, K.**, *Table of Laser Lines in Gases and Vapors*, 2nd ed., Springer-Verlag, Berlin, 1978.
582. **Coleman, C. D., Bozman, W. R., and Meggers, W. F.**, Table of Wave Numbers, Vols. 1 and 2, U.S. National Bureau Standards Monograph 3, 1960.

583. **McFarlane, R. A.,** Laser oscillation on visible and ultraviolet transitions of singly and multiply ionized oxgen, carbon and nitrogen, *Appl. Phys. Lett.*, 5, 91—93, 1964.
584. **Von Engel, A.,** *Ionized Gases*, Oxford University Press, London, 1975.
585. **Boguslovskii, A. A., Guryev, T. T., Didrikell, L. N., Novikova, V. A., Kune, V. V., and Stepanov, A. F.,** *Electronnaya Tekhn.*, Ser. 3, No. 1, 8, 1967.
586. **Beaulieu, A. J.,** Transversely excited atmospheric pressure CO_2 laser, *Appl. Phys. Lett.*, 16, 504—505, 1970.
587. **Dumanchin, R. and Rocca-Serra, J.,** High power density pulsed molecular laser, Paper 18-6, presented at the 6th Int. Quantum Electronics Conf., Kyoto, Japan, September 1970.
588. **Dumanchin, R., Michon, M., Farcy, J. C., Boudinet, G., and Rocca-Serra, J.,** Extension of TEA CO_2 laser capabilities, *IEEE J. Quantum Electron.*, QE-8, 163—165, 1972.
589. **Cobine, J. D.,** *Gaseous Conductors*, Dover, New York, 1958.
590. **Lamberton, H. M. and Pearson, P. R.,** Improved excitation techniques for atmospheric pressure CO_2 lasers, *Electron. Lett.*, 7, 141—142, 1971.
591. **Pearson, P. R. and Lamberton, H. M.,** Atmospheric pressure CO_2 lasers giving high output energy per unit volume, *IEEE J. Quantum Electron.*, QE-8, 145—149, 1972.
592. **Richardson, M. C., Alcock, A. J., Leopold, K., and Burtyn, P.,** A 300-J multigigawatt CO_2 laser, *IEEE J. Quantum Electron.*, QE-9, 236—243, 1973.
593. **Reid, R. D., McNeil, J. R., and Collins, G. J.,** New ion laser transitions in He-Au mixtures, *Appl. Phys. Lett.*, 29, 666—668, 1976.
594. **Borodin, V. S. and Kagan, Yu. M.,** Investigations of hollow-cathode discharges. I. Comparison of the electrical characteristics of a hollow cathode and a positive column, *Sov. Phys. Tech. Phys.*, 11, 131—134, 1966.
595. **Hodgson, R. T. and Dreyfus, R. W.,** Vacuum-uv laser action observed in H_2 Werner Bands: 1161-1240 Å, *Phys. Rev. Lett.*, 28, 536—539, 1972.
596. **Labuda, E. F.,** Ph.D. dissertation, Polytechnic Institute of Brooklyn, New York, 1967, (obtainable from University Microfilms, 313N First Street, Ann Arbor, Mich. 48108)
597. **Silfvast, W. T.,** Efficient CW laser oscillation at 4416 Å in Cd(II), *Appl. Phys. Lett.*, 15, 23—25, 1969.
598. **Goldsborough, J. P.,** Stable long life CW excitation of helium-cadmium lasers by dc cataphoresis, *Appl. Phys. Lett.*, 15, 159—161, 1969.
599. **Goldsborough, J. P.,** Continuous laser oscillation at 3250 Å in cadmium ion, *IEEE J. Quantum Electron.*, QE-5, 133, 1969.
600. **Gross, R. W. F. and Bott, J. F., Eds.,** *Handbook of Chemical Lasers*, John Wiley & Sons, New York, 1976.
601. **Thom, K. and Schneider, R. T.,** Nuclear pumped gas lasers, *AIAA J.*, 10, 400—406, 1972.
602. **Russell, G. R.,** Feasibility of a nuclear laser excited by fission fragments produced in a pulsed nuclear reactor, in Research on Uranium Plasmas, NASA SP-236, National Aeronautics and Space Administration, Langley Air Force Base, Va., 1971.
603. **DeShong, J. A.,** Nuclear Pumped Carbon Dioxide Gas Lasers — Model I Experiments, ANL-7310, Argonne National Lab, Argonne, Ill., 1966.
604. **Andriakin, V. M., Velikhov, E. P., Golubev, S. A., Krasil'nikov, S. S., Prokhorov, A. M., Pismennyi, V. D., and Rakhimov, A. T.,** Increase of CO_2 laser power under the influence of a beam of fast protons, *JETP Lett.*, 8, 214—216, 1968.
605. **Gancey, T., Verdeyen, J. T., and Miley, G. H.,** Enhancement of CO_2 laser power and efficiency by neutron irradiation, *Appl. Phys. Lett.*, 18, 568—569, 1971.
606. **Allario, F. and Schneider, R. T.,** Enhancement of Laser Output by Nuclear Reactions, NASA SP-236, National Aeronautics and Space Administration, Langley Air Force Base, Va., 1971.
607. **Rhoads, H. S. and Schneider, R. T.,** Nuclear enhancement of CO_2 laser output, *Trans. Am. Nucl.* Soc. 14, 429, 1971.
608. **DeYoung, R. J., Wells, W. E., and Miley, G. H.,** Enhancement of He-Ne lasers by nuclear radiation, *Trans. Am. Nucl. Soc.*, 19, 66, 1974.
609. **McArthur, D. A. and Tollefsrud, P. B.,** Observation of laser action in CO gas excited only by fission fragments, *Appl. Phys. Lett.*, 26, 187—190, 1975.
610. **Helmick, H. K., Fuller, J. L., and Schneider, R. T.,** Direct nuclear pumping of a helium-xenon laser, *Appl. Phys. Lett.*, 26, 327—328, 1975.
611. **Warner, B. E., Gerstenberger, D. C., Reid, R. D., McNeil, J. R., Solanki, R., Perrson, K. B., and Collins, G. J.,** 1 W operation of singly ionized silver and copper lasers, *IEEE J. Quantum Electron.*, QE-14, 568—570, 1978.
612. **Von Engel, A. and Steenbeck, N.,** in *Elektrische Gasentladungen*, Band II, Springer-Verlag, Berlin, 1932, 85.
613. **Dorgela, H. B., Alting, H., and Boers, J.,** Electron temperature in the positive column in mixtures of neon and argon or mercury, *Physica*, 2, 959—967, 1935.

614. **Massey, H. S. W. and Burhop, E. H. S.**, in *Electronic and Ionic Impact Phenomena,* Clarendon Press, Oxford, 1952.
615. **Condon, E. U. and Shortley, G. H.**, *The Theory of Atomic Spectra,* Cambridge University Press, New York, 1935, 304.
616. **Chen, C. H., Haberland, H., and Lee, Y. T.**, Interaction potential and reaction dynamics of He(2^1S, 2^3S) + Ne, Ar by the crossed molecular beam method, *J. Chem. Phys.,* 61, 3095—3103, 1974.
617. **Leasure, E. L. and Mueller, C. R.**, Crossed-molecular beams investigation of the excitation of ground-state neon atoms by 4.6-eV helium metastables, *J. Appl. Phys.,* 47, 1062—1064, 1976.
618. **Brion, C. E. and Olsen, L. A. R.**, Threshold electron impact excitation of the rare gases, *J. Phys. B,* 3, 1020—1033, 1970.
619. **Holstein, T.**, Imprisonment of resonance radiation in gases, *Phys. Rev.,* 72, 1212—1233, 1947.
620. **Holstein, T.**, Imprisonment of resonance radiation in gases. II, *Phys. Rev.,* 83, 1159—1168, 1951.
621. **McDaniel, E. W.**, in *Collision Phenomena in Ionized Gases,* John Wiley & Sons, New York, 1964, 584.
622. **Donohue, T. and Wiesenfeld, J. R.**, Relative yields of electronically excited iodine atoms, I ($5^2P_{1/2}$), in the photolysis of alkyl iodides, *Chem. Phys. Lett.,* 33, 176—180, 1975.
623. **Donohue, T. and Wiesenfeld, J. R.**, Photodissociation of alkyl iodides, *J. Chem. Phys.,* 63, 3130—3135, 1975.
624. **Ershov, L. S., Zalesskii, V. Yu., and Sokolov, V. N.**, Laser photolysis of perfluoroalkyl iodides, *Sov. J. Quantum Electron.,* 8, 494—501, 1978.
625. **Derwent, R. G. and Thrush, B. A.**, Excitation of iodine by singlet molecular oxygen, *Faraday Trans. II, Chem. Soc.,* 68, 720, 1972; *Faraday Disc. Chem. Soc.,* 53, 16—167, 1972.
626. **Pirkle, R. J., Wiesenfeld, J. R., Davis, C. C., Wolga, G. J., and McFarlane, R. A.**, Production of electronically excited iodine atoms, I ($^2P_{1/2}$) following injection of HI into a flow of discharged oxygen, *IEEE J. Quantum Electron.,* QE-11, 834—838, 1975.
627. **Beaty, E. C. and Patterson, P. L.**, Mobilities and reaction rates of ions in helium, *Phys. Rev.,* A137, 346—357, 1965.
628. **Deese, J. E. and Hassan, H. A.**, Analysis of nuclear induced plasmas, *AIAA J.,* 14, 1589—1597, 1976.
629. **DeYoung, R. J. and Winters, P. A.**, Power deposition in He from the volumetric ^{3}He(n,p)^{3}H reaction, *J. Appl. Phys.,* 48, 3600—3602, 1977.
630. **Wilson, J. W. and DeYoung, R. J.**, Power density in direct nuclear-pumped ^{3}He lasers, *J. Appl. Phys.,* 49, 980—988, 1978.
631. **Carter, B. D., Rowe, M. J., and Schneider, R. T.**, Nuclear-pumped CW lasing of the ^{3}He-Ne system, *Appl. Phys. Lett.,* 36, 115—117, 1980.
632. **Falcone, R. W. and Zdasiuk, G. A.**, Pair-absorption-pumped barium laser, *Opt. Lett.,* 5, 155—157, 1980.
633. **Ramsay, J. V. and Tanaka, K.**, Construction of single-mode dc operated He/Ne lasers, *Jpn. J. Appl. Phys.,* 5, 918—923, 1966.
634. **Sakurai, T.**, Dependence of He-Ne laser output power on discharge current, gas pressure and tube radius, *Jpn. J. Appl. Phys.,* 11, 1826—1831, 1972.
635. **Brunet, H.**, Laser gain measurements in a xenon-krypton discharge, *Appl. Opt.,* 4, 1354, 1965.
636. **Mash, D. I., Papulovskii, V. F., and Chirina, L. P.**, On the operation of a xenon-krypton laser, *Opt. Spectrosc.,* 17, 431—432, 1964.
637. **Horrigan, F. A.**, Estimated Lifetimes in Neon I and Xenon I, unpublished report on contract No. AF33(615)-1949, Raytheon Company, Waltham, Mass., 1966.
638. **Allen, L., Jones, D. G. C., and Schofield, D. G.**, Radiative lifetimes and collisional cross sections for XeI and II, *J. Opt. Soc. Am.,* 59, 842—847, 1969.
639. **Allen, L. and Peters, G. I.**, Superradiance, coherence brightening and amplified spontaneous emission, *Phys. Lett.,* 31A, 95—96, 1970.
640. **Peters, G. I. and Allen, L.**, Amplified spontaneous emission. I. The threshold condition, *J. Phys. A,* 4, 238—243, 1971.
641. **Allen, L. and Peters, G. I.**, Amplified spontaneous emission. II. The connection with laser theory, *J. Phys. A,* 4, 377—381, 1971.
642. **Allen, L. and Peters, G. I.**, Amplified spontaneous emission. III. Intensity and saturation, *J. Phys. A,* 4, 564—573, 1971.
643. **Peters, G. I. and Allen, L.**, Amplified spontaneous emission. IV. Beam divergence and spatial coherence, *J. Phys. A,* 5, 546—554, 1972.
644. **Willett, C. S., Gleason, T. J., and Kruger, J. S.**, unpublished work, 1970.
645. **Shelton, R. A. J.**, Vapour pressures of the solid copper (I) halides, *Trans. Faraday. Soc.,* 57, 2113—2118, 1961.
646. **Tobin, R. C.**, Rapid differential decay of metastable populations in a copper halide laser, *Opt. Commun.,* 32, 325—330, 1980.

647. **Rothe, D. E. and Tan, K. O.,** High power N_2^+ laser pumped by change transfer in a high-pressure pulsed glow discharge, *Appl. Phys. Lett.,* 30, 152—154, 1977.
648. **Olson, R. A., Grosjean, D., Sarka, B., Jr., Garscadden, A., and Bletzinger, P.,** High-repetition-rate closed-cycle rare gas electrical discharge laser, *Rev. Sci. Instrum.,* 47, 677—683, 1976.
649. **Hasle, E. K.,** Polarization properties of He-Ne lasers, *Opt. Commun.,* 31, 206—210, 1979.
650. **Shore, B. W. and Menzel, D. H.,** *Principles of Atomic Spectra,* John Wiley & Sons, New York, 1968.
651. **Ershov, L. S., Zalesskii, V. Yu., and Sokolov, V. N.,** Laser photolysis of perfluoroalkyl iodides, *Sov. J. Quantum Electron.,* 8, 494—501, 1978.
652. **Majer, J. R. and Simons, J. P.,** Photochemical process in halogenated compounds, *Adv. Photochem.,* 2, 137—182, 1964.
653. **Riley, S. J. and Wilson, K. R.,** Excited fragments from excited molecules: energy partitioning in the photodissociation of alkyl iodides, *Faraday Disc. Chem. Soc.,* 53, 132—146, 1972.
654. **Harris, G. M. and Willard, J. E.,** Photochemical reactions in the system methyl iodide-iodine-methane: the reaction $C^{14}H_3 + CH_4 - C^{14}H_4 + CH_3$, *J. Am. Chem. Soc.,* 76, 4678—4687, 1954.
655. **Jaseja, T. S., Javan, A., and Townes, C. H.,** Frequency stability of He-Ne masers and measurements of length, *Phys. Rev. Lett.,* 10, 165—167, 1963.
656. **Jaseja, T. S., Javan, A., Murray, J., and Townes, C. H.,** Test of special relativity or of the isotropy of space by use of infrared masers, *Phys. Rev. A,* 133, 1221—1225, 1964.
657. **Hochuli, U. E. and Haldemann, P.,** private communication.
658. **McFarlane, R. A., Bennett, W. R., Jr., and Lamb, W. E., Jr.,** Single mode tuning dip in the power output of an He-Ne optical maser, *Appl. Phys. Lett.,* 2, 189—190, 1963.
659. **Tobias, I., Skolnick, M., Wallace, R. A., and Polanyi, T. G.,** Deviation of a frequency-sensitive signal from a gas laser in an axial magnetic field, *Appl. Phys. Lett.,* 6, 198—201, 1965.
660. **Wallard, A. J.,** The frequency stabilization of gas lasers, *J. Phys. E,* 6, 793—807, 1973.
661. **Birnbaum, G.,** Frequency stabilization of gas lasers, *Proc. IEEE,* 55, 1015—1026, 1967.
662. **Laures, P.,** Stabilisation de la frequence des lasers a gaz, *Onde Electr.,* No. 469, 1966.
663. **White, A. D.,** Frequency stabilization of gas lasers, *IEEE J. Quantum Electron.,* QE-1, 349—357, 1965.
664. **Hochuli, U. E., Haldemann, P., and Li, H. A.,** Factors influencing the relative frequency stability of He-Ne laser structures, *Rev. Sci. Instrum.,* 45, 1378—1381, 1974.
665. **Hanes, G. R. and Dahlstrom, C. E.,** Iodine hyperfine structure observed in saturated absorption at 633 nm, *Appl. Phys. Lett.,* 14, 362—364, 1969.
666. *P. V. Seances Com. Int. Poids Mes.,* 28, 70, 1960.
667. **VanOorschot, B. D. J. and VanderHoeven, C. J.,** A recently developed iodine-stabilized laser, *J. Phys. E,* 12, 51—55, 1979.
668. **Schweitzer, W. G., Jr., Kessler, E. G, Jr., Deslattes, R. D., Layer, H. P., and Whetstone, J. R.,** Description, performance and wavelengths of iodine stabilized lasers, *Appl. Opt.,* 12, 2927—2938, 1973.
669. **Cérez, P., Brillet, A., and Hartmann, F.,** Metrological properties of the R(127) line of iodine studied by laser saturated absorption, *IEEE Trans. Inst. Meth.,* IM-23, 526—528, 1974.
670. **Helmcke, J. and Bayer-Helms, F.,** He-Ne laser stabilized by saturated absorption in I_2, *IEEE Trans. Inst. Meth.,* IM-23, 529—531, 1974.
671. **Wallard, A. J.,** The reproducibility of 633 nm lasers stabilized by $^{127}I_2$, *IEEE Trans. Inst. Meth.,* IM-23, 52—535, 1974.
672. **Cole, J. B. and Bruce, C. F.,** Iodine stabilized laser with three internal mirrors, *Appl. Opt.,* 14, 1303—1310, 1975.
673. **Cerez, P., Brillet, A., Hajdukovic, S., and Man, N.,** Iodine stabilized He-Ne laser with a hot wall iodine cell, *Opt. Commun.,* 21, 332—336, 1977.
674. **Melnikov, N. A., Privalov, V. E., and Fofanov, Ya. A.,** Experimental investigation of He-Ne lasers stabilized by saturation absorption in iodine, *Opt. Spectrosc.,* 42, 425—428, 1977.
675. **Layer, H. P.,** The iodine stabilized laser as a realization of the length unit, *Proc. Soc. Photo Opt. Instrum. Eng.,* 129, 9—11, 1977.
676. **Chartier, J. M., Helmcke, J., and Wallard, A. J.,** International intercomparison of the wavelength of iodine-stabilized lasers, *IEEE Trans. Inst. Meth.,* IM-25, 450—453, 1976.
678. **Hall, J. C.,** Saturated absorption spectroscopy with applications to the 3.39 μm methane transition, in Atomic Physics 3 (Proc. 3rd Int. Conf. on Atomic Physis, Boulder, Colo., 1972,) Smith, S. J., Walters, G. K., and Volsky, L. H., Eds., Plenum Press, New York, 1973, 615.
678. **Bagaev, S. N., Baklanov, E. V., and Chebotaev, V. O.,** Anomalous decrease of the shift of the center of the Lamb dip in low-pressure molecular gases, *JETP Lett.,* 16, 243—246, 1972.
679. **Koshelyaevskii, N. B., Tatarenkov, V. M., and Titov, A. N.,** Power shift of the frequency of an He-Ne-CH_4 laser, *Sov. J. Quantum Electron.,* 6, 222—226, 1976.

680. **Jolliffe, B. W., Kramer, G., and Chartier, J. M.**, Methane-Stabilized He-Ne laser intercomparisons 1976, *IEEE Trans. Inst. Meth.*, IM-25, 447—450, 1976.

681. **Alekseev, V. A. and Malyugin, A. V.**, Influence of the hyperfine structure on the frequency reproducibility of an He-Ne laser with a methane absorption cell, *Sov. J. Quantum Electron.*, 7, 1075—1081, 1977.

682. **Nakazawa, M., Musha, T., and Tako, T.**, Frequency-stabilized 3.39 μm He-Ne laser with no frequency modulation, *J. Appl. Phys.*, 50, 2544—2547, 1979.

683. **Evenson, K. M., Day, G. W., Wells, J. S., and Mullen, L. O.**, Extension of absolute frequency measurements to the CW He-Ne laser at 88 THz(3.39 μ), *Appl. Phys. Lett.*, 20, 133—134, 1972.

684. **Evenson, K. M., Wells, J. S., Petersen, F. R., Danielson, B. L., and Day, G. W.**, Absolute frequencies of molecular transitions used in laser stabilization: the 3.39 μm transition in CH_4 and the 9.33- and 10.18-μm transitions in CO_2, *Appl. Phys. Lett.*, 22, 192—195, 1973.

685. **Barger, R. L. and Hall, J. L.**, Wavelength of the 3.39 μm laser-saturated absorption line of methane, *Appl. Phys. Lett.*, 22, 196—199, 1973.

686. **Giacomo, P.**, *Atomic Masses and Fundamental Constants*, Vol. 4, Plenum Press, New York, 1972, 378.

687. **Jennings, D. A., Petersen, F. R., and Evenson, K. M.**, Frequency measurements of the 260-THz(1.15 μm) He-Ne laser, *Opt. Lett.*, 4, 129—130, 1979.

688. **Baird, K. M., Evenson, K. M., Hanes, G. R., Jennings, D. A., and Peterson, F. R.**, Extension of absolute-frequency measurements to the visible: frequencies of ten hyperfine components of iodine, *Opt. Lett.*, 4, 263—264, 1979.

689. **Jennings, D. A., Petersen, F. R., and Evenson, K. M.**, Extension of absolute frequency measurements to 148THz: frequencies of the 2.0- and 3.5-μ Xe laser, *Appl. Phys. Lett.*, 510—511, 1975.

690. **Baird, K. M., Smith, D. S., and Whitford, B. G.**, Confirmation of the currently accepted value 299792458 metres per second for the speed of light, *Opt. Commun.*, 31, 367—368, 1979.

691. **Evenson, K. M., Wells, J. S., Petersen, F. R., Danielson, B. L., Day, G. W., Barger, R. C., and Hall, J. C.**, Speed of light from direct frequency and wavelength measurements of the methane-stabilized laser, *Phys. Rev. Lett.*, 29, 1346—1349, 1972.

692. **Knight, T. G., Rowley, W. R. C., Shotton, K. C., and Woods, P. T.**, Measurement of the speed of light, *Nature (London)*, 251, 46, 1974.

693. **Blaney, T. G., Bradley, C. C., Edwards, G. J., Knight, D. J. E., Woods, P. T., and Jolliffe, B. W.**, Absolute frequency measurement of the R(12) transition of CO_2 at 9.3 μm, *Nature (London)*, 244, 504, 1973.

694. **Jolliffe, B. W., Rowley, W. R. C., Shotton, K. C., Wallard, A. J., and Woods, P. T.**, Accurate wavelength measurement on up-converted CO_2 laser radiation, *Nature (London)*, 251, 46—47, 1974.

695. **Woods, P. T., Shotton, K. C., and Rowley, W. R. C.**, Frequency determination of visible laser light by interferometric comparison, *Appl. Opt.*, 17, 1048—1054, 1978.

696. **Markova, S. V., Petrash, G. G., Chere zov, V. M.**, Investigation of the stimulated emission mechanism in a pulsed bismuth vapor laser, *Sov. J. Quantum Electron.*, 9, 707—711, 1979.

697. **Fill, E. E., Thieme, W. H., and Volk, R.**, A tunable iodine laser, *J. Phys. D*, 12, L41—L45, 1979.

698. **Mott, N. F. and Massey, H. S. W.**, *The Theory of Atomic Collisions*, 3rd ed., Oxford University Press, New York, 1965.

699. **Silfvast, W. T., Szeto, L. H., and Woods, O. R.**, II, Recombination lasers in expanding CO_2 laser-produced plasmas of argon, krypton and xenon, *Appl. Phys. Lett.*, 31, 334—337, 1977.

700. **Silfvast, W. T., Szeto, L. H., and Wood, O. R.**, II, Ultra-high-gain laser-produced plasma laser in xenon using periodic pumping, *Appl. Phys. Lett.*, 34, 213—215, 1979.

701. **Silfvast, W. T., Szeto, L. H., and Wood, O. R.**, II, Power output enhancement of a laser-produced Cd plasma recombination laser by plasma confinement, *Appl. Phys. Lett.*, 36, 500—502, 1980.

Section 2
Ionized Gas Lasers

2. IONIZED GAS LASERS

William B. Bridges

INTRODUCTION TO THE TABLES

The tables in this section include all known laser lines originating from transitions between energy levels of the ionized states of atoms in gas discharges. Figure 2.1 indicates by the shaded squares on the periodic chart those elements which produce ion laser lines. Nor surprisingly, most of the elements are those which occur as gases at room temperature or are easily vaporized. Some of the more refractory elements also exhibit ion laser oscillation when sputtered in a hollow-cathode discharge.

The tables following Figure 2.1 list the observed wavelengths with identifications and references.* The tables are arranged in the order in which they occur as columns in the periodic system, reading from left to right, and then within each column, in the order of increasing atomic number. An explanation of the entries and abbreviations used in each column is given below.

Wavelength

This column lists the most accurate available wavelength for the given transition, in micrometers. In most cases, this value of wavelength has been derived from energy levels given in the spectroscopic literature, assuming the indicated classification for the line. The literature sources are given in the notes to the tables; if no source is listed, the wavelength has been calculated from the energy levels listed by Moore in "Atomic Energy Levels", National Bureau of Standards Circular 467,[241] but hereafter referred to as AEL in the text. The wavelength in air has been calculated from the vacuum energy levels using the Edlen formula:

$$n(v) = 1 + 6432.8 \times 10^{-8} + \frac{2949810}{146 \times 10^{8} - v^{2}} + \frac{25540}{41 \times 10^{8} - v^{2}}$$

where

$$\lambda[\text{microns, air}] = 10^{4}/n(v)v[\text{cm}^{-1}\text{, vacuum}]$$

The calculations were made with a programmable hand calculator and are accurate to $\pm 10^{-6}$ μm. For levels known only to the nearest 1 cm^{-1} (or so), the wavelength observed in spontaneous emission is doubtless more accurate and is listed as given by Harrison,[141] or some other spectroscopic source.

Abbreviations	
*	Strong or characteristic laser line in pure gas
?	Existence of this laser line may be in doubt
v	Vacuum wavelength
[H]	Wavelength from Harrison[141]
[#]	Wavelength from reference cited

Measured Value

This column lists the most accurate measured wavelength of the actual laser output, in micrometers, along with the estimated error in the final digits as cited in the references, even though some authors seem overly optimistic in their claimed accuracy.

Identification

This column lists the emission spectrum (II for single ionized, III for doubly ionized, and so on), the upper, and then the lower level of the atomic transition believed to be

* Tables begin on p. 198.

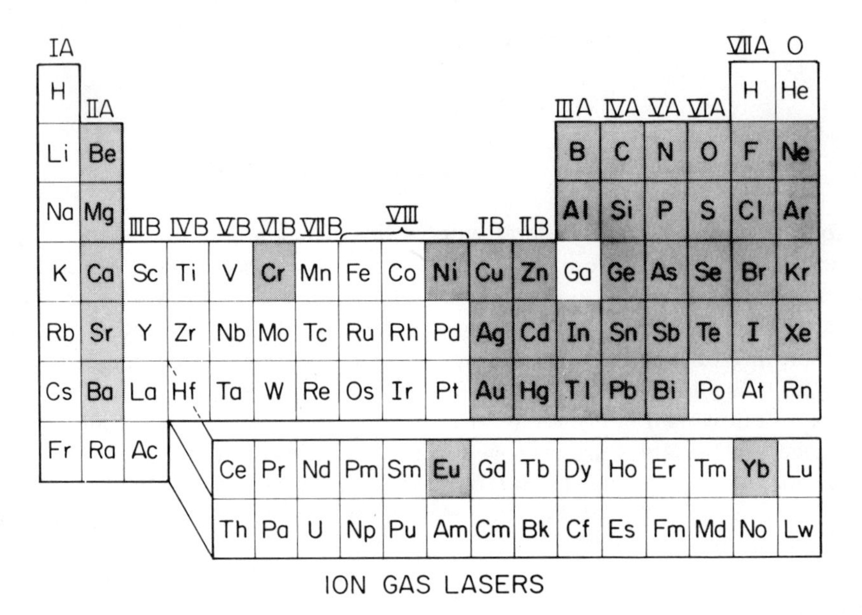

FIGURE 2.1. Periodic table of the elements. Shaded blocks indicate those elements whose ions have exhibited laser oscillation in gaseous or vapor form.

responsible for the laser emission. The level notation follows AEL, with the core configuration being given in parentheses. In cases where the configuration is completely unknown, the level energy rounded to the nearest centimeter^{-1} is given, with the J-value as a subscript. In cases where the level configuration is partly known, either the partial designation in AEL or the rounded energy designation is given. Identifications of the observed laser lines have generally been made only on the basis of the measured value of wavelength and theoretical plausibility. This can be done with some degree of confidence in the absence of other nearby spectral lines, but occasionally two equally plausible lines fall within the measurement error. In these cases, both identifications are given. Laser transitions can often be assigned to an ionization state on the basis of their behavior with discharge current relative to lines of known ionization state, even if they cannot be assigned to particular energy levels.

Abbreviations

[ ]	Classification by reference cited
?	Classification or ionization state uncertain
. . .	Classification unknown

References

The first reference listed is the first report of laser action on that line, although it may not represent the most accurate measurement of the laser wavelength. No attempt has been made to list all references in the case of lines which have received much attention; only references giving added data on gain or power or those which new excitation techniques or other new features are listed. All references are given if the line was observed by only a few investigators.

Abbreviations

**	See more complete discussion following the tables
CW	Continuous oscillation reported
[ ]	Reference cited changes previous data
A	Laser oscillation observed only in the afterglow of the discharge pulse

D	Ion obtained by dissociation of molecular compound
E	Error in classification or wavelength
G	Gain measured
H	Hollow-cathode excitation
I	Identification of line given or discussed
P	Power output reported
S	Superradiant operation reported
U	Unique or unusual excitation method
X	Existence of line discussed
hfs	Hyperfine structure investigated
λ	Accurate wavelength measurement
(M^+)	Excited by charge-exchange collisions with ground-state ion of M
(M^*)	Excited by Penning collisions with metastable
(M)	Collisions with M required for laser oscillation

ENERGY LEVEL DIAGRAMS

Figures 2.2 through 2.15 are energy level diagrams for selected laser ions. The arrangement of the levels differs from diagram to diagram according to the complexity of the coupling scheme appropriate to the particular ion. Not all levels are shown, and only laser transitions are included, rather than all levels and all observed spectral lines as in conventional Grotrian diagrams.

Individual levels and laser transitions are not shown (Figures 2.2 to 2.4) for Cu II, Ag II, and Au II. However, the range of energies over which particular groups of levels occur is indicated by the vertical extent of the boxes. The number of laser transitions observed to date is given by the number beside each arrow. Both Penning and thermal charge-exchange collisional excitation processes are known to be responsible for various laser lines in Cu, Ag, Au, Zn, Cd, and Hg; accordingly, the energies of the ions and metastables in He and Ne are indicated with respect to the neutral atom ground state on Figures 2.2 through 2.7.

Figures 2.8 through 2.15 for the noble gas ions are somewhat more complicated. Each known atomic energy state (up to the maximum energy indicated) appears as a horizontal line, and these states are classified in columns according to the orbital angular momentum, ℓ, of the excited electron (ℓ = 0,1,2 for s,p,d states). The configuration designations are given in the L-S (Russell-Saunders) coupling scheme, as listed in AEL, except as noted in the captions. Core configurations corresponding to unprimed and primed state designations follow AEL and are explicitly given at the bottom of the figures. Higher orbital states (f,g, etc.) are not shown, since very few laser lines are known to involve any of these levels. In addition, levels not yet assigned to the s,p, or d systems are omitted, even though some laser lines involve these levels. The total angular quantum number J of each state appears to the right of the horizontal line denoting that state.

This arrangement of states in columns automatically separates states of even and odd parity, the latter being denoted °. Note that the only selection rules satisfied by all the laser lines are those of parity change and orbital momentum change ($\Delta\ell = \pm 1$); L-S coupling is badly violated in many of these atoms, and changes in core configuration or total spin S are relatively common for laser transitions.

Vertical lines and dots are used to connect states having a common state designation, differing only in their total angular momentum, J.

The vertical scale represents energy of excitation above the ground state for the given ionic species. The left-hand axis is marked in thousands of centimeter^{-1}, and corresponding tic marks are indicated on the right-hand axis. In addition, the right-hand axis indicates energy in electron volts (8068 cm^{-1} = 1 eV). The energy placement of the atomic states is that given in AEL, with a few exceptions. The designation "I.P."

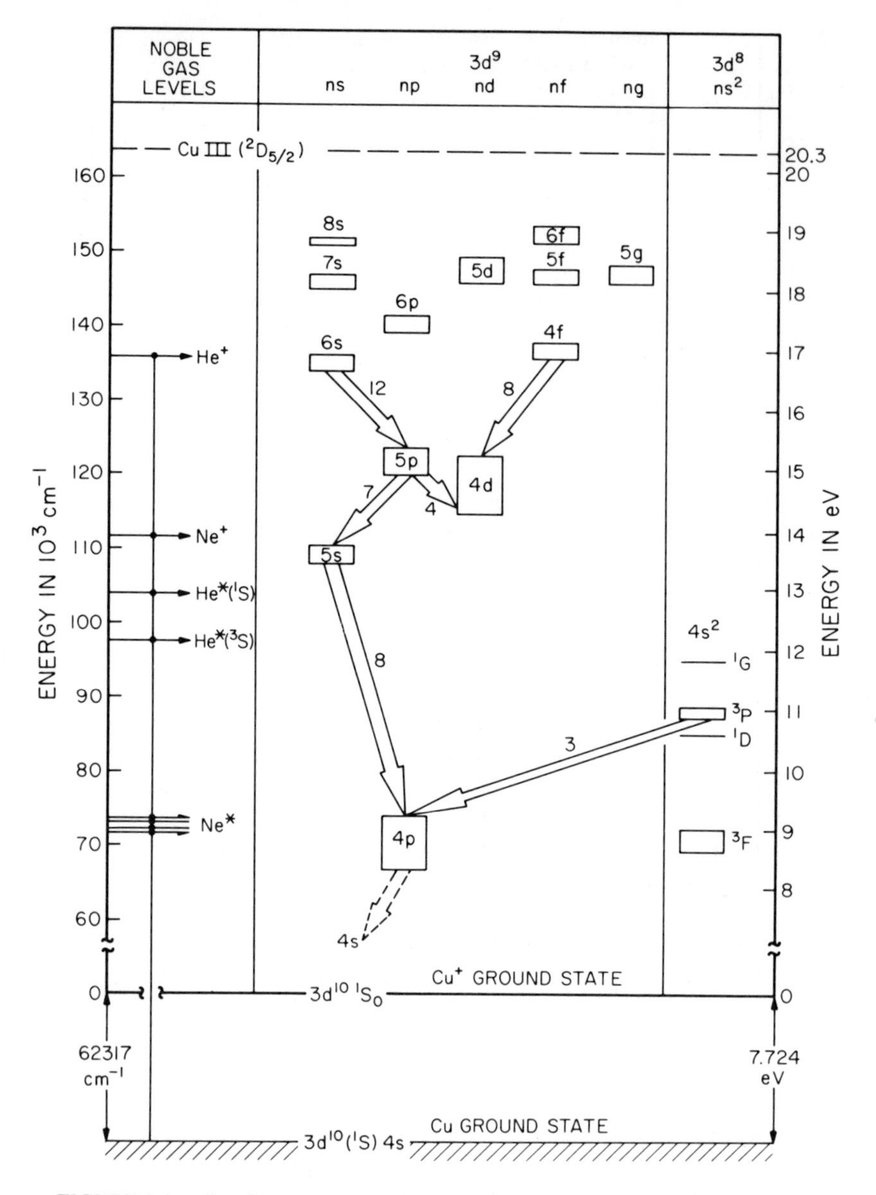

FIGURE 2.2. Simplified energy level diagram for singly-ionized copper (Cu II), showing the number of laser lines in each supermultiplet. The energies of the He and Ne metastables and ground-state ions are also shown with respect to the ground state of the neutral copper atom. (From Bridges, W. B., *Methods of Experimental Physics,* Vol. 15A, Tang, C. L., Ed., Academic Press, New York, 1979. With permission.)

in the upper corners of the figure gives the ionization potential of the particular ion considered, in centimeter^{-1} and in electronvolts.

The diagonal lines drawn on the figures represent known ion laser lines. Their wavelengths (in air) are indicated in micrometers with four-place accuracy. A dotted line means that the indicated transition is one of two that could be responsible for the observed laser line. For Ar II and Kr II, the laser lines ending on the s^2P energy states are listed in two separate columns, keyed to the corresponding upper energy levels; the transitions in these cases were so numerous that diagonal lines would have been difficult to distinguish.

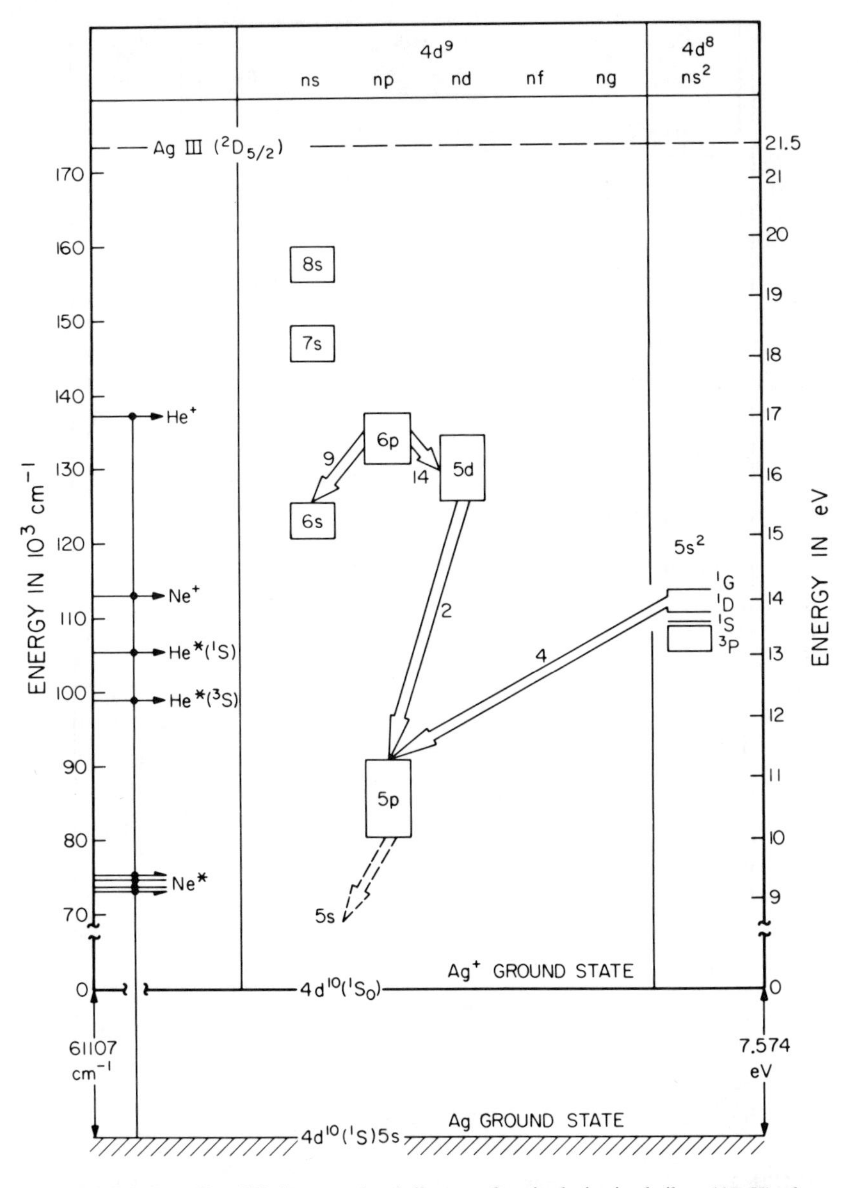

FIGURE 2.3. Simplified energy level diagram for singly-ionized silver (Ag II), showing the number of laser lines in each supermultiplet. The energies of the He and Ne metastables and ground-state ions are also shown with respect to the ground state of the neutral Ag atom.

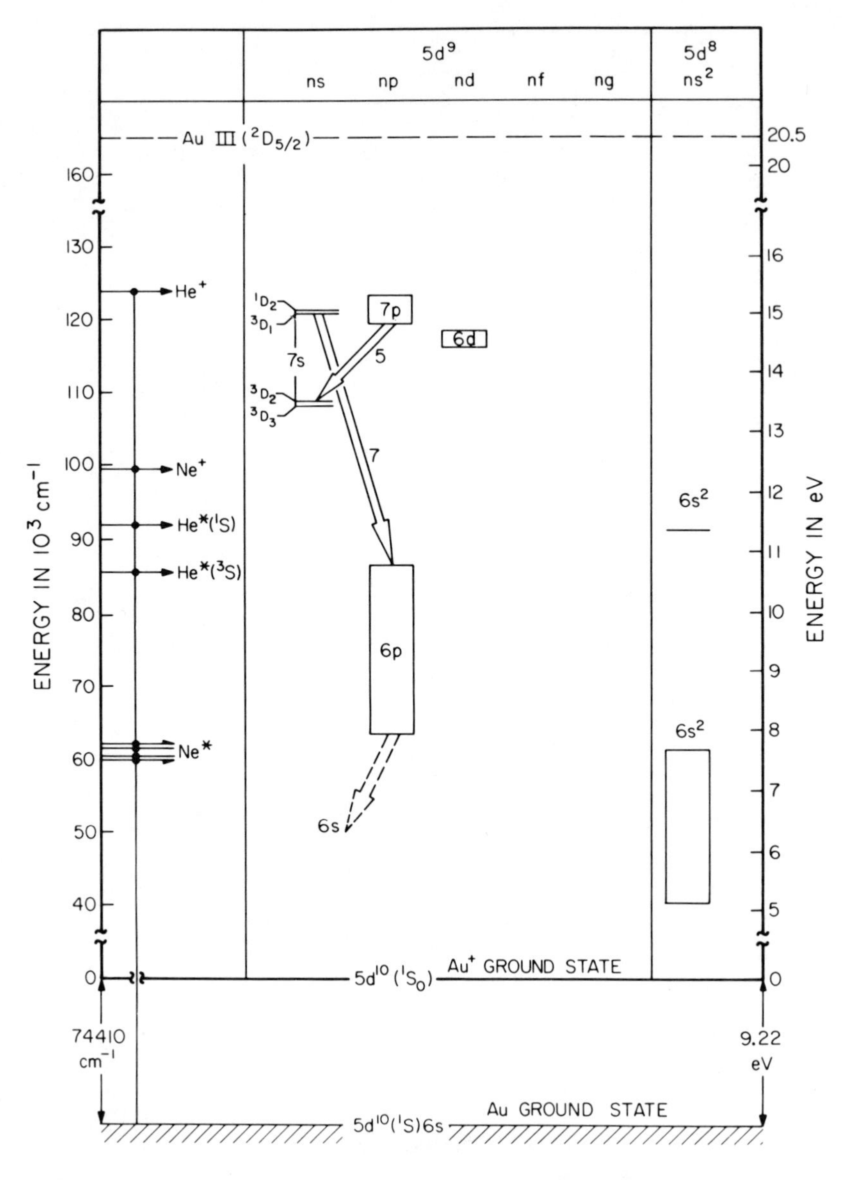

FIGURE 2.4. Simplified energy level diagram for singly-ionized gold (Au II), showing the number of laser lines in each supermultiplet. The energies of the He and Ne metastables and ground-state ions are also shown with respect to the ground state of the neutral Au atom.

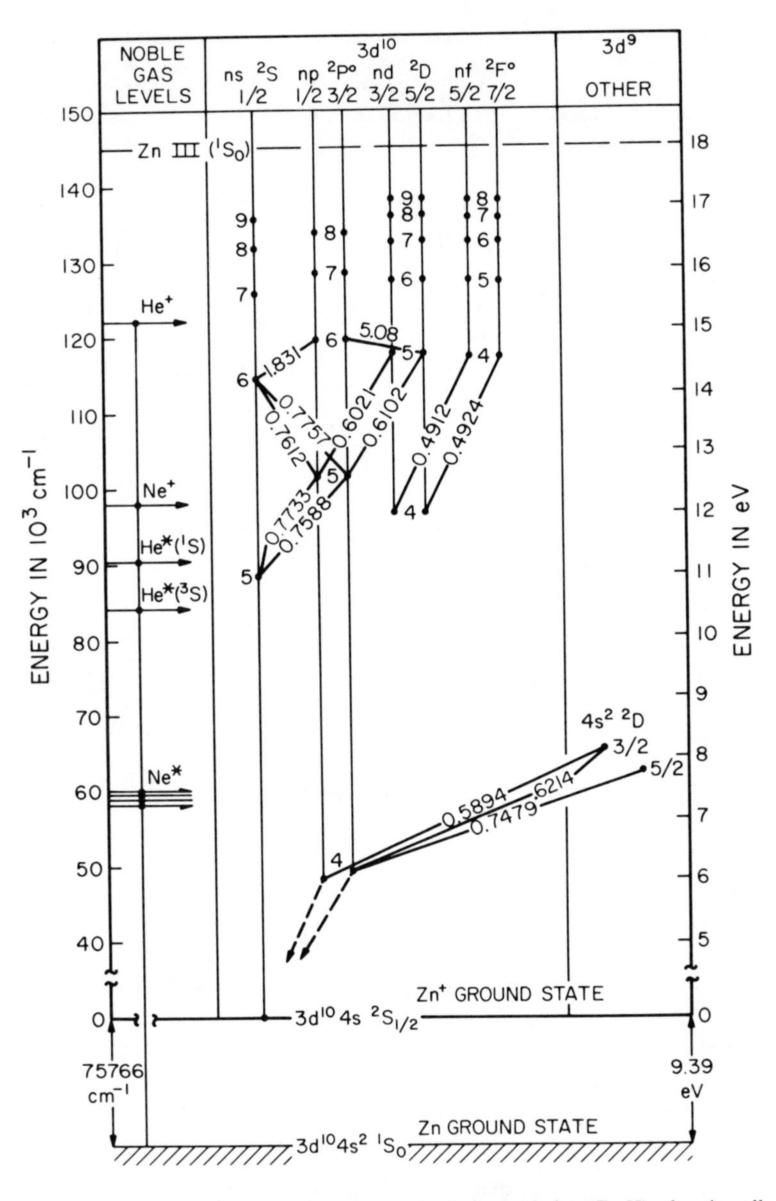

FIGURE 2.5. Energy level diagram for singly-ionized zinc (Zn II), showing all the Zn II laser lines observed to date. The energies of the He and Ne metastables and ground-state ions are also shown with respect to the ground state of the neutral Zn atom. (With revisions from Bridges, W. B., *Methods of Experimental Physics,* Vol. 15A, Tang, C. L., Ed., Academic Press, New York, 1979. With permission.)

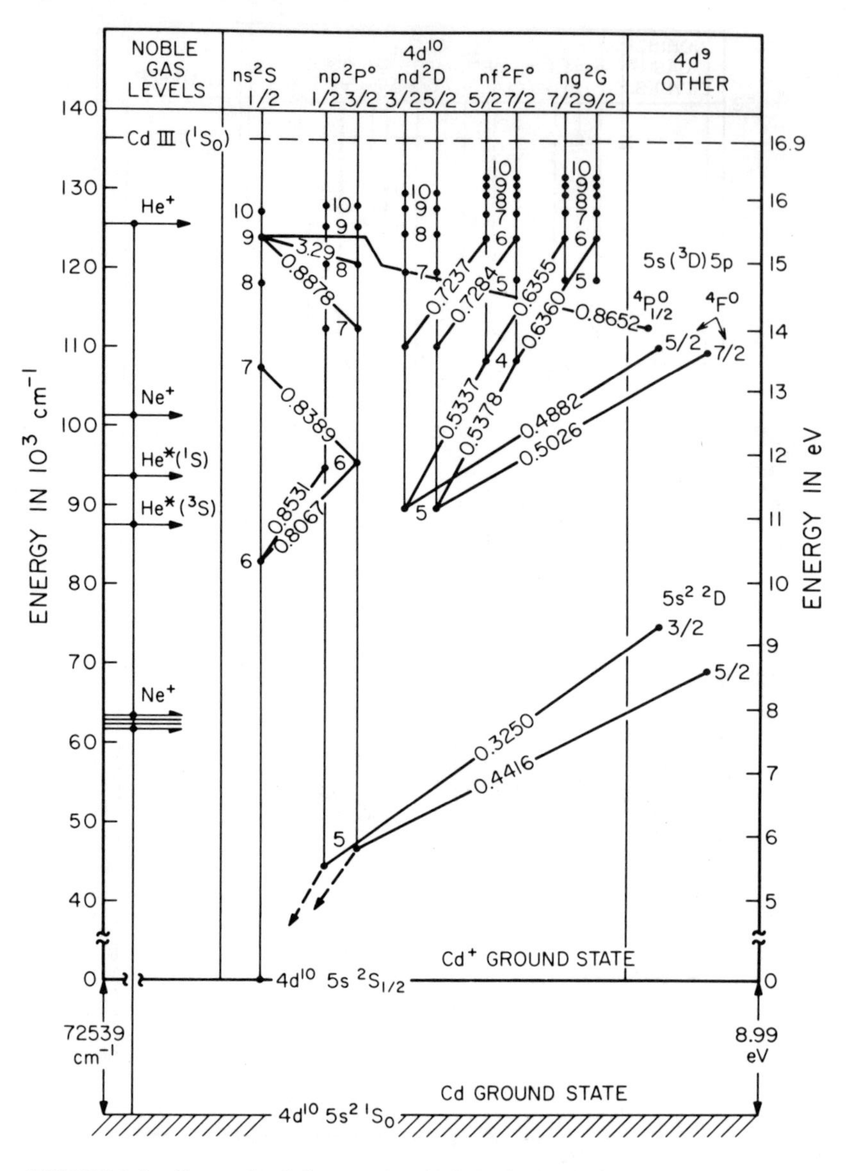

FIGURE 2.6. Energy level diagram for singly-ionized cadmium (Cd II), showing all the Cd II laser lines observed to date. The energies of the He and Ne metastables and ground-state ions are also shown with respect to the ground state of the neutral Cd atom. (With revisions from Bridges, W. B., *Methods of Experimental Physics*, Vol. 15A, Tang, C. L., Ed., Academic Press, New York, 1979. With permission.)

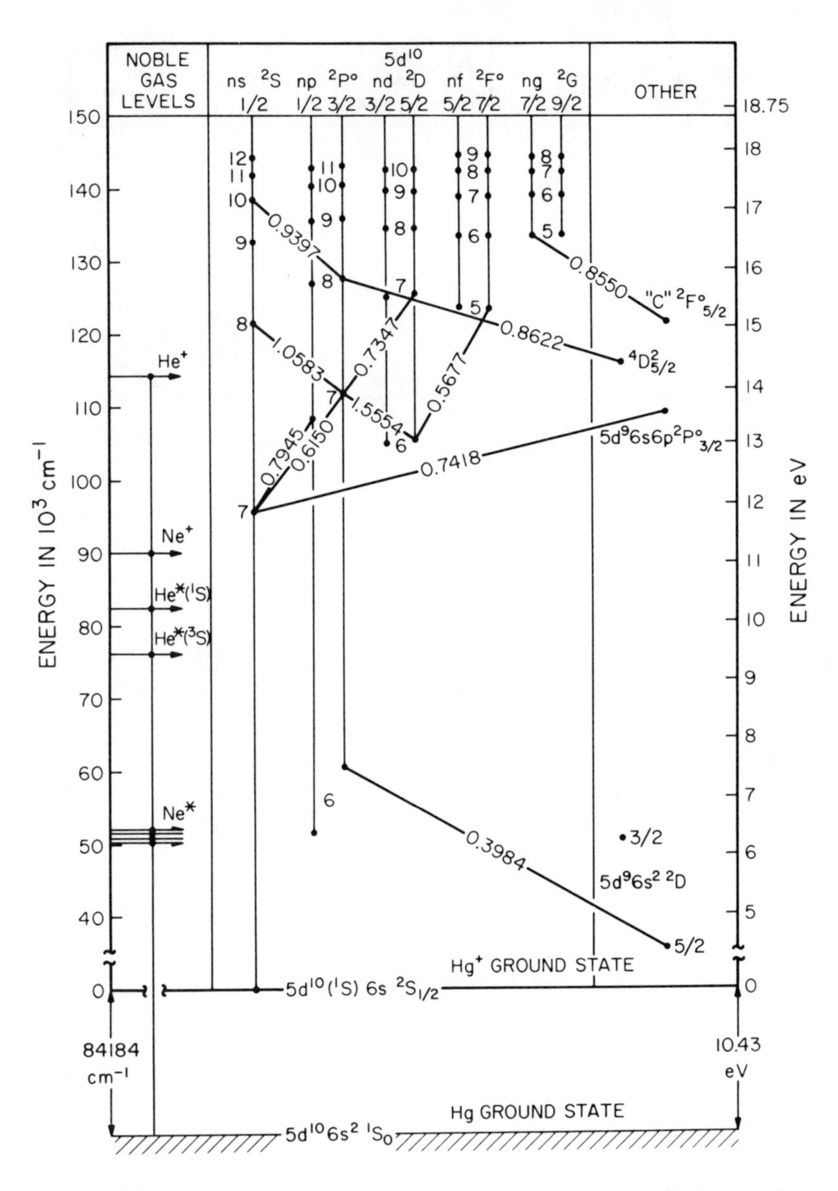

FIGURE 2.7. Energy level diagram for singly-ionized mercury (Hg II), showing all of the Hg II laser lines observed to date. The energies of the He and Ne metastables and ground-state ions are also shown with respect to the ground state of the neutral Hg atom. (With revisions from Bridges, W. B., *Methods of Experimental Physics*, Vol. 15A, Tang, C. L., Ed., Academic Press, New York, 1979. With permission.)

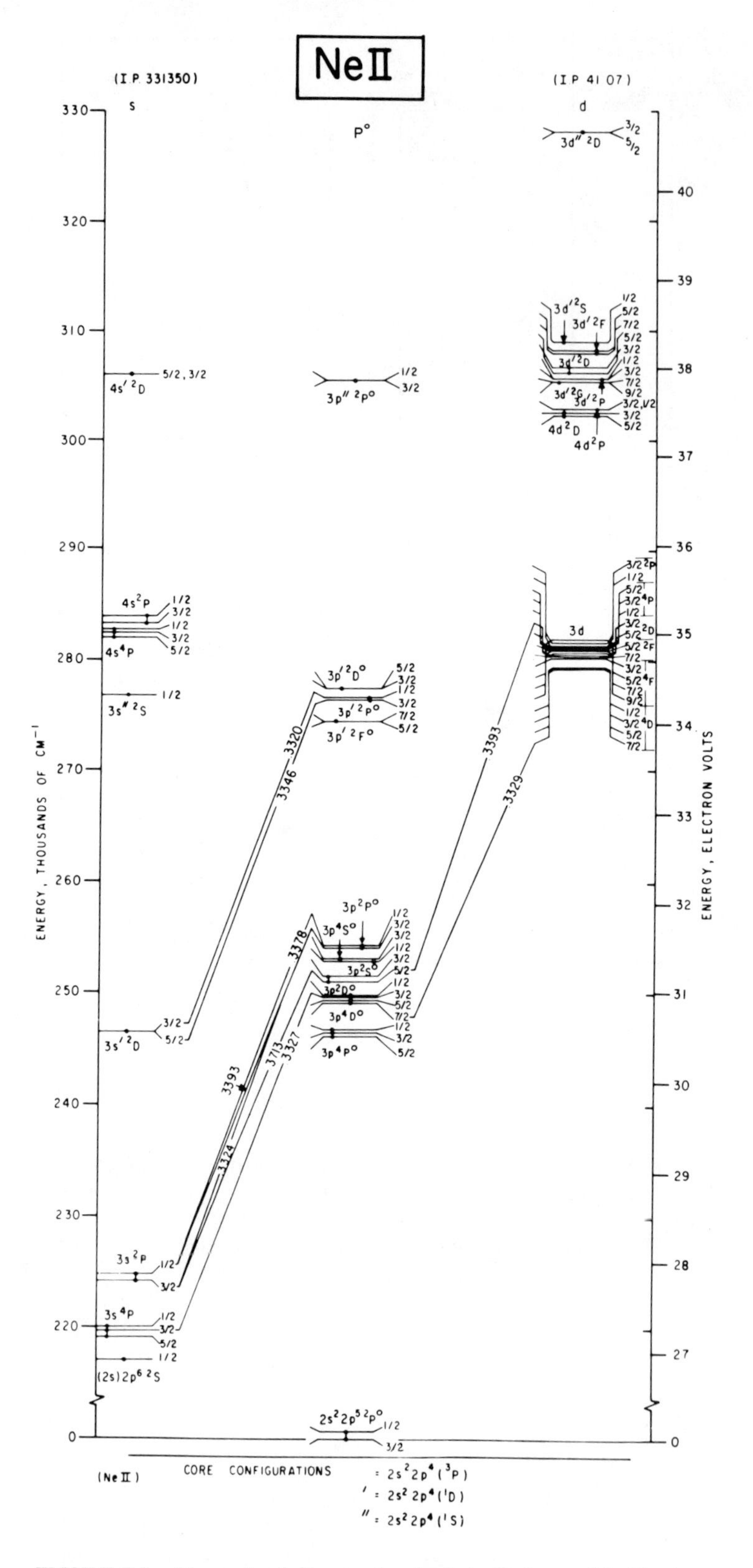

FIGURE 2.8. Energy level diagram for singly-ionized neon (Ne II), showing all classified Ne II laser lines. Energy levels taken from AEL.[241]

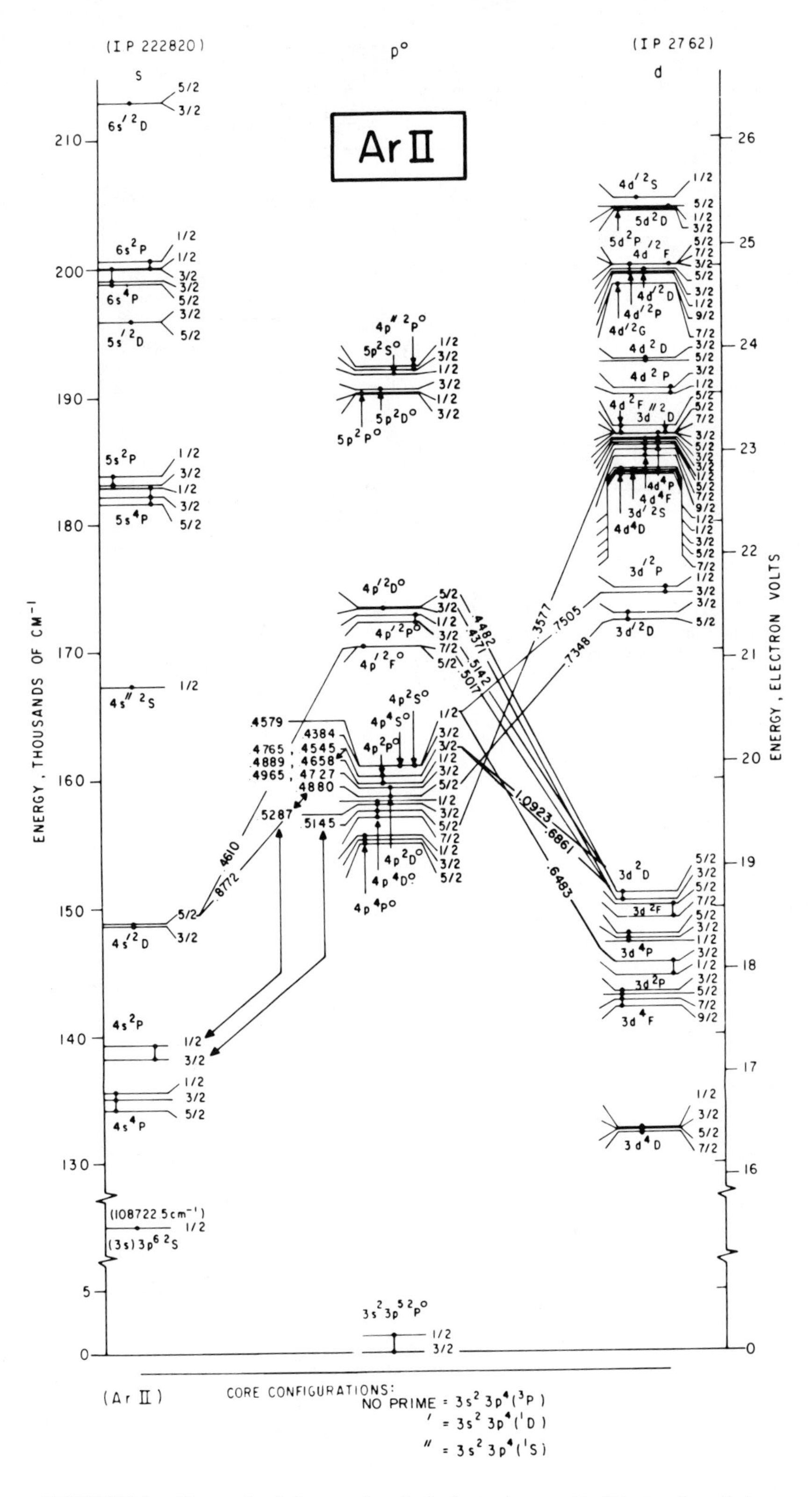

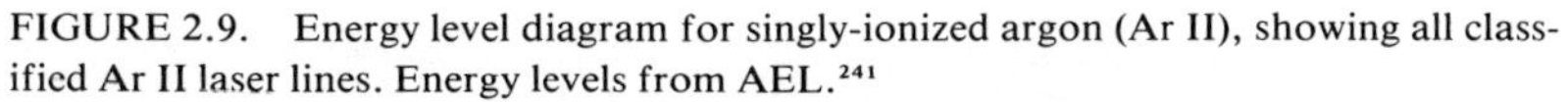

FIGURE 2.9. Energy level diagram for singly-ionized argon (Ar II), showing all classified Ar II laser lines. Energy levels from AEL.[241]

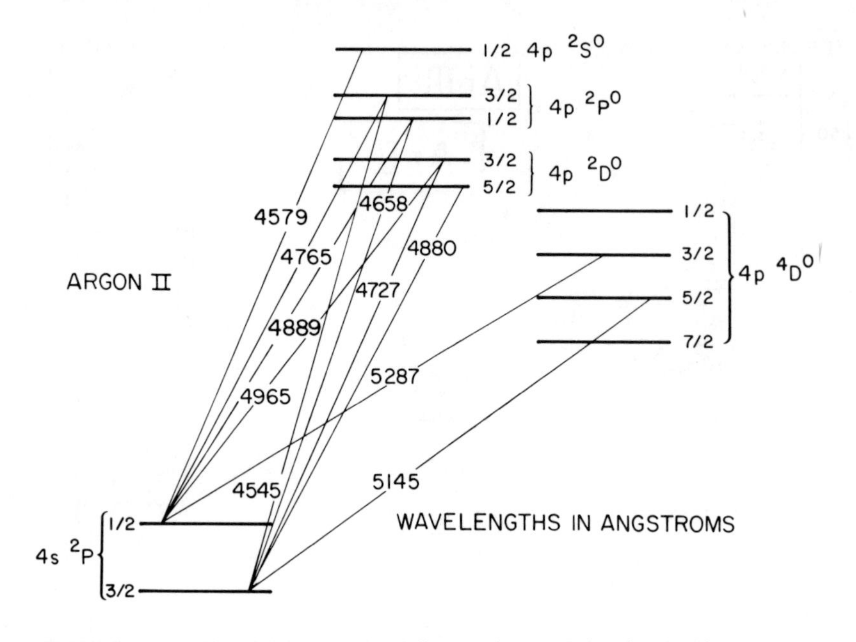

FIGURE 2.10. Simplified energy level diagram for Ar II showing the blue-green laser lines in the 4p → 4s supermultiplet. (From Bridges, W. B., *Appl. Phys. Lett.*, 4, 128—130, 1964. By courtesy of the American Institute of Physics.)

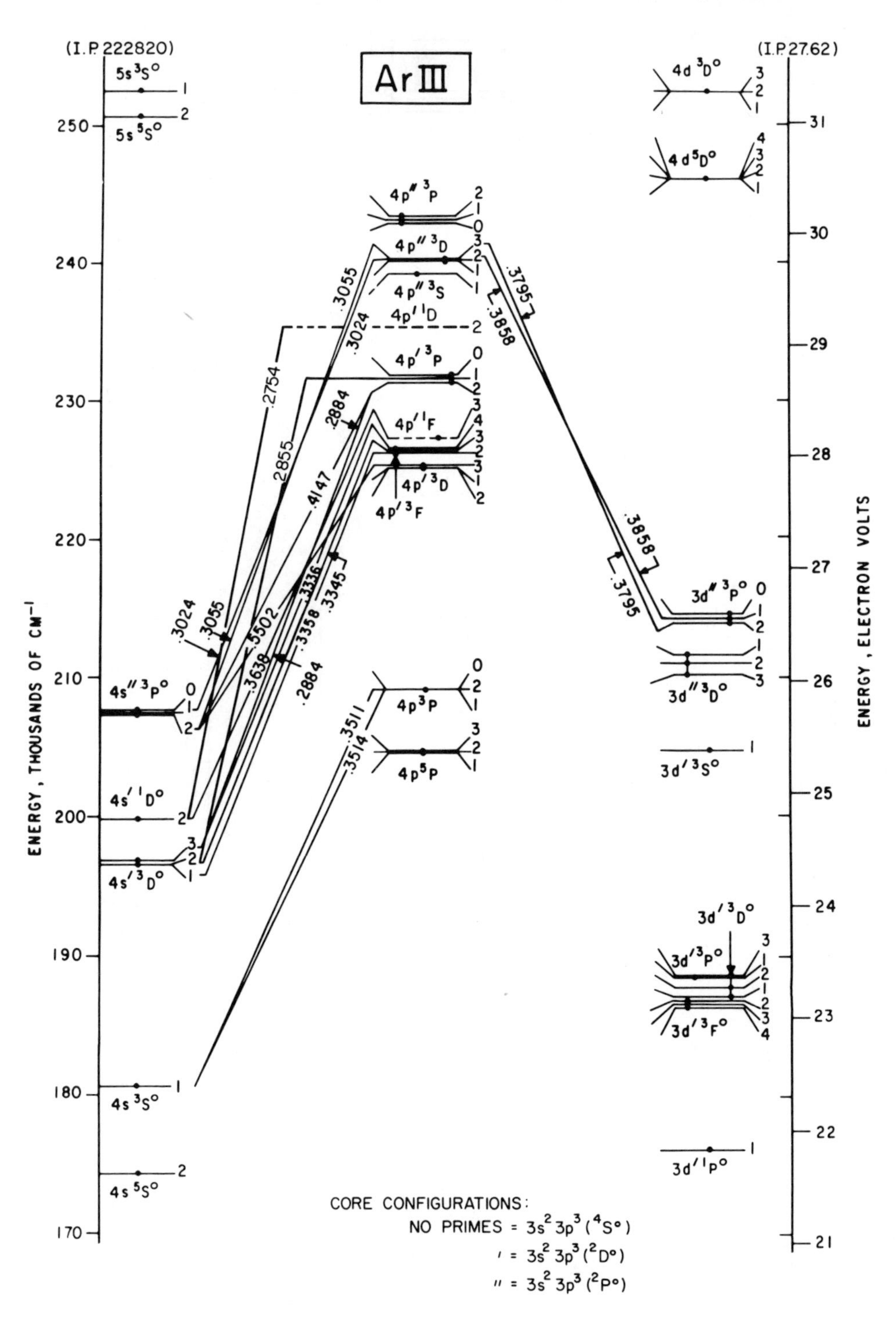

FIGURE 2.11. Energy level diagram for doubly-ionized argon (Ar III), showing all classified Ar III laser lines. Energy levels are from AEL,[241] except those shown dashed which are positioned through isoelectronic arguments by McFarlane[230] and Marling.[222]

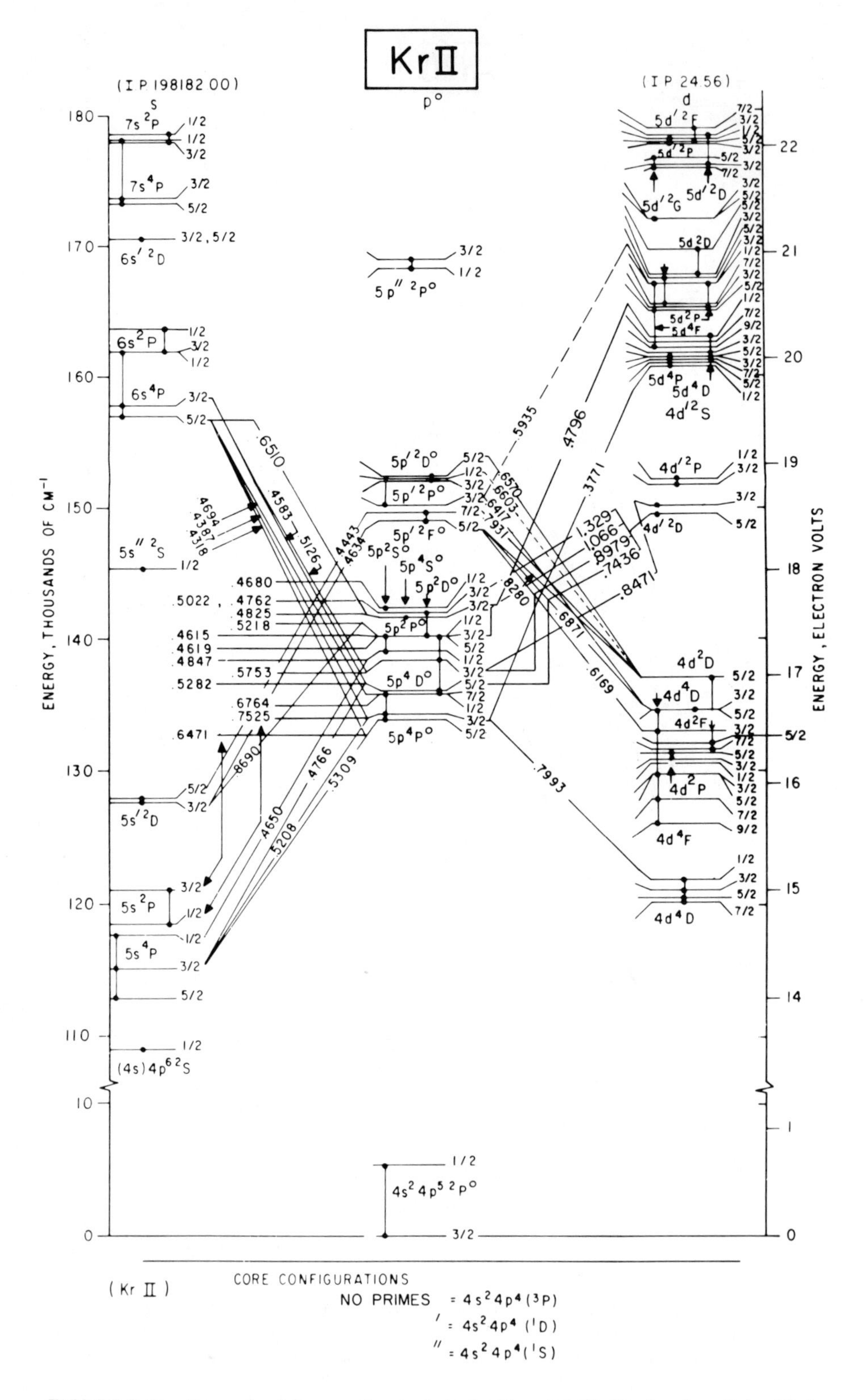

FIGURE 2.12. Energy level diagram for singly-ionized krypton (Kr II), showing all classified Kr II laser lines. Energy levels are from AEL,[241] with revisions according to Minnhagen et al.[240]

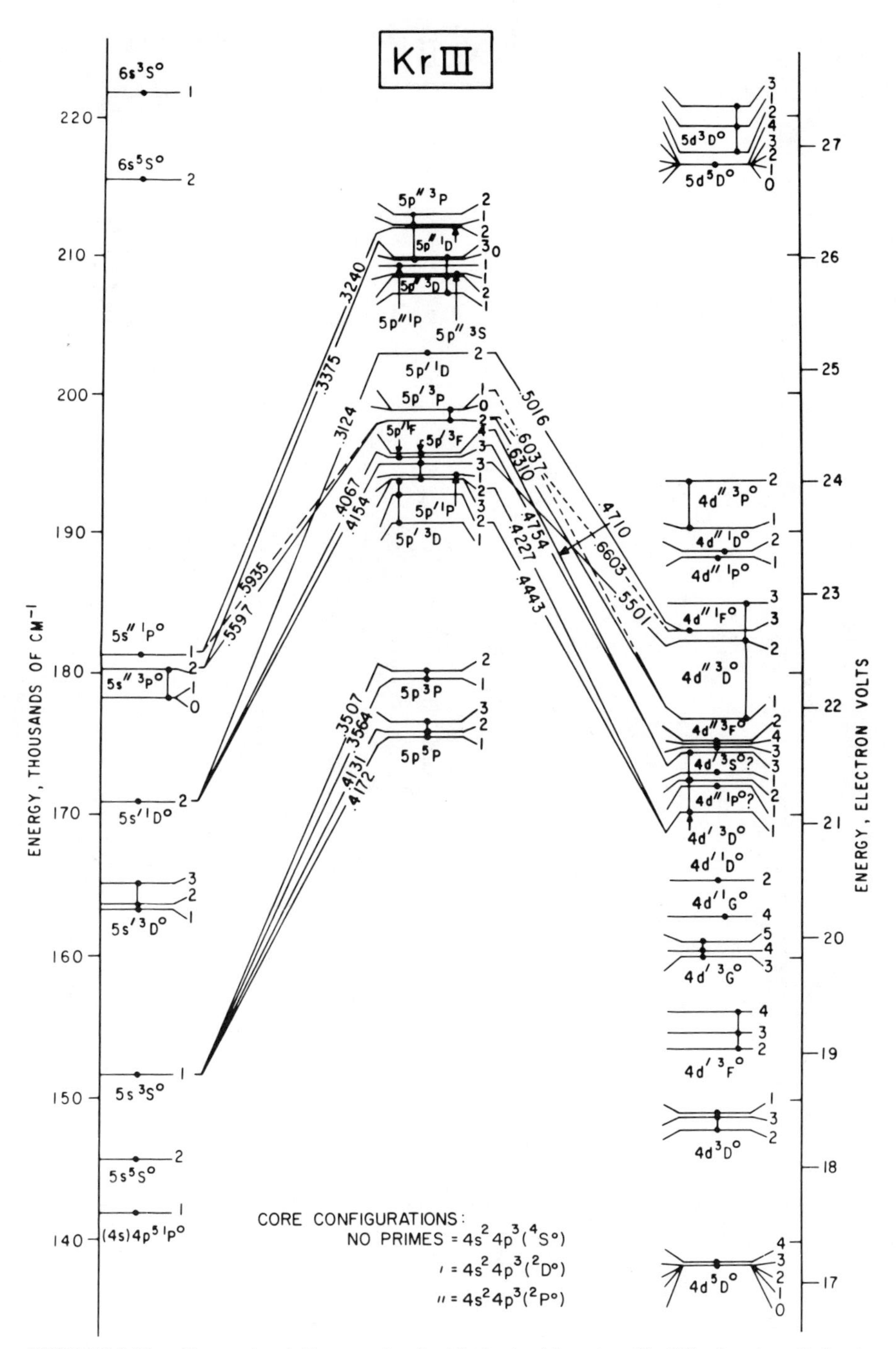

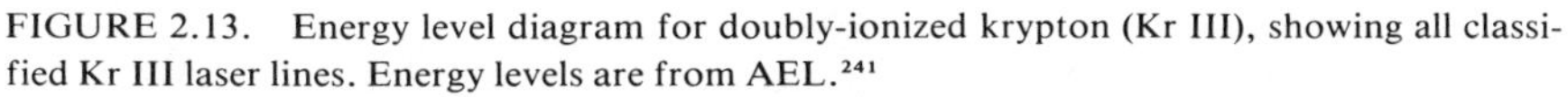

FIGURE 2.13. Energy level diagram for doubly-ionized krypton (Kr III), showing all classified Kr III laser lines. Energy levels are from AEL.[241]

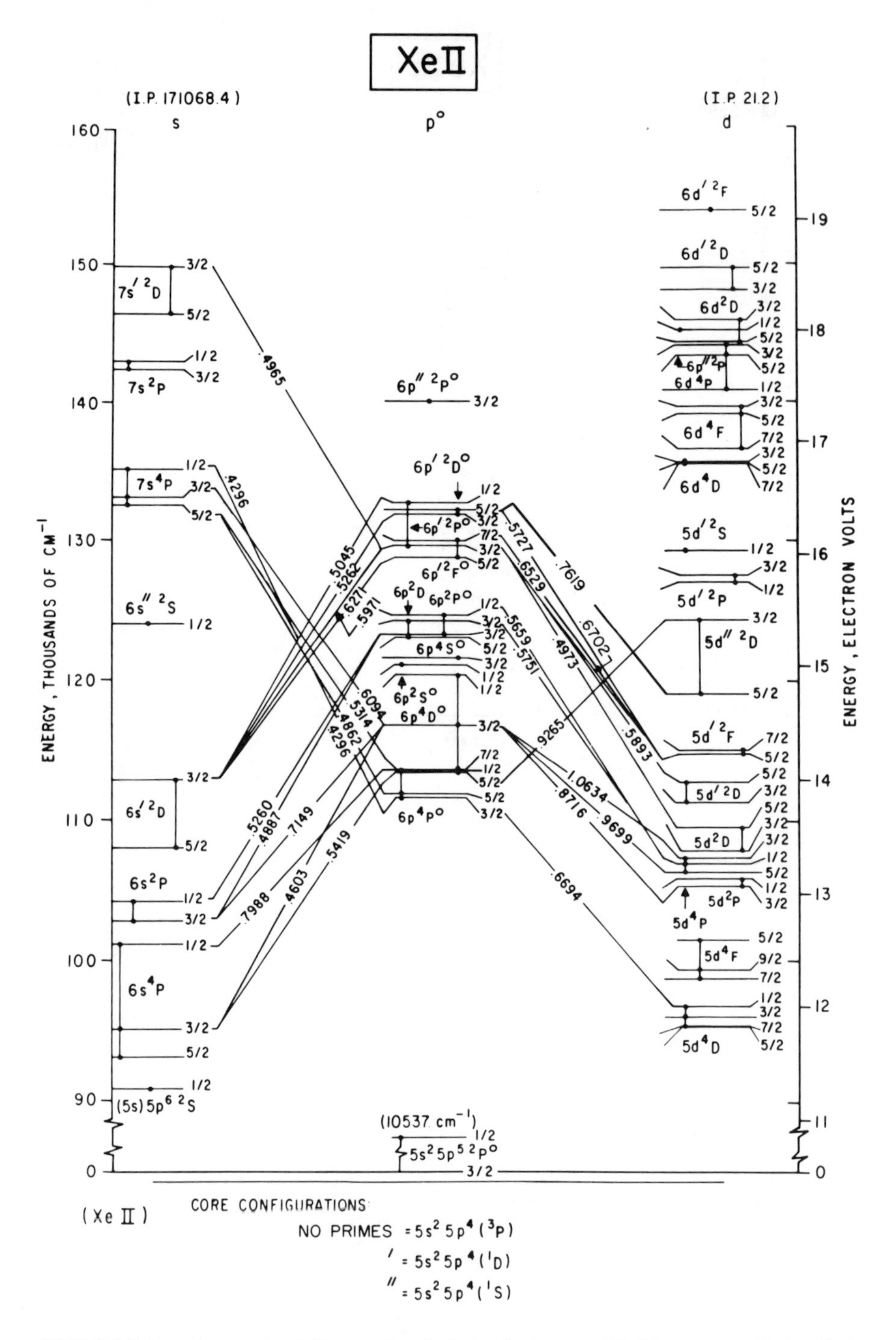

FIGURE 2.14. Energy level diagram for singly-ionized xenon (Xe II), showing all classified Xe II laser lines. Energy levels are from AEL.[241]

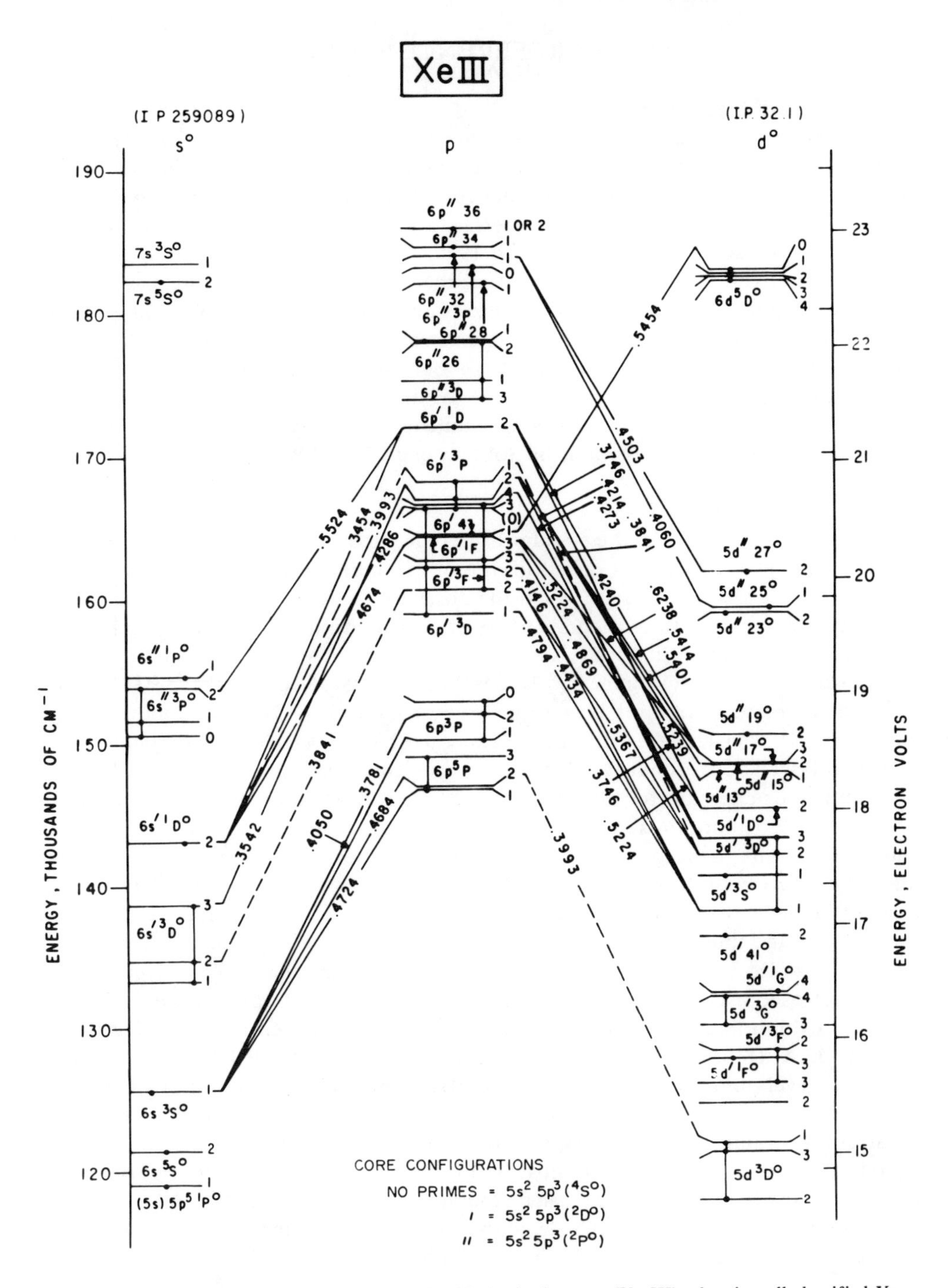

FIGURE 2.15. Energy level diagram for doubly-ionized xenon (Xe III), showing all classified Xe III laser lines. Energy levels are from AEL,[241] with revisions according to Gallardo et al.[112]

OTHER REFERENCE SOURCES

Laser Lines and Mechanisms

Since the publication of the ion laser article in the *CRC Handbook of Lasers* [45] in 1971, several other listings of ion laser lines have appeared. Willett[339] includes listings of both neutral and ion laser transitions complete through 1974. This book also includes chapters on the discharge properties and excitation mechanisms. Davis and King[83] list ion lasers transitions through 1975. This reference also contains the most extensive discussion of ion laser excitation mechanisms, plasma properties, and discharge technology to appear to date. Beck et al.[14] list all neutral, ion, and molecular transitions through 1979. Bennett[21] lists neutral and ion laser transitions observed through 1979, and includes an annotated bibliography. Bridges[49] gives brief discussions of ion laser mechanisms and performance through 1979, but does not include tables of lines. Earlier reviews of ion laser mechanisms and performance, particularly the CW noble gas ion lasers were made by Sobolev et al.[196] and Bridges et al.[46]

Wavelengths and Energy Levels

Moore's tables[241] remain the standard reference and *sine qua non* for energy levels. More recent spectroscopic work modifying or extending Moore's listings, where they relate to laser transitions, are given in the notes to Tables 2.1 to 2.13.* References to the spectroscopic literature, including lists of classified lines, Grotrian diagrams, etc., are given as references in Moore's volumes and are current through the publication dates of those volumes (Volume 1, 1949; Volume 2, 1952; Volume 3, 1958). References to more recent spectroscopic literature can be found in the bibliographies by Hagen and Martin.[137,138] There are a number of general tables of wavelengths that have proven useful to laser spectroscopists. Harrison[141] lists observed wavelengths and intensities in spontaneous emission for all elements over the wavelength range 10,000 to 2000 Å. The lines included are those known to originate from neutral and singly ionized states and those of unknown ionization state. Striganov and Sventitskii[312] list wavelengths, intensities, and classifications for all classified lines from all ionization states of the elements H, He, Li, C, N, O, Ne, Na, Ar, K, Kr, Xe, and Cs. Zaidel' et al.[344] list wavelengths and intensities similar to Harrison.

There are also several collections of Grotrian diagrams that have proven useful to laser spectroscopists. The classic volume by Grotrian[133] contains diagrams for several ions that exhibit laser oscillation: Be II, Ca II, Sr II, Ba II, Zn II, Cd II, Hg II, B II, Al II, In II, C II, III, Si II, III, IV, Ge II, Sn II, Pb II, N III, P III, IV, S IV, and V. The older level notation used by Grotrian differs somewhat from that in AEL. Moore and Merrill[242] have published selected partial Grotrian diagrams with modern level notation, including those of Be II, Ca II, Mg II, Cr II, Ni II, C II, III, Si II, N II, O II, III, and S II. In perhaps the most heroic undertaking since Moore's compilation of AEL, Bashkin and Stoner published volumes of Grotrian diagrams and energy levels for all ionization states of all the elements. To date, two volumes have appeared, H I through P XV (Volume 1)[12] and S I through Ti XXII (Volume 2).[13]

MECHANISMS AND PERFORMANCE OF SELECTED ION LASERS

Of all the laser lines listed in Tables 2.1 to 2.13, very few have received more attention than an initial demonstration of oscillation. The most well-studied and highly developed ion lasers are those using the blue-green laser lines in singly ionized argon. Commercial lasers with up to 30- to 40-W CW output are available, and laboratory demonstrations of as much as 500-W CW have been made. Other commercially available ion lasers use the visible Kr II lines, UV Ar III and Kr III lines, or the UV and blue lines of Cd II. The Se II lines and high-power, pulsed Xe IV visible lines were

* The remainder of the tables and figures may be found beginning on p. 198.

also available briefly as commercial products, but found little application. Other lasers, such as the Zn II and Hg II lasers, have received substantial laboratory study to determine the population inversion mechanisms, but they have received no commercial development.

In the brief discussions that follow, excitation mechanisms and performance characteristics are given for selected lasers. These lasers have been divided into two general groups according to excitation mechanisms: those populated by collisions with atoms or ions of another gas (usually He or Ne) and those populated by collisions with electrons.

Ion Lasers Excited by Collisions With Atoms or Ions

About half of the elements lasing as ions employ collisions with He or Ne to excite the upper laser levels. A few of these lasers have received sufficient study so that the excitation pathways are quite well understood, while for most atoms there is still some uncertainty in the details. Two basic processes are primarily responsible for the excitation: Penning, ionization,[204,261]

$$A^* + B \rightarrow A + B^{+*} + e$$

where A^* is a He or Ne atom excited to a metastable level and B^+ is the laser ion upper level; and thermal charge exchange,[90,91,221]

$$A^+ + B \rightarrow A + B^{+*}$$

where A^+ is the He or Ne ground-state ion. The thermal charge-exchange reaction is somewhat resonant, with energy defects of 0.1 to 2 eV being typically observed. The Penning process, on the other hand, prefers to excite levels lying several eV below the metastable energy rather than those in close coincidence.[131] Table 2.14 lists the lasing ions that are known to be populated by each process. Only singly ionized lasers have been produced by these means. In some cases, additional laser lines originate as a result of radiative cascade from the charge-exchange-excited levels to lower levels. In other cases, electron collision is responsible for transfer of excitation from the charge-exchange-excited levels to the upper laser level. Many of the atoms listed in Table 2.14 also exhibit laser oscillation by electron collision on other wavelength and in higher ionization states.

Mechanisms

Copper—The singly ionized copper vapor laser provides a good example of the several population pathways that can occur in the same ion. Figure 2.2 illustrates these processes. The 6s → 5p laser transitions result from charge-exchange population of the 6s upper levels by He^+, while the 5p → 5s and 5p → 4d transitions result by radiative cascade from the 6s levels. The 4f → 4d blue-green laser lines are also assumed to result from charge-exchange collisions with He^+. However, Figure 2.2 shows that the 4f levels lie a few tenths of an electron volt above the He^+ level, indicating an endothermic reaction. It may be that the Cu 4f levels are populated from the 6s levels by an additional collision with discharge electrons similar to the process described by Green et al.[130] for the Zn 4f levels. In any case, the 4f → 4d transitions have been reported by only one group (see McNeil et al.),[232] so that some discharge peculiarity may also be required for their operation. The UV 5s → 4p transitions occur only in Ne discharges and result from charge-exchange collisions with Ne^+. The $4s^2$ → 4p transitions occur in He discharges and are attributed to Penning collisions with He^*, although the definitive studies have not yet been carried out.

Silver—As Figure 2.3 indicates, the 6p levels are populated by charge exchange collisions with He^+, resulting directly in oscillation on the 6p → 6s and 6p → 5d lines and on the 5d → 5p lines by cascade. Oscillation in Ne discharges is presumed to result from charge-exchange collisional population of the $5s^2$ levels by Ne^+.

Gold—The relative positions of the energy levels in Au are somewhat different than those in Ag and Cu, as seen in Figure 2.4. The uppermost pair of 7s levels are populated by charge exchange with He^+ and produce the seven laser lines on the 7s → 6p supermultiplet. Some of the 7p levels are also populated by He^+ to produce an inversion with respect to the lower two 7s levels which are farther from resonance with He^+ than the upper two and are evidently not as strongly populated by He^+. The position of the $6s^2$ multiplet several electronvolts below the He^+ level is favorable for Penning excitation, but the 6p levels lie above the $6s^2$ levels (except for $6s^2$ 91114_o), so that oscillation analogous to the Cu II $4s^2$ → 4p lines is not possible. Cascade excitation of the 6p levels from the 7s levels may be sufficient to invert members of the 6p → $6s^2$ supermultiplet, but no laser lines have been reported to date. Likewise, no results have been reported in Ne-Au discharges; it may be that the 6p levels lie sufficiently below the Ne^+ level, so that efficient charge transfer excitation is not possible.

Zinc—Figure 2.5 shows the individual energy levels in singly ionized Zn that take part in laser oscillation. The $4s^2$ levels are populated by Penning collisions with He* as first proposed by Silfvast[298] and later confirmed by Riseberg and Schearer,[279] Collins et al.,[66] and Webb et al.[326,336] The 6s and 5d levels are excited by charge exchange from He^+, and the 5p level is excited by cascade as first suggested by Jensen et al.[180,181] The 4f levels lie below the He^+ energy and were originally thought to be populated by direct charge exchange also.[180,181,326,336] Later measurements by Green et al.[130] using flowing afterglow apparatus indicate that only the 6s and 5d levels are populated directly and that the 4f levels are populated from the 5d levels by electron collision. The study of the radial profiles of the various excited states in a Zn hollow-cathode discharge by Gill and Webb[116] is also consistent with this picture. The two laser transitions originating from the 6p level were reported by Chou and Cool[63] in a pulsed-discharge dissociation of $Zn(CH_3)_2$ and are presumably not produced in conventional positive columns or hollow-cathode He-Zn discharges. Figure 2.5 also suggests that vacuum UV oscillation on the 4d → 4p lines may be possible by charge-exchange excitation in a Ne-Zn discharge.

Cadmium—Figure 2.6 shows the energy levels of singly ionized Cd. The blue and UV transitions from the $5s^2$ levels are primarily populated by Penning collisions with He* as first proposed by Silfvast[298] and confirmed by several studies.[65,67,130,286,326,336] However, oscillation on these same two lines has also been obtained in Ne-Cd discharges by Csillag et al.[76] and in Ne-Cd, Ar-Cd, and Xe-Cd discharges by Wang and Siegman;[331] evidently electron-collisional population of the $5s^2$ levels is sufficient to produce oscillation as well. However, much greater outputs are obtained with He. Flowing afterglow studies by Turner-Smith et al.[326] show that the 9s, 6f, and 6g levels are populated directly by charge-exchange collisions with He^+ and that the $5s(^3D)5p$ ^4F levels are also probably excited by He^+. They also confirm that the 6p and 4f levels are populated by radiative cascade from higher-lying states. The 9s → 8p line at 3.29 μm was obtained by Chou and Cool[63] by dissociation of $Cd(CH_3)_2$ in a He discharge.

Mercury—The laser lines in singly ionized Hg shown in Figure 2.7 form a much less orderly pattern than those in Figures 2.2 to 2.6. The 7p levels are populated by charge exchange with He^+ as first proposed by Dyson[93] and confirmed by later studies.[3,101,187,188,216,267,343] The Hg levels lying above the He^+ energy are evidently populated by electron collision; for example, the 0.5677-μm line will oscillate in He, Ne, or Ar discharges.[15,25,34] The mechanisms responsible for the population of the 6p level in a Ne discharge is not known; it lies about 4 eV below the Ne^+ level and just slightly above the Ne* levels, so that neither charge-exchange nor Penning collisions seem favorable.

Performance of Lasers Pumped by Atoms or Ions

Performance characteristics of selected lasers excited by charge-exchange or Penning collisions are given in this section; included are those lasers that have had significant study or development. Unless otherwise stated, all power and gain characteristics are for CW operation. Both positive-column discharges (PCD) and hollow-cathode discharges (HCD) have been successfully employed with these collision lasers. In Cu, Ag, and Au HCDs, sputtering of the active material from the cooled cathode has been used; usually a trace of a heavier noble gas (Ar, Kr, Xe) is added to the He or Ne buffer in order to enhance the sputtering rate.

Copper—The highest power reported to date for Cu II is 1 W at 0.7808 μm, obtained by Warner et al.[332] in an HCD 120-cm long with a rectangular cross section of 4 × 12 mm. The electrical input was 40 A at 400 V. The same group of investigators also reports substantially higher efficiency by evaporating copper from a discharge-heated cathode than by sputtering; they obtained 500 mW at 0.7807 μm from a hollow molybdenum cathode 13 cm long and 9.5 mm in diameter with molten Cu wetting the inner wall. The electrical input for this laser was 3.8 A at 440 V, much less than that for the 1 W sputtered atom laser. McNeil and Collins[234] report 350-mW peak power (200 μsec, 0.8% duty pulses) as atotal for 5 or 6 of the UV lines in a Ne-Cu hollow cathode discharge 2 × 6 × 50 cm long. Jain[173] has recently reported 800-mW peak output from the same group of UV lines in a hollow cathode 25 cm long. Piper and Neely[271] obtained oscillation on four of the 6s → 5p lines and one of the 5p → 5s lines around 0.7 μm using CuCl, CuBr, or CuI as sources for the Cu vapor rather than sputtering or evaporating atoms from a Cu surface. No output powers were given, but the current thresholds reported were substantially lower than those reported by McNeil et al.[232,233] and Rozsa et al.[280] for sputtered metal lasers.

Silver—Warner et al.[332] report 1-W quasi-CW output (200-μsec pulses) for the group of three Ag II lines at 0.8404, 0.8255, and 0.8004 μm in a 2 × 6 × 50 cm slotted-cathode HCD in He. They also obtained a CW output of 100 mW on the same lines. The other strong Ag lines observed in the same laser structure, but with a Ne discharge, were 0.3181 (350-mW peak), 0.4086 (350-mW peak), and 0.4788 μm (100-mW peak), all at 1% duty cycle. Jain[173] has reported 1.5-mW CW and 50-mW peak output from the 0.2243-μm line in a He HCD 25 cm in length.

Gold—Reid et al.[275] report 125-mW peak output (200-μsec pulses, 0.8% duty cycle) for the UV lines 0.2823 to 0.2918 μm in a 2 × 6 × 50 cm He HCD. Jain[174] reports 20-mW CW and 600-mW peak output on the same lines in a 25-cm HCD.

Zinc—Jensen et al.[181] reported CW outputs of 12 mW at 0.7479 μm, 7 mW at 0.5894 μm, and 1 mW at 0.7588 μm in a He PCD 4 mm in diameter by 90 cm long. Piper and Gill[268] have used a He hollow cathode 3 mm in diameter by 50 cm long in both CW and quasi-CW operation (1.2 msec pulses at 2 pps repetition rate to limit the average power dissipation of the hollow-cathode structure). They obtained the following maximum powers for each line (CW, quasi-CW): 0.7588 μm (12, 25 mW), 0.7479 μm (35, 60 mW), 0.5894 μm (10, 15 mW), 0.4924 μm (25, 95 mW), 0.4912 μm (15, 50 mW). The maximum output on the 0.7494- and 0.5894-μm lines was obtained at less than the maximum current used (6 A), while the remaining lines were still increasing at this current. Piper and Brandt[270] have obtained oscillation on ten of the stronger Zn II lines by using $ZnCl_2$, $ZnBr_2$, or ZnI_2 as the source of Zn in a He HCD. The only power given was 3 mW at 0.5894 μm for $ZnBr_2$.

Cadmium—The singly ionized Cd laser is one of the few lasers that has seen commercial development as well as scientific study. The 0.4416-μm lines have been important sources for photochemical reactions, particularly photolithography and other photographic exposures on blue and UV-sensitive materials. Because of the possibility of producing blue, green, and red laser wavelengths simultaneously from the same discharge, some efforts have also been made to balance the colors to produce a ''white light'' laser, even though the applications for such a laser are not clear.

Early He-Cd PCD lasers suffered from cataphoretic separation of the He and Cd, with all the Cd ending up at the cathode end of the discharge. Goldsborough,[123] Fendley et al.,[99] and Sosnowski[310] independently conceived and demonstrated the use of cataphoresis to provide a continuous flow of cadmium from anode to cathode. A design based on cataphoretic flow was used by Goldsborough[123] to obtain 200-mW CW output at 0.4416 μm and 20 mW at 0.3250 μm in a tube 143 cm long and 2.4 mm in diameter. These figures remain the maximum published outputs to date, even though several other designs have been developed to meet particular requirements of long life and low-noise operation.[149,299,301,311,150,152,153] Johnston and Kolb[185] have published the results of an extensive parametric study, mapping the noise performance as a function of the He and Cd pressures and the discharge current in small diameter (1 to 1.6 mm) PCD lasers. Table 2.15 gives the scaling relations they determined for optimum noise performance.

Hollow-cathode discharges in He have also been used to produce He-Cd laser oscillation.[189,288,289,313,314.] Piper and Webb[266] report oscillation on ten lines in Cd II, with outputs of 10 mW each on 0.4416- and 0.5378-μm lines in a HCD 85 cm long. Fukuda and Miya[108] have obtained 12.5 mW at 0.8878 μm and lesser powers on several other Cd II lines in a HCD 45 cm long. Oscillation on the Cd II 0.5378-μm line has also been reported by Toyoda et al.[350] in a tiny plasma produced by evaporating a Cd surface in vacuum with a CO_2 laser pulse.

Kano et al.[186] have obtained Cd II oscillation in a PCD using CdI_2 as the Cd source. Piper and Brandt[270] have obtained oscillation on Cd II lines using $CdCl_2$ and/or CdI_2 in a He HCD (as well as 11 of the stronger I II lines with CdI_2; there are no known charge-exchange-excited transitions in Cl II or Br II).

Mercury—Historically the first ion laser, Hg has received much attention in laboratory studies, but no serious commercial development. The lines originating from the 7p levels, 0.6150, 0.7945, and 1.5554 μm are pumped by charge-exchange and will oscillate CW in either positive-column or hollow-cathode discharges, the remaining lines are excited by electron collision and are observed only in pulsed PCDs. Peak powers of tens of watts are typical for the green 0.5677 μm line, as reported originally by Bell.[16] CW operation on the 0.6150 μm was first obtained in a PCD by Ferrario,[101] who observed a maximum of 10 mW from a tube 100 cm long by 3 mm in diameter. Kano et al.[187] obtained 6 mW at 0.7945 μm from a 123 cm by 3.5 mm diameter tube. Table 2.16 lists the performances obtained in HCDs by several workers. The highest power reported to date is 80 mW at 0.6150 μm by Piper and Webb.[267]

Selenium—Besides the He-Cd laser, the singly ionized He-Se laser is the only other member of this class of lasers that has received commercial development. The wide range of wavelengths throughout the visible spectrum that can be obtained at the few milliwatt level from a simple discharge tube with only a few hundred watts input is indeed an attractive feature. The performances of various PCD He-Se lasers is reported by Silfvast and Klein,[199,300] Keidan et al.,[194] and Hernqvist and Pultorak[151] are summarized in Table 2.17. Piper and Webb[266] have also obtained CW oscillation in a HCD on 13 of the stronger Se II lines, but no output powers were reported.

Iodine—Oscillation on singly ionized iodine, I II, was first obtained in a pulsed discharge by Fowles and Jensen,[102,103] who were also the first to propose charge exchange as the pumping mechanism for this (or any other) laser. There have been many laboratory studies of the He-I laser, but no commercial development. The best performance for a PCD laser is that of Hattori et al.,[145] who obtained 40 mW total on the five visible lines: 0.5407, 0.5678, 0.5761, 0.6127, and 0.6585 μm, with 18 mW on the strongest, 0.5761 μm. They used a sealed-off tube 123 cm long and 3.5 mm in diameter, with cataphoretic flow. Later work by the same group, Goto et al.,[129] demonstrated 1000-hr operating life at 20 to 30 mW for the same five lines in a 100 cm by 2.5-mm diameter tube. They also obtained operation on 0.7033-, 0.7618-, 0,8170-, and 0.8804-μm lines with IR mirrors. Piper and Webb[264] have obtained CW operation in a He HCD 75 cm long and 2.5 mm in diameter with outputs of 55 mW at 0.5678 μm, 100 mW at 0.5761 μm, and 70 mW at 0.6127 μm. Output powers were tens of milliwatts for the other stronger lines, and almost 1 W was observed summed over all lines.

Lasers Excited by Collisions With Electrons

There are many ion lasers in which the population inversion is thought to occur by selective population of the upper laser level by collisions with electrons. This is the case for the remainder of the elements shown shaded in Figure 2.1 and not listed in Table 2.14 (and for a few already included in Table 2.14 as already noted). Unfortunately, for most of these elements, the assignment to the "electron-excited" category is effected in a somewhat negative way, namely, that the laser oscillates with more than one buffer gas or with no buffer gas at all, so that Penning or charge-exchange collisions are unlikely contributors. For only a few lasers have detailed studies been carried out to confirm the electron-excitation hypothesis. In this brief review, only the noble gas ion lasers will be discussed, since these are the only electron-excited lasers that have received significant study of excitation mechanisms or practical development. In the absence of evidence to the contrary, it is reasonable to assume that many other ion lasers employ the same or similar excitation pathways and that similar performance would result if they were developed. Of course, it may not be possible to carry out the development because of some inherent difficulties such as chemical reactivity in a discharge. The elements that seem to be most analogous to the ionized noble gases in demonstrated or potential laser performance are N, P, O, S, F, Cl, and Br. Some of these electron-excited lasers have already shown reasonable performance in pulsed[222,224] or r-f excited ring discharges,[2,17,18] where chemical poisoning of a cathode is tolerable or nonexistent. For summaries of what is known about these and other electron-excited lasers, see the reviews by Davis and King[83] or Willett.[339]

Noble Gas Ion Lasers

To date over 200 papers have appeared dealing with noble gas ion laser mechanisms. For a guide to this vast literature, see the reviews by Davis and King,[83] Dunn and Ross,[92] Willett,[339] or Bridges.[49] The discussion of mechanisms given below will serve as an outline and summary, but with only a few of the references cited. The detailed arguments are considerably more complicated and somewhat less certain than this outline would indicate.

Excitation Mechanisms in the Ar II Laser

Historically, the first model of ion laser excitation was that proposed by Gordon et al.[206,237] shown schematically in Figure 2.16A. In this model, the upper laser level is excited by an electron collision with an ion in its ground state (Process 2); the ion is assumed to have been created by a collision of an electron with a neutral atom in the ground state (Process 1). Cascade from higher states and destruction by electrons are neglected. The lower level is assumed to be rapidly depopulated by radiation (in the vacuum UV, approximately 740 Å for the 4s states of Ar II). The depopulation is assumed to occur much more rapidly than collisional electron excitation (dotted arrow). If the further assumptions of approximate charge neutrality in the plasma and current-independent electron temperature are made, then the upper laser level population N_2 varies as

$$N_2 \sim n_e n_i \sim n_e^2 \sim J^2$$

where n_e, n_i, and J are the electron density, ion density, and discharge current density, respectively. This quadratic current dependence has been observed in spontaneous emission measurements over a wide range of currents, gas pressures, and tube dimensions typical of CW ion laser operation.

An alternate excitation model involving a single electron collision as shown in Figure 2.16B was proposed by Bennett et al.[20] This mode of operation apparently does occur in low-pressure, short pulse-excited discharges. A very high E/p is required to create a sufficiently high electron temperature to achieve single-step excitation. This mode of operation is also characterized by a different output spectral distribution. For example, in Ar II 0.4765 μm is the strongest laser line. This distribution is never observed in CW operation, where the 0.4880 and 0.5145 μm lines are always dominant. (An energy level diagram for the 4p → 4s Ar II laser transitions as shown in Figure 2.10 for reference.) In addition, the single-step process, with constant electron temperature, would yield spontaneous emission from the upper laser level which varies linearly with discharge current.

One of the arguments advanced by Bennett et al.[20] for the single-step process involves the favorable selection rules for excitation from the $3p^6$ neutral ground state to the $3p^4\ {}^3P)4p$ excited ionic state (which does have a change in parity, even though it is a p → p transition, because of the electron lost in ionization). The highest yields are expected for the $3p^4\ ({}^3P)\ 4p\ {}^2P^o$ levels, and hence the prediction of the 0.4765-μm $4p\ {}^2P^o_{3/2} \rightarrow 4s\ {}^2P_{1/2}$ as the strongest transition. In fact, this line was observed by Bennett el al. to be the strongest in a discharge with E/p ∿1000 V/cm-torr and excitation pulse ∿20 nsec. Experiments by Kobayashi et al.,[201] Demtroder,[86] and Davis and King[81] confirm that there are two discharge regimes: (1) Short pulse, low pressure, and (2) Long pulse/CW, high pressure. The 0.4765-μm line was also the only line observed by Kulagin et al.[205] in a very high-current (15 kA) self-constricted discharge operated at $E/p > 1000$ V/cm-torr and ∿200-nsec pulses.

Arguments based on selection rules weigh against the two-step process of Figure 2.16A because it requires a $3p^5 \rightarrow 3p^4({}^3P)4p$ excitation, a violation of the dipole selection rules. In fact, it would seem that the transition to the lower laser level, $3p^5 \rightarrow 3p^4({}^3P)4s$, would be more probable. However, the dipole selection rules arise from the Born approximation which is valid only for electron energies well above threshold for the excitation. Since the electron temperature is small compared with the threshold energy for excitation (particularly under conditions of CW laser operation), most of the excitations actually take place quite near threshold. As pointed out by Massey and Burhop,[228] in this energy region, the Born approximation does not hold, and excitation cross sections for transitions with no parity change (such as the $3p^5 \rightarrow 3p^4(3P)4p$ excitation) can even exceed cross sections for "optically allowed" excitations. An alternative answer proposed by Labuda et al.[207] is that ionic metastables serve as the intermediate species, rather than the ion ground state. This is shown schematically in Figure 2.16C. Absorption measurements[207] show these metastable levels to be highly populated.

The ionic metastables are populated either by electron collision with the ionic ground state (Process 1 in Figure 2.16C) or, more probably, by cascade from higher-lying states (Process 2) which exhibit populations $N_X \sim J^2$ (from spontaneous emission data). It can be argued that the metastable population N_M will be proportional to J rather than J^2, however, because it is primarily destroyed by electron collisions (see Reference 207 or 49 for the argument). The upper laser level is then populated by a second electron collision (Process 3) which is more readily permitted by selection rules (requiring at most only a core or spin change rather than a parity violation), and the relation $N_2 \sim J^2$ is obtained as before.

The picture of the upper level excitation process remained essentially as described above until measurements made by Rudko and Tang[283] were reported. Their measurements indicated that a significant fraction of the upper laser level population was created by radiative cascade from higher-lying states (see Figure 2.16D). By summing the spontaneous emission intensities of all lines terminating on a given upper laser level (e.g., 4p $^2D^o_{5/2}$, the upper level of the 0.4880-μm transition) and then comparing this with the sum of intensities of those lines originating from that level, we may determine the population fraction due to cascade from the higher-lying states. Rudko and Tang reported that approximately 50% of the 4p $^2D^o_{5/2}$ level population is created by cascade, in a 1-mm diameter tube operated without a magnetic field. Similar measurements were made by Bridges and Halsted,[41] who obtained a value of 23% for the cascade contribution to this same level in a 3-mm diameter tube operated without a magnetic field, and 22% for the cascade contribution to the 4p $^4D^o_{5/2}$ (0.5145 μm) upper level. Since the populations N_C from which the cascade takes place also exhibit a quadratic dependence on J, the same quadratic dependence $N_2 \sim J^2$ results again. The point is that the fraction due to cascade cannot be separated from that due to two-electron-collision processes by its current dependence.

Trapping of the vacuum UV radiation which depopulates the lower laser level, N_1, can upset the simple quadratic variation predicted above. The amount of trapping in CW operation is generally thought to be small,[197,198] but it definitely shows up in pulsed ion laser operation. Figure 2.17 shows the different time variation of the usual blue-green Ar II lines in a pulsed laser when an approximately square current pulse is applied. The different shapes reflect the different time histories of the population differences $[N_2(t)-N_1(t)]$ for each line. By comparison with spontaneous emission data [which gives only $N_2(t)$], it is clear that the primary peculiarities in shape are caused by $N_1(t)$, evidently due to the complex behavior of the excitation and trapping functions for the lower levels. Figure 2.17 shows the effect of trapping on a single line (0.4880 μm) as the discharge current or the gas pressure is increased. A region of decreased inversion develop a few microseconds after the leading edge of the current pulse as the current is increased. Eventually the inversion (N_2-N_1) goes to zero or even becomes negative for a period as shown by the absorption measurements of Gordon[127] and Davis and King[81] and then recovers. The recovery is presumed to be due to a decrease in the radiation trapping coefficient caused by doppler broadening of the vacuum UV line as the gas temperature rises. The result is typically the "double-pulse" output exhibited in the lower traces of Figure 2.17. If the current pulse rises slowly enough, the "first" pulse often disappears completely, since trapping causes the lower level to fill up faster than the upper level is populated. The resulting laser output appears to be simply delayed by several microseconds. The amount of delay depends on the gas pressure (longer for higher pressures), the kind of gas (longer for Kr and Xe than Ar) and the magnetic field (longer if a magnetic field is used). All these dependencies point to gas heating (or rather, increased ion velocities) as the mechanism which reduces the trapping and allows the inversion to be reestablished.

Performance of the Ar II Laser

Whatever the details of the excitation and deexcitation mechanisms in Ar, the performance of the CW blue-green Ar II lasers is reasonably described by the scaling relation

$$\frac{P}{L} = K\,(JR)^2$$

where P is the total blue-green output on the transitions shown in Figure 2.10, J is the current density, L is the discharge length, and R is the discharge radius. The constant of proportionality K has the value of 2×10^{-3} to 5×10^{-3}, when P/L is in watts per meter and JR is in amperes per centimeter. Figure 2.18 shows a summary of Ar II laser performance from several sources, plotted as P/L vs. JR. The value of $K = 5 \times 10^{-3}$ would be a straight line of slope 2 just touching the upper left edge of the curves ("best results"), while $K = 2 \times 10^{-3}$ would put the line through the center of the data spread. Table 2.18 lists the operating parameters for the highest-power data shown on Figure 2.18. It is unfortunate that data from commercially available Ar ion lasers cannot be compared on this graph, since laser manufacturers consider operating current and tube dimensions for their "best performance" proprietary information; it would be unfair to compare the typical performance of commercial lasers on the same graph as the best laboratory data. Table 2.19 gives the data sheet performance for a medium-power vs. a high-power commercial Ar II laser from each of two manufacturers, showing the typical power outputs on each of the lines.

Typical Operating Parameters of the Ar II Laser

That the curves of P vs. J (or P vs. I) deviate from the simple quadratic dependence given above at a fixed gas pressure is well known. At low currents, the laser is nearer threshold and exhibits a much steeper variation of output with current because it behaves as an inhomogeneously broadened medium. (True inhomogeneous broadening would yield a quartic dependence of power on current.) At the highest currents, the neutral atom density in the active region is depleted by the discharge ionizing the gas to still higher states and driving it from the hot small-bore region into the cooler electrode regions. The relative importance of these two processes is not known, but they both result in a subquadratic dependence at high currents. However, if the total gas pressure in the laser is increased, the P vs. I curve becomes steeper and continues to higher currents, so that a set of P vs. I curves at different fill pressures overlap to form an envelope as shown in Figure 2.19. The envelope exhibits an I^2 dependence over a much wider range than any individual curve. This is also seen in Figure 2.18.

Figure 2.20 shows the typical distribution of the blue-green wavelengths from an Ar II laser (the same tube as Figure 2.19; this tube is described in detail in Reference 39. These curves were taken with laser mirrors that had roughly the same reflectivity across the 0.5200-4000-μm range. Different wavelengths can be emphasized by tailoring the laser mirror reflectivities; for example, the 0.4880-μm line has the highest gain and can be emphasized by using a high transmission output mirror. The 0.5145-μm line can be emphasized by using higher reflectivity mirrors biased toward the green. An intracavity dispersive element such as a prism can be used to select an individual wavelength. However, the power output on the wavelength selected will not necessarily be more than if the same wavelength were selected by a prism outside the laser, since competition effects are small or zero. The two strongest lines, 0.4880 and 0.5145 μm, for example, do not compete even though they share a common lower level. The situation is different for the strong red and yellow lines of Kr II, the 0.6471 and 0.5682 μm lines, which also share a common lower level (in fact, the level in Kr II analogous to the 0.4880/0.5145 lower level in Ar II). These two lines compete strongly in a typical CW Kr II ion laser. Evidently their lower level is not nearly so well-depleted by radiation as the corresponding Ar II lower level. A simple $P/L \sim (JR)^2$ behavior would not be expected (and, in fact, is not observed) for Kr II, unless very low gas pressures are used to reduce radiation trapping effects.

Figure 2.21 shows the typical variation with current of the optimum gas filling pressure for argon ion lasers of different bore diameters. In all cases, the optimum pressure (i.e., the gas pressure in the tube measured before the discharge is struck) increases with increasing discharge current and with decreasing tube diameter. Figure 2.21 is not universal in the sense that discharge tubes of different lengths, type of construction (smooth walls, disks, segments, etc.,) or tubes with different gas return paths will exhibit somewhat different values. (See References 41, 46, 61, and 62 for discussions of gas pumping and gas return paths.) The general trends are shown in Figure 2.21, however. Likewise, pulsed operation shows the same trends, but with approximately an order of magnitude lower pressure. The heavier noble gases (Kr, Xe) also prefer lower pressures than shown in Figure 2.21 both in CW operation and in pulsed operation. Ne pressures are generally similar to Ar.

Figure 2.22 shows the general behavior of the optimum magnetic field for CW Ar ion lasers. The data are taken from several sources. The optimum magnetic field decreases with increasing tube diameter. Again, the curve is not universal, since there is some interaction between gas pressure and magnetic field. For the heavier noble gases (Kr, Xe), the optimum magnetic field generally decreases from the corresponding Ar value and becomes more critical; also with Kr and Xe, different transitions may optimize at quite different magnetic fields, so that the laser can actually be ''color-tuned'' with magnetic field. (In Ar, all the Ar II lines optimize together.)

Performance of the Kr II Laser

Figure 2.23 gives a typical comparison between the performances obtained with Ar and Kr in the same laser. The three curves for Ar (3-, 4-, and 5-mm tubes) give the total multimode blue-green output in the usual distribution among colors (see Figure 2.20), while the two curves for Kr give the output on the red lines 0.6471 and 0.6764 μm. At the maximum output shown, about 80% of the power occurs at 0.6471 μm. The Kr blue-green output is typically equal to the red output, but spread among more lines from 0.4619 to 0.5308 μm. Table 2.20 gives the performance of three commercial high-power Kr lasers in the visible Kr II lines and the UV Kr III lines.

Performance of the Ne II, Ar III, and Kr III UV Lasers

High-power CW UV operation of the noble gas ion lasers has received much attention in the past few years. Commercial lasers are available, as indicated in Table 2.20 which gives the relative output powers at several UV and violet transitions in Kr III. Figure 2.24 shows the Ar III UV output power for a small-bore laser[43] vs. discharge current and pressure, while Figure 2.25 compares Ar III and Kr III UV outputs in the same tube.[43] This particular tube used W disks 2 in. in diameter, spaced about 1 cm with a 2.3-mm hole to define the discharge. Optimum magnetic fields for this type of tube were high, typically 1 to 1.5 kG. Commercially available UV lasers are generally of the small-bore, high magnetic field variety. However, still higher output powers have been demonstrated on the Ne II, Ar III, and Kr III transitions by using larger-diameter discharges at low magnetic fields, just as with the blue-green Ar II lines. Table 2.21 illustrates the increase in performance with the evolution from small to large bore diameters over the past decade. Specific outputs as high as 36 W/m have been observed on the Ar III UV lines.[219] Figures 2.26 and 2.27 show the variation of P/L with JR for the highest performance Ar III and Kr III UV lasers obtained by Luthi et al.[219]

Performance of the Xe IV Ion Laser

The highest pulsed output powers generated to date by ion lasers have been on the blue and green lines in Xe. These laser lines were first observed by Bridges and Chester[37] and independently by Dahlquist.[78] They are not listed in the published spectroscopic literature, but they are listed by Humphreys.[163] No classification of the energy levels involved can be made but several studies[43,47,82,136,158,159,217] indicate that triply ionized species are responsible. Table 2.22 compares the performance of the strong blue and green Xe IV lines obtained by several groups of investigators. Table 2.23 gives the line-by-line output power for most of the lasers listed in Table 2.22. Marling and Lang[222,223] has also obtained outputs over a kilowatt on UV lines near 0.2 μm in both Xe IV and Kr IV.

Table 2.1.1
BERYLLIUM

Wavelength (μm)	Measured value (μm)	Identification: Ion	Upper level	—	Lower level	Ref.
0.467346	0.4675 ± ?	Be II	4f $^2F^o_{5/2,7/2}$	—	3d $^2D_{d/2,5/2}$	347—A,G,H,(He⁺)
0.527032 0.527084	0.5272 ± ?	Be II	4s $^2S_{1/2}$ 4s $^2S_{1/2}$	— —	3p $^2P^o_{1/2}$ 3p $^2P^o_{3/2}$	347—A,(He⁺)
1.209298 1.209561	1.2096 ± ?	Be II	3p $^2P^o_{3/2}$ 3p $^2P^o_{1/2}$	— —	3s $^2S_{1/2}$ 3s $^2S_{1/2}$	347—A,H,(Ne⁺)

Table 2.1.2
MAGNESIUM

Wavelength (μm)	Measured value (μm)	Identification: Ion	Upper level	—	Lower level	Ref.
		Spectroscopic Reference: Mg II (278)				
0.921825	0.921800 ± 150	Mg II	4p $^2P^o_{3/2}$	—	4s $^2S_{1/2}$	156—CW,G,(He*); 210—A,(He⁺)
0.924427	0.924400 ± 150	Mg II	4p $^2P^o_{1/2}$	—	4s $^2S_{1/2}$	156—CW,G,(He*); 210—A,(Ne⁺)
1.091423 1.091527	1.091500 ± 150	Mg II	4p $^2P^o_{3/2}$ 4p $^2P^o_{3/2}$	— —	3d $^2D_{5/2}$ 3d $^2D_{3/2}$	156—CW,G,(He*)
1.095178	1.095200 ± 150	Mg II	4p $^2P^o_{1/2}$	—	3d $^2D_{3/2}$	156—CW,G(He*)
2.40415	2.404160 ± 300	Mg II	5p $^2P^o_{3/2}$	—	4d $^2D_{5/2}$	56
2.41246	2.412520 ± 300	Mg II	5p $^2P^o_{1/2}$	—	4d $^2D_{3/2}$	56

Table 2.1.3
CALCIUM

Wavelength (μm)	Measured value (μm)	Identification: Ion	Upper level	—	Lower level	Ref.
		Spectroscopic Reference: Ca II (94)				
0.370602	0.3706 ± ?	Ca II	(^{1}S)5s $^2S_{1/2}$	—	(^{1}S)4p $^2P^o_{1/2}$	210—A,(He⁺); 348
0.373690	0.3737 ± ?	Ca II	(^{1}S)5s $^2S_{1/2}$	—	(^{1}S)4p $^2P^o_{3/2}$	210—A,G,P,(He⁺); 348—P
0.854209	0.854180 ± 60	Ca II	(^{1}S)4p $^2P^o_{3/2}$	—	(^{1}S)3d $^2D_{5/2}$	328—G,P,S
0.866214	0.866200 ± 60	Ca II	(^{1}S)4p $^2P^o_{1/2}$	—	(^{1}S)3d $^2D_{3/2}$	328—G,P,S
0.992342	0.9940 ± ?	Ca II	(^{1}S)6s $^2S_{1/2}$	—	(^{1}S)5p $^2P^o_{3/2}$	210—A

Table 2.1.4
STRONTIUM

Wavelength (μm)	Measured value (μm)	Identification				Ref.
		Ion	Upper level	—	Lower level	
0.407771	0.4078 ± ?	Sr II	$(^1S)5p\ ^2P^o_{3/2}$	—	$(^1S)5s\ ^2S_{1/2}$	132—S,U
0.416179	0.4162 ± ?	Sr II	$(^1S)6s\ ^2S_{1/2}$	—	$(^1S)5p\ ^2P^o_{1/2}$	210—A,(He⁺); 348
0.430544	0.4305 ± ?	Sr II	$(^1S)6s\ ^2S_{1/2}$	—	$(^1S)5p\ ^2P^o_{3/2}$	210—A,G,P,S,(He⁺); 348—G,P,S
0.868930	0.8700 ± ?	Sr II	$(^1S)6d\ ^2D_{5/2}$	—	$(^1S)6p\ ^2P^o_{3/2}$	210—A,G,(He⁺)
1.03273	1.033050 ± 50	Sr II	$(^1S)5p\ ^2P^o_{3/2}$	—	$(^1S)4d\ ^2D_{5/2}$	85—A,G,S; 118
1.09149	1.091450 ± 50	Sr II	$(^1S)5p\ ^2P^o_{1/2}$	—	$(^1S)4d\ ^2D_{3/2}$	85—A,G,S; 118
1.122506	1.1230 ± ?	Sr II	$(^1S)7s\ ^2S_{1/2}$	—	$(^1S)6p\ ^2P^o_{3/2}$	210—A,G

Table 2.1.5
BARIUM

Wavelength (μm)	Measured value (μm)	Identification				Ref.
		Ion	Upper level	—	Lower level	
*0.614171	0.614170 ± 10	Ba II	$(^1S)6p\ ^2P^o_{3/2}$	—	$(^1S)5d\ ^2D_{5/2}$	167—P,S; 11—H,P
0.649690	0.649690 ± 10	Ba II	$(^1S)6p\ ^2P^o_{1/2}$	—	$(^1S)5d\ ^2D_{3/2}$	167—S; 11—H,P
1.24748	1.2478 ± ?	Ba II	$(^1S)8s\ ^2S_{1/2}$	—	$(^1S)7p\ ^2P^o_{3/2}$	210—A,(He⁺)
2.59243	2.592300 ± 150	Ba II	$(^1S)7p\ ^2P^o_{3/2}$	—	$(^1S)6d\ ^2D_{5/2}$	54, 55
2.90572	2.905900 ± 200	Ba II	$(^1S)7p\ ^2P^o_{1/2}$	—	$(^1S)6d\ ^2D_{3/2}$	54,55

Table 2.2.1
CHROMIUM

Wavelength (μm)	Measured value (μm)	Identification				Ref.
		Ion	Upper level	—	Lower level	
0.868347	0.868400 ± 100	Cr II	$(a^5D)4p\ z^4D^o_{3/2}$	—	$(a^3D)4s\ b^2D_{3/2}$	175—H,(He⁺)

Table 2.3.1
NICKEL

Wavelength (μm)	Measured value (μm)	Identification				Ref.
		Ion	Upper level	—	Lower level	
0.796213	0.796070 ± 250	Ni II	$(^1D)5p\ ^2F^o_{7/2}$	—	$(^1D)5s\ ^2D_{5/2}$	172—H,(He⁺)
0.797537	0.797480 ± 250	Ni II	$(^3P)5p\ ^4D^o_{7/2}$	—	$(^3P)5s\ ^4P_{5/2}$	172—H,(He⁺)

Table 2.4.1
COPPER

Wavelength (μm)	Measured value (μm)	Identification: Ion	Upper level	—	Lower level	Ref.
0.248579	0.248580 ± 10	Cu II	$(^2D_{3/2}5s\ ^3D_1$	—	$(^2D_{3/2})4p\ ^3F^o_2$	233—H,P,(Ne⁺); 173—H,P
0.250627	0.250650 ± 10	Cu II	$(^2D_{5/2})5s\ ^3D_2$	—	$(^2D_{5/2})4p\ ^3F^o_3$	233—H,P,(Ne⁺)
0.252930	0.252920 ± 10	Cu II	$(^2D_{3/2})5s\ ^1D_2$	—	$(^2D_{5/2})4p\ ^3D^o_3$	234—H,P,(Ne⁺); 173—H,P
0.259052	0.259060 ± 10	Cu II	$(^2D_{3/2})5s\ ^3D_1$	—	$(^2D_{5/2})4p\ ^3D^o_2$	233—CW,H,P,(Ne⁺); 154—CW,H,P; 173—H,P
0.259881	0.259900 ± 10	Cu II	$(^2D_{5/2})5s\ ^3D_2$	—	$(^2D_{3/2})4p\ ^3F^o_2$	233—CW,H,P,(Ne⁺); 154—CW,H,P; 332—H,P; 173—H,P
0.260027	0.260030 ± 10	Cu II	$(^2D_{3/2})5s\ ^1D_2$	—	$(^2D_{3/2})4p\ ^1F^o_3$	234—H,P,(Ne⁺); 332—H,P; 173—H,P
0.270097	0.270070 ± 30	Cu II	$(^2D_{3/2})5s\ ^1D_2$	—	$(^2D_{3/2})4p\ ^1D^o_2$	9—H,P,(Ne⁺)
0.270318	0.270310 ± 10	Cu II	$(^2D_{3/2})5s\ ^3D_1$	—	$(^2D_{3/2})4p\ ^3D^o_1$	234—H,P,(Ne⁺); 154—CW,H,P; 173—H,P
0.272168	0.2722 ± ?	Cu II	$(^2D_{3/2})5s\ ^3D_1$	—	$(^2D_{3/2})4p\ ^1D^o_2$	173—H,P,(Ne⁺); 174—H
0.450598	0.450660 ± 30	Cu II	$4s^2\ ^3P_1$	—	$(^2D_{5/2})4p\ ^3P^o_2$	232—H,(He*)
0.455591	0.455630 ± 30	Cu II	$4s^2\ ^3P_2$	—	$(^2D_{5/2})4p\ ^3P^o_2$	232—H,(He*)
0.467356	0.467320 ± 30	Cu II	$(^2D_{5/2})4f\ ^3P^o_0$	—	$(^2D_{5/2})4d\ ^3S_1$	232—H,(He⁺)
0.468198	0.468290 ± 30	Cu II	$(^2D_{5/2})4f\ ^3P^o_1$	—	$(^2D_{5/2})4d\ ^3S_1$	232—H,(He⁺)
0.485496	0.485580 ± 30	Cu II	$(^2D_{5/2})4f\ ^3G^o_5$	—	$(^2D_{5/2})4d\ ^3G_5$	232—H,(He⁺)
0.490972	0.490970 ± 30	Cu II	$(^2D_{5/2})4f\ ^3H^o_6$	—	$(^2D_{5/2})4d\ ^3G_5$	232—CW,H,(He⁺)
0.493169	0.493170 ± 30	Cu II	$(^2D_{5/2})4f\ ^3H^o_5$	—	$(^2D_{5/2})4d\ ^3G_4$	232—CW,E,H,(He⁺); [21]
0.501262	0.501330 ± 30	Cu II	$(^2D_{5/2})4f\ ^3G^o_4$	—	$(^2D_{5/2})4d\ ^3F_3$	232—H,(He⁺)
0.502128	0.502190 ± 30	Cu II	$(^2D_{5/2})4f\ ^3D^o_3$	—	$(^2D_{5/2})4d\ ^3D_3$	232—H,(He⁺)
0.505177	0.505210 ± 30	Cu II	$(^2D_{5/2})4f\ ^3G^o_5$	—	$(^2D_{5/2})4d\ ^3F_4$	232—H,(He⁺)
0.506062	0.506050 ± 30	Cu II	$4s^2\ ^3P_1$	—	$(^2D_{3/2})4p\ ^3P^o_0$	232—H,(He*)
0.725576	0.725600 ± 20	Cu II	$(^2D_{3/2})6s\ ^1D_2$	—	$(^2D_{5/2})5p\ ^3D^o_2$	309—H,(He⁺)
0.739988	0.740020 ± 20	Cu II	$(^2D_{5/2})5p\ ^3D^o_3$	—	$(^2D_{5/2})5s\ ^3D_3$	309—H,(Ne)
0.740434	0.740450 ± 30	Cu II	$(^2D_{5/2})6s\ ^3D_3$	—	$(^2D_{5/2})5p\ ^3P^o_2$	232—CW,H,(He⁺); 282—CW,H,P,U; 271—CW,D,H; 170—CW,H

0.743816	0.743900	±20?	Cu II	$(^2D_{3/2})6s\ ^3D_1$ — $(^2D_{3/2})\ 5p\ ^3P^o_0$	309—E,H,(He+)
0.766466	0.766470	±30	Cu II	$(^2D_{5/2})6s\ ^3D_2$ — $(^2D_{5/2})\ 5p\ ^3F^o_3$	232—CW,H,(He+); 282—CW,H,P,U; 271—CW,D,H; 170—CW,H
0.773866	0.773870	±30	Cu II	$(^2D_{3/2})6s\ ^3D_1$ — $(^2D_{3/2})5p\ ^3F^o_2$	232—CW,H,(He+); 282—CW,H,P,U; 271—CW,D,H; 170—CW,H
0.777879	0.777890	±30	Cu II	$(^2D_{5/2})6s\ ^3D_2$ — $(^2F\ )4p\ ^1D^o_2$	232—CW,H,(He+)
0.780523	0.780530	±30	Cu II	$(^2D_{5/2})6s\ ^3D_2$ — $(^2D_{5/2})5p\ ^3P^o_1$	232—CW,H,(He+); 170—CW,H
0.780767	0.780780	±30	Cu II	$(^2D_{5/2})6s\ ^3D_3$ — $(^2D_{5/2})5p\ ^3F^o_4$	77—CW,H,P, (He+); 232—CW,H,P; 282—CW,H,P,U; 332—CW,H,P; 271—CW, D,H; 170—CW,H
0.782565	0.782600	±30	Cu II	$(^2D_{5/2})5p\ ^3F^o_4$ — $(^2D_{5/2})5s\ ^3D_3$	232—CW,H,(He+); 170—CW,H
0.784508	0.784530	±30	Cu II	$(^2D_{3/2})6s\ ^1D_2$ — $(^2D_{3/2})5p\ ^1F^o_3$	232—CW, H,(He+); 282—CW,H,P,U; 170—CW,H
0.789584	0.789600	±30	Cu II	$(^2D_{3/2})5p\ ^3F^o_2$ — $(^2D_{3/2})5s\ ^3D_1$	232—CW,H,(He+;) 170—CW,H
0.790254	0.790270	±30	Cu II	$(^2D_{3/2})5p\ ^1F^o_3$ — $(^2D_{3/2})5s\ ^1D_2$	232—CW,H,(He+); 171—CW, H
0.794442	0.794480	±30	Cu II	$(^2D_{5/2})5p\ ^3P^o_1$ — $(^2D_{5/2})5s\ ^3D_2$	232—CW,H,(He+); 171—CW,H
0.798814	0.798820	±30	Cu II	$(^2D_{5/2})6s\ ^3D_3$ — $(^2F\ )4p\ ^1F^o_3$	232—CW,H,(He+); 170—CW,H
0.808862	0.808858	±?	Cu II	$(^2D_{3/2})6s\ ^3D_1$ — $(^2D_{3/2})5p\ ^3D^o_1$	171—CW,H,(He+)
0.809555	0.8096	±?	Cu II	$(^2D_{5/2})5p\ ^3F^o_3$ — $(^2D_{5/2})5s\ ^3D_2$	170—CW,H,(He+)
0.819233 } 0.819223 }	0.819228	±?	Cu II	{ $(^2D_{5/2})6s\ ^3D_2$ — $(^2D_{5/2})5p\ ^3D^o_3$; $(^2D_{3/2})6s\ ^1D_2$ — $(^2D_{3/2})5p\ ^1D^o_2$ }	170—CW,H,(He+)
0.827758	0.827700	±200	Cu II	$(^2D_{5/2})tp\ ^3P^o_2$ — $(^2D_{5/2})5s\ ^3D_3$	308—H,(He+)
0.828317	0.828321	±?	Cu II	$(^2D_{5/2})6s\ ^3D_3$ — $(^2D_{5/2})5p\ ^3D^o_3$	171—H,(He+)
0.851108	0.851104	±?	Cu II	$(^2D_{5/2})6s\ ^3D_2$ — $(^2D_{5/2})5p\ ^3D^o_2$	171—H,(He+)
1.771	1.771000	±200	Cu II?	—	308—E,H
1.91495	1.915600	±200?	Cu II	$(^2D_{5/2})5p\ ^3F^o_4$ — $(^2D_{5/2})4d\ ^3G_5$	308—H,(He,Ne)
1.9473	1.948000	±200	Cu II	$(^2D_{3/2})5p\ ^1F^o_3$ — $(^2D_{3/2})4d\ ^1G_4$	308—H,(Ne)
1.97071	1.971000	±200	Cu II	$(^2D_{3/2})5p\ ^1D^o_2$ — $(^2D_{3/2})4d\ ^1F_3$	308-H,(He)
2.00002	2.000400	±200	Cu II	$(^2D_{3/2})5p\ ^3F^o_2$ — $(^2D_{3/2})4d\ ^3G_3$	308—H,(Ne)

Table 2.4.2
SILVER

Wavelength (μm)	Measured value (μm)	Identification: Ion	Upper level	—	Lower level	Ref.
0.224357	0.224340 ±20	Ag II	$(^2D_{3/2})5d\ ^1S_0$	—	$(^2D_{3/2})5p\ ^1P^o_1$	235—CW,H,(He⁺); 173—CW,H,P
0.227743	0.227760 ±20	Ag II	$(^2D_{5/2})5d\ ^3D_2$	—	$(^2D_{5/2})5p\ ^3P^o_1$	235—CW,H,(He⁺)
0.318070	0.318060 ±20	Ag II	$5s^2\ ^1G_4$	—	$(^2D_{5/2})5p\ ^3F^o_3$	235—CW,H,P,(Ne⁺); 332—H,P
0.408590	0.408620 ±20	Ag II	$5s^2\ ^1G_4$	—	$(^2D_{3/2})5p\ ^1F^o_3$	184—H,P, (Ne⁺); 332—H,P
0.478839	0.478840 ±20	Ag II	$5s^2\ ^1D_2$	—	$(^2D_{3/2})5p\ ^1P^o_1$	184—H,P,(Ne⁺); 332—H,P
0.502733	0.502720 ±20	Ag II	$5s^2\ ^1D_2$	—	$(^2D_{3/2})5p\ ^1D^o_2$	184—H,P,(Ne⁺)
0.640529	0.640430 ±20?	Ag II	$(^2D_{3/2})6p\ ^3F^o_2$	—	$(^2D_{5/2})6s\ ^3D_2$	276—CW,H,P,(He⁺)
0.800517	0.800540 ±10	Ag II	$(^2D_{5/2})6p\ ^3D^o_3$	—	$(^2D_{5/2})6s\ ^3D_3$	190—A,(He⁺); 277—CW,H,(He⁺); 332-CW,H,P
0.825482	0.825450 ±10	Ag II	$(^2D_{5/2})6p\ ^3D^o_3$	—	$(^2D_{5/2})6s\ ^3D_2$	276—CW,H,(He⁺); 277—H; 332—H,P; 171—CW,H
0.826284	0.826300 ±200	Ag II	$(^2D_{3/2})6p\ ^3D^o_1$	—	$(^2D_{3/2})6s\ ^3D_1$	308—H,(He⁺)
0.832469	0.832480 ±10	Ag II	$(^2D_{5/2})6p\ ^3D^o_2$	—	$(^2D_{5/2})6s\ ^3D_2$	276—CW,H,(He⁺); 277—H; 171—CW,H
0.837957	0.837950 ±?	Ag II	$(^2D_{5/2})6p\ ^3P^o_1$	—	$(^2D_{5/2})6s\ ^3D_2$	171—CW,H,(He⁺); 308—H
0.840384	0.840350 ±10	Ag II	$(^2D_{5/2})6p\ ^3F^o_4$	—	$(^2D_{5/2})6s\ ^3D_3$	190—A,(He⁺); 276—CW,H,G,P,(He⁺); 277—H; 332—CW,H,P; 171—CW,H
0.874833	0.874760 ±?	Ag II	$(^2D_{3/2})6p\ ^1P^o_1$	—	$(^2D_{3/2})6s\ ^1D_2$	171—CW,H,(He⁺); 308—CW,H
0.877527	0.877300 ±200	Ag II	$(^2D_{3/2})6p\ ^3F^o_2$	—	$(^2D_{3/2})6s\ ^3D_1$	308—CW,H,(He⁺)
1.37538	1.375900 ±200	Ag II	$(^2D_{5/2})6p\ ^3P^o_1$	—	$(^2D_{5/2})5d\ ^3S_1$	308—H,(He⁺)
1.59759	1.598200 ±200	Ag II	$(^2D_{5/2})6p\ ^3D^o_3$	—	$(^2D_{5/2})5d\ ^3P_2$	308—H,(He⁺)
1.64593	1.646400 ±200	Ag II	$(^2D_{5/2})6p\ ^3P^o_1$	—	$(^2D_{5/2})5d\ ^3P_1$	308—H,(He⁺)
1.71983	1.720200 ±200	Ag II	$(^2D_{5/2})6p\ ^3D^o_3$	—	$(^2D_{5/2})5d\ ^3D_3$	308—CW,H,(He⁺)
1.73413	1.734600 ±200	Ag II	$(^2D_{5/2})6p\ ^3F^o_4$	—	$(^2D_{5/2})5d\ ^3G_5$	308—CW,H,(He⁺)
1.74749	1.748000 ±200	Ag II	$(^2D_{3/2})6p\ ^3D^o_1$	—	$(^2D_{3/2})5d\ ^3D_1$	308—H,(He⁺)
1.84051	1.841000 ±200	Ag II	$(^2D_{5/2})6p\ ^3D^o_2$	—	$(^2D_{5/2})5d\ ^3F_3$	308—CW,H,(He⁺)
1.84583	1.846400 ±200?	Ag II	$(^2D_{5/2})6p\ ^3D^o_3$	—	$(^2D_{5/2})5d\ ^3F_4$	308—CW,H,(He⁺)
1.87200	1.872400 ±200	Ag II	$(^2D_{5/2})6p\ ^3F^o_3$	—	$(^2D_{5/2})5d\ ^3G_4$	308—H,(He⁺)
1.87896	1.879600 ±200	Ag II	$(^2D_{5/2})6p\ ^3P^o_1$	—	$(^2D_{5/2})5d\ ^3D_2$	308—CW,H,(He⁺)

1.89752	1.898000 ± 200	Ag II	$(^2D_{3/2})6p\ ^3F^o_2$ — $(^2D_{3/2})5d\ ^3G_3$	308—CW,H,(He⁺)
1.97086	1.971600 ± 200	Ag II	$(^2D_{3/2})6p\ ^3D^o_1$ — $(^2D_{3/2})5d\ ^3F_2$	308—H,(He⁺)
1.98177	1.982400 ± 200?	Ag II	$(^2D_{3/2})6p\ ^1P^o_1$ — $(^2D_{3/2})5d\ ^1D_2$	308—H,(He⁺)
2.07907	2.079400 ± 200	Ag II	$(^2D_{5/2})6p\ ^3F^o_3$ — $(^2D_{5/2})5d\ ^3D_3$	308—H,(He⁺)

Table 2.4.3
GOLD

Wavelength (μm)	Measured value (μm)	Identification: Ion	Upper level	—	Lower level	Ref.
0.226362	0.226400 ±?	Au II	$(^2D_{3/2})7s\ ^3D_1$	—	76659°_{2}	174—H,(He⁺)
0.253353	0.253350 ±20	Au II	$(^2D_{3/2})7s\ ^1D_2$	—	81659°_{1}	275—CW,H,P,(He⁺)
0.261640	0.261640 ±20	Au II	$(^2D_{3/2})7s\ ^3D_1$	—	82613°_{0}	275—CW,H,P, (He⁺)
0.282254	0.282220 ±20	Au II	$(^2D_{3/2})7s\ ^1D_2$	—	85699°_{3}	275—CW,H,P,(He⁺); 332—CW, H,P; 173—CW,H,P
0.284694	0.284720 ±20	Au II	$(^2D_{3/2})7s\ ^3D_1$	—	85707°_{1}	275—CW,H,P,(He⁺); 332—CW,H,P; 173—CW,H,P
0.289328	0.289350 ±20	Au II	$(^2D_{3/2})7s\ ^1D_2$	—	86565°_{2}	275—CW,H,P,(He⁺); 332—CW,H,P; 173—CW,H,P
0.291826	0.291810 ±20	Au II	$(^2D_{3/2})7s\ ^3D_1$	—	86565°_{2}	275—CW,H,P,(He⁺); 332—CW,H,P; 173—CW,H,P
0.755581	0.755600 ±20	Au II	121862°_{1}	—	$(^2D_{5/2})7s\ ^3D_2$	275—CW,H,(He⁺)
0.760050	0.760040 ±20	Au II	121784°_{2}	—	$(^2D_{5/2})7s\ ^3D_2$	275—CW,H,(He⁺)
0.827293	0.827290 ±?	Au II	120256°_{3}	—	$(^2D_{5/2})7s\ ^3D_3$	171—H,(He⁺)
0.859927	0.859900 ±?	Au II	120256°_{3}	—	$(^2D_{5/2})7s\ ^3D_2$	174—H,(He⁺)
0.886761	0.886760 ±?	Au II	119446°_{2}	—	$(^2D_{5/2})7s^3D_3$	171—H,(He⁺)

Table 2.5.1
ZINC

Wavelength (μm)	Measured value (μm)	Identification: Ion	Upper level	—	Lower level	Ref.
			Spectroscopic Reference: Zn II (73)			
0.491166[H]	0.49116 ±?	Zn II	$(^1S)4f\ ^2F^o_{5/2}$	—	$(^1S)4d\ ^2D_{3/2}$	180—(He⁺); 181—(He⁺); 314—CW,H,(He⁺); 290; 316; 266—P; 268—P; 270—D
0.492404[H]	0.4925 ±?	Zn II	$(^1S)4f\ ^2F^o_{7/2}$	—	$(^1S)4d\ ^2D_{5/2}$	105; 26—CW; 189—CW,H,(He⁺); 314; 316; 266—P; 268—P; 270—D
*0.589435[H]	0.5894 ±?	Zn II	$4s^2\ ^2D_{3/2}$	—	$(^1S)4p\ ^2P^o_{1/2}$	181—CW,P,(He*); 316; 266—P; 268—P; 270-D
0.602126[H]	0.6021 ±?	Zn II	$(^1S)5d\ ^2D_{3/2}$	—	$(^1S)5p\ ^2P^o_{1/2}$	182—CW,H,(He)
0.610249	0.610280 ±70	Zn II	$(^1S)5d\ ^2D_{5/2}$	—	$(^1S)5p\ ^2P^o_{3/2}$	296; 314—CW,H,(He); 316
0.621459[H]	0.6214 ±?	Zn II	$4s^2\ ^2D_{3/2}$	—	$(^1S)4p\ ^2P^o_{3/2}$	268—CW,H,(He*)
*0.747879[H]	0.747830 ±160	Zn II	$4s^2\ ^2D_{5/2}$	—	$(^1S)4p\ ^2P^o_{3/2}$	296; 181—CW,P,(He*); 298—CW,G; 256—hfs; 316; 266—P; 268—P; 270—D
0.758848[H]	0.758750 ±160	Zn II	$(^1S)5p\ ^2P^o_{3/2}$	—	$(^1S)5s\ ^2S_{1/2}$	296; 181—CW,P,(He⁺); 298; 316; 189—CW,H,(He⁺); 266; 268—P; 270—D
0.761219[H]	0.761118 ±160	Zn II	$(^1S)6s\ ^2S_{1/2}$	—	$(^1S)5p\ ^2P^o_{1/2}$	296
0.773250[H]	0.773250 ±50	Zn II	$(^1S)5p^2\ P^o_{1/2}$	—	$(^1S)5s\ ^2S_{1/2}$	164—CW,H,(He)
0.775786[H]	0.7757 ±?	Zn II	$(^1S)6s\ ^2S_{1/2}$	—	$(^1S)5p\ ^2P^o_{3/2}$	105—E; 66—(He⁺)
1.83083	1.831000 ±100	Zn II	$(^1S)6p\ ^2P^o_{3/2}$	—	$(^1S)6s\ ^2S_{1/2}$	63—D,P
5.08483	5.086000 ±400	Zn II	$(^1S)6p\ ^2P^o_{3/2}$	—	$(^1S)5d\ ^2D_{5/2}$	63—D,P

Table 2.5.2
CADMIUM

Wavelength (μm)	Measured value (μm)	Identification			Ref.
		Ion	Upper level	Lower level	
0.325029	0.3250 ± ?	Cd II	$5s^2\ ^2D_{3/2}$	$(^1S)5p\ ^2P^o_{1/2}$	121—CW,P,(He*); 298—CW,G,P; 123—P; 266—CW,H,(He*)
*0.441565	0.441560 ± 70	Cd II	$5s^2\ ^2D_{5/2}$	$(^1S)5p\ ^2P^o_{3/2}$	296; 107—CW,P; 297—CW,G,P,(He*);314—CW, H,(He*);123—P;316; 256—hfs;266; 270—D; **
0.488172	0.4882 ± ?	Cd II	$(^3D)5p\ ^4F^o_{5/2}$	$(^1S)5d\ ^2D_{3/2}$	26—CW,(He)
0.502548	0.50259 ± ?	Cd II	$(^3D)5p\ ^4F^o_{7/2}$	$(^1S)5d\ ^2D_{5/2}$	26—CW,P,(He)
*0.533748	0.5337 ± ?	Cd II	$(^1S)4f\ ^2F^o_{5/2}$	$(^1S)5d\ ^2D_{3/2}$	105; 189—CW,H,P,(He); 288—CW,H; 314; 256—hfs; 289—P; 316; 108—P; 266; 270—D; **
0.537813	0.5338 ± ?	Cd II	$(^1S)4f\ ^2F^o_{7/2}$	$(^1S)5d\ ^2D_{5/2}$	Same as above
0.635480	0.63548 ± ?	Cd II	$(^1S)6g\ ^2G_{7/2}$	$(^1S)4f, ^2F^o_{5/2}$	288—CW,H,(He⁺); 315; 289—P; 316; 266
0.636004	0.63601 ± ?	Cd II	$(^1S)6g\ ^2G_{9/2}$	$(^1S)4f\ ^2F^o_{7/2}$	288—CW,H,P,(He⁺); 313;315; 289—P; 316; 108—P; 266; 270—D
0.723689	0.72369 ± ?	Cd II	$(^1S)6f\ ^2F^o_{5/2}$	$(^1S)6d\ ^2D_{3/2}$	288—CW,H,(He⁺); 313, 315; 289—P; 316; 108—P; 266
0.728423	0.72843 ± ?	Cd II	$(^1S)6f\ ^2F^o_{7/2}$	$(^1S)6d\ ^2D_{5/2}$	288—CW,H,(He⁺); 313; 315; 316; 108—P; 266; 270—D
0.806687	0.80669 ± ?	Cd II	$(^1S)6p\ ^2P^o_{3/2}$	$(^1S)6s\ ^2S_{1/2}$	189—H,P,(He⁺); 290—CW,H; 316; 108—P; 266; 270—D
0.838889	0.8389 ± ?	Cd II	$(^1S)7s\ ^2S_{1/2}$	$(^1S)6p\ ^2P^o_{3/2}$	320—A,G; 89—A, λ
0.853026	0.85309 ± ?	Cd II	$(^1S)6p\ ^2P^o_{1/2}$	$(^1S)6s\ ^2S_{1/2}$	290—CW,H,(He⁺); 316; 108—P; 266; 270—D
0.865190	0.8652 ± ?	Cd II	$(^1S)9s\ ^2S_{1/2}$	$(^3D)5p\ ^4P^o_{1/2}$	174—H,(He⁺)
0.887770	0.88778 ± ?	Cd II	$(^1S)9s\ ^2S_{1/2}$	$(^1S)7p\ ^2P^o_{3/2}$	316—CW,H,(He⁺); 108—P; 270—D
3.28825	3.289000 ± 200	Cd II	$(^1S)9s\ ^2S_{1/2}$	$(^1S)8p\ ^2P^o_{3/2}$	63—D,P

Table 2.5.3
MERCURY

Wavelength (μm)	Measured value (μm)	Identification: Ion	Upper level	—	Lower level	Ref.
0.398398[H]	0.398399 ±2	Hg II	$(^1S)6p\ ^2P^o_{3/2}$	—	$6s^2\ ^2D_{5/2}$	166—S,(Ne)
0.479701	0.479700 ±10	Hg III	$5d^86s^2\ 126468_4$	—	$5d^96p\ 105628_3$	113—G,P; 114—λ; 51—CW,P
0.521082	0.5210 ±?	Hg III	$5d^86s^2\ 122735_1$	—	$5d^96p\ 103549_2$	51—CW,P
*0.567717[H]	0.5678 ±?	Hg II	$(^1S)5f\ ^2F^o_{7/2}$	—	$(^1S)6d\ ^2D_{5/2}$	16—G,P; 17—U; **
*0.614947[93]	0.6150 ±?	Hg II	$(^1S)7p\ ^2P^o_{3/2}$	—	$(^1S)7s\ ^2S_{1/2}$	16—G,P; 3; 15—CW; 52—H,hfs,λ; 93; 343; 291—CW,H,(He$^+$); 338—G,P; 101—CW,P,(He$^+$); 187—CW,G,P; 267—CW,H,P; **
0.650138	0.6501 ±?	Hg III	$5d^86s^2\ 118926_2$	—	$5d^96p\ 103549_2$	51—CW,P
0.734637[H]	0.7346 ±?	Hg II	$(^1S)7d\ ^2D_{5/2}$	—	$(^1S)7p\ ^2P^o_{3/2}$	16
0.74181[H]	0.74181 ±?	Hg II	$5d^96s6p\ ^2P^o_{3/2}$	—	$(^1S)7s\ ^2S_{1/2}$	124—G,H
*0.794466[H]	0.7945 ±?	Hg II	$(^1S)7p\ ^2P^o_{1/2}$	—	$(^1S)7s\ ^2S_{1/2}$	52—H,(He$^+$); 3; 343; 291—CW,H,(He$^+$); 187—CW,G.P; 267—CW,H,P,(He$^+$);**
0.85482[H]	0.8547 ±?	Hg II	$(^1S)5g\ ^2G_{7/2}$	—	$121960\ ^2F^o_{5/2}$	25
0.86220	0.8628 ±?	Hg II?	$?(^1S)8p\ ^2P^o_{3/2}$	—	$116200_{5/2}$	25
0.8677	0.8677 ±?	Hg II?		—		25
0.93968	0.9396 ±?	Hg II	$(^1S)10s\ ^2S_{1/2}$	—	$(^1S)8p\ ^2P^o_{3/2}$	25
1.0584	1.0586 ±?	Hg II	$(^1S)8s\ ^2S_{1/2}$	—	$(^1S)7p\ ^2P^o_{3/2}$	16; 25
1.2545	1.2545 ±?	Hg II?		—		25
1.2981	1.2981 ±?	Hg II?		—		25
1.5555	1.5550 ±?	Hg II	$(^1S)7p\ ^2P^o_{3/2}$	—	$(^1S)6d\ ^2D_{5/2}$	25; 338

Table 2.6.1
BORON

Wavelength (μm)	Measured value (μm)	Identification				Ref.
		Ion	Upper level	—	Lower level	
0.345129	0.345132 ± ?	B II	$2p^2\ ^1D_2$	—	$(^2S)2p\ ^1P^o_1$	71

Table 2.6.2
ALUMINUM

Wavelength (μm)	Measured value (μm)	Identification				Ref.
		Ion	Upper level	—	Lower level	
0.358655	0.358660 ± 10	Al II	$(^2S)4f\ ^3F^o_4$	—	$(^2S)3d\ ^3D_3$	292—H,(Ne$^+$)
0.358744	0.358740 ± 20	Al II	$(^2S)4f\ ^3F^o_2$	—	$(^2S)3d\ ^3D_1$	115—H,P,(Ne$^+$)
0.691994	0.691998 ± 10	Al II	$(^2S)5s\ ^1S_0$	—	$(^2S)4p\ ^1P^o_1$	115—CW,H,P,(Ne$^+$); 292—G,H,λ; 281—CW,H,U
0.704205	0.704209 ± 10	Al II	$(^2S)4p\ ^3P^o_2$	—	$(^2S)4s\ ^3S_1$	115—CW,H,P,(Ne$^+$); 292—G,H,λ; 281—CW,H,U
0.705661	0.705640 ± 20	Al II	$(^2S)4p\ ^3P^o_1$	—	$(^2S)4s\ ^3S_1$	115—H,P,(Ne$^+$)
0.747145	0.747149 ± 10	Al II	$(^2S)4f\ ^1F^o_3$	—	$(^2S)3d\ ^1D_2$	115—CW,H,(Ne$^+$); 292—G,λ; 281—CW, H,U

Table 2.6.3
INDIUM

Wavelength (μm)	Measured value (μm)	Identification				Ref.
		Ion	Upper level	—	Lower level	
0.468111	0.468050 ± 70	In II	$(^2S)4f\ ^3F^o_4$	—	$(^2S)5d\ ^3D_3$	296

Table 2.6.4
THALLIUM

Wavelength (μm)	Measured value (μm)	Identification				Ref.
		Ion	Upper level	—	Lower level	
0.473705[H]	0.4737 ± ?	T1 II	$(^2S)5f\ ^1F_3^o$	—	$(^2S)6d\ ^1D_2$	168—A,G,(He)
0.498135[H]	0.4981 ± ?	T1 II	$(^2S)5f\ ^3F_2^o$	—	$(^2S)6d\ ^3D_1$	168—A,G,(He)
0.507854[H]	0.5079 ± ?	T1 II	$(^2S)5f\ ^3F_3^o$	—	$(^2S)6d\ ^3D_2$	168—A,G,(He)
0.515214[H]	0.5152 ± ?	T1 II	$(^2S)5f\ ^3F_4^o$	—	$(^2S)6d\ ^3D_3$	168—A,G,(He); 211—CW,G,P,U,(Ne^+)
0.594904[H]	0.5949 ± ?	T1 II	$(^2S)7p\ ^3P_2^o$	—	$(^2S)7s\ ^3S_1$	168—A,G,(Ne); 211—CW,G,P,U,(Ne^+); 134—CW,H,P,(Ne^+); 135—CW,D,G,H
0.69505[H]	0.6951 ± ?	T1 II	$(^2S)7p\ ^1P_1^o$	—	$(^2S)7s\ ^1S_0$	168—A,G,(Ne); 211—CW,G,P,U,(Ne^+); 135—CW,D,G,H
0.9350	0.9350 ± ?	T1 II		—		168—A,G,(Ne)
1.1350	1.1350 ± ?	T1 II		—		168—A,G,(He)
1.1788	1.1750 ± ?	T1 II	$(^2S)6f\ ^3F_3^o$	—	$(^2S)7d\ ^3D_2$	168—A,G,(He)

Table 2.7.1
CARBON

Wavelength (μm)	Measured value (μm)	Identification				Ref.
		Ion	Upper level	—	Lower level	
		Spectroscopic References: C III (27); C IV (28)				
v.154819	0.154820 ±30	C IV	2p $^2P^o_{3/2}$	—	2s $^2S_{1/2}$	334—D,G,P,U
v.155077	0.155090 ±30	C IV	2p $^2P^o_{1/2}$	—	2s $^2S_{1/2}$	334—D,G,P,U
0.464742	0.464740 ±04	C III	(^{2}S)3p $^3P^o_2$	—	(^{2}S)3s $^3S_1$	229—D; 37—λ: 111—X; 47-X
0.465025	0.465021 ±04	C III	(^{2}S)3p $^3P^o_1$	—	(^{2}S)3s $^3S_1$	229—D; 37—λ; 111—X; 47—X
0.514516	0.514570 ±50	C II	(^{3}P°)3p $^4P_{5/2}$	—	(^{3}P°)3s $^4P_{5/2}$	263—D,P
0.657803	0.657800 ±50	C II	(^{1}S)3p $^2P^o_{3/2}$	—	(^{1}S)3s $^2S_{1/2}$	263—D,P
0.678375	0.678360 ±50	C II	(^{3}P°)3p $^4D_{7/2}$	—	(^{3}P°)3s $^4P_{5/2}$	263—D,P

Table 2.7.2
SILICON

Wavelength (μm)	Measured value (μm)	Identification				Ref.
		Ion	Upper Level	—	Lower level	
		Spectroscopic References: Si II (294); Si III (324); Si IV (323)				
0.408885	0.408890 ±10	?Si IV	4p $^2P^o_{3/2}$	—	4s $^2S_{1/2}$	37: 57—X: [cf. Ar]
0.455262	0.455259 ±06	Si III	(^{2}S)4p $^3P^o_2$	—	(^{2}S)4s $^3S_1$	60—E; 253
0.456782	0.456784 ±06	Si III	(^{2}S)4p $^3P^o_1$	—	(^{2}S)4s $^3S_1$	60—E; 253
0.634710	0.634724 ±06	Si II	(^{1}S)4p $^2P^o_{3/2}$	—	(^{1}S)4s $^2S_{1/2}$	60—E; 57—D: 253
0.637136	0.637148 ±06	Si II	(^{1}S)4p $^2P^o_{1/2}$	—	(^{1}S)4s $^2S_{1/2}$	60—E; 253
0.667188	0.667193 ±06	Si II	(^{3}P°)4p $^4D_{7/2}$	—	(^{3}P°)4s $^4P^o_{5/2}$	60—E; 253

Table 2.7.3
GERMANIUM

Wavelength (μm)	Measured value (μm)	Identification				Ref.
		Ion	Upper level	—	Lower level	
		Spectroscopic reference: Ge II (295)				
0.513175	0.513150 ± 70	Ge II	$(^1S)4f\ ^2F^o_{5/2}$	—	$(^1S)4d\ ^2D_{3/2}$	296
0.517865	0.517840 ± 70	Ge II	$(^1S)4f\ ^2F^o_{7/2}$	—	$(^1S)4d\ ^2D_{5/2}$	296

Table 2.7.4
TIN

Wavelength (μm)	Measured value (μm)	Identification				Ref.
		Ion	Upper level	—	Lower level	
0.558943[H]	0.5589 ± ?	Sn II	$(^1S)4f\ ^2F^o_{5/2}$	—	$(^1S)5d\ ^2D_{3/2}$	346—A
0.579918[H]	0.579870 ± 70	Sn II	$(^1S)4f\ ^2F^o_{7/2}$	—	$(^1S)5d\ ^2D_{5/2}$	296; 346—A
0.645358[H]	0.645300 ± 70	Sn II	$(^1S)6p\ ^2P^o_{3/2}$	—	$(^1S)6s\ ^2S_{1/2}$	296; 298—CW,G,(He*)
0.657926[H]	0.657903 ± 06	Sn		—		71; 58—D,X; 346—X
0.684420[H]	0.684400 ± 70	Sn II	$(^1S)6p\ ^2P^o_{1/2}$	—	$(^1S)6s\ ^2S_{1/2}$	296; 298—CW, (He*)
1.06118	1.062 ± ?	Sn II	$(^1S)5f\ ^2F^o_{5/2}$	—	$(^1S)6d\ ^2D_{3/2}$	346—A; 317—E
1.07388	1.074 ± ?	Sn II	$(^1S)5f\ ^2F^o_{7/2}$	—	$(^1S)6d\ ^2D_{5/2}$	346—A

Table 2.7.5
LEAD

Wavelength (μm)	Measured value (μm)	Identification: Ion	Upper level	—	Lower level	Ref.
0.53721[H]	0.537210 ±70	Pb II	$(^1S)5f\ ^2F^o_{7/2}$	—	$6s6p^2\ ^4P_{5/2}$	296
0.56088[H]	0.560860 ±50	Pb II	$(^1S)7p\ ^2P^o_{3/2}$	—	$(^1S)7s\ ^2S_{1/2}$	303—CW
0.66600[H]	0.666010 ±50	Pb II	$(^1S)7p\ ^2P^o_{1/2}$	—	$(^1S)7s\ ^2S_{1/2}$	303—CW
1.1592	1.159 ±?	Pb II	$6p^2\ ^2D_{3/2}$	—	$(^1S)7p\ ^2P_{1/2}$	318—G

Table 2.8.1
NITROGEN

Spectroscopic References: N II (97); N IV (139)

Wavelength (μm)	Measured value (μm)	Identification: Ion	Upper level	—	Lower level	Ref.
0.336734	0.336732 ±06	N III	$(^3P^o)3p\ ^4P_{5/2}$	—	$(^3P^o)3s\ ^4P^o_{5/2}$	59
0.347871	0.347876 ±05	N IV	$(^2S)3p\ ^3P^o_2$	—	$(^2S)3s\ ^3S_1$	229; 143—U; 222—P
0.348299	0.348302 ±06	N IV	$(^2S)3p\ ^3P^o_1$	—	$(^2S)3s\ ^3S_1$	59
0.399501	0.399499 ±01	N II	$(^2P^o)3p\ ^1D_2$	—	$(^2P^o)3s\ ^1P^o_1$	5; 325—P
0.409732	0.409729 ±06	N III	$(^1S)3p\ ^2P^o_{3/2}$	—	$(^1S)3s\ ^2S_{1/2}$	229; 143-U; 325—P; 222—P
0.410338	0.410336 ±02	N III	$(^1S)3p\ ^2P^o_{1/2}$	—	$(^1S)3s\ ^2S_{1/2}$	229; 5—λ; 325—P; 222—P
0.451088	0.451089 ±02	N III	$(^3P^o)3p\ ^4D_{5/2}$	—	$(^3P^o)3s\ ^4P^o_{3/2}$	229; 325—P; 222—P,λ
0.451487	0.451486 ±03	N III	$(^3P^o)3p\ ^4D_{7/2}$	—	$(^3P^o)3s\ ^4P^o_{5/2}$	229; 5—λ; 325—P; 222—P
0.462140	0.462100 ±80	N II	$(^2P^o)3p\ ^3P_0$	—	$(^2P^o)3s\ ^3P^o_1$	249; 325—P
0.463055	0.463051 ±02	N II	$(^2P^o)3p\ ^3P_2$	—	$(^2P^o)3s\ ^3P^o_2$	229; 5—λ; 325—P; 222—P

0.464310	0.464390 ±80	N II	$(^2P^o)3p\ ^3P_1$	—	$(^2P^o)3s\ ^3P^o_2$	249; 325—P
0.501638	0.501639 ±?	N II	$(^2P^o)3d\ ^3F_2$	—	$(^2P^o)3p\ ^3D_2$	109
0.566663	0.566662 ±03	N II	$(^2P^o)3p\ ^3D_2$	—	$(^2P^o)3s\ ^3P^o_1$	231; 109
0.567601	0.567603 ±03	N II	$(^2P^o)3p\ ^3D_1$	—	$(^2P^o)3s\ ^3P^o_0$	231
*0.567956	0.567953 ±03	N II	$(^2P^o)3p\ ^3D_3$	—	$(^2P^o)3s\ ^3P^o_2$	148; 37; 231—λ; 109
0.568622	0.568690 ±80	N II	$(^2P^o)3p\ ^3D_1$	—	$(^2P^o)3s\ ^3P^o_1$	249
0.648209	0.648260 ±60	N II	$(^2P^o)3p\ ^1P_1$	—	$(^2P^o)3s\ ^1P^o_1$	249; 22

Table 2.8.2
PHOSPHORUS

Wavelength (μm)	Measured value (μm)	Identification: Ion	Upper level	—	Lower level	Ref.
		Spectroscopic Reference: P II (225)				
0.334769	0.334776 ±06	P IV	$(^2S)4p\ ^3P^o_2$	—	$(^2S)4s\ ^3S_1$	60
0.422208	0.422225 ±06	P III	$(^1S)4p\ ^2P^o_{3/2}$	—	$(^1S)4s\ ^2S_{1/2}$	60
*0.602418	0.602427 ±06	P II	$(^2P^o)4p\ ^3D_2$	—	$(^2P^o)4s\ ^3P^o_1$	106; 24—CW,U; 60—λ
0.603404	0.603419 ±06	P II	$(^2P^o)4p\ ^3D_1$	—	$(^2P^o)4s\ ^3P_0$	60
*0.604312	0.604322 ±06	P II	$(^2P^o)4p\ ^3D_3$	—	$(^2P^o)4s\ ^3P^o_2$	106; 24—CW,U; 60—λ
0.608782	0.608804 ±06	P II	$(^2P^o)4p\ ^3D_1$	—	$(^2P^o)4s\ ^3P^o_1$	60
0.616559	0.616574 ±06	P II	$(^2P^o)4p\ ^3D_2$	—	$(^2P^o)4s\ ^3P^o_2$	60
0.784563	0.7846 ±?	P II	$(^2P^o)4p\ ^1P_1$	—	$(^2P^o)4s\ ^1P^o_1$	106

Table 2.8.3
ARSENIC

Wavelength (μm)	Measured value (μm)	Identification Ion	Upper level — Lower level	Ref.
			Spectroscopic Reference: (215)	
0.538520	0.538510 ±40	As II	$(^2P^o)6s\ ^1P^o_1$ — $(^2P^o)5p\ ^3P_1$	265—CW,H,(He⁺); 272—D
0.549695	0.549680 ±40	As II	$(^2P^o)6s\ ^3P^o_2$ — $(^2P^o)5p\ ^3D_3$	265—CW,H,(He⁺); 272—D
0.549773	0.549760 ±40	As II	$(^2P^o)5p\ ^3D_1$ — $(^2P^o)5s\ ^3P^o_0$	18—U; 265-CW,H,λ,(He⁺); 272—D,P
0.555809	0.555820 ±40	As II	$(^2P^o)5p\ ^3D_2$ — $(^2P^o)5s\ ^3P^o_1$	18—U; 265—CW,H,λ,(He⁺); 272—D,P
0.565132	0.565200 ±100	As II	$(^2P^o)5p\ ^3D_3$ — $(^2P^o)5s\ ^3P^o_2$	18—U
0.583790	0.583800 ±40	As II	$(^2P^o)6s\ ^3P^o_2$ — $(^2P^o)5p\ ^3P_2$	265—CW,H,(He⁺); 272—D
0.617027	0.617020 ±40	As II	$(^2P^o)5p\ ^1P_1$ — $(^2P^o)5s\ ^3P^o_1$	18—U; 265—CW,H,(He⁺); 272—D
0.651174	0.651180 ±40	As II	$(^2P^o)6s\ ^1P^o_1$ — $(^2P^o)5p\ ^1D_2$	265—CW,G,H,P,(He⁺); 272—D,P
0.710272	0.710250 ±40	As II	$(^2P^o)5p\ ^3P_1$ — $4p^3\ ^3P^o_2$	265—CW,H,(He⁺); 272—D

Table 2.8.4
ANTIMONY

Wavelength (μm)	Measured value (μm)	Identification Ion	Upper level — Lower level	Ref.
0.613000	0.613000 ±100	Sb II	$(^2P^o)6p\ ^3D_3$ — $(^2P^o)6s\ ^3P^o_2$	19
0.699563	0.6996 ±?	Sb II	$(^2P^o)6p\ ^3D_1$ — $(^2P^o)6s\ ^3P^o_1$	169—G,(He⁺)

Table 2.8.5
BISMUTH

Wavelength (μm)	Measured value (μm)	Identification: Ion	Upper level — Lower level	Ref.
0.456118	0.456070 ± 10	Bi III	$(^1S)7p\ ^2P^o_{1/2}$ — $(^1S)7s\ ^2S_{1/2}$	193—G
0.571924	0.571920 ± 10	Bi II	$6p_{1/2}7p_{1/2}\,^3P_0$ — $6p_{1/2}\,7s\ ^3P^o_1$	193—G
0.75990	0.759870 ± 50	Bi III	$(^1S)6f\ ^2F^o_{5/2}$ — $(^1S)7d\ ^2D_{3/2}$	193—G
0.806881	0.806920 ± 50	Bi III	$(1S)6f\ ^2F^o_{7/2}$ — $(^1S)7d\ ^2D_{5/2}$	193—G

Table 2.9.1
OXYGEN

Wavelength (μm)	Measured value (μm)	Identification: Ion	Upper level — Lower level	Ref.
		Spectroscopic Reference: O IV (50)		
0.278104	0.278150 ± 50	O V	$(^2S)3p\ ^3P^o_2$ — $(^2S)3s\ ^3S_1$	144—U
*0.298378	0.298389 ± 06	O III	$(^2P^o)3p\ ^1D_2$ — $(^2P^o)3s\ ^1P^o_1$	37—E; 59; 144—U; 222—P
0.304713	0.304715 ± 06	O III	$(^2P^o)3p\ ^3P_2$ — $(^2P^o)3s\ ^3P^o_2$	37—E; 59; 144—U; 222—P
0.306342	0.306346 ± 06	O IV	$(^1S)3p\ ^2P^o_{3/2}$ — $(^1S)3s\ ^2S_{1/2}$	59; 144—U; 222—P
0.338130 } 0.338121 }	0.338134 ± 06	O IV	{ $(^3P^o)3p\ ^4D_{3/2}$ — $(^3P^o)3s\ ^4P^o_{1/2}$; $(^3P^o)3p\ ^4D_{5/2}$ — $(^3P^o)3s\ ^4P^o_{3/2}$ }	59
0.338554	0.338554 ± 06	O IV	$(^3P^o)3p\ ^4D_{7/2}$ — $(^3P^o)3s\ ^4P^o_{5/2}$	59; 144—U
0.372733	0.372711 ± 50	O II	$(^3P)3p\ ^4S^o_{3/2}$ — $(^3P)3s\ ^4P_{3/2}$	144—U
*0.374949	0.374947 ± 04	O II	$(^3P)3p\ ^4S^o_{3/2}$ — $(^3P)3s\ ^4P_{5/2}$	229; 37; 144—U; 2—P,U; 222—P
*0.375467	0.375468 ± 04	O III	$(^2P^o)3p\ ^3D_2$ — $(^2P^o)3s\ ^3P^o_1$	229; 37; 144—U; 259—P; 2—P,U; 222—P
0.375720	0.37572 ± ?	O III	$(^2P^o)3p\ ^3D_1$ — $(^2P^o)3s\ ^3P^o_0$	259—P;222—E,P
*0.375988	0.375989 ± 05	O III	$(^2P^o)3p\ ^3D_3$ — $(^2P^o)3s\ ^3P^o_2$	229; 37; 144—U; 259—P; 2—P,U; 222—P
0.377399	0.37738 ± ?	O III	$(^2P^o)3p\ ^3D_1$ — $(^2P^o)3s\ ^3P^o_1$	259

Table 2.9.1 (continued)
OXYGEN

Wavelength (μm)	Measured value (μm)	Identification: Ion	Upper level	—	Lower level	Ref.
*0.434738	0.434738 ±04	O II	$(^1D)3p\ ^2D^o_{3/2}$	—	$(^1D)3s\ ^2D_{3/2}$	229; 37; 144—U; 325—P
*0.435128	0.435126 ±04	O II	$(^1D)3p\ ^2D^o_{5/2}$	—	$(^3P)3s\ ^2P_{3/2}$	229; 37; 325—P; 259
*0.441488	0.441439 ±04	O II	$(^3P)3p\ ^2D^o_{5/2}$	—	$(^3P)3s\ ^2P_{3/2}$	229; 37; 325—P
*0.441697	0.441697 ±04	O II	$(^3P)3p\ ^2D^o_{3/2}$	—	$(^3P)3s\ ^2P_{1/2}$	229; 37; 325—P
0.460552	0.460552 ±09	O		—		229
0.464914	0.464908 ±10	O II	$(^3P)3p\ ^4D^o_{7/2}$	—	$(^3P)3s\ ^4P_{5/2}$	148; 37
*0.559237	0.559237 ±06	O III	$(^2P^o)3p\ ^1P_1$	—	$(^2P^o)3s\ ^1P^o_1$	229; 22—P; 37; 109; 259—G,P; 222—P
0.664099	0.664020 ±100	O II	$(^3P)3p\ ^2S^o_{1/2}$	—	$(^3P)3s\ ^2P_{1/2}$	22—P
0.666694	0.666694 ±?	O II	$(^3P)4p\ ^2P^o_{1/2}$	—	$(^3P)3d\ ^2P_{3/2}$	109; [83]
0.672136	0.672138 ±04	O II	$(^3P)3p\ ^2S^o_{1/2}$	—	$(^3P)3s\ ^2P_{3/2}$	229; 109

Table 2.9.2
SULFUR

Wavelength (μm)	Measured value (μm)	Identification: Ion	Upper level	—	Lower level	Ref.
0.263896	0.263896 ±01	S III, IV?		—		70—D; 224—D,P,λ
0.332486	0.332486 ±01	S III	$(^2P^o)4p\ ^3P_2$	—	$(^2P^o)3d\ ^3P^o_2$	70—D; 224—D,P,λ
0.349733	0.349733 ±01	S III		—		70—D; 224—D,P,λ
0.370937	0.370935 ±01	S III	$(^2P^o)4p\ ^3D_2$	—	$(^2P^o)3d\ ^3P^o_1$	70—D; 224—D,P,λ
0.492532	0.492560 ±06	S II	$(^3P)4p\ ^4P^o_{3/2}$	—	$(^3P)4s\ ^4P_{1/2}$	70—D
0.501401	0.501424 ±06	S II	$(^3P)4p\ ^2P^o_{3/2}$	—	$(^3P)4s\ ^2P_{3/2}$	70—D
0.503239	0.503262 ±06	S II	$(^3P)4p\ ^4P^o_{5/2}$	—	$(^3P)4s\ ^4P_{5/2}$	70—D
0.516032	0.516032 ±06	S		—		70—D
0.521962	0.521962 ±06	S		—		70—D
*0.532070	0.532088 ±06	S II	$(^1D)4p\ ^2F^o_{7/2}$	—	$(^1D)4s\ ^2D_{5/2}$	106; 24—CW,U; 70—D,λ
*0.534567	0.534583 ±06	S II	$(^1D)4p\ ^2F^o_{5/2}$	—	$(^1D)4s\ ^2D_{3/2}$	106; 24—CW,U; 70—D,λ
0.542864	0.542874 ±06	S II	$(^3P)4p\ ^4D^o_{3/2}$	—	$(^3P)4s\ ^4P_{1/2}$	106; 70—D,λ
*0.543274	0.543287 ±06	S II	$(^3P)4p\ ^4D^o_{5/2}$	—	$(^3P)4s\ ^4P_{3/2}$	106; 24—CW,U; 70—D,λ
*0.545380	0.545388 ±06	S II	$(^3P)4p\ ^4D^o_{7/2}$	—	$(^3P)4s\ ^4P_{5/2}$	106; 24—CW,U; 70—D,λ
0.547360	0.547374 ±06	S II	$(^3P)4p\ ^4D^o_{1/2}$	—	$(^3P)4s\ ^4P_{1/2}$	106; 70—D,λ
0.550965	0.550990 ±06	S II	$(^3P)4p\ ^4D^o_{3/2}$	—	$(^3P)4s\ ^4P_{3/2}$	70—D
0.556491	0.556511 ±06	S II	$(^3P)4p\ ^4D^o_{5/2}$	—	$(^3P)4s\ ^4P_{5/2}$	70—D
*0.563999	0.564012 ±06	S II	$(^3P)4p\ ^2D^o_{5/2}$	—	$(^3P)4s\ ^2P_{3/2}$	106; 24—CW,U; 70—D,λ
0.564698	0.564716 ±06	S II	$(^3P)4p\ ^2D^o_{3/2}$	—	$(^3P)4s\ ^2P_{1/2}$	106; 70—D,λ
0.581919	0.581935 ±06	S II	$(^3P)4p\ ^2D^o_{3/2}$	—	$(^3P)4s\ ^2P_{3/2}$	70—D

Table 2.9.3
SELENIUM

Wavelength (μm)	Measured value (μm)	Identification: Ion	Upper level	—	Lower level	Ref.
0.446760	0.446800 ± 50	Se II	$(^3P)5p\ ^2P^o_{1/2}$	—	$(^3P)5s\ ^2P_{1/2}$	199—CW,G,(He⁺)
0.460431	0.460460 ± 50	Se II	$(^3P)5p\ ^2D^o_{5/2}$	—	$(^3P)5s\ ^4P_{5/2}$	300—CW,P,(He⁺); 151—P; 266-CW,H,(He⁺)
0.461877	0.461910 ± 50	Se II	$(^3P)5p\ ^4P^o_{5/2}$	—	$(^3P)5s\ ^4P_{3/2}$	199—CW,G,(He⁺)
0.464843	0.464860 ± 50	Se II	$(^3P)5p\ ^4P^o_{3/2}$	—	$(^3P)5s\ ^4P_{1/2}$	300—CW,P,(He⁺); 151—P
0.471782	0.471850 ± 50	Se II	$?(^3P)5p\ ^4S^o_{3/2}$	—	$98118_{1/2}$	199—CW,G,(He⁺)
0.474098	0.474060 ± 50	Se II	$(^3P)5p\ ^2P^o_{3/2}$	—	$4s4p^4\ ^2P_{3/2}$	199—CW,G,(He⁺)
0.476368	0.476410 ± 50	Se II	$(^3P)5p\ ^2D^o_{3/2}$	—	$(^3P)5s\ ^4P_{3/2}$	300—CW,(He⁺)
0.476554	0.476510 ± 50	Se II	$(^3P)5p\ ^2P^o_{1/2}$	—	$4s4p^4\ ^2P_{3/2}$	199—CW,G,(He⁺); 151—P
0.484063	0.484060 ± 50	Se II	$(^3P)5p\ ^2S^o_{1/2}$	—	$(^3P)5s\ ^4P_{3/2}$	300—CW,(He⁺)
0.484499	0.485500 ± 50	Se II	$(^3P)5p^4S^o_{3/2}$	—	$(^3P)5s^4P_{5/2}$	300—CW,G,(He⁺); 194—G; 151—P; 266-CW,H,(He⁺)
*0.497572	0.497610 ± 50	Se II	$(^3P)5p^2D^o_{5/2}$	—	$4s4p^{42}P_{3/2}$	300—CW,G,P,(He⁺); 194—G; 151—P; 266-CW,H,(He⁺)
*0.499279	0.499290 ± 50	Se II	$(^3P)5p\ ^4P^o_{3/2}$	—	$(^3P)5s\ ^4P_{3/2}$	300—CW,G,P,(He⁺); 194—G; 151—P; 266—CW,H,(He⁺)
*0.506865	0.506870 ± 50	Se II	$(^3P)5p\ ^4P^o_{5/2}$	—	$(^3P)5s\ ^4P_{5/2}$	300—CW,G,P,(He⁺); 199—P; 194—G; 151—P; 266—CW,H,(He⁺)
0.509650	0.509610 ± 50	Se II	$?(^3P)5p\ ^4D^o_{7/2}$	—	$(^3P)4d\ ^4F_{9/2}$	18—U; 300—CW,(He⁺); 194—G; 151
0.514215	0.514190 ± 50	Se II	$(^3P)5p\ ^4D^o_{3/2}$	—	$(^3P)5s\ ^4P_{1/2}$	300—CW,G,(He⁺); 194—G
*0.517593	0.517600 ± 50	Se II	$(^3P)5p\ ^4D^o_{5/2}$	—	$(^3P)5s\ ^4P_{3/2}$	300—CW,G,P,(He⁺); 199—P; 194—G; 151—P; 266—CW,H,(He⁺)
*0.522749	0.522760 ± 50	Se II	$(^3P)5p\ ^4D^o_{7/2}$	—	$(^3P)5s\ ^4P_{5/2}$	18—U; 300—CW,G,P,(He⁺); 199—G,P,hfs; 194—G; 151—P; 266-Cw,H,(He⁺)
0.525309	0.525260 ± 50	Se II	$(^3P)5p\ ^2D^o_{5/2}$	—	$(^3P)5s\ ^2P_{3/2}$	300—CW,G,(He⁺); 194—G; 266—CW,H,(He⁺)

0.525367	0.525320 ±50	Se II	$(^3P)5p\ ^4D^o_{1/2}$ — $(^3P)5s\ ^4P^o_{1/2}$	300—CW,G,(He+); 194—G; 266—CW,H,(He+)
0.527115	0.527130 ±50	Se II	$(^1D)5p\ ^2D^o_{5/2}$ — $(^1D)5s\ ^2D_{3/2}$	300—CW,(He+); 194—G
0.527127			$(^3P)5p\ ^4D^o_{5/2}$ — $(^3P)4d\ ^4F_{7/2}$	
0.530539	0.530550 ±50	Se II	$(^3P)5p\ ^2D^o_{3/2}$ — $(^3P)5s\ ^2P_{1/2}$	300—CW,G,P,(He+); 194—G; 151—P
0.552266	0.552280 ±50	Se II	$(^3P)5p\ ^4P^o_{3/2}$ — $(^3P)5s\ ^4P_{5/2}$	300—CW,(He+); 194—G; 151—P; 266—CW,H,(He+)
0.552244			$(^3P)5p\ ^4P^o_{5/2}$ — $4s4p^4\ ^2P_{3/2}$	
0.556688	0.556710 ±50	Se II	$(^3P)5p\ ^4D^o_{3/2}$ — $(^3P)5s\ ^4P_{3/2}$	199—CW,G,(He+); 194—G
0.559113	0.559160 ±50	Se II	$(^3P)5p\ ^4P^o_{3/2}$ — $(^3P)5s\ ^2P_{1/2}$	300—CW,(He+); 151—P
0.562315	0.562280 ±50	Se II	$(^3P)5p\ ^4P^o_{1/2}$ — $(^3P)5s\ ^4P_{1/2}$	199—CW,G,(He+)
0.569782	0.569790 ±50	Se II	$(^3P)5p\ ^4D^o_{1/2}$ — $(^3P)5s\ ^4P_{3/2}$	300—CW,(He+); 151—P
0.574761	0.574790 ±50	Se II	$(^3P)5p\ ^4D^o_{5/2}$ — $(^3P)5s\ ^4P_{5/2}$	300—CW,(He+)
0.584261	0.584280 ±50	Se II	$(^3P)5p\ ^2S^o_{1/2}$ — $4s4p^4\ ^2P_{3/2}$	199—CW,G,(He+)
0.586623	0.586670 ±50	Se II	$(^3P)5p\ ^4P^o_{5/2}$ — $(^3P)5s\ ^2P_{3/2}$	199—CW,G,(He+); 151—P
0.605592	0.605630 ±50	Se II	$(^3P)5p\ ^2P^o_{3/2}$ — $104874_{3/2}$	300—CW,G, (He+); 194—G; 151—P; 266—CW,H, (He+)
0.606573	0.606610 ±50	Se II	$(^3P)5p\ ^4P^o_{3/2}$ — $4s4p^4\ ^2P_{3/2}$	199—CW,G,(He+)
0.610198	0.610210 ±50	Se II	$(^3P)5p\ ^2D^o_{3/2}$ — $(^3P)5s\ ^2P_{3/2}$	199—CW,G,(He+)
0.644426	0.644390 ±50	Se II	$(^3P)5p\ ^2D^o_{5/2}$ — $104874_{3/2}$	300—CW,(He+); 194—G; 151—P; 266—CW,G,(He+)
0.649049	0.649010 ±50	Se II	$(^3P)5p\ ^4D^o_{1/2}$ — $(^3P)5s\ ^2P_{1/2}$	300—CW,(He+); 194—G; 151—P; 266—CW,(He+)
0.653440	0.653460 ±50	Se II	$(^3P)5p\ ^2P^o_{1/2}$ — $105974_{1/2}$	300—CW,(He+)
0.706389	0.706420 ±50	Se II	$(^3P)5p\ ^4P^o_{1/2}$ — $(^3P)5s\ ^2P_{1/2}$	199—CW,G,(He+)
0.739206	0.739240 ±50	Se II	$(^3P)5p\ ^4P^o_{5/2}$ — $104874_{3/2}$	199—CW,G,(He+)
0.767472	0.767490 ±50	Se II	$(^3P)5p\ ^2P^o_{3/2}$ — $(^1D)5s\ ^2D_{5/2}$	199—CW,G,(He+)
0.772406	0.772360 ±50	Se II	$(^3P)5p\ ^4D^o_{1/2}$ — $(^3P)5s\ ^2P_{3/2}$	199—CW,G,(He+)
0.779610	0.779620 ±50	Se II	$(^3P)5p^2P^o_{1/2}$ — $(^1D)5s\ ^2D_{3/2}$	199—CW,G,(He+)
0.783877	0.783930 ±50	Se II	$(^3P)5p\ ^4P^o_{1/2}$ — $4s4p^4\ ^2P_{3/2}$	199—CW,G,(He+)
0.830930	0.830890 ±50	Se II	$(^3P)5p\ ^2D^o_{5/2}$ — $(^1D)5s\ ^2D_{5/2}$	199—CW,G,(He+)
0.924930	0.924930 ±100	Se II	—	199—CW,G,(He+)
0.995515	0.995470 ±100	Se II	$(^3P)5p\ ^4P^o_{5/2}$ — $(^1D)5s\ ^2D_{5/2}$	199—CW,G,(He+)
*1.04088	1.040940 ±100	Se II	$(^3P)5p\ ^4D^o_{1/2}$ — $104694_{3/2}$	199—CW,G,(He+)
*1.25867	1.258790 ±100	Se II	$(^3P)5p\ ^4P^o_{3/2}$ — $108834_{5/2}$	199—CW,G,(He+)

Table 2.9.4
TELLURIUM

Wavelength (μm)	Measured value (μm)	Identification: Ion	Upper level	—	Lower level	Ref.
		Spectroscopic Reference: Te II (140)				
0.48429	0.484330 ±40	Te II	$122888^o_{5/2}$	—	$102245_{3/2}$	304—CW,G,(He⁺)
0.502039[140]	0.502000 ±40	Te II		—		304—CW,G,(He⁺)
0.525641[140]	0.525640 ±40	Te II		—		304—CW,G,(He⁺)
0.544985	0.544980 ±40	Te II	$103936^o_{3/2}$	—	$85592_{5/2}$	304—CW,G,(Ne⁺)
0.54540	0.545400 ±50	Te II?		—		335
0.547905	0.547930 ±40	Te II	$105006^o_{3/2}$	—	$86760_{3/2}$	304—CW,G,(Ne⁺)
0.557634	0.557650 ±40	Te II	$5s^25p^2\,(^1D)6p^o_{7/2}$	—	$94860_{5/2}$	19; 335-λ; 304—CW,G,(He⁺)
0.56405	0.564050 ±50	Te II?		—		335
0.566622	0.566610 ±40	Te II	$101221^o_{3/2}$	—	$83577_{1/2}$	304—CW,G,(He⁺)
0.570815	0.570850 ±50	Te II	$103106^o_{7/2}$	—	$85592_{5/2}$	19; 335—λ; 333—CW,P; 304—G, (He,Ne)
0.574160	0.574150 ±40	Te II	$112272^o_{3/2}$	—	$94860_{5/2}$	304—CW,G,(Ne⁺)
0.575589	0.575570 ±40	Te II	$100112^o_{5/2}$	—	$82743_{3/2}$	304—CW,G,(Ne,He)
0.576528	0.576490 ±40	Te II	$112549_{5/2}$	—	$95208_{3/2}$	304—CW,G,(He⁺)
0.585105	0.585100 ±40	Te II	$111947^o_{5/2}$	—	$94860_{5/2}$	304—CW,G,(Ne,He)
0.593618	0.593650 ±50	Te II	$99585^o_{3/2}$	—	$82743_{3/2}$	335—A; 304—CW,G,(Ne⁺)
0.597267	0.597230 ±40	Te II	$111947^o_{5/2}$	—	$95208_{3/2}$	304—CW,G,(Ne,He)
0.597471	0.597430 ±40	Te II	$102324^o_{5/2}$	—	$85592_{5/2}$	304—CW,G,(He⁺)
0.601447	0.601470 ±40	Te II	$105583^o_{5/2}$	—	$88961_{3/2}$	304—CW,G,(Ne⁺)
0.60824	0.608240 ±40	Te II		—		304—CW,G,(He⁺)
0.623074	0.623040 ±40	Te II	$105006^o_{3/2}$	—	$88961_{3/2}$	304—CW,G,(Ne⁺)
0.624549	0.624550 ±50	Te II	$99585^o_{3/2}$	—	$83577_{1/2}$	335—A; 304—CW,G,(Ne,He)
0.65850	0.658500 ±40	Te II		—		304—CW,G,(He⁺)
0.664859	0.664820 ±40	Te II	$97780^o_{1/2}$	—	$82743_{3/2}$	304—CW,G,(Ne⁺)
0.667602	0.667650 ±40	Te II	$103936^o_{3/2}$	—	$88961_{3/2}$	304—CW,G,(Ne⁺)
0.688508	0.688530 ±40	Te II	$100112^o_{5/2}$	—	$85592_{5/2}$	304—CW,G,(Ne⁺)

Wavelength (μm)	Measured value (μm)	Ion	Upper level	—	Lower level	Ref.
0.703904	0.703920 ±60	Te II	$97780^{o}_{1/2}$	—	$83577_{1/2}$	335—A; 304—CW,G,(Ne^{+})
0.780162	0.780160 ±60	Te II	$105006^{o}_{3/2}$	—	$92192_{3/2}$	304—CW,G,(Ne^{+})
0.792155	0.792140 ±60	Te II	$97780^{o}_{1/2}$	—	$85160_{3/2}$	304—CW,G,(Ne^{+})
0.86044	0.860440 ±60	Te II		—		304—CW,G,(He^{+})
0.87343	0.873430 ±60	Te II		—		304—CW,G,(He^{+})
0.897212	0.897190 ±60	Te II	$103936^{o}_{3/2}$	—	$92793_{5/2}$	304—CW,G,(Ne^{+})
0.89982	0.899820 ±60	Te II		—		304—CW,G,(Ne^{+})
0.937842	0.937790 ±60	Te II	$99585^{o}_{3/2}$	—	$88925_{5/2}$	304—CW,G,(Ne^{+})

Table 2.10.1
FLUORINE

Wavelength (μm)	Measured value (μm)	Identification				Ref.
		Ion	Upper level	—	Lower level	
		Spectroscopic References: F II (254), F III (255)				
0.275963	0.275959 ±06	F III	$(^{1}D)3p\ ^{2}D^{o}_{5/2}$	—	$(^{1}D)3s\ ^{2}D_{5/2}$	60—E
0.282613	0.282608 ±06	F IV	$(^{2}P^{o})3p\ ^{3}D_{3}$	—	$(^{2}P^{o})3s\ ^{3}P^{o}_{2}$	60—E
0.312154	0.312150 ±01	F III	$(^{3}P)3p\ ^{4}D^{o}_{7/2}$	—	$(^{3}P)3s\ ^{4}P_{5/2}$	60; 224—D,P,λ
0.317417	0.317418 ±06	F III	$(^{3}P)3p\ ^{2}D^{o}_{5/2}$	—	$(^{3}P)3s\ ^{2}P_{3/2}$	60—E
0.320275	0.320274 ±06	F II	$(^{2}D^{o})3p\ ^{1}D_{2}$	—	$(^{2}D^{o})3s\ ^{1}D^{o}_{2}$	60
0.402472	0.402478 ±06	F II	$(^{4}S^{o})3p\ ^{3}P_{2}$	—	$(^{4}S^{o})3s\ ^{3}S^{o}_{1}$	60—E

Table 2.10.2
CHLORINE

Wavelength (μm)	Measured value (μm)	Identification: Ion	Upper level	—	Lower level	Ref.
0.263267	0.263269 ±01	Cl III	$(^1D)4d\ ^2D_{5/2}$	—	$(^1D)4p\ ^2F^o_{7/2}$	60; 253; 224—D,P,λ
0.319146	0.319142 ±01	Cl III	$(^3P)4p\ ^4S^o_{3/2}$	—	$(^3P)4s\ ^4P_{5/2}$	60—E; 224—D,P,λ
0.339288	0.339286 ±01	Cl III	$(^1D)4p\ ^2D^o_{3/2}$	—	$(^1D)4s\ ^2D_{3/2}$	60—E; 224—D,P,λ
0.339345	0.339344 ±01	Cl III	$(^1D)4p\ ^2D^o_{5/2}$	—	$(^1D)4s\ ^2D_{5/2}$	60—E; 224—D,P,λ
0.353003	0.353002 ±01	Cl III	$(^1D)4p\ ^2F^o_{7/2}$	—	$(^1D)4s\ ^2D_{5/2}$	60—E; 224-D,P,λ
0.356068	0.356063 ±01	Cl III	$(^1D)4p\ ^2F^o_{5/2}$	—	$(^1D)4s\ ^2D_{3/2}$	60—E; 224—D,P,λ
0.360210	0.360210 ±06	Cl III	$(^3P)4p\ ^4D^o_{7/2}$	—	$(^3P)4s\ ^4P_{5/2}$	60—E
0.361282	0.361210 ±06	Cl III	$(^3P)4p\ ^4D^o_{5/2}$	—	$(^3P)4s\ ^4P_{3/2}$	60—E
0.362268	0.362269 ±06	Cl III	$(^3P)4p\ ^4D^o_{3/2}$	—	$(^3P)4s\ ^4P_{1/2}$	60—E
0.372045	0.372044 ±01	Cl III	$(^3P)4p\ ^2D^o_{5/2}$	—	$(^3P)4s\ ^2P_{3/2}$	60—E; 224—D,P,λ
0.374880	0.374877 ±01	Cl III	$(^3P)4p\ ^2D^o_{3/2}$	—	$(^3P)4s\ ^2P_{1/2}$	60—E; 224—D,P,λ
0.413250	0.413250 ±10	Cl II	$(^2D^o)4p\ ^1D_2$	—	$(^2D^o)4s\ ^1D^o_2$	345—CW,U
0.474042	0.474040 ±10	Cl II	$(^2P^o)4p\ ^1P_1$	—	$(^2P^o)3d\ ^1D^o_2$	345—CW,U
0.476870	0.476874 ±06	Cl II	$(^2P^o)4p\ ^3D_2$	—	$(^2P^o)4s\ ^3P^o_1$	60; 345—CW,U
0.478134	0.478134 ±03	Cl II	$(^2P^o)4p\ ^3D_3$	—	$(^2P^o)4s\ ^3P^o_2$	230; 24—CW,U; 231—λ
0.489685	0.489688 ±03	Cl II	$(^2D^o)4p\ ^3F_4$	—	$(^2D^o)4s\ ^3D^o_3$	230; 24—CW,U; 231—λ; 110—P
0.490482	0.490473 ±03	Cl II	$(^2D^o)4p\ ^3F_3$	—	$(^2D^o)4s\ ^3D^o_2$	230; 24—CW,U; 231—λ
0.491780	0.491766 ±03	Cl II	$(^2D^o)4p\ ^3F_2$	—	$(^2D^o)4s\ ^3D^o_1$	230; 24—CW,U; 231—λ
0.507828	0.507830 ±03	Cl II	$(^2D^o)4p\ ^3D_3$	—	$(^2D^o)4s\ ^3D^o_3$	230; 24—CW,U; 231—λ
0.510309	0.510310 ±10	Cl II	$(^2D^o)4p\ ^3D_2$	—	$(^2D^o)4s\ ^3D^o_2$	345—CW,U
0.521792	0.521790 ±03	Cl II	$(^4S^o)4p\ ^3P_2$	—	$(^4S^o)4s\ ^3S^o_1$	230; 231—CW,λ; 110—P; 327—D,P
0.522135	0.522130 ±03	Cl II	$(^4S^o)4p\ ^3P_1$	—	$(^4S^o)4s\ ^3S^o_1$	230; 24—CW,U; 231—λ; 327—D,P
0.539216	0.539215 ±03	Cl II	$(^2D^o)4p\ ^1F_3$	—	$(^2D^o)4s\ ^1D^o_2$	230; 24—CW,U; 231—λ; 110—P
0.609472	0.609474 ±03	Cl II	$(^2D^o)4p\ ^1P_1$	—	$(^2D^o)4s\ ^1D^o_2$	230; 24—CW,U; 231—λ

Table 2.10.3
BROMINE

Wavelength (μm)	Measured value (μm)	Identification				Ref.
		Ion	Upper level	—	Lower level	
		Spectroscopic Reference: Br II (274)				
0.236246	0.236246 ±01	Br IV?		—		224—D,P
0.258125	0.258125 ±01	Br IV?		—		224—D,P
0.278762	0.278762 ±01	Br III?		—		224—D,P
0.474270	0.474266 ±03	Br II	$(^2P^o)5p\ ^3D_3$	—	$(^2P^o)5s\ ^3P^o_2$	191
0.505465	0.505463 ±05	Br II	$(^2D^o)5p\ ^3F_3$	—	$(^4S^o)4d\ ^3D^o_2$	191
0.518227	0.518238 ±02	Br II	$(^4S^o)5p\ ^3P_2$	—	$(^4S^o)5s\ ^3S^o_1$	191; 18—U; 24—CW,U
0.523823	0.523826 ±04	Br II	$(^4S^o)5p\ ^3P_1$	—	$(^4S^o)5s\ ^3S^o_1$	191; 24—CW,U
0.533205	0.533203 ±03	Br II	$(^2D^o)5p\ ^1F_3$	—	$(^2D^o)5s\ ^1D^o_2$	191; 18—U; 24—CW,U
0.611761	0.611756 ±06	Br II	$(^4S^o)5p\ ^5P_2$	—	$(^4S^o)5s\ ^3S^o_1$	192; 71—D
0.616870	0.616878 ±06	Br II	$(^4S^o)5p\ ^5P_1$	—	$(^4S^o)5s\ ^3S^o_1$	192

Table 2.10.4
IODINE

Wavelength	Measured value (μm)	Identification: Ion	Upper Level — Lower level	Ref.
		Spectroscopic Reference: I II (226)		
0.448855	0.448850 ± 20	I II	$(^2D^o)6p\ ^3D_1$ — $(^4S^o)\ 5d\ ^5D^o_1$	269—CW,H,(He⁺)
0.453379	0.453379 ± 03	I III, IV?	—	203
0.467440	0.467440 ± 03	I III, IV?	—	203
0.467553	0.467560 ± 20	I II	$(^2D^o)6p\ ^1D_2$ — $(^2D^o)6s\ ^1D^o_2$	269—CW,H,(He⁺)
0.493467	0.493467 ± 03	I III, IV?	—	203
0.498692	0.498670 ± 20	I II	$(^2D^o)6p\ ^3D_2$ — $(^4S^o)5d\ ^3D^o_1$	179—A; 273—G; 269—CW,H,λ,(He⁺)
0.52140	0.521430 ± 20	I II	$(^2D^o)6p\ ^3D_2$ — $(^4S^o)5d\ ^3D^o_3$	269—CW,H,(He⁺)
0.521626	0.521630 ± 20	I II	$(^2D^o)6p\ ^3F_2$ — $(^4S^o)5d\ ^3D^o_1$	179—A; 342—hfs; 273—G; 269—CW, H,λ,(He⁺)
*0.540737	0.540750 ± 20	I II	$(^2D^o)6p\ ^3D_2$ — $(^2D^o)6s\ ^3D^o_2$	103—(He⁺); 104—hfs; 179—A; 342—hfs; 264—CW,H,(He⁺); 273—G; 146—CW,P; 269—CW,H,λ,(He⁺)
0.55931	0.559310 ± 20	I II	$(^2D^o)6p\ ^3D_1$ — $(^4S^o)5d\ ^3D^o_1$	269—CW,H,(He⁺)
0.562569	0.5625 ± ?	I II	$(^4S^o)6p\ ^3P_2$ — $(^4S^o)6s\ ^3S^o_1$	179
*0.567807	0.567820 ± 20	I II	$(^2D^o)6p\ ^3F_2$ — $(^2D^o)6s\ ^3D^o_2$	103—(He⁺); 104—hfs; 179—A; 342—hfs; 264—CW,H,(He⁺); 273—CW,G; 146—P; 269—CW,H,λ,(He⁺)
*0.576071	0.576070 ± 20	I II	$(^2D^o)6p\ ^3D_2$ — $(^2D^o)6s\ ^3D^o_1$	103—(He⁺);104—hfs; 179—A; 340—G; 342—hfs; 264—CW,H,(He⁺); 273—CW,G; 146—P; 269—CW,H,λ,(He⁺)
0.606895	0.606900 ± 20	I II	$(^2D^o)6p\ ^3F_2$ — $(^2D^o)6s\ ^3D_1$	341—G; 342—hfs; 269—CW,H,λ,(He⁺)

*0.612750	0.612740 ±20	I II	$(^2D^o)6p\ ^3D_1$ — $(^2D^o)6s\ ^3D^o_2$	10; 104—hfs; 179—A; 342—hfs; 264—CW,H,(He⁺); 68—CW,D; 273—CW,G; 147—P; 269—CW,H,λ,(He⁺)
0.620485	0.620490 ±20	I II	$(^2D^o)6p\ ^1P_1$ — $(^2D^o)6s\ ^1D^o_2$	269—CW,H,(He⁺)
0.634001	0.633990 ±20	I II	$(^2D^o)6p\ ^3F_3$ — $(^2D^o)6s\ ^3D^o_3$	269—CW,H,(He⁺)
0.635738	0.635737 ±?	I II	$(^4S^o)8d\ ^5D^o_1$ — $(^2P^o)6p\ ^3P_1$	110
0.648899	0.648897 ±?	I II	$(^4S^o)8d\ ^5D^o_4$ — $(^4S^o)7p\ ^5P_3$	110—(He⁺)
0.651615	0.651620 ±20	I II	$(^2D^o)6p\ ^3F_2$ — $5s5p^5\ ^1P^o_1$	340; 342—hfs; 264—CW,H,(He⁺); 269—CW,H,λ,(He⁺)
*0.658520	0.658530 ±20	I II	$(^2D^o)6p\ ^3D_1$ — $(^2D^o)6s\ ^3D^o_1$	103—(He⁺); 104—hfs; 179—A; 342—hfs; 264—CW,H,(He⁺); 273—CW,G; 146—P; 269—CW,H,λ,(He⁺)
0.662236	0.662250 ±20	I II	$(^2D^o)6p\ ^3D_2$ — $(^2D^o)5d\ ^3D^o_2$	269—CW,H,(He⁺)
0.667228	0.667227 ±?	I II	$(^2P^o)6p\ ^3D_1$ — $(^2D^o)5d\ ^1D^o_2$	110
0.682520	0.682520 ±20	I II	$(^2D^o)6p\ ^3F_2$ — $(^2D^o)6s\ ^1D^o_3$	341; 269—CW, H, (He⁺)
0.690480	0.6904 ±?	I II	$(^2D^o)6p\ ^3D_2$ — $(^2D^o)6s\ ^1D^o_2$	179—A,E; 110
0.703299	0.703300 ±20	I II	$(^2D^o)6p\ ^3F_2$ — $(^2D^o)5d\ ^3D^o_2$	103—(He); 104—hfs; 179—A; 341—hfs; 264—CW,H,(He⁺); 273—G; 269—CW,H,λ,(He⁺)
0.713898	0.713900 ±20	I II	$(^2D^o)6p\ ^3D_2$ — $(^2D^o)5d\ ^3D^o_3$	341; 264—CW,H,(He⁺); 269—CW,H,λ,(He⁺)
0.761849	0.761850 ±20	I II	$(^2D^o)6p\ ^3F_2$ — $(^2D^0)5d\ ^3D^0_3$	269—CW,H,(He⁺)
0.773577	0.773580 ±20	I II	$(^2D^o)6p\ ^3D_1$ — $(^2D^o)5d\ ^3D^o_2$	269—CW,H,(He⁺)
0.817010	0.817020 ±20	I II	$(^2D^o)6p\ ^3D_2$ — $(^2D^o)5d\ ^3F^o_3$	269—CW,H,(He⁺)
0.825381	0.825390 ±20	I II	$(^2D^o)6p\ ^3D_1$ — $(^2D^o)5d\ ^3P^o_0$	179—A; 264—CW,H,(He⁺); 269—CW,H,λ
0.880428	0.880428 ±20	I II	$(^2D^o)6p\ ^3F_2$ — $(^2D^o)5d\ ^3F^o_3$	179—A; 264—CW,H,(He⁺); 269—CW,H,λ
0.887757	0.887740 ±20	I II	$(^2D^o)6p\ ^3F_3$ — $(^2D^o)5d\ ^3G^o_4$	269—CW,H,(He⁺)
1.041 72	1.041720 ±60	I III,IV?	—	203

Table 2.11.1
NEON

Wavelength (μm)	Measured value (μm)	Identification: Ion	Upper level	—	Lower level	Ref.
			Spectroscopic Reference: Ne II (262)			
0.201844[312]	0.201842 ±01	Ne IV	$(^1D)3p\ ^2D^o_{3/2}$	—	$(^1D)3s\ ^2D_{3/2}$	222—P
0.202219[312]	0.202219 ±01	Ne IV	$(^1D)3p\ ^2D^o_{5/2}$	—	$(^1D)3s\ ^2D_{5/2}$	222—P; 223—P
0.206530	0.206530 ±01	Ne III,IV		—		222—G,P; 223—P
0.217771	0.217770 ±01	Ne III	$(^2D^o)3p\ ^3P_1$	—	$(^2D^o)3s\ ^3D^o_2$	222—P
0.218088	0.218086 ±01	Ne III	$(^2D^o)3p\ ^3P_2$	—	$(^2D^o)3s\ ^3D^o_3$	222—P
0.226570	0.226570 ±?	Ne V	$(^2P^o)3p\ ^3D_3$	—	$(^2P^o)3s\ ^3P^o_2$	222—P
0.228579	0.228579 ±01	Ne IV	$(^1D)3p\ ^2F^o_{7/2}$	—	$(^1D)3s\ ^2D_{5/2}$	222—P
0.235252	0.235255 ±01	Ne IV	$(^3P)3p\ ^4D^o_{5/2}$	—	$(^3P)3s\ ^4P_{3/2}$	222—P
0.235796	0.235798 ±01	Ne IV	$(^3P)3p\ ^4D^o_{7/2}$	—	$(^3P)3s\ ^4P_{5/2}$	59; 222—G,P,λ
0.237320	0.237320 ±01	Ne IV?		—		222—P
0.247340	0.247340 ±01	Ne III	$(^2D^o)3p\ ^1D_2$	—	$(^2D^o)3s\ ^1D^o_2$	59; [222]—G,P,λ
0.261003	0.260998 ±01	Ne III	$(^2D^o)3p\ ^3F_4$	—	$(^2D^o)3s\ ^3D^o_3$	222—P
0.261341	0.261340 ±10	Ne III	$(^2D^o)3p\ ^3F_3$	—	$(^2D^o)3s\ ^3D^o_2$	222—P
0.267790	0.267792 ±01	Ne III	$(^4S^o)3p\ ^3P_{2,0}$	—	$(^4S^o)3s\ ^3S^o_1$	37; 59; 143—U; 222—G,P,λ
0.267864	0.267869 ±01	Ne III	$(^4S^o)3p\ ^3P_1$	—	$(^4S^o)3s\ ^3S^o_1$	37; 59; 143—U; 222—G,P,λ
0.277765	0.277763 ±01	Ne III	$(^2D^o)3p\ ^3D_3$	—	$(^2D^o)3s\ ^3D^o_3$	37; 222—P,λ
0.286673	0.286673 ±01	Ne III	$(^2D^o)3p\ ^1F_3$	—	$(^2D^o)3s\ ^1D^o_2$	37; 59; [222]—P,λ
0.331972	0.331975 ±01	Ne II	$(^1D)3p\ ^2P^o_{1/2}$	—	$(^1D)3s\ ^2D_{3/2}$	80; 59; 143—U; 222—P,λ
*0.332373	0.332375 ±01	Ne II	$(^3P)3p\ ^2P^o_{3/2}$	—	$(^3P)3s\ ^2P_{3/2}$	37; 64—S; 96—S; 98—CW,G,P; 251—P; 143—U; 222—P,λ; 321—CW,P; 220—CW,P
0.332715	0.332750 ±50	Ne II	$(^3P)3p\ ^4D^o_{3/2}$	—	$(^3P)3s\ ^4P_{3/2}$	37
0.332916	0.332902 ±10	Ne II	$(^3P)3d\ ^4D_{7/2}$	—	$(^3P)3p\ ^4D^o_{7/2}$	80
0.333114	0.333107 ±10	Ne III	$(^2P^o)3p\ ^3D_2$	—	$(^2P^o)3s\ ^1P^o_1$	80
*0.334545	0.334545 ±02	Ne II	$(^1D)3p\ ^2P^o_{3/2}$	—	$(^1D)3s\ ^2D_{5/2}$	80; 59; 96—S; 98—CW; 143—U; 222—P,λ
*0.337822	0.337826 ±01	Ne II	$(^3P)3p\ ^2P^o_{1/2}$	—	$(^3P)3s\ ^2P_{1/2}$	37; 96—S; 98—CW,G,P; 251; 143—U; 222—P.λ

Wavelength (μm)	Measured value (μm)	Ion	Upper level — Lower level	Ref.
*0.339280	0.339280 ±01	Ne II	$(^3P)3p\ ^2P^o_{3/2}$ — $(^3P)3s\ ^2P_{1/2}$	37; 59; 98—CW,G,P; 251; 222—P,λ
0.339318	0.339340 ±10	Ne II	$(^3P)3d\ ^2D_{3/2}$ — $(^3P)3p\ ^2D^o_{5/2}$	80
*0.371308	0.371300 ±100	Ne II	$(^3P)3p\ ^2D^o_{5/2}$ — $(^3P)3s\ ^2P_{3/2}$	40—CW,G; 98—CW,P

Table 2.11.2
ARGON

Wavelength (μm)	Measured value (μm)	Identification		Ref.
		Ion	Upper level — Lower level	
		Spectroscopic Reference: Ar II (239)		
v.184343	v.184343 ±03	Ar V?	—	223
0.211398	0.211398 ±01	Ar IV	—	222—P
0.224884	0.224884 ±01	Ar IV	—	222—P
0.251328	0.251330 ±02	Ar IV	$(^3P)4p\ ^4S^o_{3/2}$ — $(^3P)4s\ ^4P_{5/2}$	222—E,P
0.262135	0.262138 ±01	Ar IV	$(^1D)4p\ ^2D^o_{3/2}$ — $(^1D)4s\ ^2D_{3/2}$	222—P
0.262493	0.262488 ±01	Ar IV	$(^1D)4p\ ^2D^o_{5/2}$ — $(^1D)4s\ ^2D_{5/2}$	59; 222—P,λ
0.275392[H]	0.275388 ±01	Ar III	$[(^2D^o)4p\ ^1D_2$ — $(^2D^o)4s\ ^1D_2]$	37; 59; 222—CW,I,P,λ; 48—I; 219—CW,P;**
0.285521	0.285537 ±01	Ar III	$(^2D^o)4p\ ^3P_1$ — $(^2D^o)4s\ ^3D^o_2$	222—P
0.288416	0.288422 ±01	Ar III	$(^2D^o)4p\ ^3P_2$ — $(^2D^o)4s\ ^3D^o_3$	80; 59; 222—CW,P,λ
0.291300	0.291292 ±01	Ar IV	$(^3P)4p\ ^2D^o_{5/2}$ — $(^3P)4s\ ^2P_{3/2}$	37; 59; 214—G; 143—U; 222—P,λ
0.292627	0.292623 ±01	Ar IV	$(^3P)4p\ ^2D^o_{3/2}$ — $(^3P)4s\ ^2P_{1/2}$	37; 59; 222—P,λ
0.300264	0.300264 ±01	Ar III	$[(^2P^o)4p\ ^1P_1$ — $(^2P^o)3d\ ^1D^o_2]$	37—E; 59; 143—U; 222—CW,I,P,λ; 219—CW,P; **
0.302405	0.302400 ±50	Ar III	$(^2P^o)4p\ ^3D_3$ — $(^2P^o)4s\ ^3P^o_2$	37; 219—CW,P; **
0.305484	0.305480 ±50	Ar III	$(^2P^o)4p\ ^3D_2$ — $(^2P^o)4s\ ^3P^o_1$	37; 219—CW,P; **
0.333613	0.333621 ±06	Ar III	$(^2D^o)4p\ ^3F_4$ — $(^2D^o)4s\ ^3D^o_3$	37; 59; 98—CW,P; 219—CW,P; **
0.334472	0.334479 ±06	Ar III	$(^2D^o)4p\ ^3F_3$ — $(^2D^o)4s\ ^3D^o_2$	37; 59; 98—CW,P; 219—CW,P; **
0.335849	0.335852 ±06	Ar III	$(^2D^o)4p\ ^3F_2$ — $(^2D^o)4s\ ^3D^o_1$	37; 59; 98—CW,P; 219—CW,P; **

Table 2.11.2 (continued) ARGON

Wavelength (μm)	Measured value (μm)	Identification: Ion	Upper level	—	Lower level	Ref.
*0.351112	0.351112 ±06	Ar III	$(^4S^o)4p\ ^3P_2$	—	$(^4S^o)4s\ ^3S^o_1$	230; 37; 96—S; 98—CW,G,P; 251—CW,P; 165—P,U; 222—P,S; 321—CW,P; 219—CW,P; **
0.351418	0.351415 ±06	Ar III	$(^4S^o)4p\ ^3P_1$	—	$(^4S^o)4s\ ^3S^o_1$	230; 37—λ; 43—CW; 143—U; 325—P; 222—P
0.357661	0.357690 ±50	Ar II	$(^3P)4d\ ^4F_{7/2}$	—	$(^3P)4p\ ^4D^o_{5/2}$	37—E
*0.363789[H]	0.363786 ±04	Ar III	$[(^2D^o)4p\ ^1F_3$	—	$(^2D^o)4s\ ^1D^o_2]$	230—I; 37; 40—CW,P; 98—CW,G,P; 143—U; 325—P; 222—P; 321—CW,P; 219—CW,P; **
?0.37052	0.370520 ±50	Ar		—		37; 36
0.379532	0.379528 ±06	Ar III	$(^2P^o)4p\ ^3D_3$	—	$(^2P^o)3d\ ^3P^o_2$	37; 43—CW
0.385829	0.385826 ±06	Ar III	$(^2P^o)4p\ ^3D_2$	—	$(^2P^o)3d\ ^3P^o_1$	37; 43—CW
0.40886	0.408860 ±20	Ar		—		43—CW; 45—X; [cf. Si IV]
0.414671	0.414660 ±04	Ar III	$(^2D^o)4p\ ^3P_2$	—	$(^2P^o)4s\ ^3P^o_2$	37; 43—CW
*0.418298[312]	0.418292 ±06	Ar III	$[(^2D^o)4p\ ^1P_1$	—	$(^2D^o)4s\ ^1D^o_2]$	37; 98—CW; 48—I
0.437075	0.437073 ±06	Ar II	$(^1D)4p\ ^2D^o_{3/2}$	—	$(^3P)3d\ ^2D_{3/2}$	37; 38—CW
0.438375	0.438360 ±60	Ar II	$(^3P)4p\ ^4S^o_{3/2}$	—	$(^3P)4s\ ^2P_{3/2}$	249
0.448181	0.448200 ±100	Ar II	$(^1D)4p\ ^2D^o_{5/2}$	—	$(^3P)3d\ ^2D_{5/2}$	24—CW; 208—CW
0.454505	0.454504 ±10	Ar II	$(^3P)4p\ ^2P^o_{3/2}$	—	$(^3P)4s\ ^2P_{3/2}$	33; 126—CW; 37—λ; 117—CW,H,(He); **
*0.457935	0.457936 ±16	Ar II	$(^3P)4p\ ^2S^o_{1/2}$	—	$(^3P)4s\ ^2P_{1/2}$	33;20—G; 126—CW; 117—CW,H,(He); **
0.460956	0.460957 ±10	Ar II	$(^1D)4p\ ^2F^o_{7/2}$	—	$(^1D)4s\ ^2D_{5/2}$	37
0.465789	0.465795 ±02	Ar II	$(^3P)4p\ ^2P^o_{1/2}$	—	$(^3P)4s\ ^2P_{3/2}$	33; 20—G; 126—CW; **
0.472686	0.472689 ±04	Ar II	$(^3P)4p\ ^2D^o_{3/2}$	—	$(^3P)4s\ ^2P_{3/2}$	33; 126—CW; **

*0.476486	0.476488	±04	Ar II	$(^3P)4p\ ^2P^o_{3/2}$ — $(^3P)4s\ ^2P_{1/2}$	33; 20—G,P,S; 69—E; 126—CW; 205—P,U; 165—P,U; 143—U; 117—CW,H,(He); **
*0.487986	0.487986	±04	Ar II	$(^3P)4p\ ^2D^o_{5/2}$ — $(^3P)4s\ ^2P_{3/2}$	33—E,G,P; 20—G,S; 126—CW,P;**
0.488903	0.488906	±06	Ar II	$(^3P)4p\ ^2P^o_{1/2}$ — $(^3P)4s\ ^2P_{1/2}$	37; 38—CW
*0.496507	0.496509	±02	Ar II	$(^3P)4p\ ^2D^o_{3/2}$ — $(^3P)4s\ ^2P_{1/2}$	33; 20—G,S; 126—CW; **
?0.499255	0.499255	±05	Ar	—	23
*0.501716	0.501717	±02	Ar II	$(^1D)4p\ ^2F^o_{5/2}$ — $(^3P)3d\ ^2D_{3/2}$	33; 20—G; 126—CW; **
0.506204	0.506210	±25	Ar II	$(^3P)4p\ ^4P^o_{3/2}$ — $(^3P)4s\ ^4P_{1/2}$	21—U; 351—U
0.514179	0.514180	±05	Ar II	$(^1D)4p\ ^2F^o_{7/2}$ — $(^3P)3d\ ^2D_{5/2}$	37; 38—CW
*0.514532	0.514533	±02	Ar II	$(^3P)4p\ ^4D^o_{5/2}$ — $(^3P)4s\ ^2P_{3/2}$	33—G,P; 20—G; 126—CW;**
0.528690	0.528700	±100	Ar II	$(^3P)4p\ ^4D^o_{3/2}$ — $(^3P)4s\ ^2P_{1/2}$	33; 126—CW; **
0.550220	0.550220	±50	Ar III	$(^2D^o)4p\ ^3D_3$ — $(^2P^o)4s\ ^3P^o_2$	37
0.648308	0.648280	±20	Ar II	$(^3P)4p\ ^2S^o_{1/2}$ — $(^3P)3d\ ^2P_{3/2}$	177—CW,H,(He); 309—CW,H,(He)
?0.673000	0.673000	±50	Ar	—	157
0.686127	0.686110	±20	Ar II	$(^3P)4p\ ^2P^o_{3/2}$ — $(^3P)3d\ ^2P_{3/2}$	177—CW,H,(He); 309—CW,H,(He)
0.734805	0.734804	±05	Ar II	$(^1D)3d\ ^2D_{5/2}$ — $(^3P)4p\ ^2D^o_{5/2}$	96—S
0.750514	0.750508	±05	Ar II	$(^1D)3d\ ^3P_{3/2}$ — $(^3P)4p\ ^2S^o_{1/2}$	96—S
0.877186	0.878000	±300	Ar II	$[(^3P)4p\ ^2P^o_{3/2}$ — $(^1D)4s\ ^2D_{5/2}]$	306; 36—I; 309—CW,H,(He)
*1.092344	1.092300	±100	Ar II	$(^3P)4p\ ^2P^o_{3/2}$ — $(^3P)3d\ ^2D_{5/2}$	160—CW,P; 36—CW,P; 7—P

Table 2.11.3
KRYPTON

Wavelength (μm)	Measured value (μm)	Identification				Ref.
		Ion	Upper level	—	Lower level	
		Spectroscopic Reference: Kr II (240)				
0175641	0.175641 ±03	Kr IV		—		223—P
0.183243	0.183243 ±03	Kr V?		—		223—P
0.195027	0.195027 ±03	Kr IV		—		222—P; 223—P
0.196808	0.196808 ±03	Kr IV		—		223—P
0.205108	0.205108 ±01	Kr IV		—		222—P; 223—P
0.219192	0.219192 ±01	Kr IV		—		222—P; 223—P
0.225464	0.225464 ±01	Kr IV		—		222—P
0.233848	0.233848 ±01	Kr IV		—		222—P
0.241784	0.241784 ±01	Kr III,IV		—		222—P
0.264936	0.264936 ±01	Kr IV		—		59; 45—E; 222—P,λ
0.266440	0.266440 ±01	Kr IV		—		59; 45—E; 222—P,λ
0.274138	0.274138 ±01	Kr IV		—		59; 222—P,λ
0.304970	0.304970 ±02	Kr III		—		37; 59; 222—P,λ
0.312438	0.312436 ±01	Kr III	$(^2D^o)5p\ ^1D_2$	—	$(^2D^o)5s\ ^1D^o_2$	59; 222—CW,P,λ; 219—CW,P; **
0.323951	0.323951 ±01	Kr III	$(^2P^o)5p\ ^1D_2$	—	$(^2D^o)5s\ ^1P^o_1$	37; 59; 222—CW,P,λ; 219—CW, P; **
0.337496	0.337500 ±50	Kr III	$(^2P^o)5p\ ^3D_3$	—	$(^2P^o)5s\ ^3P^o_2$	37; 98—CW; 219—CW,P; **
*0.350742	0.350742 ±01	Kr III	$(^4S^o)5p\ ^3P_2$	—	$(^4S^o)5s\ ^3S^o_1$	37; 98—CW,G,P; 251—CW,P; 325—P; 222—P,λ; 321—P; 219—P; **
*0.356432	0.356420 ±06	Kr III	$(^4S^o)5p\ ^3P_1$	—	$(^4S^o)5s\ ^3S^o_1$	59; 98—CW,G,P; 222—P,λ; 321—P; 219—P; **
0.377134	0.377134 ±05	Kr II	$(^1D)4d\ ^2S_{1/2}$	—	$(^3P)5p\ ^4P^o_{3/2}$	96—P,S
*0.406737	0.406736 ±06	Kr III	$(^2D^o)5p\ ^1F_3$	—	$(^2D^o)5s\ ^1D^o_2$	37; 40—CW; 98—CW,P; 209—CW,P; 325—P; 7—P; **
*0.413133	0.413138 ±06	Kr III	$(^4S^o)5p\ ^5P_2$	—	$(^4S^o)5s\ ^3S^o_1$	37; 98—CW; 209—CW,P; 325—P; **
0.415444	0.415445 ±04	Kr III	$(^2D^o)5p\ ^3F_3$	—	$(^2D^o)5s\ ^1D^o_2$	37; 43—CW; **

0.417179	0.417181	±10	Kr III	$(^4S^o)5p\ ^5P_1$ — $(^4S^o)5s\ ^3S^o_1$	37
0.422658	0.422651	±06	Kr III	$(^2D^o)5p\ ^3F_2$ — $(^2D^o)4d\ ^3D^o_1$	37; **
0.431780	0.431800	±20	Kr II	$(^3P)6s\ ^4P_{5/2}$ — $(^3P)5p\ ^4P^o_{5/2}$	212; 79—A; 117—CW,H,(He⁺); 252—CW,H,(He⁺); 309—CW,H,(He⁺)
0.438653	0.438610	±20	Kr II	$(^3P)6s\ ^4P_{5/2}$ — $(^3P)5p\ ^4P^o_{3/2}$	212—E; 79—A,E; 117—CW,H,(He⁺) 309—CW,H,(He⁺)
0.444329	0.444328	±04	Kr III	$(^2D^o)5p\ ^3D_2$ — $(^2D^o)4d\ ^3D^o_1$	37
0.457720	0.457720	±10	Kr II	$(^1D)5p\ ^2F^o_{7/2}$ — $(^1D)5s\ ^2D_{5/2}$	35; 1—CW; 37-λ; 309—CW,H,(He⁺)
0.458285	0.458300	±100	Kr II	$(^3P)6s\ ^4P_{3/2}$ — $(^3P)5p\ ^4D^o_{5/2}$	212; 79—A; 48—H,(He⁺); 177—CW,H,(He⁺); 309—CW,H,(He⁺)
0.461528	0.461520	±10	Kr II	$(^3P)5p\ ^2P^o_{3/2}$ — $(^3P)5s\ ^2P_{3/2}$	247
0.461915	0.461917	±10	Kr II	$(^3P)5p\ ^2D^o_{5/2}$ — $(^3P)5s\ ^2P_{3/2}$	35; 37-λ; 36—CW
0.463388	0.463392	±06	Kr II	$(^1D)5p\ ^2F^o_{5/2}$ — $(^1D)5s\ ^2D_{3/2}$	35; 37-λ;38—CW
0.465016	0.465016	±10	Kr II	$(^3P)5p\ ^2P^o_{1/2}$ — $(^3P)5s\ ^4P_{1/2}$	37;[cf. C III, Xe IV]
*0.468041	0.468045	±06	Kr II	$(^3P)5p\ ^2S^o_{1/2}$ — $(^3P)5s\ ^2P_{1/2}$	35; 37-λ; 36-CW; **
0.469443	0.469410	±20	Kr II	$(^3P)6s\ ^4P_{5/2}$ — $(^3P)5p\ ^4D^o_{7/2}$	212; 79—A; 176—CW,H,(He⁺); 352—P,U; 280; 177; 252; 309
0.471046	0.471030	±60	Kr III	$(^2D^o)5p\ ^3F_4$ — $(^2D^o)4d\ ^3D^o_3$	249
0.475447	0.475450	±30	Kr III	$(^2D^o)5p\ ^1F_3$ — $(^2D^o)4d\ ^3D^o_3$	249—E
*0.476243	0.476244	±06	Kr II	$(^3P)5p\ ^2D^o_{3/2}$ — $(^3P)5s\ ^2P_{1/2}$	35; 17—U; 37—λ; 1—CW; **
0.476573	0.476571	±10	Kr II	$(^3P)5p\ ^4D^o_{5/2}$ — $(^3P)5s\ ^4P_{3/2}$	35; 37—λ; 36—CW
0.479633	0.479630	±60	Kr II	$(^3P)5d\ ^4D_{1/2}$ — $(^3P)5p\ ^4S^o_{3/2}$	249
*0.482518	0.482518	±06	Kr II	$(^3P)5p\ ^4S^o_{3/2}$ — $(^3P)5s\ ^2P_{1/2}$	35; 37—λ; 1—CW; **
0.484660	0.484666	±06	Kr II	$(^3P)5p\ ^2P^o_{1/2}$ — $(^3P)5s\ ^2P_{3/2}$	37; 36—CW
0.501645	0.501640	±10	Kr III	$(^2D^o)5p\ ^1D_2$ — $(^2P^o)4d\ ^1F^o_3$	246; 157—CW
0.502239	0.502200	±100	Kr II	$(^3P)5p\ ^4D^o_{3/2}$ — $(^3P)5s\ ^2P_{3/2}$	24; 38—CW
0.503747[84]	0.503750	±60	Kr II	—	249
0.512572	0.512600	±10	Kr II	$(^3P)6s\ ^4P_{3/2}$ — $(^3P)5p\ ^4D^o_{3/2}$	212; 79—A; 117—CW,H,(He⁺)
*0.520832	0.520832	±04	Kr II	$(^3P)5p\ ^4P^o_{3/2}$ — $(^eP)5s\ ^3P_{3/2}$	35; 17—U; 37—λ; 36—CW; **
0.521793	0.521820	±40	Kr II	$(^3P)5p\ ^4D^o$ — $(^3P)5s\ ^2P_{1/2}$	249
*0.530865	0.530868	±04	Kr II	$(^3P)5p\ ^4P^o_{5/2}$ — $(^3P)5s\ ^4P_{3/2}$	35; 37—λ; 36—CW; **
0.550143	0.550150	±50	Kr III	$(^2D^o)5p\ ^3F_3$ — $(^2P^o)4d\ ^3D^o_2$	249—E
0.559732	0.559770	±100	Kr III	$(^2D^o)5p\ ^3P_2$ — $(^2P^o)5s\ ^3P^o_2$	249

Table 2.11.3 (continued)
KRYPTON

Wavelength (μm)	Measured value (μm)	Ion	Upper level	—	Lower level	Ref.
*0.568188	0.568192 ±04	Kr II	$(^3P)5p\ ^4D^o_{5/2}$	—	$(^3P)5s\ ^2P_{3/2}$	35; 1—CW,P; 17—U; 37—λ; 36—CW; **
0.57529	0.575340 ±50	Kr II	$(^3P)5p\ ^4D^o_{3/2}$	—	$(^3P)5s\ ^2P_{1/2}$	24—CW; 157—CW,λ
0.593506		Kr III	$(^2D^o)5p\ ^3P_2$	—	$(^2P^o)5s\ ^1P^o_1$	
	0.593530 ±60	or				249
0.593529		Kr II	$(^3P)5d\ ^4P_{3/2}$	—	$(^1D)5p\ ^2P^o_{3/2}$	
0.603716		Kr III	$(^2D^o)5p\ ^3P_1$	—	$(^2P^o)4d\ ^3D^o_1$	
	0.603760 ±80	or				248
0.603811		Kr II	$(^3P_2)4f\ [2]^o_{3/2}$	—	$(^1D)4d\ ^2P_{3/2}$	
?0.6072	0.607200 ±100	Kr		—		72
0.616880	0.616880 ±50	Kr II	$(^1D)5p\ ^2F^o_{5/2}$	—	$(^3P)4d\ ^4P_{3/2}$	38—CW
0.631024	0.631030 ±80	Kr III	$(^2D^o)5p\ ^3P_2$	—	$(^2P^o)4d\ ^3D^o_1$	248
0.631276[162]	0.631260 ±80	Kr		—		248; 157—CW
0.641660	0.641700 ±100	Kr II	$(^1D)5p\ ^2P^o_{3/2}$	—	$(^3P)4d\ ^2P_{3/2}$	72
*0.647088	0.647100 ±50	Kr II	$(^3P)5p\ ^4P^o_{5/2}$	—	$(^3P)5s\ ^2P_{3/2}$	35; 1—CW; 251—CW,P; **
0.651016	0.651000 ±10	Kr II	$(^3P)6s\ ^4P_{5/2}$	—	$(^3P)5p\ ^4S^o_{3/2}$	117—CW,H; 309—CW,H,(He⁺)
0.657008	0.657000 ±50	Kr II	$(^1D)5p\ ^2D^o_{5/2}$	—	$(^3P)4d\ ^2D_{5/2}$	35; 38—CW
0.660293		Kr III	$[(^2D^o)5p\ ^3P_2$	—	$(^2P^o)4d\ ^1F^o_3]$	
	0.560280 ±80	or				248; 157—CW
0.660275		Kr II	$(^1D)5p\ ^2P^o_{1/2}$	—	$(^3P)4d\ ^2D_{5/2}$	
*0.676442	0.676457 ±10	Kr II	$(^3P)5p\ ^4P^o_{1/2}$	—	$(^3P)5s\ ^2P_{1/2}$	35; 1—CW; 37—λ; **
0.687085	0.687096 ±10	Kr II	$(^1D)5p\ ^2F^o_{5/2}$	—	$(^3P)4d\ ^2D_{3/2}$	35; 37—λ; 38—CW
0.743578	0.743576 ±01	Kr II	$(^1D)4d\ ^2D_{5/2}$	—	$(^3P)5p\ ^4D^o_{5/2}$	322; 222—P,S,λ
*0.752546	0.752550 ±10	Kr II	$(^3P)5p\ ^4P^o_{3/2}$	—	$(^3P)5s\ ^2P_{1/2}$	183—CW; **
0.793141	0.79314 ±?	Kr II	$(^1D)5p\ ^2F^o_{7/2}$	—	$(^3P)4d\ ^2F_{5/2}$	208—CW; **
0.799322	0.799300 ±50	Kr II	$(^3P)5p\ ^4P^o_{3/2}$	—	$(^3P)4d\ ^4D_{1/2}$	35; 208—CW; **
0.828034	0.828030 ±10	Kr II	$(^1D)5p\ ^2F^o_{5/2}$	—	$(^3P)4d\ ^2F_{5/2}$	208—CW; 183—λ
0.8334	0.8334 ±?	Kr II?		—		8—CW

0.847333	0.8473 ± ?	Kr II	(^{1}D)4d $^2D_{3/2}$	—	(^{3}P)5p $^4D^o_{3/2}$	322
0.8589	0.858900 ± 300	Kr III?		—		306
0.869014	0.86901 ± ?	Kr II	(^{3}P)5p $^2P^o_{1/2}$	—	(^{1}D)5s $^2D_{3/2}$	208—CW
0.897869	0.89784 ± ?	Kr II	(^{1}D)4d $^2D_{5/2}$	—	(^{3}P)5p $^4D^o_{3/2}$	322—E
1.06596	1.06596 ± ?	Kr II	(^{1}D)4d $^2D_{5/2}$	—	(^{3}P)5p $^2P^o_{3/2}$	322
1.329404	1.3295 ± ?	Kr II	(^{1}D)4d $^2D_{5/2}$	—	(^{3}P)5p $^2D^o_{3/2}$	322

Table 2.11.4
XENON

Wavelength (μm)	Measured value (μm)	Identification				Ref.
		Ion	Upper level	—	Lower level	
0.223244	0.223244 ± 01	Xe III,IV		—		222—P; **
*0.231536	0.231536 ± 01	Xe III,IV		—		222—G,P; **
0.247739	0.247739 ± 03	Xe		—		59; 111—I,λ; 217—I,λ
0.252666	0.252666 ± 01	Xe IV		—		222—P
0.269194	0.269194 ± 01	Xe IV		—		59; 111—I; 217—I; 222—P,λ
0.307974	0.307974 ± 02	Xe III		—		37—E; 59; 111—I; 217—I; 222—P,λ; **
0.324692	0.324692 ± 01	Xe IV		—		59—E; 217—I; 257; 222—P,λ; 159—I; **
0.330596	0.330596 ± 01	Xe IV		—		37—E; 59; 217—I; 222—λ; 159—I
*0.333087	0.333087 ± 02	Xe IV		—		37; 42—CW; 59; 111—I; 257—P; 217—I; 222—P,λ; **
0.334974	0.334974 ± 06	Xe III		—		59; 111—I; 217—I; 159—I
*0.345424	0.345425 ± 01	Xe III	(^{2}D°)6p $^1D_2$	—	(^{2}D°)6s $^1D^o_2$	59; 98—CW,P; 257; 222—P,λ
0.348322	0.348331 ± 03	Xe IV		—		37; 59; 111—I; 217—I; 159—I,λ
0.354233	0.354231 ± 05	Xe III	(^{2}D°)6p $^3P_2$	—	(^{2}D°)6s $^3D^o_3$	96—S
0.359661	0.359600 ± 100	Xe III		—		98—CW; 159—I,λ
*0.364548	0.364548 ± 01	Xe IV		—		80—E; 59; 111—I; 258—G,P; 217—I; 257—P; 325—P; 222—P,λ; 159—I; **
0.366920	0.366920 ± 03	Xe III		—		59; 43—CW; 111—I; 217—I; 257; 159—I
0.374571	0.374573 ± 06	Xe III	(^{2}D°)6p $^1D_2$	—	(^{2}D°)5d $^1D^o_2$	59; 98—CW; 257; 321—CW,P; **

Table 2.11.4 (continued)
XENON

Wavelength (μm)	Measured value (μm)	Identification: Ion	Upper level	—	Lower level	Ref.
0.375994	0.375994 ±03	Xe IV?		—		37—E; 36—E; 111—I; 47—I; 217—I; 159—I
*0.378097	0.378099 ±02	Xe III	$(^4S^o)6p\ ^3P_2$	—	$(^4S^o)6s\ ^3S^o_1$	37; 59; 98—CW,G,P; 257; 222—P,λ; 321—CW,P
0.380329	0.380329 ±03	Xe IV		—		59; 111—I; 258; 217—I; 257; 159—I
0.384152	0.384100 ±100	Xe III	$(^2D^o)6p\ ^3F_2$	—	$(^2D^o)6s\ ^3D^o_2$	43—CW
		or				
0.384186		Xe III	$(^2D^o)6p\ ^3P_1$	—	$(^2D^o)5d\ ^3D^o_2$	
0.397301	0.397301 ±03	Xe IV?		—		59; 111—I; 258; 217—I; 257; 159—I
0.399285	0.399300 ±100	Xe III	$(^2D^o)6p\ ^3P_1$	—	$(^2D^o)6s\ ^1D^o_2$	249; 43—CW,I
		or				
0.399255		Xe III	$[(^4S^o)6p\ ^5P_2$	—	$(^4S^o)5d\ ^3D^o_1]$	
0.405005	0.404990 ±20	Xe III	$(^4S^o)6p\ ^3P_1$	—	$(^4S^o)6s\ ^3S^o_1$	248; 325—P
*0.406041	0.406048 ±06	Xe III	$(^2P^o)6p\ 32_1$	—	$(^2P^o)5d\ 25^o_1$	37; 78; 80; 98—CW,P; 258, 257; 325—P; 222—P
0.414572	0.414530 ±60	Xe III	$(^2D^o)6p\ ^3D_2$	—	$(^2D^o)5d\ ^3D^o_1$	249; 43—CW
*0.421401	0.421405 ±06	Xe III	$(^2D^o)6p\ ^3P_2$	—	$(^2D^o)5d\ ^3D^o_3$	37; 98—CW,G,P
*0.424024	0.424026 ±10	Xe III	$(^2D^o)6p\ ^1D_2$	—	$(^2P^o)5d\ 17^o_3$	37; 98—CW,P
*0.427259	0.427260 ±06	Xe III	$(^2D^o)6p\ ^3F_4$	—	$(^2D^o)5d\ ^3D^o_3$	37; 98—CW,P
0.428588	0.428592 ±06	Xe III	$(^2D^o)6p\ ^3D_3$	—	$(^2D^o)6s\ ^1D^o_2$	37
0.429639	0.429633 ±05	Xe II	$(^3P)7s\ ^4P_{1/2}$	—	$(^3P)6p\ ^4P^o_{3/2}$	96—P,S; 325—P
0.430575	0.430575 ±03	Xe IV		—		37—E; 36—E; 78—E; 111—I; 258—G; 47—I; 217—I; 257—P; 325—P; 222—P; 159—I
0.441314	0.441300 ±60	Xe III		—		249; 159—I
0.443415	0.443422 ±10	Xe III	$(^2D^o)6p\ ^3F_2$	—	$(^2D^o)5d\ ^3D^o_1$	37; 157—CW
0.450345	0.450350 ±60	Xe III	$(^2P^o)6p\ 32_1$	—	$(^2P^o)5d\ 27^o_2$	249
0.455874	0.455874 ±06	Xe IV		—		158; 217—I; 325—P; 159—I

*0.460303	0.460302 ±04	Xe II	$(^3P)6p\ ^4D^o_{3/2}$ — $(^3P)6s\ ^4P_{3/2}$	35; 37—λ; 36—CW; 120—CW,U
0.464740	0.464740 ±04	Xe IV	—	37—X; 36—X; 111—I; 47—X; 217—I; 159—I; [cf. C III]
0.465025	0.465025 ±01	Xe III,IV?	—	37—X; 36—X; 111—I; 47—X; 217—I; 159—I; [cf. C III, Kr II]
0.467368	0.467373 ±06	Xe III	$(^2D^o)6p\ ^1F_3$ — $(^2D^o)6s\ ^1D^o_2$	37; 38—CW; 337—G,P; 325—P; 222—P
0.468354	0.468357 ±06	Xe III	$(^4S^o)6p\ ^5P_2$ — $(^4S^o)6s\ ^3S^o_1$	37; 42—CW
0.472360	0.472357 ±05	Xe III	$(^4S^o)6p\ ^5P_1$ — $(^4S^o)6s\ ^3S^o_1$	247; 217—I; 159—I,λ
0.474894	0.474894 ±01	Xe III	—	38—CW; 111—I; 217—I,λ; 159—I
0.479448	0.479450 ±60	Xe III	$(^2D^o)6p\ ^3D_1$ — $(^2D^o)5d\ ^3D^o_1$	249
0.486249	0.486200 ±100	Xe II	$(^3P)7s\ ^4P_{5/2}$ — $(^3P)6p\ ^4P^o_{5/2}$	213—(Ne); 79—A; 280—CW,H,(Ne⁺)
0.486946	0.486948 ±06	Xe III	$(^2D^o)6p\ ^3F_3$ — $(^2D^o)5d\ ^3D^o_2$	37; 38—CW
0.488730	0.488700 ±100	Xe II	$(^3P)6p\ ^2P_{3/2}$ — $(^3P)6s\ ^2P_{3/2}$	24—CW
*0.495413[163]	0.495418 ±03	Xe IV	—	37; 42—CW; 78; 111—I; 258; 305—G,P; 337—G,P; 217—I; 257—G,P; 325—P; 222—P; 159—I
0.496508	0.496508 ±06	Xe II	$(^1D)7s\ ^2D_{3/2}$ — $(^1D)6p\ ^2P^o_{3/2}$	37; 24—CW
0.497270	0.497271 ±05	Xe II	$(^1D)6p\ ^2P^o_{3/2}$ — $(^3P)5d\ ^2D_{5/2}$	147; 78; 96—S,λ
*0.500774[163]	0.500780 ±03	Xe IV	—	37; 42—CW; 78; 111—I; 258; 305—G,P; 337—P; 217—I; 257—P; 222—P; 159—I; **
0.504492	0.504489 ±06	Xe II	$(^1D)6p\ ^2P^o_{1/2}$ — $(^1D)6s\ ^2D_{3/2}$	35; 37—λ; 36—CW; 120—CW,U
0.515704	0.515704 ±06	Xe IV	—	158; 217—I; 159—I
*0.515902[163]	0.515908 ±03	Xe IV	—	37; 42—CW; 78; 111—I; 258—G; 305—G,P; 337—P; 217—I; 257; 222—P; 159—I; **
0.522364	0.522340 ±60	Xe III	$(^2D^o)6p\ ^1F_3$ — $(^2D^o)5d\ ^1D^o_2$	249
*0.523893	0.523889 ±06	Xe III	$(^2D^o)6p\ ^3P_2$ — $(^2P^o)5d\ 13^o_1$	37; 38—CW; 305—A
0.525630[163]	0.525650 ±60	Xe IV	—	249; 159—I
*0.526017[163]	0.526017 ±03	Xe IV	—	37; 36—E; 44—CW; 78; 111—I; 258—E,G,P; 305—G,P; 158—P; 337—G,P; 47—I; 243; 217—I; 257—G,P; 222—P; 159—I; **

Table 2.11.4 (continued) XENON

Wavelength (μm)	Measured value (μm)	Identification: Ion	Upper level	—	Lower level	Ref.
0.526043	0.526043 ±03	Xe II	$(^3P)6p\ ^2P^o_{3/2}$	—	$(^3P)6s\ ^2P_{1/2}$	36; 38—CW; 111—I; 47—I; 217—I; 159—I
*0.526195	0.526150 ±100	Xe II	$(^1D)6p\ ^2D^o_{3/2}$	—	$(^1D)6s\ ^2D_{3/2}$	35; 37—λ; 36—CW; 120—CW,U; 309—CW,H,(Ne)
0.531389	0.531400 ±100	Xe II	$(^3P)7s\ ^4P_{5/2}$	—	$(^3P)6p\ ^4D^o_{7/2}$	213—A,G; 79—A; 48—H,(Ne); 280—CW,H,(Ne); 309—CW,H,(Ne)
0.534334	0.534334 ±05	Xe IV		—		305; 158—λ; 217—I; 159—I
*0.535290[163]	0.535290 ±03	Xe IV		—		37; 42—CW; 78; 111—I; 178—P; 258—G,P; 305—G,P; 158—P; 337—G,P; 217—I; 257—G,P; 222—P; 159—I; **
0.536706	0.536700 ±60	Xe III	$(^2D^o)6p\ ^3F_2$	—	$(^2D^o)5d\ ^3D^o_2$	249—E
*0.539460[163]	0.539460 ±03	Xe IV		—		37; 42—CW; 78—E; 111—I; 258—G,P; 305—G,P; 158—P; 337—P; 217—I; 257—G,P; 222—P; 159—I; **
0.540100	0.540090 ±30	Xe III	$(^2D^o)6p\ ^3P_2$	—	$(^2P^o)5d\ 15^o_2$	245
0.541353	0.541350 ±60	Xe III	$(^2D^o)6p\ ^3P_2$	—	$(^2P^o)5d\ 17^o_3$	249
*0.541915	0.541916 ±06	Xe II	$(^3P)6p\ ^4D^o_{5/2}$	—	$(^3P)6s\ ^4P_{3/2}$	35; 37; 36—CW; 120—CW,U; 95—hfs
0.545433	0.545460 ±60	Xe III	$(^4S^o)6d\ ^5D^o_0$	—	$(^2D^o)6p\ 4_1$	249—E; 337—G,P
0.549942	0.549931 ±04	Xe IV		—		305; 158—λ; 217—I; 159—I
0.552437	0.552450 ±50	Xe II	$(^2D^o)6p\ ^1D_2$	—	$(^2P^o)6s\ ^3P^o_2$	38—CW
0.559227	0.559235 ±05	Xe IV		—		202; 305; 140—E,λ; 217—I; 159—I
0.565937	0.565900 ±100	Xe II	$(^3P)6p\ ^2P^o_{1/2}$	—	$(^3P)5d\ ^4P_{1/2}$	24—CW; 36—E; 217—I; 159—I
0.572690	0.572700 ±100	Xe II	$(^1D)6p\ ^2D^o_{5/2}$	—	$(^1D)5d\ ^2F_{5/2}$	24—CW; 36—I; 79—E; 309—CW,H,(Ne)
0.575102	0.575100 ±100	Xe II	$(^3P)6p\ ^2D^o_{3/2}$	—	$(^3P)5d\ ^4P_{1/2}$	24—CW; 36—E

0.589328	0.589330	±03	Xe II	$(^1D)6p\ ^2P^o_{3/2}$ — $(^1D)5d\ ^2D_{5/2}$	96—S; 111—I,λ
*0.595567	0.595567	±03	Xe IV	—	37; 78; 111—I; 305—P; 158—P; 337—P; 243; 217—I; 257—G,P; 159—I; **
0.597111	0.597112	±06	Xe II	$(^1D)6p\ ^2P_{3/2}$ — $(^1D)6s\ ^2D_{3/2}$	35; 37—λ; 36—CW; 120—CW,U
0.609361	0.609400	±100	Xe II	$(^3P)7s\ ^4P_{3/2}$ — $(^3P)6p\ ^4D^o_{3/2}$	79—A; 309—CW,H,(Ne)
0.617615	0.617619	±03	Xe III	—	248; 111—I; 157—CW; 217—I; 159—I,λ
0.623825	0.623890	±80	Xe III	$(^2D^o)6p\ ^1F_3$ — $(^2P^o)5d\ 17^o_3$	248; 157—CW
*0.627081	0.627090	±10	Xe II	$(^1D)6p\ ^2F^o_{5/2}$ — $(^1D)6s\ ^2D_{3/2}$	35; 37—λ; 36—CW; 120—CW,U; 95—hfs,λ
0.628641	0.628660	±60	Xe IV,V?	—	249; 337; 159—I
0.634343	0.634343	±05	Xe III	—	337; 217—I,λ; 159—I
0.652865	0.652850	±50	Xe II	$(^1D)6p\ ^2F^o_{7/2}$ — $(^1D)5d\ ^2F_{5/2}$	38—CW; 120—CW,U
0.669431	0.66943	±?	Xe II	$(^3P)6p\ ^4P^o_{3/2}$ — $(^3P)5d\ ^4D_{1/2}$	208—CW
0.669950	0.669950	±30	Xe IV	—	158; 159—I
0.670225	0.670200	±100	Xe II	$(^1D)6p\ ^2P^o_{3/2}$ — $(^1D)5d\ ^2F_{5/2}$	120—CW,U; 343
0.707234	0.70723	±?	Xe II	$37^o_{5/2}$ — $(^3P)6d\ ^4D_{5/2}$	208—CW
0.714903	0.714894	±60	Xe II	$(^3P)6p\ ^4D^o_{3/2}$ — $(^3P)6s\ ^2P_{3/2}$	247—E; 208—CW; 120—CW, U; 217—I; 159—I
0.761859	0.761900	±20	Xe II	$(^1D)6p\ ^2D_{5/2}$ — $(^1S)5d\ ^2D_{5/2}$	309—CW,H,(Ne)
0.782763	0.782800	±300	Xe II	$35^o_{5/2}$ — $16_{3/2}$	306
0.798800	0.798900	±300	Xe II	$(^3P)6p\ ^4P^o_{1/2}$ — $(^3P)6s\ ^4P_{1/2}$	306
0.833271	0.833000	±300	Xe II	$27^o_{5/2}$ — $(^3P)6d\ ^4D_{5/2}$	306
0.844619	0.844300	±300	Xe II	$27^o_{5/2}$ — $(^3P)6d\ ^4D_{3/2}$	306
0.8569	0.856900	±300	Xe III?	—	306; 111—I; 217—I; 159—I
0.858251	0.858200	±300	Xe II	$31^o_{3/2}$ — $10_{5/2}$	306
0.871617	0.871400	±300	Xe II	$(^3P)6p\ ^4D^o_{3/2}$ — $(^3P)5d\ ^2P_{3/2}$	306; 208—CW
0.905930	0.906300	±400	Xe II	$27^o_{5/2}$ — $16_{3/2}$	306
0.926539	0.926500	±400	Xe II	$(^1S)5d\ ^2D_{3/2}$ — $(^3P)6p\ ^4D^o_{5/2}$	306
0.928854	0.928700	±400	Xe II	$13^o_{1/2}$ — $(^1D)5d\ ^2S_{1/2}$	306
0.969859	0.969700	±200	Xe II	$(^3P)6p\ ^4D^o_{3/2}$ — $(^3P)5d\ ^4P_{5/2}$	306; 319—CW
1.063385	1.063400	±600	Xe II	$(^3P)6p\ ^4D^o_{3/2}$ — $(^3P)5d\ ^4P_{3/2}$	306
1.0950	1.095000	⁺600	Xe	—	306

Table 2.12.1
EUROPIUM

Wavelength (μm)	Measured value (μm)	Identification: Ion	Upper level — Lower level	Ref.
		Spectroscopic Reference: Eu II (285)		
0.664506	0.6645 ± ?	Eu II	$z\,^9P_5$ — $a\,^9D^\circ_6$	32—S,(He)
0.989827	0.9898 ± ?	Eu II	$z\,^7P_3$ — $a\,^7D^\circ_4$	32—P,(He)
1.00195	1.0020 ± ?	Eu II	$z\,^7P_4$ — $a\,^7D^\circ_5$	31—G,P,(He); 32—P
1.01656	1.0166 ± ?	Eu II	$z\,^7P_4$ — $a\,^7D^\circ_4$	31—G,P,(He); 32—P
1.36070	1.3610 ± ?	Eu II	$z\,^9P_4$ — $a\,^7D^\circ_5$	31—G,P,(He); 32
1.47665	1.4770 ± ?	Eu II	$z\,^9P_3$ — $a\,^7D^\circ_4$	32—P

Table 2.12.2
YTTERBIUM

Wavelength (μm)	Measured value (μm)	Identification: Ion	Upper level — Lower level	Ref.
		Spectroscopic Reference: Yb II (236)		
1.26925	1.271400 ± 100	Yb II	$(^2F^\circ_{7/2})6s6p\,(^3P_0)_{7/2}$ — $(^2F^\circ_{5/2})5d6s\,^3[9/2]^\circ_{7/2}$	200—S
1.34527	1.345300 ± 100	Yb II	$(^1S)6p\,^2P^\circ_{3/2}$ — $(^1S)5d\,^2D_{3/2}$	200—E,S
1.64984	1.649800 ± 200	Yb II	$(^1S)6p\,^2P^\circ_{3/2}$ — $(^1S)5d\,^2D_{5/2}$	53—G,S; 55—hfs
1.804031	1.8057 ± ?	Yb II	$(^2F^\circ_{7/2})6s6p\,(^3P_1)_{5/2}$ — $(^2F^\circ_{7/2})5d6s\,^3[5/2]^\circ_{5/2}$	200
2.1480	2.1480 ± ?	Yb II?	—	200
2.43775	2.437700 ± 200	Yb II	$(^1S)6p\,^2P^\circ_{1/2}$ — $(^1S)5d\,^2D_{3/2}$	53; 55—hfs

Table 2.13
MISCELLANEOUS LINES

Measured wavelength (μm)	Original identification: Species	Original identification: Ref.	Probable identification; comments
0.298370 ± 50	Xe III	36; 37; 217	O III; not seen in 222
0.332437 ± 10	Ne	80	Ne II, 0.332377; λ not seen in 222
0.657745 ± 12	Ne, He	37	?
0.840800 ± 300	Xe	6—λ; 306	Xe I
1.117682 ± 10	Hg II?	25	Hg I $7p^1P_1^o \rightarrow 7s^3S_1$ [Ref. 29]
1.1869 ± ?	Cd II	320	Cd I $6p^1P_1 \rightarrow 6s^3S_1$ [Ref. 89]
1.3655 ± ?	Hg II?	25	Hg I $7p^3P_1^o \rightarrow 7s^3S_1$ [Ref. 29]

Note: In addition to the above, several unidentified lines in Cu, Ag, and Au were reported in References 184, 235, 275, 276, and 308; these later proved to be grating ghosts.[21,277]

Table 2.14
SINGLY IONIZED LASERS EXCITED BY CHARGE-EXCHANGE OR PENNING COLLISIONS

Collision partner	Lasing ion
He^+	Be, Ca, Sr, Ba, Ni, Cu, Ag, Au, Zn, Cd, Hg, Al, Tl, As, Se, Te, I, Kr?
Ne^+	Be, Cu, Ag, Hg, Al, Tl, Te, Xe?
He*	Mg, Cu, Zn, Cd, Sn?, Pb, Kr?
Ne*	Xe?

Table 2.15
SCALING RELATIONS FOR ^4He-$^NCd^+$ LASERS OF 1 < D < 1.6 mm

Quantity		Value	Units
Helium pressure,	$p_{He}D$	1.00	torr — cm
Cadmium pressure,	$p_{Cd}D$	2.5×10^{-4}	torr — cm
Electric field,	ED	7.6	volts
Current density,	i/D	0.43	A/cm
Power output,	P_{out}/LD	4.3	mW/cm^2
Gain,	gD	1.4×10^{-4}	—
Efficiency,[a]	η/D	1.4×10^{-3}	cm^{-1}
Pressure sensitivity,[b]	$\Delta p_{He}D$	0.93	torr — cm
Noise,[c]	—	4	percent

[a] $\eta = P_{out}/iEL$.
[b] Change in $p_{He}D$ to decrease P_{out} to 0.5 its maximum value.
[c] Measured over 10 Hz to 1 MHz band width at maximum P_{out}.

From Johnston, T. F., Jr. and Kolb, W. P., *IEEE J. Quantum Electron.*, QE-12, 482, 1976. With permission.

Table 2.16
PARAMETERS OF HOLLOW-CATHODE He-Hg⁺ ION LASERS

Date	Dimensions D (mm)	Dimensions L (cm)	Output λ (μm)	Output P (W)	Excitation I (A)	Excitation V (kV)	Excitation τ (μsec)	Excitation Rate (kHz)	Gas Hg (°C)	Gas He (torr)	Comment	Ref.
1965	25	30	0.6150	2—3	—	5	2	1	30	1—2	g = 2.2 dB/m	52
			0.7945	—	—	—	2	1	30	1—2	—	
1966	25	20	0.6150	—	30	5	—	1	—	—	^{202}Hg; g = 7 dB	244
1967	25	15	0.6150	1.8	10	2	1—3	30	—	—	^{202}Hg; g = 0.5 dB/cm	338
			1.5554	—	—	—	—	—	—	—	—	
1969	24	53	0.6150	—	60	—	2	—	80—100	10	^{202}Hg, —	124
			0.7418	—	60	—	—	—		—	^{202}Hg, —	
			0.7945	7	60	—	—	—		30	^{202}Hg; g = 0.6% cm^{-1}	
1971	5.6	50	0.6150	—	0.7^{t}	—	CW	—	60—80	4—9		291
			0.7945	—	0.24^{t}	—	CW	—	60—80	4—9		
1975	3	50	0.6150	0.08	10	—	1200	—	115	16		267
			0.6150	0.02	3.5	—	CW		115	16		
			0.7945	0.005	4	—	CW		115	10		

Note: t, threshold current; power output not measured.

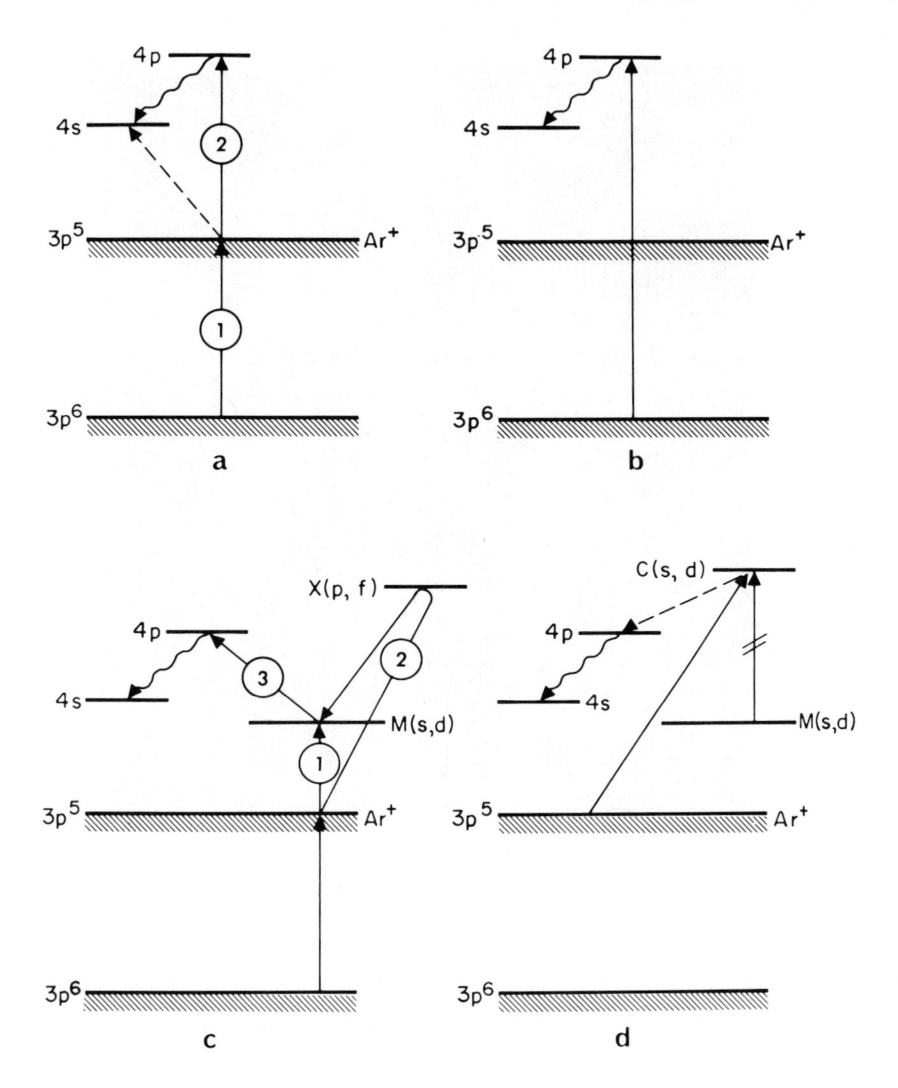

FIGURE 2.16. Schematic energy level diagrams for the 4p → 4s laser transitions in Ar II, showing various excitation models. (From Bridges, W. B. and Halsted, A. S., Gaseous Ion Laser Research, Tech. Rep. AFAL-TR-67-89 (DOC No. AD-814897), Hughes Research Laboratories, Malibu, Calif., 1967. With permission.)

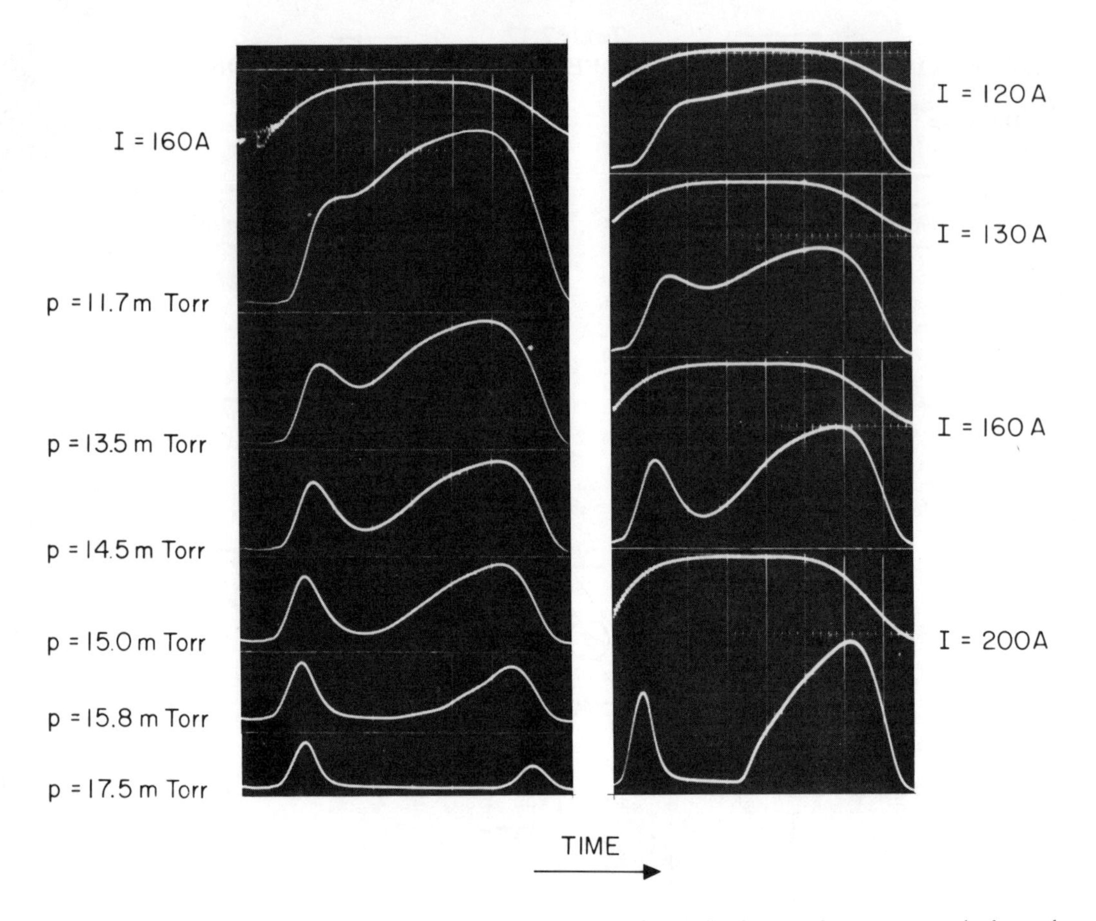

FIGURE 2.17. Variation of the output waveform of a pulsed Ar II ion laser as the gas pressure is changed at constant current (left) or the discharge current is changed at constant gas pressure (right). The current pulse is about 50 μsec in this 5-mm diameter laser. (From Bridges, W. B., *Methods of Experimental Physics*, Vol. 15A, Tang, C. L., Ed., Academic Press, New York, 1979. With permission.)

Table 2.17
OPERATING PARAMETERS AND PERFORMANCE OF He-Se⁺ ION LASERS

Discharge dimensions			Power output		Optimum values				
Diameter (mm)	Length (cm)	Total wavelengths observed	mW	λ's	I (mA)	He (torr)	Se (mtorr)	Other	Ref.
3	50	—	—	—	500	6	5	—	300
4	100	27	30	6	> 200[a]	8	5	—	300
2	100	17	20—30	7—10	200	4.5	7	—	194
3	200	46	250	?	400	7	5	—	199
1	10	5	—	—	50	20	—	V_{TUBE} = 850 V	151
1.5	50	17	28	6	115	20	—	V_{TUBE} = 2750 V	151
2	90	24[b]	—	—	175	15	—	V_{TUBE} = 3600 V	151

[a] Not optimum current; maximum value limited by glass tube walls.
[b] Some wavelengths not confirmed by other references.

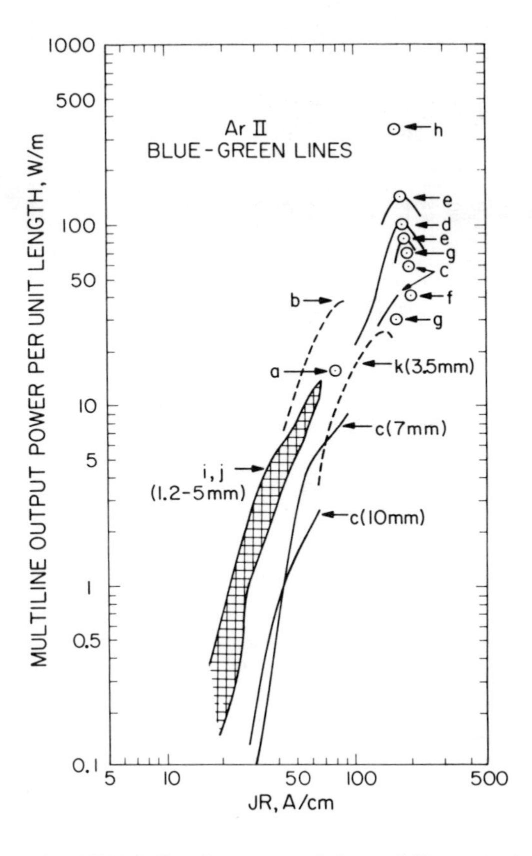

FIGURE 2.18. Summary of the multiline output powers per unit length as a function of the product of current density and discharge radius (JR) observed by various investigators. The solid curves and points are large-bore, zero magnetic field lasers. The dotted and shaded curves are smaller-bore tubes with magnetic fields of 0.5 to 1.5 kG. Sources of data are a[39]; b[100]; c[155]; d[30]; e[87]; f[329]; g[330]; h[4]; i[206]; j[41]; k[350]. (From Bridges, W. B., *Methods of Experimental Physics*, Vol. 15A, Tang, C. L., Ed., Academic Press, New York, 1979. With permission.)

Table 2.18
PERFORMANCE OF HIGH-POWER CW BLUE-GREEN Ar ION LASERS

	Operating parameters							Performance				
Date	D (mm)	L (cm)	I (A)	V (V)	B (G)	p (torr)	JR (A/cm)	P (W)	P/L (W/m)	η (%)	Bore type	Ref.
1967	4	70	50	219	900	0.5	80	10	14.3	0.09	Fused silica	34
1970	6.35	254	90	600	600	0.38	90	93	35.4	0.16	Graphite disks	100
1969	12	200	(320)	(280)	0	(0.4)	170	120	(60)	0.13	Fused silica	155
1970	12	85	340	?	0	1.1	180	90	105	?	Al_2O_3 on Al segments	30
1972	11	55	320	—	0	1	(185)	45	(82)	—	Al_2O_3 on Al segments	87
	16	135	420	255	0	0.7	(168)	175	(130)	0.16		
1972	12	85	(377)	—	0	—	200	(47)	40	—	W segments	329
1973	12	85	300	(150)	0	—	(159)	25	(29)	0.055	W segments	330
	12	185	360	(430)	0	—	(191)	125	(68)	0.08		
1973	16	150	390	(< 640)	0	0.4	(155)	500	(333)	0.2	Al_2O_3 on Al segments	4
1979	16	200	400	500	0	—	(160)	300	(150)	0.15	Al_2O_3 on Al segments	88

Table 2.19
COMPARISON OF OUTPUT POWERS OF THE BLUE-GREEN Ar II LASER LINES FROM COMMERCIAL ION LASERS

	Laser output power (W)			
λ (μm)	CR-2[a]	164-00[b]	CR-18[a]	170-03[b]
0.4545	0.05	—	0.7	0.5
0.4579	0.15	0.15	1.3	1.2
0.4658	0.05	0.05	0.7	0.5
0.4727	0.06	0.06	1.0	0.8
0.4765	0.3	0.3	2.5	1.5
0.4880	0.7	0.7	6.0	5.0
0.4889	—[c]	—[c]	—[c]	—[c]
0.4965	0.3	0.3	2.5	1.5
0.5017	0.14	0.14	1.5	0.7
0.5145	0.8	0.8	7.5	6.0
0.5287	0.1	—	1.5	1.3

[a] Power output specifications taken from Coherent Inc. ion laser brochure, dated December 1976.

[b] Power output specifications taken from Spectra-Physics data sheets, undated.

[c] This line oscillates both pulsed and CW, but is often overlooked because it is difficult to separate from 0.4880 μm with intracavity or external prisms.

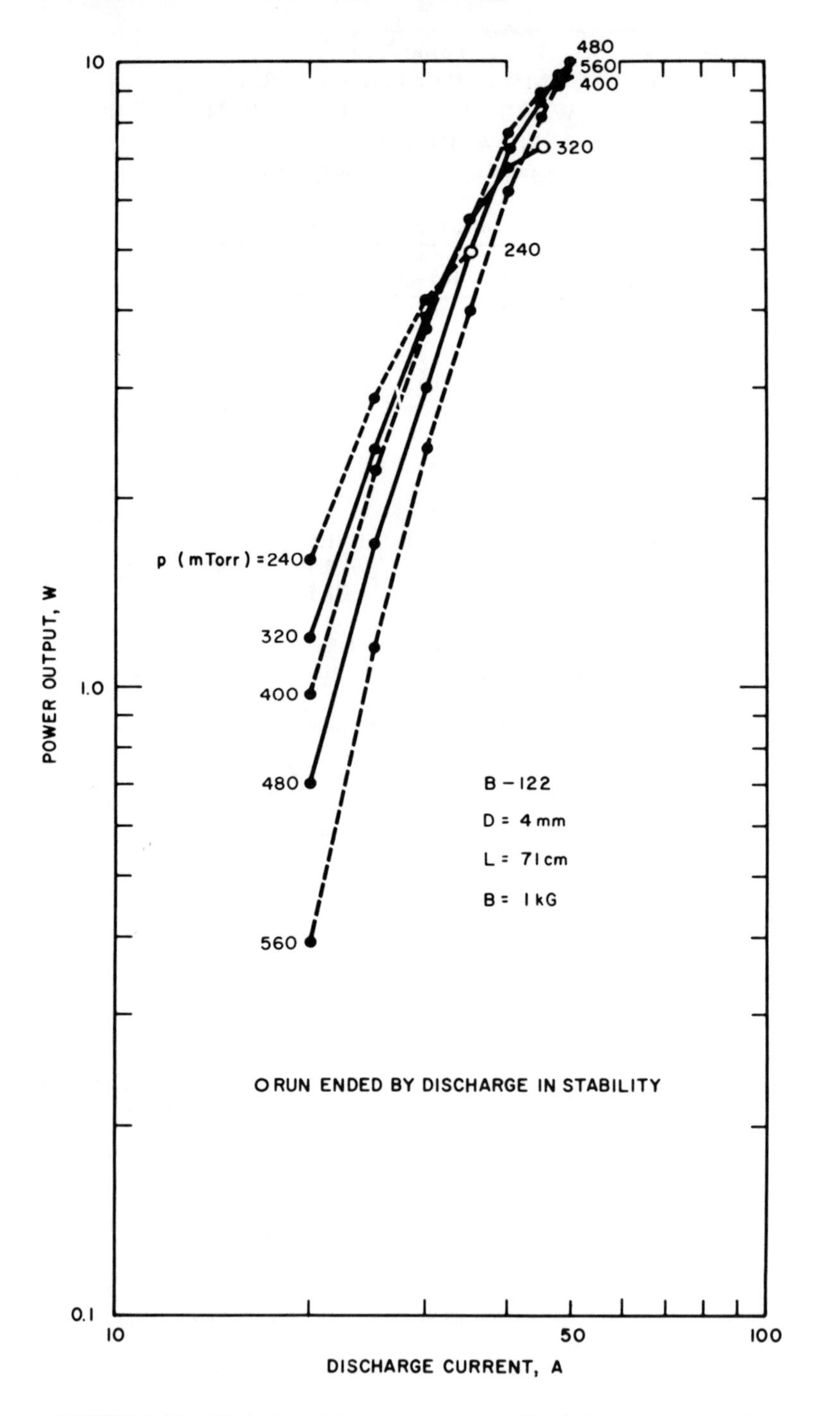

FIGURE 2.19. Variation of the output power with discharge current and gas filling pressure (cold) for a CW Ar II ion laser. (From Bridges, W. B. and Halsted, A. S., Gaseous Ion Laser Research, Tech. Ref. AFAL-TR-67-89 (DDC No. AD-814897), Hughes Research Laboratories, Malibu, Calif., 1967. With permission.)

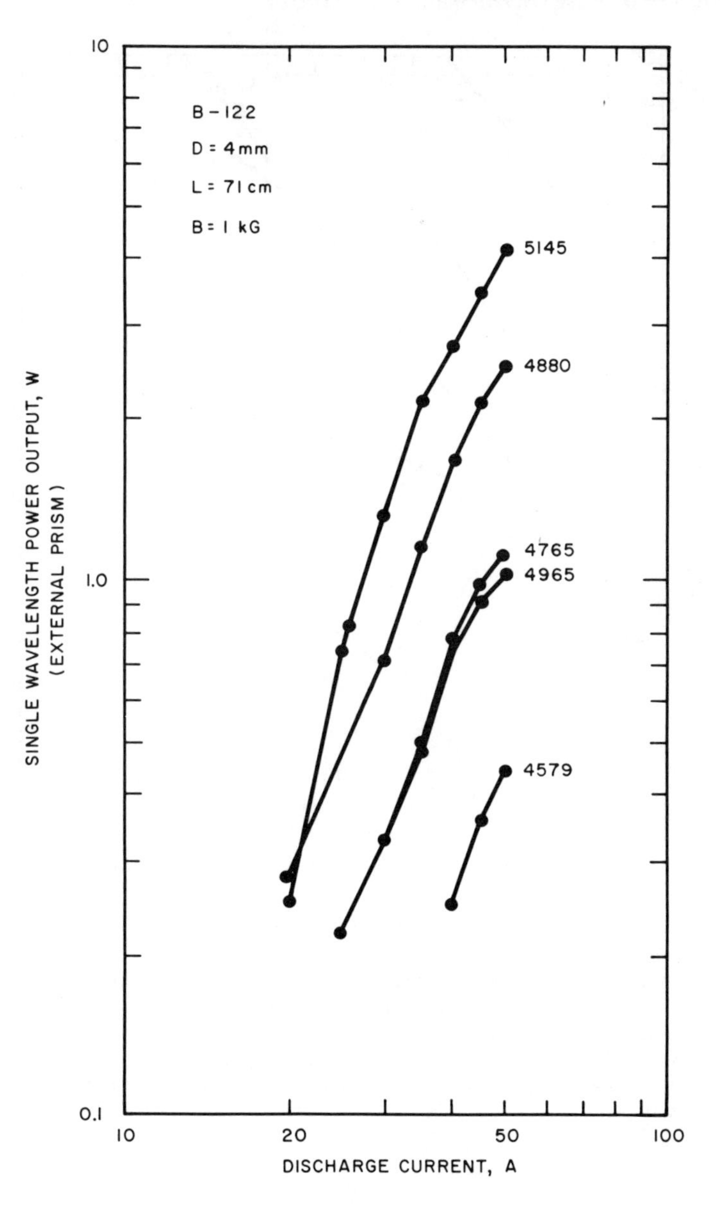

FIGURE 2.20. Variation of the output power on individual wavelengths for the stronger Ar II laser lines, using an external prism to separate the colors. This data taken with the same laser as Figure 19, with p = 480 mtorr. (From Bridges, W. B. and Halsted, A. S., Gaseous Ion Laser Research, Tech. Rep. AFAL-TR-67-89 (DDC No. AD-814897) Hughes Research Laboratories, Malibu, Calif., 1967. With permission.)

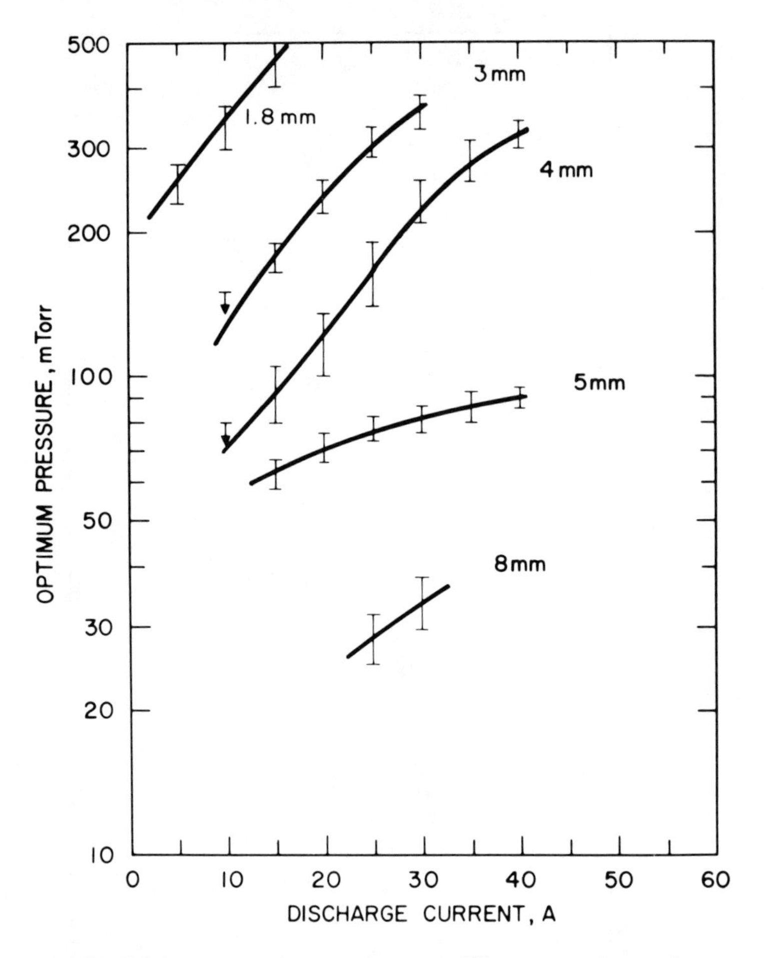

FIGURE 2.21. Approximate optimum gas filling pressure for maximum Ar II laser output, plotted as a function of discharge current for various capillary diameters. (With additions, from Bridges, W. B. and Halsted, A. S., Gaseous Ion Laser Research, Tech. Rep. AFAL-TR-67-89 (DDC No. AD-814897) Hughes Research Laboratories, Malibu, Calif., 1967. With permission.)

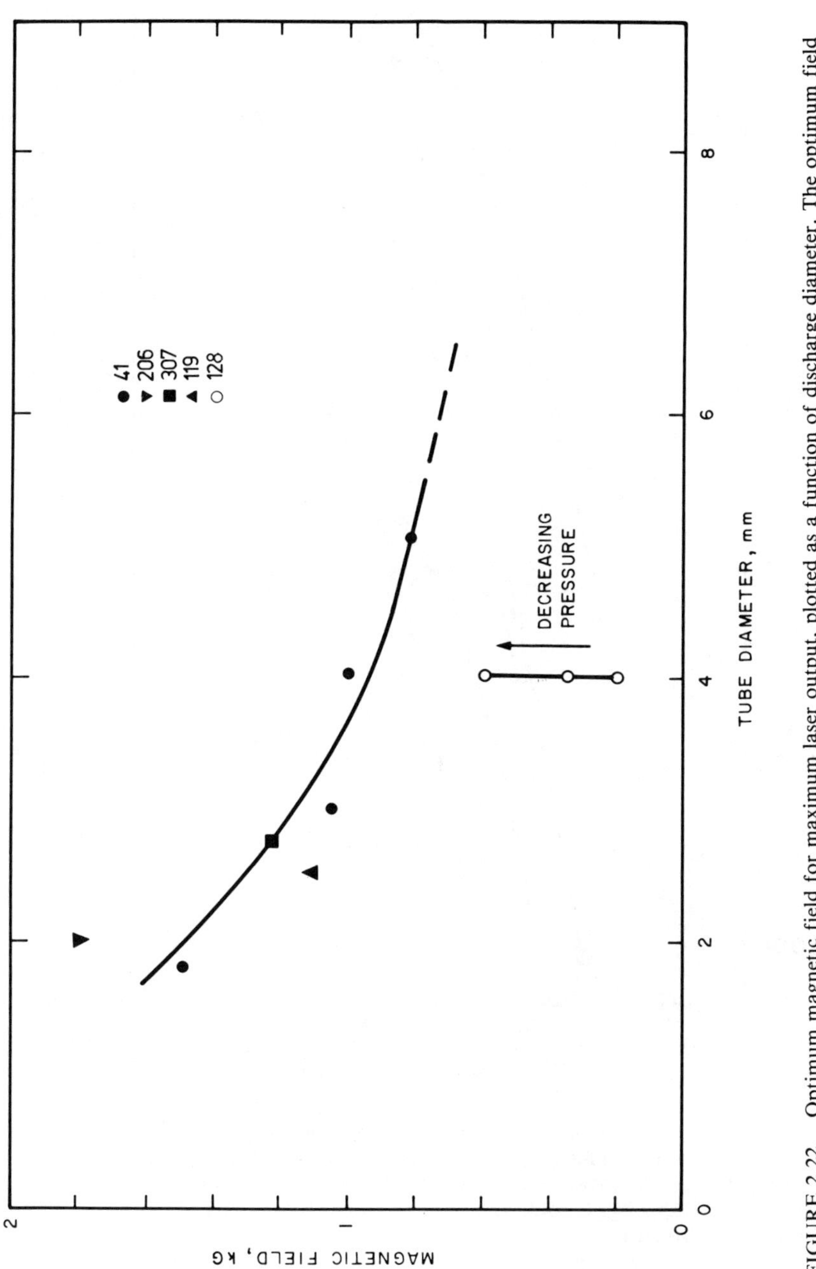

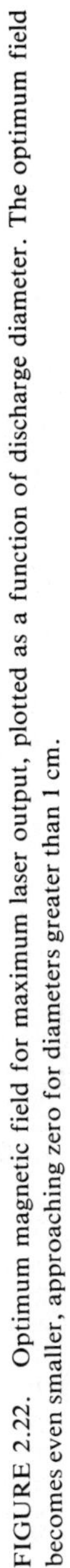

FIGURE 2.22. Optimum magnetic field for maximum laser output, plotted as a function of discharge diameter. The optimum field becomes even smaller, approaching zero for diameters greater than 1 cm.

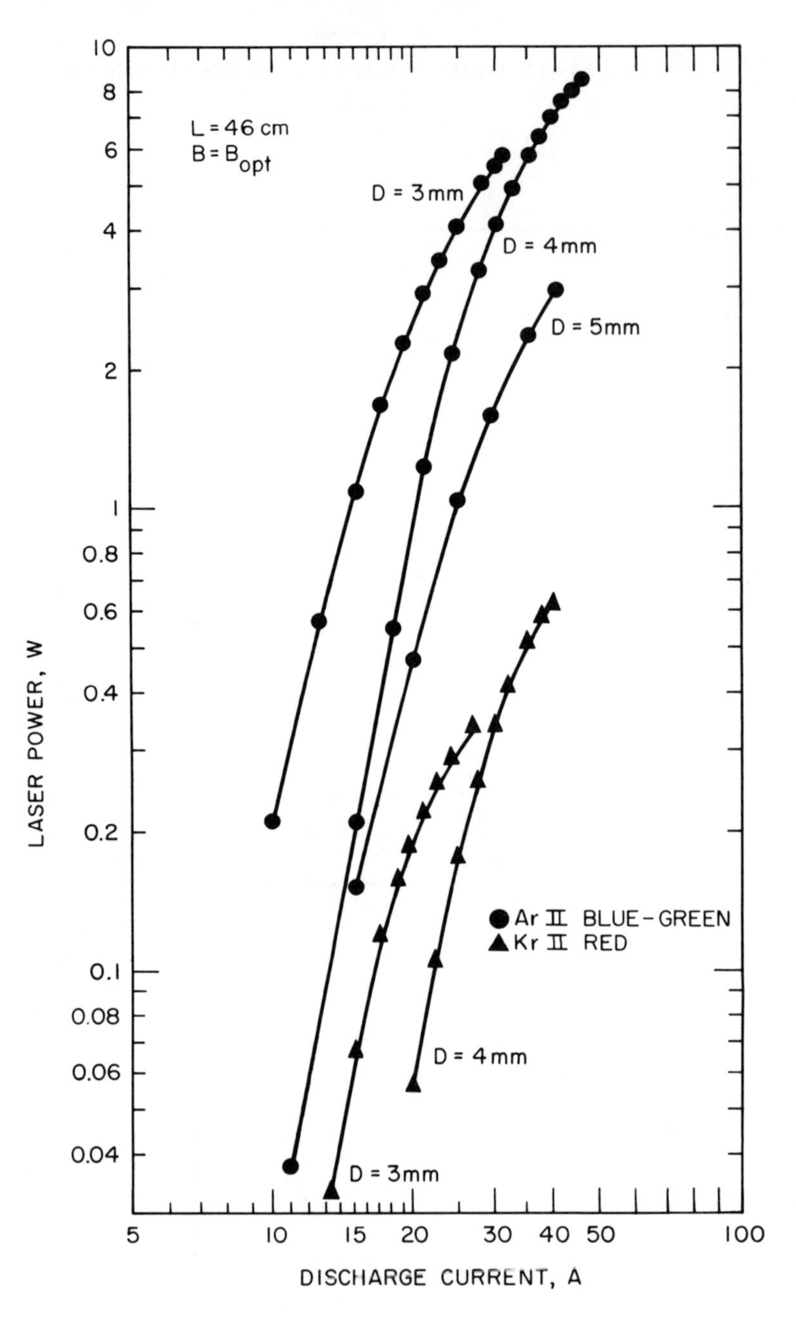

FIGURE 2.23. Comparison between the Ar II blue-green lines and the Kr II red (0.6471 + 0.6764 μm) lines for typical small-bore ion lasers. The gas filling pressure and magnetic field were optimized at the highest power shown for each curve. (From Bridges, W. B. and Chester, A. N., unpublished data, 1969 and 1970. With permission.)

Table 2.20
PERFORMANCE OF COMMERCIAL KRYPTON ION LASERS

Wavelength (μm)	Ion	Output power (W)				
		CR3000K[a]	SP171-01[b]		SP921-01[b,c]	
0.3374	III					
0.3507	III	2.0	1.1		2.5	
0.3564	III					
0.4067	III	0.9	—		—	
0.4131	III	1.5	0.5		2.4	
0.4154	III	0.1	—	1.3	—	3.0
0.4226	III	—	—		—	
0.4680	II	0.5	0.3		0.8	
0.4762	II	0.4	0.4		1.0	
0.4825	II	0.4	0.4		1.0	
0.5208	II	0.7	0.7		2.0	
0.5309	II	1.5	1.4		3.0	
0.5682	II	1.1	1.0		2.5	
0.6471	II	3.5	3.5		7.0	
0.6764	II	0.9	0.9		2.5	
0.7525	II	1.2	1.1		3.0	
0.7931	II	0.3	—		—	
0.7993	II		0.25		1.2	
Line current @ 460 V 3ϕ (A)		70	60		60 × 2	

[a] Data source: Super Graphite Ion Lasers and Accessories, Coherent Radiation, Palo Alto, Calif., December 1976.

[b] Data source: Spectra-Physics High Power Ion Lasers, Spectra-Physics, Mountain View, Calif., May 1977.

[c] This laser is essentially two SP171-01 discharge tubes in a common optical cavity.

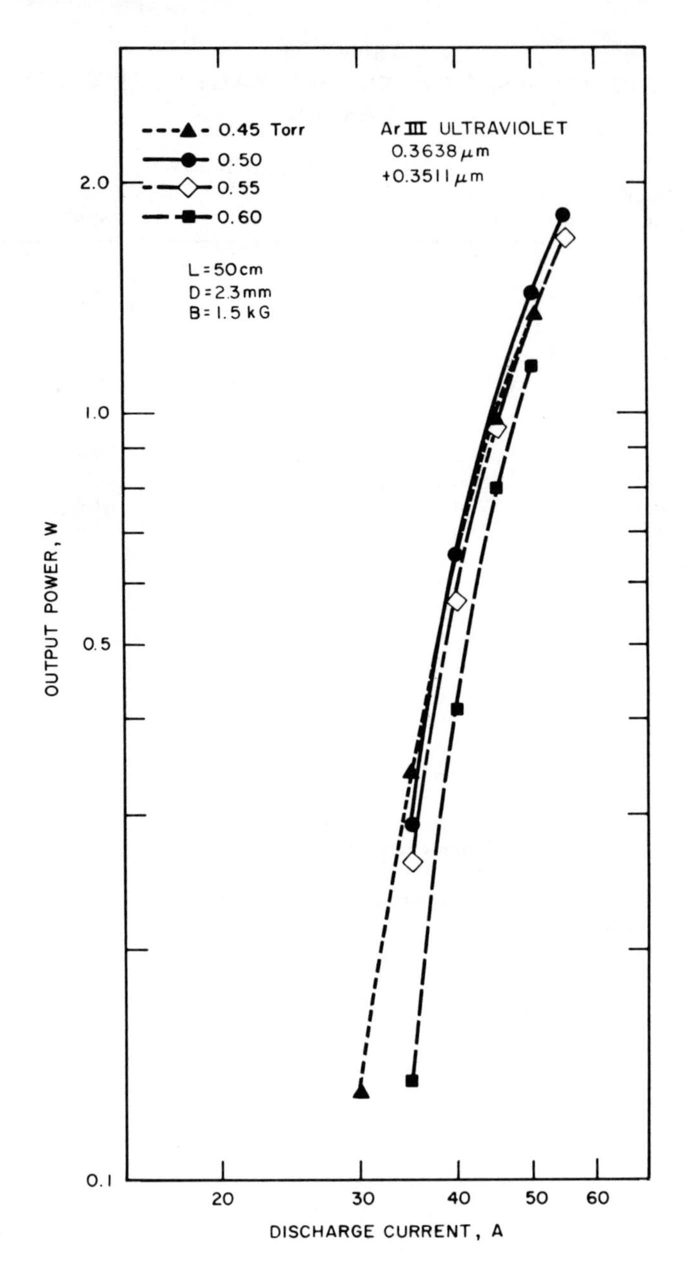

FIGURE 2.24. Output power of a disk bore Ar III UV ion laser. (From Bridges, W. B. and Mercer, G. N., Ultraviolet Ion Lasers, Tech. Rep. ECOM-0229-F (DDC No. AD-861927), Hughes Research Laboratories, Malibu, Calif., 1969. With permission.)

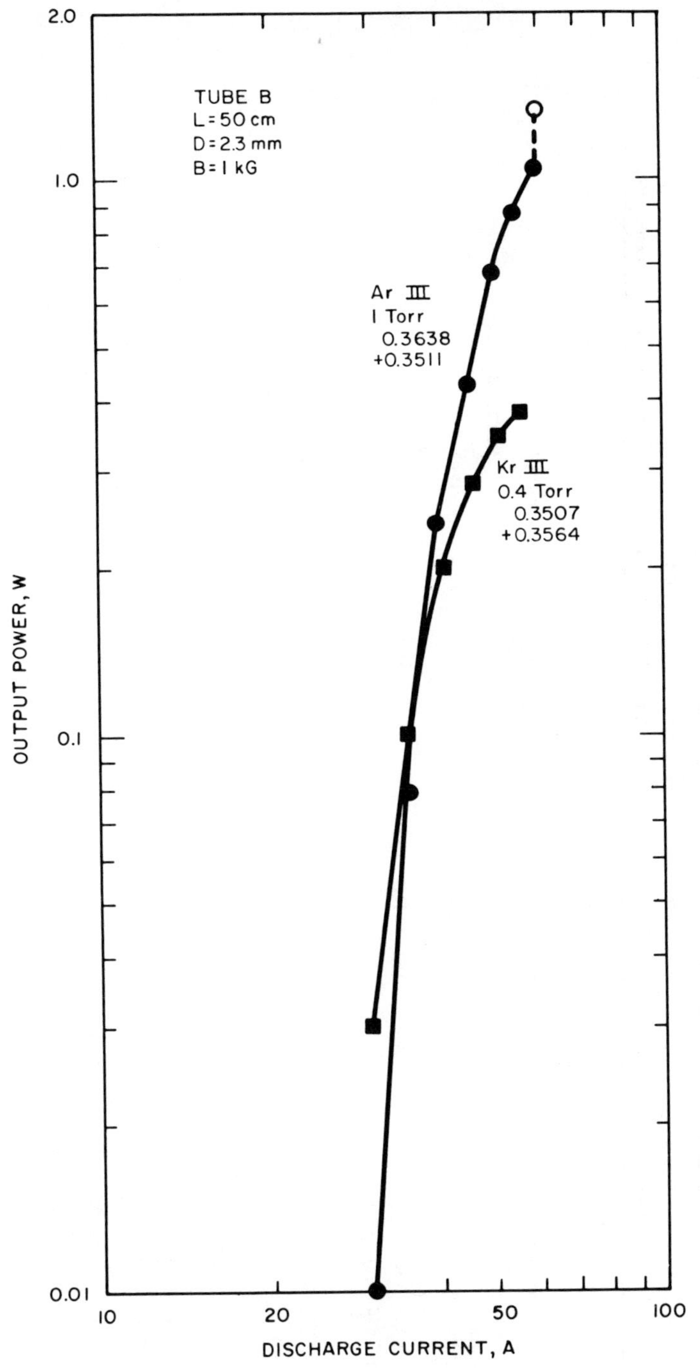

FIGURE 2.25. Comparison of the output powers for the Ar III and Kr III UV laser lines. (From Bridges, W. B. and Mercer, G. N., Ultraviolet Ion Laser, Tech. Rep. ECOM-0229-F (DDC No. AD-861927), Hughes Research Laboratories, Malibu, Calif., 1969. With permission.)

Table 2.21
PERFORMANCE OF CW UV NOBLE GAS ION LASERS

			Operating parameters							Performance				
Date	Ion	λ	D (mm)	L (cm)	I (A)	V (V)	B (G)	p (torr)	JR (A/cm)	P (W)	P/L (W/m)	η (%)	Bore type and comment	Ref.
	Ne II	0.3324 (0.75)			18	510	1100	0.7	67	0.13	0.38	0.0014		
		0.3378 (0.25)												
	Ar III	0.3511 (0.5)			20	315	1200	0.33	75	0.1	0.35	0.0016		
1968		0.3638 (0.5)	1.7	34									Graphite segments	98
	Kr III	0.3507 (0.7)			20	270	1200	0.35	75	0.14	0.41	0.0025		
		0.3564 (0.3)												
	Xe III	0.3781 (0.8)			18	240	1100	0.25	67	0.075	0.22	0.0017		
		0.3454 (0.2)												
	Ar III	0.3511								1.5	0.75	—		
1968		0.3638	12	200	280		0	0.3	150				Fused silica	10
	Kr III	0.3507								1.3	0.65	—		
	Ar III	0.3511 (0.2)			135	—			214	1.7	1.9	—		
		0.3638 (0.8)												
1968			4	90			840	0.85					W disks	209
	Kr III	0.3507			120	—			185	0.44	0.5	—		
	Ar III	0.3511			64	390	1500	0.8	177	2.3	5	0.01		
		0.3638												
1969			2.3	46									W disks	43
	Kr III	0.3507			57	—	1500	—	158	0.38	0.83	—		
		0.3564												
1970	Ar III	0.3511	12	85	(475)	—	0	—	250	3	3.5	—	Al_2O_3 on Al segments	30
		0.3638												

1976	Ne II	0.3324	12	185	400	—	0	0.8—1	212	0.15	0.08	—	W segments	321
	Ar III	0.3511 (0.45)			485	—	0	0.7—1.4	180	16	8.6	—		
		0.3638 (0.55)												
	Kr III	0.3507 (0.8)			430	—	0	0.6—1.2	228	7	3.8	—		
		0.3564 (0.2)												
	Xe III	0.3781 (0.9)			340	—	0	0.8—1.1	257	1.8	1	—		
		0.3746 (0.1)												
1977	Ar III	0.3511	12	170	490	480	25	1.2	260	40	24	0.017	Al_2O_3 on Al segments	218
		0.3638												
1977	Ar III	0.3511		170					255	61	36		Al_2O_3 on Al segments	219
		0.3638												
	Ar III	0.33 triplet								17	10			
	Ar III	0.30 triplet								4	2.3			
	Ar III	0.275								0.4	0.23			
	Kr III	0.3507								19	11.2			
		0.3564												
	Kr III	0.32 triplet								4.5	2.7			
1978	Ne II	0.33	12	75	580	—	73	1.2	350	6.3	8.4		Al_2O_3 on Al segments	220

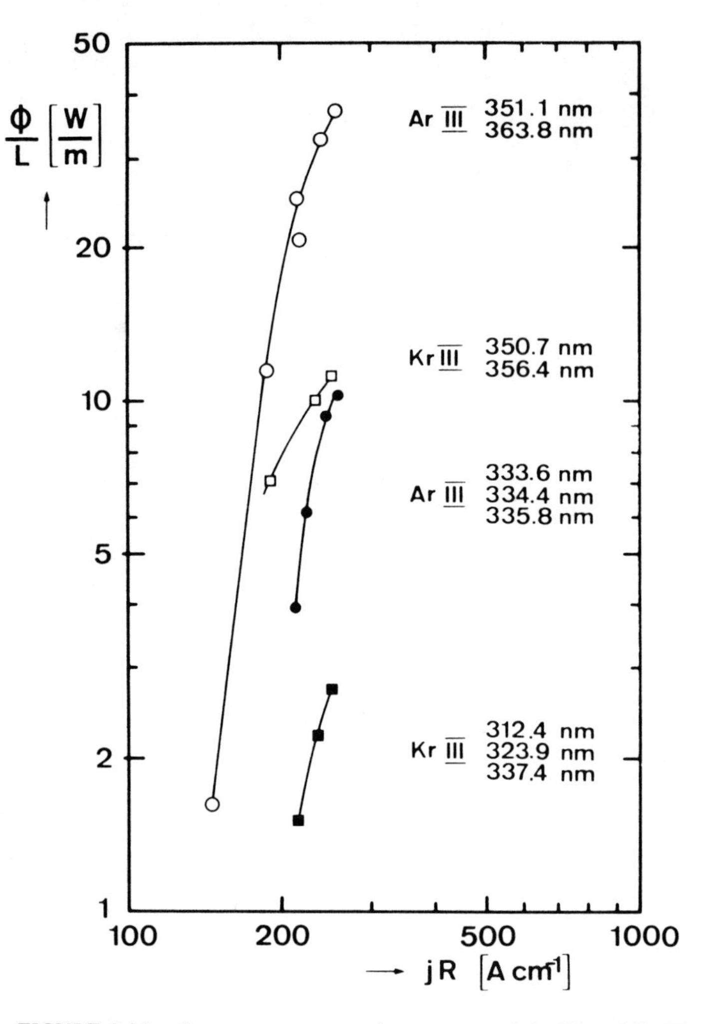

FIGURE 2.26. Power output on various groups of Ar III and Kr III UV laser lines as a function of discharge current, plotted as power per unit length [Φ/L] vs. the product of current density and discharge radius [JR]. Data is for a 12 mm diameter discharge with a 30 G magnetic field for stabilization. (From Lüthi, H. R., Seelig, W., and Steinger, J., *Appl. Phys. Lett.*, 31, 670, 1977. With permission.)

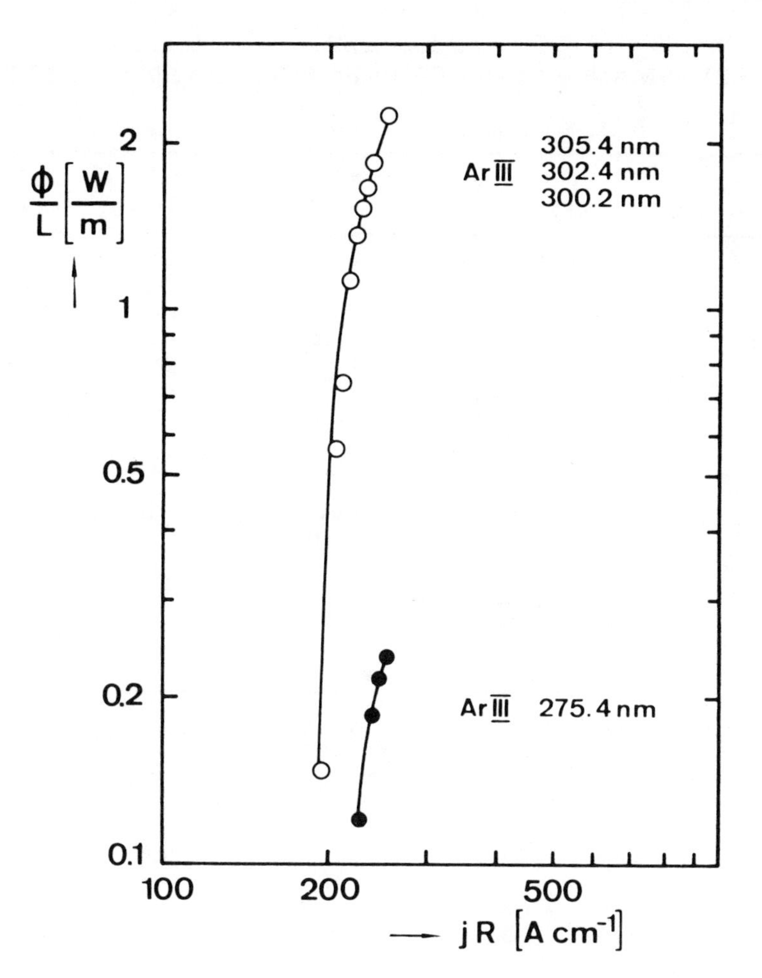

FIGURE 2.27. Power output of additional groups of Ar III laser lines for the same conditions as Figure 26. (From Lüthi, H. R., Seelig, W., and Steinger, J., *Appl. Phys. Lett.*, 31, 670, 1977. Courtesy of the American Institute of Physics.)

Table 2.22
PERFORMANCE OF BLUE-GREEN PULSED Xe ION LASERS

	Discharge parameters					Output			
Year	D (mm)	L (cm)	I (kA)	Rep. rate (Hz)	Press (mtorr)	P_{peak} (kW)	τ (μsec)	P_{avg} (W)	Ref.
1970	2.3	152	0.1—0.6	—	8—28	0.9	0.5—5	—	(305)
1970	5	175	1.8	10	—	0.4	0.6—1.5	—	(158)
1970	4	<150	1—1.2	100—200	—	2	0.3	(0.4)	(349)
1973	7—12	90	1.5—4	—	16—20	2	0.2	—	(257)
1974	17	300	2	10—36	6	82	0.3	0.1—0.25	(142)
1975	7	150	10	600	11	6.5	0.2—0.3	—	(222)
1975	4	120	—	30	—	4	0.35	0.02	(289)

Table 2.23
COMPARISON OF POWER OBSERVED ON Xe IV LINES

Ref.	(305)	(158)	(257)	(287)	(222)
Date	1970	1973	1973	1975	1975
	P(W)				
λ(μm)					
0.2232	—	—	—	—	10—50
0.2315	—	—	—	—	1400
0.3079	—	—	—	—	32
0.3247	—	—	Weak	—	—
0.3331	—	—	100	—	330
0.3645	—	—	400	—	3600
0.4306	—	—	240	50	1000
0.4954	155	—	525	250	
0.5008	150	—	180	—	
0.5159	85	—	Weak	250	6500 (0.5159—0.5956)
0.5260	160	110	350	680	
0.5353	180	60	570	800	
0.5395	170	120	560	480	
0.5956	—	100	120	110	

REFERENCES

1. **der Agobian, R., Otto, J.-L., Cagnard, R., Barthélémy, J., and Échard, R.,** Emission Stimulee en Regime Permanent dans le Spectre Visible du Krypton Ionise, *C.R. Acad. Sci. Paris,* 260, 6327—6329, 1965.
2. **Akirtava, O. S., Bogus, A. M., Dzhilkiya, V. L., and Oleinik, Yu. M.,** Quasicontinuous emission from ion lasers in electrodeless high-frequency discharges, *Sov. J. Quantum Electron.,* 3, 519-520, 1974.
3. **Aleinikov, V. S.,** Use of an electron gun to determine the nature of collisions of the second kind in a mercury-helium mixture, *Opt. Spectra,* 28, 15-17, 1970.
4. **Alferov, G. N., Donin, V. I., and Yurshin, B. Ya.,** CW argon laser with 0.5 kW output power, *JETP Lett.,* 18, 369—370, 1973.
5. **Allen, R. B., Starnes, R. B., and Dougal, A. A.,** A new pulsed ion laser transition in nitrogen at 3995 Å, *IEEE J. Quantum Electron.,* QE-2, 334, 1966.
6. **Andrade, O., Gallardo, M., and Bockasten, K.,** High-gain laser lines in noble gases, *Appl. Phys. Lett.,* 11, 99—100, 1967.

7. **Anon.** *Spectra-Phys. Laser Rev.*, 3(1), 1976.
8. **Anon.**, *Spectra-Phys. Laser Rev.*
9. **Auschwitz, B., Eichler, H. J., and Wittwer, W.**, Extension of the operating period of an UV Cu II-laser by admixture of argon, *Appl. Phys. Lett.*, 36, 804—805, 1980.
10. **Banse, K., Herziger, G., Schafer, G., and Seelig, W.**, Continuous UV-laser power in the watt range, *Phys. Lett.*, 27A, 682—683, 1968.
11. **Baron, K. U. and Stadler, B.**, New visible laser transitions in Ba I and Ba II, *IEEE J. Quantum Electron.*, QE-11, 852—853, 1975.
12. **Bashkin, S. and Stoner, J. O., Jr.**, *Atomic Energy Levels and Grotrian Diagrams*, Vol. 1, North-Holland, Amsterdam, 1975.
13. **Bashkin, S. and Stoner, J. O., Jr.**, *Atomic Energy Levels and Grotrian Diagrams*, Vol. 2, North-Holland, Amsterdam, 1978.
14. **Beck, R., Englisch, W., and Gurs, K.**, *Tables of Laser Lines in Gases and Vapors*, 3rd ed., Springer-Verlag, Berlin, 1980.
15. **Bell, W. E.**, private communication, 1964.
16. **Bell, W. E.**, Visible laser transitions in Hg^+, *Appl. Phys. Lett.*, 4, 34—35, 1964.
17. **Bell, W. E.**, Ring discharge excitation of gas ion lasers, *Appl. Phys. Lett.*, 7, 190—191, 1965.
18. **Bell, W. E., Bloom, A. L., and Goldsborough, J. P.**, Visible laser transitions in ionized selenium, arsenic, and bromine, *IEEE J. Quantum Electron.*, QE-1, 400, 1965.
19. **Bell, W. E., Bloom, A. L., and Goldsborough, J. P.**, New laser transitions in antimony and tellurium, *IEEE J. Quantum Electron.*, QE-2, 154, 1966.
20. **Bennett, W. R., Jr., Knutson, J. W., Jr., Mercer, G. N., and Detch, J. L.**, Super-radiance, excitation mechanisms, and quasi-cw oscillation in the visible Ar^+ laser, *Appl. Phys. Lett.*, 4, 180—182, 1964.
21. **Bennett, W. R., Jr.**, *Atomic Gas Laser Transition Data — A Critical Evaluation*, IFI/Plenum Press, New York, 1979.
22. **Birnbaum, M., Tucker, A. W., Gelbwachs, J. A., and Fincher, C. L.**, New O II 6640 Å laser line, *J. Quantum Electron.*, QE-7, 208, 1971.
23. **Birnbaum, M. and Stocker, T. L.**, private communication, 1965.
24. **Bloom, A. L.**, private communication.
25. **Bloom, A. L., Bell, W. E., and Lopez, F. O.**, Laser spectroscopy of a pulsed mercury-helium discharge, *Phys. Rev.*, 135, A578—A579, 1964.
26. **Bloom, A. L. and Goldsborough, J. P.**, New CW laser transitions in cadmium and zinc ion, *IEEE J. Quantum Electron.*, QE-6, 164, 1970.
27. **Bockasten, K.**, A study of C III by means of a sliding vacuum spark, *Ark. Fys.*, 9 (30), 457—481, 1955.
28. **Bockasten, K.**, A study of C IV: term values, series formulae, and stark effect, *Ark. Fys.*, 10 (40), 567—582, 1956.
29. **Bockasten, K., Garvaglia, M., Lengyel, B. A., and Lundholm, T.**, Laser lines in Hg I, *J. Opt. Soc. Am.*, 55, 1051—1053, 1965.
30. **Boersch, H., Boscher, J., Hoder, D., and Schafer, G.**, Saturation of laser power of CW ion laser with large bored tubes and high power CW UV, *Phys. Lett.*, 31A, 188—189, 1970.
31. **Bokhan, P. A., Klimkin, V. M., and Prokop'ev, V. E.**, Gas laser using ionized europium, *JETP Lett.*, 18, 44—45, 1973.
32. **Bokhan, P. A., Klimkin, V. M., and Prokop'ev, V. E.**, Collision gas-discharge laser utilizing europium vapor. I. Observation of self-terminating oscillations and transition from cyclic to quasicontinum operation, *Sov. J. Quantum Electron.*, 4, 752—754, 1974.
33. **Bridges, W. B.**, Laser oscillation in singly ionized argon in the visible spectrum, *Appl. Phys. Lett.*, 4, 128—130, 1964; Erratum: *Appl. Phys. Lett.*, 5, 39, 1964.
34. **Bridges, W. B.**, unpublished data, 1964.
35. **Bridges, W. B.**, Laser action in singly ionized krypton and xenon, *Proc. IEEE*, 52, 843—844, 1964.
36. **Bridges, W. B. and Chester, A. N.**, Spectroscopy of ion lasers, *IEEE J. Quantum Electron.*, QE-1, 66—84, 1965.
37. **Bridges, W. B. and Chester, A. N.**, Visible and uv laser oscillation at 118 wavelengths in ionized neon, argon, krypton, xenon, oxygen, and other gases, *Appl. Opt.*, 4, 573—580, 1965.
38. **Bridges, W. B. and Halsted, A. S.**, New CW laser transitions in argon, krypton, and xenon, *IEEE J. Quantum Electron.*, QE-2, 84, 1966.
39. **Bridges, W. B., Clark, P. O., and Halsted, A. S.**, High power gas laser research, Tech. Rep. AFAL-TR-66-369, DDC No. AD-807363, Hughes Research Laboratories, Malibu, Calif., 1967.
40. **Bridges, W. B., Freiberg, R. J., and Halsted, A. S.**, New continuous UV ion laser transitions in neon, argon, and krypton, *IEEE J. Quantum Electron.*, QE-3, 339, 1967.
41. **Bridges, W. B. and Halsted, A. S.**, Gaseous Ion Laser Research, Tech. Rep. AFAL-TR-67-89; DDC No. AD-814897, Hughes Research Laboratories, Malibu, Calif., 1967.

42. **Bridges, W. B. and Mercer, G. N.**, CW operation of high ionization states in a xenon laser, *IEEE J. Quantum Electron.*, QE-5, 476—477, 1969.
43. **Bridges, W. B. and Mercer, G. N.**, Ultraviolet Ion Laser, Tech. Rep. ECOM-0229-F, DDC No. AD-861927, Hughes Research Laboratories, Malibu, Calif., 1969.
44. **Bridges, W. B. and Chester, A. N.**, unpublished data, 1969 and 1970.
45. **Bridges, W. B. and Chester, A. N.**, Ionized gas lasers, in *Handbook of Lasers*, Pressley, R. J., Ed., CRC Press, Boca Raton, Fla., 1971, 242—297.
46. **Bridges, W. B., Chester, A. N., Halsted, A. S., and Parker, J. V.**, Ion laser plasmas, *Proc. IEEE*, 59, 724—737, 1971.
47. **Bridges, W. B. and Chester, A. N.**, Comments on the identification of some xenon ion laser lines, *IEEE J. Quantum Electron.*, QE-7, 471—472, 1971.
48. **Bridges, W. B.**, unpublished data, 1975.
49. **Bridges, W. B.**, Atomic and ionic gas lasers, in *Methods of Experimental Physics*, Vol. 15A, Tang, C. L., Ed., Academic Press, New York, 1979.
50. **Bromander, J.**, The spectrum of triply-ionized oxygen, O IV, *Ark. Fys.*, 40 (23), 257—274, 1969.
51. **Burkhard, P., Lüthi, H. R., and Seelig, W.**, Quasi-CW laser action from Hg-III lines, *Opt. Commun.*, 18, 485—487, 1976.
52. **Byer, R. L., Bell, W. E., Hodges, E., and Bloom, A. L.**, Laser emission in ionized mercury: isotope shift, linewidth, and precise wavelength, *J. Opt. Soc. Am.*, 55, 1598—1602, 1965.
53. **Cahuzac, Ph.**, Emissions Laser infrarouges dans les vapeurs de Terres Rares, *Phys. Lett.*, 31A, (10), 541—542, 1970.
54. **Cahuzac, Ph.**, Nouvelles raies laser infrarouges dans la vapeur de baryum, *Phys. Lett.*, 32a, 150—151, 1970.
55. **Cahuzac, Ph.**, Raies laser infrarouges dans les vapeurs de terres rares et D'Alcalino-Terreux, *J. Phys.*, 32, 499—505, 1971.
56. **Cahuzac, Ph.**, New infrared laser lines in Mg vapor, *IEEE J. Quantum Electron.*, QE-8, 500, 1972.
57. **Carr, W. C. and Grow, R. W.**, Silicon and chlorine laser oscillations in $SiCl_4$, *Proc. IEEE*, 55, 726, 1967.
58. **Carr, W. C. and Grow, R. W.**, A new laser line in tin using stannic chloride vapor, *Proc. IEEE*, 55, 1198, 1967.
59. **Cheo, P. K. and Cooper, H. G.**, Ultraviolet ion laser transitions between 2300 and 4000 Å, *J. Appl. Phys.*, 36, 1862—1865, 1965.
60. **Cheo, P. K. and Cooper, H. G.**, UV and visible laser oscillations in fluorine, phosphorus, and chlorine, *Appl. Phys. Lett.*, 7, 202—204, 1965; for corrections: see H. P. Palenius.
61. **Chester, A. N.**, Gas pumping in discharge tubes, *Phys. Rev.*, 169, 172—184, 1968.
62. **Chester, A. N.**, Experimental measurements of gas pumping in an argon discharge, *Phys. Rev.*, 169, 184—193, 1968.
63. **Chou, M. S. and Cool, T. A.**, Laser operation by dissociation of metal complexes. II. New transitions in Cd, Fe, Ni, Se, Sn, Te, V, and Zn, *J. Appl. Phys.*, 48, 1551—1555, 1977.
64. **Clunie, D. M.**, private communication, 1966.
65. **Collins, G. J.**, Cw Oscillation and Charge Exchange Excitation in the Zinc-Ion Laser, Ph.D. thesis, Yale University, New Haven, Conn., 1970.
66. **Collins, G. J., Jensen, R. C., and Bennett, W. R., Jr.**, Excitation mechanisms in the zinc ion laser, *Appl. Phys. Lett.*, 18, 282—284, 1971.
67. **Collins, G. J., Jensen, R. C., and Bennett, W. R., Jr.**, Charge-exchange excitation in the He-Cd laser, *Appl. Phys. Lett.*, 19, 125—130, 1971.
68. **Collins, G. J., Kuno, H., Hattori, S., Tokutome, K., Ishikawa, M., and Kamide, N.**, Cw laser oscillation at 6127 Å in singly ionized iodine, *IEEE J. Quantum Electron.*, QE 8, 679—680, 1972.
69. **Convert, G., Armand, M., and Martinot-Lagrade, P.**, Effet laser dans des melanges mercure-gaz rares, *C.R. Acad. Sci. Paris*, 258, 3259—3260, 1964.
70. **Cooper, H. G. and Cheo, P. K.**, Ion laser oscillations in sulfer, Physics of Quantum Electronics Conference, San Juan, P.R., 1965, *Proceedings*, McGraw-Hill, New York, 1966, 690—697.
71. **Cooper, H. G. and Cheo, P. K.**, Laser transitions in B II, Br II, and Sn, *IEEE J. Quantum Electron.*, QE-2, 785, 1966.
72. **Cottrell, T. H. E., Sinclair, D. C., and Forsyth, J. M.**, New laser wavelengths in krypton, *IEEE J. Quantum Electron.*, QE-2, 703, 1966.
73. **Crooker, A. M. and Dick, K. A.**, Extensions to the spark spectra of zinc. I. Zinc II and Zinc IV, *Can. J. Phys.*, 46, 1241—1251, 1968.
74. **Csillag, L., Janossy, M., Kantor, K., Rozsa, K., and Salamon, T.**, Investigation on a continuous wave 4416 Å cadmium ion laser, *Appl. Phys.*, 3, 64—68, 1970.
75. **Csillag, L., Janossy, M., and Salamon, T.**, Time delay of laser oscillation of the green transition of a pulsed He-Cd laser, *Phys. Lett.*, 31A, 532—533, 1970.

76. **Csillag, L., Itagi, V. V., Janossy M., and Rozsa, K.,** Laser oscillation at 4416 Å in a Ne-Cd discharge, *Phys. Lett.,* 34A, 110—111, 1971.
77. **Csillag, L., Janossy, M., Rozsa, K., and Salamon, T.,** Near infrared Cw laser oscillation in Cu II, *Phys. Lett.,* 50A, 13—14, 1974.
78. **Dahlquist, J. A.,** New line in a pulsed xenon laser, *Appl. Phys. Lett.,* 6, 193—194, 1965.
79. **Dana, L. and Laurès, P.,** Stimulated emission in krypton and xenon ions by collisions with metastable atoms, *Proc. IEEE,* 53, 78—79, 1965.
80. **Dana, L., Laures, P., and Rocherolles, R.,** Raies laser ultraviolettes dans le neon, l'argon, et le xenon, *C.R. Acad. Sci. Paris,* 260, 481—484, 1965.
81. **Davis, C. C. and King, T. A.,** Time-resolved gain measurements and excitation mechanisms of the pulsed argon ion laser, *Phys. Lett.,* 36A, 169—170, 1971.
82. **Davis, C. C. and King, T. A.,** Laser action on unclassified xenon transitions in a highly ionized plasma, *IEEE J. Quantum Electron.,* QE-8, 755—757, 1972.
83. **Davis, C. C. and King, T. A.,** Gaseous ion lasers, in *Advances in Quantum Electronics,* Vol. 3, Goodwin, D. W., Ed., Academic Press, New York, 1975, 169.
84. **deBruin, T. L., Humphreys, C. J., and Meggers, W. F.,** The second spectrum of krypton, *J. Res. Natl. Bur. Std.,* 11, 409—440, 1933.
85. **Deech, J. S. and Sanders, J. H.,** New self-terminating laser transitions in calcium and strontium, *IEEE J. Quantum Electron.,* QE-4, 474, 1968.
86. **Demtroder, W.,** Excitation mechanisms of pulsed argon ion lasers at 4880 Å, *Phys. Lett.,* 22, 436—438, 1966.
87. **Donin, V. I.,** Output power saturation with a discharge current in powerful continuous argon lasers, *Sov. Phys. JETP,* 35, 858—864, 1972.
88. **Donin, V. I., Shipilov, A. F., and Grigor'ev, V. A.,** High-power cw ion lasers with an improved service life, *Sov. J. Quantum Electron.,* 9, 210—212, 1979.
89. **Dubrovin, A. N., Tibilov, A. S., and Shevtsov, A. K.,** Lasing on Cd, Zn, and Mg lines and possible applications, *Opt. Spectrosc. (USSR),* 32, 685, 1972.
90. **Duffendack, O. S. and Thomson, K.,** Some factors affecting action cross section for collisions of the second kind between atoms and ions, *Phys. Rev.,* 43, 106—111, 1933.
91. **Duffendack, O. S. and Gran, W. H.,** Regularity along a series in the variation of the action cross section with energy discrepancy in impacts of the second kind, *Phys. Rev.,* 51, 804—809, 1937.
92. **Dunn, M. H. and Ross, J. N.,** The argon ion laser, *Progr. Quantum Electron.,* 4, 233—269, 1976.
93. **Dyson, D. J.,** Mechanism of population inversion at 6149 Å in the mercury ion laser, *Nature,* (London) 207, 361—363, 1965.
94. **Edlén, B. and Risberg, P.,** The spectrum of singly-ionized calcium, Ca II, *Ark. Fys.,* 10 (39), 553—566, 1956.
95. **Engelhard, E. and Spieweck, F.,** Ein Ionen-Laser Fur Metrologische Anwendungen, Mitteilung aus der Physikalisch-Technischen Bundesanstalt Braunschweig, *Z. Naturforsch.,* 25a, 156, 1969.
96. **Ericsson, K. G. and Lidbolt, L. R.,** Superradiant transitions in argon, krypton, and xenon, *IEEE J. Quantum Electron.,* QE-3, 94, 1967.
97. **Eriksson, K. B. S.,** The spectrum of the singly-ionized nitrogen atom, *Ark. Fys.,* 13 (25), 303—329, 1958.
98. **Fendley, J. R., Jr.,** Continuous UV lasers, *IEEE J. Quantum Electron.,* QE-4, 627—631, 1968.
99. **Fendley, J. R., Jr., Gorog, I., Hernqvist, K. G., and Sun, C.,** Characteristics of a sealed-off He^3-Cd^{114} laser, *IEEE J. Quantum Electron.,* QE-6, 8, 1970.
100. **Fendley, J. R., Jr. and O'Grady, J. J.,** Development, Construction, and Demonstration of a 100W cw Argon-Ion Laser, Tech. Rep. ECOM-0246-F, RCA, Lancaster, Pa., 1970.
101. **Ferrario, A.,** Excitation mechanism in Hg^+ ion laser, *Opt. Commun.,* 7, 375—378, 1973.
102. **Fowles, G. R. and Jensen, R. C.,** Visible laser transitions in the spectrum of singly ionized iodine, *Proc. IEEE,* 52, 851—852, 1964.
103. **Fowles, G. R. and Jensen, R. C.,** Visible laser transitions in ionized iodine, *Appl. Opt.,* 3, 1191—1192, 1964.
104. **Fowles, G. R. and Jensen, R. C.,** Laser oscillation on a single hyperfine transition in iodine, *Phys. Rev. Lett.,* 14, 347—348, 1965.
105. **Fowles, G. R. and Silfvast, W. T.,** Laser action in the ionic spectra of zinc and cadmium, *IEEE J. Quantum Electron.,* QE-1, 131, 1965.
106. **Fowles, G. R., Silfvast, W. T., and Jensen, R. C.,** Laser action in ionized sulfur and phosphorus, *IEEE J. Quantum Electron.,* QE-1, 183—184, 1965.
107. **Fowles, G. R. and Hopkins, B. D.,** CW laser oscillation at 4416 Å in cadmium, *IEEE J. Quantum Electron.,* QE-3, 419, 1967.
108. **Fukuda, S. and Miya, M.,** A metal-ceramic He-Cd II laser with sectioned hollow cathodes and output power characteristics of simultaneous oscillations, *Jpn. J. Appl. Phys.,* 13, 667—674, 1974.

109. **Gadetskii, N. P., Tkach, Yu. V., Slezov, V. V., Bessarab, Ya. Ya., and Magda, I. I.,** New mechanism of coherent-radiation generation in the visible region of the spectrum in ionized oxygen and nitrogen, *JETP Lett.,* 14, 101—105, 1971.
110. **Gadetskii, N. P., Tkach, Yu. V., Bessarab, Ya. Ya., Sidel'nikova, A. V., and Magda, I. I.,** Stimulated emission of visible light due to transitions in singly ionized chlorine and iodine atoms, *Sov. J. Quantum Electron.,* 3, 168—169, 1973.
111. **Gallardo, M., Garavaglia, M., Tagliaferri, A. A., and Gallego Lluesma, E.,** About unidentified ionized Xe laser lines, *IEEE J. Quantum Electron.,* QE-6, 745—747, 1970.
112. **Gallardo, M., Massone, C. A., Tagliaferri, A. A., and Garavaglia, M.,** $5s^25p^3(^4S)nl$ Levels of Xe III, *Phys. Scripta,* 19, 538—544, 1979.
113. **Gerritsen, H. J. and Goedertier, P. V.,** Blue gas laser using Hg^{2+}, *J. Appl. Phys.,* 35, 3060—3061, 1964.
114. **Gerritsen, H. J.,** private communication, 1965.
115. **Gerstenberger, D. C., Reid, R. D., and Collins, G. J.,** Hollow-cathode aluminum ion laser, *Appl. Phys. Lett.,* 30, 466—468, 1977.
116. **Gill, P. and Webb, C. E.,** Radial profiles of excited ions and electron density in the hollow cathode He/Zn laser, *J. Phys. D.,* 11, 245—254, 1978.
117. **Gilles, M.,** Recherches sur la Structure des Spectres du Soufre, *Ann. Phys.,* Series 10, 15, 267—410, 1931.
118. **Cem Gokay, M., Soltanolkotabi, M., and Cross, L. A.,** Single- and double-pulse experiments on the Sr^+ cyclic ion laser, *IEEE J. Quantum Electron.,* QE-14, 1004—1007, 1978.
119. **Goldsborough, J. P., Hodges, E. B., and Bell, W. E.,** RF induction excitation of CW visible laser transitions in ionized gases, *Appl. Phys. Lett.,* 8, 137—139, 1966.
120. **Goldsborough, J. P. and Bloom, A. L.,** New CW ion laser oscillation in microwave-excited xenon, *IEEE J. Quantum Electron.,* QE-3, 96, 1967.
121. **Goldsborough, J. P.,** Continuous laser oscillation at 3250 Å in cadmium ion, *IEEE J. Quantum Electron.,* QE-5, 133, 1969.
122. **Goldsborough, J. P. and Hodges, E. B.,** Stable long-life operation of helium-cadmium lasers at 4416 Å and 3250 Å, *IEEE J. Quantum Electron.,* QE-5, 361—367, 1969.
123. **Goldsborough, J. P.,** Stable, long life cw excitation of helium-cadmium lasers by dc cataphoresis, *Appl. Phys. Lett.,* 15, 159—161, 1969.
124. **Goldsborough, J. P. and Bloom, A. L.,** Near-infrared operating characteristics of the mercury ion laser, *IEEE J. Quantum Electron.,* QE-5, 459—460, 1969.
125. **Goldsmith, S. and Kaufman, A. S.,** The spectra of Ne IV, Ne V, and Ne VI: a further analysis, *Proc. Phys. Soc.,* 81, 544—552, 1963.
126. **Gordon, E. I., Labuda, E. F., and Bridges, W. B.,** Continuous visible laser action in singly ionized argon, krypton, and xenon, *Appl. Phys. Lett.,* 4, 178—180, 1964.
127. **Gordon, E. I., Labuda, E. F., Miller, R. C., and Webb, C. E.,** Excitation mechanisms of the argon-ion laser, in *Proc. Phys. of QE Conf.,* McGraw-Hill, New York, 1966, 664—673.
128. **Gorog, I. and Spong, F. W.,** High pressure, high magnetic field effects in continuous argon ion lasers, *Appl. Phys. Lett.,* 9, 61—63, 1966.
129. **Goto, T., Kano, H., Yoshino, N., Mizeraczyk, J. K., and Hattori, S.,** Construction of a practical sealed-off He-I^+ laser device, *J. Phys. B,* 10, 292—295, 1977.
130. **Green, J. M., Collins, G. J., and Webb, C. E.,** Collisional excitation and destruction of excited Zn II levels in a helium afterglow, *J. Phys. B,* 6, 1545—1555, 1973.
131. **Green, J. M. and Webb, C. E.,** The production of excited metal ions in thermal energy charge transfer and Penning reactions, *J. Phys. B,* 7 , 1698—1711, 1974.
132. **Green, W. R. and Falcone, R. W.,** Inversion of the resonance line of Sr^+ produced by optically pumped Sr atoms, *Opt. Lett.,* 2, 115—116, 1978.
133. **Grotrian, W.,** *Graphische Darstellung der Spektren von Atomen und Ionen mit ein, zwei, und drei Valenzelektronen,* Vol. II, Julius Springer, Berlin, 1928.
134. **Grozeva, M. G., Sabotinov, N. V., and Vuchkov, N. K.,** CW laser generation on Tl II in a hollow-cathode Ne-Tl discharge, *Opt. Commun.,* 29, 339—340, 1979.
135. **Grozeva, M. G., Sabotinov, N. V., Telbizov, P. K., and Vuchkov, N. K.,** CW laser oscillation on transitions of Tl in a hollow-cathode Ne-Tl halide discharge, *Opt. Commun.,* 31, 211—213, 1979.
136. **Gunderson, M. and Harper, C. D.,** A high-power pulsed xenon ion laser, *IEEE J. Quantum Electron.,* QE-9, 1160, 1973.
137. **Hagen, L. and Martin, W. C.,** Bibliography on Atomic Energy Levels and Spectra, July 1968 through June 1971, Special Publication 363, U.S. National Bureau of Standards, Washington, D.C., 1972.
138. **Hagen, L.,** Bibliography on Atomic Energy Levels and Spectra, July 1971 through June 1975, Special Publication 363, Supplement 1, U.S. National Bureau of Standards, Washington, D.C., 1977.
139. **Hallin, R.,** The spectrum of N IV, *Ark. Fys.,* 32 (11), 201—210, 1966.

140. **Handrup, M. B. and Mack, J. E.,** On the spectrum of ionized tellurium, Te II, *Physica,* 30, 1245—1275, 1964.
141. **Harrison, G. R.,** *MIT Wavelength Tables,* John Wiley & Sons, New York, 1952.
142. **Harper, C. D. and Gundersen, M.,** Construction of a high power xenon ion laser, *Rev. Sci. Instrum.,* 45, 400—402, 1974.
143. **Hashino, Y., Katsuyama, Y., and Fukuda, K.,** Laser oscillation of multiply ionized Ne, Ar, and N ions in a Z-pinch discharge, *Jpn. J. Appl. Phys.,* 11, 907, 1972.
144. **Hashino, Y., Katsuyama, Y., and Fukuda, K.,** Laser oscillation of O V in Z-pinch discharge, *Jpn. J. Appl. Phys.,* 12, 470, 1973.
145. **Hattori, S., Kano, H., Tokutome, K., Collins, G. J., and Goto, T.,** CW iodine-ion laser in a positive-column discharge, *IEEE J. Quantum Electron.,* QE-10, 530—531, 1974.
146. **Hattori, S., Kano, H., and Goto, T.,** A continuous positive-column He-I$^+$ laser using a sealed-off tube, *IEEE J. Quantum Electron.,* QE-10, 739—740, 1974.
147. **Heard, H. G. and Peterson, J.,** Orange through blue-green transitions in a pulsed-CW xenon gas laser, *Proc. IEEE,* 52, 1050, 1964.
148. **Heard, H. G. and Peterson, J.,** Visible laser transitions in ionized oxygen, nitrogen, and carbon monoxide, *Proc. IEEE,* 52, 1258, 1964.
149. **Hernqvist, K. G. and Pultorak, D. C.,** Simplified construction and processing of a helium-cadmium laser, *Rev. Sci. Instrum.,* 41, 696—697, 1970.
150. **Hernqvist, K. G.,** Stabilization of He-Cd laser, *Appl. Phys. Lett.,* 16, 464—466, 1970.
151. **Hernqvist, K. G. and Pultorak, D. C.,** Study of He-Se laser performance, *Rev. Sci. Instrum.,* 43, 290—292, 1972.
152. **Hernqvist, K. G.,** He-Cd lasers using recirculation geometry, *IEEE J. Quantum Electron.,* QE-8, 740—743, 1972.
153. **Hernqvist, K. G.,** Vented-bore He-Cd lasers, *RCA Rev.,* 34, 401—407, 1973.
154. **Hernqvist, K. G.,** Continuous laser oscillation at 2703 Å in copper ion, *IEEE J. Quantum Electron.,* QE-13, 929, 1977.
155. **Herziger, G. and Seelig, W.,** Ionenlaser hober Leistung, *Z. Phys.,* 219, 5—31, 1969.
156. **Hodges, D. T.,** CW laser oscillation in singly ionized magnesium, *Appl. Phys. Lett.,* 18, 454—456, 1971.
157. **Hodges, D. T. and Tang, C. L.,** New CW ion laser transitions in argon, krypton, and xenon, *IEEE J. Quantum Electron.,* QE-6, 757—758, 1970.
158. **Hoffmann, V. and Toschek, P.,** New laser emission from ionized xenon, *IEEE J. Quantum Electron.,* QE-6, 757, 1970.
159. **Hoffmann, V. and Toschek, P. E.,** On the ionic assignment of xenon laser lines, *J. Opt. Soc. Am.,* 66, 152—154, 1976.
160. **Horrigan, F. A., Koozekanani, S. H., and Paananen, R. A.,** Infrared laser action and lifetimes in Argon II, *Appl. Phys. Lett.,* 6, 41—43, 1965.
161. **Humphreys, C. J., Meggers, W. F., and de Bruin, T. L.,** Zeeman effect in the second and third spectra of xenon, *J. Res. Natl. Bur. Stand.,* 23, 683—699, 1939.
162. **Humphreys, C. J.,** Line List for Ionized Krypton, unpublished.
163. **Humphreys, C. J.,** Line List for Ionized Xenon, unpublished.
164. **Iijima, T. and Sugawara, Y.,** New CW laser oscillation in He-Zn hollow cathode laser, *J. Appl. Phys.,* 45, 5091—5092, 1974.
165. **Illingworth, R.,** Laser action and plasma properties of an argon Z-pinch discharge, *Appl. Phys.,* 3, 924—930, 1970.
166. **Isaev, A. A. and Petrash, G. G.,** New superradiance on the violet line of the mercury ion, *J. Appl. Spectrosc. (USSR),* 12, 835—837, 1970.
167. **Isaev, A. A., Kazaryan, M. A., and Petrash, G. G.,** Emission of laser pulses due to transitions from a resonance to a metastable level in barium vapor, *Sov. J. Quantum Electron.,* 3, 358—359, 1974.
168. **Ivanov, I. G. and Sém, M. F.,** New lasing lines in thallium, *J. Appl. Spectrosc. (USSR),* 19, 1092—1093, 1973.
169. **Ivanov, I. G., Il'yushko, V. G., and Sém, M. F.,** Dependences of the gain of cataphoretic lasers on helium pressure and discharge-tube diameter, *Sov. J. Quantum Electron.,* 4, 589—593, 1974.
170. **Jain, K.,** Cw laser oscillation at 8096 Å in Cu II in a hollow cathode discharge, *Opt. Commun.,* 28, 207—208, 1979.
171. **Jain, K.,** New ion laser transitions in copper, silver, and gold, *Appl. Phys. Lett.,* 34, 398—399, 1979.
172. **Jain, K.,** A nickel-ion laser, *Appl. Phys. Lett.,* 34, 845—846, 1979.
173. **Jain, K.,** A milliwatt-level cw laser source at 224 nm, *Appl. Phys. Lett.,* 36, 10—11, 1980.
174. **Jain, K.,** New UV and IR transitions in gold, copper, and cadmium hollow cathode lasers, *IEEE J. Quantum Electron.* , QE-16, 387—388, 1980.
175. **Jain, K.,** Laser action in chromium vapor, *Appl. Phys. Lett.,* 37, 362—364, 1980.

176. **Jánossy, M., Csillag, L., Rózsa, K., and Salamon, T.**, CW laser oscillation in a hollow cathode He-Kr discharge, *Phys. Lett.*, 46A, 379—380, 1974.
177. **Jánossy, M., Rózsa, K., Csillag, L., and Bergou, J.**, New cw laser lines in a noble gas mixture high voltage hollow cathode discharge, *Phys. Lett.*, 68A, 317—318, 1978.
178. **Jarrett, S. M. and Barker, G. C.**, High-power output at 5353 Å from a pulsed xenon ion laser, *IEEE J. Quantum Electron.*, QE-5, 166, 1969.
179. **Jensen, R. C. and Fowles, G. R.**, New laser transitions in iodine-inert gas mixtures, *Proc. IEEE*, 52, 1350, 1964.
180. **Jensen, R. C. and Bennett, W. R., Jr.**, Role of charge exchange in the zinc ion laser, *IEEE J. Quantum Electron.*, QE-4, 356, 1968.
181. **Jensen, R. C., Collins, G. J., and Bennett, W. R., Jr.**, Charge-exchange excitation and cw oscillation in the zinc-ion laser, *Phys. Rev. Lett.*, 23, 363—367, 1969.
182. **Jensen, R. C., Collins, G. J., and Bennett, W. R., Jr.**, Low-noise CW hollow-cathode zinc-ion laser, *Appl. Phys. Lett.*, 18, 50—51, 1971.
183. **Johnson, A. M. and Webb, C. E.**, New CW laser wavelength in Kr II, *IEEE J. Quantum Electron.*, QE-3, 369, 1967.
184. **Johnson, W. L., McNeil, J. R., Collins, G. J., and Persson, K. B.**, CW laser action in the blue-green spectral region from AgII, *Appl. Phys. Lett.*, 29, 101—102, 1976.
185. **Johnston, T. F., Jr. and Kolb, W. P.**, The self-heated 442-nm He-Cd laser: optimizing the power output, and the origin of beam noise, *IEEE J. Quantum Electron.*, QE-12, 482—493, 1976.
186. **Kano, H., Goto, T., and Hattori, S.**, Electron temperature and density in the He-CdI_2 positive column used for an I^+ laser, *IEEE J. Quantum Electron.*, QE-9, 776—778, 1973.
187. **Kano, H., Goto, T., and Hattori, S.**, CW laser oscillation of visible and near-infrared Hg(II) lines in a He-Hg positive column discharge, *J. Phys. Soc. Jpn.*, 38, 596, 1975.
188. **Kano, H., Shay, T., and Collins, G. J.**, A second look at the excitation mechanism of the He-Hg^+ laser, *Appl. Phys. Lett.*, 27, 610—612, 1975.
189. **Karabut, E. K., Mikhalevskii, V. S., Papakin, V. F., and Sém, M. F.**, Continuous generation of coherent radiation in a discharge in Zn and Cd vapors obtained by cathode sputtering, *Sov. Phys. Tech. Phys.*, 14, 1447—1448, 1970.
190. **Karabut, E. K., Kravchenko, V. F., and Papakin, V. F.**, Excitation of the Ag II Lines by a pulsed discharge in a mixture of silver vapor and helium, *J. Appl. Spectrosc. (USSR)*, 19, 938—939, 1973.
191. **Keeffe, W. M. and Graham, W. J.**, Laser oscillation in the visible spectrum of singly ionized pure bromine vapor, *Appl. Phys. Lett.*, 7, 263—264, 1965.
192. **Keeffe, W. M. and Graham, W. J.**, Observation of New Br II laser transitions, *Phys. Lett.*, 20, 643, 1966.
193. **Keiden, V. F. and Mikhalevskii, V. S.**, Pulsed generation in bismuth vapor, *J. Appl. Spectrosc. (USSR)*, 9, 1154, 1968.
194. **Keiden, V. F., Mikhalevskii, V. S., and Sém, M. F.**, Generation from ionic transitions of selenium, *J. Appl. Spectrosc. (USSR)*, 15, 1089—1090, 1971.
195. **Kiess, C. C. and de Bruin, T. L.**, Second spectrum of chlorine and its structure, *J. Res. Natl. Bur. Stand.*, 23, 443—470, 1939.
196. **Kitaeva, V. F., Odintsov, A. N., and Sobolev, N. N.**, Continuously operating argon ion lasers, *Sov. Phys. Usp.*, 12, 699—730, 1970.
197. **Klein, M. B.**, Radiation Trapping Processes in the Pulsed Ion Laser, Ph.D. dissertation, University of California, Berkeley, 1969.
198. **Klein, M. B.**, Time-resolved temperature measurements in the pulsed argon ion laser, *Appl. Phys. Lett.*, 17, 29—32, 1970.
199. **Klein, M. B. and Silfvast, W. T.**, New CW laser transitions in Se II, *Appl. Phys. Lett.*, 18, 482—485, 1971.
200. **Klimkin, V. M.**, Investigation of an ytterbium vapor laser, *Sov. J. Quantum Electron.*, 5, 326—329, 1975.
201. **Kobayashi, S., Izawa, T., Kawamura, K., and Kamiyama, M.**, Characteristics of a pulsed Ar II ion laser using the external spark gap, *IEEE J. Quantum Electron.*, QE-2, 699—700, 1966.
202. **Kobayashi, S., Kurihara, K., and Kamiyama, M.**, New laser oscillation in ionized xenon at 5592 Å, *Oyo Butsuri*, 38, 766—768, 1969 (in Japanese).
203. **Koval'chuk, V. M. and Petrash, G. G.**, New generation lines of a pulsed iodine-vapor laser, *JETP Lett.*, 4, 144—146, 1966.
204. **Kruithof, A. A. and Penning, F. M.**, Determination of the Townsend ionization coefficient α for mixtures of neon and argon, *Physica*, 4, 430—449, 1937.
205. **Kulagin, S. G., Likhachev, V. M., Markuzon, E. V., Rabinovich, M. S., and Sutovskii, V. M.**, States with population inversion in a self-compressed discharge, *JETP Lett.*, 3, 6—8, 1966.

206. **Labuda, E. F., Gordon, E. I., and Miller, R. C.,** Continuous-duty argon ion lasers, *IEEE J. Quantum Electron.,* QE-1, 273—279, 1965.
207. **Labuda, E. F., Webb, C. E., Miller, R. C., and Gordon, E. I.,** A study of capillary discharges in noble gases at high current densities, *Bull. Am. Phys. Soc.,* 11, 497, 1966.
208. **Labuda, E. F. and Johnson, A. M.,** Threshhold properties of continuous duty rare gas ion laser transitions, *IEEE J. Quantum Electron.,* QE-2, 700—701, 1966.
209. **Latimer, I. D.,** High power quasi-CW ultra-violet ion laser, *Appl. Phys. Lett.,* 13, 333—335, 1968.
210. **Latush, E. L. and Sém, M. F.,** Stimulated emission due to transitions in alkaline-earth metal ions, *Sov. J. Quantum Electron.,* 3, 216—219, 1973.
211. **Latush, E. L., Mikhalevskiĭ, V. S., Sém, M. F., Tolmachev, G. N., and Khasilov, V. Ya.,** Metal-ion transition lasers with transverse HF excitation, *JETP Lett.,* 24, 69—71, 1976.
212. **Laures, P., Dana, L., and Frapard, C.,** Nouvelles transitions laser dans le domaine 0.43-0.52 μ Obtenues a Partir du Spectre du Krypton Ionise, *C.R. Acad. Sci. Paris,* 258, 6363—6365, 1964.
213. **Laures, P., Dana, L., and Frapard, C.,** Nouvelles Raies Laser Visibles dans le Xenon Ionise, *C.R. Acad. Sci. Paris,* 259, 745—747, 1964.
214. **Levinson, G. R., Papulovskiy, V. F., and Tychinskiy, V. P.,** The mechanism of inversion of the populations of the various levels in multivalent argon ions, *Radio Eng. Electron. Phys.,* 13, 578—582, 1968.
215. **Li, H. and Andrew, K. L.,** First spark spectrum of arsenic, *J. Opt. Soc. Am.,* 61, 96—109, 1971.
216. **Littlewood, I. M., Piper, J. A., and Webb, C. E.,** Excitation mechanisms in CW He-Hg lasers, *Opt. Commun.,* 16, 45—49, 1976.
217. **Lluesma, E. G., Tagliaferri, A. A., Massone, C. A., Garavaglia, A., and Gallardo, M.,** Ionic assignment of unidentified xenon laser lines, *J. Opt. Soc. Am.,* 63, 362—364, 1973.
218. **Lüthi, H. R., Seelig, W., Steinger, J., and Lobsiger, W.,** Continuous 40-W UV laser, *IEEE J. Quantum Electron.,* QE-13, 404—405, 1977.
219. **Lüthi, H. R., Seelig, W., and Steinger, J.,** Power enhancement of continuous ultraviolet lasers, *Appl. Phys. Lett.,* 31, 670—672, 1977.
220. **Lüthi, H. R. and Steinger, J.,** Continuous operation of a high power neon ion laser, *Opt. Commun.,* 27, 435—438, 1978.
221. **Manley, J. H. and Duffendack, O. S.,** Collisions of the second kind between magnesium and neon, *Phys. Rev.,* 47, 56—61, 1935.
222. **Marling, J. B.,** Ultraviolet ion laser performance and spectroscopy. I. New strong noble-gas transitions below 2500 Å, *IEEE J. Quantum Electron.,* QE-11, 822—834, 1975.
223. **Marling, J. B. and Lang, D. B.,** Vacuum ultraviolet lasing from highly ionized noble gases, *Appl. Phys. Lett.,* 31, 181—184, 1977.
224. **Marling, J. B.,** Ultraviolet ion laser performance and spectroscopy for sulfur, fluorine, chlorine, and bromine, *IEEE J. Quantum Electron.,* QE-14, 4—6, 1978.
225. **Martin, W. C.,** Atomic energy levels and spectra of neutral and singly ionized phosphorus (P I and P II), *J. Opt. Soc. Am.,* 49, 1071—1085, 1959.
226. **Martin, W. C. and Corliss, C. H.,** The spectrum of singly ionized atomic iodine (I.II), *J. Res. Natl. Bur. Stand.,* 64A, 443—477, 1960.
227. **Martin, W. C. and Kaufman, V.,** New vacuum ultraviolet wavelengths and revised energy levels in the second spectrum of zinc (Zn II), *J. Res. Natl. Bur. Stand.,* 74A, 11—22, 1970.
228. **Massey, H. S. W. and Burhop, E. H. S.,** *Electronic and Ionic Impact Phenomena,* Oxford, 1952, 165.
229. **McFarlane, R. A.,** Laser oscillation on visible and ultraviolet transitions of singly and multiply ionized oxygen, carbon, and nitrogen, *Appl. Phys. Lett.,* 5, 91—93, 1964.
230. **McFarlane, R. A.,** Optical maser oscillation on iso-electronic transitions in Ar III and Cl II, *Appl. Opt.,* 3, 1196, 1964.
231. **McFarlane, R. A.,** private communication.
232. **McNeil, J. R., Collins, G. J., Persson, K. B., and Franzen, D. L.,** CW laser oscillation in Cu II, *Appl. Phys. Lett.,* 27, 595—598, 1975.
233. **McNeil, J. R., Collins, G. J., Persson, K. B., and Franzen, D. L.,** Ultraviolet laser action from Cu II in the 2500 Å region, *Appl. Phys. Lett.,* 28, 207—209, 1976.
234. **McNeil, J. R. and Collins, G. J.,** Additional ultraviolet laser transitions in Cu II, *IEEE J. Quantum Electron.,* QE-12, 371—372, 1976.
235. **McNeil, J. R., Johnson, W. L., Collins, G. J., and Persson, K. B.,** Ultraviolet laser action in He-Ag and Ne-Ag mixtures, *Appl. Phys. Lett.,* 29, 172—174, 1976.
236. **Meggers, W. F.,** The second spectrum of ytterbium (Yb II), *J. Res. Natl. Bur. Stand.,* 71A, 396—544, 1967.
237. **Miller, R. C., Labuda, E. F., and Gordon, E. I.,** paper presented at the Conference on Electron Device Research, Cornell University, Ithaca, N.Y., 1964.

238. **Minnhagen, L. and Stigmark, L.,** The excitation of ionic spectra by 100 kw high frequency pulses, *Ark. Fys.,* 13, 27—36, 1957.
239. **Minnhagen, L.,** The spectrum of singly ionized argon, Ar II, *Ark. Fys.,* 25, 203—284, 1963.
240. **Minnhagen, L., Strihed, H., and Petersson, B.,** Revised and extended analysis of singly ionized krypton, Kr II, *Ark. Fys.,* 39 (34), 471—493, 1969.
241. **Moore, C. E.,** Atomic Energy Levels, Circular 467, Vols. 1, 2, and 3, U.S. National Bureau of Standards, Washington, D.C., 1949, 1952, and 1958.
242. **Moore, C. E. and Merrill, P. W.,** Partial Grotrian Diagrams of Astrophysical Interest, NSRDS-NBS23, National Bureau of Standards, Washington, D.C., 1968.
243. **Moskalenko, V.F., Ostapchenko, E. P., Perchurina, S. V., Stepanov, V. A. and Tsukanov, Yu. M.,** Radiation of a pulsed ion laser, *Opt. Spectrosc.,* 30, 201-202, 1971.
244. **Myers, R. A., Wieder, H. and Pole, R. V.,** 9A5 - Wide field active imaging, *IEEE J. Quantum Electron.,* QE-2, 270-275, 1966.
245. **Neusel, R. H.,** A new xenon laser oscillation at 5401 Å, *IEEE J. Quantum Electron.,* QE-2, 70, 1966.
246. **Neusel, R. H.,** A new krypton laser oscillation at 5016.4 Å, *IEEE J. Quantum Electron.,* QE-2, 106, 1966.
247. **Neusel, R. H.,** New laser oscillations in krypton and xenon, *IEEE J. Quantum Electron.,* QE-2, 334, 1966.
248. **Neusel, R. H.,** New laser oscillations in xenon and kryton, *IEEE J. Quantum Electron.,* QE-2, 758, 1966.
249. **Neusel, R. H.,** New laser oscillations in Ar, Kr, Xe, and N, *IEEE J. Quantum Electron.,* QE-3, 207—208, 1967.
250. **Olme, A.,** The spectrum of singly ionized boron, B II, *Phys. Scripta,* 1, 256—260, 1970.
251. **Paananen, R.,** Continuously operated Ultraviolet Laser, *Appl. Phys. Lett.,* 9, 34—35, 1966.
252. **Pacheva, Y., Stefanova, M., and Pramatarov, P.,** Cw laser oscillations on the Kr II 4694 and Kr II 4318 A lines in a hollow-cathode He-Kr discharge, *Opt. Commun.,* 27, 121—122, 1978.
253. **Palenius, H. P.,** The identification of some Si and Cl laser lines observed by Cheo and Cooper, *Appl. Phys. Lett.,* 8, 82, 1966.
254. **Palenius, H. P.,** Spectrum and term system of singly ionized fluorine, F II, *Ark. Fys.,* 39 (3), 15—64, 1969.
255. **Palenius, H. P.,** Spectrum and term system of doubly ionized fluorine, F III, *Phys. Scripta,* 1, 113—135, 1970.
256. **Papakin, V. F. and Sém, M. F.,** Use of isotopes in cadmium and zinc vapor lasers, *Sov. Phys. J.,* 13, 230—231, 1970.
257. **Papayoanou, A., Buser, R. G., and Gumeiner, I. M.,** Parameters in a dynamically compressed xenon plasma laser, *IEEE J. Quantum Electron.,* QE-9, 580—585, 1973.
258. **Papayoanou, A. and Gumeiner, I.,** High power xenon laser action in high current pinched discharges, *Appl. Phys. Lett.,* 16, 5—8, 1970.
259. **Pappalardo, R.,** Observation of afterglow character and high gain in the laser lines of O(III), *J. Appl. Phys.,* 45, 3547—3553, 1974.
260. **Pappalardo, R.,** Some observations on multiply ionized xenon laser lines, *IEEE J. Quantum Electron.,* QE-10, 897—898, 1974.
261. **Penning, F. M.,** The starting potential of the glow discharge in neon-argon mixtures between large parallel plates, *Physica,* 1, 1028—1044, 1934.
262. **Persson, W.,** The spectrum of singly ionized neon, Ne II, *Phys. Scripta,* 3, 133—155, 1971.
263. **Petersen, A. B. and Birnbaum, M.,** The singly ionized carbon laser at 6783, 6578, and 5145 A, *IEEE J. Quantum Electron.,* QE-10, 468, 1974.
264. **Piper, J. A., Collins, G. J., and Webb, C. E.,** CW laser oscillation in singly ionized iodine, *Appl. Phys. Lett.,* 21, 203—205, 1972.
265. **Piper, J. A. and Webb, C. E.,** Continuous-wave laser oscillation in singly ionized arsenic, *J. Phys. B,* 6, L116—L120, 1973.
266. **Piper, J. A. and Webb, C. E.,** A hollow cathode device for CW helium-metal vapor laser systems, *J. Phys. D,* 6, 400—407, 1973.
267. **Piper, J. A. and Webb, C. E.,** Power limitations of the CW He-Hg laser, *Opt. Commun.,* 13, 122—125, 1975.
268. **Piper, J. A. and Gill, P.,** Output characteristics of the He-Zn laser, *J. Phys. D,* 8, 127—134, 1975.
269. **Piper, J. A. and Webb, C. E.,** High-current characteristics of the continuous-wave hollow-cathode He-I_2 laser, *IEEE J. Quantum Electron.,* QE-12, 21—25, 1976.
270. **Piper, J. A. and Brandt, M.,** Cw laser oscillation on transitions of Cd^+ and Zn^+ in He-Cd-halide and He-Zn-halide discharges, *J. Appl. Phys.,* 48, 4486—4494, 1977.
271. **Piper, J. A. and Neely, D. F.,** Cw laser oscillation on transitions of Cu^+ in He-Cu-halide gas discharges, *Appl. Phys. Lett.,* 33, 621—623, 1978.

272. **Piper, J. A.**, CW laser oscillation on transitions of As^+ in He-AsH_3 gas discharges, *Opt. Commun.*, 31, 374—376, 1979.
273. **Pugnin, V. I., Rudelev, S. A., and Stepanov, A. F.**, Laser generation with ionic transitions of iodine, *J. Appl. Spectrosc. (USSR)*, 18, 667—668, 1973.
274. **Rao, Y. B.**, Structure of the spectrum of singly ionized bromine, *Indian J. Phys.*, 32, 497—515, 1958.
275. **Reid, R. D., McNeil, J. R., and Collins G. J.**, New ion laser transitions in He-Au mixtures, *Appl. Phys. Lett.*, 29, 666—668, 1976.
276. **Reid, R. D., Johnson, W. L., McNeil, J. R., and Collins, G. J.**, New infrared laser transitions in Ag II, *IEEE J. Quantum Electron.*, QE-12, 778—779, 1976.
277. **Reid, R. D., Gerstenberger, D. C., McNeil, J. R., and Collins, G. J.**, Investigations of unidentified laser transitions in Ag II, *J. Appl. Phys.*, 48, 3994, 1977.
278. **Risberg, P.**, The spectrum of singly ionized magnesium, Mg II, *Ark. Fys.*, 9 (31), 483-494, 1955.
279. **Riseberg, L. A. and Schearer, L. B.**, On the excitation mechanism of the He-Zn laser, *IEEE J. Quantum Electron.*, QE-7, 40—41, 1971.
280. **Rózsa, K., Jánossy, M., Bergou, J., and Csillag, L.**, Noble gas mixture CW hollow cathode laser with internal anode system, *Opt. Commun.*, 23, 15—18, 1977.
281. **Rózsa, K., Jánossy, M., Csillag, L., and Bergou, J.**, CW aluminum ion laser in a high voltage hollow cathode discharge, *Phys. Lett.*, 63A, 231—232, 1977.
282. **Rózsa, K., Jánossy, M., Csillag, L., and Bergou, J.**, CW Cu II laser in a hollow anode-cathode discharge, *Opt. Commun.*, 23, 162—164, 1977.
283. **Rudko, R. I. and Tang, C. L.**, Effects of cascade in the excitation of the Ar II laser, *Appl. Phys. Lett.*, 9, 41—44, 1966.
284. **Rudko, R. I. and Tang, C. L.**, Spectroscopic studies of the Ar^+ laser, *J. Appl. Phys.*, 38, 4731—4739, 1967.
285. **Russell, H. N., Albertson, W., and Davis, D. N.**, The spark spectrum of europium, Eu II, *Phys. Rev.*, 60, 641—656, 1941.
286. **Schearer, L. D. and Padovani, F. A.**, De-excitation cross section of metastable helium by Penning collisions with cadmium atoms, *J. Chem. Phys.*, 52, 1618—1619, 1970.
287. **Schearer, L. D.**, A high-power pulsed xenon ion laser as a pump source for a tunable dye laser, *IEEE J. Quantum Electron.*, QE-11, 935—937, 1975.
288. **Schuebel, W. K.**, New cw Cd-vapor laser transitions in a hollow-cathode structure, *Appl. Phys. Lett.*, 16, 470—472, 1970.
289. **Schuebel, W. K.**, Transverse-discharge slotted hollow-cathode laser, *IEEE J. Quantum Electron.*, QE-6, 574—575, 1970.
290. **Schuebel, W. K.**, New CW laser transitions in singly-ionized cadmium and zinc, *IEEE J. Quantum Electron.*, QE-6, 654—655, 1970.
291. **Schuebel, W. K.**, Continuous visible and near-infrared laser action in Hg II, *IEEE J. Quantum Electron.*, QE-7, 39—40, 1971.
292. **Schuebel, W. K.**, Laser action in Al II and He I in a slot cathode, *Appl. Phys. Lett.*, 30, 516—519, 1977.
293. **Shay, T., Kano, H., Hattori, S., and Collins, G. J.**, Time-resolved double-probe study in a He-Hg afterglow plasma, *J. Appl. Phys.*, 48, 4449—4453, 1977.
294. **Shenstone, A. G.**, The second spectrum of silicon, *Proc. R. Soc.*, A261, 153—174, 1961.
295. **Shenstone, A. G.**, The second spectrum of germanium, *Proc. R. Soc.*, A276, 293—307, 1963.
296. **Silfvast, W. T., Fowles, G. R., and Hopkins, B. D.**, Laser action in singly ionized Ge, Sn, Pb, In, Cd, and Zn, *Appl. Phys. Lett.*, 8, 318—319, 1966.
297. **Silfvast, W. T.**, Efficient cw laser oscillation at 4416 Å in Cd(II), *Appl. Phys. Lett.*, 13, 169—171, 1968.
298. **Silfvast, W. T.**, New cw metal-vapor laser transitions in Cd, Sn, and Zn, *Appl. Phys. Lett.*, 15, 23—25, 1969.
299. **Silfvast, W. T. and Szeto, L. H.**, A simple high temperature system for CW metal vapor lasers, *Appl. Opt.*, 9, 1484—1485, 1970.
300. **Silfvast, W. T. and Klein, M. B.**, CW laser action on 24 visible wavelengths in Se II, *Appl. Phys. Lett.*, 17, 400—403, 1970.
301. **Silfvast, W. T. and Szeto, L. H.**, Simplified low-noise He-Cd laser with segmented bore, *Appl. Phys. Lett.*, 19, 445—447, 1971.
302. **Silfvast, W. T.**, Penning ionization in a He-Cd DC discharge, *Phys. Rev. Lett.*, 27, 1489—1492, 1971.
303. **Silfvast, W. T.**, private communication, 1971.
304. **Silfvast, W. T. and Klein, M. B.**, CW laser action on 31 transitions in tellurium vapor, *Appl. Phys. Lett.*, 20, 501—504, 1972.

305. **Simmons, W. W. and Witte, R. S.,** High power pulsed xenon ion lasers, *IEEE J. Quantum Electron.,* QE-6, 466—469, 1970.
306. **Sinclair, D. C.,** Near-infrared oscillation in pulsed noble-gas-ion lasers, *J. Opt. Soc. Am.,* 55, 571—572, 1965.
307. **Sinclair, D. C.,** Polarization characteristics of an ionized-gas laser in a magnetic field, *J. Opt. Soc. Am.,* 56, 1727—1731, 1966.
308. **Solanki, R., Latush, E. L., Fairbank, W. M., Jr., and Collins, G. J.,** New infrared laser transitions in copper and silver hollow cathode discharges, *Appl. Phys. Lett.,* 34, 568—570, 1979.
309. **Solanki, R., Latush, E. L., Gerstenberger, D. C., Fairbank, W. M., Jr., and Collins, G. J.,** Hollow-cathode excitation of ion laser transitions in noble-gas mixtures, *Appl. Phys. Lett.,* 35, 317—319, 1979.
310. **Sosnowski, T. P.,** Cataphoresis in the helium-cadmium laser discharge tube, *J. Appl. Phys.,* 40, 5138—5144, 1969.
311. **Sosnowski, T. P. and Klein, M. B.,** Helium cleanup in the helium-cadmium laser discharge, *IEEE J. Quantum Electron.,* QE-7, 425—426, 1971.
312. **Striganov, A. R. and Sventitskii, N. S.,** *Tables of Spectral Lines of Neutral and Ionized Atoms,* IFI/Plenum Press, New York, 1968.
313. **Sugawara, Y. and Tokiwa, Y.,** CW hollow cathode laser oscillation in Zn^+ and Cd^+, Technology Reports of Seikei University, 9, 1970, 759—760.
314. **Sugawara, Y. and Tokiwa, Y.,** CW laser oscillations in Zn II and Cd II in hollow cathode discharges, *Jpn. J. Appl. Phys.,* 9, 588—589, 1970.
315. **Sugawara, Y., Tokiwa, Y., and Iijima, T.,** Excitation mechanisms of CW laser oscillations in Zn II and Cd II in hollow cathode discharges, *Int. Quantum Electronics Conf.,* Kyoto, Japan, September 7 to 10, 1970, 320-321.
316. **Sugawara, Y., Tokiwa, Y., and Iijima, T.,** New CW laser oscillations in Cd-He and Zn-He hollow cathode lasers, *Jpn. J. Appl. Phys.,* 9, 1537, 1970.
317. **Sutton, D. G., Galvan, L., and Suchard, S. N.,** Two-electron laser transition in Sn I?, *IEEE J. Quantum Electron.,* QE-11, 312, 1975.
318. **Szeto, L. H., Silfvast, W. T., and Wood, O. R., II,** High-gain laser in Pb^+ populated by direct electron excitation from the neutral ground states, *IEEE J. Quantum Electron.,* QE-15, 1332-1334, 1979.
319. **Tell, B., Martin, R. J., and McNair, D.,** CW laser oscillation in ionized xenon at 9697 Å, *IEEE J. Quantum Electron.,* QE-3, 96, 1967.
320. **Tibilov, A. S.,** Generation of radiation in He-Cd and Ne-Cd mixtures, *Opt. Spectrosc.,* 19, 463—464, 1965.
321. **Tio, T. K., Luo, H. H., and Lin, S-C.,** High cw power ultraviolet generation from wall-confined noble gas ion lasers, *Appl. Phys. Lett.,* 29, 795—797, 1976.
322. **Tolkachev, V. A.,** Super-radiant transitions in Ar and Kr, *J. Appl. Spectrosc. (USSR),* 8, 449—451, 1968.
323. **Toresson, Y. G.,** Spectrum and term system of trebly ionized silicon, Si IV, *Ark. Fys.,* 17 (12), 179—192, 1959.
324. **Toresson, Y. G.,** Spectrum and term system of doubly ionized silicon, Si III, *Ark. Fys.,* 18 (28), 389—416, 1960.
325. **Tucker, A. W. and Birnbaum, M.,** Pulsed-ion laser performance in nitrogen, oxygen, krypton, xenon, and argon, *IEEE J. Quantum Electron.,* QE-10, 99—100, 1974.
326. **Turner-Smith, A. R., Green, J. M., and Webb, C. E.,** Charge transfer into excited states in thermal energy collisions, *J. Phys. B.,* 6, 114—130, 1973.
327. **Vuchkov, N. K. and Sabotinov, N. V.,** Pulse generation of Cl II in vapours of CuCl and $FeCl_2$, *Opt. Commun.,* 25, 199—200, 1978.
328. **Walter, W. T., Solimene, N., Piltch, M., and Gould, G.,** Efficient pulsed gas discharge lasers, *IEEE J. Quantum Electron.,* QE-2, 474—479, 1966.
329. **Wang, C. P. and Lin, S.-C.,** Experimental study of argon ion laser discharge at high current, *J. Appl. Phys.,* 43, 5068—5073, 1972.
330. **Wang, C. P. and Lin, S.-C.,** Performance of a large-bore high-power argon ion laser, *J. Appl. Phys.,* 44, 4681—4682, 1973.
331. **Wang, S. C. and Siegman, A. E.,** Hollow-cathode transverse discharge He-Ne and He-Cd^+ lasers, *Appl. Phys.,* 2, 143—150, 1973.
332. **Warner, B. E., Gerstenberger, D. C., Reid, R. D., McNeil, J. R., Solanki, R., Persson, K. B., and Collins, G. J.,** 1W operation of singly ionized silver and copper lasers, *IEEE J. Quantum Electron.,* QE-14, 568—570, 1978.
333. **Watanabe, S., Chihara, M., and Ogura, I.,** New continuous oscillation at 5700 Å in He-Te laser, *Jpn. J. Appl. Phys.,* 11, 600, 1972.

334. **Waynant, R. W.,** Vacuum ultraviolet laser emission from C IV, *Appl. Phys. Lett.,* 22, 419—420, 1973.
335. **Webb, C. E.,** New pulsed laser transitions in Te II, *IEEE J. Quantum Electron.,* QE-4, 426—427, 1968.
336. **Webb, C. E., Turner-Smith, A. R., and Green, J. M.,** Optical excitation in charge transfer and Penning ionization, *J. Phys. B,* 3, L134—L138, 1970.
337. **Wheeler, J. P.,** New xenon laser line observed, *IEEE J. Quantum Electron.,* QE-7, 429, 1971.
338. **Wieder, H., Myers, R. A., Fisher, C. L., Powell, C. G., and Colombo, J.,** Fabrication of wide bore hollow cathode Hg^+ lasers, *Rev. Sci. Instrum.,* 38, 1538—1546, 1967.
339. **Willett, C. S.,** *Introduction to Gas Lasers: Population Inversion Mechanisms,* Pergamon Press, Oxford, 1974.
340. **Willett, C. S. and Heavens, O. S.,** Laser transition at 651.6 nm in ionized iodine, *Opt. Acta,* 13, 271—274, 1966.
341. **Willett, C. S.,** New laser oscillations in singly-ionized iodine, *IEEE J. Quantum Electron.,* QE-3, 33, 1967.
342. **Willett, C. S. and Heavens, O. S.,** Laser oscillation on hyperfine transitions in ionized iodine, *Opt. Acta,* 14, 195—197, 1967.
343. **Willett, C. S.,** Note on near-infrared operating characteristics of the mercury ion laser, *IEEE J. Quantum Electron.,* QE-6, 469—471, 1970.
344. **Zaidĕl, A. N., Prokof'ev, V. K., Raiskii, S. M., Slavnyi, V. A., and Shreider, E. Ya.,** *Tables of Spectral Lines,* IFI/Plenum Press, New York, 1970.
345. **Zarowin, C. B.,** New visible CW laser lines in singly ionized chlorine, *Appl. Phys. Lett.,* 9, 241—242, 1966.
346. **Zhukov, V. V., Latush, E. L., Mikhalevskii, V. S., and Sĕm, M. F.,** New laser transitions in the spectrum of tin and population inversion mechanism, *Sov. J. Quantum Electron.,* 5, 468—469, 1975.
347. **Zhukov, V. V., Il'yushko, V. G., Latush, E. L., and Sĕm, M. F.,** Pulse stimulated emission from beryllium vapor, *Sov. J. Quantum Electron.,* 5, 757—760, 1975.
348. **Zhukov, V. V., Kucherov, V. S., Latush, E. L., and Sĕm, M. F.,** Recombination lasers utilizing vapors of chemical elements II. Laser action due to transitions in metal ions, *Sov. J. Quantum Electron.,* 7, 708—714, 1977.
349. **Anon.,** LGI-37 High Power Pulsed Optical Quantum Oscillator, Specification Sheet ca. 1970.
350. **Toyoda, K., Kobiyama, M., and Namba, S.,** Laser oscillation at 5378 Å by laser-produced cadmium plasma, *Jpn. J. Appl. Phys.,* 15, 2033—2034, 1976.
351. **Bennett, J. and Bennett, W. R., Jr.,** CW oscillation on a new argon ion laser line at 5062 Å and relation to laser Raman spectroscopy, *IEEE J. Quantum Electron.,* QE-15, 842—843, 1979.
352. **Kato, I., Nakaya, M., Satake, T., and Shimizu, T.,** Output power characteristics of microwave-pulse-excited He-Kr^+ ion laser, *Jpn. J. Appl. Phys.,* 14, 2001—2004, 1975.

Section 3
Molecular Gas Lasers

3.1 Electronic Transition Lasers
3.2 Vibrational Transition Lasers
3.3 Far Infrared Lasers

3.1 ELECTRONIC TRANSITION LASERS

Robert S. Davis and Charles K. Rhodes

INTRODUCTION

Since the publication of the last CRC Press *Handbook of Lasers* in 1971, by far the most important class of electronic transition lasers to appear is that of excimer systems. These systems combine high power and high efficiency in the UV wavelength region.

Although many excimer systems now exist, most share certain properties, particularly in regard to their molecular structure and kinetic mechanisms. The radiating transition of interest usually involves a bound upper state and a repulsive lower level. This aspect automatically assures both inversion and sufficient band width to permit significant tunability.

The properties of excimer media are generally conducive to the efficient flow of electronic energy. The origin of this characteristic can be seen from an analysis of the molecular potentials for the most elementary rare-gas dimer system He_2^*. The molecular curves of Ginter and Battino,[124] shown in Figure 3.1.1, illustrate these potentials. The existence of families of closely nested curves for the excited states and of level crossings linking specific levels is striking. These crossings provide important kinetic pathways for the population of the excimer states.

Indeed, this class of molecular excimers represents a special case in which highly organized electronically excited products develop spontaneously as the system evolves along a relatively simple path or reaction coordinate. As is to be expected, such media channel energy efficiently and are good energy funnels.

The rare-gas dimer systems (e.g., Xe_2^*, Kr_2^*, and Ar_2^*) are unusually clear examples of this kinetic ordering and provide an excellent illustration of the manner in which basic chemical reactions enter into these mechanisms. Figure 3.1.2 shows the energy flow pattern for the formation of electronically excited rare-gas dimers.[125] The top of the diagram represents energy in the form of ionized matter, the bulk of the material initially formed by the process of excitation. The bottom of Figure 3.1.2 corresponds to cold ground-state atoms, the lowest energy state available. Following excitation, the system evolves rapidly from top to bottom by relaxation mechanisms through a nested sequence of electronic states, linked by numerous curve crossings similar to those illustrated in Figure 3.1.1. This mechanism enables the selective population of the upper level to occur at a high rate.

The most important aspect of this diagram is the fact that the nature of the particle interactions is such that no single step makes a direct transition from the top to the bottom possible. Therefore, the dynamics of the excited material force the energy to flow in an orderly pathway that leads inevitably to the upper-state manifold of interest. From this level, the only available downward channel is radiation, since no thermally accessible crossings leading to the ground state are available. We should note, finally, that behavior of this nature is not limited to diatomic systems; triatomic species, such as Xe_2Cl^*, exhibit very similar properties.

Of course, a wide variety of other systems has been observed under conditions of stimulated emission. These include diatomic halogen species, CO, H_2, metallic halides, N_2, alkali dimers, molecular ions, and rare earth complexes. All these materials appear in the tabulation.

Excitation by either electical discharge or optical means is by far the most common. The former generally delivers a greater power, while the latter has much greater selectivity. In cases for which favorable kinetic pathways exist (e.g., excimers and H_2), the

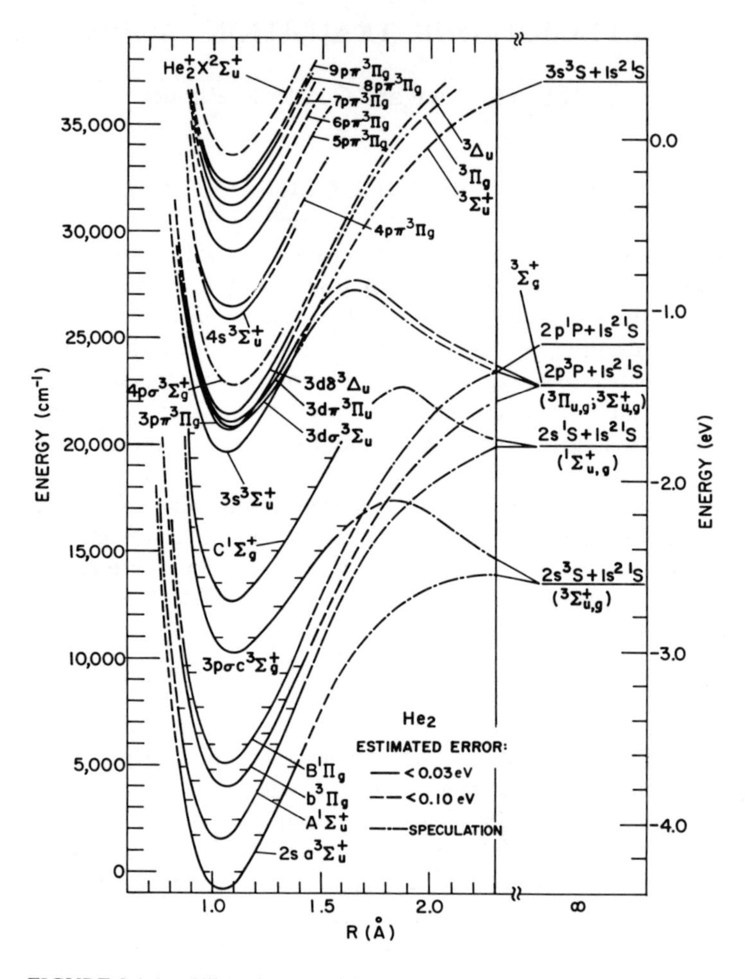

FIGURE 3.1.1. Vibronic potential-energy curves for selected electronic states of He_2. Energy in cm^{-1} is based on N = 0 and v = 0 of the $a^3\Sigma^+_u$ state, while energy in electron volts is based on the lowest level in $X^2\Sigma^+_u$ of He^+_2. When practical, the observed vibrational levels are indicated by horizontal lines at the edges of their appropriate curves. The repulsive ground $^1\Sigma^+_g$ state, which lies at ~ −22.4 eV, is not shown. (From Ginter, M. L. and Battino, R., *J. Chem. Phys.*, 52, 4469, 1970. With permission.)

electrical method can be used. In other examples for which greater specificity in deposition is needed (e.g., rare earth complexes), optical excitation has been employed.

The tables are arranged in molecular alphabetical order, similar to the compilation of Rosen[126] for diatomics. These tabulations include specific data on the wavelengths, transition, and mode of excitation. The nomenclature for the nature of the transition is that used by Herzberg.[127] In all cases, except the rare earth complexes, the molecular transitions involved represent dipole allowed amplitudes.

INTRODUCTION TO THE TABLES

The tables within this section list all molecular electronic transition laser lines known prior to March 1980. The range of wavelengths listed is 109.82 (a para-H_2 C→ X transition) to 8210.2 nm (a N_2 B → A transition). These tables are arranged by increasing number of atoms constituting the molecule. Within the tables, the ordering scheme is as follows:

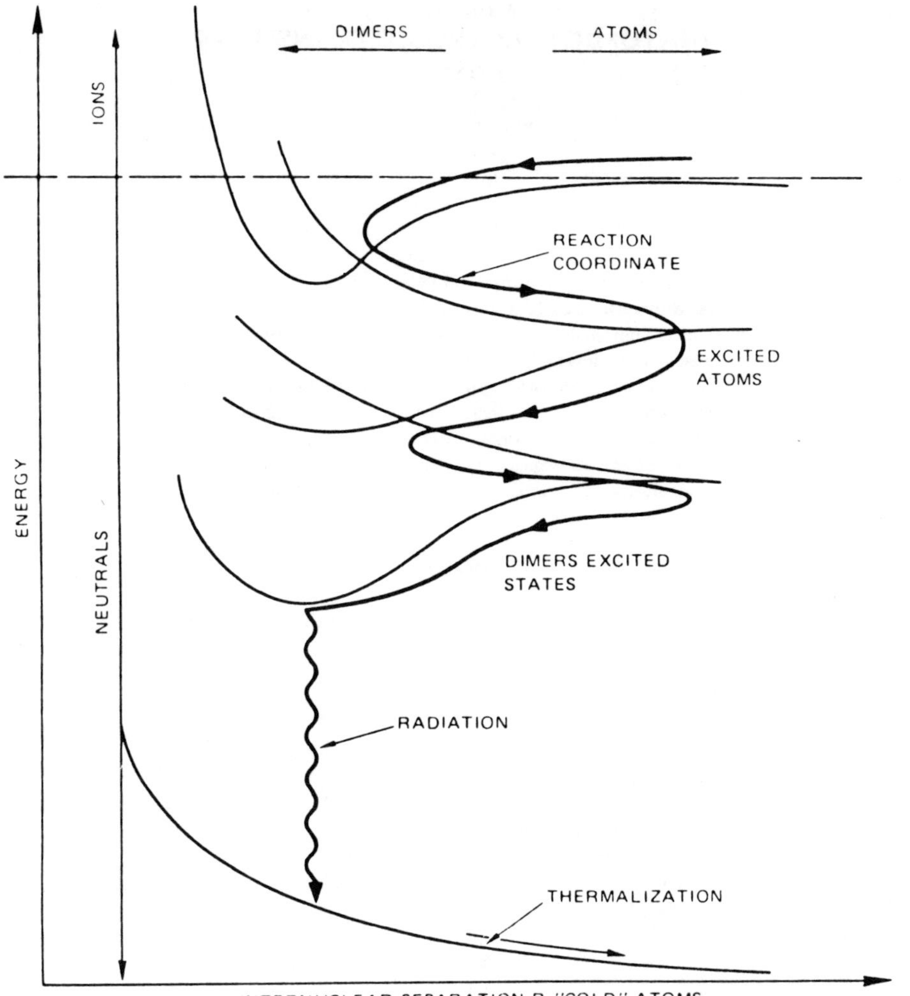

FIGURE 3.1.2. Dimer chemical pathway. (From Rhodes, C. K., Ed., *Excimer Lasers*, Springer-Verlag, New York, 1979. With permission.)

1. The alphabetical order of the chemical formulae
2. Increasing isotopic mass
3. Increasing band-center wavelength
4. Increasing lower vibrational state energy
5. Increasing transition wavelength within a given vibronic group

All pulse times are to be understood to be FWHM (full width at half maximum). The frequency values listed may have an uncertainty of five in the last digit, but in most cases the uncertainty will be less than two. The experimental conditions (pumping method, pump energy, and temperature and pressure of lasant and diluent species), peak output power, and selected references are given in the footnotes.

Table 3.1.1
DIATOMIC ELECTRONIC TRANSITION LASERS

Section	Species	Wavelength, λ air (nm)	Transition	Ref.
2.1	Ar_2[a-d]	126.1	$^1\Sigma^+_u \rightarrow X^1\Sigma^+_g$	1-9

[a] Electron-beam pumping (800 keV, 60 nsec) of reagent grade Ar at pressures of 20 to 68 atm.[1]
[b] Collisional processes in dense, electron beam-excited rare gases are reviewed.[2]
[c] Spectral and temporal characteristics of high-current density (1.7 kA/cm^2) electron beam-excited argon at 0.2 to 65 atm.[3]
[d] Microwave- and condensed discharge-excited emission continua of the rare gases were investigated with reference to photodetection methods, order separation, and wavelength standards in the VUV.[4]

Section	Species	Wavelength, λ air (nm)	Transition	Ref.
2.2	ArCl[a,b]	169.0[c] 175.0	$B^2\Sigma^+_{1/2} \rightarrow X^2\Sigma^+_{1/2}$	9-16

[a] First rare gas-halide exciplex system found to lase after the advent of the rare gas lasers.[10]
[b] ArCl exhibits predissociation as do KrCl and XeBr and is therefore an inefficient lasing material.[11]
[c] Fast discharge excitation (80 to 110 kV, 2.5 nsec rise time) of Cl_2 to Ar to He (1:15:84%) at atmospheric pressure produces a 10-nsec pulse of energy ~0.2 mJ.[12]

Section	Species	Wavelength, λ air (nm)	Transition	Ref.
2.3	ArF[a,b]	193.3	$B^2\Sigma^+_{1/2} \rightarrow X^2\Sigma^+_{1/2}$	9,11, 13-15, 17,18

[a] Axial e-beam excitation (2 MeV, 55 kA, 55 nsec) of 5-torr F_2 in Ar at ~2 atm produces a 55-nsec pulse of energy 92 J with the peak power of 1.6 GW.[15]
[b] UV-preionized transverse electric discharge (25 kV,10 J,40 nsec) excitation of F_2 to Ar to He (0.3:30:69.7%) at a pressure; $\lessapprox$ 2 atm produces a 20- to 25-nsec pulse with peak energy of 60 mJ and intrinsic efficiency of $\gtrapprox$1%.[17]

Table 3.1.1 (continued)
DIATOMIC ELECTRONIC TRANSITION LASERS

Section	Species	Wavelength, λ air (nm)	Transition	Ref.
2.4	ArO[a]	558.0[b,c]	$O(^1S_0) \rightarrow O(^1D_2)$	10,19-26

[a] This system once appeared to be an attractive candidate for high-energy storage short-pulse laser applications, e.g., thermonuclear fusion. Excitation mechanisms, kinetic processes, and collisionally stimulated emission in Group VI elements are examined.[19-22]

[b] A theoretical expression for the emission-rate coefficient of ArO is developed, interatomic potentials and transition moments are discussed, and a theory is given which explains the main features of $O(^1S)$ collision-induced emission in rare gases.[23,24]

[c] Vacuum UV photolysis of 2 torr N_2O in 12 to 41 atm Ar utilizing 192-nm Ar excimer radiation produced $O(^1S_0)$ at high densities. The gain profile and stimulated emission cross section near 558 nm were determined.[25]

Section	Species	Wavelength, λ air (nm)	Transition	Ref.
2.5	Bi_2[a]	592.9[b]	$A(0^-_u) \rightarrow X(0^-_g)$	27,28
		616.0		
		623.9		
		630.0		
		633.9		
		641.4		
		642.2		
		657.6		
		658.2		
		660.3		
		665.0		
		680.9		
		700.6		
		701.3		
		729.2		
		730.1		
		733.5		
		736.4		
		736.6		
		737.6		
		739.8		
		740.8		
		743.9		
		746.8		
		747.1		
		747.5		
		748.2		
		754.3		
		755.1		

[a] Optical pumping by a dye laser ($540 < \lambda < 580$ nm) of ~2 torr Bi_2 in 10 torr Ar at 1400 K produces a dense output

Table 3.1.1 (continued) DIATOMIC ELECTRONIC TRANSITION LASERS

spectra covering the entire region between 660 and 710 nm. Multiline output power up to 350 mW observed with several watts pumping power.[28]

[b] Optical pumping (Ar^*_2, 514.5 nm) of Bi_2 to Bi to Ar (6:22:10—55 torr) exhibits an optimum operating temperature of 1300 K.[27]

Section	Species	Wavelength, λ air (nm)	Transition	Ref.
2.6	Br_2	291.5[a,b]	$E^3\Pi_{2g} \rightarrow B^3\Pi_{2u}$	29-32

[a] Electron-beam excitation (235 keV, 5A/cm², 50 mJ/cm³) of 1 to 3 torr Br_2 in ~2 atm Ar produced maximum energy of 17 mJ with an efficiency ~0.1%.[29]

[b] Discharge pumping (11 kV/cm, 25 A/cm²) in a high-energy electron beam of 0.4% Br_2 in Ar at 4 atm produced >200-nsec pulses with energy 10 μJ.[30]

Section	Species	Wavelength, λ air (nm)	Transition	Ref.
2.7	Cl_2[a]	258.0	$E^3\Pi_{2g} \rightarrow B^3\Pi_{2u}$	33-34

[a] Longitudinal electron-beam excitation of 2 torr Cl_2 in ~12 atm He yielded 96 mJ with an efficiency ~0.4%.[33]

Section	Species	Wavelength, λ air (nm)	Transition	Ref.
2.8	ClF[a]	285.0	$3\Pi_{2g} \rightarrow {}^3\Pi_{2u}$	34

[a] Fluorescence and laser emission were observed utilizing coaxial electron beam (600 keV, 5 kA, 2 nsec) excitation of Cl_2 to F_2 to Ne (0.025:0.025:99.5%) at 20 torr. Simultaneous fluorescing of F_2 (158. nm) and Cl_2 (258. nm) with similar band intensities occurred.[34]

Section	Species	Wavelength, λ air (nm)	Transition	Ref.
2.9	CN[a,b]		$A^2\Pi_{3/2} \rightarrow X^2\Sigma^+$ (0-0) Band	35,36
		1099.63	$Q_1(3)$	
		1099.65	$Q_1(4)$	
		1099.66	$Q_1(2)$	
		1099.74	$Q_1(1)$	
		1099.74	$Q_1(5)$	
		1099.87	$Q_1(6)$	
		1100.07	$Q_1(7)$	
		1100.31	$Q_1(8)$	

Table 3.1.1 (continued)
DIATOMIC ELECTRONIC TRANSITION LASERS

Section	Species	Wavelength, λ air (nm)	Transition	Ref.
			$A^2\Pi_{3/2} \rightarrow X^2\Sigma^+$	
2.9	CN[a,b]		(0-0) Band	35,36
		1100.61	$Q_1(9)$	
		1100.96	$Q_1(10)$	
		1101.36	$Q_1(11)$	
		1101.82	$Q_1(12)$	
		1102.32	$Q_1(13)$	
		1102.88	$Q_1(14)$	
		1103.48	$Q_1(15)$	
		1104.14	$Q_1(16)$	
		1104.45	$P_1(9)$	
		1104.85	$Q_1(17)$	
		1105.21	$P_1(10)$	
		1105.60	$Q_1(18)$	
		1106.03	$P_1(11)$	
		1106.41	$Q_1(19)$	
		1106.89	$P_1(12)$	
		1107.26	$Q_1(20)$	
		1107.82	$P_1(13)$	
		1108.79	$P_1(14)$	
		1109.81	$P_1(15)$	
		1110.90	$P_1(16)$	
		1112.00	$P_1(17)$	
		1113.21	$P_1(18)$	
			(0-1) Band	
		1418.30	$Q_1(5)$	
		1418.49	$Q_1(6)$	
		1418.76	$Q_1(7)$	
		1419.11	$Q_1(8)$	
		1419.54	$Q_1(9)$	
		1420.05	$Q_1(10)$	
		1420.65	$Q_1(11)$	
		1421.32	$Q_1(12)$	
		1422.07	$Q_1(13)$	
		1422.89	$Q_1(14)$	
		1423.80	$Q_1(15)$	
		1424.78	$Q_1(16)$	
		1425.83	$Q_1(17)$	
		1426.96	$Q_1(18)$	
		1428.08	$Q_1(19)$	
		1429.45	$Q_1(20)$	
		1430.81	$Q_1(21)$	
			(0-2) Band	
		1986.16	$Q_1(4)$	
		1986.29	$Q_1(5)$	
		1986.58	$Q_1(6)$	
		1987.01	$Q_1(7)$	
		1987.59	$Q_1(8)$	
		1988.31	$Q_1(9)$	
		1989.18	$Q_1(10)$	
		1990.20	$Q_1(11)$	
		1991.35	$Q_1(12)$	

Table 3.1.1 (continued)
DIATOMIC ELECTRONIC TRANSITION LASERS

Section	Species	Wavelength, λ air (nm)	Transition $A^2\Pi_{3/2} \rightarrow X^2\Sigma^+$	Ref.
2.9	CN[a,b]		(0-2) Band	35,36
		1992.63	$Q_1(13)$	
		1994.06	$Q_1(14)$	
		1995.63	$Q_1(15)$	
		1997.33	$Q_1(16)$	
		1998.8	$P_1(8)$	
		2000.9	$P_1(9)$	
		2003.1	$P_1(10)$	
		2005.5	$P_1(11)$	
		2008.0	$P_1(12)$	
		2010.7	$P_1(13)$	
		2013.5	$P_1(14)$	
		2016.4	$P_1(15)$	
		2019.6	$P_1(16)$	

[a] Photodissociation and predissociation by flash photolysis (15 to 18 kV, 117 to 2.5 kV) of the parent molecules: HCN, ClCN, BrCN, ICN,$(CN)_2$, CH_3CN, CF_3CN, and C_2F_5CN in Ar diluent. Methyl isocyanide to Ar (1:99%) at 50 torr produces the most intense emission. Not all lines are observed with each species.[35]

[b] Open discharge, electrical explosion of W wire pumping (50 kV, 60 kJ) of $(CN)_2C_4F_8$ to Ar (5:95%) at 200 torr produced 70 mJ.[36]

Section	Species	Wavelength, λ vac (nm)	Frequency [ν(cm⁻¹)]	Transition	Ref.
2.10a	CO			$B'\Sigma$ - $A'\Pi$[b]	37,38
				(0-3) Band	
		559.21[a]	17882.2	Q(11)	
		559.49	17873.3	Q(10)	
		559.75	17864.9	Q(9) or R(13)	
		559.98	17857.6	Q(8)	
		560.19	17850.9	Q(7)	
		560.40	17844.3	Q(6)	
		560.53	17840.1	Q(5)	
				(0-4) Band	
		606.46	16489.2	Q(9)	
		606.74	16481.6	Q(8)	
		606.99	16474.8	Q(7)	
		607.22	16448.6	Q(6)	
		607.42	16463.1	Q(5)	
		607.59	16458.6	Q(4)	
				(0-5) Band	
		659.73	15157.7	Q(10)	
		660.13	15148.5	Q(9)	
		660.49[a]	15140.2	Q(8)or	

Table 3.1.1 (continued)
DIATOMIC ELECTRONIC TRANSITION LASERS

Section	Species	Wavelength, λ vac (nm)	Frequency [ν (cm⁻¹)]	Transition	Ref.
2.10a	CO			$B'\Sigma - A'\Pi$[b]	37,38
				(0-5) Band	
				P(13)	
		660.82	15132.7	Q(7)	
		661.09	15126.5	Q(6)	
		661.33	15121.0	Q(5)	
		661.53	15116.4	Q(4)	
2.10b	CO			$A'\Pi - X'\Sigma^+$	33-34
				(2,6) Band	
		181.085[d]		Q(5-13), R(2-9)	
		187.831		Q(5-13), R(2-9)	
		189.784		Q(5-12), R(2-9)	
		195.006		Q(5-11), R(2-9)	
		197.013		Q(5-11)	

[a] Observed in a pulsed (25 pps) discharge (80 A peak) in a 1.17-m long, 10-mm I.D. tube.[37]

[b] $\pm 1\ 0^{-5}\mu m$.

[c] ± 0.005 nm.

[d] Superradiant 1.5-nsec 6-W pulses were produced in a 0.05- × 1.2- × 120-cm tube containing research grade CO (60 torr) with a high-current, Blumlein-circuit, parallel-plate discharge.[38]

Section	Species	Wavelength, λ air (nm)	Transition	Ref.
2.10c	CO^{+}[a]		$B^2\Sigma^+_u \rightarrow X^2\Sigma^+_g$	39
		247.0	(0-2)	
2.10d	CO^{+}[a]		$B^2\Sigma^+_u \rightarrow A^2\Pi$	39
		395.4	(0-0)	
		421.0	(0-1)	

[a] Electron-beam excitation (0.5 MeV, 10 kA, 3 nsec) of 0.1 to 10.0 torr CO in 0.5 to 3 atm He. This laser may exhibit an efficiency approaching 12.7%.

Section	Species	Wavelength, λ air (nm)	Transition	Ref.
2.11	F_2[a-c]		$^3\Pi g \rightarrow {}^3\Pi_u$[a,b]	40-42
		156.71[a]		
		157.48		
		157.59		

[a] Electron-beam excitation (1.9 MeV, 400 J, 50 nsec) with magnetic field guidance (4.5 kG) of 10 torr F_2 in 10.2 atm He produced a 33-nsec pulse of energy 0.25 J/cm^2 with a peak power of 7.6 MW/cm^2 and 2.6% efficiency.[40]

Table 3.1.1 (continued)
DIATOMIC ELECTRONIC TRANSITION LASERS

[b] Electron-beam pumping (850 keV, 20 kA, 300 nsec) in a 3-KG field of 1 to 4 torr F_2 in 2 atm He produced a maximum energy of 22 mJ.[41]

[c] UV-preionized (20 kV, 10 J, 300 nsec) fast-discharge pumping (35, 40 nsec) produced power output with marked sensitivity towards oxygen impurity.[42]

Section	Species	Wavelength, λ air (nm)	Transition	Ref.
2.12a	H_2		$C'\Pi_u \rightarrow X'\Sigma^+_g$[a]	43-50
		110.20	(0-2) P(2)	
		111.89	(1-3) P(2)	
		114.86	(0-3) P(2)	
		116.13	(1-4) Q(1)	
		116.39	(1-4) P(2)	
		116.62	(1-4) P(3)	
		117.58	(2-5) Q(1)	
		117.83	(2-5) P(2)	
		118.05	(2-5) P(3)	
		118.93	(3-6) Q(1)	
		120.67	(1-5) Q(1)	
		120.93	(1-5) P(2)	
		121.73	(2-6) R(1)	
		121.89	(206) Q(1)	
		122.14	(2-6) P(2)	
		122.36	(2-6) P(3)	
		122.99	(3-7) Q(1)	
		123.23	(3-7) P(2)	
		123.94	(4-8) Q(1)	
		124.17	(4-8) P(2)	
2.12a	Para-H_2		$C^2\Pi_u \rightarrow X^1\Sigma^+_g$[a]	43
		109.82	(0-2) R(0)	
		111.52	(1-3) R(0)	
		114.46	(0-3) R(0)	
		116.00	(1-4) R(0)	
		117.46	(2-5) R(0)	
		120.54	(1-5) R(0)	
		121.77	(2-6) R(0)	
		122.87	(3-7) R(0)	
		123.83	(4-8) R(0)	
		124.62	(5-9) R(0)	
		125.20	(6-10) R(0)	
2.12a	HD		$C^1\Pi \rightarrow X^1\Sigma^+$[a]	43
		113.86	(1-4) R(0)	
		114.15	(1-4) P(2)	
		115.20	(2-5) R(0)	
		117.81	(1-5) R(0)	
		119.00	(2-6) R(0)	
		119.28	(2-6) P(2)	
		120.10	(3-7) R(0)	
		121.13	(4-8) R(0)	
		122.84	(6-10) R(0)	

Table 3.1.1 (continued)
DIATOMIC ELECTRONIC TRANSITION LASERS

Section	Species	Wavelength, λ air (nm)	Transition	Ref.
2.12a	D_2		$C^1\Pi_u \rightarrow X^1\Sigma^+_g$ [a]	43
		111.34	(1-4) R(0)	
		113.77	(1-5) R(0)	
		114.76	(1-5) P(2)	
		115.65	(2-6) R(0)	
		115.84	(2-6) P(2)	
		118.81	(2-7) R(0)	
		119.01	(2-7) P(2)	
		119.75	(3-8) R(0)	
		119.94	(3-8) P(2)	
		120.64	(4-9) R(0)	
		120.82	(4-9) P(2)	
		122.80	(3-9) R(0)	
		123.56	(4-10) R(0)	
		124.24	(5-11) R(0)	
		124.41	(5-11) P(2)	
		124.83	(6-12) R(0)	
		125.00	(6-12) P(2)	
		125.33	(7-13) R(0)	
2.12b	H_2		$B^1\Pi^+_u \rightarrow X^1\Sigma^+_g$	43,44, 46-50
		134.23	(0-4) P(3)[a]	
		140.26	(0-5) P(3)[a]	
		141.87	(4-9) P(3)[a]	
		143.62	(1-6) P(3)[a]	
		144.09	(3-7) P(3)[a]	
		146.38	(0-6) P(3)[a]	
		146.70	(6-9) P(3)[a]	
		148.65	(1-7) R(0)[a]	
		149.42	(11-14) P(3)[b]	
		149.52	(1-7) P(3)[a]	
		152.33	(2-8) P(3)[c]	
		154.49	(1-8) R(0)[a]	
		155.34	(1-8) P(3)[a]	
		156.55	(8-14) R(1)[b]	
		156.63	(8-14) R(0)[a]	
		156.73	(8-14) P(3)[c,d]	
		157.20	(2-9) P(1)[c,d]	
		157.43	(2-9) P(2)[b]	
		157.74	(2-9) P(3)[c,d]	
		157.77	(7-13) R(0)[a]	
		157.92	(7-13) P(1)[c]	
		158.00	(7-13) P(2)[c]	
		158.08	(7-13) P(3)[d,e]	
		158.90	(3-10) R(0)[a]	
		159.13	(3-10) P(1)[c,d]	
		159.34	(3-10) P(2)[c]	
		159.61	(3-10) P(3)[d,e]	
		160.49	(4-11) P(1)[d,e]	
		160.59	(6-13) R(0)[a]	
		160.62	(4-11) P(2)[c]	
		160.75	(6-13) P(1)[d,e]	
		160.83	(6-13) P(2)[a]	
		160.84	(4-11) P(3)[d,e]	

Table 3.1.1 (continued)
DIATOMIC ELECTRONIC TRANSITION LASERS

Section	Species	Wavelength, λ air (nm)	Transition	Ref.
2.12b	H_2		$B^1\Pi^+_u \rightarrow X^1\Sigma^+_g$	43,44, 46-50
		160.90	(6-13) P(3)[e]	
		161.03	(5-12) P(1)[e]	
		161.17	(5-12) P(2)[e]	
		161.32	(5-12) P(3)[d,e]	
		161.48	(5-12) P(4)[b]	
		161.65	(5-12) P(5)[b]	
		163.95	(4-11) R(0)[a]	
		164.15	(4-11) P(1)[a]	
		164.29	(4-11) P(2)[a]	
		164.44	(4-11) P(3)[a]	
2.12b	Para-H_2		$B^1\Sigma^+_u \rightarrow X^1\Sigma^+_g$[a]	43,45
		127.95	(0-3) P(2)	
		133.86	(0-4) P(2)	
		135.98	(4-6) P(2)	
		136.80	(6-7) P(2)	
		139.90	(0-5) P(2)	
		140.75	(0-5) P(4)	
		143.26	(1-6) P(2)	
		143.76	(3-7) P(2)	
		144.06	(7-9) P(2)	
		146.02	(0-6) P(2)	
		146.41	(6-9) P(2)	
		146.84	(0-6) P(4)	
		149.17	(1-7) P(2)	
		151.57	(4-9) P(2)	
		151.99	(2-8) P(2)	
		153.49	(5-10) P(2)	
		155.01	(1-8) P(2)	
		156.40	(8-14) P(4)	
		156.75	(8-14) P(2)	
		157.43	(2-9) P(2)	
		157.77	(7-13) R(0)[c]	
		158.00	(7-13) P(2)	
		158.11	(2-9) P(4)[c]	
		158.14	(7-13) P(4)[c]	
		158.90	(3-10) R(0)[c]	
		159.34	(3-10) P(2)	
		159.93	(3-10) P(4)	
		160.24	(4-11) R(0)[c]	
		160.59	(6-13) R(0)[c]	
		160.62	(4-11) P(2)[c]	
		160.83	(6-13) P(2)	
		160.96	(6-13) P(4)	
		161.03	(5-12) P(1)[c]	
		161.09	(4-11) P(4)[c]	
		161.17	(5-12) P(2)	
		161.32	(5-12) P(3)[c]	
		161.49	(5-12) P(4)	
		164.29	(4-11) P(2)	
		164.60	(4-11) P(4)	

Table 3.1.1 (continued)
DIATOMIC ELECTRONIC TRANSITION LASERS

2.12b	HD		$B^1\Sigma^+_u \rightarrow X^1\Sigma^+_g$	43,45
		124.57	(6-5) P(2)	
		125.28	(0-3) P(2)	
		130.33	(0-4) P(2)	
		135.51	(0-5) P(2)	
		140.77	(0-6) P(2)	
		148.84	(1-8) P(2)	
		151.36	(2-9) P(2)	
		152.99	(5-11) P(2)	
		156.20	(2-10) P(2)	
		157.13	(9-16) R(0)	
		157.27	(9-16) P(3)	
		157.43	(9-16) P(2)	
		158.01	(8-15) R(0)	
		158.09	(3-11) P(2)	
		158.19	(8-15) P(2)	
		158.25	(8-15) P(3)	
		158.31	(3-11) P(3)	
		159.38	(4-12) P(1)	
		159.55	(4-12) P(2)	
		159.74	(4-12) P(3)	
		160.23	(5-13) R(0)	
		160.37	(5-13) P(1)	
		160.46	(7-15) R(0)	
		160.52	(5-13) P(2)	
		160.57	(7-15) P(1)[c]	
		160.65	(7-15) P(2)	
		160.67	(6-14) R(0)	
		160.68	(5-13) P(3)	
		160.69	(7-15) P(3)	
		160.75	(7-15) P(4)[c]	
		160.79	(6-14) P(1)[c]	
		160.83	(5-13) P(4)[c]	
		160.91	(6-14) P(2)	
		161.03	(6-14) P(3)	
		161.13	(6-14) P(4)	
2.12b	D_2		$B^a\Sigma^+_u \rightarrow X^1\Sigma^{+\,a}_g$	43,45
		130.36	(0-5) P(2)	
		134.59	(0-6) P(2)	
		138.88	(0-7) P(2)	
		143.22	(0-8) P(2)	
		157.58	(3-13) P(2)	
		158.63	(7-16) P(1)	
		158.64	(10-19) P(1)	
		158.67	(10-19) P(2)	
		158.69	(10-19) P(3)	
		158.71	(10-19) P(4)	
		158.72	(7-16) P(2)	
		158.98	(4-14) P(2)	
		159.13	(9-18) P(2)	
		159.14	(8-17) P(2)	
		159.23	(8-17) P(3)	
		159.26	(9-18) P(4)	
		160.09	(5-15) P(2)	
		160.21	(5-15) P(3)	

Table 3.1.1 (continued)
DIATOMIC ELECTRONIC TRANSITION LASERS

Section	Species	Wavelength, λ air (nm)	Transition	Ref.
		160.35	(5-15) P(4)[c]	
		160.58	(9-19) R(0)	
		160.65	(9-19) P(1)	
		160.68	(9-19) P(2)	
		160.77	(6-16) P(1)	
		160.85	(6-16) P(2)	
		160.96	(6-16) P(3)	
		161.07	(7-17) R(0)	
		161.08	(6-16) P(4)	
		161.15	(8-18) P(1)	
		161.17	(7-17) P(1)	
		161.20	(8-18) P(2)	
		161.24	(7-17) P(2)	
		161.26	(8-18) P(3)	
		161.32	(7-17) P(3)	
		161.41	(7-17) P(4)[c]	
		161.66	(9-20) P(2)	
2.12c	H_2		$E,F^1\Sigma^+_g \rightarrow B^1\Sigma^+_u$ [f,g]	48-50
		835.19	(2-1) P(2)	
		887.87	(1-0) P(4)	
		890.13	(1-0) P(2)	
		1116.5	(0-0) P(4)	
		1122.5	(0-0) P(2)	
		1306.1	(0-1) P(4)	
		1316.6	(0-1) P(2)	
2.12c	HD		$E,F^1\Sigma^+_g \rightarrow B^1\Sigma^+_u$ [g]	
		916.3	(1-0)	
2.12c	D_2			
		827.98	(2-0) P(3)	
		953.26	(1-0) P(3)	

[a] Electron-beam excitation (600 kV, 10 kA) of hydrogen isotopes at a pressure of ∿8 torr in a liquid nitrogen-cooled stainless tube.[43]

[b] Optimum conditions in a TEA H and D laser, 0.1 torr and 5-kV discharge voltage produced 0.5 mJ.[44]

[c] Additional observations made with the same apparatus.[45]

[d] Observed in a pulsed discharge in a 1-m long, 12- × 3-mm tube. Maximum power output is several hundred kilowatts in a 1-nsec pulse.[46]

[e] Observed in a pulsed discharge in a 1.2-m long, 12- × 0.4-mm tube; pressure, 20 to 150 torr H_2; maximum power output, 1.5 kW in a 2-nsec pulse.[47]

[f] Observed in a pulsed discharge in a 1.02-m long, 7-mm I.D. tube; pressure, 3 torr H_2.[48]

[g] Observed in a pulsed (20 pps) discharge in a 1.45-m long, 15-mm I.D. tube; in H_2 or H_2 - D_2 mixtures, pressure 3 torr.[49,50]

Table 3.1.1 (continued)
DIATOMIC ELECTRONIC TRANSITION LASERS

Section	Species	Wavelength, λ air (nm)	Transition electronic vibrational	Ref.
2.13	HgBr		$B^2\Sigma_{1/2}^+ \rightarrow X^2\Sigma_{1/2}^+$	51-54
		499.0[a]	(0,21)	
		501.8[b]	(0,22)[c]	
		502.3[b]	(0,22)[d]	
		502.6[b]	(0,22)[d]	
		503.9[b]	(0,23[c])	
		504.2[b]	(0,23)[c]	
		504.6[b]	(0,23)[d]	

[a] Electron-beam sustained discharge pumping (25 kV, 2.5 A/cm², 800 nsec) of HgBr to CCl_4 to Ar (0.2:05:99.3%) at ∼2 atm produces ∼1 mJ in an unoptimized system.[51]

[b] Electron-beam excitation (240 kV, 100 A/cm², 150 nsec) of a mixture of HBr to Hg to Xe to Ar (0.8:2.0:10.8:86.4%) at an Ar density of 3 amagats produced 501.8-nm radiation of 3.2-mJ energy in a 60-nsec pulse resulting in a peak power of 50 kW and an efficiency of 0.25%.[52]

[c] $HgBr^{81}$ transition.

[d] $HgBr^{79}$ transition.

[e] Optical pumping (ArF*, 193 nm) on the photodissociation of $HgBr_2$ at a vapor pressure of 50 torr produced 0.25 mJ.[53]

[f] UV-preionized transverse-electric-discharge photodissociation of $HgBr_2$ in N_2 to He (100:900 torr) produced 7.5 mJ with an efficiency of 0.1%.[54]

Section	Species	Wavelength, λ air (nm)	Transition electronic vibrational	Ref.
2.14	HgCl[a-d]		$B^2\Sigma_{1/2}^+ \rightarrow X^2\Sigma^+_{1/2}$	51,54-59
		551.6	(0,21)	
		552.3	(0,23)	
		555.0	(0,22)	
		557.6[a]	(0,22)	
		558.4[b]	(1,23)	
		559.0	(1,23)	

[a] First report of electron-beam excitation of CCl_4 to Hg to Xe to Ar (1.1:2.1:11.1:85.7%) at an Ar density of 3 amagats on the 557.6-nm line produced 138 kW with an efficiency of 0.5%.[56]

[b] Electron-beam (300 kV, 2 A/cm², 400 nsec) controlled discharge pumping (2.2 kV, 2.5 kA) of Ar to Hg to Cl_2 (0.7:0.5:98%) at ∼2 atm lased with 0.1 mJ energy and an efficiency of 0.01%.[57]

[c] UV-preionized transverse-discharge-initiated predissociation of $HgCl_2$ to N_2 to He (1:10:89%) resulted in an order of magnitude improvement in efficiency and output producing 6 J in a 50-nsec pulse with an efficiency of 0.05%.[54]

[d] Photolyic predissociation (Xe_2, 172 nm) of $HgCl_2$ in He at 200 torr produces emission on the (0,22) and (1,23) lines of the B → X transition.[58]

Table 3.1.1 (continued) DIATOMIC ELECTRONIC TRANSITION LASERS

Section	Species	Wavelength, λ air (nm)	Transition	Ref.
2.15	HgI		$B^2\Sigma^+_{1/2} \rightarrow X^2\Sigma^+_{1/2}$	54-55
		443.0	(0,15)	
		445.0	(0,17)	

[a] UV-preionized discharge (6 J in 50 cm^3, 50 nsec) dissociation of HgI_2 (1 to 10 torr) in He (350 to 1000 torr) produces maximal energy of 0.3 mJ in a mixture of HgI_2 to N_2 to He (1:10:89%). Energy output was an order of magnitude less without the nitrogen.[54] The spectrum consists of six lines which are correlated to the (0,15) and (0,17) vibrational bands.[55]

Section	Species	Wavelength, λ air (nm)	Transition	Ref.
2.16a	I_2		$^3\Pi_{2g} \rightarrow {}^3\Pi_{2u}$[c]	60-62
		342.0[a,b]	(0-12), (2-15)	
		342.3	(3-17)	
		342.4	(1-14)	
		342.8	(2-16), (0-13)	
2.16b	I_2		$B^3\Pi_{2u} \rightarrow X^1\Sigma^+_g$	63-67
		544.3[d]	—	
		555.0[e]	—	
		556.7[d]	—	
		568.0[e]	—	
		569.7[f]	(43-9)	
		574.5[e]	—	
		576.4[d]	—	
		581.5[e]	—	
		583.0[f]	(43-11)	
		588.0[e]	—	
		590.5[d]	—	
		596.8[f]	(43-13)	
		602.5[e]	—	
		604.8[d]	—	
		611.1[f]	(43-15)	
		617.5[e]	—	
		617.520[e]	(34-13) R(84)	
		617.730[e]	(35-13) R(107)	
		617.900[e]	(33-13) R(58)	
		617.970[e]	(34-13) R(89)	
		617.990[e]	(34-13) P(86)	
		618.245[e]	(33-13) P(60)	
		618.325[e]	(35-13) P(109)	
		618.490[e]	(34-13) P(91)	
		618.580[e]	(33-13) P(65)	
		619.8[d]	—	
		625.8[g]	P(17), R(17)	
		626.0[f]	(43-17)	
		633.0[e]	—	

Table 3.1.1 (continued)
DIATOMIC ELECTRONIC TRANSITION LASERS

Section	Species	Wavelength, λ air (nm)	Transition	Ref.
2.16b	I_2		$B^3\Pi_{2u} \rightarrow X^1\Sigma^+_g$	63-67
		635.2[d]	—	
		649.0[e]	—	
		651.1[d]	—	
		659.2[d]	—	
		664.5[e]	—	
		676.3[d]	—	
		693.6[d]	—	
		711.4[d]	—	
		814.4[g]	P(38),R(38)	
		835.8[g]	P(40),R(40	
		857.8[g]	P(42),R(42)	
		857.9[f]	(43-42) P(17) R(15), R(11)	
		880.4[g]	P(44), R(44)	
		880.6[f]	(43-44) P(17), R(15)	
		881.3[d]	—	
		903.7[g]	P(46),R(46)	
		903.8[f]	(43-46)	
		904.7[d]	—	
		906.0[d]	—	
		927.4[g]	P(48),R(48)	
		927.6[f]	(43-48) P(17), P(13),R(15), R(11)	
		928.8[d]	—	
		929.5[d]	—	
		930.5[d]	—	
		951.8[g]	P(50),R(50)	
		952.0[f]	(43-50)	
		954.5[d]	—	
		955.5[d]	—	
		976.6[g]	P(52),R(52)	
		976.7[f]	(43-52) P(17), R(15)	
		996.3[d]	—	
		997.3[d]	—	
		1001.9[f]	(43-54) P(17), P(13),R(15), R(11)	
		1005.3[d]	—	
		1022.5[d]	—	
		1024.5[d]	—	
		1025.5[d]	—	
		1027.4[f]	(43-56) P(17), P(13),R(15), R(11)	
		1053.4[d]	—	
		1077.5[d]	—	
		1078.8[d]	—	
		1106.6[d]	—	
		1107.3[d]	—	
		1120.6[h]	(11-44) R(40)	
		1121.4[h]	(11-44) P(42)	

Table 3.1.1 (continued)
DIATOMIC ELECTRONIC TRANSITION LASERS

Section	Species	Wavelength, λ air (nm)	Transition	Ref.
2.16b	I_2		$B^3\Pi_{2u} \rightarrow X^1\Sigma^+_g$	63-67
		1121.6[h]	(11-44) R(58)	
		1122.6[h]	(11-44) P(60)	
		1125.5[d]	—	
		1132.8[h]	(12-45) R(127)	
		1133.4[h]	(13-46) R(84)	
		1134.7[h]	(12-45) P(129)	
		1134.8[h]	(13-46) P(86)	
		1135.0[d]	—	
		1145.3[h]	(12-46) R(61)	
		1146.4[h]	(12-46) P(63)	
		1150.2[i]	(13-47) P(57)	
		1151.0[i]	(13-47) R(57)	
		1151.5[i]	(13-47) P(82)	
		1152.2[i]	(13-47) R(75)	
		1152.9[i]	(13-47) R(82)	
		1169.8[i]	(13-48) P(75)	
		1170.3[i]	(13-48) P(82)	
		1171.1[i]	(13-48) R(75)	
		1171.8[i]	(13-48) R(82)	
		1174.0[i]	(14-49) P(62)	
		1175.0[i]	(14-49) R(63)	
		1217.0[g]	P(71), R(71)	
		1274.0[g]	P(76),R(76)	
		1287.0[d]	—	
		1292.5[d]	—	
		1294.0[g]	P(78),R(78)	
		1301.0[i]	(27-66) P(66)	
		1302.0[i]	(27-66) R(66)	
		1304.0[g]	P(79),R(79)	
		1306.9[i]	(29-68) P(64)	
		1308.0[i]	(26-68) R(64)	
		1315.3[d]	—	
		1319.2[d]	—	
		1320.0[g]	P(81),R(81)	
		1328.2[d]	—	
		1329.1[d]	—	
		1331.0[d]	—	
		1332.4[d]	—	
		1333.3[d]	—	
		1334.9[d]	—	
		1338.0[g]	P(83),R(83)	
		1340.6[i]	(42-82) P(57)	
		1341.8[i]	(42-82) R(57)	
		1342.1[i]	(45-85) P(83), (44-84) P(75), (43-83) R(53)	
		1342.9[i]	(45-85) R(74), (45-85) R(72), (44-84) R(62)	

[a] Electron-beam pumping (1 MeV, 20 kA, 20 nsec) of 30 torr CF_3I in Ar at 10 atm produced 36 mJ in a 10-nsec pulse corresponding to a peak power of 3.6 MW.[60]

Table 3.1.1 (continued)
DIATOMIC ELECTRONIC TRANSITION LASERS

[b] Electron-beam pumping of HI, CF_3I, or CH_3I in Ar at ∿4 atm produced lasing on eight vibrational bands; maximum energy obtained for 8 torr HI in Ar at 5.3 atm was 1 J in a 40-nsec pulse corresponding to a peak power of 25 MW.[61]

[c] Exploding W wire (45 kV, 6 kJ, 5 μsec) photoexcitation of 5 torr I_2 in 1.5 atm SF_6 produces a 5-μsec pulse with an energy of 0.4 J and an overall efficiency of 0.1%.[62]

[d] Optical pumping at 531.9 nm (Nd to YAG laser, 190-nsec pulse) of I_2 in a Brewster cell exhibits pumping threshold of a few microjoules.[63]

[e] Pump line is at 530.6 nm in same apparatus as in Remark d.[63]

[f] CW-optical pumping at 514.5 nm (AR^+ laser, 1 to 4 W) of I_2 vapor in a room-temperature 50-cm cell produces a maximum of 3 mW average power in ∿50 μsec pulses with a maximum conversion efficiency of 0.14%.[64]

[g] CW-oscillation produced by Ar^+ laser pumping at 514.5 nm yields up to 250 mW with a pump power of 3 W. CW-oscillation in the spectral ranges 625.8 to 814.4 and 976.6 to 1217 nm should be possible with suitable mirrors.[65]

[h] Optical pumping in the wavelength range 593.1 to 597.2 nm by a dye laser (1 to 2 kW, 1 μsec) of I_2 in a 70-cm glass tube produced 20 to 40 W in a 0.2- to 0.6-μsec pulse. At 125°C an output power of 150 W is produced.[66]

[i] Superfluorescent emission is produced via pumping with a broad-band dye laser over the range 515 to 583 nm. Threshold pump intensity is $\gtrsim$ 10 μJ at 532 nm. Rotational assignment is tentative.[67]

Section	Species	Wavelength, λ air (nm)	Transition	Ref.
2.17	KrCl	222.	$C^2\Sigma^+_{1/2} \rightarrow X^2\Sigma^+_{1/2}$	12,13,16, 68-69

[a] First demonstration of electron-beam pumping (600 keV, 760 J, 50 nsec) of a mixture of Cl_2 to Kr to Ar (0.15:2.96.95%) at 4.5 atm produces 50 mJ in a ∿30-nsec pulse.[68]

[b] Discharge pumping (80 to 110 kV) of mixtures of Cl_2 to Kr to He(1.0:10.89%) at ∿1 atm produced superfluorescent lasing with 1.3 mJ in a 10-nsec pulse corresponding to a peak power of 60 kW.[12]

[c] Discharge excitation of mixtures of HCl to Kr to He (0.15:10:89.85%) at 3.3 atm produced 100 mJ per pulse.[16]

Section	Species	Wavelength, λ air (nm)	Transition	Ref.
2.18	KrF[a-g]	248.4 249.1	${}^2\Sigma^+_{1/2} \rightarrow X^2\Sigma^+_{1/2}$	9,11,13-15,17,18, 70-76

[a] *Ab initio* configuration interaction calculations were performed; the main emission band corresponds to the $III_{1/2} \rightarrow$

Table 3.1.1 (continued)
DIATOMIC ELECTRONIC TRANSITION LASERS

$I_{1/2}$ ionic to covalent transition, while emission features at 220 and 275 nm are assigned to the $IV_{1/2} \rightarrow I_{1/2}$ and $II_{3/2} \rightarrow I_{3/2}$ transitions, respectively.[70]

[b] Axial electron-beam excitation (2 MeV, 55 kA, 55 nsec) of a mixture of F_2 to Kr to Ar (0.3:≥7.1:≤92.6%) produced 108 J of energy in a ∼55-nsec pulse corresponding to a peak power of ∼1.96 W and an intrinsic efficiency of ∼3%.

[c] Fast-discharge pumping (5 to 20 kV, 10 nsec) of NF_3 to Kr to He (0.1:1.0:98.9%) at 1 to 4 atm produced 6 to 8 mJ in a ∼100-nsec pulse with an intrinsic efficiency of 0.4%.[72]

[d] Electron-beam controlled-discharge pumping (350 kV, 2.2 kA, 100 nsec) of a mixture of F_2 to Kr to Ar (0.1:2.1:97.9%) at 1 atm produced a maximum of 6.0 mJ energy in a 90-nsec pulse corresponding to >100 kW and an efficiency of 0.2%.[73]

[e] Electron-beam excitation (250 keV, 12 A/cm^2, 0.5 to 1.0 μsec) of a mixture of F_2 to Kr to Ar (0.2:4.0:95.8%) at 1.7 atm and a volume of 8.5 ℓ produced 102 J with an efficiency of 9%. An accurate computer code is developed which explains the main features of the electron beam-produced KrF radiation.[74]

[f] Discharge-pumping of F_2 to Kr to Ar (0.7:4.2:95.1%) produces 10 W of average power at a pulse rate of 1 kHz with an efficiency of 0.13%.[75]

[g] An ultra-high spectral brightness KrF laser with a pulse energy of ∼60 mJ and spectral width of 150 ± 30 MHz is tunable over 2 cm^{-1} and has beam divergence of 50 μrad.[76]

Section	Species	Wavelength, λ air (nm)	Transition	Ref.
2.19	Kr_2[a-d]	145.7[a]	$^1\Sigma^+_u \rightarrow X^1\Sigma^+_g$	2-9,77

[a] First report of coherent oscillation in Kr_2 with a 10-nsec pulse. The 147.0-nm resonance line of Xe impurity is a prominent absorptive feature in the spectrum. Threshold pressure for oscillation is ∼250 psi.[5]

[b] The physics of electron beam-excited rare gases at high densities is reviewed.[2]

[c] Measurements were made of conversion reaction coefficients for the three-body conversion reactions of Kr and Xe to their molecular ions, Kr^+_2 and Xe^+_2, in electron afterglow plasmas.[77]

[d] An early dynamic model of high-pressure rare gas excimer lasers is given.[6]

Section	Species	Wavelength, λ air (nm)	Transition	Ref.
2.20	KrO	557.81[a]	$O(^1S_0) \rightarrow O(^1D_2)$[b]	10,19-21, 23,26

[a] Electron-beam excitation (1 MeV, 50 nsec, 2 × 10 cm) of 5 torr O_2 in 25 atm Kr produces an optimum output of 5 to 10 mJ in a 100-nsec pulse corresponding to a peak power output of ∼100 kW.[19]

Table 3.1.1 (continued)
DIATOMIC ELECTRONIC TRANSITION LASERS

[b] The metastable $^3P_2$ state of Kr strongly resembles Rb in its chemical properties.[10]

Section	Species	Wavelength, (λ air (nm)	Transition	Ref.
2.21a	N_2[a]		$C^3\Pi_u \rightarrow B^3\Pi_g$ (0-0) Band	78,79, 83-85
		336.4903	$R_1(7)$	
		336.5474	$R_3(6)$	
		336.5537	$R_1(6)$	
		336.6156	$R_1(4)$	
		336.6211	$R_3(5)$	
		336.6682	$R_1(4)$	
		336.6911	$R_3(4)$	
		336.7218	$R_1(3)$	
		336.8432	$P'_3(20)$	
		336.8917	$P_3(19)$	
		336.9250	$P_1(1)$	
		336.9361	$P'_3(18)$	
		336.9502	$P'_2(18)$	
		336.9542	$Q_3(2)$	
		336.9555	$P'_2(2)$	
		336.9575	$P'_1(18)$	
		336.9760	$P_3(17)$	
		336.9838	$P_1(3)$	
		336.9852	$P_2(17)$	
		337.9898	$P_1(17)$	
		337.0081	$P'_1(4)$	
		337.0121	$P'_3(16)$	
		337.0138	$P'_1(16)$	
		337.0161	$P_1(16)$	
		337.0169	$P'_2(16)$	
		337.0297	$P_1(5)$	
		337.0316	$P_2(3)$	
		337.0360	$P'_1(15)$	
		337.0374	$P_1(15)$	
		337.0434	$P_2(15)$, $P_3(15)$	
		337.0472	$P'_1(6)$	
		337.0529	$P'_1(14)$	
		337.0551	$P_1(14)$	
		337.0559	$P'_2(4)$	
		337.0614	$P'_1(7)$	
		337.0623	$P_1(7)$	
		337.0663	$P'_2(14)$	
		337.0682	$P_1(13)$	
		337.0716	$P'_1(8)$, $P'_3(14)$	
		337.0731	$P_1(8)$	
		337.0757	$P'_1(12)$, $P_3(3)$	
		337.0762	$P_2(5)$	
		337.0787	$P'_1(9)$	
		337.0803	$P_1(9)$	
		337.0821	$P'_1(10)$, $P_1(11)$	

Table 3.1.1 (continued)
DIATOMIC ELECTRONIC TRANSITION LASERS

Section	Species	Wavelength, λ air (nm)	Transition	Ref.
2.21a	N_2[a]		$C^3\Pi_u \rightarrow B^3\Pi_g$ (0-0) Band	78,79, 83-85
		337.0843	$P_2(13)$	
		337.0924	$P'_2(6)$	
		337.0941	$P_3(13)$	
		337.0990	$P'_2(12)$, $P'_3(4)$	
		337.1042	$P_2(7)$	
		337.1082	$P_2(11)$	
		337.1120	$P'_2(8)$	
		337.1129	$P'_3(12)$	
		337.1141	$P'_2(10)$	
		337.1147	$P_2(9)$	
		337.1179	$P_3(5)$	
		337.1271	$P_3(11)$	
		337.1312	$P'_3(6)$	
		337.1371	$P'_3(10)$	
		337.1398	$P_3(7)$	
		337.1427	$P_3(9)$	
		337.1433	$P'_3(8)$	
			(0-1) Band	
		357.5980	$P_1(6)$	
		357.6194	$P_1(11)$	
		357.6250	$P_1(9)$	
		357.6320	$P_2(5)$	
		357.6571	$P_2(7)$	
		357.6613	$P_3(11)$	
		357.6778	$P_3(5)$	
		357.6899	$P_3(9)$	
		357.6955	$P_3(7)$	
2.21b	N_2[b]		$B^3\Pi_g \rightarrow A^3\Sigma^+_u$ (4-2) Band	79
		748.2187	$^QR_{23}(1)$	
		748.5941	—	
		748.6135	$^PQ_{23}(3)$	
		748.6413	$P_2(4)$	
		748.6253	$^PQ_{23}(4)$	
		748.7409	$Q_1(9)$	
		748.8046	—	
		748.8246	$P_2(6)$	
		748.9107	$^PQ_{23}(6)$	
		748.9626	$Q_1(8)$	
		748.9809	$P_2(13)$	
		749.0096	$P_2(12)$	
		749.0317	$^PQ_{23}(8)$	
		749.1510		
		749.1705	$Q_1(7)$	
		749.2379	$^OP_{23}(4)$	
		749.3082	$^QR_{12}(6)$	
		749.3716	$Q_1(6)$	
		749.3910	—	
		749.5086	$^OP_{23}(5)$	
		749.5465	$Q_1(5)$	

Table 3.1.1 (continued)
DIATOMIC ELECTRONIC TRANSITION LASERS

Section	Species	Wavelength, λ air (nm)	Transition	Ref.
2.21b	N_2[b]		$B^3\Pi_g \rightarrow A^3\Sigma^+_u$ (4-2) Band	79
		749.5660	—	
		749.6024	—	
		749.7256	—	
		749.7524	$Q_1(4)$	
			(4-2) Band	
		749.7728	—	
		749.8898	—	
		749.9013	—	
		749.9327	$Q_1(3)$	
		749.9593	—	
		749.9825	$P_1(13)$	
		750.0071	$^OP_{23}(7)$	
		750.0646	$^PQ_{12}(12)$	
		750.0734	—	
		750.1056	$Q_1(2)$	
		750.1295	—	
		750.1404	—	
		750.1553	$P_1(11)$	
		750.2139	$^PQ_{12}(9)$	
		750.2729	$Q_1(1)$	
		750.2768	$P_1(9)$	
		750.3035	$^PQ_{12}(7)$	
		750.3371	$^PQ_{12}(6)$	
		750.3418	$^PR_{13}(8)$	
		750.3642	$^PQ_{12}(5)$	
		750.3669		
		750.3697	$P_1(7)$	
		750.3838	$^PQ_{12}(4)$	
		750.3960	$^PR_{13}(7)$	
		750.3994	$^PQ_{12}(3)$	
		750.4106	$^PQ_{12}(2)$	
		750.4160	$^PQ_{12}(1)$	
		750.4184	$^PR_{13}(6)$	
		750.4274	$Q_1(0)$	
		750.4598	$P_1(3)$	
		750.4768	$P_1(1)$	
		750.5113	$^PR_{13}(2)$	
		750.5710	—	
		750.5903	—	
		750.6063	—	
		750.6356	$^OP_{23}(10)$	
		750.8145	$^OP_{23}(11)$	
		750.9890	—	
		751.0133	$^OP_{12}(3)$	
		751.0923	—	
		751.1592	$^OP_{12}(4)$	
		751.2799	—	
		751.3003	$^OP_{12}(5)$	
		751.3569	—	
		751.4357	$^OP_{12}(6)$	
		751.5079	—	
		751.5446	—	

Table 3.1.1 (continued)
DIATOMIC ELECTRONIC TRANSITION LASERS

Section	Species	Wavelength, λ air (nm)	Transition	Ref.
2.21b	N_2[b]		$B^3\Pi_g \rightarrow A^3\Sigma^+_u$	79
			(4-2) Band	
		751.5650	${}^{O}P_{12}(7)$	
		751.7728	${}^{O}Q_{13}(8)$	
		751.8013	${}^{O}P_{12}(9)$	
			(3-1) Band	
		758.6439	$Q_3(9)$	
		758.7693	$Q_3(7)$	
		760.3477	—	
		760.6374	${}^{P}Q_{23}(2)$	
		760.7626	${}^{P}Q_{23}(3)$	
		760.8801	${}^{P}Q_{23}(4)$	
		760.9853	${}^{P}Q_{23}(5)$	
		761.0759	${}^{P}Q_{23}(6)$	
		761.1082	$Q_1(8)$	
		761.1514	${}^{P}Q_{23}(7)$	
		761.2105	${}^{P}Q_{23}(8)$	
		761.2528	${}^{P}Q_{23}(9)$	
		761.3260	$Q_1(7)$	
		761.5347	$Q_1(6)$	
		761.6994	${}^{O}P_{23}(5)$	
		761.7357	$Q_1(5)$	
		761.9288	$Q_1(4)$	
		762.0844	—	
		762.0943	—	
		762.1161	$Q_1(3)$	
		762.2235	${}^{O}P_{23}(7)$	
		762.2565	—	
		762.2959	$Q_1(2)$	
		762.3256	—	
		762.3311	—	
		762.3582	${}^{P}Q_{12}(10)$	
		762.3686	—	
		762.3918	—	
		762.4220	${}^{P}Q_{12}(9)$	
		762.4690	$Q_1(1)$	
		762.4924	$P_1(9)$	
		762.5115	${}^{P}Q_{12}(7)$	
		762.5445	${}^{P}Q_{12}(6)$	
		762.5709	${}^{P}Q_{12}(5)$	
		762.5770	—	
		762.5812	$P_1(7)$	
		762.5906	${}^{P}Q_{12}(4)$	
		762.6007	${}^{P}R_{13}(7)$	
		762.6044	${}^{P}Q_{12}(3)$	
		762.6114	$P_1(6)$	
		762.6180	${}^{P}Q_{12}(2)$	
		762.6207	${}^{P}Q_{12}(1)$	
		762.6360	$P_1(5)$	
		762.6560	$P_1(4)$	
		762.6700	$P_1(3)$	
		762.6749	$P_1(1)$	
		762.6826	${}^{O}P_{23}(9)$	
		762.8854	${}^{O}P_{23}(10)$	

Table 3.1.1 (continued)
DIATOMIC ELECTRONIC TRANSITION LASERS

Section	Species	Wavelength, λ air (nm)	Transition	Ref.
2.21b	N_2[b]		$B^3\Pi_g \rightarrow A^3\Sigma^+_u$	79
			(3-1) Band	
		762.9102	—	
		763.0305	—	
		763.1880	—	
		763.2446	$^OP_{12}(3)$	
		763.3348	—	
		763.3985	$^OP_{12}(4)$	
		763.4546	—	
		763.4779	—	
		763.5474	$^OP_{12}(5)$	
		763.6126	—	
		763.6904	$^OP_{12}(6)$	
		763.7586	—	
		763.8274	$^OP_{12}(7)$	
		763.9571	$^OP_{12}(8)$	
		763.9715	—	
		764.0383	—	
		764.0794	$^OP_{12}(9)$	
		764.1929	$^OP_{12}(10)$	
		764.2478	—	
		764.4612	—	
			(2-0) Band	
		774.3859	$Q_1(5)$	
		775.2354	$P_1(8)$	
		775.3652	$P_1(2)$	
			(2-1) Band	
		866.9223	$Q_3(9)$	
		867.1332	$Q_3(7)$	
		869.2580	$^QP_{23}(1)$	
		869.6366	$^PQ_{23}(2)$	
		869.7945	$^PQ_{23}(3)$	
		869.8263	$Q_1(9)$	
		869.9397	$^PQ_{23}(4)$	
		870.0670	$^PQ_{23}(5)$	
		870.0684	$^PQ_{23}(5)$	
		870.1481	$Q_1(8)$	
		870.1718	$^PQ_{23}(6)$	
		870.2451	$^PQ_{23}(7)$	
		870.2681	$^OP_{23}(3)$	
		870.3093	$^PQ_{23}(8)$	
		870.3457	$^PQ_{23}(9)$	
		870.4549	$Q_1(7)$	
		870.7478	$Q_1(6)$	
		871.0118	$^OP_{23}(5)$	
		871.0273	$Q_1(5)$	
		871.2956	$Q_1(4)$	
		871.3533	$^OP_{23}(6)$	
		871.5519	$Q_1(3)$	
		871.6718	$^OP_{23}(7)$	
		871.7377	$P_1(11)$	
		871.7970	$Q_1(2)$	
		871.8571	$P_1(10)$	
		871.8654	$^PQ_{12}(9)$	
		871.9537	$^PQ_{12}(8)$	

Table 3.1.1 (continued)
DIATOMIC ELECTRONIC TRANSITION LASERS

Section	Species	Wavelength, λ air (nm)	Transition	Ref.
2.21b	N_2[b]		$B^3\Pi_g \rightarrow A^3\Sigma^+_u$	79
			(2-1) Band	
		871.9562	$P_1(9)$	
		871.9791		
		872.0251	$^PQ_{12}(7)$	
		872.0284	$Q_1(1)$	
		872.0308	—	
		872.0419	$P_1(8)$	
		872.0848	$^PQ_{12}(6)$	
		872.1155	$P_1(7)$	
		872.1327	$^PQ_{12}(5)$	
		872.1718	$P_1(6)$	
		872.1971	—	
		872.2007	$^PQ_{12}(3)$	
		872.2220	$P_1(5)$	
		872.2341	$^PQ_{12}(1)$	
		872.2569	$P_1(4)$	
		872.2836	$P_1(3)$	
		872.3057	$P_1(1)$	
		872.6333	$^OP_{12}(1)$	
		872.8430	$^OP_{12}(2)$	
		873.0453	$^OP_{12}(3)$	
		873.2394	$^OP_{12}(4)$	
		873.4247	$^OP_{12}(5)$	
		873.5995	$^OP_{12}(6)$	
		873.7644	$^OP_{12}(7)$	
		873.9162	$^OP_{12}(8)$	
		874.0559	$^OP_{12}(9)$	
		874.2917	$^OP_{12}(11)$	
			(1-0) Band	
		884.5349	$^SR_{32}(1)$	
		885.6271	$Q_3(9)$	
		885.8470	$Q_3(7)$	
		888.0521	$^QR_{23}(1)$	
		888.4527	$^PQ_{23}(2)$	
		888.6204	$^PQ_{23}(3)$	
		888.6378	$Q_1(9)$	
		888.7756	$^PQ_{23}(4)$	
		888.9111	$^PQ_{23}(5)$	
		888.9738	$Q_1(8)$	
		889.0243	$^PQ_{23}(6)$	
		889.1133	$^PQ_{23}(7)$	
		889.1769	$^PQ_{23}(8)$	
		889.2149	$^PQ_{23}(9)$	
		889.2940	$Q_1(7)$	
		889.6001	$Q_1(6)$	
		889.8930	$Q_1(5)$	
		889.9078	$^OP_{23}(5)$	
		890.1733	$Q_1(4)$	
		890.2711	$^OP_{23}(6)$	
		890.4419	$Q_1(3)$	
		890.6097	$^OP_{23}(7)$	
		890.6649	$P_1(11)$	
		890.6994	$Q_1(2)$	
		890.7920	$^PQ_{12}(9)$	

Table 3.1.1 (continued)
DIATOMIC ELECTRONIC TRANSITION LASERS

Section	Species	Wavelength, λ air (nm)	Transition	Ref.
2.21b	N_2[b]		$B^3\Pi_g \rightarrow A^3\Sigma^+_u$	79
			(1-0) Band	
		890.8808	$^PQ_{12}(8)$	
		890.8878	$P_1(9)$	
		890.9451	$Q_1(1)$	
		890.9527	$^PQ_{12}(7)$	
		890.9750	$P_1(8)$	
		891.0132	$^PQ_{12}(6)$	
		891.0480	$P_1(7)$	
		891.0612	$^PQ_{12}(5)$	
		891.1001	$^PQ_{12}(4)$	
		891.1063	$P_1(6)$	
		891.1280	$^PQ_{12}(3)$	
		891.1502	$^PQ_{12}(2)$	
		891.1538	$P_1(5)$	
		891.1608	$^PQ_{12}(1)$	
		891.1898	$P_1(4)$	
		891.2139	$^OP_{23}(9)$	
		891.8033	$^PP_{12}(2)$	
		892.0184	$^OP_{12}(3)$	
		892.2249	$^OP_{12}(4)$	
		892.4223	$^OP_{12}(5)$	
		892.6099	$^OP_{12}(6)$	
		892.7865	$^OP_{12}(7)$	
		892.9509	$^OP_{12}(8)$	
		893.1019	$^OP_{12}(9)$	
		893.3580	$^OP_{12}(11)$	
			(3-3) Band	
		965.389	$P_2(11)$	
		965.846	$Q_1(7)$	
		966.599	$Q_1(5)$	
		967.270	$Q_1(3)$	
		967.758	$^PQ_{12}(7)$	
		967.943	$^PQ_{12}(5)$	
		968.061	$^PQ_{12}(3)$	
		969.552	$^OP_{12}(5)$	
		969.879	$^OP_{12}(7)$	
2.21c	N_2		$a^1\Pi_g - a'^1\Sigma^-_u$	80,81
			(2-1) Band	
		3294.63[c,d]	Q(14)	
		3301.49[c,d]	Q(12)	
		3307.34[c,d]	Q(10)	
		3309.89[c,d]	Q(9)	
		3312.21[c,d]	Q(8)	
		3314.26[d]	Q(7)	
		3316.07[c,d]	Q(6)	
		3317.60[d]	Q(5)	
		3318.89[c,d]	Q(4)	
		3320.69[d]	Q(2)	
			(1-0) Band	
		3451.84[c,d]	Q(12)	
		3458.32[c,d]	Q(10)	
		3461.14[d]	Q(9)	

Table 3.1.1 (continued) DIATOMIC ELECTRONIC TRANSITION LASERS

Section	Species	Wavelength, λ air (nm)	Transition	Ref.
2.21c	N_2		$a^1\Pi_g - a'^1\Sigma^-_u$	80,81
			(1-0) Band	
		3463.68[c,d]	Q(8)	
		3465.96[d]	Q(7)	
		3467.95[c,d]	Q(6)	
		3469.67[d]	Q(5)	
		3471.09[c,d]	Q(4)	
			(0-0) Band	
		8148.3[d]	Q(10)	
		8182.7[c,d]	Q(8)	
		8210.2[c,d]	Q(6)	
2.21d	N_2[e]		$w^1\Delta_u - a^1\Pi_g$	82
			(0-0) Band	
		3623.49	R(4)	
		3626.14	R(3)	
		3629.10	R(2)	
		3643.13	Q(4)	
		3644.72	Q(5)	
		3646.62	Q(6)	
		3648.83	Q(7)	
		3651.38	Q(8)	
		3654.24	Q(9)	
		3657.45	Q(10)	
		3660.95	Q(11)	
		3664.83	Q(12)	
		3668.99	Q(13)	
		3673.52	Q(14)	
		3678.34	Q(15)	

[a] Electrical-discharge (23 kV) in a 240- × 10- × 5-mm channel containing 70 torr of N_2 produces a 1-nsec pulse of energy 0.4 mJ and a peak power of 400 kW with no mirror. High-resolution Czerny-Turner spectrograph employed.[78]

[b] Discharge pumping (50 kV) of 0.5 to 1 torr N_2 in 2.5 to 11 mm I.D., 30- to 120-mm pyrex tubes at liquid air temperature.[79]

[c] Observed in a pulsed discharge (several hundred amperes) in a 2.25-m, 15-mm I.D. tube containing 0.15 torr N_2 and 0.5 torr Ne.[80]

[d] Observed under similar conditions as Remark C.[81]

[e] Observed in a pulsed (15 pps) discharge at pressure of 1 torr.[82]

Section	Species	Wavelength, λ air (nm)	Transition	Ref.
2.22	N^+_2[a-d]		$B^2\Sigma^+_u \rightarrow X^2\Sigma^+_g$	86-89
		391.4[a]	(0-0)	
		427.81[a-c]	(0-1)	
		470.9[c]	(0-2)	
		522.8[d]	(0-3)	

Table 3.1.1 (continued) DIATOMIC ELECTRONIC TRANSITION LASERS

[a] Preionized discharge excitation (60 kV) of N_2 to He (1:99%) at 3 to 11 atm produced superradiant power of 400 kW.[86]

[b] Electron-beam excitation (200 keV, 3kA, 50 nsec) of 0.02 to 0.2% N_2 in He at 0.5 to 7 atm.[87]

[c] Spark-preionized discharge excitation (30 kV, 27 J, 15 nsec) of 1 torr N_2 in 2.6 atm. He produces a 6-nsec pulse with energy 3 mJ and peak power of 0.5 MW.[88]

[d] Electron-beam pumping (900 kV, 14 kA, 20 nsec) of 10 torr N_2 in 7 atm He produces a 15-nsec pulse of energy 0.27 J/ℓ, power 0.5 MW, and an efficiency of 1.9%.[89]

Section	Species	Wavelength, λ air (nm)	Transition	Ref.
2.23a	Na_2		$B^1\Pi_u \rightarrow X^1\Sigma^+_g$	90-92
		524.5[a]		
		526.33[b]		
		527.4[a]		
		532.1[a]		
		533.3[a]		
		534.0[a]		
		534.90[b]		
		536.90[b]		
		537.0[a]		
		537.1[a]		
		537.6[a]		
		537.81[b]		
		538.9[a]		
		540.24[b]		
		541.31[b]		
		541.6[a]		
		544.69[b]		
		544.8[a]		
		545.4[a]		
		546.7[a]		
		546.9[a]		
		548.0[a]		
		548.5[a]		
		549.16[b]		
		549.9[a]		
		558.1[a]		
		559.1[a]		
		559.7[a]		
2.23b	Na_2		$A^1\Sigma^+_u \rightarrow X^1\Sigma^+_g$	90-91
		529.82[b]		
		529.95[b]		
		534.15[b]		
		534.28[b]		
		538.50[b]		
		538.63[b]		
		784.93[c]		
		785.69[c]		
		789.74[c]		

Table 3.1.1 (continued)
DIATOMIC ELECTRONIC TRANSITION LASERS

Section	Species	Wavelength, λ air (nm)	Transition	Ref.
2.23b	Na_2		$A^1\Sigma^+_u \rightarrow X^1\Sigma^+_g$	90-91
		789.79[c]		
		791.78[c]		
		792.95[c]		
		793.70[c]		
		797.47[c]		
		797.66[c]		
		799.09[c]		
		799.66[c]		
		800.84[c]		
		803.65[c]		
		803.93[c]		
		804.45[c]		
		805.37[c]		
		805.61[c]		
		806.94[c]		
		808.05[c]		

[a] Optical pumping with the continuous Ar^+ - laser in the range 458 to 488 nm of Na_2 in a heat pipe produces continuous oscillation with an output power of up to 3 mW.[90]

[b] Nd to YAG 473-nm radiation focused through antireflection windows into a stainless steel pipe at 605°C containing Na_2 to Na in a He buffer at 30 torr produces an average superfluorescent pulse energy of ~0.13 μJ with an intrinsic efficiency of ~0.07%.[91]

[c] A-band laser lines excited by 659-nm lines of apparatus in Remark b produces only 2.4 nJ due to nonoptimum output coupling and losses at the Na vapor-fogged windows.[91]

Section	Species	Wavelength, λ air (nm)	Transition	Ref.
2.24a	S_2	365—570[a]	$B^3\Sigma_u$ - $X^3\Sigma^-_g$[b]	93-94
2.24b	S_2	1086.0[c]	$^1\Sigma^+_g \rightarrow X^3\Sigma^-_g$	95
		1091.5		
		1091.7		
		1092.0		
		1092.3		
		1094.1		
		1094.6		
		1099.0		
		1100.0		
		1158.7		

[a] Line-tunable laser action is observed via frequency-doubled dye laser pumping of 1 to 10 torr S_2. Transitions corresponding to (3-0) to (7-0) were excited; intrinsic efficiency is ~2%.[93]

[b] Vibronic transitions of S_2 are given.[94]

[c] Photolysis of ~30 torr COS in 100 torr Xe diluent produced specific energy output of 0.5 mJ/cm³.[95]

Table 3.1.1 (continued)
DIATOMIC ELECTRONIC TRANSITION LASERS

Section	Species	Wavelength, λ air (nm)	Transition	Ref.
2.25	Te_2[a]		$B(O^+_u) \rightarrow X(O^+_g)$	27
		557.1		
		557.5		
		557.8		
		557.9		
		562.6		
		563.8		
		564.2		
		564.3		
		564.6		
		564.7		
		564.9		
		565.0		
		569.6		
		570.1		
		571.1		
		571.4		
		571.5		
		571.9		
		572.0		
		572.1		
		572.4		
		576.6		
		576.7		
		577.3		
		577.4		
		578.0		
		578.3		
		578.4		
		578.5		
		578.7		
		578.9		
		579.0		
		579.3		
		579.4		
		579.7		
		579.8		
		584.1		
		584.9		
		585.7		
		585.9		
		586.5		
		586.9		
		587.0		
		587.4		
		592.4		
		592.7		
		593.4		
		593.6		
		600.2		
		600.4		
		600.5		

Table 3.1.1 (continued)
DIATOMIC ELECTRONIC TRANSITION LASERS

Section	Species	Wavelength, λ air (nm)	Transition	Ref.
2.25	Te_2[a]		$B(O^+_u) \rightarrow X(O^+_g)$	27
		600.8		
		600.9		
		608.2		
		608.3		
		608.5		
		608.7		
		608.9		
		616.2		
		616.5		
		616.8		
		617.0		
		620.4		
		627.8		
		628.7→→→		
		628.8		
		629.5		
		637.1		
		637.9		
		638.1		
		638.8		
		647.7		
		648.4		
		656.1		
		657.4		
		658.1		

[a] CW-optical pumping with the 476.5-nm Ar laser line of Te_2 vapor in a heat pipe at 1000 K produces CW-output having maximum multiline power output of 20 mW with a pump power of 1 W.[27]

Section	Species	Wavelength, λ air (nm)	Transition	Ref.
2.26	XeBr		$B^1\Sigma^+_{1/2} \rightarrow X^1\Sigma^+_{1/2}$	9,13-15,96-99
		281.8[a,b]		

[a] Electron-beam pumping (433 keV, 36 J, 50 nsec) of 0.10 to 4% analytical grade Br_2 in Xe at 500 to 1500 torr produced peak power of 200 W.[96]

[b] Electrical-discharge pumping (26 kV, ≃100 kW, ∿35 nsec) of HBr to Xe to He (0.1:5:94.9%) at 3.3 atm overcomes the problem of Xe^+_2 absorption and produces 60 mJ per pulse. Simultaneous lasing of Br_2 at 291.3 nm is observed.[97]

Table 3.1.1 (continued) DIATOMIC ELECTRONIC TRANSITION LASERS

Section	Species	Wavelength, λ air (nm)	Transition	Ref.
2.27	XeCl[a-c]		$B^2\Sigma^+_{1/2} \rightarrow X^2\Sigma^+_{1/2}$	9,13-16,99-103
		307.0[b]	(1-5)	
		307.3[b]	(1-6)	
		307.65[b,c]	(0-0)	
		307.92[b,c]	(0-1)	
		308.17[b,c]	(0-2)	
		308.43[b,c]	(0-3)	

[a] Electron-beam pumping (300 keV, 14 A/cm², 1.2 μsec) of HCl to Xe to Ne (0.07:1.0:98.93%) at 4 atm produces 3.0 J/ℓ with an efficiency of 5%.[100]

[b] UV-preionized-discharge pumped HCl to Xe to He (0.2:3.0:96.8%) at 2.7 atm produced maximum output energy of 1110 mJ in a 30 nsec pulse with an efficiency of >0.8%.[101]

[c] Electric-discharge excitation (48 kV, 150 kA) of HCl to Xe to He (0.2:5.0:94.8%) at ∿3.3 atm produces peak energy of 180 mJ in a ∿30 nsec pulse and is amenable to continuous operation.[16]

Section	Species	Wavelength, λ air (nm)	Transition	Ref.
2.28a	XeF[a]		$B^2\Sigma^+_{1/2} \rightarrow X^2\Sigma^+_{1/2}$	17,104-106,109
		348.8[b,c]	(2-5)	
		351.1[b,c]	(0-2), (1-4)	
		353.1[b,c]	(0-3), (1-6)	
		354.0[d]	(0-4), (1-7)	
2.28b	XeF		$C^2\Pi_u \rightarrow A^2\Pi_g$	107,108
		483.0[e]		
		486.0[f]		

[a] Fast-discharge Blumlein-type circuit electrical discharge with preionization (20 to 33 kV, ∿1 kA, 8 nsec) of NF_3 to Xe to He (1.2:97.6%) at 750 torr produced a 4-nsec pulse of energy 100 mJ corresponding to peak power of 25 MW.[104]

[b] UV-preionized discharge-pumped (25 kV, 10 J, 40 nsec) NF_3 to Xe to He (0.3:1.0:98.7%) at ∿2 atm produced 65 mJ.[17]

[c] Electron-beam pumping (1 MeV, 20 kA, 20 nsec) of NF_3 to Xe to Ar (0.36:10.0:89.4%) at 1.7 atm produces a 10-nsec pulse of 5 mJ corresponding to 500 kW with an efficiency >1%. Coaxial excitation produced a 100-nsec pulse of 80 mJ with an efficiency of 3%.[105]

[d] Electron-beam excitation of F_2 to Xe to Ar (0.1:0.3:99.6%) at ≲4 atm produces 6 kW.[106]

[e] Photodissociation (Xe_2^*, 172 nm, 10^5 W/cm²) of ∿2 torr

Table 3.1.1 (continued)
DIATOMIC ELECTRONIC TRANSITION LASERS

XeF_2 in Xe at 6 atm produces output on the 353- and 483-nm bands of slightly more than 1 mJ.[107]

[f] Electron-beam pumping (1 MeV, 20 kA, 8 nsec) of NF_3 to Xe to Ar (0.17:0.33:99.5%) at 5.9 atm produced a 20-nsec pulse with peak power of 5 kW.[108]

Section	Species	Wavelength, λ air (nm)	Transition	Ref.
2.29	XeO[a]		$2^1\Sigma^+ \rightarrow 1^1\Sigma^{+b}$	19,20, 110
		537.6	(0-5)	
		544.2	(0-6)	

[a] Pulsed electron-beam pumping (1 MeV, 50 kA, 50 nsec) of 10 torr O_2 in 12 atm Xe produces a 60-nsec pulse of energy ∿10 mJ corresponding to a peak power of ∿80 kW.[19]

[b] Open high-current discharge (45 kV, 20 μsec) optical pumping ($130 < \lambda < 145$ nm) of 3.5 torr N_2O in Xe at 0.7 atm at a temperature of 170 K produced 10 μJ/cm³ in a 4- to 5-μsec pulse.[110]

Section	Species	Wavelength, λ air (nm)	Transition	Ref.
2.30	Xe_2	171.6	$^1\Sigma^+_u \rightarrow X^1\Sigma^+_g$	2-9,15, 68,77, 111-116

[a] First report of laser action on liquified Xe by electron-beam pumping (300 A/cm²) produces peak power of ∿1 kW on a line near 176 nm.[111]

[b] First report of lasing on high pressure (1 to 30 atm) gaseous Xe using electron-beam excitation (0.6 MeV, 7 kA, 2 nsec) showed a threshold pressure of 13.6 atm and produced a 3-nsec pulse with an intrinsic efficiency of ∿20%.[116]

Table 3.1.2
TRIATOMIC ELECTRONIC TRANSITION LASERS

Section	Species	Wavelength, λ air (nm)	Transition	Ref.
3.1	Kr_2F	430.0[a]	(?)	117

[a] First observation of lasing produced via e-beam-excited Ar/Kr/NF_3 (4 torr, 400 torr, 6 to 8 atm) yields ∿10 kW in a 25-nsec pulse.[117]

Section	Species	Wavelength, λ air (nm)	Transition	Ref.
3.2	Xe_2Cl	518.0[a]	(?)[b]	14, 118-19

[a] First observation of laser action produced from e-beam-excited Ar/Xe/CCl_4 (4 to 9 atm, 100 to 750 torr, 0.5 to 10 torr). Peak laser power measured to be ∿2 kW.[118]

[b] Gain was measured on the blue-green band of Xe_2Cl and its characteristics were shown to be similar to those of the XeF(C → A) transition.[119]

Table 3.1.3
POLYATOMIC ELECTRONIC TRANSITION LASERS

Section	Species	Wavelength, λ air (nm)	Transition	Ref.
4.1	$NdAl_3Cl_{12}$[a] ↔$NdAl_4Cl_{15}$	1060	$^4F_{3/2} \rightarrow {}^4I_{11/2}$	120

[a] First observation of optical gain for trivalent rare earth molecular vapors. Purified sample of $NdCl_3/AlCl_3$ in a Brewster-angle quartz cell at ∿350°C is excited by a pulsed dye laser (587.2 nm, ∿1.5 torr). The existence of optical gain, etching, impurities, and Schlieren effects prevent demonstration of laser oscillation.[120]

Section	Species	Wavelength, λ air (nm)	Transition	Ref.
4.2	$Nd(thd)_3$[a]	1060.0	$^4F_{3/2} \rightarrow {}^4I_{4/2}$	121

[a] Fluorescence decays were studied using the vapor-phase chelate 2,2,2,6-tetramethyl-3,5-heptanediove.

Table 3.1.1 (continued) DIATOMIC ELECTRONIC TRANSITION LASERS

Section	Species	Wavelength, λ air (nm)	Transition	Ref.
4.3	$TbAl_3Cl_{12}$ ↔ $TbAl_4Cl_{15}$	435	$^5D_3 \rightarrow {}^7F_4$	122-123
		545	$^5D_4 \rightarrow {}^7F_5$	

[a] A double Brewster-angle quartz cell containing the $TbCl_3$/$AlCl_3$ (55.8 mg/354.8 mg) starting materials was situated in a cylindrical oven. The driver, a 15- to 20-mJ KrF* laser directed axially along the cell, produced excited-state densities of ∼10J/ℓ sustained for ∼10 μsec.[122]

Section	Species	Wavelength, λ air (nm)	Transition	Ref.
4.4	POPOP[a]	385	$S_1 \rightarrow S_0$	128

[a] Stimulated emission from POPOP [*p*-phenylene-bis (5-phenyl-2-oxazole)] has been observed under electron beam excitation in a mixture involving a 4 to 5 atm argon buffer.

REFERENCES

1. **Hughes, W. M., Shannon, J., and Hunter, R.,** 126.1 nm argon laser, *Appl. Phys. Lett.*, 24, 488, 1974.
2. **Lorents, D. C.,** The physics of electron-beam excited rare-gases at high densities, *Physica*, 82C, 19, 1976.
3. **Koehler, H. A., Ferderber, L. J., Redhead, D. L., and Ebert, P. J.,** Vacuum-ultraviolet emission from high-pressure xenon and argon excited by high-current relativistic electron-beams, *Phys. Rev. A.*, 9, 768, 1974.
4. **Wilkinson, P. G. and Byram, E. T.,** Rare-gas light sources for the vacuum untraviolet, *Appl. Opt.*, 4, 581, 1965.
5. **Hoff, P. W., Swingle, J. C., and Rhodes, C. K.,** Observations of stimulated emission from high-pressure krypton and argon/xenon mixtures, *Appl. Phys. Lett.*, 23, 245, 1973.
6. **Werner, C. W., George, E. V., Hoff, P. W., and Rhodes, C. K.,** Dynamic model of high-pressure rare-gas excimer lasers, *Appl. Phys. Lett.*, 25, 235, 1974.
7. **Gedanken, A., Jortner, J., Raz, B., and Szöke, A.,** Electronic energy transfer phenomena in rare gases, *J. Chem. Phys.*, 57, 3456, 1972.
8. **Verkovtseva, E. T., Ovechkin, A. E., and Fogel, Y. M.,** The vacuum-uv spectra of supersonic jets of Ar-Kr-Xe mixtures excited by an electron beam, *Chem. Phys. Lett.*, 30, 120, 1975.
9. **McDaniel, E. W., Flannery, M. R., Ellis, H. W., Eisele, F. L., and Pope, W.,** Compilation of Data Relevant to Rare Gas - Rare Gas and Rare Gas - Monohalide Excimer Lasers, Vols. 1 and 2, Tech. Rep. H-78-1, U.S. Army Missile Research and Development Command, Redstone Arsenal, Ala., 1978.
10. **Golde, M. F. and Thrush, B. A.,** Vacuum UV emission from reactions of metastable inert-gas atoms: chemiluminescence of ArO and ArCl, *Chem. Phys. Lett.*, 29, 486, 1974.

11. **Velazco, J. E., Kolts, J. H., and Setser, D. W.,** Quenching rate constants for metastable argon, krypton, and xenon atoms by fluorine containing molecules and branching ratios for XeF* and KrF* formation, *J. Chem. Phys.,* 65, 3468, 1976.
12. **Waynant, R. W.,** A discharge-pumped ArCl superfluorescent laser at 175.0 nm, *Appl. Phys. Lett.,* 30, 234, 1977.
13. **Golde, M. F.,** Interpretation of the oscillatory spectra of the inert-gas halides, *J. Mol. Spectrosc.,* 58, 261, 1975.
14. **Lorents, D. C., Huestis, D. L., McCusker, R. V., Nakano, H. H., and Hill, R. M.,** Optical emissions of triatomic rare gas halides, *J. Chem. Phys.,* 68, 4657, 1978.
15. **Hoffman, J. M., Hays, A. K., and Tisone, G. C.,** High-power noble-gas-halide lasers, *Appl. Phys. Lett.,* 28, 538, 1976.
16. **Sze, R. C. and Scott, P. B.,** Intense lasing in discharge-excited noble-gas monochlorides, *Appl. Phys. Lett.,* 33, 419, 1978.
17. **Burnham, R. and Djeu, N.,** Ultraviolet-preionized discharge-pumped lasers in XeF, KrF, and ArF, *Appl. Phys. Lett.,* 29, 707, 1976.
18. **Rokni, M., Jacob, J. H., and Mangano, J. H.,** Dominant formation and quenching processes in e-beam pumped ArF* and KrF* lasers, *Phys. Rev. A,* 16, 2216, 1977.
19. **Powell, H. T., Murray, J. R., and Rhodes, C. K.,** Laser oscillation on the greenbands of XeO and KrO, *Appl. Phys. Lett.,* 25, 730, 1974.
20. **Murray, J. R. and Rhodes, C. K.,** The possibility of high-energy-storage lasers using the auroral and transauroral transitions of column-VI elements, *J. Appl. Phys.,* 47, 5041, 1976.
21. **Rockwood, S. D.,** Mechanisms for Achieving Lasing on the 5577 Å Line of Atomic Oxygen, LAUR 73-1031, Los Alamos Scientific Laboratory, Los Alamos, N.M., 1973.
22. **Murray, J. R., Powell, H. T., Schlitt, L. G., and Toska, J.,** Laser Program Annual Report, UCRL-50021-76, Lawrence Livermore Laboratory, Livermore, Calif., 1976.
23. **Julienne, P. S., Krauss, M., and Stevens, W.,** Collision-induced O^1D^2 - ^1S_o emission near 5577 Å in argon, *Chem. Phys. Lett.,* 38, 374, 1976.
24. **Krauss, M. and Mies, F. H.,** Electronic structure and radiative transitions of excimer systems, in *Excimer Lasers,* Rhodes, C. K., Ed., Springer-Verlag, New York, 1979.
25. **Hughes, W. M., Olson, N. T., and Hunter, R.,** Experiments on the 558 nm argon oxide laser system, *Appl. Phys. Lett.,* 28, 81, 1976.
26. **Cunningham, D. L. and Clark, K. C.,** Rates of collision-induced emission from metastable $O(^1S)$ atoms, *J. Chem. Phys.,* 61, 1118, 1974.
27. **Wellegehausen, B., Friede, D., and Steger, G.,** Optically pumped continuous Bi_2 and Te_2 lasers, *Opt. Commun.,* 26, 391, 1978.
28. **West, W. P. and Broida, H. P.,** Optically pumped vapor phase Bi_2 laser, *Chem. Phys. Lett.,* 56, 283, 1978.
29. **Murray, J. R., Swingle, J. C., and Turner, C. E., Jr.,** Laser oscillation of the 292 nm band system of Br_2, *Appl. Phys. Lett.,* 28, 530, 1976.
30. **Ewing, J. J., Jacob, J. H., Mangano, J. A., and Brown, H. A.,** Discharge pumping of the Br_2^* laser, *Appl. Phys. Lett.,* 28, 656, 1976.
31. **Hunter, R. O.,** ARPA Review Meeting, Stanford Research Institute, Menlo Park, Calif., 1975 (unpublished).
32. **Veukateswarlu, P. and Verma, R. D.,** Emission spectrum of bromine excited in the presence of argon — Part I, *Proc. Indian Acad. Sci.,* 46, 251, 1957.
33. **Hays, A. K.,** Cl_2 laser emitting 96 mJ at 258 nm, seen promising for iodine-laser pump, *Laser Focus,* 14, 28, 1978.
34. **Diegelmann, M., Hohla, K., and Kompa, K. L.,** Interhalogen UV laser on the 285 nm line band of ClF*, *Opt. Commun.,* 29, 334, 1979.
35. **Baboshii, V. N., Dobychin, S. L., Zuev, V. S., Mikheev, L. D., Pavlov, A. B., Startsev, A. V., and Fokanov, V. P.,** Laser utilizing an electronic transition in CN radicals pumped by radiation from an open high current discharge, *Sov. J. Quantum Electron.,* 7, 1183, 1977.
36. **West, G. A. and Berry, M. J.,** CN photodissociation and predissociation chemical lasers: molecular electronic and vibrational laser emissions, *J. Chem. Phys.,* 61, 4700, 1974.
37. **Mathias, L. E. S. and Crocker, A.,** Visible laser oscillations from carbon monoxide, *Phys. Lett.,* 7, 194, 1963.
38. **Hodgson, R. T.,** Vacuum-ultraviolet lasing observed in CO: 1800-2000 Å, *J. Chem. Phys.,* 55, 5378, 1971.
39. **Waller, R. A., Collins, C. B., and Cunningham, A. J.,** Stimulated emission from CO^+ pumped by charge transfer from He_2^+ in the afterglow of an e-beam discharge, *Appl. Phys. Lett.,* 27, 323, 1975.
40. **Rice, J. K., Hays, A. K., and Woodworth, J. R.,** Vacuum-UV emissions from mixtures of F_2 and the noble gases — a molecular F_2 laser at 1575 Å, *Appl. Phys. Lett.,* 31, 31, 1977.

41. **Pummer, H., Hohla, H., Diegelmann, M., and Reilly, J. P.,** Discharge pumped F_2 laser at 1580 Å, *Opt. Commun.*, 28, 104, 1979.
42. **Woodworth, J. K. and Rice, J. K.,** An efficient high-power F_2 laser near 157 nm, *J. Chem. Phys.*, 69, 2500, 1978.
43. **Dreyfus, R. W.,** Molecular hydrogen laser: 1098-1613 Å, *Phys. Rev. A*, 9, 2635, 1974.
44. **Knyazev, I. N., Letokhov, V. S., and Movshev, V. G.,** Efficient and practical hydrogen vacuum ultraviolet laser, *IEEE J. Quantum Electron.*, QE-11, 805, 1975.
45. **Waynant, R. W., Ali, A. W., and Julienne, P. S.,** Experimental observations and calculated bond strengths for the D_2 Lyman band laser, *J. Appl. Phys.*, 42, 3406, 1971.
46. **Waynant, R. W., Shipman, J. D., Jr., Elton, R. C., and Ali, A. W.,** VUV laser emission from molecular hydrogen, *Appl. Phys. Lett.*, 17, 383, 1970.
47. **Hodgson, R. T.,** VUV laser action observed in the Lyman bands of molecular hydrogen, *Phys. Rev. Lett.*, 25, 494, 1970.
48. **Bockasten, K., Lundholm, T., and Andrede, D.,** Laser lines in atomic and molecular hydrogen, *J. Opt. Soc. Am.*, 56, 1260, 1966.
49. **Bazhulin, P. A., Knyazev, I. N., and Petrash, G. G.,** Pulsed laser action in molecular hydrogen, *Sov. Phys. JETP*, 20, 1068, 1965.
50. **Bazhulin, P. A., Knyazev, I. N., and Petrash, G. G.,** Stimulated emission from hydrogen and deuterium in the infrared, *Sov. Phys. JETP*, 22, 11, 1966.
51. **Whitney, W. T.,** Sustained discharge excitation of HgCl and HgBr $B^2\Sigma^+_{1/2} \rightarrow X^2\Sigma^+_{1/2}$ lasers, *Appl. Phys. Lett.*, 32, 239, 1978.
52. **Parks, J. H.,** Laser action on the $B^2\Sigma^+{}_{1/2} \rightarrow X^2\Sigma^+{}_{1/2}$ band of HgBr at 5018 Å, *Appl. Phys. Lett.*, 31, 297, 1977.
53. **Schimitschek, E. J., Celto, J. E., and Trias, J. A.,** Mercuric bromide photodissociation laser, *Appl. Phys. Lett.*, 31, 608, 1977.
54. **Burnham, R.,** Discharge pumped mercuric halide dissociation lasers, *Appl. Phys. Lett.*, 33, 156, 1978.
55. **Maya, J.,** Ultraviolet absorption cross-sections of HgI_2, $HgBr_2$, and tin (II) halide vapors, *J. Chem. Phys.*, 67, 4976, 1977.
56. **Parks, J. H.,** Laser action on the $B^2\Sigma^+_{1/2} \rightarrow X^2\Sigma^+_{1/2}$ band of HgCl at 5576 Å, *Appl. Phys. Lett.*, 31, 192, 1977.
57. **Tang, K. Y., Hunter, R. O., Oldenettel, J., Howton, C., Huestis, D., Eckstrom, D., Perry, B., and McCusker, M.,** Electron-beam controlled HgCl* laser, *Appl. Phys. Lett.*, 32, 226, 1978.
58. **Eden, J. G.,** VUV-pumped HgCl laser, *Appl. Phys. Lett.*, 33, 495, 1978.
59. **Eden, J. G.,** Green HgCl ($B^2\Sigma^+ \rightarrow X^2\Sigma^+$) laser, *Appl. Phys. Lett.*, 31, 448, 1977.
60. **Bradford, R. S., Jr., Ault, E. R., and Bhaumik, M. L.,** High-power I_2 laser in the 342 nm band system, *Appl. Phys. Lett.*, 27, 546, 1975.
61. **Hays, A. K., Hoffman, J. M., and Tisone, G. C.,** Molecular iodine laser, *Chem. Phys. Lett.*, 39, 353, 1976.
62. **Mikheev, L. D., Shirokikh, A. P., Startsev, A. V., and Zuev, V. S.,** Optically pumped molecular iodine laser on the 342 nm band, *Opt. Commun.*, 26, 237, 1978.
63. **Byer, R. L., Herbst, R. L., and Kildal, H.,** Optically pumped molecular iodine vapor-phase laser, *Appl. Phys. Lett.*, 20, 463, 1972.
64. **Koffend, J. B. and Field, R. W.,** CW optically pumped molecular iodine laser, *J. Appl. Phys.*, 48, 4468, 1977.
65. **Wellegehausen, B., Stephan, K. H., Friede, D., and Welling, H.,** Optically pumped continuous I_2 molecular laser, *Opt. Commun.*, 23, 157, 1977.
66. **Hartmann, B., Kleman, B., and Steinvall, O.,** Quasi-tunable I_2-laser for absorption measurements in the near infrared, *Opt. Commun.*, 21, 33, 1977.
67. **Hanko, L., Benard, D. J., and Davis, S. J.,** Observation of super-fluorescent emission of the B-X system in I_2, *Opt. Commun.*, 30, 63, 1979.
68. **Murray, J. R. and Powell, H. T.,** KrCl laser oscillation at 222 nm, *Appl. Phys. Lett.*, 29, 252, 1976.
69. **Eden, J. G. and Searles, S. K.,** Observation of stimulated emission in KrCl, *Appl. Phys. Lett.*, 29, 350, 1976.
70. **Hay, P. J. and Dunning, T. H., Jr.,** The electronic states of KrF, *J. Chem. Phys.*, 66, 1306, 1977.
71. **Tisone, G. C., Hays, A. K., and Hoffman, J. M.,** 100 mW, 248.8 nm KrF laser excited by an electron beam, *Opt. Commun.*, 15, 188, 1975.
72. **Ewing, J. J. and Brau, C. A.,** Laser action on the ${}^2\Sigma^+{}_{1/2} \rightarrow {}^2\Sigma^+{}_{1/2}$ bands of KrF and XeCl, *Appl. Phys. Lett.*, 27, 350, 1975.
73. **Mangano, J. A. and Jacob, J. H.,** Electron-beam-controlled discharge pumping of the KrF laser, *Appl. Phys. Lett.*, 27, 495, 1975.
74. **Jacob, J. H., Hsia, J. C., Mangano, J. A., and Rokni, M.,** Pulse shape and laser-energy extraction from e-beam-pumped KrF, *J. Appl. Phys.*, 50, 5130, 1979.

75. **Fahlen, T. S.,** High-pulse-rate 10-W KrF laser, *J. Appl. Phys.*, 49, 455, 1978.
76. **Hawkins, R. T., Egger, H., Bokor, J., and Rhodes, C. K.,** A tunable, ultrahigh spectral brightness KrF* excimer laser source, *Appl. Phys. Lett.*, 36, 391, 1980.
77. **Smith, D., Dean, A. G., and Plumb, I. C.,** Three-body conversion reactions in pure rare gases, *J. Phys. B*, 5, 2134, 1972.
78. **Petit, A., Launay, F., and Rostas, J.,** Spectroscopic analysis of the transverse excited $C^3\Pi_u \rightarrow B^3\Pi_g$ (0,0) UV laser band of N_2 at room temperature, *Appl. Opt.*, 17, 3081, 1978.
79. **Massone, C. A., Garavaglia, M., Gallardo, M., Calatroni, J. A. E., and Tagliaferri, A. A.,** Investigation of a pulsed molecular nitrogen laser at low temperature, *Appl. Opt.*, 11, 1317, 1972.
80. **McFarlane, R. A.,** Observation of $a'^1\Pi$ - $^1\Sigma^-$ transition in the nitrogen molecule, *Phys. Rev.*, 140, 1070, 1965.
81. **McFarlane, R. A.,** unpublished work.
82. **McFarlane, R. A.,** Precision spectroscopy of new infrared emission systems of molecular nitrogen, *IEEE J. Quantum Electron.*, QE-2, 229, 1966.
83. **Ault, E. R., Bhaumik, M. L., and Olson, N. T.,** High-power Ar-N_2 transfer laser at 3577 Å, *IEEE J. Quantum Electron.*, QE-10, 624, 1974.
84. **Black, G., Sharpless, R. L., Slanger, T. G., and Lorents, D. C.,** Quantum yields for the production of $O(^1S)$, $N(^2D)$, and $N_2(A^3\Sigma^+_u)$ from VUV photolysis of N_2O, *J. Chem. Phys.*, 62, 4266, 1975.
85. **Searles, S. K. and Hart, G. A.,** Laser emission at 3577 and 3805 Å in electron-beam-pumped Ar-N_2 mixtures, *Appl. Phys. Lett.*, 25, 79, 1974.
86. **Ischenko, V. N., Lisitsyn, V. N., Razhev, A. M., Starinskii, V. N., and Chapovsky, P. L.,** The N^+_2 laser, *Opt. Commun.*, 13, 231, 1975.
87. **Basov, N. G. Vasil'ev, L. A., Danilychev, V. A., Dolgov-Saval'ev, G. G., Dolgikh, V. A., Kerimov, O. M., Kozorovitskii, L. L., Orlov, V. K., and Khodkevitch, D. D.,** High-pressure N^+_2 laser emitting violet radiation, *Sov. J. Quantum Electron.*, 5, 869, 1975.
88. **Rothe, D. E. and Tan, K. O.,** High-power N^+_2 laser pumped by charge transfer in a high-pressure pulsed glow discharge, *Appl. Phys. Lett.*, 30, 152, 1977.
89. **Collins, C. B., Cunningham, A. J., and Stockton, M.,** A nitrogen ion laser pumped by charge transfer, *Appl. Phys. Lett.*, 25, 344, 1974.
90. **Wellegehauser, B., Shahdin, S., Friede, D., and Welling, H.,** Continuous laser oscillation in alkali dimers, *IEEE J. Quantum Electron.*, QE-13, 65D, 1977.
91. **Henesian, M. A., Herbst, R. L., and Byer, R. L.,** Optically pumped superfluorescent Na_2 molecular laser, *J. Appl. Phys.*, 47, 1515, 1976.
92. **Itoh, H., Uchiki, H., and Matsuoka, M.,** Stimulated emission from molecular sodium, *Opt. Commun.*, 18, 271, 1976.
93. **Leone, S. R. and Kosnik, K. G.,** A tunable visible and ultraviolet laser on S_2, *Appl. Phys. Lett.*, 30, 346, 1977.
94. **Rosen, B.,** *Selected Constants — Spectroscopic Data Relative to Diatomic Molecules,* Pergamon Press, Oxford, 1970.
95. **Zuev, V. S., Mikheev, L. D., and Yalovi, V. I.,** Photochemical laser utilizing the $^1\Sigma^+_g$ - $^3\Sigma^-_g$ vibronic transition in S_2, *Sov. J. Quantum Electron.*, 5, 442, 1975.
96. **Searles, S. K. and Hart, G. A.,** Stimulated emission at 281.8 nm from XeBr, *Appl. Phys. Lett.*, 27, 243, 1975.
97. **Sze, R. C. and Scott, P. B.,** High-energy lasing of XeBr in an electric discharge, *Appl. Phys. Lett.*, 32, 479, 1978.
98. **Tellinghuisen, J., Hays, A. K., Hoffman, J. M., and Tisone, G. C.,** Spectroscopic studies of diatomic noble gas halides. II. Analysis of bound-free emission from XeBr, XeI, and KrF, *J. Chem. Phys.*, 65, 4473, 1976.
99. **Velazco, J. E. and Setser, D. W.,** Bound-free emission spectra of diatomic xenon halides, *J. Chem. Phys.*, 62, 1990, 1975.
100. **Champagne, L. F.,** Efficient operation of the electron-beam-pumped XeCl laser, *Appl. Phys. Lett.*, 33, 523, 1978.
101. **Burnham, R.,** Improved performance of the discharge-pumped XeCl laser, *Opt. Commun.*, 24, 161, 1978.
102. **Bichkov, Y. I., Gorbatenko, A. I., Mesyats, G. A., and Tarasenko, V. F.,** Effective XeCl-laser performance conditions with combined pumping, *Opt. Commun.*, 30, 224, 1979.
103. **Sur, A., Hui, A. K., and Tellinghuisen, J.,** Noble gas halides, *J. Mol. Spectrosc.*, 74, 465, 1979.
104. **Burnham, R., Powell, F. X., and Djeu, N.,** Efficient electric discharge lasers in XeF and KrF, *Appl. Phys. Lett.*, 29, 30, 1976.
105. **Ault, E. R., Bradford, R. S., Jr., and Bhaumik, M. L.,** High-power xenon fluoride laser, *Appl. Phys. Lett.*, 27, 413, 1975.

106. **Brau, C. A. and Ewing, J. J.,** 354-nm laser action on XeF, *Appl. Phys. Lett.*, 27, 435, 1975.
107. **Bischel, W. K., Nakano, H. H., Eckstrom, D. J., Hill, R. M., Huestis, D. L., and Lorents, D. C.,** A new blue-green excimer laser in XeF, *Appl. Phys. Lett.*, 34, 565, 1979.
108. **Ernst, W. E. and Tittel, F. K.,** A new electron-beam pumped XeF laser at 486 nm, *Appl. Phys. Lett.*, 35, 36, 1979.
109. **Tellinghuisen, J., Tellinghuisen, P. C., Tisone, G. C., Hoffman, J. M., and Hays, A. K.,** Spectroscopic studies of diatomic noble gas halides. III. Analysis of XeF 3500 Å band system, *J. Chem. Phys.*, 68, 5177, 1978.
110. **Basov, N. G., Babeiko, Y. A., Zuev, V. S., Mikheev, L. D., Orlov, V. K., Pogorel'skii, I. V., Staurovskii, D. B., Startsev, A. V., and Yalovoi, V. I.,** Laser emission from the XeO molecule under optical pumping conditions, *Sov. J. Quantum Electron.*, 6, 505, 1976.
111. **Basov, N. G., Danilychev, V. A., and Popov, Y. M.,** Stimulated emission in the vacuum-ultraviolet region, *Sov. J. Quantum Electron.*, 1, 18, 1971.
112. **Turner, C. E., Jr.,** Near-atmospheric-pressure xenon excimer laser, *Appl. Phys. Lett.*, 31, 659, 1977.
113. **Hughes, W. M., Shannon, J., Kolb, A., Ault, E., and Bhaumik, M.,** High-power ultraviolet laser radiation from molecular xenon, *Appl. Phys. Lett.*, 23, 385, 1973.
114. **Hoff, P. W., Swingle, J. C., and Rhodes, C. K.,** Demonstration of temporal coherence, spatial coherence, and threshold effects in the molecular xenon laser, *Opt. Commun.*, 8, 128, 1973.
115. **Bradley, D. J., Hull, D. R., Hutchinson, M. H. R., and McGeoch, M. W.,** Co-axially pumped, narrow band, continuously tunable, high power VUV xenon laser, *Opt. Commun.*, 14, 1, 1975.
116. **Koehler, H. A., Ferderber, L. J., Redhead, D. L., and Ebert, P. J.,** Stimulated VUV emission in high-pressure xenon excited by relativistic electron beams, *Appl. Phys. Lett.*, 21, 198, 1972.
117. **Tittel, F. K.,** private communication.
118. **Tittel, F. K., Wilson, W. L., Stickel, R. E., Marowsky, G., and Ernst, W. E.,** A triatomic Xe_2Cl excimer laser in the visible, *Appl. Phys. Lett.*, 36, 405, 1980.
119. **Tang, K. Y., Lorents, D. C., and Huestis, D. L.,** Gain measurements on the triatomic excimer Xe_2Cl, *Appl. Phys. Lett.*, 36, 347, 1980.
120. **Jacobs, R. R. and Krupke, W. F.,** Optical gain at 1.06 μm in the neodymium chloride-aluminum chloride vapor complex, *Appl. Phys. Lett.*, 32, 31, 1978.
121. **Jacobs, R. R. and Krupke, W. F.,** Excited state kinetics for $Nd(thd)_3$ and $Tb(thd)_3$ chelate vapors and prospects as fusion laser media, *Appl. Phys. Lett.*, 34, 497, 1979.
122. **Jacobs, R. R. and Krupke, W. F.,** Kinetics and fusion laser potential for the terbium aluminum chloride vapor complex, *Appl. Phys. Lett.*, 35, 126, 1979.
123. **Jacobs, R. R., Weber, M. J., and Pearson, R. K.,** Nonradiative intramolecular deactivation of Tb^{+3} fluorescence in a vapor phase terbium (III) complex, *Chem. Phys. Lett.*, 34, 80, 1975.
124. **Ginter, M. L. and Battino, R.,** Potential-energy curves for the He_2 molecule, *J. Chem. Phys.*, 52, 4469, 1970.
125. **Rhodes, C. K., Ed.,** *Excimer Lasers*, Springer-Verlag, New York, 1979.
126. **Rosen, B., Ed.,** *Données Spectroscopiques Relatives aux Molécules Diatomiques*, Pergamon Press, New York, 1970.
127. **Herzberg, G.,** *Spectra of Diatomic Molecules*, D. van Nostrand, Princeton, 1950.
128. **Marowsky, G.,** *IEEE J. Quantum Electron.*, QE-16, 49, 1980.

3.2 VIBRATIONAL TRANSITION LASERS

T. Y. Chang

DIATOMIC MOLECULES

The CO molecule and all the H halide molecules have a singlet ground electronic state $X'\Sigma^+$. This state contains a nearly harmonic ladder of vibrational states which are designated by the vibrational quantum number v (= 0, 1, 2 . . .). Each vibrational state is further subdivided into rotational levels which are designated by the rotational quantum number J (= 0, 1, 2, . . .). The energies of these levels can be expressed as a double power series

$$E(v,J) = \sum_{m,n} Y_{mn} \left(v + \frac{1}{2}\right)^m [J(J+1)]^n$$

The interdependence among the coefficients Y_{mn} and their isotope shifts have been derived theoretically by Dunham.[141] For several isotopic CO molecules, the values of 15 leading Dunham coefficients are now quite accurately known.[23]

The principal radiative transitions among the vibrational states are of the type v + 1 ↔ v. The overtone transitions (v + Δv ↔ v, where Δv = 2, 3, . . .) are also weakly allowed, but their strength decreases rapidly with the increasing value of Δv. Each vibrational transition actually consists of a large number of vibrational-rotational transitions of types (v + 1, J + 1) ↔ (v,J) and (v + 1, J − 1) ↔ (v,J). The former is known as the R(J) transition, and the latter is known as the P(J) transition. These transitions are illustrated by means of an energy-level diagram in Figure 3.2.1.

The molecules CN, NO, and OH have a doublet ground electronic state of one type or the other due to an unpaired electronic spin. In these molecules, there is a varying degree of coupling among the components of angular momenta associated with electronic spin, electronic orbit, and nuclear rotation. For the details of their energy levels and selection rules, the reader should consult a standard reference, e.g., Herzberg.[4] Suffice it to say that in these molecules J takes on half integer values and Q(J) transitions, i.e., (v + 1, J) ↔ (v, J), are also allowed.

Vibrationally excited molecules in a diatomic laser are often generated in an exothermic chemical reaction[142] which is initiated either by light (photolysis), electronic impact (glow discharge), or by rapid mixing of reactant gases. In these reactions, the molecules are formed predominantly in the high v states, resulting in an inversion of population with respect to the lower v states. The population inversion is sustained by continuous replenishment of the reactants and rapid removal of the reaction products after their radiative decay.

The CO and NO lasers are also excited by thermal heating or electronic impact in a discharge followed by preferential cooling of the lower v states by rapid expansion[20] or with the help of a cold tube wall.[143] The faster relaxation of lower v states arises from closer resonance of lower v + 1 ↔ v transitions with the v = 0 ↔ 1 transition of the background molecules.[144] The time evolution of population distribution among the vibrational states after pulse heating is illustrated in Figure 3.2.2. The distribution of population in a CW CO (0.07 torr)-O_2 (0.03 torr)-He (7.3 torr) laser discharge is shown in Figure 3.2.3. Notice that in the latter case no real inversion of population exists between any pair of vibrational states. Yet, the laser exhibited P-branch gain on bands v = 7 → 6 through v = 16 → 15. The explanation for this apparent paradox

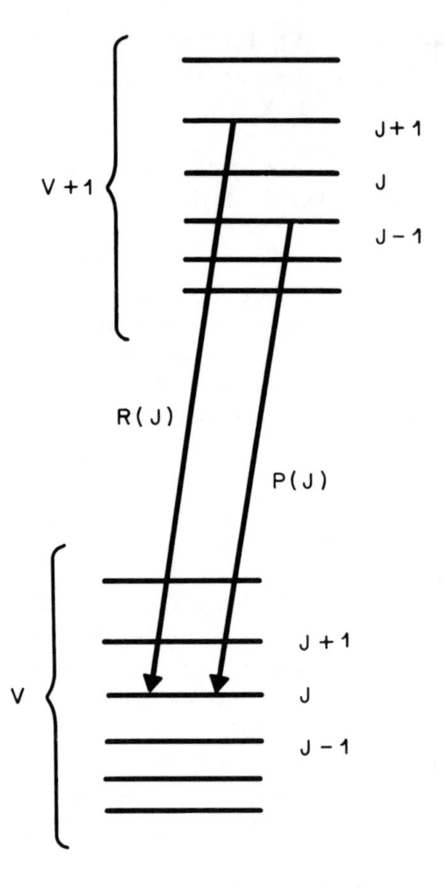

FIGURE 3.2.1. Partial energy-level diagram for a diatomic molecule showing P(J) and R(J) transitions.

lies in the concept of partial inversion. Consider a case in which two vibrational states, v + 1 and v, happen to have the same total population. Since rotational relaxation is always much faster than vibrational relaxation, the populations in both states will be distributed among the respective rotational levels according to the same Boltzmann distribution with the same rotational temperature. This means the Boltzmann factor for the J + 1 level in either vibrational state will be smaller than that for the J level in either vibrational state. Consequently a P(J) transition would exhibit gain while an R(J) transition would exhibit absorption. A P(J) transition could exhibit gain even when the total population of the v + 1 state is slightly less than that of the v state. A calculated example is shown in Figure 3.2.4.[14]

Another important feature of diatomic molecular lasers is cascading of laser action. The change of population due to laser action on the v + 1 → v band directly affects the gain or absorption on the v + 2 → v + 1 and the v → v − 1 bands. As a result, output pulses due to successive bands on the cascade often appear successively in time, as is illustrated in Figure 3.2.5.[14]

LINEAR TRIATOMIC AND FOUR-ATOMIC MOLECULES

CO_2, COS, CS_2, HCN, and N_2O are all linear triatomic molecules. They all have two stretching vibrational models, ν_1 and ν_3, and a bending mode, ν_2, which can give rise to a nonzero vibrational angular momentum. An excited vibrational state can be

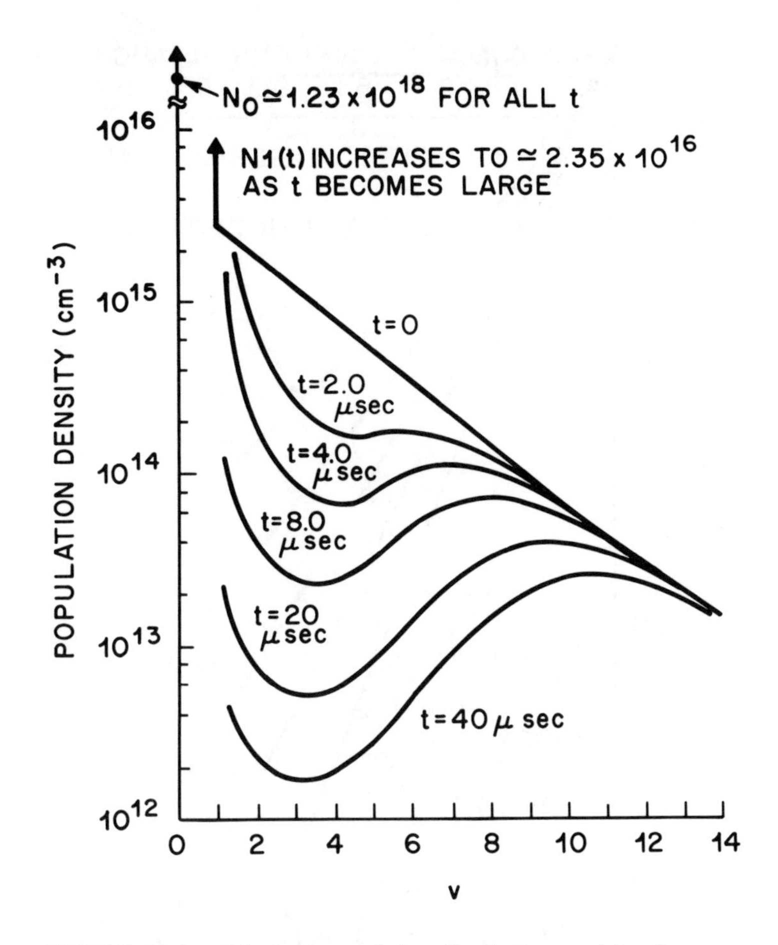

FIGURE 3.2.2. Calculated population distributions evolving from a Boltzmann distribution at t = 0 in CO. Initial distribution: $N_0 = 1.24 \times 10^{18}$ cm^{-3}, with 1% of N_0 excited to a vibrational temperature of 7000 K. (From Jeffers, W. Q. and Wiswall, C. E., *J. Appl. Phys.*, 42, 5059, 1971. With permission.)

specified by four vibrational quantum numbers in the form (v_1, v_2^ℓ, v_3), where ℓ specifies the vibrational angular momentum.

For CO_2, which is the most important laser molecule in this group, states with even values of v_2, v_3, and $\ell = 0$ are symmetric, while those with an odd value of v_3, an even value of v_2, and $\ell = 0$ are antisymmetric. For CO_2^{16} and CO_2^{18}, where the spins of the identical oxygen nuclei are zero, all the odd J levels are missing in the symmetric states and all the even J levels are missing in the antisymmetric states. For further details, the reader is referred to Herzberg.[3] The (1,0°,0) state and the (0,2°,0) state are mixed and shifted due to a strong resonant interaction (Fermi resonance). It has been found that the conventional designations for these two states (as used in Herzberg[3] and in the present table) are theoretically unsatisfactory.[147] Therefore, a different set of notations, $[10°0, 02°0]_I$ for (10°0) and $[10°0, 02°0]_{II}$ for (02°0), is often used. This type of nomenclature also applies to other sets of states involving ($v_1 + 1$, v_2^ℓ, v_3) and (v_1, $[v_2 + 2]_\ell$, v_3), and in some cases also ($v_1 - 1$, $[v_2 + 4]_\ell$, v_3).

A partial energy-level diagram for CO_2 is shown in Figure 3.2.6. Laser transitions are indicated by solid arrows along with their band-center wavelengths. Excitation energy transfer and relaxation processes are indicated by dotted arrows. By far the most important laser band is the (00°1) → (10°0) band centered at 10.4 μm. For CW

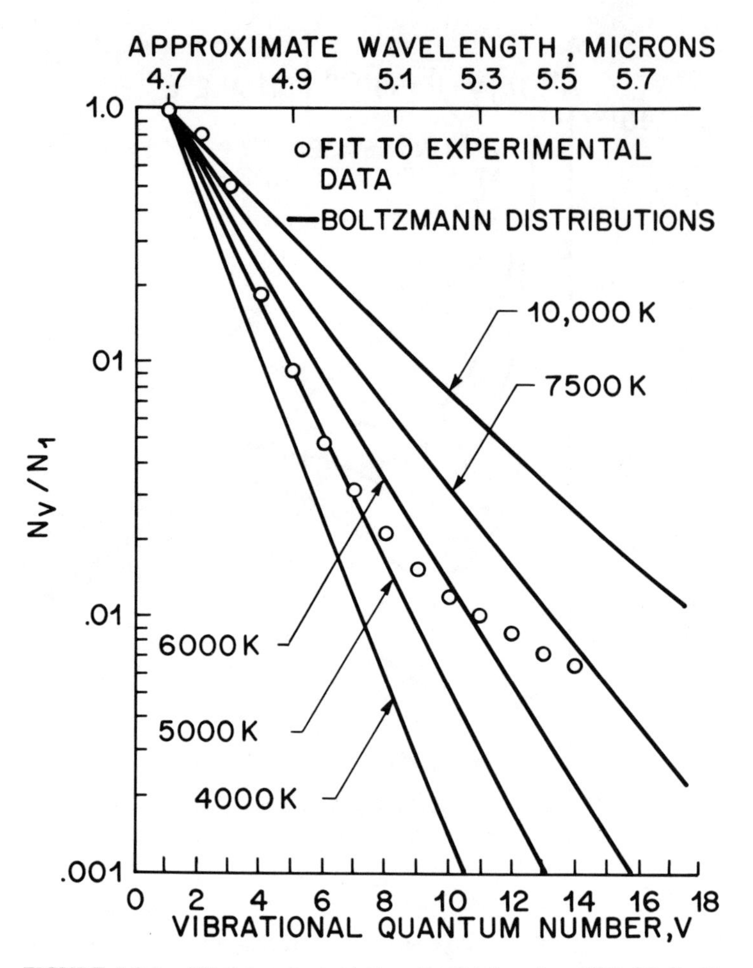

FIGURE 3.2.3. Vibrational population distributions in a CW CO-O_2-He discharge. (From Rich, J. W. and Thompson, H. M., *Appl. Phys. Lett.*, 19, 3, 1971. With permission.)

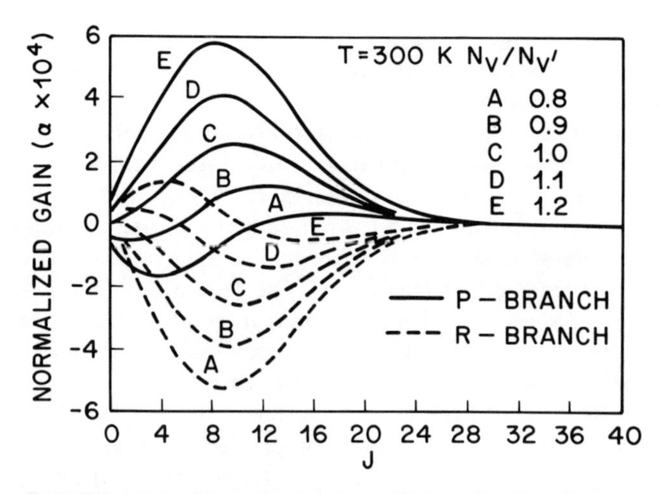

FIGURE 3.2.4. Normalized gain coefficients for 7 → 6 vibrational-rotational band of CO, plotted as a function of upper level J for T = 300 K and N_v/N_v' = 0.8, 0.9, 1.0, 1.1, and 1.2, where N_v and N_v' are the population densities for the upper- and lower-vibrational states, respectively. (From Patel, C. K. N., *Phys. Rev.*, 141, 71, 1966. With permission.)

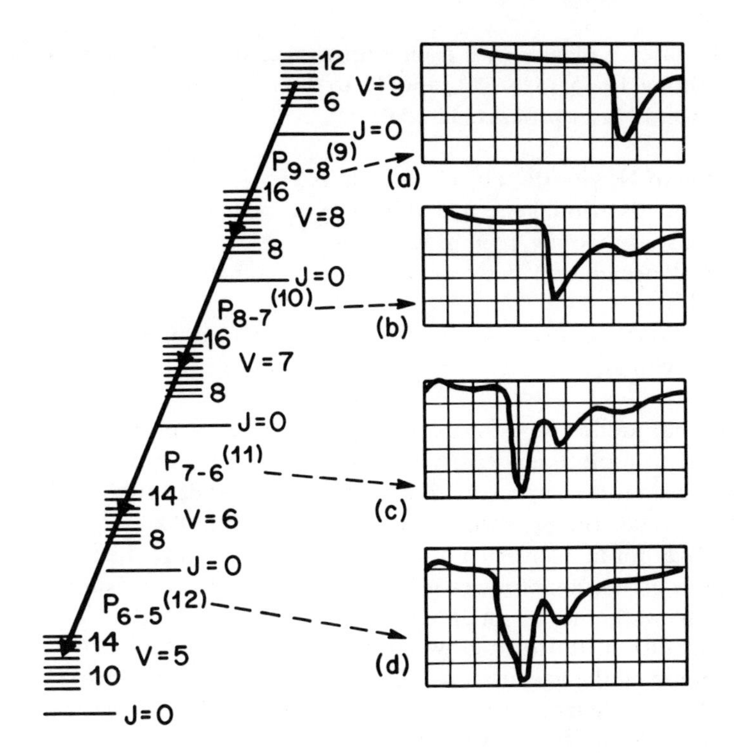

FIGURE 3.2.5. Laser pulses showing cascade nature of CO laser transitions. Energy-level diagram (not to scale) is shown on the left. In each of the oscilloscope traces, the current pulse occurs at the left-hand edge. (From Patel, C. K. N., *Phys. Rev.*, 141, 71, 1966. With permission.)

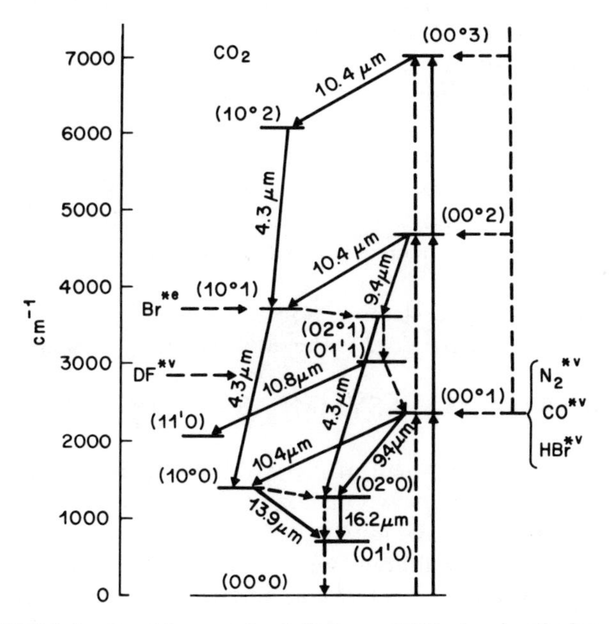

FIGURE 3.2.6. Partial energy-level diagram of CO_2 showing the laser transitions (solid downward arrows) and excitation, energy-transfer, relaxation pathways (dotted arrows). $N_2{}^{*v}$ denotes vibrationally excited nitrogen molecule, and Br^{*e} denotes electronically excited Br atom. Solid upward arrows indicate direct optical pumping.

operation, the laser typically uses a glow discharge in 6 to 15 torr of CO_2 - N_2 - He mixture containing 10 to 15% CO_2 and 10 to 20% N_2. The principal mechanisms involved in the overall laser process include[148,149]

1. Excitation of N_2 vibration by electron impact
2. Transfer of vibrational energy from N_2 to the nearly resonant ν_3 mode of CO_2
3. Laser transition from ν_3 to ν_1 mode
4. Sharing of population between ν_1 and $2\nu_2^l$ modes and relaxation within the ν_2 manifold
5. The vibrational energy in the ν_2 manifold converted into translational energy by collisions with He

The vibrational modes of CO_2 are also excited by electron impact. By maintaining an electric field to molecular density ratio (E/N) of 2 to 3 $\times$ 10^{-16} V cm^2 in the glow discharge, the excitation can be maximized for the ν_3 mode relative to other modes.[150] However, even under the optimum condition, more than 80% of power dissipated in the discharge will not be converted into IR output. Therefore, the attainment of a high volumetric laser efficiency requires efficient removal of this waste heat by a fast gas flow,[67] cold tube wall, and/or small tube diameter.[151]

In pulsed operation, uniform glow discharge with a suitable E/N value can be obtained by preionizing the gas mixture with a broad-area electron beam or UV radiation from an auxiliary discharge.[152] By these techniques, the CO_2-N_2-He laser can be operated at pressures from 10^2 to over 10^4 torr. Since approximately the same degree of population inversion can be achieved over the entire pressure range, the output energy per unit volume per pulse increases approximately linearly with pressure.[152] The width of each laser line also increases linearly with pressure. Above 4 atm, the laser lines begin to show significant spectral overlap. By 20 atm all the lines in each laser branch have merged into a nearly continuous gain band.[153] The width of the main output pulse is, on the other hand, approximately inversely proportional to the pressure. As a result, the peak output power increases quadratically with pressure.[152]

The CO_2 laser can also be excited by energy transfer from chemically excited DF,[142] by energy transfer from photodissociated Br or optically pumped CO, HBr, DF, or by direct optical pumping.[154] An optically pumped CO_2-N_2O energy transfer laser has been operated at pressures up to 42 atm in a 2-mm cavity, giving a continuous tuning range for a single longitudinal mode of approximately 100 GHz.[154] It should be noted that many vibrational transition lasers cannot be operated at multiatmosphere pressures due to the presence of a ground state absorption band that closely overlaps the laser band.

C_2H_2 and C_2D_2 are linear four-atomic molecules. Both have five different modes of vibration. Otherwise, the spectroscopy is the same as linear three-atomic molecules.

SYMMETRIC AND ASYMMETRIC TOP MOLECULES

For a linear molecule, the moment of inertia about the molecular axis is essentially zero, while those about the other two orthogonal directions are finite and equal. For a symmetric top molecule, the axial moment of inertia can be smaller (prolate) or larger (oblate) than the other two equal moments of inertia. Due to this fact, it requires two quantum numbers to specify the rotational angular momentum of a symmetric top molecule. These quantum numbers are denoted by J, which is related to quantization of the total angular momentum, and K, which is related to quantization of the axial component of the angular momentum.

There are two types of vibrational transitions in symmetric top molecules: one in

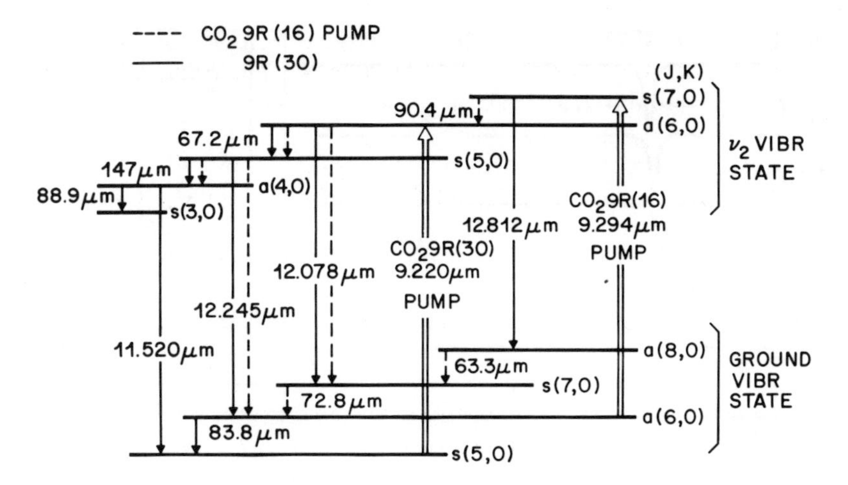

FIGURE 3.2.7. Relevant energy levels of $N^{14}H_3$ showing optical pumping by 9.220- R(30) and 9.294-μm R(16) CO_2 laser lines. The solid and broken lines indicate observed IR and FIR laser emissions for the R(30) and R(16) pumping cases, respectively. (From Yoshida, T., Yamabayashi, N., Miyazaki, K., and Fujisawa, K., *Opt. Commun.*, 26, 410, 1978. With permission.)

which the transition dipole moment is parallel to the molecular axis and the other in which the dipole moment is perpendicular to the molecular axis. Only the first type (parallel band) is encountered in this section. Radiative transitions of this type obey the selection rules $\Delta K = 0$, $\Delta J = \pm 1$, or 0 (except when $K = 0$). This means that for each value of K, the vibrational-rotational transitions form P, Q, and R branches like in a linear molecule (except that each branch starts with $J = K$). P(J,K) lines of the same J value are usually spaced much closer together than those of different J values.

The symmetric-top molecules encountered in this section include BCl_3, NH_3, CF_3I, and CH_3F. For NH_3, there is an added complication due to the constant tunneling of the N atom through the H_3 triangle. Each rotational level is split into a symmetric level and a slightly higher antisymmetric level (the inversion splitting). Radiative transitions occur only between levels of opposite symmetry. A P-branch transition whose lower level is antisymmetric is denoted by aP(J,K).

Most vibrational-transition lasers in symmetric-top molecules are excited by optical pumping which is much more specific and selective than electron impact. The energy-level diagram for two optically pumped NH_3 lasers along with some of their associated cascade transitions is shown in Figure 3.2.7. The spectral line shapes of cascade transitions are expected to be similar to those of discharge excited laser lines. The spectral properties of directly optically pumped laser transitions, such as the 12.8121 and the 12.078-μm lines in Figure 3.2.7, on the other hand, are strongly modified by the dynamic Stark effect.[155] If the frequency of the pump laser is exactly on resonance with the molecular transitions being pumped, then the resulting laser line will usually exhibit a double peaked spectral profile. In most cases, the pump frequency is several line widths off resonance. The spectral profile for the resulting laser line is illustrated in Figure 3.2.8. Gain appears on a Raman-like component which is shifted from the original spectral line. The latter remains as a somewhat weakened absorption line. The two components shift away from each other with increasing pump intensity due to the dynamic Stark effect. There are two interesting consequences of this effect:

1. There is no self absorption in the gas due to the shift of the gain peak.
2. Laser gain can be obtained without inverting the population.

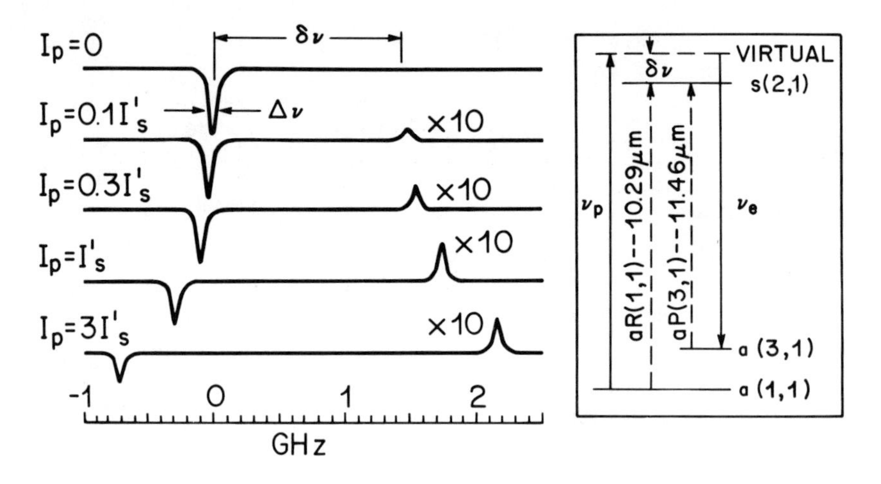

FIGURE 3.2.8. Calculated line profiles for the 11.46-μm line in NH_3 with pumping at 10.29 μm. The pump frequency is assumed to be off resonant by $\delta\nu = 1440$ MHz. The Lorentz full width is $\Delta\nu = 59.14$ MHz (at 2 torr) and Doppler broadening is ignored. $I'_s = (4/3)I_s$, where I_s is the saturation pump intensity for the given $\delta\nu$ as calculated from rate equations. The relevant energy levels are shown in the inset. (From Chang, T. Y. and McGee, J. D., *Appl. Phys. Lett.*, 29, 725, 1976. With permission.)

When the moments of inertia about all three orthogonal directions are equal, we have a spherical top, a special case of symmetric tops. CF_4, SiF_4, SiH_4, and SF_6 are spherical top molecules. An illustrative energy-level diagram for CF_4 with allowed radiative transitions is shown in Figure 3.2.9. The upper laser level in this example is split into three Coriolis sublevels due to the Coriolis interactions among three degenerate vibrations.

The moments of inertia about three orthogonal molecular axes are all different in H_2O, COF_2, HCOOH, and C_2H_4. These are asymmetric top molecules. The rotational levels of an asymmetric top molecule are labeled by the notation (J,τ) where τ takes integer values from $-J$ to $+J$. The selection rules for radiative transition include the basic rule:

$$\Delta J = \pm 1, \text{ or } 0 \text{ (except when } J = 0)$$

Further details are beyond the scope of this résumé. The reader may consult Herzberg's book[3] for further details on the spectroscopy of polyatomic molecules.

INTRODUCTION TO THE TABLES

The tables in this section cover all known laser lines between 1.8 and 33 μm that are based on the "usual" vibrational-rotational transitions in molecular gases. The spectral coverage overlaps partly with that of the following section on far IR lasers. Some of the far-IR transitions not covered in this section include transitions between resonantly mixed vibrational states (as in H_2O and HCN), inversion transitions (as in NH_3), transitions associated with torsional vibrations and internal rotations, and transitions due to ring puckering modes. These transitions are distinctly different from pure rotational transitions, but occur in the same spectral region as the latter. Consequently, the far-IR laser section is considered to be the proper section for their inclusion.

The tables in this section and their contents are organized according to the following ordering scheme:

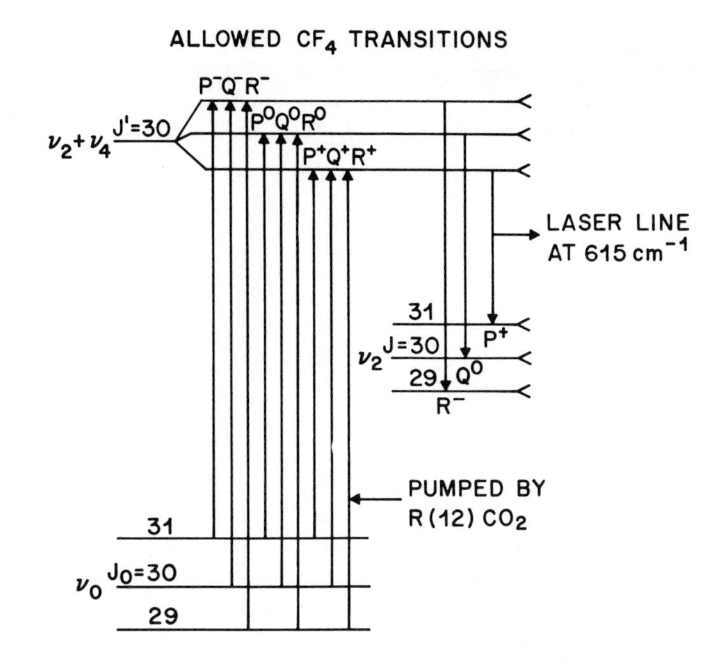

FIGURE 3.2.9. Typical energy-level diagram for the CF_4 laser with the total angular momentum quantum numbers in the vibrational ground state and in $\nu_2 + \nu_4$ and ν_2 denoted by J_0, J', and J, respectively. The arrows indicate allowed transitions to the three Coriolis sublevels of $J' = 30$ of $\nu_2 + \nu_4$ and allowed laser transition down to ν_2. R(12) of $C^{12}O^{16}_2$ pumps the transition $R^+(29)$, populating the + level of $J' = 31$ of ν_2, producing the strong laser line P(31) at 615 cm^{-1}. (From McDowell, R. S., Patterson, C. W., Jones, C. R., Buchwald, M. I., and Telle, J. M., *Opt. Lett.*, 4, 274, 1979. With permission.)

a. The increasing number of atoms in the laser molecule
b. The alphabetical order of common chemical formulas (with D treated as H^2)
c. The increasing isotopic mass
d. The increasing band-center wavelength
e. The increasing lower vibrational state energy
f. The increasing transition wavelength within a given vibrational rotational band

A further explanation of the columns in the tables is given below.

Wavelength and Frequency

The estimated uncertainty of any listed value is always less than 20 in the last two digits, and in most cases it is less than 10. When values of comparable accuracy are available from several different sources, the value used is chosen according to the following priority order, namely,

1. The direct laser measurement
2. Other spectroscopic measurements
3. The calculated value

Calculated values are used wherever highly accurate molecular constants from extensive polynomial fitting are available. Where accurate frequencies are known in units of hertz, the values are converted to centimeter^{-1} unit by using the value of $c = 2.99792458 \times 10^{10}$ cm/sec.

All wavelengths given are vacuum values. To obtain wavelengths in dry air, the reader may consult the monograph by Coleman et al.[1] Wavelength values are not given beyond 10^{-5} μm. In the rare event that more accuracy is needed for a truncated wavelength value, the full accuracy can be recovered by a simple inversion of the frequency value given. A wavelength value given inside parentheses indicates a group of unresolved laser lines.

Transition

All vibrational transitions encountered in this section occur within the ground electronic state of the laser molecules. The spectroscopic notations used conform to the convention used in Herzberg's books.[2,3] Brief explanations are to be found at the end of this section and in some footnotes.

Footnotes

Selected references, the experimental conditions used for observation, and the peak output power obtained are given in footnotes. The number of footnotes for a given line roughly reflects the relative importance of that line within a given band. For some laser lines that are optically pumped by another laser line, additional information is given in this column to identify the transition of the pump laser.

A short discussion of basic principles and operating characteristics of vibrational transition lasers is given at the end of this section.

Table 3.2.1
DIATOMIC VIBRATIONAL TRANSITION LASERS

Wavelength (μm)vac	Frequency (cm^{-1})	Transition P(N)[a]	Remarks
3.2.1.1 CN Laser			
		v = 2 → 1 Band	b
5.0372	1985.22	P(8)	
5.0474	1981.20	P(9)	
5.0578	1977.15	P(10)	
5.0683	1973.06	P(11)	
5.0789	1968.94	P(12)	
5.0896	1964.79	P(13)	
		v = 3 → 2 Band	b
5.0838	1967.05	P(6)	
5.0939	1963.13	P(7)	
5.1042	1959.18	P(8)	
5.1146	1955.19	P(9)	
5.1251	1951.17	P(10)	
5.1358	1947.12	P(11)	
5.1466	1943.03	P(12)	
5.1575	1938.91	P(13)	
5.1686	1934.75	P(14)	
		v = 4 → 3 Band	b
5.1422	1944.69	P(5)	
5.1524	1940.84	P(6)	
5.1627	1936.96	P(7)	
5.1732	1933.04	P(8)	
5.1838	1929.08	P(9)	
5.1945	1925.10	P(10)	
5.2054	1921.08	P(11)	
5.2164	1917.02	P(12)	
5.2276	1912.94	P(13)	
5.2389	1908.81	P(14)	
5.2503	1904.66	P(15)	
		v = 5 → 4 Band	b
5.2337	1910.71	P(7)	
5.2443	1906.83	P(8)	
5.2551	1902.91	P(9)	
5.2661	1898.95	P(10)	
5.2771	1894.97	P(11)	
5.2884	1890.94	P(12)	
5.2997	1886.89	P(13)	
5.3112	1882.81	P(14)	
5.3229	1878.69	P(15)	

[a] The values of P(N) given here represent the calculated mean values of closely spaced spin doublets P_1 (J − ½) and P_2 (J + ½)[4] which were not resolved in the experiment.[5]

[b] Observed during flash photolysis of $(CN)_2$, CF_3CN, BrCN, or C_2F_5CN in up to 50 torr of Ar. All transitions listed here are in the ground electronic state. Electronic transitions give rise to laser lines in the 1- to 2-μm region.[5]

Table 3.2.1 (continued)
DIATOMIC VIBRATIONAL TRANSITION LASERS

3.2.1.2a $C^{12}O^{16}$ Laser

Wavelength (μm)vac	Frequency (cm^{-1})	Transition P(N)[a]	Remarks
		v = 12 → 10 Band	
2.68864	3719.3550	P(5)	a
2.69142	3715.5124	P(6)	a
		v = 13 → 11 Band	
2.72619	3668.1177	P(5)	a
2.72903	3664.3102	P(6)	a
2.73192	3660.4336	P(7)	a,b
2.73486	3656.4882	P(8)	b
2.73787	3652.4740	P(9)	b
2.74093	3648.3912	P(10)	b
2.74406	3644.2399	P(11)	b
2.74724	3640.0204	P(12)	b
2.75048	3635.7327	P(13)	b
2.75378	3631.3770	P(14)	b
2.75714	3626.9534	P(15)	b
2.76055	3622.4621	P(16)	b
2.76403	3617.9032	P(17)	b
2.76757	3613.2769	P(18)	b
2.77117	3608.5833	P(19)	b
		v = 14 → 12 Band	
2.76469	3617.0445	P(5)	a
2.76757	3613.2720	P(6)	a
2.77052	3609.4305	P(7)	a
2.77659	3601.5410	P(9)	b
2.77971	3597.4933	P(10)	b
2.78290	3593.3771	P(11)	b
2.78614	3589.1927	P(12)	b
2.78945	3584.9401	P(13)	b
2.79281	3580.6195	P(14)	b
2.79624	3576.2310	P(15)	b
2.79973	3571.7749	P(16)	b
2.80328	3567.2511	P(17)	b
2.80689	3562.6600	P(18)	b
2.81057	3558.0015	P(19)	b
2.81430	3553.2759	P(20)	b
2.81811	3548.4833	P(21)	b
2.82197	3543.6239	P(22)	b
2.82590	3538.6978	P(23)	b
2.82989	3533.7051	P(24)	b
		v = 15 → 13 Band	
2.80415	3566.1373	P(5)	a
2.80710	3562.3998	P(6)	a
2.81010	3558.5933	P(7)	a
—	—	—	
2.82272	3542.6802	P(11)	b
2.82603	3538.5309	P(12)	b

Table 3.2.1 (continued)
DIATOMIC VIBRATIONAL TRANSITION LASERS

Wavelength (μm)vac	Frequency (cm^{-1})	Transition P(N)[a]	Remarks
3.2.1.2a $C^{12}O^{16}$ Laser			
		v = 15 → 13 Band	
2.82940	3534.3134	P(13)	b
2.83284	3530.0279	P(14)	b
2.83634	3525.6746	P(15)	b
2.83990	3521.2536	P(16)	b
2.84352	3516.7650	P(17)	b
2.84721	3512.2090	P(18)	b
2.85096	3507.5858	P(19)	b
2.85478	3502.8954	P(20)	b
2.85866	3498.1380	P(21)	b
2.86261	3493.3138	P(22)	b
2.86662	3488.4229	P(23)	b
2.87070	3483.4655	P(24)	b
		v = 16 → 14 Band	
2.84463	3515.3976	P(5)	a
2.84763	3511.6952	P(6)	a
2.85069	3507.9237	P(7)	a
		—	
2.87038	3483.8543	P(13)	b
2.87389	3479.6039	P(14)	b
2.87746	3475.2857	P(15)	b
2.88110	3470.8999	P(16)	b
2.88480	3466.4465	P(17)	b
2.88857	3461.9257	P(18)	b
2.89240	2457.3376	P(19)	b
2.89630	3452.6825	P(20)	b
2.90027	3447.9603	P(21)	b
2.90430	3443.1713	P(22)	b
2.90840	3438.3157	P(23)	b
		v = 17 → 15 Band	
2.88921	3461.1592	P(6)	a
2.89233	3457.4228	P(7)	a
		—	
2.91965	3425.0656	P(15)	b
2.92337	3420.7149	P(16)	b
2.92715	3416.2967	P(17)	b
2.93099	3411.8111	P(18)	b
2.93491	3407.2582	P(19)	b
2.93890	3402.6383	P(20)	b
2.94295	3397.9514	P(21)	b
2.94707	3393.1976	P(22)	b
2.95127	3388.3772	P(23)	b
2.95553	3383.4903	P(24)	b
2.95986	3378.5370	P(25)	b

Table 3.2.1 (continued)
DIATOMIC VIBRATIONAL TRANSITION LASERS

Wavelength (μm)vac	Frequency (cm^{-1})	Transition P(N)[a]	Remarks
3.2.1.2a $C^{12}O^{16}$ Laser			
		v = 18 → 16 Band	
2.92875	3414.4251	P(5)	a
2.93187	3410.7927	P(6)	a
2.93505	3407.0913	P(7)	a
		—	
2.97061	3366.3163	P(17)	b
2.97454	3361.8659	P(18)	b
2.97854	3357.3482	P(19)	b
2.98261	3352.7635	P(20)	b
2.98676	3348.1118	P(21)	b
2.99097	3343.3934	P(22)	b
2.99526	3338.6082	P(23)	b
2.99962	3333.7566	P(24)	b
3.00405	3328.8386	P(25)	b
3.00856	3323.8543	P(26)	b
3.01313	3318.8039	P(27)	b
		v = 19 → 17 Band	
2.97248	3364.1932	P(5)	a
2.97566	3360.5957	P(6)	a
2.97891	3356.9294	P(7)	a
—	—	—	
3.02750	3303.0584	P(20)	b
3.03173	3298.4419	P(21)	b
3.03605	3293.7587	P(22)	b
3.04043	3289.0089	P(23)	b
3.04489	3284.1925	P(24)	b
3.04942	3279.3098	P(25)	b
3.05403	3274.3609	P(26)	b
3.05872	3269.3458	P(27)	b
3.06348	3264.2649	P(28)	b
3.06831	3259.1181	P(29)	b
3.07323	3253.9056	P(30)	b
		v = 20 → 18 Band	
3.01738	3314.1304	P(5)	a
3.02063	3310.5680	P(6)	a
—	—	—	
3.07793	3248.9413	P(21)	b
3.08234	3244.2933	P(22)	b
3.08682	3239.5787	P(23)	b
3.09138	3234.7977	P(24)	b
3.09602	3229.9503	P(25)	b
3.10074	3225.0367	P(26)	b
3.10554	3220.0570	P(27)	b
3.11041	3215.0114	P(28)	b
3.11536	3209.9000	P(29)	b
3.12039	3204.7229	P(30)	b
3.12551	3199.4804	P(31)	b

Table 3.2.1 (continued)
DIATOMIC VIBRATIONAL TRANSITION LASERS

Wavelength (μm)vac	Frequency (cm^{-1})	Transition P(N)[a]	Remarks
3.2.1.2a $C^{12}O^{16}$ Laser			
		v = 21 → 19 Band	
3.06682	3260.7084	P(6)	a
		—	
3.13448	3190.3169	P(23)	b
3.13915	3185.5711	P(24)	b
3.14390	3180.7590	P(25)	b
3.14873	3175.8807	P(26)	b
3.15364	3170.9364	P(27)	b
3.15863	3165.9262	P(28)	b
3.16371	3160.8502	P(29)	b
3.16886	3155.7085	P(30)	b
3.17410	3150.5014	P(31)	b
		v = 22 → 20 Band	
3.18347	3141.2215	P(23)	b
3.18826	3136.5111	P(24)	b
3.19312	3131.7344	P(25)	b
3.19806	3126.8915	P(26)	b
3.20309	3121.9825	P(27)	b
3.20821	3117.0076	P(28)	b
3.21340	3111.9670	P(29)	b
3.21868	3106.8608	P(30)	b
3.22405	3101.6891	P(31)	b
		v = 1 → 0 Band	
4.74513[i]	2107.4236[i]	P(9)	c
4.75450	2103.2701	P(10)	c
4.76398	2099.0831	P(11)	c,d
4.77358	2094.8627	P(12)	c,d
4.78330	2090.6091	P(13)	c
4.79312	2086.3223	P(14)	c
4.80307	2082.0026	P(15)	c
4.81313	2077.6501	P(16)	c
4.82331	2073.2650	P(17)	c
4.83361	2068.8473	P(18)	c
4.84403	2064.3973	P(19)	c
4.85457	2059.9150	P(20)	c
4.86523	2055.4007	P(21)	c
		v = 2 → 1 Band	
4.77689[i]	2093.4110[i]	P(6)	c
4.78608	2089.3940	P(7)	c,d
4.79537	2085.3431	P(8)	c,d
4.80479	2081.2583	P(9)	c,d
4.81431	2077.1399	P(10)	c,d
4.82395	2072.9879	P(11)	c,d
4.83371	2068.8025	P(12)	c,d
4.84359	2064.5839	P(13)	c,d
4.85359	2060.3322	P(14)	c

Table 3.2.1 (continued)
DIATOMIC VIBRATIONAL TRANSITION LASERS

3.2.1.2a $C^{12}O^{16}$ Laser

Wavelength (μm)vac	Frequency (cm^{-1})	Transition P(N)[a]	Remarks
		v = 2 → 1 Band	
4.86370	2056.0475	P(15)	c
4.87394	2051.7301	P(16)	c
4.88429	2047.3799	P(17)	c
4.89477	2042.9973	P(18)	c
4.90537	2038.5823	P(19)	c
4.91609	2034.1351	P(20)	c
		v = 3 → 2 Band	
4.85623[i]	2059.2089[i]	P(8)	c
4.86580	2055.1591	P(9)	c,d
4.87549	2051.0757	P(10)	c,d
4.88530	2046.9588	P(11)	c,d
4.89522	2042.8084	P(12)	c,d
4.90527	2038.6248	P(13)	c,d
4.91543	2034.4082	P(14)	c,d
4.92572	2030.1586	P(15)	c
4.93614	2025.8761	P(16)	c
4.94667	2021.5611	P(17)	c
4.95733	2017.2135	P(18)	c
4.96812	2012.8335	P(19)	c
4.97903	2008.4214	P(20)	c
4.99008	2003.9771	P(21)	c
5.00125	1999.5010	P(22)	c
5.01255	1994.9931	P(23)	c
5.02398	1990.4535	P(24)	c
5.03554[i]	1985.8824[i]	P(25)	c
5.04724	1981.2800	P(26)	c
5.05907	1976.6465	P(27)	c
5.07104	1971.9818	P(28)	c
5.08314	1967.2863	P(29)	c
5.09539	1962.5599	P(30)	c
5.10777	1957.8030	P(31)	c
5.12029	1953.0156	P(32)	c
5.13295	1948.1979	P(33)	c
		v = 4 → 3 Band	
4.90888[i]	2037.1233[i]	P(7)	d
4.91849	2033.1424	P(8)	c,d
4.92823	2029.1277	P(9)	c,d
4.93808	2025.0793	P(10)	c,d
4.94805	2020.9973	P(11)	c,d
4.95815	2016.8820	P(12)	c,d
4.96837[i]	2012.7335[i]	P(13)	c,d,e
4.97871	2008.5518	P(14)	c,d,e
4.98918	2004.3373	P(15)	c,d,e
4.99978	2000.0899	P(16)	c
5.01050	1995.8099	P(17)	c
5.02135	1991.4973	P(18)	c

Table 3.2.1 (continued)
DIATOMIC VIBRATIONAL TRANSITION LASERS

Wavelength (μm)vac	Frequency (cm^{-1})	Transition P(N)[a]	Remarks
3.2.1.2a $C^{12}O^{16}$ Laser			
		v = 4 → 3 Band	
5.03233	1987.1524	P(19)	c
5.04344	1982.7753	P(20)	c
5.05468	1978.3662	P(21)	c
5.06605	1973.9251	P(22)	c
		v = 5 → 4 Band	
4.96277[i]	2015.0029[i]	P(6)	d
4.97243	2011.0910	P(7)	d
4.98220	2007.1452	P(8)	c,d
4.99210	2003.1655	P(9)	c,d
5.00212	1999.1521	P(10)	c,d,e
5.01227[i]	1995.1052[i]	P(11)	c,d,e
5.02254	1991.0249	P(12)	c,d,e
5.03294	1986.9114	P(13)	c,d,e
5.04346	1982.7648	P(14)	c,d,e
5.05412	1978.5853	P(15)	c,d,e
5.06490	1974.3730	P(16)	c,e
5.07581	1970.1280	P(17)	c,e
5.08686	1965.8505	P(18)	c,e,g
5.09803	1961.5407	P(19)	c,g
5.10934	1957.1986	P(20)	c,g
5.12079	1952.8245	P(21)	c,g
5.13237	1948.4185	P(22)	c,g
5.14408	1943.9807	P(23)	c,g
5.15594	1939.5112	P(24)	c,g
5.16793	1935.0103	P(25)	c,g
5.18006	1930.4780	P(26)	c,g
5.19234	1925.9146	P(27)	c,g
5.20475[i]	1921.3201[i]	P(28)	c
5.21732	1916.6946	P(29)	c
5.23002	1912.0385	P(30)	c
5.24287	1907.3517	P(31)	c
5.25587	1902.6344	P(32)	c
5.26902	1897.8868	P(33)	c
		v = 6 → 5 Band	
5.01794[i]	1992.8494[i]	P(5)	d
5.02764	1989.0066	P(6)	d
5.03745	1985.1298	P(7)	d,f
5.04740	1981.2189	P(8)	c,d,f
5.05747	1977.2742	P(9)	c,d,f
5.06766	1973.2959	P(10)	c,d,e,f
5.07799	1969.2840	P(11)	c,d,e,f
5.08844	1965.2388	P(12)	c,d,e,f
5.09902	1961.1603	P(13)	c,d,e,f
5.10974[i]	1957.04860[i]	P(14)	c,d,e,f,h
5.12058	1952.90444	P(15)	c,d,e
5.13156	1948.7270	P(16)	c,d,e,g

Table 3.2.1 (continued)
DIATOMIC VIBRATIONAL TRANSITION LASERS

3.2.1.2a $C^{12}O^{16}$ Laser

Wavelength (μm)vac	Frequency (cm^{-1})	Transition P(N)[a]	Remarks
		v = 6 → 5 Band	
5.14267	1944.5171	P(17)	c,d,e,g
5.15391	1940.2747	P(18)	c,g
5.16529	1935.9999	P(19)	c,g
5.17681	1931.6929	P(20)	c,g
5.18846	1927.3538	P(21)	c,g
5.20025	1922.9829	P(22)	c,g
5.21219	1918.5801	P(23)	c,g
5.22426	1914.1458	P(24)	c,g
5.23648	1909.6799	P(25)	c,g
5.24884	1905.1827	P(26)	c,g
5.26135	1900.6543	P(27)	c,g
5.27400	1896.0949	P(28)	c,g
5.28680	1891.5046	P(29)	c
5.29974	1886.8835	P(30)	c
5.31284[i]	1882.2318[i]	P(31)	c
5.32609	1877.5496	P(32)	c
5.33949	1872.8371	P(33)	c
		v = 7 → 6 Band	
5.08417[i]	1966.8907[i]	P(5)	d
5.09403	1963.0830	P(6)	d
5.10402	1959.2411	P(7)	c,d,f
5.11413	1955.3653	P(8)	c,d,f
5.12438	1951.4556	P(9)	c,d,e,f
5.13476	1947.5123	P(10)	c,d,e,f
5.14526	1943.5355	P(11)	c,d,e,f
5.15590	1939.5253	P(12)	c,d,e,f,h
5.16667	1935.48165	P(13)	c,d,e,f,h
5.17758	1931.4054	P(14)	c,d,e,f,h
5.18862	1927.2959	P(15)	c,d,e,f
5.19979	1923.1537	P(16)	c,d,f
5.21110[i]	1918.9788[i]	P(17)	c,d,e,g
5.22256	1914.7714	P(18)	c,d,g
5.23414	1910.5317	P(19)	c,g
5.24587	1906.2598	P(20)	c,g
5.25775	1901.9558	P(21)	c,g
5.26976	1897.6199	P(22)	c,g
5.28192	1893.2523	P(23)	c,g
5.29422	1888.8530	P(24)	c,g
5.30667	1884.4222	P(25)	c,g
5.31926	1879.9601	P(26)	c,g
5.33201	1875.4668	P(27)	c,g
5.34490	1870.9425	P(28)	c,g
5.35794	1866.3872	P(29)	c
5.37114	1861.8012	P(30)	c
5.38449	1857.1846	P(31)	c
5.39800	1852.5375	P(32)	c

Table 3.2.1 (continued)
DIATOMIC VIBRATIONAL TRANSITION LASERS

Wavelength (μm)vac	Frequency (cm^{-1})	Transition P(N)[a]	Remarks
3.2.1.2a $C^{12}O^{16}$ Laser			
		v = 8 → 7 Band	
5.15197[i]	1941.0063[i]	P(5)	d
5.16200	1937.2336	P(6)	d
5.17216	1933.4267	P(7)	c,d,f
5.18246	1929.5859	P(8)	c,d,f
5.19289	1925.7113	P(9)	c,d,e,f
5.20345	1921.8030	P(10)	c,d,e,f
5.21414	1917.8612	P(11)	c,d,e,f
5.22497	1913.8861	P(12)	c,d,e,f,h
5.23594	1909.87722	P(13)	c,d,e,f
5.24704	1905.83608	P(14)	c,d,e,f,g,h
5.25828	1901.7618	P(15)	c,d,e,g
5.26966	1897.6546	P(16)	c,d,e,g
5.28118	1893.51508	P(17)	c,d,g
5.29285	1889.3425	P(18)	c,g
5.30465	1885.13791	P(19)	c,g
5.31660	1880.90096	P(20)	c,g
5.32869	1876.6321	P(21)	c,g
5.34094	1872.3313	P(22)	c,g
5.35332	1867.9987	P(23)	c,g
5.36586	1863.6345	P(24)	c,g
5.37855	1859.2388	P(25)	c,g
5.39138	1854.8118	P(26)	c,g
5.40437	1850.3536	P(27)	c,g
5.41752	1845.8643	P(28)	c,g
5.43082	1841.3442	P(29)	c
5.44427	1836.7933	P(30)	c
5.45788	1832.2118	P(31)	c
5.47166	1827.5998	P(32)	c
5.48559	1822.9575	P(33)	c
		v = 9 → 8 Band	
5.09200[i]	1963.8640[i]	R(8)	c
5.10043	1960.6190	R(7)	c
5.10898[i]	1957.3379[i]	R(6)	c
—	—		
5.21132	1918.9013	P(4)	d
5.22139	1915.1978	P(5)	d
5.23160	1911.4601	P(6)	c,d
5.24195	1907.6882	P(7)	c,d,f
5.25243	1903.8824	P(8)	c,d,e,f
5.26304	1900.0428	P(9)	c,d,e,f
5.27379	1896.1696	P(10)	c,d,e,f
5.28468	1892.26240	P(11)	c,d,e,f,h
5.29571	1888.32225	P(12)	c,d,e,f,h
5.30687	1884.3493	P(13)	c,d,e,f
5.31818	1880.3429	P(14)	c,d,e,f,h
5.32963	1876.3036	P(15)	c,d,e,g
5.34122	1872.2315	P(16)	c,d,e,g
5.35295	1868.12715	P(17)	c,d,g
5.36484	1863.9895	P(18)	c,d,g

Table 3.2.1 (continued)
DIATOMIC VIBRATIONAL TRANSITION LASERS

3.2.1.2a $C^{12}O^{16}$ Laser

Wavelength (μm)vac	Frequency (cm^{-1})	Transition P(N)[a]	Remarks
		v = 9 → 8 Band	
5.37686	1859.82039	P(19)	c,g
5.38904	1855.6181	P(20)	c,g
5.40136	1851.3843	P(21)	c,g
5.41384	1847.1185	P(22)	c,g
5.42646	1842.8210	P(23)	c,g
5.43924	1838.4919	P(24)	c,g
5.45217	1834.1313	P(25)	c,g
5.46526	1829.7394	P(26)	c,g
5.47850	1825.3163	P(27)	c,g
5.49190	1820.8621	P(28)	c,g
5.50546	1816.3771	P(29)	c
5.51919	1811.8613	P(30)	c
5.53307	1807.3149	P(31)	c
5.54712	1802.7380	P(32)	c
5.56133	1798.1309	P(33)	c
		v = 10 → 9 Band	
5.09683[i]	1962.0047[i]	R(16)	c
5.10441	1959.0892	R(15)	c
5.11212	1956.1363	R(14)	c
5.11994	1953.1463	R(13)	c
5.12789	1950.1192	R(12)	c
5.13596	1947.0552	R(11)	c
—	—		
5.28224	1893.1353	P(4)	d
5.29250	1889.4668	P(5)	d
5.30289	1885.7640	P(6)	c,d
5.31342	1882.0272	P(7)	c,d
5.32409	1878.2564	P(8)	c,d,e,f
5.33489	1874.4518	P(9)	c,d,e,f
5.34584	1870.61378	P(10)	c,d,e,f
5.35693[i]	1866.7418[i]	P(11)	c,d,e,f
5.36816	1862.83705	P(12)	c,d,e,f,h
5.37953	1858.8984	P(13)	c,d,e,f,h
5.39105	1854.92760	P(14)	c,d,e,g
5.40271	1850.9228	P(15)	c,d,e,g
5.41452	1846.8857	P(16)	c,d,e,g
5.42648	1842.8160	P(17)	c,d,g
5.43858	1838.7138	P(18)	c,d,g
5.45084	1834.5793	P(19)	c,g
5.46325	1830.4126	P(20)	c,g
5.47581	1826.2139	P(21)	c,g
5.48852	1821.9832	P(22)	c,g
5.50139	1817.7208	P(23)	c,g
5.51442	1813.4268	P(24)	c,g
5.52761	1809.1013	P(25)	c,g
5.54095	1804.7445	P(26)	c,g
5.55446	1800.3565	P(27)	g

Table 3.2.1 (continued)
DIATOMIC VIBRATIONAL TRANSITION LASERS

Wavelength (μm)vac	Frequency (cm^{-1})	Transition P(N)[a]	Remarks
3.2.1.2a $C^{12}O^{16}$ Laser			
		v = 11 → 10 Band	
5.12275[i]	1952.0755[i]	R(22)	c
5.12973	1949.4220	R(21)	c
5.13682	1946.7302	R(20)	c
5.14403	1944.0003	R(19)	c
5.15137	1941.2325	R(18)	c
5.15882	1938.4270	R(17)	c
5.16640	1935.5838	R(16)	c
5.17410	1932.7032	R(15)	c
5.18192	1929.7852	R(14)	c
5.18987	1926.8301	R(13)	c
—	—		
5.35490	1867.4481	P(4)	d
5.36534	1863.8146	P(5)	d
5.37592	1860.1468	P(6)	d
5.38664	1856.4450	P(7)	c,d
5.39750	1852.7092	P(8)	c,d,e
5.40851	1848.9396	P(9)	c,d,e
5.41965	1845.1364	P(10)	c,d,e
5.43095	1841.2997	P(11)	c,d,e,h
5.44239	1837.42954	P(12)	c,d,e
5.45397	1833.5264	P(13)	c,d,e,g
5.46571	1829.5901	P(14)	c,d,e,g
5.47759	1825.6208	P(15)	c,d,e,g
5.48962	1821.6188	P(16)	c,d,g
5.50181	1817.5842	P(17)	c,d,g
5.51415	1813.5171	P(18)	c,d,g
5.52664	1809.4176	P(19)	c,g
5.53929	1805.2860	P(20)	c,g
5.55209	1801.1223	P(21)	c,g
5.56506	1796.9268	P(22)	c,g
5.57818	1792.6995	P(23)	c,g
5.59146[i]	1788.4405[i]	P(24)	c,g
5.60491	1784.1501	P(25)	c,g
5.61852	1779.8284	P(26)	c
5.63229	1775.4756	P(27)	c
		v = 12 → 11 Band	
5.17961[i]	1930.6480[i]	R(24)	c
5.18644	1928.1062	R(23)	c
5.19339	1925.5258	R(22)	c
5.20046	1922.9071	R(21)	c
5.20765	1920.2502	R(20)	c
5.21497	1917.5552	R(19)	c
5.22242	1914.8223	R(18)	c
5.22998	1912.0516	R(17)	c
5.23768	1909.2434	R(16)	c
5.24550	1906.3976	R(15)	c
—	—	—	—
5.43998[i]	1838.2426[i]	P(5)	d
5.45075	1834.6099	P(6)	d

Table 3.2.1 (continued)
DIATOMIC VIBRATIONAL TRANSITION LASERS

3.2.1.2a $C^{12}O^{16}$ Laser

Wavelength (μm)vac	Frequency (cm^{-1})	Transition P(N)[a]	Remarks
		v = 12 → 11 Band	
5.46167	1830.9430	P(7)	d
5.47273	1827.2423	P(8)	c,d,e
5.48394	1823.5077	P(9)	c,d,e
5.49529	1819.7395	P(10)	c,d,e,h
5.50680	1815.9378	P(11)	c,d,e,h
5.51845	1812.1028	P(12)	c,d,e
5.53026	1808.2346	P(13)	c,d,e
5.54221	1804.33320	P(14)	c,d,e
5.55432	1800.39913	P(15)	c,d,e
5.56659	1796.4322	P(16)	c,d
5.57901	1792.4326	P(17)	c,g
5.59159	1788.4006	P(18)	c,d,g
5.60433	1784.3362	P(19)	c,g
5.61722	1780.2396	P(20)	c,g
5.63028	1776.1110	P(21)	c,g
5.64350	1771.9506	P(22)	c,g
5.65688	1767.7584	P(23)	c,g
5.67043	1763.5345	P(24)	c,g
5.68415	1759.2792	P(25)	c,g
5.69803	1754.9926	P(26)	c
		v = 13 → 12 Band	
5.34326[i]	1871.5177[i]	R(12)	c
5.35172	1868.5584	R(11)	c
5.36031	1865.5624	R(10)	c
5.36904	1862.5298	R(9)	c
5.37790	1859.4608	R(8)	c
5.38690	1856.3555	R(7)	c
5.39603	1853.2141	R(6)	c
5.40530	1850.0367	R(5)	c
5.41470[i]	1846.8234[i]	R(4)	c
5.42425	1843.5745	R(3)	c
5.43393	1840.2900	R(2)	c
—	—		
5.51647	1812.7521	P(5)	d
5.52744	1809.1544	P(6)	d
5.53856	1805.5225	P(7)	d,e
5.54983	1801.8568	P(8)	d,e
5.56125	1798.1572	P(9)	c,d,e
5.57282	1794.4241	P(10)	c,d,e
5.58454	1790.65726	P(11)	c,d,e
5.59642	1786.85747	P(12)	c,d,e,h
5.60845	1783.02431	P(13)	c,d,e,h
5.62063	1779.15855	P(14)	c,d,e
5.63298	1775.2588	P(15)	c,d,e,g
5.64549	1771.3269	P(16)	c,g
5.65815	1767.3624	P(17)	c,d,g
5.67097	1763.3655	P(18)	c,d,g
5.68396	1759.3362	P(19)	c,g
5.69711	1755.2747	P(20)	c,g

Table 3.2.1 (continued)
DIATOMIC VIBRATIONAL TRANSITION LASERS

Wavelength (μm)vac	Frequency (cm^{-1})	Transition P(N)[a]	Remarks
3.2.1.2a $C^{12}O^{16}$ Laser			
		v = 13 → 12 Band	
5.71043	1751.1812	P(21)	c,g
5.72392	1747.0558	P(22)	c,g
5.73757	1742.8987	P(23)	c,g
5.75139	1738.7100	P(24)	c,g
5.76538	1734.4898	P(25)	c
5.77955	1730.2383	P(26)	c
5.79389	1725.9557	P(27)	c
5.80841	1721.6420	P(28)	c
		v = 14 → 13 Band	
5.35465[i]	1867.5363[i]	R(20)	c
5.36219	1864.9111	R(19)	c
5.36985	1862.2479	R(18)	c
5.37765	1859.5469	R(17)	c
5.38559	1856.8084	R(16)	c
5.39365	1854.0324	R(15)	c
5.40185	1851.2190	R(14)	c
5.41018	1848.3685	R(13)	c
5.41864	1845.4810	R(12)	c
5.42724	1842.5566	R(11)	c
5.43598	1839.5955	R(10)	c
5.44485	1836.5978	R(9)	c
—	—		
5.59489	1787.3442	P(5)	d
5.60607	1783.7814	P(6)	d
5.61740	1780.1846	P(7)	d,e
5.62888	1776.5538	P(8)	d,e
—	—		
5.65230	1769.1911	P(10)	c,d,e
5.66425	1765.4595	P(11)	c,d,e,h
5.67635[i]	1761.6945[i]	P(12)	c
5.68862	1757.8964	P(13)	c,d,e
5.70104	1754.0652	P(14)	c,d,e
5.71363	1750.2011	P(15)	c,d,e,g
5.72638	1746.3042	P(16)	c,d,g
5.73929	1742.3748	P(17)	c,d,g
5.75237	1738.4129	P(18)	c,d,g
5.76562	1734.4187	P(19)	g
5.77904	1730.3923	P(20)	g
5.79262	1726.3339	P(21)	g
5.80638	1722.2436	P(22)	g
5.82031	1718.1216	P(23)	g
5.83441	1713.9680	P(24)	g
5.84870	1709.7829	P(25)	g
		v = 15 → 14 Band	
5.37448[i]	1860.6453[i]	R(28)	c
5.38108	1858.3635	R(27)	c
5.38781	1856.0426	R(26)	c

Table 3.2.1 (continued)
DIATOMIC VIBRATIONAL TRANSITION LASERS

3.2.1.2a $C^{12}O^{16}$ Laser

Wavelength (μm)vac	Frequency (cm^{-1})	Transition P(N)[a]	Remarks
		v = 15 → 14 Band	
5.39467	1853.6828	R(25)	c
5.40166	1851.2841	R(24)	c
5.40878	1848.8468	R(23)	c
5.41603	1846.3710	R(22)	c
5.42341	1843.8568	R(21)	c
5.43093	1841.3044	R(20)	c
5.43858	1838.7140	R(19)	c
5.44637	1836.0857	R(18)	c
5.45429	1833.4196	R(17)	c
5.46234	1830.7158	R(16)	c
5.47054	1827.9747	R(15)	c
—	—		
5.68669[i]	1758.4918[i]	P(6)	d,e
5.69823	1754.9300	P(7)	d,e
—	—		
5.72179	1747.7047	P(9)	d,e
5.73381	1744.0416	P(10)	d,e,h
5.74599	1740.3449	P(11)	d,e
5.75833	1736.6150	P(12)	c,d,e
5.77083	1732.8519	P(13)	c,d,e
5.78350	1729.0557	P(14)	c,d,e,g
5.79634	1725.2267	P(15)	c,d,g
5.80934	1721.3649	P(16)	c,d,g
5.82252	1717.4705	P(17)	g
5.83586	1713.5437	P(18)	g
5.84937	1709.5846	P(19)	g
5.86306	1705.5933	P(20)	g
5.87693	1701.5699	P(21)	g
5.89096	1697.5148	P(22)	g
5.90518	1693.4279	P(23)	g
5.91958	1689.3094	P(24)	g
		v = 16 → 15 Band	
5.75778[i]	1736.7792[i]	P(5)	e
—	—		
5.78115	1729.7595	P(7)	d
5.79308	1726.1988	P(8)	d,e
5.80516	1722.6042	P(9)	c,d,e
5.81742	1718.9761	P(10)	d,e
5.82983	1715.3145	P(11)	d,e
5.84242	1711.6196	P(12)	d,e,h
5.85517	1707.8915	P(13)	d,e
5.86810	1704.1304	P(14)	d
—	—		
5.89446	1696.5097	P(16)	g
5.90789	1692.6503	P(17)	g
5.92151	1688.7586	P(18)	d,g
5.93530	1684.8345	P(19)	g
5.94927	1680.8783	P(20)	g
5.96342	1676.8901	P(21)	g
5.97775	1672.8700	P(22)	g

Table 3.2.1 (continued)
DIATOMIC VIBRATIONAL TRANSITION LASERS

Wavelength (μm)vac	Frequency (cm^{-1})	Transition P(N)[a]	Remarks
3.2.1.2a $C^{12}O^{16}$ Laser			
		v = 17 → 16 Band	
5.85423[i]	1708.1655[i]	P(6)	e
5.86623	1704.6736	P(7)	e
5.87838	1701.1478	P(8)	d
5.89071	1697.5883	P(9)	d,e
5.90320	1693.9952	P(10)	d,e,h
5.91587	1690.3686	P(11)	d,e
5.92871	1686.7087	P(12)	d,e
—	—		
5.95490	1679.2896	P(14)	d,e
5.96826	1675.5306	P(15)	d
5.98180	1671.7390	P(16)	d,g
5.99551	1667.9147	P(17)	g
6.00941	1664.0580	P(18)	g
6.02348	1660.1690	P(19)	g
6.03774	1656.2479	P(20)	g
		v = 18 → 17 Band	
5.95354[i]	1679.6725[i]	P(7)	d,e
5.96594	1676.1817	P(8)	d,e
5.97851	1672.6572	P(9)	d,e
—	—		
6.00418	1665.5075	P(11)	d,e
6.01727	1661.8826	P(12)	d,e
6.03055	1658.2246	P(13)	d,e
—	—		
6.05763	1650.8096	P(15)	d,g
6.07145	1647.0530	P(16)	d,g
6.08545[i]	1643.2638[i]	P(17)	g
6.09964	1639.4422	P(18)	g
6.11401	1635.5883	P(19)	g
6.12857	1631.7022	P(20)	g
		v = 19 → 18 Band	
6.04319[i]	1654.7561[i]	P(7)	d
6.05583	1651.3003	P(8)	d,e
6.06866	1647.8108	P(9)	d,e
6.08166	1644.2876	P(10)	d,e
6.09484	1640.7311	P(11)	d,e
6.10821	1637.1412	P(12)	d,e
6.12176	1633.5182	P(13)	d,e
6.13549	1629.8622	P(14)	d
6.14941	1626.1733	P(15)	d
		v = 20 → 19 Band	
6.10999[i]	1636.6641[i]	P(5)	d
6.12253	1633.3111	P(6)	d,e
6.13525	1629.9242	P(7)	d,e
6.14816	1626.5033	P(8)	d,e

Table 3.2.1 (continued)
DIATOMIC VIBRATIONAL TRANSITION LASERS

Wavelength (μm)vac	Frequency (cm^{-1})	Transition P(N)[a]	Remarks
3.2.1.2a $C^{12}O^{16}$ Laser			
		v = 20 → 19 Band	
6.16124	1623.0488	P(9)	d,e
6.17451	1619.5606	P(10)	d,e
6.18797	1616.0391	P(11)	d,e
6.20161	1612.4842	P(12)	d,e
6.21544	1608.8962	P(13)	d,e
6.22946	1605.2752	P(14)	d
6.24367	1601.6214	P(15)	d
		v = 21 → 20 Band	
6.20407[i]	1611.8460[i]	P(5)	d
6.21686	1608.5280	P(6)	d,e
6.22985	1605.1759	P(7)	d,e
6.24302	1601.7901	P(8)	d,e
6.25637	1598.3705	P(9)	d,e
6.26992	1594.9173	P(10)	d,e
6.28365	1591.4307	P(11)	d,e
6.29758	1587.9109	P(12)	d,e
6.31171	1584.3579	P(13)	d
6.32602	1580.7720	P(14)	d
6.34054	1577.1532	P(15)	d
6.35525	1573.5017	P(16)	d
6.37017	1569.8177	P(17)	d
		v = 22 → 21 Band	
6.30076[i]	1587.1106[i]	P(5)	d,e
6.31382	1583.8275	P(6)	d,e
6.32707	1580.5105	P(7)	d,e
6.34051[i]	1577.1595[i]	P(8)	d,e
6.35415	1573.7749	P(9)	d,e
6.36798	1570.3568	P(10)	d,e
6.38201	1566.9052	P(11)	d,e
6.39623	1563.4203	P(12)	d
—	—		
6.42528	1556.3514	P(14)	d
6.44011	1552.7677	P(15)	d
6.45515	1549.1512	P(16)	d
6.47039	1545.5023	P(17)	d
		v = 23 → 22 Band	
6.40018[i]	1562.4566[i]	P(5)	d,e
6.41351	1559.2085	P(6)	d
6.42704	1555.9264	P(7)	d,e
6.44077	1552.6104	P(8)	d,e
6.45469	1549.2608	P(9)	d,e
6.46882	1545.8776	P(10)	d,e
6.48315	1542.4610	P(11)	d
—	—		
6.51242	1535.5282	P(13)	d

Table 3.2.1 (continued)
DIATOMIC VIBRATIONAL TRANSITION LASERS

Wavelength (μm)vac	Frequency (cm^{-1})	Transition P(N)[a]	Remarks
3.2.1.2a $C^{12}O^{16}$ Laser			
		v = 23 → 22 Band	
6.52736	1532.0123	P(14)	d
6.54252	1528.4635	P(15)	d
6.55788	1524.8821	P(16)	d
		v = 24 → 23 Band	
6.50245[i]	1537.8824[i]	P(5)	d,e
6.51606	1534.6691	P(6)	d,e
6.52988	1531.4220	P(7)	d,e
6.54390	1528.1410	P(8)	d,e
6.55812	1524.8263	P(9)	d,e
—	—		
6.58720[i]	1518.0964[i]	P(11)	d,e
6.60205	1514.6816	P(12)	d,e
6.61711	1511.2336	P(13)	d,e
—	—		
6.64788	1504.2390	P(15)	d
		v = 25 → 24 Band	
6.60770[i]	1513.3856[i]	P(5)	d,e
6.62161	1510.2073	P(6)	d,e
6.63572	1506.9951	P(7)	d
6.65005	1503.7490	P(8)	d,e
6.66458	1500.4693	P(9)	d,e
6.67933	1497.1560	P(10)	d,e
6.69429	1493.8093	P(11)	d,e
6.70948	1490.4295	P(12)	d,e
6.72487	1487.0165	P(13)	d
6.74049	1483.5706	P(14)	d
6.75634[i]	1480.0919[i]	P(15)	d
6.77240	1476.5806	P(16)	d
		v = 26 → 25 Band	
6.70208[i]	1492.0731[i]	P(4)	d,e
—	—		
6.73029	1485.8205	P(6)	d,e
6.74471	1482.6431	P(7)	d,e
6.75935	1479.4320	P(8)	d,e
—	—		
6.78929	1472.9089	P(10)	d,e
6.80459	1469.5972	P(11)	d,e
6.82011	1466.2523	P(12)	d,e
6.83586	1462.8743	P(13)	d,e
6.85183	1459.4634	P(14)	d
6.86804	1456.0198	P(15)	d
6.88448	1452.5435	P(16)	d

Table 3.2.1 (continued)
DIATOMIC VIBRATIONAL TRANSITION LASERS

3.2.1.2a $C^{12}O^{16}$ Laser

Wavelength (μm)vac	Frequency (cm^{-1})	Transition P(N)[a]	Remarks
		v = 27 → 26 Band	
6.84226[i]	1461.5055[i]	P(6)	d,e
—	—		
6.87197	1455.1869	P(8)	d
6.88716	1451.9770	P(9)	d,e
6.90258	1448.7337	P(10)	d,e
6.91823	1445.4569	P(11)	d,e
6.93411	1442.1470	P(12)	d,e
6.95022	1438.8040	P(13)	d
6.96656	1435.4282	P(14)	d
6.98315	1432.0195	P(15)	d
		v = 28 → 27 Band	
6.94284[i]	1440.3326[i]	P(5)	d,e
—	—		
6.97276	1434.1515	P(7)	d,e
6.98807	1431.0102	P(8)	d,e
7.00361	1427.8353	P(9)	d,e
7.01938	1424.6268	P(10)	d,e
7.03539	1421.3851	P(11)	d,e
7.05164	1418.1101	P(12)	d
7.06813	1414.8021	P(13)	d
7.08486	1411.4612	P(14)	d
7.10183	1408.0876	P(15)	d
7.11905	1404.6813	P(16)	d
		v = 29 → 28 Band	
7.07676[i]	1413.0769[i]	P(6)	d,e
7.09218	1410.0043	P(7)	d,e
7.10784	1406.8979	P(8)	d,e
7.12374	1403.7579	P(9)	d,e
7.13988	1400.5844	P(10)	d,e
7.15626	1397.3776	P(11)	d,e
7.17289[i]	1394.1376[i]	P(12)	d
7.18977	1390.8646	P(13)	d
7.20690	1387.5587	P(14)	d
7.22428	1384.2200	P(15)	d
7.24192	1380.8488	P(16)	d
		v = 30 → 29 Band	
7.19966[i]	1388.9548[i]	P(6)	d,e
7.21544	1385.9171	P(7)	d,e
7.23147	1382.8456	P(8)	d,e
7.24774	1379.7405	P(9)	d,e
7.26426	1376.6019	P(10)	d,e
7.28104	1373.4300	P(11)	d
7.29807	1370.2250	P(12)	d
7.31536	1366.9869	P(13)	d
7.33291	1363.7160	P(14)	d

Table 3.2.1 (continued)
DIATOMIC VIBRATIONAL TRANSITION LASERS

3.2.1.2a $C^{12}O^{16}$ Laser

Wavelength (μm)vac	Frequency (cm⁻¹)	Transition P(N)[a]	Remarks
		v = 30 → 29 Band	
7.35071	1360.4124	P(15)	d
7.36878	1357.0761	P(16)	d
		v = 31 → 30 Band	
7.31071[i]	1367.8565[i]	P(5)	d
7.32661	1364.8875	P(6)	d,e
7.34277	1361.8847	P(7)	d,e
7.35917	1358.8481	P(8)	d,e
7.37584	1355.7779	P(9)	d,e
7.39276	1352.6742	P(10)	d,e
7.40995	1349.5373	P(11)	d
7.42739	1346.3672	P(12)	d
7.44511	1343.1641	P(13)	d
		v = 32 → 31 Band	
7.44156[i]	1343.8036[i]	P(5)	d
7.45785	1340.8695	P(6)	d,e
7.47439	1337.9015	P(7)	d,e
7.49120	1334.8997	P(8)	d,e
7.50827	1331.8644	P(9)	d,e
7.52561	1328.7957	P(10)	d,e
7.54322	1325.6937	P(11)	d,e
		v = 33 → 32 Band	
7.59362[i]	1316.8944[i]	P(6)	d
7.61057	1313.9613	P(7)	d,e
7.62780	1310.9944	P(8)	d,e
7.64529	1307.9940	P(9)	d,e
7.66307	1304.9602	P(10)	d,e
7.68112	1301.8931	P(11)	e
		v = 34 → 33 Band	
7.73422[i]	1292.9555[i]	P(6)	e
7.75159	1290.0573	P(7)	e
7.76925[i]	1287.1253[i]	P(8)	e
7.78719	1284.1597	P(9)	e
7.80542	1281.1608	P(10)	e
7.82394	1278.1286	P(11)	e
		v = 35 → 34 Band	
7.87994[i]	1269.0454[i]	P(6)	e
7.89776	1266.1819	P(7)	e
7.91587	1263.2848	P(8)	e
—	—		
7.95298	1257.3901	P(10)	e

Table 3.2.1 (continued)
DIATOMIC VIBRATIONAL TRANSITION LASERS

Wavelength (μm)vac	Frequency (cm^{-1})	Transition P(N)[a]	Remarks
3.2.1.2a $C^{12}O^{16}$ Laser			
		v = 36 → 35 Band	
8.03112[i]	1245.1558[i]	P(6)	e
8.04941	1242.3272	P(7)	e
8.06800	1239.4649	P(8)	e
8.08689	1236.5691	P(9)	e
8.10609	1233.6399	P(10)	e
		v = 37 → 36 Band	
8.20692[i]	1218.4842[i]	P(7)	e
8.22601	1215.6567	P(8)	e
8.24541	1212.7958	P(9)	e
8.26514	1209.9014	P(10)	e

[a] Observed in a supersonic flow laser which uses an electric-discharge-excited CO-He-O_2 mixture.[6] The transition frequencies are calculated from the best spectroscopic data available.[7] A similar output extending to 2.35 μm has also been reported for a CS_2-O_2 chemical laser.[8] However, the reported spectrum is inconsistent with the well-established spectroscopic data on CO.

[b] Observed in an electron-beam-controlled discharge in CO-N_2-He mixtures cooled to 100 K.[9] The transition frequencies are calculated from the best spectroscopic data available.[7]

[c] Observed during flash photolysis of CS_2-O_2 mixtures.[10]

[d] Observed in a liquid-nitrogen cooled longitudinal discharge in CO-N_2-He mixtures.[11]

[e] Observed in a liquid-nitrogen cooled longitudinal discharge in CO-He-O_2 mixtures with Q-switching.[12]

[f] Observed in a pulsed longitudinal discharge in CO.[13]

[g] Observed in a CW discharge in CO-N_2 mixtures with the tube jacket cooled to either −78°C or 15°C.[15] A multiline CW output of 25 W has been reported for a water-cooled tube using a He-CO-N_2-Hg-Xe mixture.[16] A multiline pulsed output of 5 J has been reported for a transverse discharge tube using a CO-N_2-He mixture at −20°C.[17]

[h] Observed in a liquid-nitrogen cooled longitudinal discharge in CO-N_2-O_2-He mixtures.[18] A multiline peak output power of 7.7 kW and a greater number of laser lines were obtained in Q-switched operation.[18] In CW mode, a multiline output of 70 W has been obtained with the addition of Xe.[19] A discharge excited, supersonic expansion CO-He-O_2 laser has produced a 940 W CW output with a similar output spectrum.[20] An output energy of 1600 J per pulse has been reported for an electron-beam sustained discharge CO-N_2 laser.[21]

[i] The transition frequencies given are either calculated values (eight digits) or observed values (nine digits) based on the best data available.[7]

Table 3.2.1 (continued)
DIATOMIC VIBRATIONAL TRANSITION LASERS

Wavelength (μm)vac	Frequency (cm^{-1})	Transition P(N)[a]	Remarks
3.2.1.2b $C^{13}O^{16}$ Laser			
		v = 7 → 6 Band	a
5.30328	1885.6249	P(16)	
5.31451	1881.6403	P(17)	
5.32588	1877.6253	P(18)	
5.33738	1873.5800	P(19)	
		v = 8 → 7 Band	a
5.32843	1876.7267	P(12)	
5.33932	1872.8979	P(13)	
—	—		
5.36151	1865.1475	P(15)	
5.37280	1861.2262	P(16)	
5.38423	1857.2743	P(17)	
5.39580	1853.2920	P(18)	
5.40751	1849.2795	P(19)	
		v = 9 → 8 Band	a
5.37702	1859.7664	P(10)	
5.38784	1856.0326	P(11)	
5.39879	1852.2676	P(12)	
5.40988	1848.4715	P(13)	
—	—		
5.43246	1840.7866	P(15)	
5.44396	1836.8980	P(16)	
5.45560	1832.9710	P(17)	
5.46738	1829.0295	P(18)	
—	—		
5.49137	1821.0397	P(20)	
5.50358	1816.9998	P(21)	
		v = 10 → 9 Band	a
5.43779	1838.9835	P(9)	
5.44866	1835.3138	P(10)	
5.45967	1831.6128	P(11)	
5.47082	1827.8805	P(12)	
5.48210	1824.1171	P(13)	
5.49353	1820.3229	P(14)	
5.50510	1816.4978	P(15)	
5.51681	1812.6420	P(16)	
5.52866	1808.7557	P(17)	
5.54066	1804.8389	P(18)	
5.55280	1800.8919	P(19)	
		v = 11 → 10 Band	a
5.50001	1818.1772	P(8)	
5.51094	1814.5717	P(9)	
5.52201	1810.9347	P(10)	
5.53322	1807.2664	P(11)	

Table 3.2.1 (continued)
DIATOMIC VIBRATIONAL TRANSITION LASERS

Wavelength (μm)vac	Frequency (cm^{-1})	Transition P(N)[a]	Remarks
3.2.1.2b $C^{13}O^{16}$ Laser			
		v = 11 → 10 Band	a
5.54457	1803.5668	P(12)	
5.55606	1799.8362	P(13)	
5.56770	1796.0747	P(14)	
5.57948	1792.2824	P(15)	
5.59140	1788.4594	P(16)	
5.60348	1784.6058	P(17)	
—	—		
5.62807	1776.8076	P(19)	
—	—		
5.66609	1764.8846	P(22)	
		v = 12 → 11 Band	a
5.57473	1793.8073	P(8)	
5.58586	1790.2345	P(9)	
5.59713	1786.6302	P(10)	
5.60854	1782.9947	P(11)	
5.62010	1779.3279	P(12)	
—	—		
5.64366	1771.9012	P(14)	
—	—		
5.66781	1764.3514	P(16)	
5.68011	1760.5306	P(17)	
—	—		
5.70516	1752.7980	P(19)	
—	—		
5.73084	1744.9448	P(21)	
		v = 13 → 12 Band	a
5.65127	1769.5132	P(8)	
5.66260	1765.9731	P(9)	
5.67408	1762.4016	P(10)	
5.68570	1758.7988	P(11)	
5.69747	1755.1647	P(12)	
—	—		
5.72147	1747.8036	P(14)	
5.73369	1744.0768	P(15)	
5.74607	1740.3193	P(16)	
5.75861	1736.5313	P(17)	
5.77130	1732.7129	P(18)	
5.78414	1728.8642	P(19)	
5.79715	1724.9854	P(20)	
5.81032	1721.0766	P(21)	
		v = 14 → 13 Band	a
5.72969	1745.2960	P(8)	
5.74122	1741.7886	P(9)	
5.75291	1738.2499	P(10)	
—	—		
5.77675	1731.0784	P(12)	

Table 3.2.1 (continued)
DIATOMIC VIBRATIONAL TRANSITION LASERS

Wavelength (μm)vac	Frequency (cm^{-1})	Transition P(N)[a]	Remarks
3.2.1.2b $C^{13}O^{16}$ Laser			
		v = 14 → 13 Band	a
5.78889	1727.4461	P(13)	
5.80119	1723.7828	P(14)	
5.81365	1720.0887	P(15)	
5.82627	1716.3640	P(16)	
5.83905	1712.6087	P(17)	
5.85198	1708.8231	P(18)	
5.86508	1705.0072	P(19)	
		v = 15 → 14 Band	a
5.79845	1724.5995	P(7)	
5.81005	1721.1565	P(8)	
—	—		
5.83371	1714.1758	P(10)	
5.84577	1710.6384	P(11)	
5.85799	1707.0698	P(12)	
5.87037	1703.4702	P(13)	
—	—		
5.90847	1692.4864	P(16)	
5.92149	1688.7639	P(17)	
—	—		
5.94803	1681.2279	P(19)	
5.96155	1677.4147	P(20)	
5.97525	1673.5715	P(21)	
		v = 16 → 15 Band	a
5.89242	1697.0954	P(8)	
5.90440	1693.6535	P(9)	
—	—		
5.92882	1686.6755	P(11)	
—	—		
5.95390	1679.5728	P(13)	
5.96668	1675.9750	P(14)	
5.97962	1672.3464	P(15)	
5.99273	1668.6872	P(16)	
6.00602	1664.9975	P(17)	
6.01946	1661.2774	P(18)	
6.03308	1657.5271	P(19)	
		v = 17 → 16 Band	a
5.96484	1676.4908	P(7)	
5.97688	1673.1132	P(8)	
—	—		
6.00145	1666.2634	P(10)	
—	—		
6.02668	1659.2883	P(12)	
6.03954	1655.7542	P(13)	
—	—		
6.06578	1648.5933	P(15)	

Table 3.2.1 (continued)
DIATOMIC VIBRATIONAL TRANSITION LASERS

Wavelength (μm)vac	Frequency (cm^{-1})	Transition P(N)[a]	Remarks
3.2.1.2b $C^{13}O^{16}$ Laser			
		v = 17 → 16 Band	a
6.07915	1644.9669	P(16)	
6.09269	1641.3099	P(17)	
—	—		
6.12031	1633.9051	P(19)	
		v = 18 → 17 Band	a
6.06351	1649.2102	P(8)	
—	—		
6.08856	1642.4258	P(10)	
6.10133	1638.9866	P(11)	
—	—		
6.12740	1632.0148	P(13)	
6.14069	1628.4825	P(14)	
6.15415	1624.9194	P(15)	
6.16779	1621.3257	P(16)	
6.18161	1617.7015	P(17)	
6.19561	1614.0470	P(18)	
6.20978	1610.3622	P(19)	
		v = 19 → 18 Band	a
6.13987	1628.6987	P(7)	
6.15238	1625.3865	P(8)	
6.16507	1622.0427	P(9)	
6.17792	1618.6674	P(10)	
6.19095	1615.2609	P(11)	
6.20415	1611.8233	P(12)	
6.21753	1608.3546	P(13)	
6.23109	1604.8550	P(14)	
6.24483	1601.3247	P(15)	
6.25875	1597.7638	P(16)	
6.27285	1594.1724	P(17)	
—	—		
6.30160	1586.8986	P(19)	
		v = 20 → 19 Band	a
6.23084	1604.9213	P(7)	
6.24359	1601.6417	P(8)	
6.25653	1598.3306	P(9)	
6.26964	1594.9881	P(10)	
6.28293	1591.6143	P(11)	
6.29640	1588.2094	P(12)	
6.31005	1584.7734	P(13)	
6.32388	1581.3066	P(14)	
6.33790	1577.8090	P(15)	
6.35211	1574.2809	P(16)	
6.36650	1570.7222	P(17)	
6.38108	1567.1332	P(18)	

Table 3.2.1 (continued)
DIATOMIC VIBRATIONAL TRANSITION LASERS

3.2.1.2b $C^{13}O^{16}$ Laser

Wavelength (μm)vac	Frequency (cm^{-1})	Transition P(N)[a]	Remarks
		v = 21 → 20 Band	a
6.33723	1577.9756	P(8)	
6.35043	1574.6972	P(9)	
6.36380	1571.3874	P(10)	
6.37736	1568.0463	P(11)	
—	—		
6.40504	1561.2708	P(13)	
—	—		
6.44797	1550.8765	P(16)	
6.46266	1547.3506	P(17)	
		v = 22 → 21 Band	a
6.44686	1551.1416	P(9)	
6.46051	1547.8644	P(10)	
6.47435	1544.5560	P(11)	
6.48838	1541.2165	P(12)	
6.50260	1537.8460	P(13)	
6.51702	1534.4447	P(14)	
6.53162	1531.0126	P(15)	
—	—		
6.56144	1524.0568	P(17)	
		v = 23 → 22 Band	a
6.55988	1524.4182	P(10)	
—	—		
6.60285	1514.4979	P(13)	
6.61757	1511.1293	P(14)	

[a] Observed in a CW longitudinal discharge in $C^{13}O$-N-Xe-He mixtures cooled to −78°C.[22] The frequencies and wavelengths are calculated from best available values of Dunham coefficients.[23]

3.2.1.3a HBr Laser

Wavelength (μm)vac	Frequency (cm^{-1})	Transition HBr^{79}	Transition HBr^{81}	Remarks
		v = 1 → 0 Band		
4.0170	2489.40	P(4)		a,b
4.0176	2489.05		P(4)	a,b
4.0470	2470.97	P(5)		a,b
4.0475	2470.63		P(5)	a,b
4.0783	2452.03	P(6)		a,b
4.0788	2451.68		P(6)	a,b
4.1107	2432.70	P(7)		a,b
4.1112	2432.36		P(7)	a,b
4.1442	2412.99	P(8)		a

Table 3.2.1 (continued)
DIATOMIC VIBRATIONAL TRANSITION LASERS

Wavelength (μm)vac	Frequency (cm^{-1})	Transition HBr^{79}	Transition HBr^{81}	Remarks
3.2.1.3a HBr Laser				
		v = 1 → 0 Band		
4.1448	2412.68		P(8)	a
4.1796	2392.56		P(9)	a
		v = 2 → 1 Band		
4.1653	2400.78	P(4)		a,b
4.1658	2400.47		P(4)	a,b
4.1970	2382.68	P(5)		a,b
4.1975	2382.35		P(5)	a,b
4.2295	2364.36	P(6)		a,b
4.2633	2345.58	P(7)		a,b
4.2639	2345.26		P(7)	a,b
4.2988	2326.23	P(8)		a,b
4.2994	2325.92		P(8)	a,b
4.3354	2306.60	P(9)		a,b
4.3359	2306.30		P(9)	a,b
		v = 3 → 2 Band		
4.3250	2312.15	P(4)		a,b
4.3255	2311.85		P(4)	a,b
4.3579	2294.68	P(5)		a,b
4.3585	2294.39		P(5)	a,b
4.3925	2276.61	P(6)		a,b
4.3931	2276.32		P(6)	a,b
4.4281	2258.29	P(7)		a,b
4.4307	2258.00		P(7)	a,b
4.4652	2239.52	P(8)		a,b
4.4658	2239.26		P(8)	a,b
4.5041	2220.20	P(9)		a
4.5047	2219.92		P(9)	a
		v = 4 → 3 Band		
4.5330	2206.07	P(5)		a,b
4.5335	2205.81		P(5)	a,b
4.5691	2188.61	P(6)		a,b
4.5696	2188.35		P(6)	a,b
4.6070	2170.61	P(7)		a,b
4.6076	2170.35		P(7)	a,b
4.6463	2152.27	P(8)		a
4.6467	2152.04		P(8)	a

[a] Observed in a pulsed longitudinal discharge in H_2-Br_2 mixtures.[24]

[b] Observed in a pulsed transverse discharge in H_2-Br_2-He mixtures.[25] A maximum energy of 550 mJ per pulse has been obtained by using a mixture of H_2, Br_2, Ar, and benzene.[26]

Table 3.2.1 (continued)
DIATOMIC VIBRATIONAL TRANSITION LASERS

Wavelength (μm)vac	Frequency (cm^{-1})	Transition DBr^{79}	Transition DBr^{81}	Remarks
3.2.1.3b DBr Laser				
		v = 1 → 0 Band		
5.5689	1795.70	P(5)		a
5.5704	1795.21		P(5)	a
5.5979	1786.40	P(6)		a
5.5994	1785.91		P(6)	a
5.6276	1776.95	P(7)		a
5.6291	1776.47		P(7)	a
5.6582	1767.34	P(8)		a
5.6597	1766.87		P(8)	a
5.6896	1757.59	P(9)		a
5.6911	1757.12		P(9)	a
		v = 2 → 1 Band		
5.711	1751.0	P(5)		a
5.713	1750.5		P(5)	a
5.741	1741.9	P(6)		a
5.743	1741.1		P(6)	a
5.772	1732.6	P(7)		a
5.773	1732.1		P(7)	a
5.803	1723.1	P(8)		a
5.8049	1722.57		P(8)	a,b
5.836	1713.5	P(9)		a
5.837	1713.1		P(9)	a
		v = 3 → 2 Band		
5.802	1723.4	P(3)		a
5.804	1722.9		P(3)	a
5.832	1714.7	P(4)		a
5.834	1714.2		P(4)	a
5.8620	1705.91	P(5)		a,b
5.8626	1705.43		P(5)	a,b
5.8928	1696.98	P(6)		a,b
5.8944	1696.52		P(6)	a,b
5.9246	1687.89	P(7)		a,b
5.9261	1687.45		P(7)	a,b
5.9573	1678.60	P(8)		a,b
5.9590	1678.13		P(8)	a,b
5.991	1669.2	P(9)		a
5.993	1668.7		P(9)	a
6.026	1659.6	P(10)		a
6.027	1659.1		P(10)	a
		v = 4 → 3 Band		
6.0209	1660.88	P(5)		a,b
6.0225	1660.45		P(5)	a,b
6.0529	1652.10	P(6)		a,b
6.0544	1651.69		P(6)	a,b

Table 3.2.1 (continued)
DIATOMIC VIBRATIONAL TRANSITION LASERS

3.2.1.3b DBr Laser

Wavelength (μm)vac	Frequency (cm^{-1})	Transition DBr^{79}	Transition DBr^{81}	Remarks
		v = 4 → 3 Band		
6.0858	1643.18	P(7)		a,b
6.0873	1642.76		P(7)	a,b
6.1200	1634.00	P(8)		a,b
6.1216	1633.57		P(8)	a,b
6.1546	1624.80	P(9)		a,b
6.1562	1624.39		P(9)	a,b
6.1903	1615.42	P(10)		a,b
6.1918	1615.03		P(10)	a,b
6.2272	1605.85	P(11)		a,b
6.2289	1605.43		P(11)	a,b
		v = 5 → 4 Band		
6.2237	1606.75		P(6)	a,b
6.2566	1598.32	P(7)		a,b
6.2581	1597.93		P(7)	a,b
6.2916	1589.41	P(8)		a,b
6.2932	1589.01		P(8)	a,b
6.3279	1580.31	P(9)		a,b
6.3294	1579.92		P(9)	a,b
6.291	1589.6	P(10)		a
6.292	1589.2		P(10)	a

[a] Observed in a pulsed transverse discharge in D_2-Br_2-He mixtures; peak output power ~6 kW.[25] Frequencies and wavelengths are calculated from spectroscopic data.[27]

[b] Observed in a pulsed longitudinal discharge in D_2-Br_2 mixtures.[24]

3.2.1.4a HCl Laser

Wavelength (μm)vac	Frequency (cm^{-1})	Transition HCl^{35}	Transition HCl^{37}	Remarks
		v = 1 → 0 Band		
3.57278	2798.940	P(4)		a
3.60262	2775.760	P(5)		a
3.63367	2752.036	P(6)		a
3.63619	2750.131		P(6)	a
3.66599	2727.777	P(7)		a
3.66848	2725.921		P(7)	a
3.69958	2703.007	P(8)		a,b
3.70207	2701.192		P(8)	a
3.73450	2677.732	P(9)		b
3.73698	2675.961		P(9)	a
3.77079	2651.966	P(10)		b
3.80847	2625.727	P(11)		b

Table 3.2.1 (continued)
DIATOMIC VIBRATIONAL TRANSITION LASERS

Wavelength (μm)vac	Frequency (cm^{-1})	Transition HCl^{35}	Transition HCl^{37}	Remarks
3.2.1.4a HCl Laser				
		v = 2 → 1 Band		
3.7071	2697.52	P(4)		a,b,c,d
3.7097	2695.63		P(4)	a
3.7383	2675.01	P(5)		a,b,c,d
3.7408	2673.23		P(5)	a,c
3.7710	2651.82	P(6)		a,b,c,d
3.7735	2650.08		P(6)	a,c
3.8050	2628.13	P(7)		a,b,c,d
3.8074	2626.45		P(7)	a,c
3.8401	2604.09	P(8)		a,b,c,d
3.8425	2602.48		P(8)	a,c
3.8768	2579.42	P(9)		a,c,d
3.9149	2554.34	P(10)		a,c,d
		v = 3 → 2 Band		
3.8509	2596.79	P(4)		a,c,d
3.8536	2594.97		P(4)	a
3.8840	2574.70	P(5)		a,c,d
3.8864	2573.07		P(5)	a
3.9181	2552.26	P(6)		a,c,d
3.9205	2550.70		P(6)	a,c
3.9536	2529.31	P(7)		a,c,d
3.9560	2527.79		P(7)	c
3.9909	2505.68	P(8)		a,c,d
4.0295	2481.69	P(9)		a,c,d
		v = 4 → 3 Band		
4.0059	2496.32	P(4)		e
4.0404	2474.99	P(5)		e
4.0764	2453.12	P(6)		e
4.1140	2430.73	P(7)		e

[a] Observed in a pulsed transverse discharge in H_2-Cl_2-He mixtures; peak output power ~2.4 kW.[25] Frequencies and wavelengths for the v = 1 → 0 band are taken from absorption spectroscopy.[28,29]

[b] Observed during a flash initiated reaction of H_2 and Cl_2.[30,31]

[c] Observed in a pulsed longitudinal discharge in H_2-Cl_2 mixtures.[24]

[d] Observed in a pulsed longitudinal discharge in H_2-Cl_2 or H_2-NOCl mixtures.[32]

[e] Observed in a pulsed transverse discharge in H_2-ICl-He mixtures.[25] Frequencies and wavelengths are calculated from spectroscopic data.[28]

Table 3.2.1 (continued)
DIATOMIC VIBRATIONAL TRANSITION LASERS

Wavelength (μm)vac	Frequency (cm^{-1})	Transition DCl^{35}	Transition DCl^{37}	Remarks
3.2.1.4b DCl Laser				
		v = 1 → 0 Band		
5.0031	1998.76	P(8)		a
5.0098	1996.07		P(8)	a
5.0344	1986.34	P(9)		a
5.0412	1983.64		P(9)	a
5.0667	1973.66	P(10)		a
5.0736	1971.00		P(10)	a
5.1000	1960.78	P(11)		a
5.1067	1958.20		P(11)	a
5.1341	1947.77	P(12)		a
5.1407	1945.26		P(12)	a
5.1691	1934.58	P(13)		a
		v = 2 → 1 Band		
5.0445	1982.35	P(5)		b
5.0514	1979.65		P(5)	b
5.0743	1970.72	P(6)		a,b
5.0811	1968.08		P(6)	a,b
5.1049	1958.90	P(7)		a,b,c
5.1118	1956.25		P(7)	a,b
5.1363	1946.94	P(8)		a,b,c
5.1431	1944.35		P(8)	a,b
5.1688	1934.67	P(9)		a,b,c
5.176	1932.1		P(9)	a
5.202	1922.3	P(10)		a
5.209	1919.8		P(10)	a
		v = 3 → 2 Band		
5.122	1952.3	P(3)		a
5.1511	1941.35	P(4)		a,b,c
5.158	1938.8		P(4)	a
5.1811	1930.10	P(5)		a,b,c
5.1879	1927.56		P(5)	a,b
5.2118	1918.73	P(6)		a,b
5.2186	1916.24		P(6)	a,b
5.2435	1907.13	P(7)		a,b
5.2503	1904.67		P(7)	a,b
5.2760	1895.37	P(8)		a,b
5.2829	1892.91		P(8)	a,b
5.3097	1883.35	P(9)		a,b
5.316	1881.0		P(9)	a
5.3443	1871.15	P(10)		a,b
5.3799	1858.77	P(11)		b
		v = 4 → 3 Band		
5.3244	1878.13	P(5)		a,b
5.3562	1867.01	P(6)		a,b
5.3629	1864.65		P(6)	b
5.3889	1855.66	P(7)		a,b

Table 3.2.1 (continued)
DIATOMIC VIBRATIONAL TRANSITION LASERS

3.2.1.4b DCl Laser

Wavelength (μm)vac	Frequency (cm^{-1})	Transition DCl^{35}	Transition DCl^{37}	Remarks
		v = 4 → 3 Band		
5.3956	1853.36		P(7)	b
5.423	1844.1	P(8)		a
5.4295	1841.79		P(8)	b
5.4577	1832.27	P(9)		a,b
5.4935	1820.34	P(10)		a,b
5.5304	1808.20	P(11)		a,b
		v = 5 → 4 Band		
5.5084	1815.38	P(6)		b
5.5423	1804.31	P(7)		b
5.5776	1792.89	P(8)		b
5.6137	1781.36	P(9)		b

[a] Observed in a pulsed transverse discharge in D_2-Cl_2-He mixtures; peak output power ~5 kW.[25] Frequencies and wavelengths for the v = 1 → 0 band are taken from absorption spectroscopy.[33]
[b] Observed in a pulsed longitudinal discharge in D_2-Cl_2 mixtures.[24]
[c] Observed during a flash initiated reaction of D_2 and Cl_2 mixtures.[31]

3.2.1.5a HF Laser

Wavelength (μm) vac	Frequency (cm^{-1})	Transition	Remarks
		v = 1 → 0 Band	
2.4138[a]	4142.83[a]	R(4)	b
2.4331	4109.95	R(3)	b
2.4538	4075.30	R(2)	b
2.4759	4038.97	R(1)	b
—	—		
2.5508	3920.29	P(1)	c
2.5789	3877.70	P(2)	c
2.6085	3833.65	P(3)	b,c,d
2.6398	3788.23	P(4)	b,c,d,e
2.6727	3741.48	P(5)	b,c,d,e
2.70752	3693.41171	P(6)	b,c,d,e,f,g
2.7441	3644.16	P(7)	b,c,d,e,f
2.7826[a]	3593.80[a]	P(8)	b,d,f
2.8231	3542.20	P(9)	f
2.8657	3489.59	P(10)	f
2.9103	3436.12	P(11)	f
2.9573	3381.50	P(12)	f,h
3.0064	3326.21	P(13)	f,h
3.0582	3269.90	P(14)	f,h
3.1125	3212.80	P(15)	f,h
3.1696	3154.95	P(16)	h
3.2293	3096.62	P(17)	h
3.2920	3037.70	P(18)	h

Table 3.2.1 (continued)
DIATOMIC VIBRATIONAL TRANSITION LASERS

Wavelength (μm) vac	Frequency (cm^{-1})	Transition	Remarks
3.2.1.5a HF Laser			
		v = 2 → 1 Band	
2.6668[a]	3749.83[a]	P(1)	d
2.6962	3708.86	P(2)	c,d,f,i
2.7275	3666.38	P(3)	c,d,f,i
2.7604	3622.71	P(4)	c,d,e,f,i
2.7953	3577.47	P(5)	c,d,e,f,i
2.8318	3531.31	P(6)	c,d,e,f,i
2.8706	3483.63	P(7)	c,d,e,f,i
2.9111	3435.17	P(8)	c,f
2.9539	3385.34	P(9)	c,f
2.9989	3334.55	P(10)	f,h
3.0461	3282.86	P(11)	f,h
3.0958	3230.18	P(12)	f,h
3.1480	3176.60	P(13)	f,h
3.2029	3122.14	P(14)	f,h
3.2603	3067.22	P(15)	f,h
3.3206	3011.51	P(16)	h
		v = 3 → 2 Band	
2.7902[a]	3583.93[a]	P(1)	d
2.8213	3544.51	P(2)	d,f
2.8540	3503.80	P(3)	d,f
2.8889	3461.54	P(4)	d,f,h
2.9256	3418.16	P(5)	d,f
2.9643	3373.46	P(6)	d,f
3.0051	3327.73	P(7)	d,f
3.0482	3280.64	P(8)	d,f,h
3.0935	3232.58	P(9)	h
3.1411	3183.60	P(10)	h
3.1912	3133.65	P(11)	h
3.2438	3082.79	P(12)	h,j
3.2991	3031.10	P(13)	h
		v = 4 → 3 Band	
2.9221[a]	3422.22[a]	P(1)	d
2.9549	3384.24	P(2)	d
2.9896	3344.92	P(3)	d,k
3.0264	3304.31	P(4)	k
3.0652	3262.46	P(5)	k
3.1062	3219.41	P(6)	j
3.1494	3175.22	P(7)	j
		v = 5 → 4 Band	
3.0982[a]	3227.67[a]	P(2)	i,j
3.1350	3189.84	P(3)	i
3.1739	3150.73	P(4)	i
3.2150	3110.41	P(5)	i
3.2585	3068.91	P(6)	i,k
3.3044	3026.29	P(7)	i

Table 3.2.1 (continued)
DIATOMIC VIBRATIONAL TRANSITION LASERS

3.2.1.5a HF Laser

Wavelength (μm) vac	Frequency (cm^{-1})	Transition	Remarks
		v = 6 → 5 Band	
3.3335[a]	2999.87[a]	P(4)	i,k
3.3772	2961.03	P(5)	i,k

[a] The laser wavelengths and frequencies given for lines covered by footnotes f and g are observed values. For all other lines, the values given are either taken or calculated from absorption spectroscopy.[34]

[b] Observed in a pulsed transverse discharge in H_2-SF_6 mixtures.[35,36]

[c] Observed in a pulsed longitudinal discharge in H_2-SF_6 mixtures.[37]

[d] Observed during flash photolysis of H_2-F_2-He mixtures.[38]

[e] Observed in a continuous mixing flow of arc-heated N_2, SF_6, and H_2, with supersonic expansion.[39] A maximum output power of 12.4 kW was obtained in a system using He, F_2, and H_2.[40]

[f] Observed in a pulsed longitudinal discharge in H_2-Freon mixtures.[41] A similar spectrum was observed in a transverse discharge laser using He-SF_6-H_2 mixtures.[25] A maximum output energy of 4.2 kJ per pulse has been obtained by electron-beam excitation of F_2-O_2-H_2 mixtures.[42]

[g] Frequency accurately measured against a CO laser line.[43]

[h] Observed in HF optically pumped by a pulsed HF laser.[44]

[i] Observed during flash photolysis of IF_5-H_2 mixtures.[45]

[j] Observed during flash photolysis of H_2-F_2 mixtures.[46]

[k] Observed in a pulsed transverse discharge in HI-He-SF_6 (or SO_2F_2) mixtures.[47]

3.2.1.5b DF Laser

Wavelength (μm) vac	Frequency (cm^{-1})	Transition	Remarks
		v = 3 → 1 Band	
1.836	5447	P(4)	a
1.844	5423	P(5)	a
1.854	5394	P(6)	a
		v = 1 → 0 Band	
3.4933	2862.65	P(2)	b
3.5214	2839.78	P(3)	b
3.5507	2816.36	P(4)	b
3.5811	2792.44	P(5)	b
3.6128	2767.91	P(6)	b
3.6456	2743.03	P(7)	b
3.6798	2717.54	P(8)	b,c
3.7155	2691.41	P(9)	c,d
3.7520	2665.25	P(10)	c,d
3.7902	2611.10	P(11)	c,d
3.8298	2583.7	P(12)	c,d,e
3.8704	2555.7	P(13)	d
3.9128		P(14)	d

Table 3.2.1 (continued)
DIATOMIC VIBRATIONAL TRANSITION LASERS

3.2.1.5b DF Laser

Wavelength (μm) vac	Frequency (cm^{-1})	Transition	Remarks
		v = 1 → 0 Band	
3.9572	2527.06	P(15)	d,e
4.0032	2498.02	P(16)	d,e
4.0491	2469.7	P(17)	d
		v = 2 → 1 Band	
3.6363	2750.05	P(3)	d,e
3.6665	2727.38	P(4)	d,e
3.6983	2703.98	P(5)	d,e
3.7310	2680.28	P(6)	d,e
3.7651	2655.97	P(7)	d,e
3.8007	2631.09	P(8)	c,d,e
3.8375	2605.87	P(9)	c,d,e
3.8757	2580.16	P(10)	c,d,e
3.9155	2553.97	P(11)	c,d,e
3.9565	2527.47	P(12)	c,d,e
3.9995	2500.32	P(13)	d,e
4.0434	2473.2	P(14)	d
4.0892	2445.5	P(15)	d
4.1369	2417.27	P(16)	d,e
4.1862	2388.79	P(17)	e
		v = 3 → 2 Band	
3.7563	2662.17	P(3)	d,e
3.7878	2640.04	P(4)	d,e
3.8206	2617.41	P(5)	d,e
3.8547	2594.23	P(6)	d,e
3.8903	2570.51	P(7)	d,e

3.2.1.5b DF Laser

Wavelength (μm) vac	Frequency (cm^{-1})	Transition	Remarks
3.9272	2546.37	P(8)	c,d,e,
3.9654	2521.81	P(9)	c,d,e
4.0054	2496.61	P(10)	c,d,e
4.0464	2471.34	P(11)	c,d,e
4.0895	2445.29	P(12)	d,e
4.1337	2419.2	P(13)	d
4.1798	2392.46	P(14)	d,e
		v = 4 → 3 Band	
3.8501	2597.3	P(2)	d
3.8816	2576.2	P(3)	d
3.9145	2554.6	P(4)	d
3.9487	2532.50	P(5)	d,e

Table 3.2.1 (continued)
DIATOMIC VIBRATIONAL TRANSITION LASERS

Wavelength (μm) vac	Frequency (cm⁻¹)	Transition	Remarks
3.2.1.5b DF Laser			
		v = 4 → 3 Band	
3.9843	2509.86	P(6)	d,e
4.0212	2486.83	P(7)	d,e
4.0594	2463.4	P(8)	d

[a] Observed during flash photolysis of N_2F_4-CD_4 mixtures.[48]
[b] Observed in a pulsed longitudinal discharge in D_2-SF_6-He mixtures.[49]
[c] Observed in a continuous mixing flow of arc-heated N_2, SF_6, and D_2, with supersonic expansion.[39] A maximum output power of 340 W was obtained.[49]
[d] Observed in a pulsed transverse discharge in He-SF_6-D_2 mixtures.[25] Maximum output power ~24 kW.
[e] Observed in a pulsed longitudinal discharge in D_2-Freon mixtures.[41]

Wavelength (μm)vac	Frequency (cm⁻¹)	Transition P(J-½) Substate $^2\Pi_{1/2}$	$^2\Pi_{3/2}$	Remarks
3.2.1.6. NO Laser				
[5.50]	[1817]	v = 2 → 1 Band		a
		v = 6 → 5 Band		
5.8462	1710.52		P(7)	b
5.8549	1707.98	P(8)		b
5.8584	1706.95		P(8)	b
5.8706	1703.41		P(9)	b
5.8789	1700.99	P(10)		b
5.9036	1693.88	P(12)		b
5.9083	1692.53		P(12)	b
5.9550	1679.3	P(16)		c
		v = 7 → 6 Band		
5.9423	1682.84		P(7)	b
5.9546	1679.37		P(8)	b
5.9632	1676.96	P(9)		b,c
5.9673	1675.81		P(9)	b
5.9756	1673.46	P(10)		b,c
5.9799	1672.27		P(10)	b
5.9882	1669.96	P(11)		b
5.9931	1668.59		P(11)	b
6.0010	1666.39	P(12)		b,c
6.0054	1665.18		P(12)	b

Table 3.2.1 (continued)
DIATOMIC VIBRATIONAL TRANSITION LASERS

Wavelength (μm)vac	Frequency (cm⁻¹)	Transition P(J-½)		Remarks
3.2.1.6. NO Laser				
		v = 7 → 6 Band		
6.0192	1661.36		P(13)	b
6.0267	1659.29	P(14)		b,c
6.0324	1657.71		P(14)	b,c
6.0402	1655.58	P(15)		b,c
		v = 8 → 7 Band		
6.0386	1656.0	P(7)		c
6.0419	1655.12		P(7)	b
6.0543	1651.72		P(8)	b
6.0628	1649.40	P(9)		b,c
6.0673	1648.19		P(9)	b
6.0801	1644.71		P(10)	b
6.0884	1642.47	P(11)		b,c
6.0934	1641.13		P(11)	b,c
6.1015	1638.95	P(12)		b
6.1204	1633.87		P(13)	b
6.1417	1628.22	P(15)		b,c
6.1546	1624.8	P(16)		c
6.1973	1613.6	P(19)		c
		v = 9 → 8 Band		
6.1538	1625.02	P(8)		b
6.1576	1624.00		P(8)	b
6.1663	1621.72	P(9)		b
6.1792	1618.34	P(10)		b
6.1838	1617.12		P(10)	b
6.1921	1614.95	P(11)		b
6.1972	1613.63		P(11)	b
6.2055	1611.48	P(12)		b
6.2110	1610.06		P(12)	b
6.2191	1607.94	P(13)		b
6.2249	1606.46		P(13)	b
		v = 10 → 9 Band		
6.2381	1603.06		P(6)	b
6.2511	1599.71		P(7)	b
6.2602	1597.39	P(8)		b
6.2645	1596.30		P(8)	b
6.2778	1592.91		P(9)	b
6.2865	1590.71	P(10)		b
6.2913	1589.50		P(10)	b
6.2998	1587.36	P(11)		b
6.3051	1586.02		P(11)	b
6.3136	1583.88	P(12)		b
6.3191	1582.50		P(12)	b
6.3274	1580.53	P(13)		b
6.3336	1578.89		P(13)	b

Table 3.2.1 (continued)
DIATOMIC VIBRATIONAL TRANSITION LASERS

Wavelength (μm)vac	Frequency (cm⁻¹)	Transition P(J-½)		Remarks
3.2.1.6. NO Laser				
		v = 11 → 10 Band		
6.3764	1568.29		P(8)	b
6.3894	1565.09		P(9)	b
6.3980	1562.99	P(10)		b
6.4031	1561.74		P(10)	b
6.4262	1556.14	P(12)		b
6.4321	1554.71		P(12)	b

[a] Observed in Br_2-NO-He mixtures during flash photolysis of Br_2. NO molecules are vibrationally excited by energy transfer from electronically excited Br atoms.[51]
[b] Observed in a pulsed longitudinal discharge in NOCl-He mixtures.[52]
[c] Observed during flash photolysis of NOCl-He mixtures,[53,54] output power 10 W.[53]

Wavelength (μm) vac	Frequency (cm⁻¹)	Transition $P_1(K)$[a]	Remarks
3.2.1.7 OH Laser			
		v = 1 → 0 Band	
2.935	3407	$P_1(4)$	b
2.969	3368	$P_1(5)$	b
		v = 2 → 1 Band	
3.078	3249	$P_1(4)$	b,c
3.115	3210	$P_2(5)$	b,c
3.157	3168	$P_1(6)$	b,c
		v = 3 → 2 Band	
3.234	3092	$P_1(4)$	b,c
3.274	3054	$P_1(5)$	b,c

[a] The subscript 1 refers to the F_1 component of the spin doublet which is expected to dominate over the F_2 component. For F_1 components, $J = K + ½$.[5]
[b] Observed during flash photolysis of O_3-H_2 mixtures.[55]
[c] Observed in a pulsed transverse discharge in O_3-H_2 mixtures.[56]

3.2.1.8 Metal-Oxide and Metal-Fluoride Lasers

IR laser action corresponding to vibrational transitions in metal oxides (MO) and metal fluorides (MF) has been observed by exploding wires or films in F_2, NF_3, or O_2 gas.[57,58] The metals used include Li, Mg, Al, Ti, V, Fe, Ni, Cu, Zn, Zr, Mo, Ag, Ta, W, Pt, Au, U, and graphite. Laser wavelengths are only roughly estimated with interference filters and generally fall in the 5- to 24-μm region. IR laser action due to AlF and BF has also been observed in quasicontinuous supersonic mixing flames.[59]

Table 3.2.2
TRIATOMIC VIBRATIONAL TRANSITION LASERS

Wavelength (μm) vac	Frequency (cm^{-1})	Transition	Remarks
2.3.2.2.1a $C^{12}O_2^{16}$ Laser			
		02°1 → 02°0 Band	a
4.31787	2315.9589	P(14)	
4.32110	2314.2250	P(16)	
4.32439	2312.4675	P(18)	
4.32772	2310.6865	P(20)	
4.33110	2308.8819	P(22)	
4.33453	2307.0538	P(24)	
4.33801	2305.2023	P(26)	
4.34155	2303.3272	P(28)	
4.34513	2301.4288	P(30)	
		10°1 → 10°0 Band	a,b
4.31311	2318.5134	P(10)	
4.31626	2316.8218	P(12)	
4.32271	2313.3638	P(16)	
4.32601	2311.5975	P(18)	
4.32937	2309.8065	P(20)	
4.33277	2307.9907	P(22)	
4.33623	2306.1501	P(24)	
4.33974	2304.2848	P(26)	
4.34331	2302.3948	P(28)	
4.35059	2298.5409	P(32)	
		10°2 → 10°1 Band	c
4.3203	2314.65	R(17)	
4.3249	2312.18	R(13)	
4.3276	2310.73	R(11)	
—	—	—	
4.3549	2296.25	P(7)	
4.3580	2294.62	P(9)	
4.3612	2292.94	P(11)	
4.3644	2291.25	P(13)	
4.3677	2289.51	P(15)	
4.3711	2287.76	P(17)	
4.3745	2286.00	P(19)	
4.3779	2284.20	P(21)	
4.3814	2282.38	P(23)	
4.3849	2280.53	P(25)	
		00°1 → 02°0 Band	
9.09349[d]	1099.6872[d]	R(62)	e
9.09976	1098.9301	R(60)	e
9.10623	1098.14940	R(58)	e
9.11291	1097.344886	R(56)	e
9.11979	1096.516356	R(54)	e
9.12689	1095.663612	R(52)	e
9.13420	1094.786462	R(50)	e
9.14173	1093.884721	R(48)	e

Table 3.2.2 (continued)
TRIATOMIC VIBRATIONAL TRANSITION LASERS

Wavelength (μm) vac	Frequency (cm^{-1})	Transition	Remarks
3.2.2.1a $C^{12}O_2^{16}$ Laser			
		00°1 → 02°0 Band	
9.14948	1092.958211	R(46)	e
9.15745	1092.006758	R(44)	e
9.16565[d]	1091.030196[d]	R(42)	e
9.17407	1090.028367	R(40)	e
9.18273	1089.001119	R(38)	e
9.19161	1087.948306	R(36)	e
9.20073	1086.869791	R(34)	e
9.21009	1085.765445	R(32)	e
9.21969	1084.635145	R(30)	e
9.22953	1083.478778	R(28)	e
9.23961	1082.296237	R(26)	e
9.24995	1081.087426	R(24)	e
9.26053	1079.852255	R(22)	e
9.27136	1078.590644	R(20)	e
9.28244	1077.302520	R(18)	e
9.29379	1075.987820	R(16)	e
9.30539	1074.646490	R(14)	e
9.31725	1073.278484	R(12)	e
9.32937	1071.883766	R(10)	e
9.34176	1070.462308	R(8)	e
9.35441	1069.014093	R(6)	e
9.36734	1067.539110	R(4)	e
9.38053	1066.037360	R(2)	e
9.39400	1064.508853	R(0)	e
—	—		
9.41472	1062.165965	P(2)	e
9.42889	1060.570666	P(4)	e
9.44333	1058.948714	P(6)	e
9.45805	1057.300161	P(8)	e
9.47306	1055.625068	P(10)	e
9.48835	1053.923503	P(12)	e
9.50394	1052.195545	P(14)	e
9.51981	1050.441282	P(16)	e
9.53597	1048.660810	P(18)	e
9.55243	1046.854234	P(20)	e
9.56918	1045.021670	P(22)	e,f
9.58623	1043.163239	P(24)	e,f
9.60357	1041.279074	P(26)	e,f
9.62122	1039.369315	P(28)	e,f
9.63917	1037.434110	P(30)	e,f
9.65742	1035.473616	P(32)	e,f
9.67597	1033.487999	P(34)	e,f
9.69483	1031.477430	P(36)	e
9.71400	1029.442092	P(38)	e
9.73348	1027.382171	P(40)	e
9.75326	1025.297865	P(42)	e
9.77336	1023.189375	P(44)	e
9.79377	1021.056912	P(46)	e
9.81450	1018.900693	P(48)	e
9.83554	1016.720942	P(50)	e
9.85690	1014.517888	P(52)	e

Table 3.2.2 (continued)
TRIATOMIC VIBRATIONAL TRANSITION LASERS

3.2.2.1a $C^{12}O_2^{16}$ Laser

Wavelength (μm) vac	Frequency (cm⁻¹)	Transition	Remarks
		00°1 → 02°0 Band	
9.87858	1012.291767	P(54)	e
9.90057	1010.042823	P(56)	e
9.92289[d]	1007.771302[d]	P(58)	e
9.94552	1005.47746	P(60)	e
9.96849	1003.1615	P(62)	e
9.99177	1000.8238	P(64)	e
10.01538	998.4646	P(66)	e
		00°2 → 02°1 Band	g
9.20917	1085.8740	R(39)	
9.21777	1084.8607	R(37)	
9.22662	1083.8211	R(35)	
9.23570	1082.7550	R(33)	
9.24503	1081.6623	R(31)	
9.25461	1080.5428	R(29)	
9.26444	1079.3965	R(27)	
9.27452	1078.2232	R(25)	
9.28485	1077.0229	R(23)	
9.29545	1075.7954	R(21)	
9.30630	1074.5406	R(19)	
9.31742	1073.2586	R(17)	
9.32880	1071.9492	R(15)	
9.34045	1070.6124	R(13)	
9.35237	1069.2481	R(11)	
9.36456	1067.8563	R(9)	
9.37702	1066.4370	R(7)	
9.38976	1064.9902	R(5)	
9.40277	1063.5159	R(3)	
—	—		
9.45055	1058.1390	P(3)	
9.46485	1056.5410	P(5)	
9.47943	1054.9155	P(7)	
9.49431	1053.2627	P(9)	
9.50948	1051.5827	P(11)	
9.52494	1049.8754	P(13)	
9.54070	1048.1410	P(15)	
9.55676	1046.3797	P(17)	
9.57312	1044.5913	P(19)	
9.58979	1042.7762	P(21)	
9.60675	1040.9344	P(23)	
9.62403	1039.0661	P(25)	
9.64161	1037.1713	P(27)	
9.65950	1035.2503	P(29)	
9.67770	1033.3031	P(31)	
9.69622	1031.3300	P(33)	
9.71505	1029.3312	P(35)	
9.73419	1027.3067	P(37)	
9.75365	1025.2569	P(39)	
9.79353	1021.0819	P(43)	
9.81395	1018.9572	P(45)	

Table 3.2.2 (continued)
TRIATOMIC VIBRATIONAL TRANSITION LASERS

3.2.2.1a $C^{12}O_2^{16}$ Laser

Wavelength (μm) vac	Frequency (cm^{-1})	Transition	Remarks
		00°1 → 10°0 Band	
10.02591[d]	997.41550[d]	R(62)	e
10.03347	996.66441	R(60)	e
10.04132	995.884686	R(58)	e
10.04948	995.076610	R(56)	e
10.05793	994.240442	R(54)	e
10.06668	993.376427	R(52)	e
10.07572	992.484803	R(50)	e
10.08506	991.565748	R(48)	e
10.09469	990.619630	R(46)	e
10.10462	989.646506	R(44)	e
10.11484	988.646626	R(42)	e
10.12535	987.620181	R(40)	e
10.13616	986.567352	R(38)	e
10.14725	985.488312	R(36)	e
10.15865[d]	984.383226[d]	R(34)	e
10.17033	983.252249	R(32)	e
10.18231	982.095531	R(30)	e
10.19458	980.913211	R(28)	e
10.20715	979.705421	R(26)	e
10.22001	978.472286	R(24)	e
10.23317	977.213922	R(22)	e
10.24663	975.930439	R(20)	e
10.26039	974.621939	R(18)	e
10.27445	973.288517	R(16)	e
10.28880	971.930258	R(14)	e
10.30347	970.547244	R(12)	e
10.31843	969.139547	R(10)	e
10.33370	967.707233	R(8)	e
10.34928	966.250361	R(6)	e
10.36518	964.768982	R(4)	e
10.38138	963.263140	R(2)	e
10.39790	961.732874	R(0)	e
—	—		
10.42327	959.391745	P(2)	e
10.44059	957.800537	P(4)	e
10.45823	956.184982	P(6)	e
10.47619	954.545087	P(8)	e
10.49449	952.880850	P(10)	e
10.51312	951.192264	P(12)	e,f
10.53209	949.479314	P(14)	e,f
10.55140	947.741979	P(16)	e,f
10.57105	945.980230	P(18)	e,f
10.59104	944.194030	P(20)	e,f
10.61139	942.383336	P(22)	e,f
10.63210	940.548098	P(24)	e,f
10.65316	938.688257	P(26)	e,f
10.67459	936.803747	P(28)	e,f
10.69639	934.894496	P(30)	e,f
10.71857[d]	932.960421[d]	P(32)	e,f
10.74112	931.001434	P(34)	e,f
10.76406	929.017437	P(36)	e,f

Table 3.2.2 (continued)
TRIATOMIC VIBRATIONAL TRANSITION LASERS

Wavelength (μm) vac	Frequency (cm^{-1})	Transition	Remarks
3.2.2.1a $C^{12}O_2^{16}$ Laser			
		00°1 → 10°0 Band	
10.78739	927.008325	P(38)	e,f
10.81111	924.973985	P(40)	e
10.83524	922.914294	P(42)	e
10.85978	920.829123	P(44)	e
10.88473	918.718331	P(46)	e
10.91010	916.581770	P(48)	e
10.93590	914.419283	P(50)	e
10.96214	912.230703	P(52)	e
10.98882	910.015853	P(54)	e
11.01595	907.774549	P(56)	e
11.04354	905.50659	P(58)	e
11.07160	903.21177	P(60)	e
11.10014	900.88992	P(62)	e
11.12915	898.54082	P(64)	e
11.15867	896.1643	P(66)	e
11.18868	893.7602	P(68)	e
		01¹1 → 11¹0 Band	h
10.59103	944.195629	R(23)	
—	—		
10.78908	926.863273	Q(11)	
—	—		
10.89018	918.258073	P(11)	
10.90096	917.350002	P(12)	
10.92147	915.627737	P(14)	
10.93071	914.853825	P(15)	
10.94235	913.880366	P(16)	
10.95149	913.118028	P(17)	
10.97261	911.359785	P(19)	
10.98527	910.310203	P(20)	
10.9944	909.55	P(21)	
11.00731	908.487360	P(22)	
11.01593	907.775947	P(23)	
11.02974	906.639294	P(24)	
11.0385	905.92	P(25)	
11.0529	904.74	P(26)	
11.0610	904.08	P(27)	
11.0762	902.84	P(28)	
11.08363	902.231482	P(29)	
11.0999	900.91	P(30)	
11.1073	900.31	P(31)	
11.1238	898.97	P(32)	
11.1309	898.40	P(33)	
11.1483	897.00	P(34)	
		00°2 → 10°1 Band	g
10.12615	987.5419	R(45)	
10.13624	986.5588	R(43)	
10.14662	985.5494	R(41)	
10.15730	984.5140	R(39)	

Table 3.2.2 (continued)
TRIATOMIC VIBRATIONAL TRANSITION LASERS

Wavelength (μm) vac	Frequency (cm^{-1})	Transition	Remarks
3.2.2.1a $C^{12}O_2^{16}$ Laser			
		00°2 → 10°1 Band	g
10.16826	983.4527	R(37)	
10.17951	982.3657	R(35)	
10.19105	981.2531	R(33)	
10.20288	980.1150	R(31)	
10.21501	978.9517	R(29)	
10.22742	977.7633	R(27)	
10.24013	976.5498	R(25)	
10.25314	975.3114	R(23)	
10.26643	974.0482	R(21)	
10.28002	972.7604	R(19)	
10.29391	971.4480	R(17)	
10.30810	970.1111	R(15)	
10.32258	968.7499	R(13)	
10.33737	967.3643	R(11)	
10.35246	965.9544	R(9)	
10.36785	964.5204	R(7)	
10.38354	963.0622	R(5)	
10.39955	961.5800	R(3)	
10.41587	960.0737	R(1)	
—	—		
10.45803	956.2031	P(3)	
10.47545	954.6130	P(5)	
10.49319	952.9989	P(7)	
10.51126	951.3608	P(9)	
10.52965	949.6988	P(11)	
10.54838	948.0129	P(13)	
10.56744	946.3030	P(15)	
10.58684	944.5691	P(17)	
10.60658	942.8111	P(19)	
10.62666	941.0291	P(21)	
10.64710	939.2229	P(23)	
10.66789	937.3926	P(25)	
10.68904	935.5380	P(27)	
10.71055	933.6591	P(29)	
10.73243	931.7558	P(31)	
10.75468	929.8281	P(33)	
10.77731	927.8757	P(35)	
10.80032	925.8987	P(37)	
10.82372	923.8969	P(39)	
10.84751	921.8702	P(41)	
10.87171	919.8185	P(43)	
10.89631	917.7417	P(45)	
10.92133	915.6396	P(47)	
10.94676	913.5120	P(49)	
		00°3 → 10°2 Band	g
10.30243	970.6452	R(20)	
10.31616	969.3531	R(18)	
10.33018	968.0370	R(16)	

Table 3.2.2 (continued)
TRIATOMIC VIBRATIONAL TRANSITION LASERS

Wavelength (μm) vac	Frequency (cm^{-1})	Transition	Remarks
3.2.2.1a $C^{12}O_2^{16}$ Laser			
		00°3 → 10°2 Band	g
10.34451	966.6968	R(14)	
10.35912	965.3326	R(12)	
10.37404	963.9446	R(10)	
10.38926	962.5328	R(8)	
10.40477	961.0973	R(6)	
—	—		
10.68565	935.8348	P(24)	
10.70652	934.0103	P(26)	
10.72775	932.1620	P(28)	
10.74934	930.2897	P(30)	
10.77130	928.3934	P(32)	
10.79362	926.4731	P(34)	
10.81632	924.5285	P(36)	
10.83941	922.5596	P(38)	
13.144	760.78	14°0 → 05¹0,Q?	i
13.154	760.21	14°0 → 05¹0,Q?	i
13.159	759.89	14°0 → 05¹0,Q?	i
13.541	738.51	21¹0 → 12²0,Q?	i
		10°0 → 01¹0 Band	
13.87	721	Q(?)	j
14.1681	705.813	P(19)	j
14.2001	704.218	P(21)	j
14.2325	702.619	P(23)	j
		02°0 → 01¹0 Band	
15.59359	641.289	R(29)	k
15.63106	639.752	R(27)	k
15.66877	638.212	R(25)	k
15.70670	636.671	R(23)	k
15.74488	635.127	R(21)	k
15.78330	633.581	R(19)	k
15.82196	632.033	R(17)	k
15.86086	630.483	R(15)	k
15.89997	628.932	R(13)	k
15.93935	627.378	R(11)	k
15.97896	625.823	R(9)	k
—	—		
16.18191	617.974	Q(8)	k
16.18267	617.945	Q(10)	k
16.18364	617.908	Q(12)	k
16.18474	617.866	Q(14)	k
16.18605	617.816	Q(16)	k
16.18744	617.763	Q(18)	k
16.18906	617.701	Q(20)	k
16.19082	617.634	Q(22)	k
16.19278	617.559	Q(24)	k

Table 3.2.2 (continued)
TRIATOMIC VIBRATIONAL TRANSITION LASERS

Wavelength (μm) vac	Frequency (cm^{-1})	Transition	Remarks
3.2.2.1a $C^{12}O_2^{16}$ Laser			
		$02°0 \rightarrow 01^10$ Band	
16.19494	617.477	Q(26)	k
16.19724	617.389	Q(28)	k
—	—		
16.36699	610.986	P(9)	k
16.40907	609.419	P(11)	k
16.45148	607.848	P(13)	k
16.49408	606.278	P(15)	k
16.53696	604.706	P(17)	k
16.58012	603.132	P(19)	k
16.62350	601.558	P(21)	k
16.66719	599.981	P(23)	k
16.71112	598.404	P(25)	k
16.75533	596.825	P(27)	k
16.79983	595.244	P(29)	k
14.1	711	$10°1 \rightarrow 01^11$,Q	e
16.586	602.96	$14°0 \rightarrow 13^10$,Q?	i
16.597	602.51	$14°0 \rightarrow 13^10$,Q?	i
17.023	587.43	$03^11 \rightarrow 02^21$,Q?	i
17.029	587.25	$03^11 \rightarrow 02^21$,Q?	i
17.036	587.00	$03^11 \rightarrow 02^21$,Q?	i
17.048	586.59	$03^11 \rightarrow 02^21$,Q?	i
17.370	575.71	$24°0 \rightarrow 23^10$,Q?	i
17.376	575.49	$24°0 \rightarrow 23^10$,Q?	i
17.390	575.05	$24°0 \rightarrow 23^10$,Q?	i

[a] Observed in pulsed transverse discharge in CO_2-N_2-He mixtures which is simultaneously stimulated by the output from a separate CO_2 laser operating on 00°2 → 02°1 or 00°2 → 10°1 sequence band.[60] Frequencies and wavelengths are calculated from spectroscopic data.[61]

[b] Observed in Br_2-CO_2-He mixtures during flash photolysis of Br_2. CO_2 molecules are excited by energy transfer from electronically excited Br atoms.[62] Frequencies and wavelengths are calculated from spectroscopic data.[61]

[c] Observed in a longitudinal discharge in He-N_2-CO_2 mixtures following Q-switched oscillation near 10.6 μm.[63]

[d] All wavelengths and frequencies given are calculated from accurately determined molecular constants.[64]

[e] Observed in a continuous longitudinal discharge in a He-N_2-CO_2 mixture.[65]

[f] Observed in a continuous longitudinal discharge in CO_2.[66] A continuous output power of 27.2 kW has been reported for a transverse-flow, transverse-discharge He-N_2-CO_2 laser,[67] and a quasicontinuous output of 60 kW has been reported for an N_2-CO_2 gas dynamic laser.[68] In the pulse mode, an output power of 3 GW has been obtained from a UV preionized transverse discharge laser,[69] while a 2-TW output beam has been produced by an electron-beam controlled transverse discharge laser.[70]

Table 3.2.2 (continued)
TRIATOMIC VIBRATIONAL TRANSITION LASERS

[g] Observed in a longitudinal-discharge He-N_2-CO_2 laser with an intracavity hot CO_2 cell.[71]

[h] Observed in a longitudinal-discharge He-N_2-CO_2 laser with an intracavity CO_2 absorption cell.[72]

[i] Observed in a pulsed longitudinal discharge in CO_2.[73]

[j] Observed in a pulsed longitudinal discharge in cryogenically cooled CO_2-N_2-He mixtures which are also irradiated by a 10.6-μm beam from a TEA CO_2 laser.[74] Frequencies and wavelengths are based on absorption spectroscopy.[75]

[k] Observed in CO_2-HBr-Ar mixtures cooled to −80°C which are optically pumped by the outputs from a pulsed HBr and a pulsed CO_2 (10- or 9-μm) laser.[76] Frequencies and wavelengths are based on absorption spectroscopy.[75]

3.2.2.lb $C^{12}O^{16}O^{18}$ Laser

		02°1 → 02°0 Band	
4.314	2318	R(8)	a
4.340	2304	P(10)	a
4.354	2297	P(19)	a
		00°1 → 02°0 Band	b
9.13784	1094.350133	R(33)	
9.14239	1093.805671	R(32)	
9.14699	1093.255670	R(31)	
9.15164	1092.700119	R(30)	
9.15634	1092.139010	R(29)	
9.16110	1091.572331	R(28)	
9.16590	1091.000075	R(27)	
9.17076	1090.422232	R(26)	
9.17567	1089.838793	R(25)	
9.18063	1089.249751	R(24)	
9.18565	1088.655097	R(23)	
9.19071	1088.054825	R(22)	
9.19583	1087.448927	R(21)	
9.20101	1086.837398	R(20)	
9.20624	1086.220230	R(19)	
9.21152	1085.597418	R(18)	
9.21685	1084.968957	R(17)	
9.22224	1084.334843	R(16)	
9.22769	1083.695069	R(15)	
9.23319	1083.049633	R(14)	
9.23874	1082.398531	R(13)	
9.24435	1081.741759	R(12)	
9.25002	1081.079315	R(11)	
9.25574	1080.411196	R(10)	
—			
9.41738	1061.865901	P(14)	

Table 3.2.2 (continued)
TRIATOMIC VIBRATIONAL TRANSITION LASERS

3.2.2.1b $C^{12}O^{16}O^{18}$ Laser

Wavelength (μm) vac	Frequency (cm^{-1})	Transition	Remarks
		00°1 → 02°0 Band	b
9.42462	1061.050831	P(15)	
9.43191	1060.230186	P(16)	
9.43927	1059.403976	P(17)	
9.44669	1058.572211	P(18)	
9.45416	1057.734903	P(19)	
9.46170	1056.892061	P(20)	
9.46931	1056.043698	P(21)	
9.47697	1055.189825	P(22)	
9.48469	1054.330454	P(23)	
9.49248	1053.465599	P(24)	
9.50033	1052.595271	P(25)	
9.50824	1051.719485	P(26)	
9.51621	1050.838254	P(27)	
9.52425	1049.951592	P(28)	
9.53235	1049.059515	P(29)	
9.54051	1048.162036	P(30)	
9.54873	1047.259171	P(31)	
9.55702	1046.350937	P(32)	
9.56537	1045.437348	P(33)	
9.57379	1044.518422	P(34)	
9.58227	1043.594175	P(35)	
9.59081	1042.664624	P(36)	
9.59942	1041.729787	P(37)	
9.60809	1040.789681	P(38)	
		00°1 → 10°0 Band	b
10.08761	991.31477	R(41)	
10.09229	990.85569	R(40)	
10.09704	990.38971	R(39)	
10.10186	989.91685	R(38)	
10.10676	989.437134	R(37)	
10.11173	988.950581	R(36)	
10.11678	988.457207	R(35)	
10.12190	987.957029	R(34)	
10.12709	987.450065	R(33)	
10.13237	986.936332	R(32)	
10.13771	986.415844	R(31)	
10.14313	985.888618	R(30)	
10.14863	985.354669	R(29)	
10.15420	984.814012	R(28)	
10.15985	984.266661	R(27)	
10.16557	983.712631	R(26)	
10.17137	983.151934	R(25)	
10.17724	982.584585	R(24)	
10.18319	982.010596	R(23)	
10.18921	981.429979	R(22)	
10.19531	980.842745	R(21)	
10.20149	980.248908	R(20)	
10.20774	979.648477	R(19)	
10.21407	979.041464	R(18)	

Table 3.2.2 (continued) TRIATOMIC VIBRATIONAL TRANSITION LASERS

Wavelength (μm) vac	Frequency (cm^{-1})	Transition	Remarks
3.2.2.lb $C^{12}O^{16}O^{18}$ Laser			
		00°1 → 10°0 Band	b
10.22048	978.427879	R(17)	
10.22696	977.807731	R(16)	
—	—		
10.43179	958.608605	P(10)	
10.44052	957.807139	P(11)	
10.44933	956.999226	P(12)	
10.45823	956.184866	P(13)	
10.46721	955.364055	P(14)	
10.47629	954.536793	P(15)	
10.48544	953.703074	P(16)	
10.49469	952.862897	P(17)	
10.50402	952.016257	P(18)	
10.51344	951.163150	P(19)	
10.52295	950.303571	P(20)	
10.53255	949.437516	P(21)	
10.54224	948.564979	P(22)	
10.55202	947.685953	P(23)	
10.56189	946.800433	P(24)	
10.57185	945.908413	P(25)	
10.58190	945.009884	P(26)	
10.59204	944.104840	P(27)	
10.60228	943.193273	P(28)	
10.61261	942.275174	P(29)	
10.62304	941.350534	P(30)	
10.63355	940.419345	P(31)	
10.64417	939.481597	P(32)	
10.65488	938.537280	P(33)	
10.66568	937.586384	P(34)	
10.67659	936.628897	P(35)	
10.68759	935.664808	P(36)	
10.69869	934.694106	P(37)	
10.70989	933.716779	P(38)	
10.72118	932.73281	P(39)	
10.73258	931.74220	P(40)	
		02°0 → 01¹0 Band	
16.596	602.6	R(6)	a
16.927	590.8	P(9)	a
16.970	589.3	P(11)	a
		02°1 → 01¹1 Band	
16.76	596.7	R(8)	a

[a] Observed in $C^{12}O^{16}O^{18}$ optically pumped by the output from an HF laser.[77]

[b] Observed in a longitudinal discharge in a mixture of $C^{12}O^{16}O^{18}$-N_2-Xe-He-H_2.[64]

Table 3.2.2 (continued)
TRIATOMIC VIBRATIONAL TRANSITION LASERS

Wavelength (μm) vac	Frequency (cm^{-1})	Transition	Remarks
3.2.2.1c $C^{12}O_2^{18}$ Laser			
		02°1 → 02°0 Band	a
4.346	2301	R(8)	
4.371	2288	P(10)	
4.377	2285	P(14)	
4.382	2282	P(18)	
4.385	2281	P(20)	
4.392	2277	P(24)	
4.398	2274	P(28)	
		00°1 → 02°0 Band	b
8.98767	1112.635004	R(50)	
8.99495	1111.735484	R(48)	
9.00238	1110.817288	R(46)	
9.00998	1109.880340	R(44)	
9.01775	1108.924564	R(42)	
9.02568	1107.949890	R(40)	
9.03378	1106.956250	R(38)	
9.04205	1105.943579	R(36)	
9.05050	1104.911817	R(34)	
9.05911	1103.860906	R(32)	
9.06790	1102.790794	R(30)	
9.07687	1101.701429	R(28)	
9.08601	1100.592768	R(26)	
9.09533	1099.464767	R(24)	
9.10484	1098.317390	R(22)	
9.11452	1097.150603	R(20)	
9.12438	1095.964378	R(18)	
9.13443	1094.758688	R(16)	
9.14467	1093.533515	R(14)	
9.15509	1092.288842	R(12)	
9.16570	1091.024658	R(10)	
9.17649	1089.740957	R(8)	
9.18748	1088.437736	R(6)	
9.19866	1087.114998	R(4)	
9.21003	1085.772750	R(2)	
—	—		
9.23931	1082.331864	P(2)	
9.25137	1080.921457	P(4)	
9.26362	1079.491631	P(6)	
9.27607	1078.042418	P(8)	
9.28873	1076.573857	P(10)	
9.30158	1075.085991	P(12)	
9.31464	1073.578866	P(14)	
9.32790	1072.052534	P(16)	
9.34137	1070.507051	P(18)	
9.35504	1068.942477	P(20)	
9.36892	1067.358878	P(22)	
9.38301	1065.756323	P(24)	
9.39730	1064.134886	P(26)	
9.41181	1062.494644	P(28)	
9.42653	1060.835680	P(30)	

Table 3.2.2 (continued)
TRIATOMIC VIBRATIONAL TRANSITION LASERS

Wavelength (μm) vac	Frequency (cm^{-1})	Transition	Remarks
3.2.2.1c $C^{12}O_2^{18}$ Laser			
		00°1 → 02°0 Band	b
9.44146	1059.158080	P(32)	
9.45661	1057.461932	P(34)	
9.47196	1055.747333	P(36)	
9.48754	1054.014378	P(38)	
9.50333	1052.263168	P(40)	
9.51933	1050.493809	P(42)	
9.53556	1048.706409	P(44)	
9.55200	1046.901078	P(46)	
9.56866	1045.077932	P(48)	
9.58555	1043.237087	P(50)	
9.60265	1041.378663	P(52)	
9.61998	1039.502785	P(54)	
9.63754	1037.609577	P(56)	
9.65531	1035.699167	P(58)	
		00°2 → 02°1 Band	c
9.07506	1101.92111	R(39)	
9.08317	1100.93743	R(37)	
9.09145	1099.93430	R(35)	
9.09991	1098.91164	R(33)	
9.10855	1097.86940	R(31)	
9.11737	1096.80751	R(29)	
9.12637	1095.72591	R(27)	
	—	—	
9.14492	1093.50343	R(23)	
9.15447	1092.36245	R(21)	
9.16421	1091.20160	R(19)	
9.17414	1090.02085	R(17)	
9.18425	1088.82016	R(15)	
9.19456	1087.59953	R(13)	
9.20506	1086.35892	R(11)	
9.21575	1085.09834	R(9)	
9.22664	1083.81777	R(7)	
9.23773	1082.51722	R(5)	
	—	—	
9.32725	1072.12716	P(9)	
9.34006	1070.65737	P(11)	
9.35307	1069.16784	P(13)	
9.36629	1067.65861	P(15)	
9.37972	1066.12974	P(17)	
9.39336	1064.58129	P(19)	
9.40722	1063.01333	P(21)	
9.42129	1061.42593	P(23)	
9.43557	1059.81917	P(25)	
9.45007	1058.19313	P(27)	
9.46479	1056.54789	P(29)	
9.47972	1054.88354	P(31)	
9.49487	1053.20018	P(33)	
9.51024	1051.49791	P(35)	
9.52583	1049.77683	P(37)	

Table 3.2.2 (continued)
TRIATOMIC VIBRATIONAL TRANSITION LASERS

Wavelength (μm) vac	Frequency (cm^{-1})	Transition	Remarks
3.2.2.1c $C^{12}O_2^{18}$ Laser			
		00°2 → 02°1 Band	c
9.54165	1048.03706	P(39)	
9.55768	1046.27869	P(41)	
9.57394	1044.50185	P(43)	
		00°1 → 10°0 Band	b
10.08328	991.274098	R(44)	
10.09604	990.487703	R(42)	
10.10435	989.673232	R(40)	
10.11295	988.830811	R(38)	
10.12186	987.960562	R(36)	
10.13107	987.062600	R(34)	
10.14058	986.137035	R(32)	
10.15039	985.183973	R(30)	
10.16050	984.203513	R(28)	
10.17091	983.195749	R(26)	
10.18163	982.160770	R(24)	
10.19265	981.098661	R(22)	
10.20398	980.009499	R(20)	
10.21562	978.893358	R(18)	
10.22756	977.750307	R(16)	
10.23981	976.580410	R(14)	
10.25238	975.383724	R(12)	
10.26525	974.160302	R(10)	
10.27844	972.910195	R(8)	
10.29195	971.633444	R(6)	
10.30577	970.330089	R(4)	
—	—		
10.38759	962.687339	P(6)	
10.40354	961.211656	P(8)	
10.41982	959.709502	P(10)	
10.43644	958.180873	P(12)	
10.45341	956.625761	P(14)	
10.47072	955.044153	P(16)	
10.48838	953.436031	P(18)	
10.50639	951.801372	P(20)	
10.52476	950.140149	P(22)	
10.54349	948.452326	P(24)	
10.56259	946.737867	P(26)	
10.58205	944.996728	P(28)	
10.60188	943.228860	P(30)	
10.62209	941.434209	P(32)	
10.64268	939.612716	P(34)	
10.66366	937.764316	P(36)	
10.68503	935.888939	P(38)	
10.70679	933.986510	P(40)	
10.72896	932.056949	P(42)	
10.75153	930.100167	P(44)	
10.77451	928.116074	P(46)	
10.79792	926.104570	P(48)	

Table 3.2.2 (continued)
TRIATOMIC VIBRATIONAL TRANSITION LASERS

Wavelength (μm) vac	Frequency (cm^{-1})	Transition	Remarks
3.2.2.1c $C^{12}O_2^{18}$ Laser			
		00°2 → 10°1 Band	c
10.20203	980.19662	R(23)	
10.21315	979.12939	R(21)	
10.22458	978.03540	R(19)	
10.23631	976.91470	R(17)	
10.24834	975.76735	R(15)	
10.26069	974.59339	R(13)	
10.27334	973.39290	R(11)	
—	—		
10.44178	957.69128	P(11)	
10.45848	956.16167	P(13)	
10.47553	954.60579	P(15)	
10.49292	953.02362	P(17)	
10.51066	951.41515	P(19)	
10.52875	949.78034	P(21)	
10.54720	948.11915	P(23)	
10.56600	946.43155	P(25)	
10.58518	944.71749	P(27)	
10.60471	942.97692	P(29)	
10.62462	941.20978	P(31)	
		02° 0 → 01¹ 0 Band	a
17.730	564.0	P(13)	
17.775	562.6	P(15)	
17.821	561.1	P(17)	
17.915	558.2	P(21)	
17.962	556.7	P(23)	
18.010	555.2	P(25)	
18.053	553.9	P(27)	
		02°1 → 01¹1 Band	a
17.463	572.6	R(8)	

[a] Observed in $C^{12}O_2^{18}$ optically pumped by the output from an HF laser.[77]

[b] Observed in a longitudinal discharge in a mixture of $C^{12}O_2^{18}$-N_2-Xe-He-H_2 cooled to −60°C.[64,78]

[c] Observed in a longitudinal-discharge $C^{12}O_2^{18}$-N_2-Xe-He laser with an intracavity hot $C^{12}O_2^{18}$ cell.[156]

Wavelength (μm) vac	Frequency (cm^{-1})	Transition	Remarks
3.2.2.1d $C^{13}O_2^{16}$ Laser			
		00°1 → 02°0 Band	
9.60169	1041.483334	R(36)	a
9.61126	1040.446675	R(34)	a
9.62110	1039.381840	R(32)	a
9.63123	1038.288700	R(30)	a

Table 3.2.2 (continued)
TRIATOMIC VIBRATIONAL TRANSITION LASERS

Wavelength (μm) vac	Frequency (cm^{-1})	Transition	Remarks
3.2.2.1d $C^{13}O_2^{16}$ Laser			
		00°1 → 02°0 Band	
9.64165	1037.167135	R(28)	a
9.65235	1036.017032	R(26)	a
9.66335	1034.838287	R(24)	a
9.67463	1033.630806	R(22)	a
9.68622	1032.394502	R(20)	a
9.69810	1031.129298	R(18)	a
9.71029	1029.835128	R(16)	a
9.72278	1028.511931	R(14)	a
9.73558	1027.159658	R(12)	a
9.74870	1025.778270	R(10)	a
9.76212	1024.367737	R(8)	a
9.77586	1022.928037	R(6)	a
9.78992	1021.459160	R(4)	a
—	—		
9.87304	1012.859224	P(6)	a
9.88923	1011.201098	P(8)	a
9.90576	1009.514024	P(10)	a
9.92262	1007.798072	P(12)	a
9.93983	1006.053323	P(14)	a
9.95738	1004.279869	P(16)	a
9.97528	1002.477810	P(18)	a
9.99353	1000.647256	P(20)	a
10.01213	998.788325	P(22)	a
10.03108	996.901145	P(24)	a
10.05039	994.985854	P(26)	a
10.07006	993.042598	P(28)	a
10.09009	991.071531	P(30)	a
10.11048	989.072816	P(32)	a
10.13123	987.046625	P(34)	a
10.15235	984.993138	P(36)	a
10.17385	982.912542	P(38)	a
		00°2 → 02°1 Band	
9.98394	1001.60906	P(17)	b
10.00202	999.79840	P(19)	b
		00°1 → 10°0 Band	
10.60063	943.340303	R(44)	a
10.61310	942.231411	R(42)	a
10.62585	941.101238	R(40)	a
10.63886	939.949924	R(38)	a
10.65215	938.777604	R(36)	a
10.66571	937.584403	R(34)	a
10.67953	936.370443	R(32)	a
10.69363	935.135838	R(30)	a
10.70801	933.880697	R(28)	a
10.72265	932.605121	R(26)	a
10.73757	931.309207	R(24)	a
10.75277	929.993046	R(22)	a

Table 3.2.2 (continued)
TRIATOMIC VIBRATIONAL TRANSITION LASERS

3.2.2.1d $C^{13}O_2^{16}$ Laser

Wavelength (μm) vac	Frequency (cm⁻¹)	Transition	Remarks
		00°1 → 10°0 Band	
10.76824	928.656723	R(20)	a,c
10.78399	927.300318	R(18)	a,c
10.80002	925.923906	R(16)	a,c
10.81634	924.527554	R(14)	a,c
10.83293	923.111328	R(12)	a
10.84981	921.675286	R(10)	a
10.86697	920.219482	R(8)	a
10.88443	918.743964	R(6)	a
10.90217	917.248777	R(4)	a
—	—		
10.98566	910.277955	P(4)	a
11.00503	908.675151	P(6)	a
11.02472	907.052844	P(8)	a
11.04471	905.411040	P(10)	a
11.06501	903.749742	P(12)	a,c
11.08563	902.068947	P(14)	a,c
11.10656	900.368647	P(16)	a,c
11.12782	898.648830	P(18)	a,c
11.14940	896.909477	P(20)	a,c
11.17131	895.150565	P(22)	a,c
11.19354	893.372066	P(24)	a,c
11.21612	891.573944	P(26)	a,c
11.23903	889.756160	P(28)	a,c
11.26229	887.918669	P(30)	a
11.28590	886.061419	P(32)	a
11.30986	884.184353	P(34)	a
11.33418	882.287407	P(36)	a
11.35885	880.370512	P(38)	a
11.38390	878.433591	P(40)	a
11.40932	876.476562	P(42)	a
11.43511	874.49933	P(44)	a
11.46129	872.50181	P(46)	a
11.48786	870.48389	P(48)	a
		00°2 → 10°1 Band	
10.65317	938.68797	R(41)	b
10.66605	937.55386	R(39)	b
10.67921	936.39893	R(37)	b
10.69263	935.22332	R(35)	b
10.70633	934.02715	R(33)	b
10.72029	932.81054	R(31)	b
10.73452	931.57361	R(29)	b
10.74903	930.31645	R(27)	b
10.76381	929.03916	R(25)	b
10.77886	927.74184	R(23)	b
10.79419	926.42457	R(21)	b
10.80979	925.08744	R(19)	b
10.82567	923.73053	R(17)	b
10.84183	922.35389	R(15)	b
10.85826	920.95760	R(13)	b

Table 3.2.2 (continued)
TRIATOMIC VIBRATIONAL TRANSITION LASERS

Wavelength (μm) vac	Frequency (cm^{-1})	Transition	Remarks
3.2.2.1d $C^{13}O_2^{16}$ Laser			
		00°2 → 10°1 Band	
10.87498	919.54173	R(11)	b
10.89198	918.10632	R(9)	b
10.90927	916.65142	R(7)	b
—	—		
11.00954	908.30302	P(3)	b
11.02873	906.72238	P(5)	b
11.04823	905.12249	P(7)	b
11.06803	903.50335	P(9)	b
11.08813	901.86496	P(11)	b
11.10855	900.20734	P(13)	b
11.12928	898.53047	P(15)	b
11.15033	896.83434	P(17)	b
11.17170	895.11894	P(19)	b
11.19339	893.38426	P(21)	b
11.21541	891.63025	P(23)	b
11.23776	889.85690	P(25)	b
11.26045	888.06417	P(27)	b
11.28347	886.25201	P(29)	b
11.30684	884.42038	P(31)	b
11.33056	882.56922	P(33)	b
11.35462	880.69848	P(35)	b
11.37905	878.80808	P(37)	b
11.40384	876.89795	P(39)	b
11.42899	874.96801	P(41)	b
11.45452	873.01818	P(43)	b
11.48042	871.04836	P(45)	b
11.50671	869.05845	P(47)	b

[a] Observed in a longitudinal discharge in a mixture of $C^{13}O_2^{16}$-N_2-Xe-He-H_2.[64,78]

[b] Observed in a longitudinal-discharge $C^{13}O_2^{16}$-N_2-Xe-He laser with an intracavity hot $C^{13}O_2^{16}$-cell.[157]

[c] Observed in a longitudinal discharge in $C^{13}O_2^{16}$-He mixtures.[79]

Wavelength (μm) vac	Frequency (cm^{-1})	Transition	Remarks
3.2.2.1e $C^{13}O_2^{18}$ Laser			
		00°1 → 02°0 Band	a
9.49996	1052.636154	R(44)	
9.50826	1051.716981	R(42)	
9.51676	1050.777466	R(40)	
9.52547	1049.817506	R(38)	
9.53437	1048.837000	R(36)	
9.54348	1047.835856	R(34)	
9.55280	1046.813984	R(32)	
9.56232	1045.771301	R(30)	
9.57206	1044.707730	R(28)	
9.58200	1043.623198	R(26)	
9.59216	1042.517640	R(24)	
9.60254	1041.390994	R(22)	
9.61314	1040.243207	R(20)	

Table 3.2.2 (continued)
TRIATOMIC VIBRATIONAL TRANSITION LASERS

Wavelength (μm) vac	Frequency (cm^{-1})	Transition	Remarks
3.2.2.1e $C^{13}O_2^{18}$ Laser			
		00°1 → 02°0 Band	a
9.62395	1039.074229	R(18)	
9.63499	1037.884020	R(16)	
9.64625	1036.672542	R(14)	
9.65773	1035.439766	R(12)	
9.66944	1034.185668	R(10)	
9.68138	1032.910232	R(8)	
9.69355	1031.613446	R(6)	
9.70596	1030.295307	R(4)	
—	—		
9.76467	1024.099642	P(4)	
9.77838	1022.664273	P(6)	
9.79233	1021.207682	P(8)	
9.80652	1019.729912	P(10)	
9.82095	1018.231016	P(12)	
9.83564	1016.711049	P(14)	
9.85057	1015.170077	P(16)	
9.86575	1013.608169	P(18)	
9.88117	1012.025402	P(20)	
9.89686	1010.421858	P(22)	
9.91279	1008.797626	P(24)	
9.92898	1007.152800	P(26)	
9.94542	1005.487482	P(28)	
9.96213	1003.801777	P(30)	
9.97909	1002.095799	P(32)	
9.99630	1000.369665	P(34)	
10.01378	998.623498	P(36)	
10.03152	996.857429	P(38)	
10.04953	995.071590	P(40)	
10.06780	993.266121	P(42)	
10.08633	991.441167	P(44)	
10.10512	989.596876	P(46)	
10.12419	987.733404	P(48)	
10.14352	985.850907	P(50)	
		00°1 → 10°0 Band	a
10.50271	952.135368	R(40)	
10.51288	951.214452	R(38)	
10.52333	950.269581	R(36)	
10.53407	949.300882	R(34)	
10.54509	948.308479	R(32)	
10.55640	947.292486	R(30)	
10.56800	946.253013	R(28)	
10.57988	945.190164	R(26)	
10.59205	944.104038	R(24)	
10.60451	942.994725	R(22)	
10.61726	941.862312	R(20)	
10.63030	940.706879	R(18)	
10.64364	939.528501	R(16)	
10.65726	938.327247	R(14)	
10.67118	937.103179	R(12)	

Table 3.2.2 (continued)
TRIATOMIC VIBRATIONAL TRANSITION LASERS

3.2.2.1e $C^{13}O_2^{18}$ Laser

Wavelength (μm) vac	Frequency (cm^{-1})	Transition	Remarks
		00°1 → 10°0 Band	a
10.68540	935.856355	R(10)	
10.69992	934.586828	R(8)	
10.71473	933.294644	R(6)	
—	—		
10.83559	922.884278	P(8)	
10.85304	921.400600	P(10)	
10.87081	919.894429	P(12)	
10.88891	918.365754	P(14)	
10.90733	916.814558	P(16)	
10.92609	915.240819	P(18)	
10.94518	913.644507	P(20)	
10.96460	912.025589	P(22)	
10.98438	910.384024	P(24)	
11.00449	908.719766	P(26)	
11.02496	907.032764	P(28)	
11.04578	905.322961	P(30)	
11.06696	903.590293	P(32)	
11.08851	901.834691	P(34)	
11.11042	900.056078	P(36)	
11.13270	898.254374	P(38)	
11.15537	896.429491	P(40)	
11.17841	894.581333	P(42)	
11.20185	892.709801	P(44)	
11.22568	890.814787	P(46)	
11.24991	888.896178	P(48)	

[a] Observed in a longitudinal discharge in a mixture of $C^{13}O_2^{18}$ - N_2-Xe-He-H_2 cooled to −60°C.[64,78]

3.2.2.1f $C^{14}O^{16}$ Laser

Wavelength (μm) vac	Frequency (cm^{-1})	Transition	Remarks
		00°1 → 02°0 Band	a
9.91788	1008.280282	R(40)	
9.92733	1007.320030	R(38)	
9.93709	1006.330912	R(36)	
9.94715	1005.312772	R(34)	
9.95753	1004.265463	R(32)	
9.96821	1003.188845	R(30)	
9.97922	1002.082785	R(28)	
9.99054	1000.947161	R(26)	
10.00218	999.781858	R(24)	
10.01415	998.586771	R(22)	
10.02645	997.361804	R(20)	
10.03908	996.106870	R(18)	
10.05205	994.821893	R(16)	
10.06536	993.506806	R(14)	
10.07900	992.161553	R(12)	
10.09300	990.786087	R(10)	
10.10734	989.380373	R(8)	
10.12203	987.944385	R(6)	

Table 3.2.2 (continued)
TRIATOMIC VIBRATIONAL TRANSITION LASERS

Wavelength (μm) vac	Frequency (cm^{-1})	Transition	Remarks
2.2.2.1f $C^{14}O^{16}$ Laser			
		00°1 → 02°0 Band	a
—	—		
10.24370	976.209763	P(8)	
10.26149	974.516934	P(10)	
10.27967	972.794124	P(12)	
10.29822	971.041421	P(14)	
10.31716	969.258921	P(16)	
10.33649	967.446731	P(18)	
10.35620	965.604971	P(20)	
10.37631	963.733766	P(22)	
10.39681	961.833254	P(24)	
10.41771	959.903583	P(26)	
10.43901	957.944909	P(28)	
10.46072	955.957396	P(30)	
10.48283	953.941220	P(32)	
10.50534	951.896562	P(34)	
10.52827	949.823614	P(36)	
		00°1 → 10°0 Band	
11.10699	900.33358	R(50)	a
11.12097	899.20226	R(48)	a
11.13520	898.05318	R(46)	a
11.14968	896.88643	R(44)	a
11.16443	895.70211	R(42)	a
11.17943	894.50031	R(40)	a
11.19468	893.28113	R(38)	a
11.21020	892.04463	R(36)	a
11.22598	890.79092	R(34)	a
11.24202	889.52005	R(32)	a
11.25832	888.23212	R(30)	a
11.27488	886.92718	R(28)	a
11.29171	885.60530	R(26)	a
11.30881	884.26654	R(24)	a
11.32617	882.91098	R(22)	a,b
11.34380	881.53866	R(20)	a,b
11.36170	880.14964	R(18)	a,b
11.37988	878.74397	R(16)	a,b
11.39833	877.32170	R(14)	a,b
11.41705	875.88288	R(12)	a
11.43605	874.42754	R(10)	a
11.45533	872.95574	R(8)	a
11.47490	871.46751	R(6)	a
11.49474	869.96288	R(4)	a
—	—		
11.60907	861.39566	P(6)	a
11.63081	859.78513	P(8)	a
11.65286	858.15839	P(10)	a
11.67521	856.51545	P(12)	a
11.69787	854.85631	P(14)	a,b
11.72084	853.18100	P(16)	a,b
11.74413	851.48950	P(18)	a,b

Table 3.2.2 (continued) TRIATOMIC VIBRATIONAL TRANSITION LASERS

Wavelength (μm) vac	Frequency (cm^{-1})	Transition	Remarks

3.2.2.1f $C^{14}O_2^{16}$ Laser

Wavelength (μm) vac	Frequency (cm^{-1})	Transition	Remarks
		00°1 → 10°0 Band	
11.76773	849.78182	P(20)	a,b
11.79165	848.05797	P(22)	a,b
11.81589	846.31794	P(24)	a,b
11.84046	844.56172	P(26)	a,b
11.86536	842.78930	P(28)	a
11.89060	841.00067	P(30)	a
11.91617	839.19581	P(32)	a
11.94209	837.37471	P(34)	a
11.96835	835.53734	P(36)	a
11.99496	833.68367	P(38)	a
12.02192	831.81368	P(40)	a
12.04925	829.92733	P(42)	a
12.07694	828.02458	P(44)	a
12.10499	826.10540	P(46)	a
12.13342	824.16974	P(48)	a

[a] Observed in a longitudinal discharge in a mixture of $C^{14}O_2^{16}$-N_2-Xe-He-H_2.[64]
[b] Observed in a longitudinal discharge in $C^{14}O_2^{16}$- He mixtures.[79]

3.2.2.1g $C^{14}O_2^{18}$ Laser

Wavelength (μm) vac	Frequency (cm^{-1})	Transition	Remarks
		00°1 → 02°0 Band	a
9.89074	1011.04715	R(52)	
9.89843	1010.260827	R(50)	
9.90636	1009.452744	R(48)	
9.91451	1008.622757	R(46)	
9.92289	1007.770725	R(44)	
9.93151	1006.896516	R(42)	
9.94945	1005.081061	R(40)	
9.95877	1004.139576	R(38)	
9.96835	1003.175439	R(36)	
9.97816	1002.188545	R(34)	
9.98823	1001.178798	R(32)	
9.99854	1000.146106	R(30)	
10.00910	999.090384	R(28)	
10.01992	998.011557	R(26)	
10.03100	996.909551	R(24)	
10.04234	995.784305	R(22)	
10.05393	994.635759	R(20)	
10.06579	993.463865	R(18)	
10.07792	992.268578	R(16)	
10.09031	991.049864	R(14)	
10.10297	989.807692	R(12)	
10.10297	989.807692	R(10)	
10.11591	988.542042	R(8)	
10.12912	987.252899	R(6)	

Table 3.2.2 (continued)
TRIATOMIC VIBRATIONAL TRANSITION LASERS

3.2.2.1g $C^{14}O_2^{18}$ Laser

Wavelength (μm) vac	Frequency (cm^{-1})	Transition	Remarks
		00°1 → 02°0 Band	a
10.14260	985.940255	R(4)	
10.15637	984.604111	R(2)	
—	—	—	
10.20676	979.742713	P(4)	
10.22180	978.301014	P(6)	
10.23713	976.835945	P(8)	
10.25276	975.347555	P(10)	
10.26867	973.835897	P(12)	
10.28488	972.301035	P(14)	
10.30139	970.743036	P(16)	
10.31819	969.161978	P(18)	
10.33530	967.557944	P(20)	
10.35271	965.931024	P(22)	
10.37042	964.281315	P(24)	
10.38843	962.608923	P(26)	
10.40676	960.913958	P(28)	
10.42539	959.196538	P(30)	
10.44433	957.456787	P(32)	
10.46359	955.694836	P(34)	
10.48316	953.910824	P(36)	
10.50304	952.104893	P(38)	
10.52325	950.277194	P(40)	
10.54376	948.427882	P(42)	
10.56460	946.557120	P(44)	
10.58576	944.665076	P(46)	
10.60724	942.751923	P(48)	
10.62905	940.817840	P(50)	
10.65118	938.863011	P(52)	
		00°1 → 10°0 Band	a
10.90027	917.40868	R(48)	
10.91123	916.48668	R(46)	
10.92246	915.54439	R(44)	
10.93396	914.58195	R(42)	
10.94572	913.59949	R(40)	
10.95774	912.59713	R(38)	
10.97002	911.57501	R(36)	
10.98258	910.53323	R(34)	
10.99539	909.47191	R(32)	
11.00847	908.39115	R(30)	
11.02182	907.29104	R(28)	
11.03544	906.17169	R(26)	
11.04932	905.03319	R(24)	
11.06347	903.87560	R(22)	
11.07789	902.69902	R(20)	
11.09258	901.50351	R(18)	
11.10754	900.28914	R(16)	
11.12278	899.05598	R(14)	
11.13829	897.80408	R(12)	

Table 3.2.2 (continued)
TRIATOMIC VIBRATIONAL TRANSITION LASERS

Wavelength (μm) vac	Frequency (cm^{-1})	Transition	Remarks
3.2.2.1g $C^{14}O_2^{18}$ Laser			
		00°1 → 10°0 Band	a
11.15407	896.53349	R(10)	
—	—		
11.29963	884.98457	P(6)	
11.31813	883.53817	P(8)	
11.33693	882.07336	P(10)	
11.35602	880.59011	P(12)	
11.37542	879.08844	P(14)	
11.39512	877.56831	P(16)	
11.41514	876.02973	P(18)	
11.43546	874.47266	P(20)	
11.45610	872.89707	P(22)	
11.47706	871.30294	P(24)	
11.49835	869.69023	P(26)	
11.51996	868.05890	P(28)	
11.54189	866.40889	P(30)	
11.56417	864.74015	P(32)	
11.58678	863.05263	P(34)	
11.60973	861.34626	P(36)	
11.63303	859.62096	P(38)	
11.65669	857.87668	P(40)	
11.68070	856.11331	P(42)	
11.70507	854.33079	P(44)	
11.72981	852.52900	P(46)	

[a] Observed in a longitudinal discharge in a mixture of $C^{14}O_2^{18}$-N_2-Xe-He-H_2.[64]

Wavelength (μm) vac	Frequency (cm^{-1})	Transition	Remarks
3.2.2.2a $C^{12}OS$ Laser			
3.428	2917	10°1 → 00°0 Band?	a
		00°1 → 10°0 Band	
8.189	1221.1	R(47)	b
8.191	1220.8	R(46)	b
8.194	1220.5	R(45)	b
8.196	1220.1	R(44)	b
8.198	1219.8	R(43)	b
8.200	1219.4	R(42)	b
8.203	1219.1	R(41)	b
8.205	1218.7	R(40)	b
8.207	1218.4	R(39)	b
8.210	1218.0	R(38)	b
8.212	1217.7	R(37)	b
8.215	1217.3	R(36)	b
8.217	1217.0	R(35)	b
8.219	1216.6	R(34)	b
8.222	1216.3	R(33)	b
8.224	1215.9	R(32)	b
8.227	1215.5	R(31)	b

Table 3.2.2 (continued)
TRIATOMIC VIBRATIONAL TRANSITION LASERS

Wavelength (μm) vac	Frequency (cm^{-1})	Transition	Remarks
3.2.2.2a $C^{12}OS$ Laser			
		00°1 → 10°0 Band	
8.229	1215.2	R(30)	b
8.232	1214.8	R(29)	b
8.234	1214.5	R(28)	b
8.237	1214.1	R(27)	b
8.2388	1213.76	R(26)	b,c
8.2416	1213.35	R(25)	b,c
8.2439	1213.02	R(24)	b,c
8.247	1212.6	R(23)	b
8.249	1212.2	R(22)	b
8.2518	1211.86	R(21)	b,c
8.2543	1211.48	R(20)	b,c
8.2571	1211.08	R(19)	b,c
8.2595	1210.73	R(18)	b,c
8.2623	1210.32	R(17)	b,c
8.2645	1209.99	R(16)	b,c
8.2673	1209.59	R(15)	b,c
8.270	1209.2	R(14)	b
8.273	1208.8	R(13)	b
8.275	1208.4	R(12)	b
8.278	1208.0	R(11)	b
8.281	1207.6	R(10)	b
8.283	1207.2	R(9)	b
8.286	1206.9	R(8)	b
—	—		
8.328	1200.8	P(6)	b
8.330	1200.4	P(7)	b
8.333	1200.0	P(8)	b
8.336	1199.6	P(9)	b
8.339	1199.2	P(10)	b
8.342	1198.8	P(11)	b
8.345	1198.3	P(12)	b
8.348	1197.9	P(13)	b
8.351	1197.5	P(14)	b
8.354	1197.1	P(15)	b
8.357	1196.7	P(16)	b
8.360	1196.2	P(17)	b
8.3625	1195.82	P(18)	b,c
8.3654	1195.40	P(19)	b,c
8.3685	1194.95	P(20)	b,c
8.3715	1194.52	P(21)	b,c
8.3746	1194.09	P(22)	b,c
8.3779	1193.62	P(23)	b,c
8.3809	1193.19	P(24)	b,c
8.3839	1192.76	P(25)	b,c
8.3870	1192.32	P(26)	b,c
8.3900	1191.89	P(27)	b,c
8.3930	1191.46	P(28)	b,c
8.3962	1191.02	P(29)	b,c
8.3999	1190.49	P(30)	b,c
8.4024	1190.14	P(31)	b,c
8.4055	1189.70	P(32)	b,c

Table 3.2.2 (continued)
TRIATOMIC VIBRATIONAL TRANSITION LASERS

Wavelength (μm) vac	Frequency (cm^{-1})	Transition	Remarks
3.2.2.2a $C^{12}OS$ Laser			
		00°1 → 10°0 Band	
8.4085	1189.27	P(33)	b,c
8.4117	1188.82	P(34)	b,c
8.4146	1188.40	P(35)	b,c
8.4178	1187.95	P(36)	b,c
8.4213	1187.46	P(37)	b,c
9.4243	1187.04	P(38)	b,c
8.428	1186.6	P(39)	b
8.431	1186.1	P(40)	b
8.434	1185.6	P(41)	b
8.437	1185.2	P(42)	b
8.441	1184.7	P(43)	b
8.444	1184.3	P(44)	b
8.447	1183.8	P(45)	b
8.451	1183.3	P(46)	b
8.454	1182.9	P(47)	b
8.457	1182.4	P(48)	b
		02°0 → 01¹0 Band	
18.983	526.8	Q(4)	d
19.057	524.7	P(5)	d
3.2.2.2b $C^{13}OS$ Laser			
(8.6)		00°1 → 10°0 Band	e,f

[a] Observed in a pulsed longitudinal discharge in COS-He mixtures.[80]
[b] Observed in COS and COS-CO mixtures optically pumped by a frequency doubled CO_2 laser.[81] The frequencies and wavelengths are calculated from existing spectroscopic data.[82,83]
[c] Observed in a pulsed longitudinal discharge in COS-He mixtures.[84]
[d] Observed in COS optically pumped by a 9.57-μm CO_2 laser.[85]
[e] Observed in C^{13} OS-CO mixtures optically pumped by a frequency doubled CO_2 laser.[81]
[f] Observed in C^{13} OS optically pumped by a CO laser.[86]

Wavelength (μm) vac	Frequency (cm^{-1})	Transition	Remarks
3.2.2.3a $C^{12}S_2$ Laser			
		10°1 → 10°0 Band	
6.6		P(60)	a
		00°1 → 10°0 Band	
11.476	871.37	P(26)	b
11.4823	870.90	P(28)	b,c
11.4893	870.38	P(30)	b,c,d
11.4962	869.85	P(32)	c
11.5031	869.33	P(34)	b,c,d

Table 3.2.2 (continued)
TRIATOMIC VIBRATIONAL TRANSITION LASERS

Wavelength (μm) vac	Frequency (cm^{-1})	Transition	Remarks
3.2.2.3a $C^{12}S_2$ Laser			
		$00°1 \rightarrow 10°0$ Band	
11.5099	868.82	P(36)	c,d
11.5166	868.31	P(38)	b,c
11.5237	867.80	P(40)	c
11.5307	867.27	P(42)	b,c
11.5376	866.73	P(44)	c,d
11.5446	866.20	P(46)	b,c
11.553	865.57	P(48)	b,c
11.560	865.08	P(50)	b,c
11.568	864.46	P(52)	b,c
11.582	863.38	P(56)	b,c

[a] Observed in CS_2 optically pumped by the second harmonic of the 9.2-μm CO_2 R(30) laser line.[81]

[b] Observed in optically pumped CO-CS_2 mixtures. A frequency doubled TEA CO_2 laser is used to excite CO molecules which in turn transfer energy to CS_2.[87]

[c] Observed in continuously flowing mixtures of preexcited N_2 and CS_2.[88]

[d] Observed in an electron-beam-stabilized electric discharge in He-CO-CS_2 mixtures.[89]

Wavelength (μm) vac	Frequency (cm^{-1})	Transition	Remarks
3.2.2.3b $C^{13}S_2$ Laser			
6.9		$10°2 \rightarrow 10°1$?	a
	or	$02°2 \rightarrow 02°1$? Band	
		$00°1 \rightarrow 10°0$? Band	
11.959	836.2	R(40)	b
11.963	835.9	R(38)	b
11.983	834.5	R(30)	b
		$01^10 \rightarrow 11^10$? Band	
12.214	818.7	P(23)	b
12.237	817.2	P(30)	b
12.247	816.5	P(32)	b

[a] Observed in C^{13} S_2 vapor optically pumped by a pulsed HF laser.[90]

[b] Observed in an electron-beam-stabilized electric discharge in He-CO-CS_2 mixtures.[89]

Wavelength (μm) vac	Frequency (cm^{-1})	Transition	Remarks
3.2.2.4 HCN Laser			
3.85		$00°1 \rightarrow 01^10$ Band	a
7.25		$10°0 \rightarrow 01^10$? Band	a
8.48		$00°1 \rightarrow 10°0$ Band P(15)?	a

Table 3.2.2 (continued)
TRIATOMIC VIBRATIONAL TRANSITION LASERS

3.2.2.4 HCN Laser

Wavelength (μm) vac	Frequency (cm^{-1})	Transition	Remarks
12.85	778	$01^10 \rightarrow 00^00$ Band R(22)?	b

[a] Observed in Br_2-HCN-He mixtures during flash photolysis of Br_2. HCN molecules are vibrationally excited by energy transfer from electronically excited Br atoms.[91]

[b] Observed in a pulsed longitudinal discharge in CH_4-N_2 mixtures.[92]

3.2.2.5 H_2O Laser

Wavelength (μm) vac	Frequency (cm^{-1})	Transition	Remarks
2.2792	4387.6	?	a
4.77	2096	$030(8_{-1}) \rightarrow 020,(7_{-4})$?	b
		$020 \rightarrow 010$ Band	
7.093	1410	$(2_2) \rightarrow (3_2)$	c
7.204	1388	$(3_1) \rightarrow (4_1)$	c
7.285	1373	—	c
7.297	1371	$(3_3) \rightarrow (4_3)$	c
7.390	1353	—	c
7.425	1347	$(4_2) \rightarrow (5_2)$	c
7.453	1342	—	c
7.4588	1340.70	$(4_3) \rightarrow (5_5)$	d
7.543	1326	—	c
7.590	1317	—	c
7.5966	1316.38	$(5_5) \rightarrow (6_5)$	d,e
7.7090	1297.19	$(6_5) \rightarrow (7_7)$	c,d,e
7.740	1292	$(6_4) \rightarrow (7_4)$	c
9.3938	1064.53	?	e
9.4747	1055.44	?	e
9.5674	1045.22	?	e
11.83	845.3	?	b
11.96	836.1	$100\,(7_1) \rightarrow 020\,(6_{-3})$	b

[a] Actual line at this location may be in question.[93]

[b] Observed in a pulsed longitudinal discharge in H_2O vapor.[92]

[c] Observed in Br_2-H_2O-He mixtures during flash photolysis of Br_2. H_2O molecules are vibrationally excited by energy transfer from electronically excited Br atoms.[51]

[d] Observed in a pulsed longitudinal discharge in H_2O-He mixtures.[93,94]

[e] Observed in a pulsed transverse discharge in H_2O-He mixtures.[95]

Table 3.2.2 (continued)
TRIATOMIC VIBRATIONAL TRANSITION LASERS

Wavelength (μm) vac	Frequency (cm^{-1})	Transition	Remarks
3.2.2.6a $N_2{}^{14}O$ Laser			
		00°1 → 10°0 Band	
10.30830[a]	970.091800[a]	R(40)	b
10.31565	969.400607	R(39)	b
10.32305	968.705789	R(38)	b
10.33050	968.007355	R(37)	b
10.33800	967.305314	R(36)	b
10.34555	966.599670	R(35)	b,c
10.35314	965.890435	R(34)	b,c
10.36079	965.177614	R(33)	b,c
10.36848	964.461213	R(32)	b,c
10.37623	963.741241	R(31)	b,c
10.38403	963.017705	R(30)	b,c
10.39187	962.290611	R(29)	b,c
10.39977	961.559966	R(28)	b,c
10.40771	960.825777	R(27)	b,c
10.41571	960.088050	R(26)	b,c
10.42376	959.346792	R(25)	b,c
10.43186	958.602009	R(24)	b,c
10.44001	957.853708	R(23)	b,c
10.44821	957.101893	R(22)	b,c
10.45646	956.346572	R(21)	b,c
10.46476	955.587750	R(20)	b,c,d
10.47312	954.825434	R(19)	b,c,d
10.48153	954.059628	R(18)	b,c,d,e
10.48998	953.290339	R(17)	b,c,d,e
10.49849	952.517571	R(16)	b,c,d,e
10.50706	951.741331	R(15)	b,c,d,e
10.51567	950.961625	R(14)	b,c,d,e
10.52434	950.178456	R(13)	b,c,d,e
10.53306	949.391831	R(12)	b,c,d,e
10.54183	948.601755	R(11)	b,c,d,e
10.55066	947.808232	R(10)	b,c,e
10.55954	947.011268	R(9)	b,c
10.56847[a]	946.210868[a]	R(8)	b,c
10.57746	945.407036	R(7)	b,c
10.58649	944.599778	R(6)	b,c
10.59559	943.789097	R(5)	b,c
10.60474	942.974999	R(4)	b,c
10.61394	942.157488	R(3)	b,c
10.62319	941.336569	R(2)	b,c
10.63250	940.512245	R(1)	b,c
10.64187	939.684522	R(0)	f
10.66077	938.018895	P(1)	f
10.67030	937.180999	P(2)	c
10.67988	936.339719	P(3)	b,c
10.68953	935.495062	P(4)	b,c
10.69923	934.647028	P(5)	b,c
10.70898	933.795623	P(6)	b,c
10.71879	932.940853	P(7)	b,c
10.72866	932.082716	P(8)	b,c
10.73859	931.221221	P(9)	b,c
10.74857	930.356369	P(10)	b,c

Table 3.2.2 (continued)
TRIATOMIC VIBRATIONAL TRANSITION LASERS

Wavelength (μm) vac	Frequency (cm^{-1})	Transition	Remarks
3.2.2.6a $N_2{}^{14}O$ Laser			
		00°1 → 10°0 Band	
10.75861	929.488164	P(11)	b,c
10.76871	928.616611	P(12)	b,c,e,g
10.77886	927.741710	P(13)	b,c,e,g
10.78908	926.863466	P(14)	b,c,e,g
10.79935	925.981884	P(15)	b,c,e,g
10.80968	925.096964	P(16)	b,c,e,g
10.82007	924.20871	P(17)	b,c,e,g
10.83052	923.31713	P(18)	b,c,e,g
10.84102	922.42221	P(19)	b,c,e,g
10.85159	921.52398	P(20)	b,c,e,g,h
10.86222	920.62242	P(21)	b,c,e,g,h
10.87290	919.71754	P(22)	b,c,g,h
10.88365	918.80935	P(23)	b,c,g,h
10.89446	917.8978	P(24)	b,c,g,h
10.90533	916.9830	P(25)	b,c,g,h
10.91626	916.0649	P(26)	b,c,g,h
10.92725	915.1434	P(27)	b,c,g,h
10.93830	914.2187	P(28)	b,c,g,h
10.94942	913.2906	P(29)	b,c,g
10.96059	912.3593	P(30)	b,c,g
10.97183	911.425	P(31)	b,c,g
10.98314	910.487	P(32)	b,c,g
10.99450	909.545	P(33)	b,c,g
11.00593	908.601	P(34)	b,c,g
11.01743	907.653	P(35)	b,c,g
11.02898	906.702	P(36)	c,g
11.04061	905.747	P(37)	c,g
		00°2 → 10°1 Band	i
10.41105	960.518	R(30)	
10.41887	959.797	R(29)	
10.42674	959.072	R(28)	
10.43466	958.345	R(27)	
10.44263	957.613	R(26)	
10.45065	956.879	R(25)	
10.45871	956.140531	R(24)	
10.46684	955.398	R(23)	
10.47501	954.653	R(22)	
10.48324	953.904	R(21)	
10.49151	953.152	R(20)	
10.49983	952.396	R(19)	
10.50821	951.637	R(18)	
10.51664	950.874	R(17)	
10.52512	950.108	R(16)	
10.53366	949.338	R(15)	
10.54224	948.565	R(14)	
10.55088	947.788633	R(13)	
10.56830	946.226	R(11)	
10.57710	945.439	R(10)	
10.58595	944.649	R(9)	

Table 3.2.2 (continued)
TRIATOMIC VIBRATIONAL TRANSITION LASERS

Wavelength (μm) vac	Frequency (cm^{-1})	Transition	Remarks
3.2.2.6a $N_2{}^{14}O$ Laser			
10.59485	943.855	R(8)	
10.60380	943.058	R(7)	
10.61281	942.258	R(6)	
10.62187	941.454	R(5)	
—	—		
10.73488	931.543	P(6)	
10.74466	930.695	P(7)	
10.75449	929.844	P(8)	
10.76438	928.989601	P(9)	
10.77433	928.132	P(10)	
10.78434	927.271	P(11)	
10.79440	926.407	P(12)	
10.80452	925.539	P(13)	
10.81470	924.668	P(14)	
10.82494	923.793	P(15)	
10.83523	922.915636	P(16)	
10.84559	922.034	P(17)	
10.85600	921.150	P(18)	
10.86647	920.262	P(19)	
10.87700	919.371	P(20)	
10.88759	918.477	P(21)	
10.89824	917.579	P(22)	
10.90895	916.678	P(23)	
10.91972	915.775	P(24)	
10.93055	914.867	P(25)	
10.94144	913.956	P(26)	
10.95240	913.042	P(27)	
10.96341	912.125	P(28)	
10.97449	911.204	P(29)	
10.98563	910.280	P(30)	

[a] All values of frequencies and wavelengths given are calculated from accurately determined molecular constants.[96]

[b] Observed in a continuous longitudinal discharge in N_2O-N_2-He mixtures.[96]

[c] Observed in a continuous longitudinal discharge in N_2O-N_2 mixtures.[97]

[d] Observed in a pulsed radio-frequency discharge in N_2O-N_2 mixtures.[98]

[e] Observed in a pulsed transverse discharge in N_2O-He mixtures;[95] output power ~6 kW.

[f] Observed in a continuous longitudinal discharge in N_2O-N_2-He mixtures.[99]

[g] Observed in continuous flowing mixtures of N_2O and discharge-excited N_2.[100]

[h] Observed in a pulsed discharge in N_2O.[101]

[i] Observed in a CW, longitudinal-discharge, N_2O-N_2-He laser with an intracavity hot N_2O cell.[102]

Table 3.2.2 (continued)
TRIATOMIC VIBRATIONAL TRANSITION LASERS

Wavelength (μm) vac	Frequency (cm^{-1})	Transition	Remarks
3.2.2.6b $N^{14}N^{15}O$ Laser			
		$10°1 \rightarrow 10°0$ Band	a
4.6204	2164.3	R(15)	
4.6812	2136.2	P(17)	
3.2.2.6c $N^{15}N^{14}O$ Laser			
		$10°1 \rightarrow 10°0$ Band	a
4.5851	2181.0	R(8)	
4.6189	2165.0	P(10)	
3.2.2.6d N^{15}_2O Laser			
		$10°1 \rightarrow 10°0$ Band	a
4.6			

[a] Observed in isotopic N_2O gas optically pumped by a pulsed HF laser.

3.2.2.7 NOCl Laser

Wavelength (μm) vac	Frequency (cm^{-1})	Transition	Remarks
		$\nu_2 + \nu_3 \rightarrow \nu_3$ Band	a
16.4	608 ± 5		10P(26)
16.52	605.5		10P(28)
16.57	603.4		10P(30)
16.69	599.1		10P(34)
16.7	598 ± 2		10P(36)
16.7	598 ± 2		10P(38)
16.75	597.0		10P(40)
16.86	593.2		10P(42)
16.9	590 ± 5		10P(44)
16.99	588.5		10P(34)

[a] Observed in NOCl at 220 K when optically pumped by various 10.7-μm lines of a TEA CO_2 laser.[103,104] The CO_2 laser transitions are identified in the last column.

3.2.2.8 NSF Laser

Wavelength (μm) vac	Frequency (cm^{-1})	Transition	Remarks
		$\nu_2 + \nu_3 \rightarrow \nu_3$ Band	a
15.19	658.3		9P40
15.26	655.1		9P36
15.34	651.7		9P46
15.42	648.6		9P48
15.47	646.5		9P36
15.61	640.7		9P44
15.61	640.6		9P36
15.838	631.41		9P44
15.89	629.2		9P46
15.98	625.9		9P42

Table 3.2.2 (continued)
TRIATOMIC VIBRATIONAL TRANSITION LASERS

Wavelength (μm) vac	Frequency (cm^{-1})	Transition	Remarks
3.2.2.8 NSF Laser			
		$\nu_2 + \nu_3 \rightarrow \nu_3$ Band	a
16.03	623.9		9P42
16.15	619.1		9P40
16.19	617.6		9P36

[a] Observed in NSF between 220 and 250 K when optically pumped by various 9.7-μm lines of a TEA CO_2 laser.[158] The CO_2 laser transitions are identified in the last column.

Table 3.2.3
FOUR-ATOMIC VIBRATIONAL TRANSITION LASERS

Wavelength (μm)vac	Frequency (cm^{-1})	Transition	Remarks
3.2.3.1 BCl_3 Laser			
18.3			a
18.8			a
19.1			a
19.4			a
20.2			a
20.6			a
22.4			a
23.0			a

[a] Observed in a longitudinal discharge in CO_2-N_2-He-BCl_3 mixtures.[105] However, this result is yet to be corroborated. The results of a subsequent study[106] cast strong doubts to the claim that these emission lines are associated with the BCl_3 molecule.

Wavelength (μm)vac	Frequency (cm^{-1})	Transition	Remarks
3.2.3.2a C_2H_2 Laser			
		$\nu_2 \rightarrow \nu_5^1$ Band	
8.0341	1244.70	Q(5) ?	a,c
8.0352	1244.52	Q(7) ?	b,c
8.0356	1244.46	Q ?	a
8.0378	1244.12	Q ?	b
8.0380	1244.09	Q ?	a
8.0402	1243.75	Q(11)	b,c
8.0409	1243.64	Q ?	a
8.0443	1243.12	Q(13)	a,c
8.0445	1243.08	Q ?	b
8.20	1220	P(10)	c
8.30	1205	P(16)	c

[a] Observed in a pulsed longitudinal discharge in H_2-C_2H_2-He mixtures.[107]
[b] Observed in CO-C_2H_2 mixtures in which CO is optically pumped by a frequency-doubled CO_2 TEA laser.[81,108]
[c] Observed in electron beam-controlled discharges in CO-C_2H_2 mixtures.[109]

Table 3.2.3 (continued)
FOUR-ATOMIC VIBRATIONAL TRANSITION LASERS

Wavelength (μm)vac	Frequency (cm^{-1})	Transition	Remarks
3.2.3.2b C_2D_2 Laser			
		$(\nu_4 + \nu_5)\ \Sigma_u^+ \rightarrow \nu_4$ Band	
17.498	571.50	R(22)	a
17.56	569.6	R(21)	a
17.610	567.87	R(20)	a
17.665	566.08	R(19)	a
17.722	564.28	R(18)	a
17.778	562.48	R(17)	a,b-9R(12)
17.835	560.68	R(16)	a
18.79	532.3	?	a
18.84	530.7	?	a
18.85	530.6	Q(1)	b-9P(24)
18.960	527.43	P(2)	a,b-9P(24)
19.03	525.6	P(3)	a
19.081	524.07	P(4)	a
19.20	520.7	P(6)	b-9P(36)
19.947	501.33	P(18)	a
20.010	499.75	P(19)	a,b-9R(12)
20.073	498.17	P(20)	a
20.13	496.8	P(21)	a
20.202	495.00	P(22)	a
20.267	493.41	P(23)	a
20.332	491.83	P(24)	a
		$(\nu_4 + \nu_5)\Delta_u \rightarrow \nu_4$ Band	
17.61	567.8	R(16)	b-9R(20)
19.67	508.4	P(18)	b-9R(20)
		Hot Band ?	
17.45	573.1	R(23) ?	b-9R(14)
18.67	535.7	Q(24)?	b-9R(14)
20.44	489.2	P(25)?	b-9R(14)
		$(2\nu_5 + \nu_4)\pi_g \rightarrow (\nu_5 + \nu_4)\ \Sigma_u$ Band	
18.79	532.2	Q?	b-9P(38)
19.03	525.6	P(4)	b-9P(38)
		$(2\nu_5 + \nu_4)\pi_g \rightarrow (\nu_5 + \nu_4)\Delta_u$ Band	
19.03	525.6	Q?	b-9P(38)
19.27	518.9	P(4)	b-9P(38)

Table 3.2.3 (continued)
FOUR-ATOMIC VIBRATIONAL TRANSITION LASERS

3.2.3.2b C_2D_2 Laser

Wavelength (μm)vac	Frequency (cm⁻¹)	Transition	Remarks
		$(2\nu_5 + \nu_4)\pi_u \rightarrow 2\nu_4\ \Sigma_g$ Band	
18.97	527.1	Q?	b-9P(26)
19.13	522.7	P(3)	b-9P(26)

[a] Observed in C_2D_2 optically pumped in the 9.4-μm R-branch of the $\nu_4 + \nu_5$ band by a high-pressure continuously tunable CO_2 laser.[159] The transition frequencies given are either observed values (4 digits) or calculated values (5 digits).[160]

[b] Observed in C_2D_2 optically pumped by a TEA CO_2 laser in the 9-μm band.[110] The CO_2 laser lines are identified in the remarks column.

3.2.3.3 COF_2 Laser

Wavelength (μm)vac	Frequency (cm⁻¹)	Transition	Remarks
		0200 → 0100 Band	a
~10			

[a] Observed in COF_2 optically pumped by a CO laser.[86]

3.2.3.4a $N^{14}H_3$ Laser

Wavelength (μm)vac	Frequency (cm⁻¹)	Transition	Remarks
		$\nu_4 \rightarrow 0$ Band	
6.27	1595		o
6.69	1495		o
		$\nu_2 \rightarrow 0$ Band	
9.3	1070	aR(6,K)	a
9.6	1037	aR(4,K) or sR(3,K)	a
9.7	1027	aR(3,1), aR(2,2), aR(2,1), aR(4,4), or aR(4,3)	a
9.9	1008	aR(2,0), sR(1,1), sR(1,0), aR(3,3), or aR(3,2)	a
10.29016	971.882	aR(1,1)	a
10.54594	948.232	sP(1,0)	a
10.6	941	aQ(9,3), aQ(10,4), aQ(9,2), aQ(10,3), or aQ(9,1)	a
10.7	931	aQ(8,6), aQ(9,7), aQ(6,5), aQ(3,3), aQ(5,4), aQ(5,5), aQ(7,6), aQ(4,4), or aQ(2,2)	a
10.73729	931.334	aQ(2,2)	b
10.74394	930.757	aQ(3,3)	b
10.75387	929.898	aQ(4,4)	b
10.76710	928.755	aQ(5,5)	b
11.01080	908.199	sP(3,0)	b
11.01	908	sP(3,1 or 2)	c
11.20879	892.157	aP(2,0)	b,d
11.261	887.99	sP(4,1-2)	d,e
11.46044	872.567	aP(3,1)	a,b,d,e,f
11.47135	871.737	aP(3,2)	d

Table 3.2.3 (continued)
FOUR-ATOMIC VIBRATIONAL TRANSITION LASERS

Wavelength (μm)vac	Frequency (cm⁻¹)	Transition	Remarks
3.2.3.4a $N^{14}H_3$ Laser			
		$\nu_2 \rightarrow 0$ Band	
11.52074	868.000	sP(5,0)	a,b,d,e,g
11.52446	867.719	sP(5,3)	v,w
11.71208	853.819	aP(4,0)	b,d,e
11.71582	853.547	aP(4,1)	a,b,e
11.72712	852.724	aP(4,2)	b,d
11.74637	851.327	aP(4,3)	b,d
11.796	847.78	sP(6,1-2)	d
11.80167	847.338	sP(6,4)	d,g
11.97859	834.823	aP(5,1)	a,b,d,e
11.99025	834.011	aP(5,2)	a,b,d,e
12.01008	832.634	aP(5,3)	a,b,d,e,i
12.03872	830.653	aP(5,4)	b,d,e
12.07912	827.875	sP(7,0)	a,b,d,e,g,j,k,w
12.24521	816.646	aP(6,0)	a,b,d,e,g,k
12.24911	816.386	aP(6,1)	a,b,d,e
12.26105	815.591	aP(6,2)	a,b,d,e
12.28136	814.242	aP(6,3)	a,b,d,e,n
12.31072	812.300	aP(6,4)	a,b,d,e
12.35002	809.715	aP(6,5)	a,b,d,e
12.37821	807.871	sP(8,1)	d
12.38425	807.477	sP(8,3)	d
12.39560	806.738	sP(8,5)	d
12.52781	798.224	aP(7,1)	a,b,d,e,v,w
12.53999	797.449	aP(7,2)	a,b,d,e,w
12.56068	796.135	aP(7,3)	a,b,d,e,w
12.59059	794.244	aP(7,4)	a,b,d,e,w
12.63063	791.726	aP(7,5)	a,b,d,e,w
12.68213	788.511	aP(7,6)	a,b,d,e
12.69716	787.578	sP(9,3)	d
12.81145	780.552	aP(8,0)	a,b,d,e,m,o,w
12.81532	780.316	aP(8,1)	a,b,d,e
12.82765	779.566	aP(8,2)	b,d,w
12.84863	778.293	aP(8,3)	a,b,d,e,w
12.87890	776.464	aP(8,4)	a,b,d,e,w
12.91946	774.026	aP(8,5)	a,b,d,e,w
12.97163	770.913	aP(8,6)	a,b,d,e,w
13.03715	767.039	aP(8,7)	a,b
13.11233	762.641	aP(9,1)	a
13.12477	761.918	aP(9,2)	a
13.14593	760.692	aP(9,3)	a,b
13.17643	758.931	aP(9,4)	a
13.21725	756.587	aP(9,5)	a
13.26978	753.592	aP(9,6)	a,b
13.33580	749.861	aP(9,7)	a
13.411	745.7	aP(9,8), aP(10,1), or aP(10,0)	a
13.57749	736.513	aP(10,6)	a
13.82608	723.271	aP(10,9)	a
		$2\nu_2 \rightarrow \nu_2$ Band	
11.55466	865.452	sP(4,1)	n

Table 3.2.3 (continued)
FOUR-ATOMIC VIBRATIONAL TRANSITION LASERS

3.2.3.4a $N^{14}H_3$ Laser

Wavelength (μm)vac	Frequency (cm⁻¹)	Transition	Remarks
		$2\nu_2 \rightarrow \nu_2$ Band	
12.11418	825.479	sP(6,4)	o
12.15575	822.656	sP(6,3)	p
12.18444	820.719	sP(6,2)	q
12.80959	780.665	sP(8,5)	r
12.98352	770.207	aR(6,5)	r
13.23390	755.635	aR(5,3)	i
13.65533	732.315	aR(4,3)	i
13.72555	728.568	aR(4,4)	o
14.78	677	—	s
15.04	665	—	s
15.08	663	—	s
15.41	649	—	s
15.47	646	—	s
15.78148	633.654	aQ(5,3)	i
15.81600	632.271	aQ(7,5)	r
15.85726	630.626	aQ(4,3)	i
15.87758	629.819	aQ(5,4)	i
15.91292	628.420	aQ(6,5)	r
15.94637	627.102	aQ(4,4)	o
18.21	549	—	s
18.92674	528.353	aP(5,4)	o
19.55019	511.504	aP(6,4)	o
20.38798	490.485	aP(7,5)	r
21.05409	474.967	aP(8,5)	r
		$3\nu_2 \rightarrow 2\nu_2$ Band	
21.471	465.74	aP(2,0)?	t
22.542	443.62	aP(3,2)	t
22.563	443.20	aP(3,1)	t
22.71	440	—	s
23.675	422.39	aP(4,0 or 1)	t
23.86	419	—	s
24.918	401.32	aP(5,1)	t
25.12	398	—	s
26.282	380.49	aP(6,0)	t
30.69	326	—	s
31.47	318	—	s
31.951	312.98	—	t
32.13	311	—	s

[a] Observed in optically pumped $N^{14}H_3$-N_2 mixtures.[111] The pump source is the 9.22-μm, R(30) line of a transverse-discharge, pulsed CO_2 laser which excites the sR(5,0) transition of the $0 \rightarrow \nu_2$ band of $N^{14}H_3$. See footnote u.

[b] Same as a.[112] See footnote u.

[c] Observed in NH_3 optically pumped by the 10.33-μm, R(8) line of a pulsed, transverse-discharge CO_2 laser.[104]

[d] Observed in NH_3-He mixtures optically pumped by the 9.22-μm, R(30) line of a pulsed, transverse-discharge CO_2 laser.[113] See footnote u.

[e] Observed in NH_3-N_2 mixtures optically pumped by the 9.22-μm, R(30) line of a pulsed, transverse-discharge CO_2 laser;[114] maximum conversion efficiency = 20%. See footnote u.

Table 3.2.3 (continued)
FOUR-ATOMIC VIBRATIONAL TRANSITION LASERS

[f] Observed in NH_3 optically pumped by the 10.29-μm, R(14) line of a pulsed, transverse-discharge CO_2 laser. The absorbing transition in NH_3 is the aR(1,1) line of the 0 → ν_2 band.[115-117]

[g] Observed in NH_3 optically pumped by the 9.22-μm, R(30) line of a transverse-discharge, pulsed CO_2 laser.[117]

[h] Observed in NH_3 optically pumped by the 10.35-μm, R(6) line of a transverse-discharge, pulsed CO_2 laser. The absorbing transition in NH_3 is the sQ(5,4) line of the 0 → ν_2 band.[116] See footnote u.

[i] Observed in NH_3 optically pumped by two pulsed transverse-discharge CO_2 lasers, one of which is tuned to the 10.72-μm P(32) line and the other to the 9.59-μm P(32) line.[118]

[j] Same as footnote g.[115,116] Peak output powers of up to 1 MW[117] and energy conversion efficiencies of up to 10%[119] have been reported.

[k] Observed in NH_3 optically pumped by the 9.29-μm, R(16) line of a transverse-discharge, pulsed CO_2 laser. The absorbing transition in NH_3 is the aR(6,0) line of the 0 → ν_2 band.[120]

[l] Observed in NH_3 optically pumped by the 10.72-μm, P(32) line of a transverse-discharge, pulsed CO_2 laser. The absorbing transition in NH_3 is the aQ(5,3) line of the 0 → ν_2 band.[116,120]

[m] Same footnote as k.[115,116,121] Peak output powers of up to 1 MW[122] and energy conversion efficiencies of up to 28%[123] have been reported.

[n] Observed in NH_3 optically pumped simultaneously by the 10.29-μm, R(14) line of a TEA C^{12} O_2 laser and the 9.94-μm, P(14) line of a TEA C^{13} O_2 laser.[124]

[o] Observed in NH_3 optically pumped simultaneously by the 10.74-μm, P(34) line and the 10.57-μm, P(18) line of TEA CO_2 lasers.[124,125]

[p] Observed in NH_3 optically pumped simultaneously by the 9.89-μm, P(8) line of a TEA C^{13} O_2 laser and the 9.59-μm, P(24) line of a TEA C^{12} O_2 laser.[124]

[q] Observed in NH_3 optically pumped simultaneously by the 9.68-μm, P(34) line and the 10.63-μm, P(24) line of TEA CO_2 lasers.[124]

[r] Observed in NH_3 optically pumped simultaneously by the 9.34-μm, R(8) line and the 10.63-μm, P(24) line of TEA CO_2 lasers.[126]

[s] Observed in a pulsed longitudinal discharge in NH_3.[127]

[t] Observed in a pulsed longitudinal discharge in NH_3.[128,129]

[u] Wavelengths and frequencies of all positively identified transitions are based on the absorption spectrum.[130]

[v] Observed in NH_3 optically pumped by a TE CO_2 laser operating in the 9.5-μm sequence Band.[161] See footnote u.

[w] Observed in NH_3 optically pumped by a high-pressure, continuously tunable CO_2 laser.[162]

Wavelength (μm)vac	Frequency (cm^{-1})	Transition	Remarks
3.2.3.4b $N^{15}H_3$ Laser			
14.3			a
14.8			a
15.2			a
15.7			a
16.0			a
17.8			a

[a] Observed in $N^{15}H_3$ optically pumped by the P(6) line of the 3 → 2 band of a pulsed HF laser.[131]

Table 3.2.4
FIVE-ATOMIC VIBRATIONAL TRANSITION LASERS

Wavelength (μm)vac	Frequency (cm⁻¹)	Transition	Remarks (J) of CO_2
3.2.4.1a $C^{12}F_4$ Laser			
		$\nu_2 + \nu_4 \rightarrow \nu_2$ Band	
15.306	653.32	R(41)	a,b(10)
15.40	649.3	R(33)	a,b,c(8)
15.46	647.0	—	d(8)
15.48	646.0	R(27)	a,b,c(6); d,e,f(24)
15.49	645.5	R(26)	g(32)
15.547	643.23	R(22)	d,e,f,g,h(20)
15.57	642.4	R(21)	a,b(4),c
15.58	641.9	R(20)	f(18)
15.60	640.9	R(18)	f(16)
15.607	640.73	R(17)	d,e,f,h(16)
15.71	636.7	R(9)	d,e,f(10)
15.71	636.6	—	d(6)
15.74	635.2	R(7)	g(26)
15.83	631.8	Q(7)?	f(6)
15.84	631.5	Q(28)	f(18)
15.84	631.3	Q(14)	b(4)
15.844	631.15	Q(36)	f,h(22)
15.845	631.12	Q(31)	a(12)
15.847	631.05	Q(29)	f,h(18)
15.85	631.0	Q(11)	g(18)
15.85	630.8	Q(35)	b(14)
15.85	630.8	?	b(4)
15.85	630.8	Q(37)	f(22)
15.91	628.5	P(5)	g(24)
15.92	628.2	P(6)	g(24)
15.94	627.3	P(8)	g(28)
16.00	624.9	P(12)	g(20)
16.02	624.4	P(14)	c(6)
16.06	622.8	P(16)	g(26)
16.07	622.4	P(17)	g(26)
16.07	622.4	P(18)	c(8)
16.09	621.4	P(19)	g(18)
16.11	620.6	P(21)	b,c(10)
16.178	618.11	P(25)	d,e,f,h(10)
16.21	617.0	P(27)	b(14)
16.24	615.7	P(30)	f(12)
16.25	615.4	P(30)?	g(28)
16.259	615.06	P(31)	d,e,f,h,i
16.27	614.7	P(32)?	f(8)
16.29	613.7	P(35)?	f(14)
16.340	611.99	P(37)	d,e,f,h(14)
16.40	609.6	P(42)	b(4)

[a] Observed in $C^{12}F_4$ at ~155 K when optically pumped by one of the 9.4-μm P(J) lines of a TEA $C^{12}O^{16}_2$ laser, where J = 4 to 12.[104]

[b] Same as footnote a except that J = 4 to 14.[132]

[c] Same as footnote a except that J = 4 to 10.[133]

[d] Observed in $C^{12}F_4$ at 100 to 200 K when optically pumped by one of the 9.3-μm P(J) lines of a TEA $C^{12}O^{18}_2$ laser, where J = 4 to 12.[104]

[e] Same as footnote d except that J = 10 to 24.[133]

[f] Same as footnote d.[132]

[g] Observed in $C^{12}F_4$ at ~155 K when optically pumped by one of the 9.4-μm P(J) lines of a TEA $C^{12}O^{18}_2$ laser, where J = 18 to 32.[133,135]

Table 3.2.4 (continued)
FIVE-ATOMIC VIBRATIONAL TRANSITION LASERS

Wavelength (μm)vac	Frequency (cm^{-1})	Transition	Remarks (J) of CO_2

3.2.4.1a $C^{12}F_4$ Laser

[h] Same as footnote d except that J = 10 to 22 and the CF_4 temperature is ∼155 K.[104]

[i] Same as footnote h except that J = 12[103] which excites the $R^+(29)$ line of the $\nu_2 + \nu_4$ band of $C^{12}F_4$.[132] An output energy of 65 mJ per pulse at an energy conversion efficiency of ∼3% has been reported.[123]

3.2.4.1b $C^{13}F_4$ Laser

$\nu_2 + \nu_4 \rightarrow \nu_2$ Band

Wavelength (μm)vac	Frequency (cm^{-1})	Transition	Remarks (J) of CO_2
15.29	654.2	R	a(8)
15.32	652.9	R	a(8)
15.42	648.4	R	a(6)
15.44	647.8	R	a(6)
15.45	647.1	R	b(10)
15.47	646.6	R	a(6)
15.53	644.0	R	c(8)
15.54	643.3	R	b,c(8)
15.59	641.3	R	a(4)
15.62	640.4	R	a(4)
15.62	640.3	R	c(6)
15.74	635.5	R	d(8)
15.75	635.0	R	e(30)
15.78	633.9	R	a(2)
15.80	632.8	R	f(28)
15.82	632.2	R	e(28)
	—		
15.89	629.5	Q	a(18)
15.89	629.5	Q	a(14)
15.89	629.5	Q	c(12)
16.00	625.0	P	c(4)
16.09	621.4	P	f(28)
16.10	621.0	P	f(28)
16.11	620.7	P	f(28)
16.15	619.3	P	b(12)
16.15	619.2	P	d(6)
16.16	618.7	P	c(12)
16.20	617.1	P	c(14)
16.23	616.0	P	a(8),d
16.25	615.4	P	f(30)
16.26	615.0	P	a(12)
16.27	614.6	P	f(30)
16.29	613.7	P	c(18)
16.30	613.4	P	c(4)
16.32	612.9	P	a(10)
16.36	611.1	P	a(12)
16.48	606.8	P	a(14)

[a] Observed in $C^{13}F_4$ optically pumped by one of the 9.3-μm R(J) lines of a TEA $C^{12}O^{16}_2$ laser, where J = 2 to 18.[136,137]

[b] Same as footnote a except pumped by one of the 9.4-μm P(J) $C^{12}O^{16}_2$ laser lines, where J = 8 to 12.[133]

Table 3.2.4 (continued)
FIVE-ATOMIC VIBRATIONAL TRANSITION LASERS

Wavelength (μm)vac	Frequency (cm^{-1})	Transition	Remarks (J) of CO_2

3.2.4.1b $C^{13}F_4$ Laser

[c] Same as footnote a except pumped by one of the 9.4-μm P(J) $C^{12}O^{16}_2$ laser lines, where J = 4 to 18.[136,137]

[d] Same as footnote a except that J = 6, 8.[133]

[e] Same as footnote a except pumped by one of the 9.4-μm $C^{12}O;^{18}_2$ laser lines, where J = 28, 30.[133]

[f] Same as footnote e.[135]

3.2.4.1c $C^{14}F_4$ Laser

$\nu_2 + \nu_4 \rightarrow \nu_2$ Band

Wavelength (μm)vac	Frequency (cm^{-1})	Transition	Remarks (J) of CO_2
15.39	649.8	R	a(16)
15.46	647.0	R	b(4)
15.51	644.9	R	b(22)
15.54	643.7	R	a(12)
15.55	643.5	R	b(20)
15.54	643.4	R	b(2)
15.58	642.0	R	b(18)
15.60	641.2	R	a(10)
15.61	640.5	R	a(10)
15.62	640.1	R	c(24)
15.64	639.2	R	b(14)
15.65	638.8	R	c(24)
15.71	636.5	R	b(10)
15.79	633.5	R	b(6)
15.79	633.4	R	b(6)
15.82	632.1	R	c(26)
15.93	627.9	Q	b(18)
15.93	627.8	Q	b(14)
15.94	627.5	Q	b(6)
15.94	627.5	Q	a(10)
15.94	627.4	Q	b(22)
15.94	627.3	Q	b(6)
15.94	627.3	Q	a(14)
15.94	627.3	Q	a(18)
15.94	627.2	Q	b(14)
15.94	627.2	Q	c(28)
15.95	627.1	Q	a(8)
	—		
15.99	625.3	P	c(30)
16.04	623.6	P	c(26)
16.08	621.7	P	a(8)
16.14	619.7	P	c(30)
16.15	619.3	P	c(30)
16.21	616.8	P	a(14)
16.22	616.6	P	b(4)
16.25	615.2	P	a(16)
16.28	614.3	P	c(20)
16.30	613.6	P	b(6)
16.33	612.2	P	a(6)

Table 3.2.4 (continued)
FIVE-ATOMIC VIBRATIONAL TRANSITION LASERS

3.2.4.1c $C^{14}F_4$ Laser

Wavelength (μm)vac	Frequency (cm^{-1})	Transition	Remarks (J) of CO_2
		$\nu_2 + \nu_4 \rightarrow \nu_2$ Band	
16.35	611.8	P	a(20)
16.44	608.2	P	b(10)

[a] Observed in $C^{14}F_4$ optically pumped by one of the 9.4-μm P(J) lines of a TEA $C^{12}O^{16}_2$ laser, where J = 4 to 20.[136,137]

[b] Same as footnote a except pumped by one of the 9.3-μm R(J) $C^{12}O^{16}_2$ laser lines, where J = 2 to 24.[136,137]

[c] Same as footnote a except pumped by one of the 9.4-μm P(J) $C^{12}O^{18}_2$ laser lines, where J = 24 to 30.[135]

3.2.4.2 CF_3I Laser

Wavelength (μm)vac	Frequency (cm^{-1})	Transition	Remarks (J) of CO_2
		$\nu_2 + \nu_3 \rightarrow \nu_3$ Band	a
13.54	738.4		9P(36)
13.57	736.8		9P(34)
13.63	733.6		9P(30)

[a] Observed in CF_3I at 170 K when optically pumped by various 9.6-μm lines of a TEA CO_2 laser.[104] The CO_2 laser lines are identified in the last column.

3.2.4.3 CH_3F Laser

Wavelength (μm)vac	Frequency (cm^{-1})	Transition	Remarks (J) of CO_2
		$2\nu_3 \rightarrow \nu_3$ Band	
9.75	1026	P(4,1)	a

[a] Observed in CH_3F optically pumped simultaneously by the 9.5-μm P(14) and the 9.6-μm P(30) lines of CO_2 TEA lasers.[138]

3.2.4.4 $FClO_3$ Laser

Wavelength (μm)vac	Frequency (cm^{-1})	Transition	Remarks (J) of CO_2
		$\nu_5 + \nu_6 \rightarrow \nu_6$ Band	a
16.32	612.6		10R(22)
16.34	612.2		10R(34)
16.35	611.7		10R(40)
16.45	607.9		10R(26)
16.49	606.3		10R(30)
16.49	606.3		10R(38)
16.50	606.1		10R(42)
16.52	605.4		10R(36)
16.52	605.3		10R(22)
16.56	603.8		10R(30)
16.56	603.8		10R(34)
16.61	601.9		10R(36)
16.66	600.3		10R(38)
16.73	597.8		10R(32)
16.75	596.9		10R(32)
16.76	596.8		10R(28)

Table 3.2.4 (continued)
FIVE-ATOMIC VIBRATIONAL TRANSITION LASERS

3.2.4.4 $FClO_3$ Laser

Wavelength (μm)vac	Frequency (cm^{-1})	Transition	Remarks (J) of CO_2
		$\nu_5 + \nu_6 \rightarrow \nu_6$ Band	a
16.76	596.5		10R(34)
16.77	596.3		10R(42)
16.79	595.6		10R(34)
16.82	594.6		10R(36)
16.93	590.7		10R(34)
17.00	588.3		10R(26)
17.15	583.1		10R(40)
17.19	581.6		10R(38)
17.22	580.7		10R(34)
17.26	579.5		10R(32)
17.28	578.8		10R(42)
17.32	577.3		10R(28)
17.36	576.1		10R(26)
17.44	573.4		10R(20)
17.46	572.8		10R(20)
17.58	568.8		10R(36)
17.71	564.8		10R(32)

[a] Observed in perchloryl fluoride ($FClO_3$) cooled to 160 K and optically pumped by a TEA CO_2 laser operating in the 10.2-μm R branch.[163] The CO_2 laser transitions are identified in the remarks column.

3.2.4.5 HCOOH Laser

Wavelength (μm)vac	Frequency (cm^{-1})	Transition	Remarks
[5.7]			a

[a] A group of lines observed in HCOOH vapor when optically pumped by the P(5) line of the v = 3 → 2 band of a pulsed HF laser.[90]

3.2.4.6 SiF_4 Laser

Wavelength (μm)vac	Frequency (cm^{-1})	Transition	Remarks (J) of CO_2
			a
24.78	403.5		P(8)
25.31	395.1		P(18)
25.36	394.3		P(24)
25.40	393.7		P(20)
25.67	389.5		P(24)
25.67	389.5		P(34)
25.68	389.4		P(18)
25.77	388.0		P(26)
25.79	387.7		P(18)
26.01	384.4		P(30)
26.14	382.5		P(18)

[a] Observed in SiF_4 when cooled to 125 K and optically pumped by various 9.6-μm lines of a TEA CO_2 laser.[139] The CO_2 laser lines are identified in the last column.

3.2.4.6 SiH_4 Laser

Wavelength (μm)vac	Frequency (cm^{-1})	Transition	Remarks
7.9020	1265.50		a
7.9220	1262.31		a

Table 3.2.4 (continued)
FIVE-ATOMIC VIBRATIONAL TRANSITION LASERS

Wavelength (μm)vac	Frequency (cm⁻¹)	Transition	Remarks (J) of CO_2
3.2.4.6 SiH_4 Laser			
7.9490	1258.02		a
7.9548	1257.10		a
7.9690	1254.86		a
7.9920	1251.25		a

[a] Observed in CO-SiH_4 mixtures in which CO is optically pumped by a frequency-doubled CO_2 TEA laser.[109]

Table 3.2.5
SIX-ATOMIC VIBRATIONAL TRANSITION LASERS

Wavelength (μm) vac	Frequency (cm⁻¹)	Transition	Remarks
3.2.5.1 C_2H_4 Laser			
10.53	950		a, R(10)
10.98	911		a, R(16)

[a] Observed in C_2H_4 optically pumped by one of the 10.3-μm R(J) lines (identified in the remarks column) of a TEA CO_2 laser.[116]

Table 3.2.6
SEVEN-ATOMIC VIBRATIONAL TRANSITION LASERS

Wavelength (μm)vac	Frequency (cm⁻¹)	Transition	Remarks
3.2.6.1 CH_3CCH Laser			a
		$\nu_9 + \nu_{10} \rightarrow \nu_{10}$ Band?	
15.711	636.5		10P14
15.716	636.3		10R4
15.721	636.1		10R16
16.121	620.3		10P14
16.39	610		10P14
16.42	609		10R16

[a] Observed in CH_3CCH (propyne, methylacetylene) when cooled to 180 K and optically pumped by various 10.4-μm lines of a TEA CO_2 laser.[164] The CO_2 laser lines are identified in the remarks column.

Table 3.2.6 (continued)
SEVEN-ATOMIC VIBRATIONAL TRANSITION LASERS

Wavelength (μm)vac	Frequency (cm^{-1})	Transition	Remarks
3.2.6.2 SF_6 Laser			
		$2\nu_3 \rightarrow \nu_2 + \nu_4$ Band?	
15.905	628.74		a

[a] Observed in SF_6 optically pumped simultaneously by 10.5-μm P(12) and P(14)lines of a TEA CO_2 laser.[140]

REFERENCES

1. **Coleman, C. D., Bozman, W. R., and Meggers, W. F.,** Table of Wavenumbers II, National Bureau of Standards Monograph 3, U.S. Government Printing Office, Washington, D.C., 1960.
2. **Herzberg, G.,** *Molecular Spectra and Molecular Structure, I, Spectra of Diatomic Molecules,* 2nd ed., Van Nostrand Reinhold, New York, 1950.
3. **Herzberg, G.,** *Molecular Spectra and Molecular Structure, Vol. 2, Infrared and Raman Spectra of Polyatomic Molecules,* Van Nostrand, Princeton, N.J., 1945.
4. **Herzberg, G.,** *Molecular Spectra and Molecular Structure, Vol. 2, Infrared and Raman Spectra of Polyatomic Molecules,* Van Nostrand, Princeton, N.J., 1945, chap. 5.
5. **West, G. A. and Berry, M. J.,** CN photodissociation and predissociation chemical lasers: molecular electronic and vibrational laser emissions, *J. Chem. Phys.,* 61, 4700, 1974.
6. **Bergman, R. C. and Rich, J. W.,** Overtone bands lasing at 2.7-3μm in electrically excited CO, *Appl. Phys. Lett.,* 31, 597, 1977.
7. **Kildal, H., Eng, R. S., and Ross, A. H. M.,** Heterodyne measurements of $^{12}C^{16}O$ laser frequencies and improved Dunham coefficients, *J. Mol. Spectrosc.,* 53, 479, 1974.
8. **Sadie, F. G., Büger, P. A., and Malan, O. G.,** Continuous-wave overtone bands in a CS_2-O_2 chemical laser, *J. Appl. Phys.,* 43, 2906, 1972.
9. **Basov, N. G., Kazakevich, V. S., and Kovsh, I. B.,** Electron-beam-controlled laser utilizing the first overtones of the vibrational-rotational transitions in the CO molecule, I and II, *Sov. J. Quantum Electron.,* 10, 1131, 1980.
10. **Gregg, D. W. and Thomas, S. J.,** Analysis of the CS_2-O_2 chemical laser showing new lines and selective excitation, *J. Appl. Phys.,* 39, 4399, 1968.
11. **Puerta, J., Herrmann, W., Bourauel, G., and Urban, W.,** Extended spectral distribution of lasing transitions in a liquid-nitrogen cooled CO-laser, *Appl. Phys.,* 19, 439, 1979.
12. **Roh, W. B. and Rao, K. N.,** CO laser spectra, *J. Mol. Spectrosc.,* 49, 317, 1974.
13. **Patel, C. K. N. and Kerl, R. J.,** Laser oscillation on $X'\Sigma^+$ vibrational rotational transitions of CO, *Appl. Phys. Lett.,* 5, 81, 1964.
14. **Patel, C. K. N.,** Vibrational-rotational laser action in carbon monoxide, *Phys. Rev.,* 141, 71, 1966.
15. **Patel, C. K. N.,** CW laser on vibrational-rotational transitions of CO, *Appl. Phys. Lett.,* 7, 246, 1965.
16. **Bhaumik, M. L.,** High efficiency CO laser at room temperature, *Appl. Phys. Lett.,* 17, 188, 1970.
17. **Schmid, W. E.,** A simple high energy TEA CO laser, in *High-Power Lasers and Applications,* Kompa, K.-L., and Walther, H., Eds., Springer-Verlag, Berlin, 1979, 148.
18. **Osgood, R. M., Jr., Nichols, E. R., Eppers, W. C., Jr., and Petty, R. D.,** Q switching of the carbon monoxide laser, *Appl. Phys. Lett.,* 15, 69, 1969.

19. **Bhaumik, M. L., Lacina, W. B., and Mann, M. M.,** Enhancement of CO laser efficiency by addition of Xenon, *IEEE J. Quantum Electron.,* QE-6, 576, 1970.
20. **Daiber, J. W. and Thompson, H. M.,** Performance of a large, cw, preexcited CO supersonic laser, *IEEE J. Quantum Electron.,* QE-13, 10, 1977.
21. **Boness, M. J. W. and Center, R. E.,** High-pressure pulsed electrical CO laser, *J. Appl. Phys.,* 48, 2705, 1977.
22. **Johns, J. W. C., McKellar, A. R. W., and Weitz, D.,** Wavelength measurements of $^{13}C^{16}O$ laser transitions, *J. Mol. Spectrosc.,* 51, 539, 1974.
23. **Ross, A.H. M., Eng, R. S., and Kildal, H.,** Heterodyne measurements of $^{12}C^{18}O$, $^{13}C^{16}O$, and $^{13}C^{18}O$ laser frequencies; mass dependence of Dunham coefficients, *Opt. Commun.,* 12, 433, 1974.
24. **Deutsch, T. F.,** New infrared laser transitions in HCl, HBr, DCl, and DBr, *IEEE J. Quantum Electron.,* QE-3, 419, 1967.
25. **Wood, O. R. and Chang, T. Y.,** Transverse-discharge hydrogen halide lasers, *Appl. Phys. Lett.,* 20, 77, 1972.
26. **Rutt, H. N.,** A high-energy hydrogen bromide laser, *J. Phys. D,* 12, 345, 1979.
27. **Keller, F. L. and Nielsen, A. H.,** The infrared spectrum and molecular constants of DBr, *J. Chem. Phys.,* 22, 294, 1954.
28. **Rank, D. H., Rao, B. S., and Wiggins, T. A.,** Molecular constants of HCl^{35}, *J. Mol. Spectrosc.,* 17, 122, 1965.
29. **Webb, D. U. and Rao, K. N.,** A heated absorption cell for studying infrared absorption bands, *Appl. Opt.,* 5, 1461, 1966.
30. **Kasper, J. V. V. and Pimentel, G. C.,** HCl chemical laser, *Phys. Rev. Lett.,* 14, 352, 1965.
31. **Corneil, P. H. and Pimentel, G. C.,** Hydrogen-chloride explosion laser II DCl, *J. Chem. Phys.,* 49, 1379, 1968.
32. **Henry, A., Bourcin, F., Arditi, I., Charneau, R., and Menard, J.,** Effect laser par reaction chimique de l'hydrogene sur du chlore ou du chlorure de nitrosyle, *C. R. Acad. Sci. Ser. B,* 267, 616, 1968.
33. **Pickworth, J. and Thompson, H. W.,** The fundamental vibration-rotation band of deuterium chloride, *Proc. R. Soc. London, Ser. A,* 218, 37, 1953.
34. **Webb, D. U. and Rao, K. N.,** Vibration rotation bands of heated hydrogen halides, *J. Mol. Spectrosc.,* 28, 121, 1968.
35. **Skribanowitz, N., Herman, I. P., Osgood, R. M., Jr., Feld, M. S., and Javan, A.,** Anisotropic ultrahigh gain emission observed in rotational transitions in optically pumped HF gas, *Appl. Phys. Lett.,* 20, 428, 1972.
36. **Skribanowitz, N., Herman, I. P., and Feld, M. S.,** Laser oscillation and anisotropic gain in the $1 \rightarrow 0$ vibrational band of optically pumped HF gas, *Appl. Phys. Lett.,* 21, 466, 1972.
37. **Ultee, C. J.,** Compact pulsed HF lasers, *Rev. Sci. Instrum.,* 42, 1174, 1971.
38. **Suchard, S. N., Gross, R. W. F., and Whittier, J. S.,** Time-resolved spectroscopy of a flash-imitated H_2-F_2 laser, *Appl. Phys. Lett.,* 19, 411, 1971.
39. **Kwok, M. A., Giedt, R. R., and Gross, R. W. F.,** Comparison of HF and DF continuous chemical lasers. II. Spectroscopy, *Appl. Phys. Lett.,* 16, 386, 1970.
40. **Kwok, M. A., Giedt, R. R., and Varwig, R. L.,** Medium diagnostics for a 10-kW cw HF chemical laser, *AIAA J.,* 14, 1318, 1976.
41. **Deutsch, T. F.,** Molecular laser action in hydrogen and deuterium halides, *Appl. Phys. Lett.,* 10, 234, 1967.
42. **Gerber, R. A. and Patterson, E. L.,** Studies of a high-energy HF laser using an electron-beam-excited mixture of high-pressure F_2 and H_2, *J. Appl. Phys.,* 47, 3524, 1976.
43. **Eng, R. S. and Spears, D. L.,** Frequency stabilization and absolute frequency measurements of a cw HF/DF laser, *Appl. Phys. Lett.,* 27, 650, 1975.
44. **Pummer, H., Proch, D., Schmailzl, U., and Kompa, K. L.,** The generation of partial and total vibrational inversion in colliding molecular systems initiated by ir-laser absorption, *Opt. Commun.,* 19, 273, 1976.
45. **Gensel, P., Kompa, K. L., and Wanner, J.,** IF_5-H_2 hydrogen fluoride chemical laser involving a chain reaction, *Chem. Phys. Lett.,* 5, 179, 1970.
46. **Dolgov-Savel'ev, G. G., Zharov, V. F., Neganov, Yu. S., and Chumak, G. M.,** Vibrational-rotational transitions in an $H_2 + F_2$ chemical laser, *Sov. Phys. JETP,* 34, 34, 1972.
47. **Mayer, S. W., Taylor, D., and Kwok, M. A.,** HF chemical lasing at higher vibrational levels, *Appl. Phys. Lett.,* 23, 434, 1973.
48. **Suchard, S. N. and Pimentel, G. C.,** Deuterium fluoride vibrational overtone chemical laser, *Appl. Phys. Lett.,* 18, 530, 1971.
49. **Ultee, C. J.,** Compact pulsed deuterium fluoride laser, *IEEE J. Quantum Electron.,* QE-8, 820, 1972.
50. **Spencer, D. J., Mirels, H., and Jacobs, T. A.,** Comparison of HF and DF continuous chemical lasers. I. Power, *Appl. Phys. Lett.,* 16, 384, 1970.

51. **Petersen, A. B., Braverman, L. W., and Wittig, C.,** H_2O, NO, and N_2O infrared lasers pumped directly and indirectly by electronic-vibrational energy transfer, *J. Appl. Phys.,* 48, 230, 1977.
52. **Deutsch, T. F.,** NO molecular laser, *Appl. Phys. Lett.,* 9, 295, 1966.
53. **Pollack, M. A.,** Molecular laser action in nitric oxide by photodissociation of NOCl, *Appl. Phys. Lett.* 9, 94, 1966.
54. **Giuliano, C. R. and Hess, L. D.,** Chemical reversibility and solar excitation rates of the nitrosyl chloride photodissociative laser, *J. Appl. Phys.,* 38, 4451, 1967.
55. **Callear, A. B. and Van den Bergh, H. E.,** An hydroxyl radical infrared laser, *Chem. Phys. Lett.,* 8, 17, 1971.
56. **Wanchop, T. S., Schiff, H. I., and Welge, K. H.,** Pulsed-discharge infrared OH laser, *Rev. Sci. Instrum.,* 45, 653, 1974.
57. **Rice, W. W. and Jensen, R. J.,** Aluminum fluoride exploding-wire laser, *Appl. Phys. Lett.,* 22, 67, 1973.
58. **Jensen, R. J.,** Metal-atom oxidation lasers, in *Handbook of Chemical Lasers,* Gross, R. W. F. and Bott, J. F., Eds., John Wiley & Sons, New York, 1976, 703.
59. **Rice, W. W., Beattie, W. H., Oldenborg, R. C., Johnson, S. E., and Scott, P. B.,** Boron fluoride and aluminum fluoride infrared lasers from quasicontinuous supersonic mixing flames, *Appl. Phys. Lett.,* 28, 444, 1976.
60. **Znotis, T. A., Reid, J., Garside, B. K., and Ballik, E. A.,** 4.3-μm TE CO_2 lasers, *Opt. Lett.,* 4, 253, 1979.
61. **Guelachvili, G.,** High-resolution Fourier spectra of carbon dioxide and three of its isotopic species near 4.3 μm, *J. Mol. Spectrosc.,* 79, 72, 1980.
62. **Petersen, A. B. and Wittig, C.,** Line-tunable CO_2 laser operating in the region 2280-2360 cm^{-1} pumped by energy transfer from $Br(4^2P_{1/2})$, *J. Appl. Phys.,* 48, 3665, 1977.
63. **Rao, D. R., Hocker, L. O., Javan, A., and Knable, K.,** Spectroscopic studies of 4.3 μ transient laser oscillation in CO_2, *J. Mol. Spectrosc.,* 25, 410, 1968.
64. **Freed, C., Bradley, L. C., and O'Donnell, R. G.,** Absolute frequencies of lasing transitions in seven CO_2 isotopic species, *IEEE J. Quantum Electron.,* QE-16, 1195, 1980.
65. **Ernst, G. J. and Witteman, W. J.,** Transition selection with adjustable outcoupling for a laser device applied to CO_2, *IEEE J. Quantum Electron.,* QE-7, 484, 1971.
66. **Patel, C. K. N.,** Continuous-wave laser action on vibrational-rotational transitions of CO_2, *Phys. Rev.,* 136, 1187, 1964.
67. **Brown, C. O. and Davis, J. W.,** Closed-cycle performance of a high-power electric discharge laser, *Appl. Phys. Lett.,* 21, 480, 1972.
68. **Gerry, E. T.,** The gas dynamic laser, *Laser Focus,* 27, 1970.
69. **Richardson, M. C., Alcock, A. J., Leopold, K., and Burtyn, P.,** A 300-J multigigawatt CO_2 laser, *IEEE J. Quantum Electron.,* QE-9, 236, 1973.
70. **Schappert, G. T., Singer, S., Ladish, J., and Montgomery, M. D.,** Comparison of theory and experiment on the performance of the Los Alamos Scientific Laboratory eight-beam 10 k J CO_2 laser, *J. Opt. Soc. Am.,* 68, 668, 1978.
71. **Siemsen, K. J. and Whitford, B. G.,** Heterodyne frequency measurements of CO_2 laser sequence-band transitions, *Opt. Commun.,* 22, 11, 1977.
72. **Whitford, B.G., Siemsen, K. J., and Reid, J.,** Heterodyne frequency measurements of CO_2 laser hot-band transitions, *Opt. Commun.,* 22, 261, 1977.
73. **Hartmann, B. and Kleman, B.,** Laser lines from CO_2 in the 11 to 18 micron region, *Can. J. Phys.,* 44, 1609, 966.
74. **Kasner, W. H. and Pleasance, L. D.,** Laser emission from the 13.9-μm $10^00 \rightarrow 01^10$ CO_2 transition in pulsed electrical discharges, *Appl. Phys. Lett.,* 31, 82, 1977.
75. **Paso, R., Kauppinen, J., and Anttila, R.,** Infrared spectrum of CO_2 in the region of the bending fundamental ν_2, *J. Mol. Spectrosc.,* 79, 236, 1980.
76. **Osgood, R. M., Jr.,** 1-mJ line-tunable optically pumped 16-μm laser, *Appl. Phys. Lett.,* 32, 564, 1978.
77. **Buchwald, M. I., Jones, C. R., Fetterman, H. R., and Schlossberg, H. R.,** Direct optically pumped multi-wavelength CO_2 laser, *Appl. Phys. Lett.,* 29, 300, 1976.
78. **Freed, C., Ross, A. H. M., and O'Donnell, R. G.,** Determination of laser line frequencies and vibrational-rotational constants of the $^{12}C^{18}O_2$, $^{13}C^{16}O_2$, and $^{13}C^{18}O_2$ isotopes from measurements of CW beat frequencies with fast HgCdTe photodiodes and microwave frequency counters, *J. Mol. Spectrosc.,* 49, 439, 1974.
79. **Siddoway, J. C.,** Calculated and observed laser transitions using $C^{14}O^{16}{}_2$, *J. Appl. Phys.,* 39, 4854, 1968.
80. **DePoorter, G. L. and Balog, G.,** New infrared laser lines in OCS and new method for C atom lasing, *IEEE J. Quantum Electron.,* QE-8, 917, 1972.

81. **Kildal, H. and Deutsch, T. F.**, Optically pumped gas lasers, in *Tunable Lasers and Applications,* Mooradian, A., Jaeger, T., and Stokseth, P., Eds., Springer-Verlag, Berlin, 1976, 367.
82. **Maki, A. G., Plyler, E. K., and Tidwell, E. D.**, Vibration-rotation bands of carbonyl sulfide, *J. Res. Natl. Bur. Stand. Sect. A,* 66, 163, 1962.
83. **Cord, M. S., Peterson, J. D., Lojko, M. S., and Haas, R. H.**, Microwave spectral tables, *Natl. Bur. of Stand. Monogr.,* 70, 4, 101, 1968.
84. **Deutsch, T. F.**, OCS molecular laser, *Appl. Phys. Lett.,* 8, 334, 1966.
85. **Schlossberg, H. R. and Fetterman, H. R.**, Optically pumped vibrational transition laser in OCS, *Appl. Phys. Lett.,* 26, 316, 1975.
86. **Nelson, L. Y., Hoverson, S. J., Newton, J. F., and Byron, S. R.**, *Math. Sci. Northwest Rep.,* 77-1064-4, 1978; **Jones, C. R.**, Optically pumped mid-ir lasers, *Laser Focus,* 70, 1978.
87. **Deutsch, T. F. and Kildal, H.**, A reexamination of the CS_2 laser, *Chem. Phys. Lett.,* 40, 484, 1976.
88. **Patel, C. K. N.**, CW laser oscillation in an N_2-CS_2 system, *Appl. Phys. Lett.,* 7, 273, 1965.
89. **Nelson, L. Y., Fisher, C. H., and Byron, S. R.**, *Appl. Phys. Lett.,* 25, 517, 1974.
90. **Bushnell, A. H., Jones, C. R., Buchwald, M. I., and Gundersen, M.**, New HF laser pumped molecular lasers in the middle infrared, *IEEE J. Quantum Electron,* QE-15, 208, 1979; **Jones, C. R.**, Optically pumped mid-ir lasers, *Laser Focus,* 68, 1978.
91. **Petersen, A. B., Wittig, C., and Leone, S. R.**, Infrared molecular lasers pumped by electronic-vibrational energy transfer from Br($4^2P_{1/2}$): CO_2, N_2O, HCN, and C_2H_2, *Appl. Phys. Lett.* 27, 305, 1975.
92. **Turner, R. and Poehler, T. O.**, Emission from HCN and H_2O lasers in the 4- to 13-μm region, *Phys. Lett. A,* 27, 479, 1968.
93. **Benedict, W. S., Pollack, M. A., and Tomlinson, W. J., III**, The water-vapor laser, *IEEE J. Quantum Electron.,* QE-5, 108, 1969.
94. **Hartmann, B., Kleman, B., and Spangstedt, G.**, Water vapor laser lines in the 7-μm region, *IEEE J. Quantum Electron.,* QE-4, 296, 1968.
95. **Wood, O. R., Burkhardt, E. G., Pollack, M. A., and Bridges, T. J.**, High-pressure laser action in 13 gases with transverse excitation, *Appl. Phys. Lett.,* 18, 112, 1971.
96. **Whitford, B. G., Siemsen, K. J., Riccius, H. D., and Hanes, G. R.**, Absolute frequency measurements of N_2O laser transitions, *Opt. Commun.,* 14, 70, 1975.
97. **Moeller, G. and Rigden, J. D.**, Observation of laser action in the R-branch of CO_2 and N_2O vibrational spectra, *Appl. Phys. Lett.,* 8, 69, 1966.
98. **Howe, J. A.**, R-branch laser action in N_2O, *Phys. Lett.,* 17, 252, 1965.
99. **Djeu, N. and Wolga, G. J.**, Observation of new laser transitions in N_2O, *IEEE J. Quantum Electron.,* QE-5, 50, 1969.
100. **Patel, C. K. N.**, CW laser action in N_2O(N_2-N_2O system), *Appl. Phys. Lett.,* 6, 12, 1965.
101. **Mathias, L. E. S., Crocker, A., and Wills, M. S.**, Laser oscillations from nitrous oxide at wavelengths around 10.9 micrometers, *Phys. Lett.,* 13, 303, 1964.
102. **Siemsen, K. and Reid, J.**, New N_2O laser band in the 10-μm wavelength region, *Opt. Commun.,* 20, 284, 1977.
103. **Tiee, J. J. and Wittig, C.**, CF_4 and NOCl molecular lasers operating in the 16-μm region, *Appl. Phys. Lett.,* 30, 420, 1977.
104. **Tiee, J. J. and Wittig, C.**, Optically pumped molecular lasers in the 11- to 17-μm region, *J. Appl. Phys.,* 49, 61, 1978.
105. **Karlov, N. V., Konev, Yu. B., Petrov, Yu. N., Prokhorov, A. M., and Stel'makh, O. M.**, Laser based on boron trichloride, *JETP Lett.,* 8, 12, 1968; *ZhETF Pis. Red.,* 8, 22, 1968.
106. **Akitt, D. P. and Yardley, J. T.**, Far-infrared laser emission in gas discharges containing boron trihalides, *IEEE J. Quantum Electron.,* QE-6, 113, 1970.
107. **Shelton, C. F. and Byrne, F. T.**, Laser emission near 8 μ from a H_2-C_2H_2-He mixture, *Appl. Phys. Lett.,* 17, 436, 1970.
108. **Kildal, H. and Deutsch, T. F.**, Opitcally pumped infrared V-V transfer lasers, *Appl. Phys. Lett.,* 27, 500, 1975.
109. **Nelson, L. Y., Fisher, C. H., Hoverson, S. J., Byron, S. R., O'Neill, F., and Whitney, W. T.**, Electron-beam-controlled discharge excitation of a CO-C_2H_2 energy transfer laser, *Appl. Phys. Lett.,* 30, 192, 1977.
110. **Rutt, H. N. and Green, J. M.**, Optically pumped laser action in dideuteroacetylene, *Opt. Commun.,* 26, 422, 1978.
111. **Fry, S. M.**, Optically pumped multiline NH_3 laser, *Opt. Commun.,* 19, 320, 1976.
112. **Yamabayashi, N., Yoshida, T., Myazaki, K., and Fujisawa, K.**, Infrared multiline NH_3 laser and its application for pumping an InSb laser, *Opt. Commun.,* 30, 245, 1979.
113. **Tashiro, H., Suzuki, K., Toyoda, K., and Namba, S.**, Wide-range line-tunable oscillation of an optically pumped NH_3 laser, *Appl. Phys.,* 21, 237, 1980.

114. **Grasiuk, A. Z.**, High-power tunable ir Raman and optically pumped molecular lasers for spectroscopy, *Appl. Phys.*, 21, 173, 1980.
115. **Danielewicz, E. J., Malk, E. G., and Coleman, P. D.**, High-power vibration-rotation emission from $^{14}NH_3$ optically pumped off reasonance, *Appl. Phys. Lett.*, 29, 557, 1976.
116. **Chang, T. Y. and McGee, J. D.**, Off-resonant infrared laser action in NH_3 and C_2H_4 without population inversion, *Appl. Phys. Lett.*, 29, 725, 1976.
117. **Mochizuki, T., Yamanaka, M., Morikawa, M., and Yamanaka, C.**, *Jpn J. Appl. Phys.*, 17, 1295, 1978.
118. **Bobrovskii, A. N., Vedenov, A. A., Kozhevnikov, A. V., and Sobolenko, D. N.**, NH_3 laser pumped by two CO_2 lasers, *JETP Lett.*, 29, 537, 1979.
119. **Baranov, V. Yu., Kazakov, S. A., Pis'menny, V. D., Starodubtsev, A. I., Velikhov, E. P., Gorokhov, Yu. A., Letokhov, V. S., Dyad'kin, A. P., Grasiuk, A. Z., and Vasil'yev, B. I.**, Multiwatt optically pumped ammonia laser operation in the 12- to 13-μm, *Appl. Phys.*, 17, 317, 1978.
120. **Yoshida, T., Yamabayashi, N., Miyazaki, K., and Fujisawa, K.**, Infrared and far-infrared laser emissions from a TE CO_2 laser pumped NH_3 gas, *Opt. Commun.*, 26, 410, 1978.
121. **Chang, T. Y. and McGee, J. D.**, Laser action at 12.812 μm in optically pumped NH_3, *Appl. Phys. Lett.*, 28, 526, 1976.
122. **Shaw, E. D. and Patel, C. K. N.**, Improved pumping geometry for high-power NH_3 lasers, *Opt. Commun.*, 27, 419, 1978.
123. **Gupta, P. K., Kar, A. K., Taghizadeh, M. R., and Harrison, R. G.**, 12.8-μ NH_3 laser emission with 40-60% power conversion and up to 28% energy conversion efficiency, *Appl. Phys. Lett.*, 39, 32, 1981.
124. **Lee, W., Kim, D., Malk, E., and Leap, J.**, Hot-band lasing in NH_3, *IEEE J. Quantum Electron.*, QE-15, 838, 1979.
125. **Jacobs, R. R., Prosnitz, D., Bischel, W. K., and Rhodes, C. K.**, Laser generation from 6 to 35 μm following two-photon excitation of ammonia, *Appl. Phys. Lett.*, 29, 710, 1976.
126. **Eggleston, J., Dallarosa, J., Bischel, W. K., Bokor, J., and Rhodes, C. K.**, Generation of 16-μm radiation in $^{14}NH_3$ by two-quantum excitation of the $2\nu_2^-$ (7.5) state, *J. Appl. Phys.*, 50, 3867, 1979.
127. **Akitt, D. P. and Wittig, C. F.**, Laser emission in ammonia, *J. Appl. Phys.*, 40, 902, 1969.
128. **Mathias, L. E. S., Crocker, A., and Wills, M. S.**, Laser oscillations at wavelengths between 21 and 32 μ from a pulsed discharge through ammonia, *Phys. Lett.*, 14, 33, 1965.
129. **Lide, D. R., Jr.**, Interpretation of the far-infrared laser oscillation in ammonia, *Phys. Lett. A*, 24, 599, 1967.
130. **Urban, S., Spirko, V., Papousek, D., McDowell, R S., Nereson, N. G., Belov, S. P., Gershstein, L. I., Maslovskij, A. V., Krupnov, A. F., Curtis, J., and Rao, K. N.**, Coriolis and 1-type interactions in the ν_2, $2\nu_2$, and ν_4 states of $^{14}NH_3$, *J. Mol. Spectrosc.*, 79, 455, 1980.
131. **Jones, C. R., Buchwald, M. I., Gundersen, M., and Bushnell, A. H.**, Ammonia laser optically pumped with an HF laser, *Opt. Commun.*, 24, 27, 1978.
132. **McDowell, R. S., Patterson, C. W., Jones, C. R., Buchwald, M. I., and Telle, J. M.**, Spectroscopy of the CF_4 laser, *Opt. Lett.*, 4, 274, 1979.
133. **Averim, V. G., Alimpiev, S. S., Baronov, G. S., Karlov, N. V., Karchevskii, A. I., Martsynkyan, V. L., Nabiev, Sh. Sh., Sartakov, B. G., and Khokhlov, E. M.**, Spectroscopic characteristics of an optically pumped carbon tetrafluoride laser, *Sov. Tech. Phys. Lett.*, 4, 527, 1978.
134. **Alimpiev, S. S., Baronov, G. S., Karlov, N. V., Karchevskii, A. I., Martsynk'yam, V. L., Nabiev, Sh. Sh., and Khokhlov, E. M.**, Tuning and stabilization of an optically pumped carbon tetrafluoride laser, *Sov. Tech. Phys. Lett.*, 4, 69, 1978.
135. **Eckardt, R., Telle, J., and Haynes, L.**, Isotopically pumped CF_4, Paper WCC 19, Conference on Laser and Electro-Optical Systems, San Diego, February 26 to 28, 1980, Tech. Dig., 54.
136. **Jones, C. R., Telle, J. M., and Buchwald, M. I.**, Optically pumped isotopic CF_4 lasers, *J. Opt. Soc. Am.*, 68, 671, 1978 (and private communication).
137. **Knyazev, I. N., Letokhov, V. S., and Lobko, V. V.**, Weakly forbidden vibration-rotation transitions $\Delta R \neq 0$ in CF_4 laser, *Opt. Commun.*, 29, 73, 1979.
138. **Prosnitz, D., Jacobs, R. R., Bischel, W. K., and Rhodes, C. K.**, Stimulated emission at 9.75 μm following two-photon excitation of methyl fluoride, *Appl. Phys. Lett.*, 32, 221, 1978.
139. **Green, J. M. and Rutt, H. N.**, Optically pumped laser action in silicon tetrafluoride, Proc. 2nd Int. Conf. on Infrared Physics, Zurich, March 5 to 9, 1979, 205.
140. **Barch, W. E., Fetterman, H. R., and Schlossberg, H. R.**, Optically pumped 15.90 μm SF_6 laser, *Opt. Commun.*, 15, 358, 1975.
141. **Dunham, J. L.**, The energy levels of a rotating vibrator, *Phys. Rev.*, 41, 721, 1932.
142. **Gross, R. W. F. and Bott, J. F., Eds.**, *Handbook of Chemical Lasers,* John Wiley & Sons, New York, 1976.
143. **Osgood, R. M., Jr. and Eppers, W. C., Jr.**, High-power CO-N_2-He laser, *Appl. Phys. Lett.*, 13, 409, 1968.

144. **Treanor, C. E., Rich, J. W., and Rehm, R. G.,** *J. Chem. Phys.,* 48, 1798, 1968.
145. **Jeffers, W. Q. and Wiswall, C. E.,** Analysis of pulsed CO lasers, *J. Appl. Phys.,* 42, 5059, 1971.
146. **Rich, J. W. and Thompson, H. M.,** Infrared sidelight studies in the high-power carbon monoxide laser, *Appl. Phys. Lett.,* 19, 3, 1971.
147. **Amat, G.,** Discussion following a paper by C. K. N. Patel, *J. Chim. Phys.,* 64, 91, 1967.
148. **Tyte, D. C.,** Carbon dioxide lasers, in *Advances in Quantum Electronics,* Vol. 1, Goodwin, D. W., Ed., Academic Press, London, 1970.
149. **Cheo, P. K.,** CO_2 lasers, in *Lasers,* Vol. 3, Levine, A. K. and DeMaria, A. J., Eds., Marcel Dekker, New York, 1971.
150. **Nigham, W. L.,** Electron energy distribution and collision rates in electrically excited N_2, CO, and CO_2, *Phys. Rev. A,* 2, 1989, 1970.
151. **Burkhardt, E. G., Bridges, T. J., and Smith, P. W.,** BeO capillary CO_2 waveguide laser, *Opt. Commun.,* 6, 193, 1972.
152. **Wood, O. R., II,** High-pressure pulsed molecular lasers, *Proc. IEEE,* 62, 355, 1974
153. **Taylor, R. S., Alcock, A. J., Sarjeant, W. J., and Leopold, K. E.,** Electrical and gain characteristics of a multiatmosphere uv-preionized CO_2 laser, *IEEE J. Quantum Electron.,* QE-15, 1131, 1979.
154. **Chang, T. Y. and Wood, O. R., II,** Optically pumped continuously tunable high-pressure molecular lasers, *IEEE J. Quantum Electron.,* QE-13, 907, 1977.
155. **Javan, A.,** Theory of a three-level maser, *Phys. Rev.,* 107, 1579, 1957.
156. **Siemsen, K. J.,** Sequence bands of the isotope $^{12}C^{18}O_2$ laser, *Opt. Lett.,* 6, 114, 1981.
157. **Siemsen, K. J.,** The sequence bands of the carbon-13 isotope CO_2 laser, *Opt. Commun.,* 34, 447, 1980.
158. **Fischer, T. A., Tiee, J. J., and Wittig, C.,** Optically pumped NSF molecular laser, *Appl. Phys. Lett.,* 37, 592, 1980.
159. **Deka, B. K., Dyer, P. E., and Winfield, R. J.,** New 17-21-μm laser lines in C_2D_2 using a continuously tunable CO_2 laser pump, *Opt. Lett.,* 5, 194, 1980.
160. **Baldacci, A., Ghersetti, S., Hurlock, S. C., and Rao, K. N.,** Spectrum of dideuteroacetylene near 18.6 microns, *J. Mol. Spectrosc.,* 42, 327, 1972.
161. **Znotins, T. A., Reid, J., Garside, B. K., and Ballik, E. A.,** 12-μm NH_3 laser pumped by a sequence CO_2 laser, *Opt. Lett.,* 5, 528, 1980.
162. **Deka, B. K., Dyer, P. E., and Winfield, R. J.,** Optically pumped NH_3 laser using a continuously tunable CO_2 laser, *Opt. Commun.,* 33, 206, 1980.
163. **Rutt, H. N.,** Optically pumped laser action in perchloryl fluoride, *Opt. Commun.,* 34, 434, 1980.
164. **Fisher, T. A. and Wittig, C.,** 16-μm laser oscillation in propyne, *Appl. Phys. Lett.,* 39, 6, 1981.

3.3 FAR INFRARED LASERS

3.3.1 INTRODUCTION

Paul D. Coleman

The basic laser problem in any range of the spectrum is to find a quantum system with appropriate energy level spacings and an efficient excitation scheme to drive it. The quantum systems explored to date for the far IR have all been gaseous atoms or molecules, and the most efficient excitation scheme has been photon pumping with CO_2 or N_2O laser. Glow discharge pumping has been used for noble gases (Ne, Ar, Xe, etc.) and a very limited number of molecules (H_2O, HCN, NH_3, SO_3, H_2S) that can survive dissociation in the discharge.

Noble gases all have far IR transitions in their highly excited states near their ionization limit, i.e., 8p → 7d, 9p → 8d, 10p → 9d, etc. and they can readily sustain glow discharges. Patel[1] in 1962-63 was the first to report far IR lasing in Ne. The laser output power was weak since it is difficult to obtain appreciable gain on these highly excited states. Glow discharges are not a very selective excitation scheme and one typically has to rely on selective relaxation for population inversion.

A partial energy level diagram of Ne, typical of the noble gases, is shown in Figure 3.3.1.1. The ns, np, nd, etc. (n on integer) levels crowd together for large values of n near the ionization limit, causing the wavelengths with the associated s ↔ p, p ↔ d, d ↔ f, etc. transitions to become longer as one goes up the energy scale. For example, 50.70 μm transition in Ne is identified as a $7p_2 \rightarrow 6d_2$.

The vibrational-rotational and pure rotational transitions in molecules[2] will typically lie in the IR to far IR spectral region. There are probably hundreds of volatile molecules at temperatures below 400°C that are potential candidates for far IR lasers. As the energy levels in these molecules become more closely spaced, relaxation rates typically increase and it becomes more difficult to obtain inversion. To obtain lasing more state selective excitation must be employed. Efficient lasing will ordinarily require a state-selective excitation mechanism.

The first far IR molecular lasers were obtained by Mathias and Parker[3] in 1963 with the glow discharge excitation of the asymmetric top molecule H_2O and the linear molecule HCN. Since glow discharge excitation of these molecules was not state specific, it was difficult to explain the lasing mechanism and identify the laser transitions.

In 1967 Lide[4] explained the basic laser process in HCN. He postulated that HCN was formed with an abnormal excitation of the 11^10 mode to give a population inversion relative to the 04^00 mode of the molecule. Irregular perturbation via Coriolis mixing gave selected 11^10 to 04^00 transitions sufficient dipole moments to achieve gain and lasing. A partial energy level diagram showing the principle lasing transitions is given in Figure 3.3.1.2. The Coriolis perturbation is between the $J = 10$ states of 11^10 and 0.4^00.

In 1968 Benedict et al.[5] made the laser assignments in H_2O. In the case of H_2O, the flow discharge creates a population inversion between the ν_1, ν_3 stretch modes and the $2\nu_2$ bending modes of the ground electronic state. Normally, the vibrational transition probabilities are very small. However, where there is an irregular perturbation between ν_1 and $2\nu_2$ or ν_3 and $2\nu_2$, the consequent mixing of states greatly increases the matrix elements to the point where lasing can occur. An energy level diagram for H_2O showing most of the laser transitions and irregularly perturbed energy levels is shown in Figure 3.3.1.3. The strong 28.054 μm lasing line is seen to be a $2\nu_2$ 6_{61} to $2\nu_2$ 5_{50} transition which is associated with irregular perturbations on both the upper and lower laser energy levels.

Photon pumping of molecules, first reported by Chang and Bridges[6] in 1970 in CH_3F, has proven to be the most versatile method of generating far IR radiation. Strong specific laser excitation makes it possible to achieve population inversion on

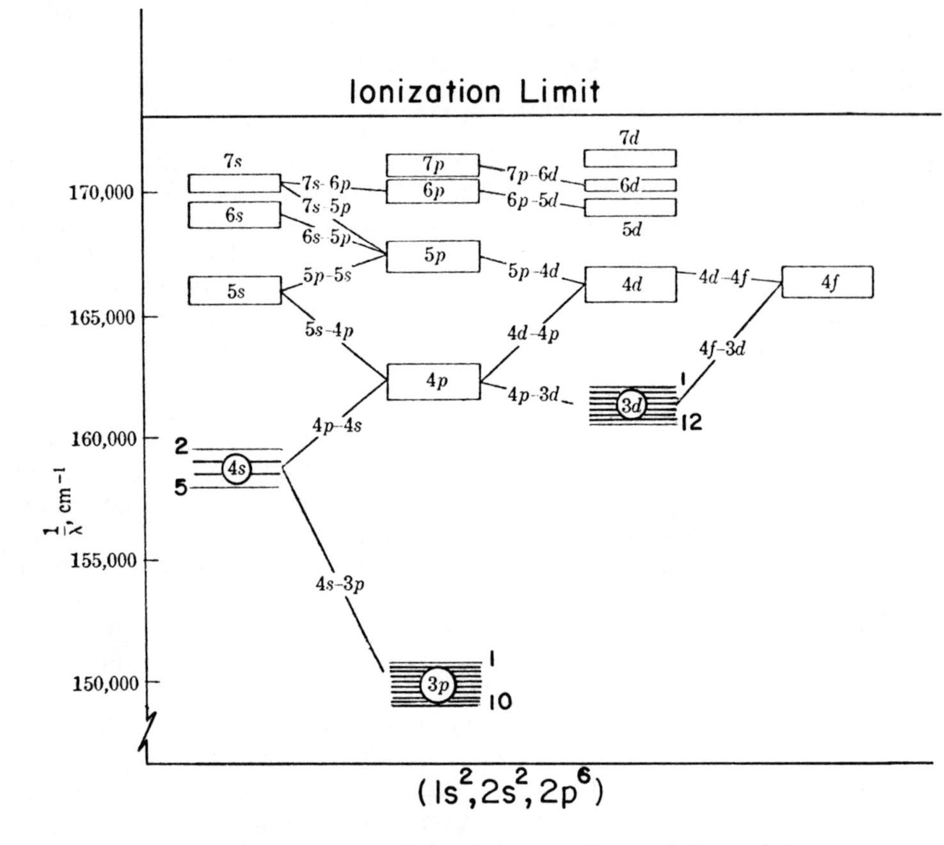

FIGURE 3.3.1.1. Laser transitions between highly-excited states in neon.

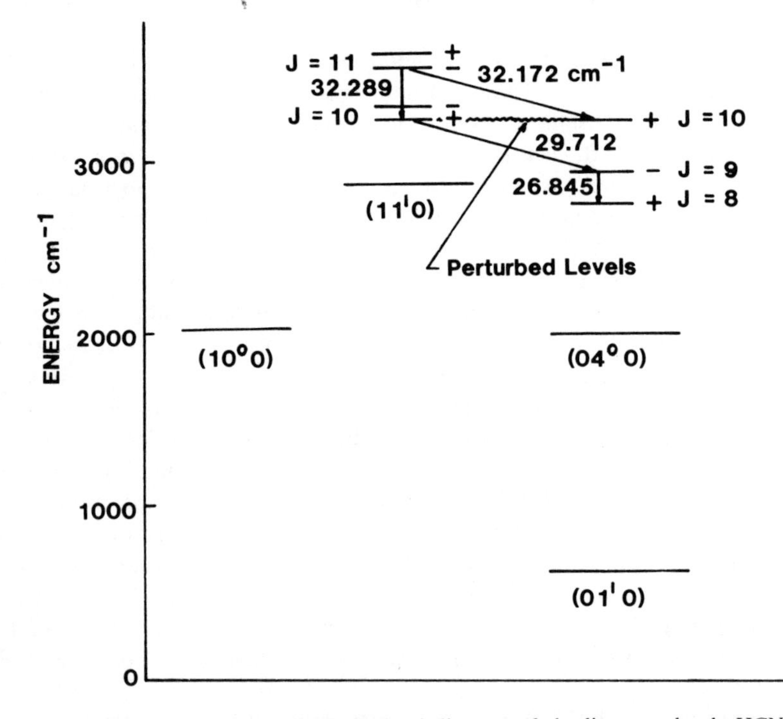

FIGURE 3.3.1.2. Partial energy level diagram of the linear molecule HCN showing laser transitions. The ℓ-doubling splits the J levels of (11¹0).

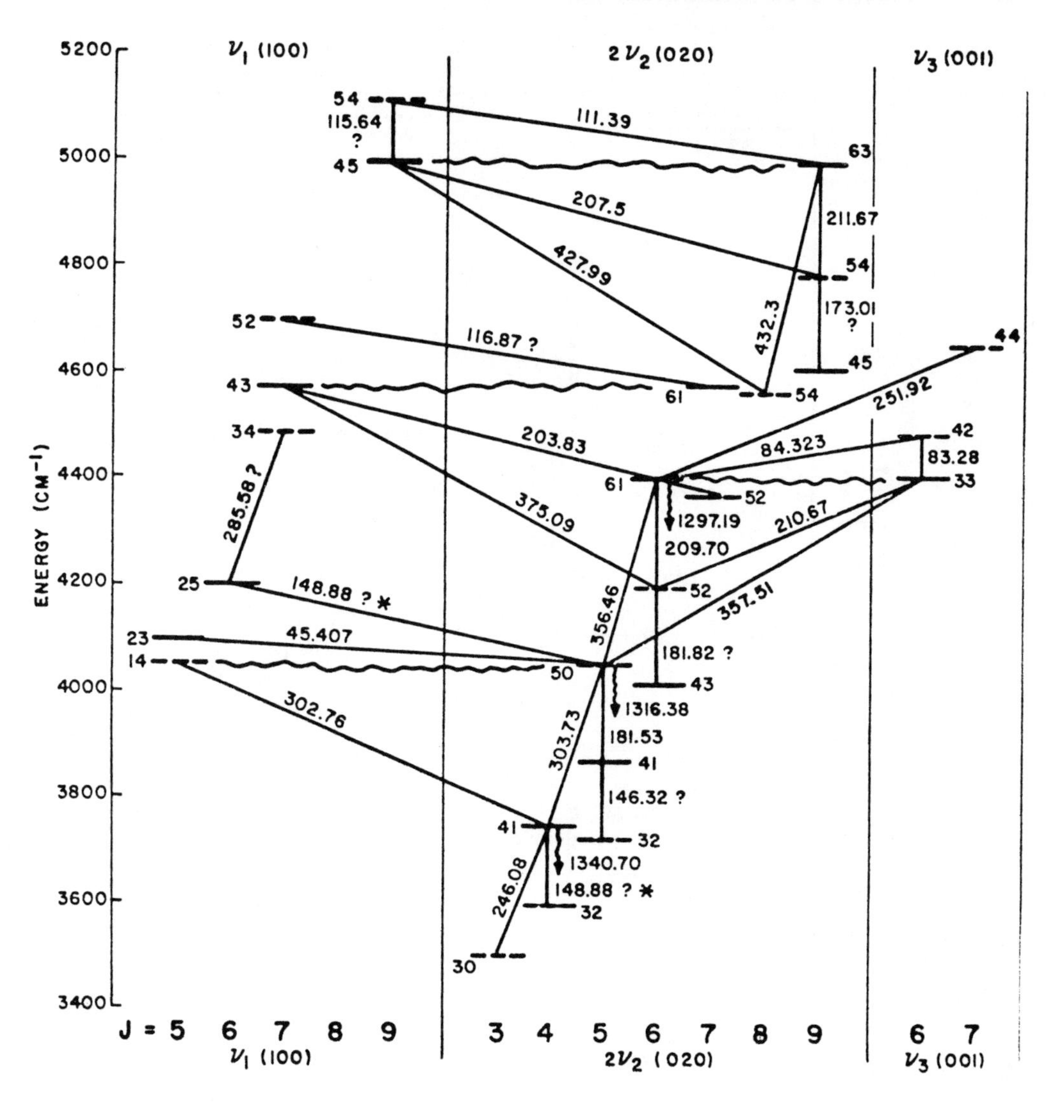

FIGURE 3.3.1.3. Partial energy level diagram for H_2O showing irregular perturbations of energy levels and laser transitions.

rotational lines as closely spaced as 1 cm^{-1} or less. State-of-the-art reviews on optical pumping have been written by Chang in 1974,[7] Coleman in 1977,[8] Hodges in 1978,[9] and DeTemple in 1979.[10] General theory papers on optical pumping have been contributed by Tucker in 1974,[11] Henningsen in 1975,[12] DeTemple in 1976,[13] Temkin in 1977,[14] and Kim in 1980.[15]

The pump source used in most of the far IR laser work is the CO_2 laser with its many powerful lasing lines around 10 μm wavelength. This requires the molecules to be pumped to have a 10 μm absorption spectrum. Also it requires that the CO_2 pump line match the molecular absorption line to within several absorption linewidths, typically less than 0.2 GHz. It turns out in practice that the probability of this spectral coincidence is rather large thanks to the many CO_2 laser lines and rich absorption spectra of the molecules.

By 1981, over 50 molecules had been optically pumped to yield over 1200 lasing lines in the far IR. These molecules ranged from two-atom diatomics to over ten-atom asymmetric tops. It is convenient to classify the molecules into the following categories:

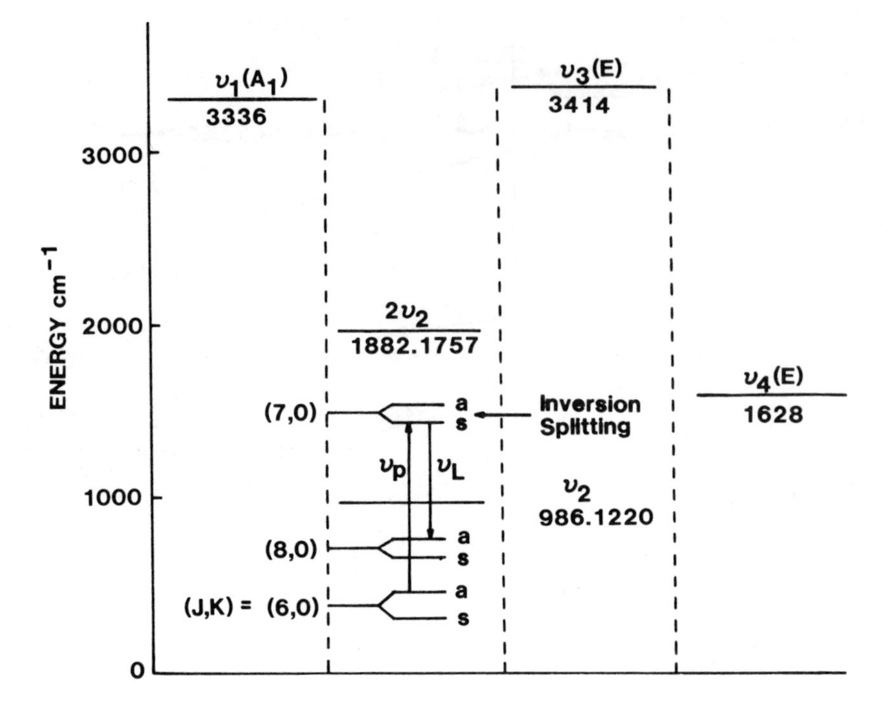

FIGURE 3.3.1.4. Partial energy level diagram for NH_3 showing the CO_2 $R_9(16)$ pump transition ν_p given by Ga(6,0) → ν_2s(7,0) and the 12.81 μm laser transition ν_2s(7,0) → Ga(8,0). The selection rules are $\Delta v = 0, \pm 1$, $\Delta J = 0, \pm 1$, $\Delta K = 0$, a ↔ s. NH_3 is a symmetric pyramidal oblate top with inversion splitting.

1. Linear molecules
2. Symmetric tops
   - 2.1 Without internal rotation
   - 2.2 With inversion
   - 2.3 With internal rotation
3. Asymmetric tops
   - 3.1 Without internal rotation
   - 3.2 With internal rotation (OH)
   - 3.3 With NH_2 groups
   - 3.4 With internal rotation

The most prolific FIR laser molecules are the symmetric tops with inversion, e.g., NH_3 and the asymmetric tops with internal rotation of the OH group, e.g., CH_3OH and HCOOH.

A partial energy level diagram of NH_3 is shown in Figure 3.3.1.4. Excellent spectroscopic data on NH_3 can be obtained from the Ph.D. thesis of Curtis.[16] This molecule has four normal modes of vibration with the ν_2 mode at 986.1220 cm^{-1} the one of interest for CO_2 laser pumping. Each J, K level will be split by the inversion into a symmetric s and antisymmetric state a, with the added selection rule for the symmetric top ($\Delta\nu = 0, \pm 1$, $\Delta J = 0, \pm 1$, $\Delta K = 0$) being a ↔ s. This inversion splitting will result in NH_3 being able to yield lasing lines from the IR to the microwave region (∿ 19 GHz). The lasing example shown in Figure 4 is the CO_2 $R_9(16)$ pump transition Ga(6,0) → ν_2s(7,0) at 9.29 m and the lasing transition ν_2s(7,0) → Ga(8,0) at 12.81 μm wavelength.

The CH_3OH molecule[17] is a weakly asymmetric top with eleven normal vibrational modes plus the torsional or hindred internal rotational mode of the OH radical. The

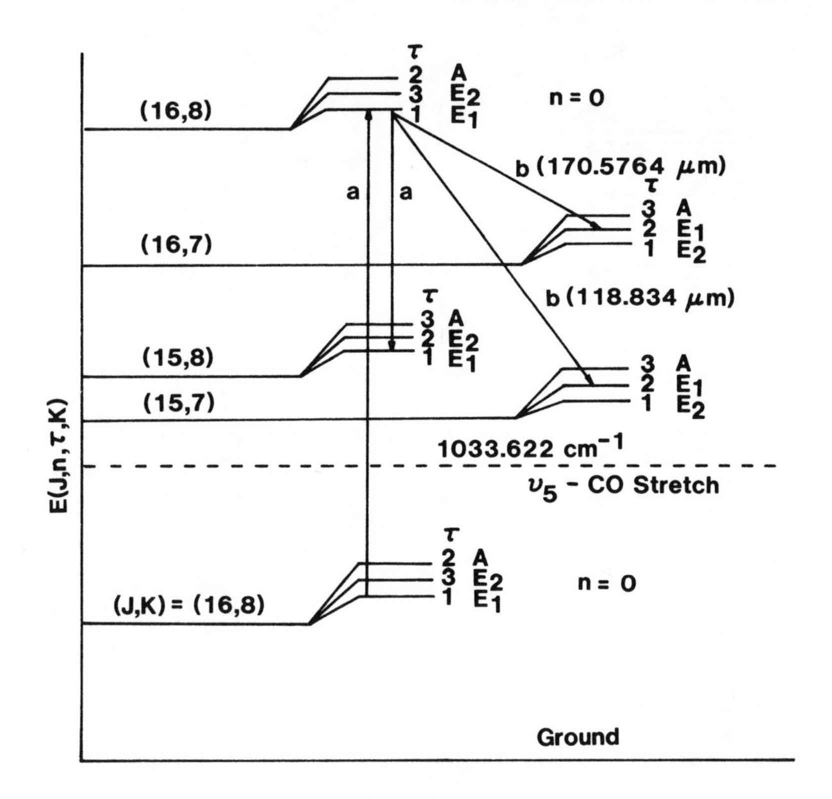

FIGURE 3.3.1.5. Partial energy level diagram for the weakly asymmetric top CH_3OH with hindered internal rotation, showing the CO_2 P_{10} (36) pump transition G(16,0,1,8) → ν_5(16,0,1,8) and 118.834 μm lasing transition ν_5(16,0,1,8) → ν_t(15,0,2,7). The state designation can be determined from the relations E_1 → $\tau + K = 3N$, A → $\tau + K = 3N + 1$, and E_2 → $\tau + K = 2N + 2$ where N is an integer. Selection rules are E_1 → F_1, A → A, E_2 → E_2.

mode of interest is ν_5, the CO stretch mode at 1033.622 cm^{-1}. An expression for the energy in this molecule requires four quantum numbers J, K, n, τ; i.e.,

$$E(J,n,\tau,K) = \nu_5 + E(J,K)^{\text{ext rot}} + E(n,\tau,K)^{\text{int rot}} \quad (1)$$

where $n = 0$, k, 2, 3, ... is the quantum number for the torsional mode and $\tau = 1, 2, 3$ is associated with the threefold potential seen by the OH group. If the state designation E_1, A and E_2 is used, where $E_1;\tau + K = 3N$, $A;\tau + K = 3N+1$ and $E_2;\tau + K = 3N+2$ with N an integer, then the selection rules are simply $E_1 \leftrightarrow E_1$, $A \leftrightarrow A$ and $E_2 \leftrightarrow E_2$. The internal rotation permits two types of transitions, the usual a-type symmetric top condition $\Delta K = 0$ plus the b-type condition $\Delta K = \pm 1$ which makes the spectra complex and extensive, leading to many far IR laser line possibilities. A partial energy diagram for CH_3OH showing the strong 118.834 μm lasing transition is given in Figure 3.3.1.5. The CO_2 pump transition $P_{10}(36)$ is given by G(16,0,1,8) → ν_5(16,0.1,8) an a-type, while the 118 μm lasing transition is on ν_5(16,0,1,8) → ν_5(15,0,2,7) a b-type transition. The Ph.D. thesis of Woods[18] is an excellent source for identifying lasing transitions in CH_3OH.

Asymmetric top molecules are important candidates for IR molecular lasers again because of the richness of the spectrum introduced by the asymmetry. A partial hybrid energy level diagram for D_2O, illustrating the 66 μm transition is shown in Figure 3.3.1.6. Here the energy levels for a prolate symmetric top is given on the left axis and that for an oblate symmetric top on the right axis. An asymmetric top energy level, i.e., G 7_{70}, represents the limiting case which connect $K = 7$ on the left with $K = 0$ on

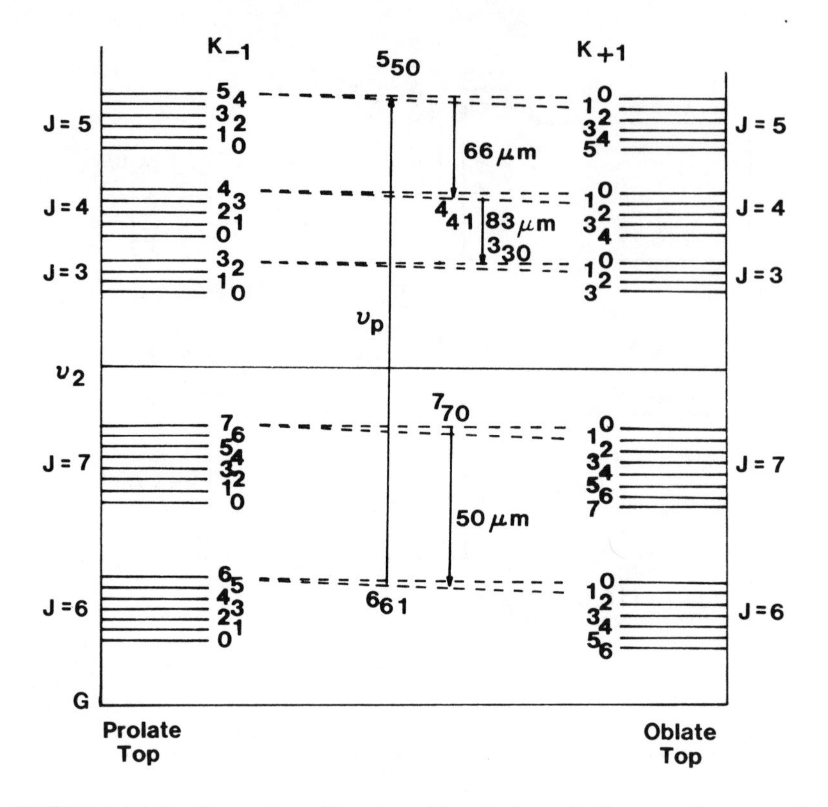

FIGURE 3.3.1.6. Illustration of laser transitions in the optically pumped asymmetric top D_2O and the energy levels $J_{K-1,K+1}$ as they are related to the symmetric prolate and oblate top. The selection rules for the $\nu_2 \rightarrow$ G transitions are $\Delta J = 0,\pm1$ with ΔK_{-1} and $\Delta K_{+1} = \pm1,\pm3,\pm5,$ ... The CO_2 P_9 (32) pump transition is seen to be G $6_{61} \rightarrow \nu_2\ 5_{50}$, while the strong 66 μm laser transition is $\nu_2\ 5_{50} \rightarrow \nu_2\ 4_{41}$.

the right. The selection rules for $\nu_2 \leftrightarrow$ G transitions are $J = 0,\pm1$, $\Delta K_{-1} = \pm1,\pm3,\pm5,$ ... and $\Delta K_{+1} = \pm1,\pm3,\pm5,$ ... The important point to be observed is the increased complexity of the asymmetric top over the symmetric top molecule. In the symmetric top there are $J + 1$ energy sub-levels for each J while for the asymmetric top there are $2J + 1$ energy levels for each J. The richer spectrum leads to more far IR laser line possibilities.

Making pump and laser transition line assignments depends directly on the availability of relevant spectroscopic data. Of the more than 50 molecules and more than 1200 laser lines that have been found, only a fraction of the transition assignments have been identified. Most far IR laser transitions will occur in the ground state of the molecule or the first vibration state where the bulk of the molecular populations will be located at room temperatures. This implies that available spectroscopic data is often useful, but that additional measurements will usually be required to solve the transition assignment problem.

Precision frequency measurements have made great progress in the far IR beginning with the early work of Hocker[20] on HCN and Frenkel[21] on H_2O in 1967, followed by the work of Evenson[22] on H_2O, CO_2 and He-Ne. Many molecular laser lines have now been measured from 337 to 3.39 μm with an accuracy of a few parts in 10^8 or better. This data can greatly improve the accuracy of the spectroscopic constants of such molecules and provide benchmarks for future precision frequency measurements. The technique of making absolute frequency measurements is illustrated in Figure 3.3.1.7. Here, the unknown frequency f_1 is mixed with an accurately known klystron frequency

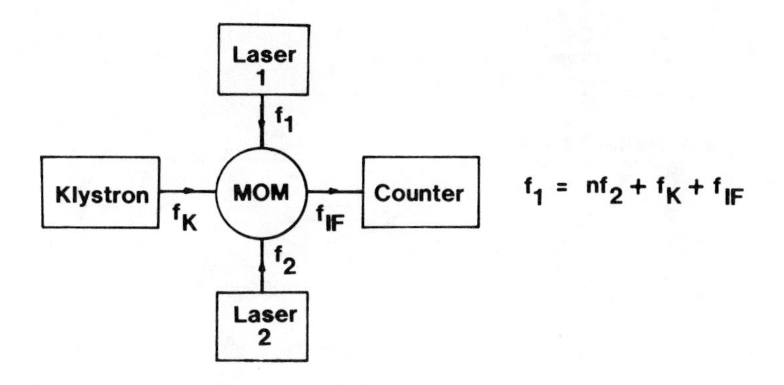

FIGURE 3.3.1.7. Absolute frequency measurement method.

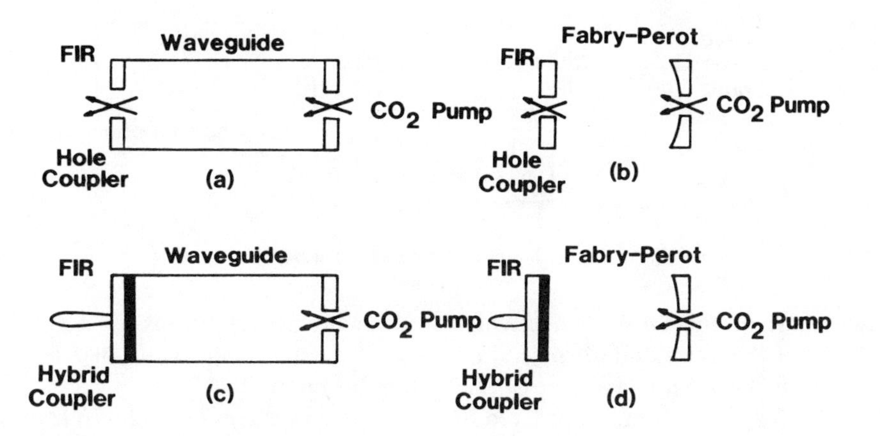

FIGURE 3.3.1.8. Waveguide (metal or glass) and Fabry-Perot resonators commonly employed in optically pumped molecular lasers. The CO_2 pump radiation is focused through a small hole in the input mirror while the far IR radiation is coupled out either through a hole coupler or hybrid coupler. The hybrid coupler is a combination dielectri-metal mesh mirror which reflects the CO_2 radiation but passes the far IR radiation.

f_K and the harmonic frequency nf_2 of a second laser in a metal-oxide-metal detector and the beat note f_{IF} determined with a frequency counter. The unknown frequency f_1 is then the sum of the three known frequencies.

Far IR optically pumped molecular lasers can employ either Fabry-Perot or waveguide resonators[23] for feedback. Four commonly used configurations are shown in Figure 3.3.1.8. In each case the CO_2 pump radiation is focused into the resonator through a small (a few millimeters in diameter) hole in the input mirror, while the output mirror is either hole coupled or a hybrid dielectric-metal mesh type.[24] A partially transmitting mirror obviously will give better beam quality and a diffraction limited beam.

For high powered pulsed operation, the intense CO_2 pump can be coupled into the resonator via a Brewster plate as shown in Figure 3.3.1.9. A Fox-Smith interferometric coupler is also employed as a mode selector. Again a hybrid dielectric-metal mesh output coupler is used for better efficiency and beam quality.

The measurement of the spectral quality of a high power pulsed optically pumped far IR molecular laser is difficult using normal methods because of the low repetition rates of 1 to 10 seconds. However, Fetterman et al.,[25] using a surface acoustic wave (SAW) dispersive delay line, reported a novel technique for spectral analysis, shown in Figure 3.3.1.10. The pulsed laser signal is down shifted via a mixer into the fre-

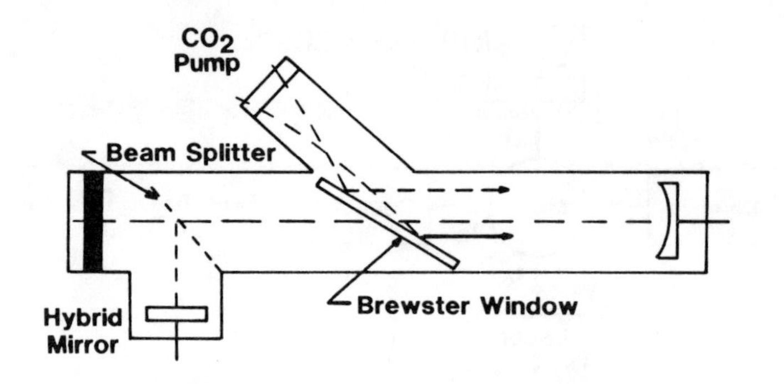

FIGURE 3.3.1.9. Pulsed laser oscillator configuration.

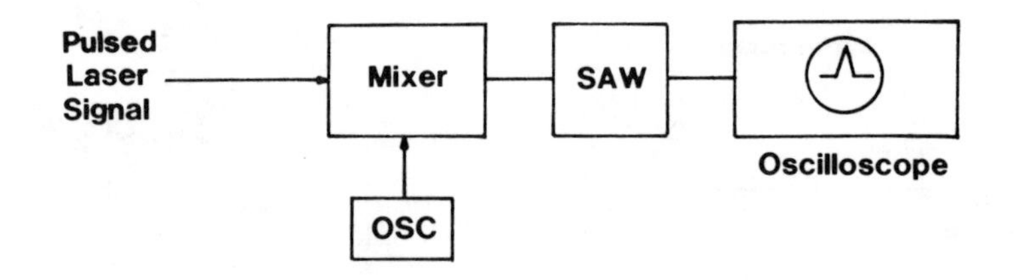

FIGURE 3.3.1.10. Simplified diagram of a SAW spectral analyzer.

quency range of the SAW device which under suitable conditions takes the Fourier transform of the pulse and displays it on an oscilloscope. A linewidth of 7.5 MHz was measured for the 385 μm laser line of D_2O using this method.

In the following section, an ordered list of far IR laser lines ($\lambda > 12$ μm) is presented with references, power levels, pump information and laser line assignment where available.

REFERENCES

1. **Patel, C. K. N., et al.**, Laser action up to 57.355 microns in gaseous discharges (Ne, He-Ne), *Appl. Phys. Lett.*, 4, 18, 1964.
2. **Herzberg, G.**, *Molecular Spectra and Molecular Structure II*, Van Nostrand, New York, 1945.
3. **Mathis, L. E. S. and Parker, J. T.**, Stimulated emission in bond spectrum of nitrogen, *Appl. Phys. Lett.*, 3, 16, 1963.
4. **Lide, D. R.**, On the explanation of the so called CN laser, *Appl. Phys. Lett.*, 11, 62, 1967.
5. **Benedict, W. S., et al.**, The water vapor laser, *IEEE J. Quantum Electron.*, QE-5, 108, 1969.
6. **Chang, T. Y. and Budgee, T.**, Laser action at 452, 496, and 541 μm in optically pumped CH_3F, *Opt. Commun.*, 1, 423, 1970.
7. **Chang, T. Y.**, Optically pumped submillimeter-wave sources, *IEEE Trans. Microwave Theory Tech.*, MTT-22, 983, 1974.
8. **Coleman, P. D.**, Present and future problems concerning lasers in the far IR spectra range, *J. Opt. Soc.*, 2567, 894, 1977.
9. **Hodges, D. T.**, A review of advances in optically pumped far IR lasers, *Inf. Phys.*, 18, 375, 1978.
10. **DeTemple, T. A.**, Pulsed optically pumped FIR lasers, in *Infrared and Millimeter Waves, Vol. I, Sources of Radiation*, Button, K. J., Ed., Academic Press, New York, 1979.
11. **Tucker, J. R.**, Theory of a FIR gas laser, in *Conference Digest 17, Int. Conf. on Submillimeter Waves and Their Applications*, Cat. no. 74-CHO-856-5-MTT, IEEE, Piscataway, N. J., 1974.
12. **Henningsen, J. O. and Jensen, H. G.**, The optically pumped FIR laser, *IEEE J. Quantum Electron.*, QE-11, 248, 1975.
13. **DeTemple, T. A. and Danielewicz, E. J.**, CW CH_3F waveguide laser at 496 μm: theory and experiment, *IEEE J. Quantum Electron.*, QE-12, 40, 1976.
14. **Temkin, R. J.**, Theory of optically pumped submillimeter lasers, *IEEE J. Quantum Electron.*, QE-13, 450, 1977.
15. **Kim, K. J. and Coleman, P. D.**, Calculated-experimental evaluation of the gain/absorption spectra of several optically pumped NH_3 systems, *IEEE J. Quantum Electron.*, QE-16, 1341, 1980.
16. **Curtis, J.**, Vibration-Rotation Bands of NH_3 in the Region 670-1860 cm^{-1}, Ph.D. thesis, The Ohio State University, Columbus, 1974.
17. **Henningsen, J. O.**, Spectroscopy of molecules by far IR laser emission, in *Infrared and Millimeter Waves*, Button, K. J., Ed., Academic Press, New York, 1981.
18. **Woods, D. R.**, The High Resolution Infrared Spectra of Normal and Deuterated Methanol Between 400 and 1300 cm^{-1}, Ph.D. Thesis, University of Michigan, Ann Arbor, 1970.
19. **Townes, C. H. and Schawlow, A. L.**, *Microwave Spectroscopy*, McGraw-Hill, New York, 1955, 86.
20. **Hocker, L. O., et al.**, Absolute frequency measurement and spectroscopy of gas laser transitions in the far IR, *Appl. Phys. Lett.*, 10, 147, 1967.
21. **Frenkel, L.**, Absolute frequency measurement of the 118.6 μm water vapor laser transition, *Appl. Phys. Lett.*, 11, 344, 1967.
22. **Evenson, K. M.**, Extension of absolute frequency measurements to the CW He-Ne laser at 88 THz (3.39 μm), *Appl. Phys. Lett.*, 20, 133, 1972.
23. **Hodges, D. T. and Hartwick, T. S.**, Waveguide lasers for the far infrared pumped by a CO_2 laser, *Appl. Phys. Lett.*, 23, 252, 1973.
24. **Danielewicz, E. J. and Coleman, P. D.**, Hybrid metal mesh-dielectric mirrors for optically pumped far IR lasers, *Appl. Opt.*, 15, 761, 1976.
25. **Fetterman, H. R., et al.**, Real-time spectral analysis of far IR laser pulses using a SAW dispersive delay line, *Appl. Phys. Lett.*, 34, 123, 1979.

3.3.2 TABLES OF CW GAS LASER EMISSIONS

D. J. E. Knight

INTRODUCTION AND COVERAGE

These tables of far-IR CW gas laser lines (with wavelengths longer than 12 μm) present experimentally observed data. The characteristics listed provide information bearing first on the operation and usefulness as a coherent radiation source and second on the distinction of one laser emission from another and on identification of the lasing transition.

Historically, the list was started in 1967 with discharge-pumped laser emissions to search for "chains" of lasers approximately harmonically related in frequency (each near a (low) harmonic of the frequency of the laser below it) for transferring absolute frequency measurement to the 10-μm region and beyond. The search for laser emissions suited to laser frequency synthesis remains the main stimulus to this work. Thus, the most accurate wavelength measurement or a good frequency one is listed, together with its uncertainty, and an approximate indication of the power output. Since the first report of optically pumped laser action in the far infrared in 1970,[CH2] a great many of these lines have been discovered and now comprise the bulk of the list. In the tables, the pump laser is identified, together with the relative polarization of the pump and laser radiations, an indication of the pump threshold power, and an indication of the optimum frequency offset of the pump laser where known.

The tabulation of 1350 radiations is broken down into separate tables for 62 different gases. The gases are arranged (following the section prefix) by increasing number of atoms and then by atomic weight, except for local rearrangements to keep isotopically related species together. The first integer of the table number represents the number of atoms in the molecule. The discharge-pumped lines (distinguishable by the total absence of pump information) are entirely from monatomic or triatomic gases, and the optically pumped lines make use of rotation-vibration transitions in triatomic or more complex molecules. Only in D_2O is lasing listed for both methods of pumping.

Contents of Tables

The tables of far IR CW gas laser lines, both discharge- and optically pumped, are arranged by increasing number of atoms (first integer(s) of table number), then approximately by molecular weight. The following tables are included:

Table number: lasing gas	Number of entries
3.3.1.1: He, helium	2
3.3.1.2: Ne, neon	45
3.3.1.3: Ar, argon	7
3.3.1.4: Xe, xenon	3
3.3.3.1: H_2O, water	12
3.3.3.2: D_2O, heavy water	6
3.3.3.3: HCN, hydrogen cyanide	7
3.3.3.4: DCN, deuterium cyanide	5
3.3.3.5: OCS, carbonyl sulphide	1
3.3.3.6: SO_2, sulfur dioxide	2
3.3.4.1: NH_3, ammonia	105
3.3.4.2: $^{15}NH_3$, ammonia with ^{15}N isotope	4

3.3.4.3: NH_2D, mono-deutero ammonia	6
3.3.4.4: ND_3, fully deuterated ammonia	1
3.3.4.5: HDCO, mono-deutero formaldehyde	2
3.3.4.6: D_2CO, di-deutero formaldehyde	13
3.3.4.7: $^{10}BCl_3$, boron trichloride with ^{10}B isotope	4
3.3.4.8: $^{11}BCl_3$, boron trichloride with ^{11}B isotope	4
3.3.5.1: CH_3F, methyl fluoride	6
3.3.5.2: $^{13}CH_3F$, methyl fluoride with ^{13}C isotope	1
3.3.5.3: CD_3F, fully deuterated methyl fluoride	15
3.3.5.4: HCOOH, formic acid	47
3.3.5.5: $H^{13}COOH$, formic acid with ^{13}C isotope	6
3.3.5.6: HCOOD, O-deutero formic acid	64
3.3.5.7: DCOOH, C-deutero formic acid	20
3.3.5.8: DCOOD, di-deutero formic acid	57
3.3.5.9: CH_3Cl, methyl chloride	20
3.3.5.10: CD_3Cl, fully deuterated methyl chloride	20
3.3.5.11: CH_2F_2, difluoromethane	48
3.3.5.12: CH_2Cl_2, methylene chloride (dichloromethane)	1
3.3.5.13: CD_2Cl_2, fully deuterated methylene chloride	7
3.3.5.14: $CH_3{}^{79}Br$, methyl bromide with ^{79}Br	9
3.3.5.15: $CH_3{}^{81}Br$, methyl bromide with ^{81}Br	7
3.3.5.16: CH_3Br, methyl bromide with mixed ^{79}Br and ^{81}Br	20
3.3.5.17: CD_3Br, fully deuterated methyl bromide	10
3.3.5.18: CF_2Cl_2, di-fluoro di-chloro methane (fluorocarbon-12)	11
3.3.5.19: CH_3I, methyl iodide	21
3.3.5.20: CD_3I, fully deuterated methyl iodide	32
3.3.5.21: $^{13}CD_3I$, fully deuterated methyl iodide with ^{13}C isotope	
3.3.5.22: CF_3Br, bromo 3-fluoromethane (fluorocarbon-1,3)	10
3.3.6.1: N_2H_4, hydrazine	29
3.3.6.2: CH_3OH, methanol (methyl alcohol)	122
3.3.6.3: $^{13}CH_3OH$, methanol with ^{13}C isotope	40
3.3.6.4: CH_2DOH, C-deutero methanol	68
3.3.6.5: CH_3OD, O-deutero methanol	47
3.3.6.6: CHD_2OH, C-dideutero methanol	11
3.3.6.7: CD_3OH, C-trideutero methanol	110
3.3.6.8: CD_3OD, fully deuterated methanol	17
3.3.6.9: CH_3CN, methyl cyanide (acetonitrile)	29
3.3.6.10: CH_3SH, methyl mercaptan	23
3.3.6.11: CH_2CHCl (or H_2CCl), vinyl chloride	19
3.3.6.12: CHF_2CH_2 (or 1,1-$C_2H_2F_2$), 1,1-difluoroethylene	16
3.3.6.13: CHFF (or *cis*-1,2-$C_2H_2F_2$), *cis*-1,2-difluoroethylene	12
3.3.6.14: C_2H_3Br, vinyl bromide, with ^{79}Br, ^{81}Br mixed 51%, 49%	45
3.3.7.1: CH_3NH_2, methylamine	35
3.3.7.2: CH_3CCH, methyl acetylene	9
3.3.7.3: CH_2CHCN, vinyl cyanide	19
3.3.8.1: CH_3CH_2F (or C_2H_5F), ethyl fluoride	16
3.3.8.2: CH_3CH_2Cl (or C_2H_5Cl), ethyl chloride	4
3.3.8.3: CH_3CHF_2 (or $C_2H_4F_2$), 1,1-difluoroethane	2
3.3.9.1: CH_3CH_2OH (or C_2H_5OH), ethyl alcohol	1
3.3.12.1: $(H_2CO)_3$, trioxane (cyclic trimer of formaldehyde)	13
62 tables — Total entries:	1350

For each gas, the observed emissions are arranged in order of increasing frequency (decreasing wavelength). The vacuum wavenumber (nu) in centimeters $^{-1}$ and frequency in gigahertz are listed in addition to the (vacuum) wavelength (lambda). (There is an arbitrary cutoff at 12 μm for the listing of Ne and Xe lines.) These quantities are interrelated by c_o = 299 792 458 $msec^{-1}$.

The numbers of significant figures shown for lambda, nu, and frequency of each laser line is computed so that the least significant decimal place corresponds to the most significant place of the uncertainty, which here is usually a SD. Where the (modulus of the) uncertainty is greater than about 10^{-5} and often as large as 2%, the measurement will have been a wavelength one; where it is 10^{-6} or less, it will have been a frequency one. The limit of resolution in the tables is five places after the decimal point for lambda, six places for nu, and four places for frequency (in gigahertz). Further small discrepancies (less than the uncertainty) may be expected from the rounding applied to each to make the tables easier to read.

The first reference given is usually the source of the wavelength or frequency data; subsequent references give backup data such as CW power, polarization, or pump threshold. The references do not necessarily contain that announcing the discovery of the laser emission nor do they necessarily give all relevant measurements, since the data base is limited to three entries. The transition assignments (for those transitions in the simpler molecules that have been unambiguously assigned) are not listed because of the size of the task, involving as it does both the pump and the lasing transitions. A list of those references (if any) which discuss transition assignments has been placed at the head of each table. Extra references which are listed only in connection with assignments have their codes suffixed with an asterisk (*). These do not appear in the table proper. (The codes for the references, consisting of two or three letters with an optional added digit, or one letter plus one digit, were necessary for computer handling. The space for the codes was limited to eight characters; therefore if the codes are two or three in number, they are run together. A glance through the reference list should minimize ambiguities). Some papers which discuss transition assignments may have been missed, e.g., where these are associated with pulsed experimental work. Molecular spectroscopy papers not specifically identifying laser transitions are excluded. Pulsed-only laser transitions are not included because of the size of the task (the larger number both of laser lines and of molecules) and the need to list a different set of parameters, such as pulse duration, energy output, and efficiency to reflect the main practical interest in such lasers.

READING THE TABLES

The contents of the columns so identified are as follows:

Lambda (μm) — Vacuum wavelength in micrometers to a maximum of five decimal places.

O/P— Laser power output (best reported) is expressed on an approximate 1, 2, 5 log scale in bels with respect to 1 mW. The power reported depends very much on the geometry and pump power. Since power measurements in the far-IR may be inaccurate by a factor of two, an approximate scale is appropriate. The sign and digit indicate the decade of milliwatts, e.g., +2 refers to 100 mW, 0 refers to 1 mW, and −2 refers to 0.01 mW. For sharper resolution, the suffixes −, blank, + represent the coefficient ranges 0.31 to 0.69,0.7 to 1.4, and 1.5 to 3.0 about the means 0.5, 1, and 2.Thus, "−1+" refers to 0.2 mW and, "+2−" refers to 50 mW. An "L" in place of the sign (mainly for discharge-pumped lasers) denotes that the laser has a Lamb dip which can act as a frequency reference. Brackets [()] around the power or a single bracket [")"] in place of the suffix indicate a cascade transition.

Pol. — A single letter denotes the relative polarization of the far-IR output radiation with respect to the pump radiation — "P", parallel; "N", normal (perpendicular); and "E", "either", meaning

1. That reports conflict
2. That the polarization is reported at a substantially intermediate angle
3. For Stark-shifted pump absorption (in ammonia) that the values of applied Stark field (listed under "OFF") are closely the same for both pump polarizations

Pump Line — Two "words" describe the gas and output wavelength and transition of the pump laser. The first word of four characters denotes one of six pump "gases":

CO2 = CO_2, carbon dioxide
CO2A = $C^{18}O_2$, carbon dioxide with ^{18}O isotope
CO2B = $^{13}CO_2$, carbon dioxide with ^{13}C isotope
CO2C = $^{13}C^{18}O$, carbon dioxide with ^{13}C and ^{18}O isotopes
CO2S = CO_2 sequence band transition
N20 = N_2O, nitrous oxide

The second word contains the wavelength in micrometers to 1 part in 10^3 and an abbreviated transition label, e.g., "10.32R10" stands for the 10.32-μm R(10) transition. In a few instances where the pump line identification was originally stated to be uncertain, this is noted with the reference.

Thres. — This is an indication of pump threshold power using the same notation as for the power output, i.e., bels with respect to 1 mW and with "−" or "+" after the digit indicating a coefficient of 0.5 or 2, e.g., "+3+" represents 2 W, "+4−" represents 5 W, "+4" represents 10 W, and so on.

Off. — Up to four characters denote the offset of the pump-absorbing transition from the center of the pump-laser transition. Except for Stark-shifted absorbing transitions in NH_3, this is given with a sign and no decimal point and denotes an offset in megahertz, e.g., "+39" indicates that the absorbing transition lies 39 MHz above the center of the pump-laser tuning profile. (Such data may be a rough estimate, in error by perhaps 20%.) In the special case of Stark-tuned pump-absorbing transitions (in NH_3), a number with decimal point and no sign indicates the Stark field in kilovolts per centimeter, e.g., "23.1" means that the laser worked at 23.1 kV/cm for the pump laser line and polarization listed. Special notes are given at Reference RE3.

Uncert. — A coefficient and exponent indicate the fractional uncertainty of lambda, nu, and frequency, e.g., "1.5E-6" means (±) 1.5 parts in 10^6, being usually a single SD.

Nu(1/cm) The vacuum wavenumber (nu) to a maximum of six decimal places.

Freq.(GHz) — The frequency, in gigahertz, to a maximum resolution of 0.1 MHz, four decimal places.

Ref. — Up to three literature references in an eight-character word. Each reference can be denoted by two to three letters followed by an optional digit, e.g., "CH4" or "ZD", or one letter and one digit, e.g., "T3".

The reference list is given alphanumerically by code. Each code is necessarily unique at the expense of not having a very regular derivation or pattern. Some merged codes can best be disentangled by looking at adjacent entries in the tables or table headings (assignment reference lists) where they may appear individually or rearranged, e.g., in Table 6.5 "NH1LDLA2" comprises NH1, LD, and LA2.

If an erroneously long reference symbol (here LDL) is searched for in the list, the truncated form (LD) should be noticed instead of that sought.

FURTHER NOTES AND EXPLANATIONS

The absence of information in the tables, except for the discharge-pumped lines in the monatomic or triatomic gases (which necessarily have no pump information), means only that no information has been found.

The use of approximate representations for output power and pump threshold has permitted (1) "translation" of authors' qualitative or roughly calibrated output powers and (2) incorporation of a pump threshold power derived, if necessary, very roughly from a single statement of notional pump laser power such as "10 W", by taking, for example, half of this. (This is still valuable, since available pump powers can range from 1 to 100 W or more.) Notes are given with the references as to how such interpolations or conventional values were arrived at. After doing this, reports from different sources usually agree within one step on the scale, i.e., within a factor two or three. Another point which gives difficulty in compiling lists, particularly of optically pumped laser lines, is that of deciding whether reports from different sources refer to the same or to different radiations. The main criteria used here were (for the same molecule) that the wavelengths and pump radiations should be the same, within the wavelength uncertainties. An exception is when the (original) source reports separately two otherwise similar radiations, implying that a critical but unpublished parameter (such as pump offset) distinguished them. The problems of distinction are worse when the wavelength data is accurate only to the order of 1%. When the same frequency is reported for two different isotopic species of the same molecule, it is likely that one was a contaminant of a supposed sample of the other, certainly if it was much more common. Polarization is an unreliable distinguishing factor, and occasionally judgments (that lines were the same) have been made contrary to the opinion of a second source when information from a third source (unknown to the second) has been found.

OTHER LISTS OF FAR-IR LASER RADIATIONS

In preparing these tables, the data was taken from the original publications listed as references, but lists prepared by other authors have been valuable to check for missing publications. Some of these lists contain information on far-IR lasers not presented here, such as transition assignments or laser emission observed only under pulsed conditions. Although the optically pumped data in particular has expanded very recently, some older lists remain valuable. These include, given broadly in order of date,

1. "*Handbook of Lasers*" (1971) — This lists mainly discharge-pumped radiations, pulsed and CW, with information on the transitions, over the whole spectrum and has a "Tabulation of gas lasers by wavelength" at the end. Chapters relevant to the far-IR are Chapter 6, "Neutral Gas Lasers" by C. S. Willett, and Chapter 8, "Molecular Gas Lasers" by M. A. Pollack (Pressley, R. J., Ed., *Handbook of Lasers*, CRC Press, Inc., Boca Raton, Fla., 1971).

2. Yamanaka, M., "Optically- Pumped Gas Lasers" — A very thorough listing and analysis (mostly in English), covering both pulsed and CW radiations, with some information on transition assignments, set out in order of wavelength and covering the whole spectrum. It is complete to the end of 1975, with 1090 entries (Yamanaka, M., Optically-pumped gas lasers — a wavelength table of laser lines, *Rev. Laser Eng. (Japan)*, 3(4), 253, 1976.

3. Rosenbluh, M., et al., "Submillimeter Laser Wavelength Tables" — Optically pumped far-IR laser lines, both pulsed and CW, are listed (1) for each gas separately and (2) in overall order of wavelength. There is much subsidiary information. The article was received March 3, 1976 (549 entries) (Rosenbluh, M., Temkin, R. J., and Button, K. J., Submillimeter wavelength tables, *Appl. Opt.*, 15 (11), 2635, 1976.)

4. Gallagher, J. J., et al., "Tabulation of Optically Pumped Far Infrared Laser Lines and Applications to Atmospheric Transmission" — Optically pumped radiations are given in order of wavelength, with separate tables for (1) CW, (2) super-radiant, and (3) pulsed emissions, and with atmospheric-attenuation charts showing spectral positions of the laser radiations. The article was received April 5, 1976 (over 450 entries) (Gallagher, J. J., Blue, M. D., Bean, B., and Perkowitz, S., Tabulation of optically pumped far infrared laser lines and applications to atmospheric transmission, *Infrared Phys.*, 17, 43, 1977).

5. Knight, D. J. E., "Ordered List of Optically-Pumped Laser Lines (Continuous, $\lambda > 9$ μm)" — Optically-pumped laser information, CW only, is presented in order of wavelength, with the frequency also listed. This differs from other tables in listing the uncertainty of the wavelength or frequency, but remains a list of radiation sources rather than of frequency "standards". The data is incorporated in the list here, with (1) separate listing for each gas, and (2) additional listing of the wavenumber. The February 1981 list had 1350 entries (Knight, D. J. E., Ordered List of Optically-Pumped Laser Lines (Continuous, $\lambda > 9$ μm), NPL Rep. QU-45, (1st revision), National Physical Laboratory, Teddington, Middlesex, U.K., February 1981).

6. Beck, R., et al., "Table of Laser Lines in Gases and Vapors" — The 1976 list covered the whole spectrum for both pulsed and CW radiations, with separate lists, of wavelengths only, for each gas. There were 4347 entries to April 1976; revised editions have been published in 1978 and in 1980 (Beck, R., Englisch, W., and Gürs, K., *Table of Laser Lines in Gases and Vapors*, Vol. 2, Springer Series in Optical Sciences, Springer-Verlag, Berlin, 1976; 2nd revised ed., 1978; 3rd revised ed., 1980).

DISCUSSION OF ERRORS AND ACCURACY

There is always risk of faults in lists such as this from both errors in transcription and errors in the original experimental observations, and it is part of a compiler's duty to keep watch for evidence of them. The reader has some protection against the former kind of error by referring to the original reference (if published), although in some cases data for this list were taken from preprints where typesetting errors occurred afterwards on publication. (The references here refer first to the actual source of the data, and notes on specific differences from the published data are sometimes added, see for example those at Reference PE2.)

As regards errors of experimental observation, we have heard of, but rarely read of, a few instances where lines reported once have not been repeated. Usually the wavelength was not precisely measured, and contrary evidence may take the form that the stated pump radiation is not absorbed in the stated gas or that some other wavelength (not overlapping in uncertainty) was seen and not that first reported. Firm published evidence, preferably from the original authors, is desirable to "clean up" such data, and nothing has been attempted here, except (more positively) to note a second reference confirming any line, especially if this was a frequency measurement. (Between one third and one half of the lasers in this list have had their frequencies measured, and while this is necessary to properly locate the radiation in the spectrum, it also confirms the existence of the line as stated.) Where frequency measurements have been made, some kinds of error show up, since each frequency is (with high probability) unique and can be accurately checked. For example, a 5-MHz error in the first frequency measurement of (the center of) the 70.5-μm methanol line (originating in an unusually large distortion of the gain profile with the pump laser not tuned to the absorption center) has been corrected. Also, the 693.7885-GHz line first ascribed to HCOOD has since been observed from HCOOH, so that it is listed now only under the latter because of likely sample contamination, as mentioned earlier.

Apart from checking for errors, it was beyond the scope of this work to use several results to compute, for example, a "best" result for frequency: a satisfactory frequency or the least uncertain wavelength was taken, and its source is the first reference quoted.

ACKNOWLEDGMENTS

I wish to thank those authors who have sent me preprints and other communications, the Director of the Bureau International des Poids et Mesures (BIPM), F-92310, Sèvres, France for the opportunity to put together the data, and F. Lesueur of BIPM for help with computing the tables. I also thank G. W. Chantry for advice on describing molecules and for reading the text.

Table 3.3.1.1
He, HELIUM

For line assignments see Reference LEV

Lambda (μm)	o/p	Pol.	Pump	Line	Thres.	Off.	Uncert.	Nu(1/cm)	Freq. (GHz)	Ref.
216.3	-1						2.0E-3	46.23	1386.	LEV
95.788	0						1.9E-5	104.397	3129.75	LEVMA2

Table 3.3.1.2
Ne, NEON

For line assignments see References PAT3, PAT4, PAT5

Lambda(μm)	o/p	Pol.	Pump	Line	Thres.	Off.	Uncert.	Nu(1/cm)	Freq. (GHz)	Ref.
132.8	-7						2.4E-3	75.3	2257.	PAT3,5
126.1	-6						2.4E-3	79.3	2377.	PAT3,5
124.4	-6						2.4E-3	80.4	2410.	PAT3,5
106.07	-6						5.0E-4	94.28	2826.	PAT3,5
93.0	-8						2.4E-3	107.5	3223.	PAT3
89.859	-6						1.0E-4	111.29	3336.3	PAT3,5
88.471	-5						1.0E-4	113.03	3388.6	PAT3,5
86.962	-7						1.0E-4	114.99	3447.4	PAT3,5
85.047							1.0E-4	117.58	3525.0	PAT3
72.108	-7						1.0E-4	138.68	4157.5	PAT3,5
68.329							1.0E-4	146.35	4387.5	PAT3
57.355	-4						1.0E-4	174.35	5227.0	PAT3,4
55.537	-6						1.0E-4	180.06	5398.1	PAT3,5
54.117	-4						1.0E-4	184.78	5539.7	PAT3,4
54.019	-3						1.0E-4	185.12	5549.8	PAT3,4
53.486	-3						1.0E-4	186.96	5605.1	PAT3,4
52.425	-7						1.0E-4	190.75	5718.5	PAT3,5
50.705	-7						1.0E-4	197.22	5912.5	PAT3,5

41.741		1.0E-4	239.57	7182.2	PAT3
37.231	-4	1.5E-4	268.59	8052.	PAT3,4
35.602	-4	1.5E-4	280.88	8421.	PAT3,4
34.679		1.5E-4	288.36	8645.	PAT3
34.552		1.5E-4	289.42	8677.	PAT3
32.830		2.0E-4	304.60	9132.	PAT3
32.516		1.5E-4	307.54	9220.	PAT3
32.016		1.5E-4	312.34	9364.	PAT3
31.928		1.5E-4	313.20	9390.	PAT3
31.553		1.5E-4	316.93	9501.	PAT3
28.053		2.0E-4	356.47	10687.	PAT3
25.423		2.0E-4	393.34	11792.	PAT3
22.836		2.5E-4	437.9	13128.	PAT3
21.752		2.5E-4	459.7	13782.	PAT3
20.480		2.5E-4	488.3	14638.	PAT3
18.396		3.0E-4	543.6	16297.	PAT3
17.888		3.0E-4	559.0	16759.	PAT3
17.841		3.0E-4	560.5	16804.	PAT3
17.804		3.0E-4	561.7	16838.	PAT3
17.189		3.0E-4	581.8	17441.	PAT3
17.158		3.0E-4	582.8	17472.	PAT3
16.947		3.0E-4	590.1	17690.	PAT3
16.893		3.0E-4	592.0	17747.	PAT3
16.668		3.0E-4	600.0	17986.	PAT3
16.638		3.0E-4	601.0	18019.	PAT3
13.759		3.0E-4	726.8	21789.	PAT3
12.835		3.0E-4	779.1	23357.	PAT3

Table 3.3.1.3
Ar, ARGON

For line assignments see Reference PAT3

Lambda(μm)	o/p	Pol.	Pump	Line	Thres.	Off.	Uncert.	Nu(1/cm)	Freq. (GHz)	Ref.
26.936							1.0E-4	371.25	11130.	PAT3
26.933							1.0E-4	371.29	11131.	PAT3
15.039							1.0E-4	664.94	19934.	PAT3
15.037							1.0E-4	665.00	19936.	PAT3
15.022							1.0E-4	665.69	19957.	PAT3
12.147							1.0E-4	823.28	24681.	PAT3
12.140							1.0E-4	823.72	24695.	PAT3

Table 3.3.1.4
Xe, XENON

For line assignments see Reference PAT3

Lambda (μ m)	o/p	Pol.	Pump	Line	Thres.	Off.	Uncert.	Nu (1/cm)	Freq. (GHz)	Ref.
18.506							3.0E-4	540.4	16200.	PAT3
12.917							3.0E-4	774.2	23209.	PAT3
12.266							3.0E-4	815.3	24441.	PAT3

Table 3.3.3.1
H_2O, WATER

For line assignments see Reference BEN1

Lambda (μm)	o/p	Pol.	Pump	Line	Thres.	Off.	Uncert.	Nu (1/cm)	Freq. (GHz)	Ref.
220.2279	-2L						1.0E-6	45.40750	1361.283	POL1
118.5910	-1L						1.0E-6	84.32343	2527.953	POL1

Lambda (μm)	o/p	Pol.	Pump	Line	Thres.	Off.	Uncert.	Nu (1/cm)	Freq. (GHz)	Ref.
115.32	-3						1.0E-4	86.715	2599.7	BEN1
79.09101	+1L						3.0E-7	126.43662	3790.475	PET
78.44333	+1L						3.0E-7	127.48057	3821.771	PET
73.401	-3						1.0E-4	136.24	4084.3	BEN1
55.088	-3						1.0E-4	181.53	5442.1	BEN1
47.687	-1						1.0E-4	209.70	6286.7	BEN1
47.46315							2.0E-6	210.6898	6316.32	DAN
47.244	-2L						1.0E-4	211.67	6345.6	BEN1
33.029							1.0E-4	302.76	9076.6	BEN1
27.97075	+2L						0.9E-8	357.516290	10718.0687	BL,EV1

Table 3.3.3.2
D_2O, HEAVY WATER

For line assignments see References BEN1, DW4

Lambda (μm)	o/p	Pol.	Pump	Line	Thres.	Off.	Uncert.	Nu (1/cm)	Freq. (GHz)	Ref.
171.67	L						1.0E-4	58.251	1746.3	BEN1
112.6	+1	N	CO2S	9.32R17	+3	-39	4.9E-3	88.8	2660.	DW4
107.72019	+1L						4.0E-7	92.83311	2783.067	PET
94.5	-1	P	CO2S	9.32R17	+3	-39	4.9E-3	105.8	3170.	DW4
84.27890	+1L						3.0E-7	118.65367	3557.147	PET
72.74778	+1L						3.0E-7	137.46124	4120.984	PET

Table 3.3.3.3
HCN, HYDROGEN CYANIDE

For line assignments see References HOC2, LM1, MA1, MK*, MK1, MK2

Lambda (μm)	o/p	Pol.	Pump	Line	Thres.	Off.	Uncert.	Nu (1/cm)	Freq. (GHz)	Ref.
372.5282							1.0E-6	26.84360	804.7509	HOC2
336.5578	+2+						1.0E-6	29.71258	890.7608	HOC2BV1
335.1831							1.0E-6	29.83444	894.4141	HOC2
310.8870	0						1.0E-6	32.16603	964.3134	HOC2
309.7140							1.0E-6	32.28786	967.9656	HOC2
211.00	-1						7.9E-5	47.393	1420.8	MF1MA1
128.629	-1						4.9E-5	77.743	2330.7	MF1MA1

Table 3.3.3.4
DCN, DEUTERIUM CYANIDE

For line assignments see References HOC3, MK*

Lambda (μm)	o/p	Pol.	Pump	Line	Thres.	Off.	Uncert.	Nu (1/cm)	Freq. (GHz)	Ref.
204.3872							2.0E-6	48.92675	1466.787	HOC3
194.7644	+2						2.0E-6	51.3441	1539.257	HOC3BV2
194.7027	+2+						2.0E-6	51.3604	1539.745	HOC3BV2
190.0090	+2						2.0E-6	52.6291	1577.781	HOC3BV2
189.9490	+2+						2.0E-6	52.6457	1578.279	HOC3BV2

Table 3.3.3.5
OCS, CARBONYL SULFIDE

For line assignments see Reference LA2

Lambda (μm)	o/p	Pol.	Pump	Line	Thres.	Off.	Uncert.	Nu (1/cm)	Freq. (GHz)	Ref.
378.4	-2+	P	CO2	9.34R08	+4-		2.0E-3	26.43	792.	LA2

Table 3.3.3.6
SO_2, SULFUR DIOXIDE

For line assignments see References HU*, STE*

Lambda (μm)	o/p	Pol.	Pump	Line	Thres.	Off.	Uncert.	Nu (1/cm)	Freq. (GHz)	Ref.
192.72	-1						3.0E-4	51.89	1555.6	DYHA
140.89	0						3.0E-4	70.98	2127.8	DYHA

Table 3.3.4.1
NH_3, AMMONIA

For line assignments see References DW4, RE1, RE2, RE3, RE4. Where the pump offset is given with a decimal point and no sign it indicates an applied Stark field in kV/cm, not detuning in MHz

Lambda (μm)	o/p	Pol.	Pump	Line	Thres.	Off.	Uncert.	Nu (1/cm)	Freq. (GHz)	Ref.
404.7		E	CO2	10.09R48	+4	20.6	1.1E-3	24.71	740.8	RE4
404.7		E	CO2	10.72P32	+4	24.5	1.1E-3	24.71	740.8	RE4
404.7	0+	E	N2O	10.73P08	+3	7.7	1.1E-3	24.71	740.8	RE3RE1
309.5		E	CO2	9.22R30	+4	12.3	9.3E-4	32.31	968.6	RE4
306.3		P	CO2	9.49P12	+4	21.0	9.2E-4	32.65	978.8	RE4
301.3	-1		CO2	10.35R06	+4	14.2	9.0E-4	33.19	995.0	RE1
290.9		E	CO2	9.55P20	+4	42.2	8.7E-4	34.37	1030.4	RE4
290.9		E	CO2	9.57P22	+4	47.5	8.7E-4	34.37	1030.4	RE4
290.9	-2-	E	CO2	10.72P32	+4	12.4	8.7E-4	34.37	1030.4	RE4RE1
290.9		E	N2O	10.35R34	+3	46.5	8.7E-4	34.37	1030.4	RE5RE3
290.4	0+	E	CO2	9.22R30	+4	8.5	8.7E-4	34.43	1032.2	RE4RE1
290.	-1-	N	CO2S	9.56P17	+3	-130	4.9E-3	34.5	1033.	DW4
289.3		E	CO2	9.73P40	+4	38.2	8.7E-4	34.56	1036.1	RE4
289.3		E	CO2	10.33R08	+4	45.4	8.7E-4	34.56	1036.1	RE4
289.3		E	CO2	10.35R06	+4	55.3	8.7E-4	34.56	1036.1	RE4
288.5		E	CO2	10.37R04	+4	21.6	8.5E-4	34.66	1039.1	RE4
288.5		N	N2O	10.38R31	+3	15.2	8.5E-4	34.66	1039.1	RE3

Table 3.3.4.1 (continued)
NH_3, AMMONIA

For line assignments see References DW4, RE1, RE2, RE3, RE4. Where the pump offset is given with a decimal point and no sign it indicates an applied Stark field in kV/cm, not detuning in MHz

Lambda (μm)	o/p	Pol.	Pump	Line	Thres.	Off.	Uncert.	Nu (1/cm)	Freq. (GHz)	Ref.
279.3		E	CO2	10.33R08	+4	34.7	8.3E-4	35.80	1073.3	RE4
279.3		E	CO2	10.35R06	+4	31.3	8.3E-4	35.80	1073.3	RE4
279.3		E	N2O	10.35R34	+3	53.6	8.3E-4	35.80	1073.3	RE3
276.8		E	N2O	10.35R34	+3	17.0	8.3E-4	36.13	1083.1	RE3
273.4		E	CO2	10.35R06	+4	14.6	8.1E-4	36.58	1096.7	RE4
273.4		P	N2O	10.36R33	+3	47.0	8.1E-4	36.58	1096.7	RE3
268.8		N	N2O	10.35R34	+3	40.1	8.0E-4	37.20	1115.2	RE3
268.8		E	N2O	10.37R32	+3	10.8	8.0E-4	37.20	1115.2	RE3
268.8		N	N2O	10.38R31	+3	54.3	8.0E-4	37.20	1115.2	RE3
263.4		N	CO2	10.38R02	+4	28.0	7.9E-4	37.96	1138.0	RE4
263.4	-2-	E	N2O	10.37R32	+2+	8.4	7.9E-4	37.96	1138.0	RE3RE1
263.40	0-	P	N2O	10.78P13	+2-		2.0E-4	37.965	1138.2	CH4T3
257.1		N	N2O	10.39R29	+3	20.3	7.6E-4	38.89	1165.9	RE3
250.1		N	N2O	10.39R29	+3	15.8	7.5E-4	39.99	1198.9	RE3
250.1		P	N2O	10.41R27	+3	52.4	7.5E-4	39.99	1198.9	RE3
225.1	-1	P	N2O	10.44R23	+3	21.3	6.8E-4	44.43	1332.0	RE3RE1
223.9		P	CO2	9.49P12	+4	26.3	6.6E-4	44.66	1338.9	RE4
223.9		E	CO2	10.72P32	+4	48.8	6.6E-4	44.66	1338.9	RE4
223.9		E	N2O	10.72P07	+3	47.3	6.6E-4	44.66	1338.9	RE3
223.9		E	N2O	10.74P09	+3	56.7	6.6E-4	44.66	1338.9	RE3
218.3		E	CO2	10.72P32	+4	42.1	6.5E-4	45.81	1373.4	RE4
218.3		E	N2O	10.72P07	+3	34.3	6.5E-4	45.81	1373.4	RE3
215.0		E	N2O	10.71P06	+3	37.7	6.5E-4	46.51	1394.3	RE3
155.28		P	CO2	9.22R30	+4	15.4	4.6E-4	64.40	1930.7	RE4
155.28		E	CO2	9.69P36	+4	26.3	4.6E-4	64.40	1930.7	RE4
155.28		E	CO2	9.71P38	+4	66.7	4.6E-4	64.40	1930.7	RE4
155.28		E	CO2	10.74P34	+4	24.4	4.6E-4	64.40	1930.7	RE4
155.28		N	N2O	10.73P08	+3	22.6	4.6E-4	64.40	1930.7	RE3

155.28		E	N2O	10.75P10	+3	68.6	4.6E-4	64.40	1930.7	RE3
151.49		E	CO2	9.68P34	+4	62.0	4.5E-4	66.01	1979.0	RE4
151.49	0+	E	CO2	10.72P32	+4	12.4	4.5E-4	66.01	1979.0	RE4FE1
151.49		E	N2O	10.71P06	+3	22.3	4.5E-4	66.01	1979.0	RE3
151.49	-1	E	N2O	10.72P07	+3	16.0	4.5E-4	66.01	1979.0	RE3RE1
147.15	-1)	P	CO2	9.22R30	+4	0.0	4.4E-4	67.96	2037.3	RE1LA3
119.02		P	CO2	9.52P16	+4	27.1	3.5E-4	84.02	2518.8	RE4
119.02		P	CO2	9.54P18	+4	48.7	3.5E-4	84.02	2518.8	RE4
119.02		N	N2O	10.74P09	+3	21.8	3.5E-4	84.02	2518.8	RE3
116.27		N	N2O	10.71P06	+3	52.7	3.5E-4	86.01	2578.4	RE3
116.27		E	N2O	10.73P08	+3	58.0	3.5E-4	86.01	2578.4	RE3
114.29	0+		CO2	9.49P12	+4	21.0	3.5E-4	87.50	2623.1	RE2FE1
112.22		P	CO2	9.47P10	+4	46.6	3.4E-4	89.11	2671.5	RE2RE5
105.35		E	CO2	9.73P40	+4	38.2	3.2E-4	94.92	2845.7	RE4
105.35		E	CO2	10.33R08	+4	45.4	3.2E-4	94.92	2845.7	RE4
105.35		E	CO2	10.35R06	+4	55.3	3.2E-4	94.92	2845.7	RE4
105.35		E	CO2	10.72P32	+4	42.1	3.2E-4	94.92	2845.7	RE4
96.67		E	CO2	9.35R06	+4	22.0	2.9E-4	103.44	3101.2	RE4
96.67		E	CO2	9.37R04	+4	34.9	2.9E-4	103.44	3101.2	RE4
96.67		E	CO2	10.76P36	+4	19.5	2.9E-4	103.44	3101.2	RE4
96.67		E	N2O	10.75P10	+3	39.4	2.9E-4	103.44	3101.2	RE3
96.67		E	N2O	10.77P12	+3	45.1	2.9E-4	103.44	3101.2	RE3
94.45		E	CO2	9.33R10	+4	47.9	2.8E-4	105.88	3174.1	RE4
94.45		E	CO2	9.34R08	+4	19.1	2.8E-4	105.88	3174.1	RE4
94.45		E	CO2	10.72P32	+4	37.9	2.8E-4	105.88	3174.1	RE4
92.88	-1	E	CO2	10.70P30	+4	8.8	2.8E-4	107.67	3227.7	RE4RE1
90.93		E	CO2	9.29R16	+4	49.5	2.7E-4	109.97	3297.0	RE4
88.06	0+	E	CO2	10.35R06	+4	7.1	2.0E-4	113.56	3404.4	RE4FE1
87.4	0	P	CO2S	9.56P17	+3	-130	4.9E-3	114.4	3430.	DW4
87.09		E	CO2	9.55P20	+4	42.2	2.5E-4	114.82	3442.3	RE4
87.09		E	CO2	9.57P22	+4	47.5	2.5E-4	114.82	3442.3	RE4
87.09		N	CO2	10.72P32	+4	12.3	2.5E-4	114.82	3442.3	RE4
81.48		P	CO2	9.19R36	+4	53.7	2.4E-4	122.73	3679.3	RE4
81.48		P	CO2	9.21R32	+4	39.8	2.4E-4	122.73	3679.3	RE4
81.50	+2-	N	N2O	10.78P13	+1+		6.0E-4	122.70	3678.	CH4T3
81.48		N	N2O	10.77P12	+3	15.6	2.4E-4	122.73	3679.3	RE3
79.62		E	CO2	9.17R42	+4	67.2	2.4E-4	125.60	3765.3	RE4

Table 3.3.4.1 (continued)
NH_3, AMMONIA

For line assignments see References DW4, RE1, RE2, RE3, RE4. Where the pump offset is given with a decimal point and no sign it indicates an applied Stark field in kV/cm, not detuning in MHz

Lambda (μm)	o/p	Pol.	Pump	Line	Thres.	Off.	Uncert.	Nu (1/cm)	Freq. (GHz)	Ref.
79.62		E	CO2	9.18R38	+4	34.6	2.4E-4	125.60	3765.3	RE4
79.62	-2-	E	CO2	10.74P34	+4	18.9	2.4E-4	125.60	3765.3	RE4RE1
79.62		N	N2O	10.73P08	+3	31.3	2.4E-4	125.60	3765.3	RE3
79.62		E	N2O	10.75P10	+3	62.8	2.4E-4	125.60	3765.3	RE3
78.27	-2-	E	N2O	10.71P06	+3	10.2	2.3E-4	127.76	3830.2	RE3RE1
67.75		N	CO2	10.72P32	+4	26.7	2.0E-4	147.60	4425.0	RE4
67.24	-1	P	CO2	9.22R30	+3+	0.0	2.0E-4	148.72	4458.5	RE1LA3
65.51		E	CO2	9.22R30	+4	12.3	2.0E-4	152.65	4576.3	RE4
65.51		E	CO2	10.37R04	+4	28.8	2.0E-4	152.65	4576.3	RE4
65.51		N	N2O	10.35R34	+3	33.1	2.0E-4	152.65	4576.3	RE3
64.50	0+	E	CO2	9.22R30	+4	8.5	1.9E-4	155.04	4647.9	RE4RE1
64.50		N	N2O	10.35R34	+3	45.0	1.9E-4	155.04	4647.9	RE3
64.50		N	N2O	10.37R32	+3	31.1	1.9E-4	155.04	4647.9	RE3
59.37		E	N2O	10.36R33	+3	17.2	1.7E-4	168.44	5049.6	RE3
58.010		N	N2O	10.36R33	+3	41.3	1.7E-4	172.38	5167.9	RE3
58.010		N	N2O	10.38R31	+3	40.2	1.7E-4	172.38	5167.9	RE3
57.040		E	CO2	10.37R04	+4	21.6	1.7E-4	175.32	5255.8	RE4
57.040		N	CO2	10.38R02	+4	48.1	1.7E-4	175.32	5255.8	RE4
57.040		N	N2O	10.38R31	+3	15.2	1.7E-4	175.32	5255.8	RE3
54.460	0+	E	N2O	10.79P14	+3	7.0	1.6E-4	183.62	5504.8	RE3RE1
52.800		N	CO2	10.38R02	+4	25.2	1.6E-4	189.39	5677.9	RE4
52.030		N	N2O	10.38R30	+3	16.0	1.6E-4	192.20	5761.9	RE3
47.910		N	N2O	10.38R30	+3	11.4	1.4E-4	208.72	6257.4	RE3

Table 3.3.4.2
$^{15}NH_3$, AMMONIA WITH ^{15}N ISOTOPE

For line assignments see Reference DW1

Lambda (μm)	o/p	Pol.	Pump	Line	Thres.	Off.	Uncert.	Nu (1/cm)	Freq. (GHz)	Ref.
375.	+1-	P	CO2	10.11R42	+3		4.9E-3	26.7	799.	DW1WO1
218.	-3+	N	CO2S	10.78P35	+3		4.9E-3	45.9	1375.	DW1
152.9	+2+		CO2B	10.78R18	+1	16	2.0E-3	65.4	1961.	WO1WO2
111.9	-3+	N	CO2S	10.73P31	+3		4.9E-3	89.4	2680.	DW1

Table 3.3.4.3
NH_2D, MONO-DEUTERO AMMONIA

For line assignments see Reference LA3

Lambda (μm)	o/p	Pol.	Pump	Line	Thres.	Off.	Uncert.	Nu (1/cm)	Freq. (GHz)	Ref.
124.	-1-	P	CO2	10.29R14	+3+		2.0E-2	81.	2420.	LA3
113.	-1-	P	CO2	10.81P40	+3+		2.0E-2	88.	2650.	LA3
108.	-2+	N	CO2	10.21R26	+4		2.0E-2	93.	2780.	LA3
86.	-1	P	CO2	10.81P40	+3-		2.0E-2	116.	3490.	LA3
77.	-1-	P	CO2	10.29R14	+3+		2.0E-2	130.	3890.	LA3
77.	-1-	P	CO2	10.18R30	+4-		2.0E-2	130.	3890.	LA3

Table 3.3.4.4
ND_3, FULLY DEUTERATED AMMONIA

For line assignments see Reference LA3

Lambda (μm)	o/p	Pol.	Pump	Line	Thres.	Off.	Uncert.	Nu(1/cm)	Freq. (GHz)	Ref.
87.	-1-	P	CO2	9.17R40	+3+		2.0E-2	115.	3450.	LA3

Table 3.3.4.5
HDCO, MONO-DEUTERO FORMALDEHYDE

For line assignments see Reference DA4

Lambda (μm)	o/p	Pol.	Pump	Line	Thres.	Off.	Uncert.	Nu (1/cm)	Freq. (GHz)	Ref.
196.0	-1-		CO2	9.46P08	+3		4.9E-3	51.0	1530.	DA2
195.0	-1-		CO2	9.24R26	+3		4.9E-3	51.3	1537.	DA2

Table 3.3.4.6
D_2CO, DI-DEUTERO FORMALDEHYDE

For line assignments see Reference DA4

Lambda (μm)	o/p	Pol.	Pump	Line	Thres.	Off.	Uncert.	Nu (1/cm)	Freq. (GHz)	Ref.
752.6808	-1-		CO2	9.21R32	+3		1.0E-6	13.28584	398.2996	DA2
733.5740	-1-		CO2	9.66P32	+3		1.0E-6	13.63189	408.6738	DA2
346.	-2)	N	CO2	10.61P22			2.0E-2	28.9	870.	LA4
341.	-2)	P	CO2	9.68P34			2.0E-2	29.3	880.	LA4
324.	-1	N	CO2	10.61P22	+3		2.0E-2	30.9	930.	LA4
320.	-1	P	CO2	9.68P34	+3+		2.0E-2	31.2	940.	LA4
294.	-2)	P	CO2	10.48P08			2.0E-2	34.0	1020.	LA4
279.	-1-		CO2	10.48P08	+3		4.9E-3	35.8	1075.	DA2
256.	-2)	P	CO2	9.25R24			2.0E-2	39.1	1170.	LA4
245.	-2)	P	CO2	9.31R14			2.0E-2	40.8	1220.	LA4
245.	-1+		CO2	9.25R24	+3		4.9E-3	40.8	1224.	DA2
244.	-1-	N	CO2	10.55P16	+3+		2.0E-2	41.0	1230.	LA4
233.	-1-		CO2	9.31R14	+3		4.9E-3	42.9	1287.	DA2

Table 3.3.4.7
$^{10}BCl_3$, BORON TRICHLORIDE WITH ^{10}B ISOTOPE

For line assignments see Reference KAR

Lambda (μm)	o/p	Pol.	Pump	Line	Thres.	Off.	Uncert.	Nu (1/cm)	Freq. (GHz)	Ref.
22.4	+1+		CO2		+4		4.9E-3	446.	13380.	KAR
19.40	+2-		CO2		+4		4.9E-3	515.	15450.	KAR
19.10	+2-		CO2		+4		4.9E-3	524.	15700.	KAR
18.30	+2-		CO2		+4		4.9E-3	546.	16380.	KAR

Table 3.3.4.8
$^{11}BCl_3$, BORON TRICHLORIDE WITH ^{11}B ISOTOPE

For line assignments see Reference KAR

Lambda (μm)	o/p	Pol.	Pump	Line	Thres.	Off.	Uncert.	Nu (1/cm)	Freq. (GHz)	Ref.
23.0	+2-		CO2		+4		4.9E-3	435.	13030.	KAR
20.6	+2		CO2		+4		4.9E-3	485.	14550.	KAR
20.2	+2		CO2		+4		4.9E-3	495.	14840.	KAR
18.80	+1+		CO2		+4		4.9E-3	532.	15950.	KAR

Table 3.3.5.1
CH_3F, METHYL FLUORIDE

For line assignments see References AR1*, CH2, CH5, DW4

Lambda (μm)	o/p	Pol.	Pump	Line	Thres.	Off.	Uncert.	Nu (1/cm)	Freq. (GHz)	Ref.
992.	0	N	CO2S	9.54P15	+3	+30	4.9E-3	10.08	302.	DW4
496.1009	0+	N	CO2	9.55P20	+3-	+44	8.0E-7	20.15719	604.2973	KW1CH5
496.072	0	N	CO2	9.55P20		-50	4.9E-6	20.1584	604.333	CH2,3
372.68	+1	P	CO2	9.84P50	+3	-50	1.2E-4	26.833	804.4	CH5
251.91	-1	P	CO2	10.16R34	+4-	+25	2.0E-4	39.697	1190.1	CH5
192.78	+1	P	CO2	10.17R32	+4-	0	2.5E-4	51.87	1555.1	CH5

Table 3.3.5.2
$^{13}CH_3F$, METHYL FLUORIDE WITH ^{13}C ISOTOPE

For line assignments see References CH5, RE (pump)

Lambda (μm)	o/p	Pol.	Pump	Line	Thres.	Off.	Uncert.	Nu (1/cm)	Freq. (GHz)	Ref.
1221.79	0	P	CO2	9.66P32	+4	-26	3.9E-5	8.1847	245.372	CH5RE1

Table 3.3.5.3
CD_3F, FULLY DEUTERATED METHYL FLUORIDE

For line assignments see Reference TO1

Lambda (μm)	o/p	Pol.	Pump	Line	Thres.	Off.	Uncert.	Nu (1/cm)	Freq. (GHz)	Ref.
1490.	-2		CO2	10.88P46	+4		6.9E-3	6.73	202.	TO1
1450.	-3		CO2	9.19R36	+4		1.3E-2	6.90	207.	TO1
384.7	-2		CO2	10.67P28	+4		1.2E-3	25.99	779.	TO1
368.4	-2	N	CO2	10.09R48	+4		1.3E-3	27.14	814.	TO1
349.0	-2		CO2	9.68P34	+4		1.3E-3	28.65	859.	TO1
336.6	-2	N	CO2	10.94P50	+4		1.3E-3	29.71	891.	TO1
323.3	-2		CO2	10.48P08	+4		1.4E-3	30.93	927.	TO1
265.	-2	P	CO2	9.86P52	+4		1.9E-2	37.7	1130.	TO1
247.5	0	P	CO2	9.33R10	+4		2.0E-3	40.40	1211.	TO1
247.3	-2	P	CO2	10.88P46	+4		2.0E-3	40.44	1212.	TO1
206.0	0+	P	CO2	9.52P16	+4		2.4E-3	48.5	1455.	TO1
201.5	-1	P	CO2	10.51P12	+4		2.4E-3	49.6	1488.	TO1
200.0	-2		CO2	9.62P28	+4		2.4E-3	50.0	1499.	TO1
172.8	-2		CO2	10.35R06	+4		2.9E-3	57.9	1735.	TO1
155.6	-2	P	CO2	10.49P10	+4		2.9E-3	64.3	1927.	TO1

Table 3.3.5.4 (continued)
HCOOH, FORMIC ACID

For line assignments see References BA, DA3, DE1 (pump), LA4

Lambda (μm)	o/p	Pol.	Pump	Line	Thres.	Off.	Uncert.	Nu (1/cm)	Freq. (GHz)	Ref.
1213.362	-1	P	CO2	9.62P28	+3+		1.0E-6	8.241561	247.0758	DY8
930.	-3		CO2	10.29R14	+3		2.0E-2	10.7	321.	RA2
930.	-3		CO2	10.17R32	+3		2.0E-2	10.8	324.	RA2
789.8396	-1	P	CO2	9.19R36	+3+		1.0E-6	12.66080	379.5612	DY8
786.9419	-1+	P	CO2	9.21R32	+3+		1.0E-6	12.70742	380.9588	DY8
786.1617	-1-		CO2	9.17R40	+3-		1.0E-6	12.72003	381.3369	DE1RA1
760.	-3-		CO2	9.25R24	+3+		2.0E-2	13.1	394.	RA1
744.050	-1+	P	CO2	9.25R24	+3		2.0E-6	13.43995	402.9196	DA1DY8
742.572	0-	P	CO2	9.17R40	+2+		2.0E-6	13.46670	403.7216	DA1DY8
705.	-1	P	CO2	9.35R06	+3+		2.0E-3	14.18	425.0	DY8
669.531	-1+	P	CO2	9.22R30	+3		1.5E-6	14.93583	447.7650	DA1DY8
580.8010	0	N	CO2	9.71P38	+3+		1.0E-6	17.21760	516.1707	DY8
580.3872	-1-		CO2	9.26R22	+3+		1.0E-6	17.22988	516.5387	DE1
535.	-1+		CO2	9.25R24	+3+		2.0E-3	18.70	561.	DY5
535.	-2+		CO2	9.54P18	+3+		2.0E-3	18.71	561.	DY5
533.7006	0	P	CO2	9.23R28	+3+		1.0E-6	18.73710	561.7240	DY8
533.6783	0	P	CO2	9.52P16	+3+		1.0E-6	18.73788	561.7475	DY8
515.1695	0	N	CO2	9.52P16	+3+		1.0E-6	19.41109	581.9297	DY8
513.0157	+1-	P	CO2	9.23R28	+3+		1.0E-6	19.49258	584.3729	DY8
513.0022	0	P	CO2	9.23R28	+3-		1.5E-6	19.49309	584.3882	DA1DY8
458.5229	0-	N	CO2	9.18R38	+3-		1.5E-6	21.80916	653.8222	DA1DY8
446.8730	-1+	P	CO2	9.29R16	+3+		1.0E-6	22.37772	670.8672	DY8
446.5054	-1		CO2	9.26R22	+3+		7.0E-7	22.39614	671.4195	RA2DY5
445.8996	-1+	P	CO2	9.27R20	+3+		1.0E-6	22.42657	672.3318	DY8
444.8	-1+	N	CO2	10.53P14	+3+		2.0E-3	22.48	674.	DY8
437.4510	-1+	N	CO2	9.52P16	+3+		1.0E-6	22.85970	685.3166	DY8
432.6665	0+	P	CO2	9.27R20	+3+		1.0E-6	23.11249	692.8950	DY8
432.6313	0+	P	CO2	9.27R20	+3		3.0E-7	23.114371	692.9514	KW1DY8
432.1094	0	P	CO2	9.26R22	+3+		1.0E-6	23.14229	693.7884	DA3DY8
420.3911	-1+	P	CO2	9.34R08	+3+		1.0E-6	23.78738	713.1276	DY8

Table 3.3.5.4
HCOOH, FORMIC ACID

For line assignments see References BA, DA3, DE1 (pump), LA4

Lambda (μm)	o/p	Pol.	Pump	Line	Thres.	Off.	Uncert.	Nu (1/cm)	Freq. (GHz)	Ref.
418.6129	0+	P	CO2	9.26R22	+3-		1.0E-6	23.88842	716.1568	DA1DY8
418.1	-1-	P	CO2	9.25R24	+3+		2.0E-3	23.92	717.	DY8
405.5848	0	P	CO2	9.28R18	+3-		9.5E-7	24.65576	739.1610	RA2DY8
405.0	-1+	P	CO2	9.60P26	+3+		2.0E-3	24.69	740.	DY5,8
404.0	-1+	N	CO2	10.11R42	+3+		2.0E-3	24.75	742.	DY8
394.2	0+		CO2	9.29R16	+3+		2.0E-3	25.37	761.	DY5
393.6311	0+	P	CO2	9.28R18	+3		4.0E-7	25.40450	761.6077	KW1DA1
359.9	-1	P	CO2	9.20R34	+3+		2.0E-3	27.79	833.	DY8
336.3	-1+	P	CO2	9.31R14	+3+		2.0E-3	29.74	891.	DY5,8
319.9	-2+	P	CO2	10.22R24	+3+		2.0E-3	31.26	937.	DY8
311.554	0-	P	CO2	10.23R22	+3+		4.9E-6	32.0972	962.249	EP1DY5,8
309.5	-2)	P	CO2	9.37R04			2.0E-3	32.31	969.	LA4
302.2781	+1-	P	CO2	9.37R04	+3-		1.0E-6	33.08212	991.7769	DA1DY8
278.5	-1+	P	CO2	9.64P30	+3+		2.0E-3	35.91	1076.	DY5,8
254.5	-1+	P	CO2	9.55P20	+3+		2.0E-3	39.29	1178.	DY5,8
196.5	-2+	N	CO2	9.26R22	+4		2.0E-3	50.9	1526.	LA4
133.9	-1-	P	CO2	9.26R22	+3+		2.0E-3	74.7	2239.	LA4

Table 3.3.5.5
$H^{13}COOH$, FORMIC ACID WITH ^{13}C ISOTOPE

Lambda (μm)	o/p	Pol.	Pump	Line	Thres.	Off.	Uncert.	Nu (1/cm)	Freq. (GHz)	Ref.
1030.378	-1-		CO2	9.22R30	+3+		1.0E-6	9.705174	290.9538	DE1
788.9192	-1+		CO2	9.49P12	+3+		1.0E-6	12.67557	380.0040	DE1
480.	-1-		CO2	10.09R46	+3+		4.9E-3	20.8	625.	DE1
448.5335	-1+		CO2	9.26R22	+3+		1.0E-6	22.29488	668.3836	DE1
313.	-1-		CO2	9.44P06	+3+		4.9E-3	31.9	958.	DE1
260.	-1-		CO2	9.52P16	+3+		4.9E-3	38.5	1153.	DE1

Table 3.3.5.6 (continued)
HCOOD, O-DEUTERO FORMIC ACID

For line assignments see Reference DE1 (pump)

Lambda (μm)	o/p	Pol.	Pump	Line	Thres.	Off.	Uncert.	Nu (1/cm)	Freq. (GHz)	Ref.
1730.833	-1+	P	CO2	10.22R24	+3+		1.0E-6	5.777564	173.2070	DY8
1541.750	-1+	P	CO2	9.64P30	+3+		1.0E-6	6.486134	194.4494	DY8
1161.676	-1	P	CO2	10.25R20	+3+		1.0E-6	8.608249	258.0688	DY8
1157.318	-1-	N	CO2	10.14R38	+3+		1.0E-6	8.640668	259.0407	DY8
986.3125	0-	P	CO2	10.17R32	+3+		1.0E-6	10.13877	303.9528	DY8
926.2087	+1	N	CO2	10.29R14	+3+		1.0E-6	10.79670	323.6770	DY8
919.9355	+1	P	CO2	10.17R32	+3+		1.0E-6	10.87033	325.8842	DY8
826.	-1	P	CO2	9.49P12	+3+		2.0E-3	12.11	363.0	DY8
819.	-3+	P	CO2	10.76P36	+3+		2.0E-3	12.21	366.0	DY8
813.7572	-1-	P	CO2	9.49P12	+3+		1.0E-6	12.28868	368.4053	DY8
733.	-1	P	CO2	10.15R36	+3+		2.0E-3	13.64	409.0	DY8
727.9491	-1+	P	CO2	10.11R42	+3+		1.0E-6	13.73722	411.8316	DY8
695.6720	0-	P	CO2	10.15R36	+3+		1.0E-6	14.37459	430.9394	DY8
692.	-1-	N	CO2	10.30R12	+3+		2.0E-3	14.44	433.0	DY8
689.9981	0+	P	CO2	10.21R26	+3+		1.0E-6	14.49279	434.4830	DY8
668.	-1	P	CO2	9.17R40	+3+		2.0E-3	14.98	449.0	DY8
660.	-1-	N	CO2	10.30R12	+3+		2.0E-3	15.14	454.0	DY8
657.	-1-	P	CO2	9.57P22	+3+		2.0E-3	15.21	456.0	DY8
630.1661	0-	N	CO2	10.32R10	+3+		1.0E-6	15.86883	475.7356	DY8
594.	-1	N	CO2	9.22R30	+3+		2.0E-3	16.85	505.	DY8
590.	-1	P	CO2	10.53P14	+3+		2.0E-3	16.95	508.	DY8
582.5536	-1+	N	CO2	9.54P18	+3+		1.0E-6	17.16580	514.6178	DY8
567.1065	0	P	CO2	10.53P14	+3+		1.0E-6	17.63337	528.6352	DY8
531.	-2+	P	CO2	9.73P40	+3+		2.0E-3	18.85	565.	DY8
498.0	-1-	P	CO2	9.77P44	+3+		2.0E-3	20.08	602.	DY8
493.1562	-1+	P	CO2	10.13R40	+3+		1.0E-6	20.27755	607.9057	DY8
477.4	-2+	P	CO2	9.50P14	+3+		2.0E-3	20.95	628.	DY8
472.9	-1	N	CO2	9.18R38	+3+		2.0E-3	21.15	634.	DY8
472.1	0	P	CO2	10.55P16	+3+		2.0E-3	21.18	635.	DY8
461.2610	+1	P	CO2	10.55P16	+3+		1.0E-6	21.67970	649.9410	DY8

Table 3.3.5.6 (continued)
HCOOD, O-DEUTERO FORMIC ACID

For line assignments see Reference DE1 (pump)

Lambda (μm)	o/p	Pol.	Pump	Line	Thres.	Off.	Uncert.	Nu (1/cm)	Freq. (GHz)	Ref.
450.9799	0-	N	CO2	10.51P12	+3+		1.0E-6	22.17394	664.7579	DY8
450.1	-1	N	CO2	10.21R26	+3+		2.0E-3	22.22	666.	DY8
446.8	-3+	P	CO2	9.62P28	+3+		2.0E-3	22.38	671.	DY8
433.2	-2+	P	CO2	10.27R16	+3+		2.0E-3	23.08	692.	DY8
430.4380	0	P	CO2	10.46P06	+3+		1.0E-6	23.23215	696.4823	DY8
429.6898	0	P	CO2	10.63P24	+3+		1.0E-6	23.27260	697.6951	DY8
417.0	0	P	CO2	9.26R22	+3+		2.0E-3	23.98	719.	DY8
411.2	-1	P	CO2	9.73P40	+3+		2.0E-3	24.32	729.	DY8
398.1	-1-	P	CO2	10.23R22	+3+		2.0E-3	25.12	753.	DY8
395.7124	0+	N	CO2	10.30R12	+3+		1.0E-6	25.27088	757.6019	DY8
395.0	-1	P	CO2	9.52P16	+3+		2.0E-3	25.32	759.	DY8
395.0	-1	P	CO2	9.47P10	+3+		2.0E-3	25.32	759.	DY8
392.9	-1+	P	CO2	9.29R16	+3+		2.0E-3	25.45	763.	DY8
391.6886	-1+	N	CO2	10.14R38	+3+		1.0E-6	25.53048	765.3846	DY8
387.8	-1+	N	CO2	10.37R04	+3+		2.0E-3	25.79	773.	DY8
373.8	-2	P	CO2	9.68P34	+3+		2.0E-3	26.75	802.	DY8
372.0	-1-	N	CO2	10.15R36	+3+		2.0E-3	26.88	806.	DY8
372.0	-1	P	CO2	10.65P26	+3+		2.0E-3	26.88	806.	DY8
369.9678	0	P	CO2	10.19R28	+3+		1.0E-6	27.02938	810.3205	DY8
361.2	0-	P	CO2	9.71P38	+3+		2.0E-3	27.69	830.	DY8
358.2	-2-	P	CO2	10.27R16	+3+		2.0E-3	27.92	837.	DY8
356.0	-1+	P	CO2	10.70P30	+3+		2.0E-3	28.09	842.	DY8
355.2	-1+	P	CO2	9.71P38	+3+		2.0E-3	28.15	844.	DY8
353.1	0-	P	CO2	10.67P28	+3+		2.0E-3	28.32	849.	DY8
353.1	0+	P	CO2	10.35R06	+3+		2.0E-3	28.32	849.	DY8
351.9	-1+	P	CO2	10.23R22	+3+		2.0E-3	28.42	852.	DY8
351.0	-1	P	CO2	9.69P36	+3+		2.0E-3	28.49	854.	DY8
347.0	-1	N	CO2	10.33R08	+3+		2.0E-3	28.82	864.	DY8
339.9	-1+	P	CO2	9.62P28	+3+		2.0E-3	29.42	882.	DY8
325.9	-1-	P	CO2	10.18R30	+3+		2.0E-3	30.68	920.	DY8

Lambda (μm)	o/p	Pol.	Pump	Line	Thres.	Off.	Uncert.	Nu (1/cm)	Freq (GHz)	Ref.
324.1	-1-	P	CO2	10.32R10	+3+		2.0E-3	30.85	925.	DY8
304.1	-1-	P	CO2	9.35R06	+3+		2.0E-3	32.89	986.	DY8
291.9	0-	P	CO2	10.72P32	+3+		2.0E-3	34.26	1027.	DY8
240.0	-1+	N	CO2	10.29R14	+3+		2.0E-3	41.67	1249.	DY8

Table 3.3.5.7
DCOOH, C-DEUTERO FORMIC ACID

Lambda (μm)	o/p	Pol.	Pump	Line	Thres.	Off.	Uncert.	Nu (1/cm)	Freq (GHz)	Ref.
1237.966	-1-	P	CO2	10.22R24	+3+		1.0E-6	8.077768	242.1654	DY8
1047.579	0-	N	CO2	10.30R12	+3+		1.0E-6	9.545824	286.1766	DY8
971.8064	-1-	P	CO2	10.19R28	+3+		1.0E-6	10.29012	308.4899	DY8
752.7485	0	P	CO2	10.16R34	+3+		1.0E-6	13.28465	398.2638	DY8
713.1056	0+	P	CO2	10.16R34	+3+		1.0E-6	14.02317	420.4040	DY8
710.	-2	N	CO2	10.46P06	+3+		2.0E-3	14.08	422.0	DY8
697.4552	-1+	P	CO2	10.15R36	+3+		1.0E-6	14.33784	429.8376	DY8
647.3485	-2+	P	CO2	10.18R30	+3+		1.0E-6	15.44763	463.1083	DY8
639.1282	0	N	CO2	10.48P08	+3+		1.0E-6	15.64631	469.0647	DY8
479.9040	0	P	CO2	10.53P14	+3+		1.0E-6	20.83750	624.6926	DY8
466.5461	0-	N	CO2	10.53P14	+3+		1.0E-6	21.43411	642.5784	DY8
433.2353	0-	N	CO2	10.29R14	+3+		1.0E-6	23.08215	691.9853	DY8
433.2	-2+	P	CO2	10.55P16	+3+		2.0E-3	23.08	692.	DY8
365.2	-2+	P	CO2	10.30R12	+3+		2.0E-3	27.38	821.	DY8
362.1	-1	P	CO2	10.59P20	+3+		2.0E-3	27.62	828.	DY8
341.8	-1	N	CO2	10.27R16	+3+		2.0E-3	29.26	877.	DY8
328.4570	0	P	CO2	10.61P22	+3+		1.0E-6	30.44539	912.7297	DY8
312.0	-1	P	CO2	10.22R24	+3+		2.0E-3	32.05	961.	DY8
272.0	-1+	P	CO2	10.70P30	+3+		2.0E-3	36.76	1102.	DY8
265.1	-2	N	CO2	10.25R20	+3+		2.0E-3	37.72	1131.	DY8

Table 3.3.5.8
DCOOD, DI-DEUTERO FORMIC ACID

For line assignments see Reference DE1 (pump)

Lambda (μm)	o/p	Pol.	Pump	Line	Thres.	Off.	Uncert.	Nu (1/cm)	Freq. (GHz)	Ref.
1281.649	-1-	P	CO2	9.71P38	+3+		1.0E-6	7.802451	233.9116	DY8
1158.	-2+	P	CO2	10.48P08	+3+		2.0E-3	8.64	259.0	DY8
1070.231	-2-	P	CO2	9.49P12	+3+		1.0E-6	9.343781	280.1195	DY8
1009.409	0	P	CO2	10.26R18	+3+		1.0E-6	9.906784	296.9979	DY8
998.5140	-1+	P	CO2	9.49P12	+3+		1.0E-6	10.01488	300.2386	DY8
936.6023	-1-	P	CO2	10.65P26	+3+		1.0E-6	10.67689	320.0851	DY8
935.0095	0	P	CO2	9.52P16	+3+		1.0E-6	10.69508	320.6304	DY8
927.9814	-1	P	CO2	10.59P20	+3+		1.0E-6	10.77608	323.0587	DY8
877.5481	0+	P	CO2	10.65P26	+3+		1.0E-6	11.39539	341.6251	DY8
843.2369	-2	P	CO2	9.49P12	+3+		1.0E-6	11.85906	355.5258	DY8
835.	-1-	P	CO2	10.25R20	+3+		2.0E-3	11.98	359.0	DY8
812.	-1+	P	CO2	10.44P04	+3+		2.0E-3	12.31	369.0	DY8
795.	-2+	P	CO2	10.44P04	+3+		2.0E-3	12.58	377.0	DY8
789.4203	0+	P	CO2	10.25R20	+3+		1.0E-6	12.66752	379.7628	DY8
779.8744	-1-	P	CO2	10.65P26	+3+		1.0E-6	12.82258	384.4112	DY8
761.7617	-1	N	CO2	10.49P10	+3+		1.0E-6	13.12747	393.5515	DY8
737.	-2+	P	CO2	10.44P04	+3+		2.0E-3	13.58	407.0	DY8
726.9203	0-	N	CO2	10.49P10	+3+		1.0E-6	13.75667	412.4145	DY8
666.	-1	P	CO2	10.70P30	+3+		2.0E-3	15.01	450.0	DY8
645.	-1-	X	CO2	10.22R24	+3+		2.0E-3	15.51	465.0	DY8
593.	-1+	N	CO2	10.57P18	+3+		2.0E-3	16.88	506.	DY8
591.6157	0	P	CO2	10.21R26	+3+		1.0E-6	16.90286	506.7351	DY8
567.8683	+1-	P	CO2	10.21R26	+3+		1.0E-6	17.60972	527.9260	DY8
561.2939	0+	N	CO2	10.59P20	+3+		1.0E-6	17.81598	534.1096	DY8
527.2146	0+	P	CO2	10.74P34	+3+		1.0E-6	18.96761	568.6346	DY8
526.4856	0+	P	CO2	10.74P34	+3+		1.0E-6	18.99387	569.4219	DY8
514.9507	0+	P	CO2	10.74P34	+3+		1.0E-6	19.41933	582.1770	DY8
508.7911	0	N	CO2	10.48P08	+3+		1.0E-6	19.65443	589.2250	DY8
508.	-2	P	CO2	10.19R28	+3+		2.0E-3	19.68	590.	DY8
491.8906	0	N	CO2	10.48P08	+3+		1.0E-6	20.32972	609.4698	DY8

478.9	-2+	P	CO2	10.37R04	+3+	2.0E-3	20.88	626.	DY8
469.2	-1	N	CO2	10.51P12	+3+	2.0E-3	21.31	639.	DY8
457.3410	-1+	P	CO2	10.70P30	+3+	1.0E-6	21.86552	655.5119	DY8
452.2	-1	P	CO2	10.32R10	+3+	2.0E-3	22.11	663.	DY8
442.8	-1-	N	CO2	10.53P14	+3+	2.0E-3	22.58	677.	DY8
425.2	0-	N	CO2	10.57P18	+3+	2.0E-3	23.52	705.	DY8
415.2	-1	P	CO2	10.29P14	+3+	2.0E-3	24.08	722.	DY8
414.1	-1+	P	CO2	10.86P44	+3+	2.0E-3	24.15	724.	DY8
397.1	0-	P	CO2	10.23R22	+3+	2.0E-3	25.18	755.	DY8
396.0	0	P	CO2	10.25R20	+3+	2.0E-3	25.25	757.	DY8
395.1488	0	P	CO2	10.32R10	+3+	1.0E-6	25.30692	758.6825	DY8
389.9070	0	P	CO2	10.30R12	+3+	1.0E-6	25.64714	768.8820	DY8
380.5654	+1	P	CO2	10.30R12	+3+	1.0E-6	26.27669	787.7555	DY8
366.9	-1	P	CO2	9.37R04	+3+	2.0E-3	27.26	817.	DY8
351.9	0	P	CO2	10.29R14	+3+	2.0E-3	28.42	852.	DY8
350.2	0	P	CO2	10.13R40	+3+	2.0E-3	28.56	856.	DY8
335.7087	0-	P	CO2	9.34R08	+3+	1.0E-6	29.78773	893.0136	DY8
325.2	-2+	P	CO2	10.19R28	+3+	2.0E-3	30.75	922.	DY8
323.1	0-	P	CO2	10.18R30	+3+	2.0E-3	30.95	928.	DY8
310.0	0+	P	CO2	10.22R24	+3+	2.0E-3	32.26	967.	DY8
304.0832	+1-	P	CO2	10.22R24	+3+	1.0E-6	32.88574	985.8897	DY8
298.0	-1+	P	CO2	10.35R06	+3+	2.0E-3	33.56	1006.	DY8
283.1	-1+	P	CO2	9.29R16	+3+	2.0E-3	35.32	1059.	DY8
276.1	0-	P	CO2	9.52P16	+3+	2.0E-3	36.23	1086.	DY8
266.1	-1+	P	CO2	10.17R32	+3+	2.0E-3	37.58	1127.	DY8
241.2	-1	P	CO2	10.15R36	+3+	2.0E-3	41.46	1243.	DY8
218.0	+1-	P	CO2	10.25R20	+3+	2.0E-3	45.87	1375.	DY8

Table 3.3.5.9
CH_3Cl- METHYL CHLORIDE

For line assignments see Reference DER*

Lambda (μm)	o/p	Pol.	Pump	Line	Thres.	Off.	Uncert.	Nu (1/cm)	Freq. (GHz)	Ref.
1886.87	0+	N	CO2	9.60P26	+4	+20	4.9E-5	5.2998	158.883	CH6
958.25	0-	P	CO2	9.71P38	+4+	-50	1.0E-4	10.436	312.85	CH6
944.0	+1	P	CO2	9.32R12	+3+	-30	1.1E-4	10.594	317.59	CH6JEN
870.80	0	P	CO2	9.86P52	+4-	+30	1.1E-4	11.484	344.27	CH6
568.8	-2+	P	CO2	10.21R26	+5-	+5	1.7E-4	17.581	527.05	CH6
511.9	-1-	N	CO2	10.07R52	+4-	0	2.0E-4	19.535	585.6	CH6
461.2	-2+	N	CO2	9.17R42	+3+	-25	2.2E-4	21.683	650.0	CH6
378.57	-1+	N	CO2	9.29R16	+4+	+50	2.5E-4	26.415	791.9	CH6
349.3	+1	P	CO2	10.26R18	+3+	-5	2.9E-4	28.625	858.2	CH6JEN
334.0	+2-	N	CO2	9.75P42	+3	+40	3.0E-4	29.944	897.7	CH6
307.65	0-	P	CO2	10.97P19	+3+	-50	3.2E-4	32.50	974.5	CH6
286.8	-1+	N	CO2	10.16R34	+4+	-35	3.5E-4	34.87	1045.3	CH6
281.7	)	P	CO2	9.31R14	+4-	+45	3.5E-4	35.50	1064.3	CH6
275.09	-1-	P	CO2	9.19R36	+4-	+10	3.5E-4	36.35	1089.8	CH6
275.00	0+	P	CO2	9.31R14	+4-	+45	3.5E-4	36.36	1090.2	CH6
271.3	0-	N	CO2	10.59P20	+4	-35	3.7E-4	36.86	1105.1	CH6
261.03	-1-	P	CO2	10.74P34	+4+	-20	3.8E-4	38.31	1148.5	CH6
240.98	0	P	CO2	10.49P10	+4	-30	4.0E-4	41.50	1244.1	CH6
236.25	-1	N	CO2	9.38R02	+3+	-10	4.1E-4	42.33	1269.0	CH6
227.15	-1	P	CO2	9.81P48	+4-	+50	4.4E-4	44.02	1319.8	CH6

Table 3.3.5.10
CD_3Cl, FULLY DEUTERATED METHYL CHLORIDE

For line assignments see Reference DH

Lambda (μm)	o/p	Pol.	Pump	Line	Thres.	Off.	Uncert.	Nu (1/cm)	Freq. (GHz)	Ref.
1990.757	0	P	CO2	9.50P14	+3+		2.0E-6	5.02322	150.5922	DY6
1239.480	0	P	CO2	9.49P12	+3+		2.0E-6	8.06790	241.8696	DY6
883.598	0	P	CO2	9.68P34	+3+		2.0E-6	11.31736	339.2860	DY6
792.	0+		CO2	9.62P28	+3	-35	4.9E-3	12.63	379.	DH
735.130	-1	N	CO2	9.44P06	+3+		2.0E-6	13.60304	407.8089	DY6
698.555	0	N	CO2	9.44P06	+3+		2.0E-6	14.31526	429.1607	DY6
519.303	0	P	CO2	9.69P36	+3+		2.0E-6	19.25657	577.298	DY6
480.3101	0	N	CO2	9.69P36	+3+		2.0E-6	20.81988	624.164	DY6
464.7567	+1	P	CO2	10.25R20	+3+		2.0E-6	21.51663	645.052	DY6
449.7997	+1	P	CO2	10.25R20	+3+		2.0E-6	22.23212	666.502	DY6
443.2645	+1	P	CO2	9.47P10	+3+		2.0E-6	22.55989	676.329	DY6
383.2845	+1	P	CO2	9.20R34	+3+		2.0E-6	26.09028	782.167	DY6
318.	-1	P	CO2	10.19R28	+3+		4.9E-3	31.4	943.	DY6
293.6480	+1	N	CO2	9.59P24	+3+		2.0E-6	34.05438	1020.925	DY6
288.	-1	P	CO2	9.52P16	+3+		4.9E-3	34.7	1041.	DY6
288.	-1	P	CO2	10.26R18	+3+		4.9E-3	34.7	1041.	DY6
249.	-1	N	CO2	9.71P38	+3+		4.9E-3	40.2	1204.	DY6
246.	0	P	CO2	10.29R14	+3+		4.9E-3	40.7	1219.	DY6
245.	0	N	CO2	9.66P32	+3+		4.9E-3	40.8	1224.	DY6
224.	-1	P	CO2	9.23R28	+3+		4.9E-3	44.6	1338.	DY6

Table 3.3.5.11
CH_2F_2, DIFLUOROMETHANE

For line assignments see References DW3 (pump), DW5*

Lambda (μm)	o/p	Pol.	Pump	Line	Thres.	Off.	Uncert.	Nu (1/cm)	Freq. (GHz)	Ref.
1448.0958	-2-	E	CO2	9.16R44	+4-		5.0E-7	6.905621	207.0253	PE3SC
724.9203	0-	N	CO2	9.43P04	+4		5.0E-7	13.794620	413.5523	PE3SC
657.2391	0-	N	CO2	9.47P10	+4		5.0E-7	15.215163	456.1391	PE3SC
642.5999	-1+	P	CO2	9.16R44	+4-		5.0E-7	15.561782	466.5305	PE3SC
588.0276	-1	E	CO2	9.15R46	+4-		5.0E-7	17.006005	509.8272	PE3SC
567.5316	-2+	E	CO2	9.23R28	+4		5.0E-7	17.620163	528.2392	PE3SC
540.9864	-1	P	CO2	9.17R42	+4		5.0E-7	18.484755	554.1590	PE3SC
511.4451	-1	N	CO2	9.23R28	+3+		5.0E-7	19.552440	586.1674	PE3SCDW3
503.0567	0-	N	CO2	9.35R06	+4		5.0E-7	19.878475	595.9417	PE3SC
464.4123	-1+	P	CO2	9.44P06	+4		5.0E-7	21.53259	645.5309	PE3SC
434.9514	0-	N	CO2	9.35R06	+3+		5.0E-7	22.99108	689.2551	PE3SCDW3
418.2703	-2	P	CO2	9.32R12	+3+		5.0E-7	23.90798	716.7433	PE3SCDW3
394.7009	0	P	CO2	9.44P06	+4		5.0E-7	25.33564	759.5433	PE3SC
382.6392	0	P	CO2	9.47P10	+4		5.0E-7	26.13428	783.4860	PE3SC
381.9956	-1	P	CO2	9.19R36	+4		5.0E-7	26.17831	784.8060	PE3SC
355.1261	-2+	P	CO2	9.46P08	+4		5.0E-7	28.15901	844.1859	PE3SC
326.4230	-2+	N	CO2	9.31R14	+3+		5.0E-7	30.63509	918.4170	PE3SCDW3
298.2910	-2+	P	CO2	9.19R36	+4		5.0E-7	33.52431	1005.0337	PE3SC
293.9015	-2	P	CO2	9.55P20	+4+		5.0E-7	34.02501	1020.0440	PE3SC
289.4999	0+	P	CO2	9.43P04	+4		5.0E-7	34.54232	1035.5527	PE3SC
287.6672	+1	P	CO2	9.20R34	+4		5.0E-7	34.76240	1042.1504	PE3SCDG
272.3389	0-	N	CO2	9.47P10	+4		5.0E-7	36.71896	1100.8067	PE3SCDG
270.0055	-1	N	CO2	9.26R22	+4+		5.0E-7	37.03629	1110.3199	PE3SC
261.7292	-1+	P	CO2	9.71P38	+4		5.0E-7	38.20743	1145.4301	PE3SCDG
256.0270	0+	N	CO2	9.59P24	+4		5.0E-7	39.05839	1170.9410	PE3SCDG

Lambda (μm)	o/p	Pol.	Pump	Line	Thres.	Off.	Uncert.	Nu (1/cm)	Freq. (GHz)	Ref.
236.6008	+1	P	CO2	9.35R06	+3+		5.0E-7	42.26529	1267.0815	PE3SCDW3
236.5915	+1	P	CO2	9.35R06	+3+		5.0E-7	42.26694	1267.1310	PE3SCDW3
235.6541	+1-	P	CO2	9.21R32	+4		5.0E-7	42.43507	1272.1714	PE3SCDG
230.1059	0-	N	CO2	9.17R42	+4		5.0E-7	43.45826	1302.8458	PE3SC
227.6570	-2	N	CO2	9.54P18	+4+		5.0E-7	43.92574	1316.8605	PE3SC
214.5791	+2+	N	CO2	9.20R34	+4		5.0E-7	46.60286	1397.1186	PE3SCDG
202.4649	0-	P	CO2	9.35R06	+3+		5.0E-7	49.39127	1480.7129	PE3SCDW3
194.44761	-2+	N	CO2	9.32R12	+3+		5.0E-7	51.42773	1541.7647	PE3SCDW3
193.90445	0-	P	CO2	9.26R22	+4+		5.0E-7	51.57179	1546.0834	PE3SC
191.84803	0	P	CO2	9.57P22	+4		5.0E-7	52.12459	1562.6559	PE3SC
184.30590	+2+	N	CO2	9.21R32	+4		5.0E-7	54.25762	1626.6026	PE3SCDG
166.67665	0	P	CO2	9.26R22	+3+		5.0E-7	59.99641	1798.6470	PE3SCDW3
166.63105	+2-	P	CO2	9.27R20	+3+		5.0E-7	60.01283	1799.1393	PE3SCDW3
158.96020	-2	P	CO2	9.55P20	+4+		5.0E-7	62.90883	1885.9593	PE3SC
158.51348	+1+	P	CO2	9.47P10	+4		5.0E-7	63.08612	1891.2742	PE3SCDG
135.26932	+1-	P	CO2	9.59P24	+4		5.0E-7	73.92659	2216.263	PE3SCDG
133.99765	0	N	CO2	9.57P22	+4		5.0E-7	74.62817	2237.296	PE3SC
122.46581	+1	N	CO2	9.26R22	+3+		5.0E-7	81.65544	2447.969	PE3SCDW3
122.46551	-2	P	CO2	9.46P08	+4		5.0E-7	81.65564	2447.975	PE3SC
117.72748	+2	N	CO2	9.27R20	+4+		5.0E-7	84.94193	2546.495	PE3SCDG
109.29579	0+	P	CO2	9.59P24	+4		5.0E-7	91.49483	2742.946	PE3SCDG
105.51827	-1+	P	CO2	9.52P16	+4+		5.0E-7	94.77032	2841.143	PE3SC
95.55106	-1-	N	CO2	9.32R12	+3+		5.0E-7	104.65609	3137.511	PE3SCDW3

Table 3.3.5.12
CH_2Cl_2, METHYLENE CHLORIDE (DICHLOROMETHANE)

Lambda (μm)	o/p	Pol.	Pump	Line	Thres.	Off.	Uncert.	Nu (1/cm)	Freq. (GHz)	Ref.
258.	-1		CO2	10.65P26	+4		1.0E-2	38.8	1160.	JEN

Table 3.3.5.13
CD_2Cl_2, FULLY DEUTERATED METHYLENE CHLORIDE

Lambda (μm)	o/p	Pol.	Pump	Line	Thres.	Off.	Uncert.	Nu (1/cm)	Freq. (GHz)	Ref.
829.	0-		CO2	10.46P06	+3+		4.9E-3	12.06	362.	ZD
631.	0-		CO2	10.26R18	+3+		4.9E-3	15.85	475.	ZD
520.	0-		CO2	10.30R12	+3+		4.9E-3	19.23	577.	ZD
469.	-1+		CO2	10.37R04	+3+		4.9E-3	21.3	639.	ZD
342.	0-		CO2	10.55P16	+3+		4.9E-3	29.2	877.	ZD
254.	-1+		CO2	10.15R36	+3+		4.9E-3	39.4	1180.	ZD
249.	0+		CO2	10.27R16	+3+		4.9E-3	40.2	1204.	ZD

Table 3.3.5.14
$CH_3{}^{79}Br$, METHYL BROMIDE WITH ^{79}Br

For line assignments see Reference MO

Lambda (μm)	o/p	Pol.	Pump	Line	Thres.	Off.	Uncert.	Nu (1/cm)	Freq. (GHz)	Ref.
1965.34	0	P	CO2	10.67P28	+4+	0	4.9E-5	5.0882	152.540	CH6MO
1572.64	0+	P	CO2	10.44P04	+4-	0	5.9E-5	6.3587	190.63	CH6MO
1310.4	+1-	N	CO2	10.37R04	+4	-30	7.9E-5	7.6314	228.78	CH6MO
925.5	+1	P	CO2	10.09R46	+4-	-35	1.1E-4	10.805	323.92	CH6MO
749.36	( )	N	CO2	10.29R14	+4	-15	1.2E-4	13.345	400.06	CH6MO
749.29	0+	N	CO2	10.53P14	+4	+5	1.2E-4	13.346	400.10	CH6MO
715.4	0+	N	CO2	10.29R14	+4	-15	1.4E-4	13.978	419.06	CH6MO
414.98	+1+	P	CO2	10.38R02	+4	+45	2.4E-4	24.098	722.4	CH6MO
264.1	+1-	P	CO2	10.32R10	+4+	+10	3.8E-4	37.87	1135.4	CH6MO

Table 3.3.5.15
$CH_3{}^{81}Br$, METHYL BROMIDE WITH ^{81}Br

For line assignments see Reference MO

Lambda (μm)	o/p	Pol.	Pump	Line	Thres.	Off.	Uncert.	Nu (1/cm)	Freq. (GHz)	Ref.
2650.			CO2	10.49P10			4.9E-3	3.77	113.1	MO
831.13	+1-	P	CO2	10.67P28	+4+	+10	1.2E-4	12.032	360.70	CH6MO
658.53	+1-	P	CO2	9.90P56	+4	0	1.5E-4	15.185	455.25	CH6MO
585.72	+1-	P	CO2	9.73P40	+4-	+25	1.7E-4	17.073	511.84	CH6MO
564.7	( )		CO2	10.79P38	+3+	-5	1.7E-4	17.709	530.91	CH6MO
545.21	+1	P	CO2	10.79P38	+3+	-5	1.7E-4	18.342	549.87	CH6MO
311.20	0	N	CO2	10.81P40	+4-	+45	3.2E-4	32.13	963.3	CH6MO

Table 3.3.5.16 (continued)
CH_3Br, METHYL BROMIDE WITH MIXED ^{79}Br AND ^{81}Br

Lambda (μm)	o/p	Pol.	Pump	Line	Thres.	Off.	Uncert.	Nu (1/cm)	Freq. (GHz)	Ref.
990.51	0	P	CO2	10.49P10	+5-	0	1.0E-4	10.096	302.66	CH6
660.70	+1+	N	CO2	10.25R20	+4	+25	1.5E-4	15.135	453.75	CH6
632.0	-2+	N	CO2	10.61P22	+4	0	1.6E-4	15.823	474.36	CH6
631.9	-3+	P	CO2	10.55P16	+5-	+20	1.6E-4	15.825	474.41	CH6
545.39	+1	P	CO2	10.17R32	+4-	+20	1.7E-4	18.336	549.68	CH6
531.1	+1-	P	CO2	10.63P24	+4+	-5	1.9E-4	18.830	564.5	CH6
508.5	0+	P	CO2	10.11R42	+4-	+30	2.0E-4	19.666	589.6	CH6
422.8	-2	P	CO2	10.21R26	+5-	+50	2.4E-4	23.653	709.1	CH6
418.3	0-	P	CO2	10.65P26	+4+	+10	2.4E-4	23.906	716.7	CH6
407.7	-2	P	CO2	9.62P28	+5-	-40	2.5E-4	24.527	735.3	CH6
380.02	+1-	N	CO2	10.26R18	+4	+50	2.5E-4	26.314	788.9	CH6
352.75	0+	N	CO2	9.54P18	+4+	+25	2.8E-4	28.349	849.9	CH6
333.15	-2	P	CO2	10.48P08	+5-	+35	3.0E-4	30.017	899.9	CH6

Table 3.3.5.16
CH_3Br, METHYL BROMIDE WITH MIXED ^{79}Br AND ^{81}Br

Lambda (μm)	o/p	Pol.	Pump	Line	Thres.	Off.	Uncert.	Nu (1/cm)	Freq. (GHz)	Ref.
332.86	0-	N	CO2	10.35R06	+4	0	3.0E-4	30.043	900.7	CH6
311.21	0-	N	CO2	10.08R50	+3+	+50	3.2E-4	32.13	963.3	CH6
311.10	0	P	CO2	10.59P20	+4	+10	3.2E-4	32.14	963.7	CH6
311.07	-2+	N	CO2	10.30R12	+4	+30	3.2E-4	32.15	963.7	CH6
294.3	0	P	CO2	10.19R28	+4+	-40	3.4E-4	33.98	1018.7	CH6
279.8	-2	P	CO2	10.07R52	+4	0	3.5E-4	35.74	1071.4	CH6
245.0	0	P	CO2	9.62P28	+4	-40	4.0E-4	40.81	1223.4	CH6

Table 3.3.5.17
CD_3Br, FULLY DEUTERATED METHYL BROMIDE

For line assignments see Reference LA4

Lambda (μm)	o/p	Pol.	Pump	Line	Thres.	Off.	Uncert.	Nu (1/cm)	Freq. (GHz)	Ref.
560.	-1-	P	CO2	9.22R30	+3+		2.0E-2	18.0	540.	LA4
550.	-2+	N	CO2	9.66P32	+4		2.0E-2	18.1	540.	LA4
530.	-1-	P	CO2	10.32R10	+3+		2.0E-2	18.9	570.	LA4
440.	-1-	P	CO2	9.28R18	+3+		2.0E-2	22.7	680.	LA4
430.	-1-	N	CO2	9.54P18	+3+		2.0E-2	23.3	700.	LA4
428.	-1-	P	CO2	10.38R02	+2+		2.0E-2	23.4	700.	LA4
367.	-1-	P	CO2	9.50P14	+3		2.0E-2	27.2	820.	LA4
341.	-2-	N	CO2	9.69P36	+4-		2.0E-2	29.3	880.	LA4
297.	-1-	P	CO2	9.66P32	+3+		2.0E-2	33.7	1010.	LA4
290.	-1-	P	CO2	10.57P18	+3		2.0E-2	34.5	1030.	LA4

Table 3.3.5.18
CF_2Cl_2, DI-FLUORO DI-CHLORO METHANE (FLUOROCARBON-12)

For line assignments see Reference LO1

Lambda (μm)	o/p	Pol.	Pump	Line	Thres.	Off.	Uncert.	Nu (1/cm)	Freq. (GHz)	Ref.
1205.	-2+		CO2	10.76P36	+4	-31	2.0E-3	8.30	248.8	LO1
1164.	-2+		CO2	10.76P36	+4	-31	2.0E-3	8.59	257.6	LO1
1025.	-2+		CO2	10.74P34	+4	+17	1.0E-3	9.756	292.5	LO1
980.	-2+		CO2	10.76P36	+4		1.0E-2	10.2	306.	LO1
858.7	-1-		CO2	10.74P34	+4	+12	6.0E-4	11.645	349.1	LO1
765.2	-1		CO2	10.84P42	+4-	+28	7.0E-4	13.068	391.8	LO1
751.4	-1		CO2	10.70P30	+4+	+10	8.0E-4	13.31	399.0	LO1
684.7	-2+		CO2	10.74P34	+4	-22	1.1E-3	14.60	437.8	LO1
684.7	-1+		CO2	10.84P42	+4-	-13	7.0E-4	14.60	437.8	LO1
638.4	-1-		CO2	10.76P36	+4	+35	9.0E-4	15.66	469.6	LO1
614.3	-1-		CO2	10.72P32	+4	+33	1.2E-3	16.28	488.0	LO1

Table 3.3.5.19 (continued)
CH_3I, METHYL IODIDE

For line assignments see Reference GR*

Lambda (μm)	o/p	Pol.	Pump	Line	Thres.	Off.	Uncert.	Nu (1/cm)	Freq. (GHz)	Ref.
1253.738	+1	P	CO2	10.72P32	+4-	+25	2.0E-6	7.97615	239.1189	DY6CH6
1063.29	+1-	P	CO2	10.78P38	+4+	+5	8.9E-5	9.4048	281.95	CH6
964.	-1	P	CO2	10.61P22	+3+		4.9E-3	10.37	311.	DY6
719.3	0-	P	CO2	10.61P22	+4+	+25	1.4E-4	13.902	416.78	CH6
671.0	0	P	CO2	10.67P28	+4+	+20	1.5E-4	14.903	446.79	CH6
639.7	-1+	P	CO2	9.44P06	+4+	+20	1.6E-4	15.632	468.62	CH6
583.87	-1	P	CO2	9.42P04	+4-	0	1.7E-4	17.127	513.46	CH6
578.90	0	P	CO2	10.16R34	+5-	-30	1.7E-4	17.274	517.87	CH6
576.17	0+	P	CO2	10.55P16	+ +	-30	1.7E-4	17.356	520.32	CH6
542.99	-3+	P	CO2	10.65P26	+4	+10	1.7E-4	18.417	552.11	CH6

Table 3.3.5.19
CH_3I, METHYL IODIDE

For line assignments see Reference GR*

Lambda (μm)	o/p	Pol.	Pump	Line	Thres.	Off.	Uncert.	Nu (1/cm)	Freq. (GHz)	Ref.
529.3	+1-	P	CO2	10.76P36	+4+	0	1.9E-4	18.894	566.4	CH6
525.32	-1	P	CO2	9.42P04	+4-	-25	1.9E-4	19.036	570.7	CH6
517.33	0-	P	CO2	10.53P14	+5-	+30	1.9E-4	19.330	579.5	CH6
508.4	+1	P	CO2	9.68P34	+4	0	2.0E-4	19.671	589.7	CH6
477.9	-2+	P	CO2	9.60P26	+5-	+10	2.0E-4	20.926	627.4	CH6
459.2	-1+	N	CO2	10.47P08	+5-	0	2.2E-4	21.778	652.9	CH6
457.2	( )	P	CO2	10.57P18	+3	0	2.2E-4	21.870	655.6	CH6
447.1421	+1+	P	CO2	10.57P18	+3	0	2.0E-6	22.36425	670.463	DY6CH6
392.48	-2+	P	CO2	9.30R14	+4+	+5	2.5E-4	25.479	763.8	CH6
390.5	0+	P	CO2	10.84P42	+4	+25	2.5E-4	25.606	767.7	CH6
377.45	+1-	P	CO2	9.29R16	+4	+5	2.5E-4	26.494	794.3	CH6DY4

Table 3.3.5.20
CD_3I, FULLY DEUTERATED METHYL IODIDE

For line assignments see Reference DY6

Lambda (μm)	o/p	Pol.	Pump	Line	Thres.	Off.	Uncert.	Nu (1/cm)	Freq. (GHz)	Ref.
1549.505	-1	P	CO2	9.33R10	+3+		2.0E-6	6.45367	193.4763	DY6
1099.544		P	CO2	10.61P22	+3+		2.0E-6	9.09468	272.6516	DY6
1005.348		P	CO2	10.74P34	+3+		2.0E-6	9.94681	298.1978	DY6
981.709		P	CO2	10.61P22	+3+		2.0E-6	10.18631	305.3780	DY6
953.880	0	P	CO2	9.23R28	+3+		2.0E-6	10.48350	314.2874	DY6
918.610	+1	P	CO2	9.23R28	+3+		2.0E-6	10.88601	326.3544	DY6
895.		P	CO2	10.70P30	+3+		4.9E-3	11.17	335.	DY6
788.482		P	CO2	10.51P12	+3+		2.0E-6	12.68260	380.2149	DY6
745.		P	CO2	10.48P08	+3+		4.9E-3	13.42	402.	DY6
734.262	0	P	CO2	9.57P22	+3+		2.0E-6	13.61911	408.2906	DY6

Lambda (μm)	o/p	Pol.	Pump	Line	Thres.	Off.	Uncert.	Nu (1/cm)	Freq. (GHz)	Ref.
730.323	0	P	CO2	9.23R28	+3+		2.0E-6	13.69256	410.4927	DY6
691.119	-1	P	CO2	9.27R20	+3+		2.0E-6	14.46928	433.7782	DY6
670.114		P	CO2	10.33R08	+3+		2.0E-6	14.92283	447.3751	DY6
670.094		P	CO2	10.33R08	+3+		2.0E-6	14.92328	447.3887	DY6
667.232		P	CO2	10.49P10	+3+		2.0E-6	14.98729	449.3075	DY6
660.582		P	CO2	10.88P46	+3+		2.0E-6	15.13816	453.8307	DY6
644.		N	CO2	10.55P16	+3+		4.9E-3	15.53	466.	DY6
640.		N	CO2	10.26R18	+3+		4.9E-3	15.62	468.	DY6
614.110	+1	P	CO2	10.23R22	+3+		2.0E-6	16.28373	488.1740	DY6
599.550	+1	P	CO2	10.23R22	+3+		2.0E-6	16.67918	500.029	DY6
569.477		P	CO2	10.76P36	+3+		2.0E-6	17.55996	526.434	DY6
556.876		P	CO2	10.76P36	+3+		2.0E-6	17.95733	538.347	DY6
540.	-1	N	CO2	9.35R06	+3+		4.9E-3	18.52	555.	DY6
523.406		P	CO2	10.79P38	+3+		2.0E-6	19.10562	572.772	DY6
490.3909	+1	P	CO2	9.26R22	+3+		2.0E-6	20.39189	611.334	DY6
487.2260	0	P	CO2	9.47P10	+3+		2.0E-6	20.52435	615.305	DY6
460.5619	+1	N	CO2	9.32R12	+3+		2.0E-6	21.71261	650.928	DY6
444.3862	+1	N	CO2	9.21R32	+3+		2.0E-6	22.50295	674.621	DY6
433.1036	-1	P	CO2	9.62P28	+3+		2.0E-6	23.08916	692.196	DY6
390.	-2	P	CO2	9.60P26	+3+		4.9E-3	25.6	769.	DY6
301.	0	P	CO2	9.24R26	+3+		4.9E-3	33.2	996.	DY6
272.	-1	N	CO2	9.49P12	+3+		4.9E-3	36.8	1102.	DY6

Table 3.3.5.21
$^{13}CD_3I$, FULLY DEUTERATED METHYL IODIDE WITH ^{13}C ISOTOPE

For line assignments see Reference DH (pump)

Lambda (μm)	o/p	Pol.	Pump	Line	Thres.	Off.	Uncert.	Nu (1/cm)	Freq. (GHz)	Ref.
806.	0		CO2	10.51P12	+3		4.9E-3	12.41	372.	DH
690.	+1-		CO2	10.49P10	+3		4.9E-3	14.49	434.	DH

Table 3.3.5.22
CF_3Br, BROMO 3-FLUOROMETHANE (FLUOROCARBON-1,3)

For line assignments see References PO1, LO2

Lambda (μm)	o/p	Pol.	Pump	Line	Thres.	Off.	Uncert.	Nu (1/cm)	Freq. (GHz)	Ref.
2140.			CO2	9.20R34	+4		9.0E-3	4.67	140.	LO2
1890.			CO2	9.27R20	+4		4.9E-3	5.29	158.6	LO2
1692.			CO2	9.17R40	+4		2.9E-3	5.91	177.2	LO2
1550.			CO2	9.18R38	+4		2.9E-3	6.45	193.4	LO2
1530.			CO2	9.29R16	+4		6.9E-3	6.55	196.	LO2
1151.			CO2	9.20R34	+4		4.0E-3	8.69	260.	LO2
1083.			CO2	9.23R28	+4		4.0E-3	9.23	277.	LO2
1040.			CO2	9.32R12	+4		1.1E-2	9.6	287.	LO2
883.	-2+		CO2	9.33R10	+4-		2.7E-3	11.33	339.5	PO1
824.	-1		CO2	9.34R08	+4-		1.7E-3	12.14	364.0	PO1

Table 3.3.6.1
N_2H_4, HYDRAZINE

Lambda (μm)	o/p	Pol	Pump	Line	Thres.	Off.	Uncert.	Nu (1/cm)	Freq. (GHz)	Ref.
1007.	-1	N	CO2	10.61P22			2.0E-3	9.93	297.7	DY3
802.	-1	P	CO2	10.22R24			2.0E-3	12.46	373.6	DY3
795.	-1	N	CO2	10.72P32			2.0E-3	12.58	377.1	DY3
734.1616	-2		CO2	10.14R38	+3		1.2E-6	13.62098	408.3467	RA2
721.	-1	P	CO2	10.51P12			2.0E-3	13.87	415.8	DY3
533.655	0	N	CO2	10.33R08			9.0E-6	18.7387	561.772	RA2DY3
527.8730	-2		CO2	9.49P12	+3		1.7E-6	18.94395	567.925	RA2
483.5	0	P	CO2	9.55P20			2.0E-3	20.68	620.	DY3
461.0718	-1	N	CO2	10.55P16	+3		1.5E-6	21.68859	650.2077	RA2DY3
435.7718	0	N	CO2	10.63P24	+3+		7.0E-7	22.94779	687.9574	RA2DY3
373.0	-1	N	CO2	10.30R12			2.0E-3	26.81	804.	DY3
372.5	-2	N	CO2	10.57P18			2.0E-3	26.85	805.	DY3
368.862	-1	N	CO2	9.28R18			6.1E-6	27.1104	812.750	RA2DY3

336.0	-1	N	CO2	10.63P24			2.0E-3	29.76	892.	DY3
331.6694	-1	N	CO2	9.49P12	+3+		1.1E-6	30.15050	903.8894	RA2DY3
331.5	-2	P	CO2	9.64P30			2.0E-3	30.17	904.	DY3
327.0	-1	P	CO2	9.26R22			2.0E-3	30.58	917.	DY3
311.0747	0	P	CO2	9.55P20	+3		1.0E-6	32.14662	963.7314	RA2DY3
301.2754	0	P	CO2	10.30R12	+3		1.5E-6	33.19222	995.078	RA2DY3
271.5	-2	P	CO2	10.57P18			2.0E-3	36.83	1104.	DY3
265.0	-2	P	CO2	10.19R28			2.0E-3	37.74	1131.	DY3
264.8014	0	P	CO2	10.25R20	+3		9.0E-7	37.76415	1132.141	RA2DY3
262.0	-2	N	CO2	10.67P28			2.0E-3	38.17	1144.	DY3
250.5	-1	P	CO2	9.34R08			2.0E-3	39.92	1197.	DY3
246.5	-1	P	CO2	10.46P06			2.0E-3	40.57	1216.	DY3
234.0	0	P	CO2	10.16R34			2.0E-3	42.74	1281.	DY3
233.9157	-1	P	CO2	10.33R08	+2+		8.0E-7	42.75044	1281.626	RA2DY3
192.9072	0	N	CO2	10.63P24	+3-		1.8E-6	51.83840	1554.076	RA2DY3
181.92643	-2	P	CO2	10.46P06	+3		3.0E-7	54.96727	1647.8774	RA2

Table 3.3.6.2 (continued)
CH_3OH, METHANOL (METHYL ALCOHOL)

For line assignments see References HEN1, HEN3*, HEN5*, HEN6*, IN1, IN2, W1*

Lambda (μm)	o/p	Pol.	Pump	Line	Thres.	Off.	Uncert.	Nu (1/cm)	Freq. (GHz)	Ref.
1223.658	-1	P	CO2	9.52P16	+3+		2.0E-6	8.17222	244.9969	RA2TA1
699.4226	-1	N	CO2	9.68P34			1.2E-6	14.29751	428.6285	PET/CH3
695.3499	-2+	P	CO2	10.27R16	+4	0	5.0E-7	14.381249	431.1390	PE2HEN1
694.1893	-1-	P	CO2	9.59P24	+4	0	5.0E-7	14.405292	431.8598	PE2HEN1
694.	-2		CO2	10.27R16	+3+		4.0E-3	14.41	432.	TA2
624.4301	-2		CO2	9.71P38	+4		5.0E-7	16.014602	480.1057	PE2
614.2851	-2+	P	CO2	9.59P24	+4	0	5.0E-7	16.279085	488.0347	PE2HEN1
602.4870	-1	P	CO2	9.59P24	+4	0	5.0E-7	16.597869	497.5916	PE2DY2
570.5687	0+	P	CO2	9.52P16			9.6E-7	17.52637	525.4275	PET/CH3
495.	-2	N	CO2	10.37R04	+3+		5.9E-3	20.2	606.	TA2PE2
486.1	+1-		CO2S	9.68P31	+3+		2.0E-3	20.57	617.	WE1
469.0233	-1+		CO2	10.14R38	+3		7.8E-7	21.32090	639.1846	PET/RA1
418.0827	(0)	N	CO2	9.69P36	+4		5.0E-7	23.91871	717.0650	PE2CH3
416.5223	-2		CO2	9.50P14	+3	-15	1.0E-6	24.00832	719.7512	HEN1TA2
392.0687	+1-	N	CO2	9.69P36	+2+		6.6E-7	25.50573	764.6426	PETCH3T3

Table 3.3.6.2 (continued)
CH_3OH, METHANOL (METHYL ALCOHOL)

For line assignments see References HEN1, HEN3*, HEN5*, HEN6*, IN1, IN2, W1*

Lambda (μm)	o/p	Pol.	Pump	Line	Thres.	Off.	Uncert.	Nu (1/cm)	Freq. (GHz)	Ref.
390.1	0-		CO2S	9.52P13	+3+		2.0E-3	25.63	769.	WE1
386.3392	-1-	P	CO2	9.50P14	+4	-15	5.0E-7	25.88399	775.9824	PE2HEN1
369.1137	+0+	P	CO2	9.52P16			6.2E-7	27.09192	812.1954	PET/CH3
346.4875	-1-	P	CO2	9.57P22	+4		5.0E-7	28.86107	865.2331	PE2DY2
311.2	-1+	N	CO2	9.59P24			2.0E-3	32.13	963.	DY2
301.9943	-1-	N	CO2	9.50P14	+4	-15	5.0E-7	33.11321	992.7091	HEN1PE2
293.8217	-1	P	CO2	10.32R10	+4		5.0E-7	34.03425	1020.3211	PE2HEN1
292.50	-1	N	CO2	9.68P34			1.7E-4	34.188	1024.9	CH3
292.1415	-1	P	CO2	9.71P38	+4		5.0E-7	34.22999	1026.1893	PE2CH3
290.62		N	CO2	9.49P12	+4	-5	3.0E-4	34.41	1031.6	HEN1
286.	0		CO2	10.09R48	+3-		1.0E-2	35.0	1050.	TA2
280.9341	-1-	P	CO2	9.28R18	+4	+5	5.0E-7	35.59553	1067.1272	PE2HEN1
278.8048	-1	P	CO2	9.71P38	+4		5.0E-7	35.86738	1075.2771	PE2CH3
274.	-1-		CO2	10.10R46	+4		1.0E-2	36.5	1090.	TA2
270.	-1-		CO2	9.66P32	+4+		4.9E-3	37.0	1110.	IN1
267.4432		N	CO2	10.16R34	+4	-10	5.0E-7	37.39112	1120.9576	HEN1PE2
264.5359	0	P	CO2	9.68P34	+4		5.0E-7	37.80205	1133.2770	PE2CH3
263.70	0	P	CO2	9.68P34			2.0E-4	37.922	1136.9	CH3
261.	-1	N	CO2	9.38R02	+3		2.0E-2	38.3	1150.	LA2
253.5530	0	P	CO2	9.68P34	+4		5.0E-7	39.43949	1182.3662	PE2CH3
253.60	0-	P	CO2	9.68P34			2.0E-4	39.432	1182.1	CH3
251.4324	-1-	P	CO2	9.28R18	+4	+5	5.0E-7	39.77212	1192.3383	PE2HEN1
251.1398	0-		CO2	10.14R38	+3-		4.0E-7	39.81846	1193.7273	PET/RA1
251.	-2		CO2	10.11R44	+4-		1.1E-2	39.8	1190.	TA2 7
250.7813	-1+		CO2	10.16R34	+4-		4.0E-7	39.87538	1195.4339	PET/RA1
242.79	0+	P	CO2	10.17R32	+4+	-45	1.0E-4	41.188	1234.8	IN1WA
242.5	0	P	CO2	9.2 R			2.0E-3	41.24	1236.	DY2
242.4727	-1+	P	CO2	10.16R34	+4	-10	5.0E-7	41.24176	1236.3969	HEN1PE2
237.60	0-	P	CO2	9.68P34			2.0E-4	42.088	1261.8	CH3
232.93906	-1-	P	CO2	9.33R10			3.9E-7	42.92968	1286.9995	PETDY2
232.7884	-2+	P	CO2	9.26R22	+4	+15	5.0E-7	42.95746	1287.8322	PE2HEN1
232.7	-1+	P	CO2	9.2 R			2.0E-3	42.97	1288.	DY2

225.5159	-2+	P	CO2	9.34R08	+4		5.0E-7	44.34277	1329.3629	PE2DY2
223.50	0	P	CO2	9.52P16			2.5E-4	44.74	1341.4	CH3
218.22	-3+	P	CO2	9.47P10	+4	-20	3.0E-4	45.83	1373.8	HEN1
214.35	-3+	P	CO2	9.47P10	+4	-20	3.0E-4	46.65	1398.6	HEN1PE2
213.4625	-1	P	CO2	9.57P22	+4		5.0E-7	46.84664	1404.4269	PE2DY2
211.3148		P	CO2	9.49P12	+4	-5	5.0E-7	47.32277	1418.7010	PE2HEN1
211.2629	-1+	P	CO2	10.37R04	+3+		5.0E-7	47.33439	1419.0493	PE2TA2
209.9302	-1-	P	CO2	9.31R14	+4		5.0E-7	47.63487	1428.0576	PE2HEN1
208.		E	CO2	9.68P34		+140	1.0E-2	48.1	1440.	IN2
206.90	-2-	P	CO2	9.49P12	+4	-5	3.0E-4	48.33	1449.0	HEN1PE2
205.		E	CO2	9.68P34		+120	4.9E-3	48.8	1462.	IN2
202.40	0	P	CO2	9.69P36			2.5E-4	49.41	1481.2	CH3
198.66433	0-	N	CO2	9.71P38	+4		5.0E-7	50.33616	1509.0402	PE2CH3
194.06320	-1	N	CO2	9.31R14	+3+		5.0E-7	51.52960	1544.8187	PE2HEN1
193.14158	-1	N	CO2	9.71P38	+4		5.0E-7	51.77549	1552.1901	PE2CH3
191.61960		P	CO2	10.32R10	+4	-15	5.0E-7	52.18673	1564.5187	PE2HEN1
191.5	-1	P	CO2	9.2 R			2.0E-3	52.2	1565.	DY2
190.72590	-1	N	CO2	9.68P34	+4		5.0E-7	52.43126	1571.8497	PE2CH3
186.04219	-1+	P	CO2	9.28R18	+4	+5	5.0E-7	53.75125	1611.4219	PE2DY2
185.9	-1+	P	CO2	9.2 R			2.0E-3	53.8	1613.	DY2
185.50040	-1+	N	CO2	9.68P34	+4		5.0E-7	53.90824	1616.1284	PE2CH3
180.4	0	N	CO2	9.68P34			2.0E-3	55.4	1662.	DY2
179.72791	-1+	P	CO2	10.37R04	+4		5.0E-7	55.63966	1668.0350	PE2
178.	-2-	P	CO2	10.38R02	+3+		2.0E-2	56.	1680.	LA2
176.	-1	N	CO2	9.38R02	+3		2.0E-2	57.	1700.	LA2
171.3	0-		CO2S	9.59P21	+3+		4.9E-3	58.4	1750.	WE1
170.57637	+1	P	CO2	9.69P36	+2+		2.8E-7	58.62477	1757.5263	PETDY2T3
167.58700	-1	P	CO2	10.13R40	+4		5.0E-7	59.67050	1788.8766	PE2
164.7832	+1	N	CO2	9.33R10	+3		1.0E-6	60.68581	1819.315	HEN1DY2
164.69747	-1	N	CO2	9.59P24	+4	0	5.0E-7	60.71739	1820.2615	PE2HEN1
164.60038	0	N	CO2	9.52P16	+4		5.0E-7	60.75320	1821.3352	PE2CH3
164.56421	-1+	N	CO2	9.50P14	+4		5.0E-7	60.76656	1821.7355	PE2
164.5076	-1-		CO2	9.50P14	+3+		1.0E-6	60.78747	1822.363	HEN1TA2
164.0	0-	P	CO2	10.09R48	+4+		4.9E-3	61.0	1828.	IN1
163.03353	0		CO2	10.14R38	+3-		2.8E-7	61.33708	1838.8393	PET/RA1
159.67569	0	P	CO2	9.24R26	+4		5.0E-7	62.62694	1877.5085	PE2DY2
159.2	0		CO2S	9.68P31	+3+		4.9E-3	62.8	1883.	WE1
152.	-1	N	CO2	9.38R02	+3-		2.0E-2	66.	1970.	LA2

Table 3.3.6.2 (continued)
CH_3OH, METHANOL (METHYL ALCOHOL)

For line assignments see References HEN1, HEN3*, HEN5*, HEN6*, IN1, IN2, W1*

Lambda (μm)	o/p	Pol.	Pump	Line	Thres.	Off.	Uncert.	Nu (1/cm)	Freq. (GHz)	Ref.
151.25369	0	P	CO2	9.24R26	+4		5.0E-7	66.11409	1982.0506	PE2DY2WA
145.5	0+	N	CO2	10.17R32	+4+	-45	4.9E-3	68.7	2060.	IN1
133.1196	0	P	CO2	9.59P24		0	1.0E-6	75.12043	2252.054	HEN1DY2
129.5497	-1	P	CO2	10.16R34	+3	-10	1.0E-6	77.19046	2314.112	HEN1KON2
121.	-2		CO2	10.11R44	+4		2.3E-2	83.	2480.	TA2
118.83409	+1+		CO2	9.69P36	+2+		2.0E-7	84.15094	2522.7816	PETDY2T3
117.95948	-1+	P	CO2	9.50P14	+4	-15	5.0E-7	84.77487	2541.487	HEN1PE2
113.73188	-2+	N	CO2	9.34R08	+4		5.0E-7	87.92609	2635.958	PE2
100.80647	-1+	P	CO2	9.31R14	+4		5.0E-7	99.19998	2973.941	PE2
98.	-1-		CO2	10.13R40	+4		3.0E-2	102.	3060.	TA2
97.51854	-1+	N	CO2	10.13R40	+4	0	5.0E-7	102.54461	3074.210	PE2HEN1
96.52239	+1+	P	CO2	9.33R10			1.6E-7	103.60290	3105.9368	PET/DY2
92.66428	-2+	P	CO2	10.16R34	+4	-10	5.0E-7	107.91644	3235.254	PE2HEN1
92.54391	-1	P	CO2	9.59P24	+4	0	5.0E-7	108.05681	3239.462	PE2HEN1
86.23938	-2+	P	CO2	9.34R08	+4		5.0E-7	115.95630	3476.283	PE2DY2
85.60093	-3+	P	CO2	9.73P40	+4	-10	5.0E-7	116.82116	3502.210	PE2HEN1
80.6	+1		CO2S	9.59P21	+3+		4.9E-3	124.1	3720.	WE1
80.	-1		CO2	9.68P34	+3		3.0E-2	125.	3700.	HO1
77.90489	-1+	P	CO2	10.27R16	+4	0	5.0E-7	128.36165	3848.186	PE2HEN1
77.40565	-1	P	CO2	9.28R08	+4		5.0E-7	129.18954	3873.005	PE2
73.30642	-3+	P	CO2	9.73P40	+4	-10	5.0E-7	136.41370	4089.580	PE2HEN1
70.51163	+2-	N	CO2	9.68P34		+26	2.0E-7	141.82058	4251.6740	BL2PE2
69.67956	-1	N	CO2	10.27R16	+4	0	5.0E-7	143.51411	4302.445	PE2HEN1
67.49536	-2	N	CO2	9.28R18	+4		5.0E-7	148.15833	4441.675	PE2
63.36954	0-		CO2	9.68P34	+3		5.0E-7	157.80452	4730.861	PE2HO1
62.96597	-1+	P	CO2	10.27R16	+4		5.0E-7	158.81595	4761.182	PE2
61.61330	-1-	N	CO2	9.28R18	+4		5.0E-7	162.30262	4865.710	PE2
60.17327	-2	P	CO2	9.73P40	+4	-10	5.0E-7	166.18674	4982.153	PE2HEN1
55.37004	-2+	N	CO2	9.73P40	+4	-10	5.0E-7	180.60308	5414.344	PE2HEN1
53.5	0+	N	CO2	10.15R36	+4+	+23	4.9E-3	186.9	5600.	IN1
43.47		P	CO2	10.16R34	+4	-10	3.0E-4	230.04	6897.	HEN1
43.1	0+		CO2	10.15R36	+4+	+23	4.9E-3	232.	6960.	IN1

42.31	-1		CO2	9.68P34	+3	+38	1.0E-3	236.4	7086.	HE1HO1
42.15908	+1	P	CO2	9.66P32	+3	-16	5.0E-7	237.1968	7110.981	PE2HEN1
39.92423	0	N	CO2	9.68P34	+3	10	5.0E-7	250.4745	7509.036	PE2HE1
37.85421	0		CO2	9.66P32	+3	-16	5.0E-7	264.1714	7919.660	PE2HO1
37.0	-3+		CO2	9.50P14	+4		2.0E-2	270.	8100.	PE2

Table 3.3.6.3

$^{13}CH_3OH$, METHANOL WITH ^{13}C ISOTOPE

For line assignments see References HEN2, HEN4*

Lambda (μm)	o/p	Pol.	Pump	Line	Thres.	Off.	Uncert.	Nu (1 /cm)	Freq. (GHz)	Ref.
629.8442		P	CO2	9.49P12	+4	+25	5.0E-7	15.876942	475.9787	HEN2PE2
461.3847	0+	N	CO2	9.49P12	+4	-20	5.0E-7	21.67389	649.7667	HEN2PE2
358.9		N	CO2	9.73P40	+4	+35	3.0E-4	27.861	835.3	HEN2
338.9638		P	CO2	9.57P22	+4	+15	5.0E-7	29.50168	884.4381	HEN2PE2
332.6034			CO2	10.27R16			5.0E-7	30.06584	901.3512	PE2
325.17		P	CO2	9.69P36	+4	-20	3.0E-4	30.753	922.0	HEN2
307.78	0-	P	CO2	9.57P22	+4	-15	3.0E-4	32.491	974.0	HEN2PE2
291.61		P	CO2	9.69P36	+4	-20	3.0E-4	34.29	1028.1	HEN2
280.2397			CO2	10.27R16			5.0E-7	35.68373	1069.7714	PE2
280.2183			CO2	10.27R16			5.0E-7	35.68647	1069.8534	PE2
268.5722	0-	P	CO2	10.27R16	+4	+20	5.0E-7	37.23394	1116.2455	HEN2PE2
238.5227	0+	P	CO2	9.49P12	+4	-20	5.0E-7	41.92474	1256.8720	HEN2PE2
237.5230		P	CO2	9.49P12	+4	+25	5.0E-7	42.10119	1262.1620	HEN2PE2
236.5303			CO2	9.49P10			5.0E-7	42.27788	1267.4590	PE2
208.4121		E	CO2	9.47P10	+4	+25	5.0E-7	47.98187	1438.4601	HEN2PE2
203.6358	0+	E	CO2	10.27R16	+4	+20	5.0E-7	49.10729	1472.1994	HEN2PE2
171.75758	0	N	CO2	10.26R18	+4	+25	5.0E-7	58.22159	1745.4395	HEN2PE2
168.84		P	CO2	9.73P40	+4	+35	3.0E-4	59.23	1775.6	HEN2
157.92848	0	P	CO2	9.49P12	+4	-20	5.0E-7	63.31980	1898.2799	HEN2PE2
152.07569			CO2	10.27R16			5.0E-7	65.75673	1971.3372	PE2
149.27228	-1	N	CO2	9.57P22	+4	+15	5.0E-7	66.99167	2008.360	HEN2PE2
148.59041			CO2	10.27R16			5.0E-7	67.29909	2017.576	PE2
147.97		N	CO2	9.64P30	+4	-25	3.0E-4	67.58	2026.0	HEN2
146.09738		N	CO2	9.47P10	+4	+25	5.0E-7	68.44750	2052.004	HEN2PE2
123.26		P	CO2	10.55P16	+4	+40	3.0E-4	81.13	2432.2	HEN2

Table 3.3.6.3 (continued)
$^{13}CH_3OH$, METHANOL WITH ^{13}C ISOTOPE
For line assignments see References HEN2, HEN4*

Lambda (μm)	o/p	Pol.	Pump	Line	Thres.	Off.	Uncert.	Nu (1 /cm)	Freq. (GHz)	Ref.
121.20		P	CO2	10.19R28	+4	-5	3.0E-4	82.51	2473.5	HEN2
118.01308	0+	N	CO2	9.57P22	+4	-15	5.0E-7	84.73637	2540.332	HEN2PE2
115.82318	0	P	CO2	10.27R16	+4	+20	5.0E-7	86.33850	2588.363	HEN2PE2
110.43238	0+	E	CO2	10.26R18	+4	+25	5.0E-7	90.55315	2714.715	HEN2PE2
105.14719	0	E	CO2	10.26R18	+4	+25	5.0E-7	95.10478	2851.170	HEN2PE2
103.58629		E	CO2	10.21R26	+4	+40	5.0E-7	96.53788	2894.133	HEN2PE2
103.48079	0	P	CO2	9.57P22	+4	+15	5.0E-7	96.63630	2897.083	HEN2PE2
87.90		N	CO2	9.46P08	+4		3.0E-4	113.77	3411.	HEN2
86.11179		P	CO2	9.47P10	+4	+25	5.0E-7	116.12812	3481.433	HEN2PE2
85.79		N	CO2	10.19R28	+4	-5	3.0E-4	116.56	3494.	HEN2
85.31729	0	P	CO2	9.57P22	+4	-15	5.0E-7	117.20954	3513.854	HEN2PE2
77.48939		P	CO2	10.21R26	+4	+40	5.0E-7	129.04993	3868.819	HEN2PE2
63.09639		P	CO2	9.49P12	+4	+25	5.0E-7	158.48767	4751.341	HEN2PE2
41.90		N	CO2	10.55P16	+4	-10	3.0E-4	238.66	7155.	HEN2
34.79		P	CO2	10.23R22	+4		3.0E-4	287.44	8617.	HEN2

Table 3.3.6.4
CH_2DOH, C-DEUTERO METHANOL

Lambda (μm)	o/p	Pol.	Pump	Line	Thres.	Off.	Uncert.	Nu (1/cm)	Freq. (GHz)	Ref.
762.50	-1-	P	CO2	9.54P18	+4+		1.2E-4	13.115	393.17	SC2
682.6	-2+	N	CO2	9.25R24	+4+		1.5E-4	14.650	439.19	SC2
616.3351	0	P	CO2	9.60P26	+4+		5.0E-7	16.224941	486.4115	SC2ZD
523.0914	-1+	P	CO2	9.73P40	+4		5.0E-7	19.117119	573.1168	SC2
509.3717	-2-		CO2	10.88P46	+3+		5.0E-7	19.632028	588.5534	SC2
468.2359	0-	P	CO2	9.60P26	+4+		5.0E-7	21.35676	640.2595	SC2ZD
452.40	0-	N	CO2	9.79P46	+3+		2.2E-4	22.104	662.7	SC2
451.4754	-1	N	CO2	9.66P32	+4+		5.0E-7	22.14960	664.0284	SC2
427.2	-2-	N	CO2	10.76P36	+4		3.0E-4	23.408	701.8	SC2
422.1512	-2+	N	CO2	9.34R08	+4		5.0E-7	23.68820	710.1543	SC2

396.00	-1-	P	CO2	9.54P18	+3+	2.5E-4	25.253	757.1	SC1ZD
387.5591	-1+	P	CO2	9.73P40	+4	5.0E-7	25.80251	773.5399	SC2
374.0861	-1	P	CO2	10.88P46	+3+	5.0E-7	26.73181	801.3996	SC2ZD
340.3566	0	P	CO2	10.17R32	+4+	5.0E-7	29.38095	880.8186	SC2
336.2461	0-	P	CO2	9.69P36	+4	5.0E-7	29.74012	891.5863	SC2
322.4522	0-	P	CO2	9.49P12	+4	5.0E-7	31.01235	929.7268	SC2ZD
308.2957	0-		CO2	10.16R34	+4	5.0E-7	32.43640	972.4187	SC2
308.0405	0-	P	CO2	9.50P14	+4	5.0E-7	32.46327	973.2243	SC2ZD
300.0	-2	P	CO2	10.27R16	+4+	1.6E-3	33.33	999.	SC2
295.6394	0	P	CO2	10.16R34	+4+	5.0E-7	33.82499	1014.0477	SC2ZD
295.3967	0-	P	CO2	9.47P10	+4	5.0E-7	33.85279	1014.6810	SC2ZD
273.0037	-2+	P	CO2	9.44P06	+4	5.0E-7	36.62954	1098.1259	SC2
272.2516	0	P	CO2	9.25R24	+4+	5.0E-7	36.73072	1101.1594	SC2ZD
266.7352	-1	P	CO2	9.66P32	+4+	5.0E-7	37.49036	1123.9327	SC2
249.7204	0-	P	CO2	10.74P34	+4	5.0E-7	40.04479	1200.5127	SC2ZD
248.1220	0	N	CO2	10.74P34	+4	5.0E-7	40.30275	1208.2460	SC2
226.2974	0-	N	CO2	9.79P46	+4	5.0E-7	44.18963	1324.7719	SC2
224.2256	-1-	P	CO2	10.76P36	+4	5.0E-7	44.59794	1337.0125	SC2
219.0960	-1+	P	CO2	9.25R24	+4+	5.0E-7	45.64209	1368.3154	SC2
218.0	-2+	P	CO2	9.26R22	+4+	5.0E-4	45.87	1375.2	SC2
216.8	-1-	P	CO2	9.29R16	+4+	5.0E-4	46.13	1382.8	SC2
212.5		P	CO2	10.27R16	+4+	5.0E-4	47.06	1410.8	SC2
206.6874	0-	P	CO2	9.50P14	+4	5.0E-7	48.38224	1450.4631	SC2ZD
200.0	-2+	N	CO2	9.71P38	+4	2.4E-3	50.0	1499.	SC2
196.10	-2+	P	CO2	10.67P28	+4+	5.0E-4	50.99	1528.8	SC2
195.49558	0-	P	CO2	9.69P36	+4	5.0E-7	51.15205	1533.4999	SC2
189.30	-2+	P	CO2	10.67P28	+4+	5.0E-4	52.83	1583.7	SC2
188.41111	0-	P	CO2	10.65P26	+4+	5.0E-7	53.07543	1591.1612	SC2
183.62132	0	P	CO2	9.47P10	+4	5.0E-7	54.45991	1632.6669	SC2
182.10	-1-	P	CO2	9.26R22	+4+	5.0E-4	54.91	1646.3	SC2
172.84620	-1+	N	CO2	9.49P12	+4	5.0E-7	57.85490	1734.4464	SC2ZD
171.8	-2	P	CO2	9.26R22	+4+	6.0E-4	58.21	1745.	SC2
167.54117	0-	P	CO2	9.54P18	+4	5.0E-7	59.68682	1789.3659	SC2ZD
167.35235	0-	P	CO2	9.66P32	+4+	5.0E-7	59.75417	1791.3848	SC2
164.74645	-1-	P	CO2	9.34R08	+4	5.0E-7	60.69934	1819.7203	SC2
162.70	-3+	N	CO2	10.70P30	+4+	6.0E-4	61.46	1843.	SC2
159.21794	0	P	CO2	10.16R34	+4+	5.0E-7	62.80699	1882.9063	SC2

Table 3.3.6.4 (continued)
CH_2DOH, C-DEUTERO METHANOL

Lambda (μm)	o/p	Pol.	Pump	Line	Thres.	Off.	Uncert.	Nu (1/cm)	Freq. (GHz)	Ref.
152.7	-1+	P	CO2	9.25R24	+4+		7.0E-4	65.49	1963.	SC2
150.81629	0+	P	CO2	10.16R34	+4+		5.0E-7	66.30584	1987.7989	SC2ZD
150.57167	0-	N	CO2	10.65P26	+4+		5.0E-7	66.41356	1991.0283	SC2
149.61284	-1-	P	CO2	10.17R32	+4+		5.0E-7	66.83918	2003.788	SC2
149.38792	-2+	P	CO2	10.76P36	+4		5.0E-7	66.93982	2006.805	SC2
140.30	-2+	N	CO2	9.55P20	+4+		7.0E-4	71.28	2137.	SC2
135.83350	-1-	N	CO2	9.34R08	+4		5.0E-7	73.61954	2207.058	SC2
135.17256	-1+	P	CO2	10.17R32	+4+		5.0E-7	73.97951	2217.850	SC2
135.17175	-1	P	CO2	10.17R32	+4+		5.0E-7	73.97995	2217.863	SC2
124.43170	0+	P	CO2	10.74P34	+4		5.0E-7	80.36537	2409.293	SC2ZD
117.08507	-1-	N	CO2	9.66P32	+4+		5.0E-7	85.40798	2560.467	SC2
112.53224	0	P	CO2	9.49P12	+4		5.0E-7	88.86342	2664.058	SC2
108.94124	-1-	N	CO2	9.66P32	+4+		5.0E-7	91.79260	2751.873	SC2
108.81775	0-	P	CO2	9.49P12	+4		5.0E-7	91.89677	2754.996	SC2ZD
102.02349	-1	N	CO2	9.52P16	+4+		5.0E-7	98.01665	2938.465	SC2
100.00	-1+	P	CO2	9.54P18	+4+		1.0E-3	100.00	2998.	SC2
90.40	-1+	P	CO2	10.70P30	+4+		1.1E-3	110.6	3316.	SC2
87.90	-1	N	CO2	9.73P40	+4		1.1E-3	113.8	3411.	SC2
87.10	-1+	N	CO2	9.54P18	+4+		1.1E-3	114.8	3442.	SC2
44.0	-1	P	CO2	9.64P30	+4+		1.1E-2	227.	6810.	SC2
42.50	-2+	N	CO2	9.71P38	+4		2.2E-3	235.3	7050.	SC2

Table 3.3.6.5
CH_3OD, O-DEUTERO METHANOL

Lambda (μm)	o/p	Pol.	Pump	Line	Thres.	Off.	Uncert.	Nu (1/cm)	Freq. (GHz)	Ref.
498.0	-1	P	CO2	9.66P32			2.0E-3	20.08	602.	DY2
417.	-1+	N	CO2	9.44P06	+3+		4.9E-3	24.0	719.	BE1

352.5	-1	N	CO2	9.64P30		2.0E-3	28.37	850.	DY2
330.	-1+	N	CO2	9.37R04	+3+	4.9E-3	30.3	908.	BE1
320.0	-1	P	CO2	9.64P30		2.0E-3	31.25	937.	DY2
305.72611	+1	P	CO2	9.34R08	+3+	1.0E-7	32.709015	980.5916	BL1KON2
294.81098	+1	P	CO2	9.34R08	+3+	1.0E-7	33.920039	1016.8972	BL1KON2
280.	-2-	P	CO2	10.57P18	+4-	4.9E-3	35.7	1071.	NH1
279.4	-1	N	CO2	9.66P32	+3+	2.0E-3	35.79	1073.	DY2LA2
241.	-1-	P	CO2	10.10R44	+3+	4.9E-3	41.5	1244.	NH1LA2
238.	-1	P	CO2	9.31R14	+4-	1.0E-2	42.0	1260.	LDLA2
238.	-2-	P	CO2	10.16R34	+4-	4.9E-3	42.0	1260.	NH1
234.	-2	P	CO2	9.31R14	+4-	4.9E-3	42.7	1281.	NH1
229.	0-	P	CO2	9.44P06	+3+	4.9E-3	43.6	1309.	BE1
225.	0-	P	CO2	9.35R06	+4-	1.0E-2	44.4	1330.	LDLA2
224.	-2+	P	CO2	10.23R22	+4	2.0E-2	44.6	1340.	LA2
215.37246	-1	P	CO2	9.31R14	+3+	1.5E-7	46.431191	1391.9721	BL1KON2
212.	0	P	CO2	9.37R04	+4-	1.0E-2	47.2	1410.	LDLA2
186.	-2+	P	CO2	9.23R28	+4-	2.0E-2	54.	1610.	LA2
182.1			CO2	9.44P06	+4-	4.9E-3	54.9	1646.	NH1
182.	-2+	P	CO2	9.24R26	+4-	2.0E-2	55.	1650.	LA2
179.0	0-	N	CO2	9.66P32	+3+	2.0E-3	55.9	1675.	DY2LA3
169.	-1-	P	CO2	9.26R22	+4-	2.0E-2	59.	1770.	LA2
168.1	-1	N	CO2	9.64P30		2.0E-3	59.5	1783.	DY2
145.6	-2-	P	CO2	9.66P32	+4-	4.9E-3	68.7	2060.	NH1
145.66171	-1	P	CO2	9.64P30	+3+	1.5E-7	68.65222	2058.1418	BL1KON2
141.	-2+	N	CO2	10.11R42	+3+	2.0E-2	71.	2130.	LA2
137.	-1-	P	CO2	10.23R22	+4-	2.0E-2	73.	2190.	LA2
136.	-1-	P	CO2	9.59P24	+3+	1.0E-2	73.5	2200.	HP1LA2
135.	-1+	N	CO2	9.44P06	+3+	1.0E-2	74.2	2230.	BE1
134.0	0+	E	CO2	9.47P10	+4-	4.0E-3	74.6	2237.	LDHP1LA2
128.0	-1+	N	CO2	9.57P22		2.0E-3	78.1	2342.	DY2
117.22707	+1+	P	CO2	9.60P26	+3	1.9E-7	85.30453	2557.3654	RA2DY2
113.8	-2	P	CO2	9.66P32	+4-	4.9E-3	87.9	2630.	NH1
110.7	-1-	E	CO2	9.66P32	+4-	4.9E-3	90.3	2710.	NH1LA2
106.	-2+	P	CO2	9.24R26	+4-	2.0E-2	94.	2830.	LA2
106.	-1-	P	CO2	9.38R02	+3+	2.0E-2	94.	2830.	LA2
110.	-1	P	CO2	10.11R44	+3	1.0E-2	90.9	2730.	KON2LA2

Table 3.3.6.5 (continued)
CH_3OD, O-DEUTERO METHANOL

Lambda (μm)	o/p	Pol.	Pump	Line	Thres.	Off.	Uncert.	Nu (1/cm)	Freq. (GHz)	Ref.
103.12463	+1	N	CO2	9.64P30	+3-		1.7E-7	96.97005	2907.0889	RA2DY2
101.6	0	E	CO2	9.60P26	+3+		4.9E-3	98.4	2950.	NH1LDLA2
89.6	0	P	CO2	9.64	+4-		4.9E-3	111.6	3350.	NH1
81.9	0	N	CO2	9.64P30	+4-		4.9E-3	122.1	3660.	NH1
80.0	0	N	CO2	9.66P32	+3+		2.0E-3	125.0	3747.	DY2LA2
70.3	-1+	P	CO2	9.29R16	+4-		4.0E-3	142.2	4260.	LD
69.5	-1		CO2	9.35R06	+4-		1.0E-2	144.	4310.	LD
57.0	0	N	CO2	9.34R08	+3		1.0E-2	175.	5260.	KON2
46.7	0-		CO2	9.34R08	+3+		1.0E-2	214.	6420.	BE1BL1

Table 3.3.6.6
CHD_2OH, C-DIDEUTERO METHANOL

Lambda (μm)	o/p	Pol.	Pump	Line	Thres.	Off.	Uncert.	Nu (1/cm)	Freq. (GHz)	Ref.
518.	0-		CO2	9.64P30	+3+		4.9E-3	19.31	579.	ZD
483.	-1+		CO2	9.44P06	+3+		4.9E-3	20.7	621.	ZD
426.	0-		CO2	10.14R38	+3+		4.9E-3	23.5	704.	ZD
363.	0+		CO2	10.27R16	+3+		4.9E-3	27.5	826.	ZD
355.	0-		CO2	10.57P18	+3+		4.9E-3	28.2	844.	ZD
346.	-1+		CO2	9.55P20	+3+		4.9E-3	28.9	866.	ZD
260.	0+		CO2	10.25R20	+3+		4.9E-3	38.5	1153.	ZD
238.	0+		CO2	10.57P18	+3+		4.9E-3	42.0	1260.	ZD
179.0	0-		CO2	10.27R16	+3+		4.9E-3	55.9	1675.	ZD
168.0	0-		CO2	10.14R38	+3+		4.9E-3	59.5	1784.	ZD
165.0	0-		CO2	9.28R18	+3+		4.9E-3	60.6	1817.	ZD

Table 3.3.6.7
CD_3OH, C-TRIDEUTERO METHANOL

Lambda (μm)	o/p	Pol.	Pump	Line	Thres.	Off.	Uncert.	Nu (1/cm)	Freq. (GHz)	Ref.
1290.	-1	P	CO2	10.25R20	+3+		4.9E-3	7.75	232.	DY7
1146.	-1	P	CO2	9.59P24	+3+		4.9E-3	8.73	262.	DY7
1100.	-1+	P	CO2	9.49P12	+3+		4.9E-3	9.09	273.	DY7
968.	-1	P	CO2	9.27R20	+3+		4.9E-3	10.33	310.	DY7
871.5850	+1-	P	CO2	10.26R18	+3+		5.0E-7	11.473351	343.9624	PE2DY7
774.	-2	P	CO2	9.64P30	+3+		4.9E-3	12.92	387.	DY7
760.	-1	P	CO2	10.57P18	+3+		4.9E-3	13.16	394.	DY7
745.	-1	P	CO2	9.24R26	+3+		4.9E-3	13.42	402.	DY7
722.	-1-	N	CO2	10.59P20	+3+		4.9E-3	13.85	415.	DY7
711.	-1	P	CO2	9.46P08	+3+		4.9E-3	14.06	422.	DY7
703.	0	N	CO2	10.15R36	+3+		4.9E-3	14.22	426.	DY7
702.	0-	P	CO2	9.59P24	+3+		4.9E-3	14.25	427.	DY7
695.	0	P	CO2	9.47P10	+3+		4.9E-3	14.39	431.	DY7
685.	-1	P	CO2	10.16R34	+3+		4.9E-3	14.60	438.	DY7
680.	-2+	P	CO2	9.44P06	+3+		4.9E-3	14.71	441.	DY7
648.	0	P	CO2	10.33R08	+3+		4.9E-3	15.43	463.	DY7
646.	0	N	CO2	10.15R36	+3+		4.9E-3	15.48	464.	DY7
599.	-1	P	CO2	10.27R16	+3+		4.9E-3	16.69	500.	DY7
583.	0	P	CO2	9.26R22	+3+		4.9E-3	17.15	514.	DY7
554.	0-	P	CO2	9.21R32	+3+		4.9E-3	18.05	541.	DY7
553.	0+	P	CO2	10.33R08	+3+		4.9E-3	18.08	542.	DY7
553.	+1-	P	CO2	9.31R14	+3+		4.9E-3	18.08	542.	DY7
551.	-1+	N	CO2	9.57P22	+3+		4.9E-3	18.15	544.	DY7
517.	-1	P	CO2	10.84P42	+3+		4.9E-3	19.34	580.	DY7
508.	-1	P	CO2	9.46P08	+3+		4.9E-3	19.69	590.	DY7
498.	-1	P	CO2	10.16R34	+3+		4.9E-3	20.1	602.	DW2
495.	0	P	CO2	10.26R18	+3+		4.9E-3	20.2	606.	DY7
483.	-1	P	CO2	10.65P26	+3+		4.9E-3	20.7	621.	DY7
483.	0-	P	CO2	9.26R22	+3+		4.9E-3	20.7	621.	DY7
480.	-2+	N	CO2	9.52P16	+3+		4.9E-3	20.8	625.	DY7
472.	-2+	P	CO2	9.29R16	+3+		4.9E-3	21.2	635.	DY7
455.	-1+	P	CO2	9.54P18	+3+		4.9E-3	22.0	659.	DY7

Table 3.3.6.7 (continued)
CD_3OH, C-TRIDEUTERO METHANOL

Lambda (μm)	o/p	Pol.	Pump	Line	Thres.	Off.	Uncert.	Nu (1/cm)	Freq. (GHz)	Ref.
435.	-3	N	CO2	9.62P28	+3+		4.9E-3	23.0	689.	GW1
422.	0-	P	CO2	9.55P20	+3+		4.9E-3	23.7	710.	DY7
421.	0-	N	CO2	10.17R32	+3+		4.9E-3	23.8	712.	DY7
418.7118	+1-	N	CO2	10.15R36	+3+		5.0E-7	23.88278	715.9876	PE2DY7
412.	-1	P	CO2	10.30R12	+3+		4.9E-3	24.3	728.	DY7
410.	-1+	P	CO2	9.66P32	+3+		4.9E-3	24.4	731.	DY7
409.	-1-	N	CO2	9.20R34	+3+		4.9E-3	24.4	733.	DY7
407.	-1+	P	CO2	9.16R44	+3+		4.9E-3	24.6	737.	DY7
398.	0-	P	CO2	10.19R28	+3+		4.9E-3	25.1	753.	DY7
386.	0-	N	CO2	9.52P16	+3+		4.9E-3	25.9	777.	DY7
385.	-2	N	CO2	9.64P30	+3+		4.9E-3	26.0	779.	DY7
370.	0+	P	CO2	9.24R26	+3+		4.9E-3	27.0	810.	DY7
370.	-1+	P	CO2	9.62P28	+3+		4.9E-3	27.0	810.	DY7
353.	-1+	P	CO2	9.35R06	+3+		4.9E-3	28.3	849.	DY7
352.	-1	P	CO2	9.23R28	+3+		4.9E-3	28.4	852.	DY7
352.	0	P	CO2	9.31R14	+3+		4.9E-3	28.4	852.	DY7
351.	-1	N	CO2	10.18R30	+3+		4.9E-3	28.5	854.	DY7
350.	0-	P	CO2	9.66P32	+3+		4.9E-3	28.6	857.	DY7
346.	-1	P	CO2	9.31R14	+3+		4.9E-3	28.9	866.	GW1
336.	-2+	N	CO2	9.21R32	+3+		4.9E-3	29.8	892.	DY7
336.	-1	P	CO2	10.18R30	+3+		4.9E-3	29.8	892.	DY7
321.	-1-	N	CO2	9.29R16	+3+		4.9E-3	31.2	934.	DY7
310.	0-	N	CO2	10.19R28	+3+		4.9E-3	32.3	967.	DY7
309.	-1-	P	CO2	10.59P20	+3+		4.9E-3	32.4	970.	DY7
299.	-2+	P	CO2	9.35R06	+3+		4.9E-3	33.4	1003.	DY7
297.	-1	N	CO2	9.28R18	+3+		4.9E-3	33.7	1009.	DY7
297.	0-	P	CO2	10.16R34	+3+		4.9E-3	33.7	1009.	DY7
290.	-1	N	CO2	10.57P18	+3+		4.9E-3	34.5	1034.	DW2
287.3076	+1-	P	CO2	10.57P18	+3+		5.0E-7	34.80590	1043.4545	PE2DY7
286.7242	0+	P	CO2	10.63P24	+3+		5.0E-7	34.87673	1045.5780	PE2DY7
286.1974	-1+	P	CO2	9.73P40	+3+		5.0E-7	34.94092	1047.5023	PE2DY7

278.	-1	P	CO2	10.22R24	+3+	4.9E-3	36.0	1078.	DY7
277.	-1	P	CO2	9.24R26	+3+	4.9E-3	36.1	1082.	DY7
276.7157	0	P	CO2	10.67P28	+3+	5.0E-7	36.13817	1083.3951	PE2DY7
268.	-1	P	CO2	9.50P14	+3+	4.9E-3	37.3	1119.	DY7
267.	-1	P	CO2	10.29R14	+3+	4.9E-3	37.5	1123.	DY7
266.	-1+	P	CO2	9.55P20	+3+	4.9E-3	37.6	1127.	DY7
265.	0-	P	CO2	10.16R34	+3+	4.9E-3	37.7	1131.	DY7
258.4356	0	P	CO2	10.61P22	+3+	5.0E-7	38.69436	1160.0278	PE2DY7
253.7196	0-	N	CO2	10.15R36	+3+	5.0E-7	39.41359	1181.5898	PE2DW2
238.	-2	N	CO2	10.63P24	+3+	4.9E-3	42.0	1258.	DW2
236.	-1	P	CO2	9.31R14	+3+	4.9E-3	42.4	1270.	GW1
232.	0-	P	CO2	9.20R34	+3+	4.9E-3	43.1	1292.	DY7
223.	-2	N	CO2	9.46P08	+3+	4.9E-3	44.8	1344.	GW1
222.	-1	P	CO2	9.44P06	+3+	4.9E-3	45.0	1350.	DY7
220.	-2	P	CO2	10.26R18	+3+	4.9E-3	45.5	1363.	DW2
215.0812			CO2	10.72P52		5.0E-7	46.49406	1393.8569	PE2
201.	-1	P	CO2	9.73P40	+3+	4.9E-3	49.8	1492.	GW1
191.9	-1	P	CO2	10.16R34	+3+	4.9E-3	52.1	1562.	DW2
188.42390			CO2	10.84P42		5.0E-7	53.07182	1591.0532	PE2
184.0	-1+	N	CO2	9.34R08	+3+	4.9E-3	54.3	1629.	DY7
182.56629			CO2	9.31R14		5.0E-7	54.77462	1642.1019	PE2
180.74051	-2	N	CO2	10.16R34	+3+	5.0E-7	55.32794	1658.6899	PE2DW2
179.0	0-	N	CO2	9.31R14	+3+	4.9E-3	55.9	1675.	GW1
158.0	0-	P	CO2	9.23R28	+3+	4.9E-3	63.3	1897.	GW1
144.11787	-1	P	CO2	10.57P18	+3+	5.0E-7	69.38765	2080.189	PE2DW2
136.62721			CO2	10.29R14		5.0E-7	73.19186	2194.237	PE2
131.56276			CO2	10.17R32		5.0E-7	76.00935	2278.703	PE2
128.7	-3	P	CO2	10.16R34	+3+	4.9E-3	77.7	2330.	DW2
112.3	-3	N	CO2	10.16R34	+3+	4.9E-3	89.0	2670.	DW2
108.66842			CO2	10.49P10		5.0E-7	92.02305	2758.782	PE2
102.6	-2	N	CO2	10.16R34	+3+	4.9E-3	97.5	2920.	DW2
86.4	-2	N	CO2	10.27R16	+3+	4.9E-3	115.7	3470.	DW2
81.55710	-1	P	CO2	10.27R16	+3+	5.0E-7	122.61348	3675.860	PE2DW2
76.1	-1	P	CO2	10.72P32	+3+	4.9E-3	131.4	3940.	DW2
71.0	-1	P	CO2	10.33R08	+3+	4.9E-3	140.8	4220.	DW2
67.47941			CO2	10.18R30		5.0E-7	148.19335	4442.725	PE2
60.10	0	P	CO2	9.20R34	+3+	5.0E-4	166.39	4988.	EDGW1

Table 3.3.6.7 (continued)
CD_3OH, C-TRIDEUTERO METHANOL

Lambda (μm)	o/p	Pol.	Pump	Line	Thres.	Off.	Uncert.	Nu (1/cm)	Freq. (GHz)	Ref.
55.56		N	CO2	9.23R28	+3+		5.0E-4	179.99	5396.	ED
53.82	0	N	CO2	9.20R34	+3+		5.0E-4	185.80	5570.	EDGW1
49.8	0	N	CO2	9.23R28	+3+		4.9E-3	201.	6020.	GW1
43.69729	0	P	CO2	10.26R18	+3+		5.0E-7	228.8471	6860.664	PE2DW2ED
41.40	-1	N	CO2	10.33R08	+3+		5.0E-4	241.5	7241.	DW2ED
41.35487	0	N	CO2	10.26R18	+3+		5.0E-7	241.8095	7249.265	PE2DW2ED
41.25		P	CO2	10.33R08	+3+		5.0E-4	242.4	7268.	ED
40.1	-1	N	CO2	10.61P22	+3+		4.9E-3	249.	7480.	DW2
37.6	-1	P	CO2	10.16R34	+3+		4.9E-3	266.	7970.	DW2
34.8	-1	P	CO2	10.61P22	+3+		4.9E-3	287.	8610.	DW2

Table 3.3.6.8
CD_3OD, FULLY DEUTERATED METHANOL

For line assignments see Reference HP1

Lambda (μm)	o/p	Pol.	Pump	Line	Thres.	Off.	Uncert.	Nu (1/cm)	Freq. (GHz)	Ref.
869.	-1		CO2	10.26R18	+3+	-20	4.9E-3	11.51	345.	DH
495.	0	P	CO2	10.22R24	+4		1.0E-2	20.2	606.	HP1
414.	0	N	CO2	10.30R12	+4		1.0E-2	24.2	724.	HP1KON
406.	0L	N	CO2	10.30R12	+3		2.0E-2	24.6	740.	KON
354.	0-		CO2	10.27R16	+4		1.0E-2	28.2	847.	HP1
339.	0L	N	CO2	10.37R04	+3		2.0E-2	29.5	880.	KON
312.		N	CO2	10.32R10	+3		2.0E-2	32.1	960.	KON
299.	0L	N	CO2	10.22R24	+3		2.0E-2	33.4	1000.	KON
255.		N	CO2	10.15R36	+3		2.0E-2	39.2	1180.	KON
229.		N	CO2	10.32R10	+3		2.0E-2	43.7	1310.	KON
184.	0L	N	CO2	10.22R24	+3		2.0E-2	54.	1630.	KON
165.	-1-		CO2	10.23R22	+4		1.0E-2	60.6	1820.	HP1
150.	-1		CO2	10.18R30	+4		1.0E-2	66.7	2000.	HP1
119.	-1+		CO2	10.21R26	+4		1.0E-2	84.0	2520.	HP1
78.0	0+		CO2	10.49P10	+4		1.0E-2	128.	3840.	HP1

Lambda (μm)	o/p	Pol.	Pump	Line	Thres.	Off.	Uncert.	Nu (1/cm)	Freq. (GHz)	Ref.
41.0		N	CO2	10.26R18	+3		2.0E-2	244.	7300.	KON
35.0	+1-		CO2	10.19R28	+4		2.0E-2	286.	8600.	HP1

Table 3.3.6.9

CH₃CN, METHYL CYANIDE (ACETONITRILE)

For line assignments see References AR2*, CH5 (pump)

Lambda (μm)	o/p	Pol.	Pump	Line	Thres.	Off.	Uncert.	Nu (1/cm)	Freq. (GHz)	Ref.
1814.37	0	P	CO2	10.88P46	+4	0	2.7E-5	5.5116	165.232	CH5
1351.8	-2+	N	CO2	9.27R20	+5-	-15	7.9E-5	7.3977	221.78	CH6
1164.8	-1-	P	CO2	9.47P10	+4+	+45	8.9E-5	8.5849	257.37	CH6
1086.89	-1+	N	CO2	9.73P40	+4	+40	8.9E-5	9.2006	275.83	CH6
1016.3	-1-	P	CO2	9.46P08	+4-	-40	1.0E-4	9.8393	294.98	CH6
1014.9	-2+	N	CO2	9.31R14	+4+	-20	1.0E-4	9.8533	295.39	CH6
854.4	-1	P	CO2	9.52P16	+4+	-10	1.2E-4	11.704	350.88	CH6
750.	-3		CO2	9.52P16	+3+		2.0E-2	13.3	400.	RA2
741.62	-1+	N	CO2	9.34R08	+4-	+35	1.2E-4	13.484	404.24	CH6
713.72	0	P	CO2	10.72P32	+4	-40	7.0E-5	14.0111	420.04	CH5
704.53	-1	P	CO2	9.20R34	+4	-15	1.4E-4	14.194	425.52	CH6
652.68	-1+	P	CO2	9.64P30	+4+	-15	1.5E-4	15.321	459.33	CH6RA1
561.4	-1+	P	CO2	9.34R08	+4+	+40	1.7E-4	17.812	534.00	CH6
510.2	-1)	N	CO2	9.44P06	+4-	-10	2.0E-4	19.602	587.6	CH6
494.6461	+1-	N	CO2	9.44P06	+3	-10	1.5E-6	20.21648	606.0747	RA2,1CH6
480.0	-1)	N	CO2	9.29R16	+4-	0	2.0E-4	20.833	624.6	CH6
466.25	0-)	N	CO2	9.29R16	+4-	0	2.0E-4	21.448	643.0	CH6
453.3974	0	N	CO2	9.29R16	+3+	0	1.5E-6	22.05570	661.2134	RA2,1CH6
441.1	-1)	N	CO2	9.29R16	+4-	0	2.3E-4	22.668	679.6	CH6
430.55	-1	P	CO2	10.57P18	+4	-15	1.2E-4	23.226	696.30	CH5
427.04	-2+	N	CO2	9.60P26	+4	+15	2.3E-4	23.417	702.0	CH6
388.4	-2-	P	CO2	9.57P22	+5-	+30	2.5E-4	25.747	771.9	CH6
387.3	0-	N	CO2	9.32R12	+4	+15	2.5E-4	25.819	774.0	CH6RA1
386.4	-3+	P	CO2	9.79P46	+4+	-15	2.5E-4	25.879	775.8	CH6
372.87	-1+		CO2	10.59P20	+3	-30	1.2E-4	26.819	804.0	RA1CH5T3
346.3	0-	P	CO2	9.52P16	+4	-10	2.9E-4	28.875	865.7	CH6
286.9	-1)	P	CO2	9.84P50	+3+	+50	3.5E-4	34.86	1045.0	CH6
281.98	0-	P	CO2	9.84P50	+3+	+50	3.5E-4	35.46	1063.2	CH6
281.18	-2+	N	CO2	9.68P34	+4	-20	3.5E-4	35.56	1066.2	CH6

Table 3.3.6.10
CH_3SH, METHYL MERCAPTAN

Lambda (μm)	o/p	Pol.	Pump	Line	Thres.	Off.	Uncert.	Nu (1/cm)	Freq. (GHz)	Ref.
456.	-1-	P	CO2	9.77P44	+4-		2.0E-2	21.9	660.	LA1
403.	-1-	N	CO2	9.49P12	+4-		2.0E-2	24.8	740.	LA1
384.	-1	P	CO2	9.52P16	+4-		2.0E-2	26.0	780.	LA1
379.	-1	P	CO2	9.57P22	+3+		2.0E-2	26.4	790.	LA1
370.	-1-	P	CO2	10.16R34	+4		2.0E-2	27.0	810.	LA1
351.	-1	N	CO2	9.23R28	+4-		2.0E-2	28.5	850.	LA1
341.	-2-	N	CO2	9.52P16	+4		2.0E-2	29.3	880.	LA1
324.	-2+	N	CO2	10.27R16	+4		2.0E-2	30.9	930.	LA1
319.	-1	P	CO2	9.71P38	+4-		2.0E-2	31.3	940.	LA1
316.	-2-		CO2	9.77P44	+4+		2.0E-2	31.6	950.	LA1
298.	-1-	P	CO2	9.64P30	+4-		2.0E-2	33.6	1010.	LA1
262.	-2+	P	CO2	9.71P38	+4		2.0E-2	38.2	1140.	LA1
234.	-2-	N	CO2	10.16R34	+4		2.0E-2	42.7	1280.	LA1
224.	-2-	P	CO2	9.77P44	+4		2.0E-2	44.6	1340.	LA1
205.	-2+	P	CO2	9.28R18	+4-		2.0E-2	48.8	1460.	LA1
185.	-1	N	CO2	9.28R18	+3+		2.0E-2	54.	1620.	LA1
161.	-2+	P	CO2	9.22R30	+4		2.0E-2	62.	1860.	LA1
147.	-2-		CO2	10.16R34	+4+		2.0E-2	68.	2040.	LA1
128.	-2-	P	CO2	9.59P24	+4-		2.0E-2	78.	2340.	LA1
127.	-2-	N	CO2	9.54P18	+4		2.0E-2	79.	2360.	LA1
124.	-2+	E	CO2	10.22R24	+4		2.0E-2	81.	2420.	LA1
117.	-2-	P	CO2	10.16R34	+4-		2.0E-2	85.	2560.	LA1
116.	-2-	P	CO2	9.20R34	+4		2.0E-2	86.	2580.	LA1

Table 3.3.6.11
CH_2CHCl (OR H_2CCl), VINYL CHLORIDE

Lambda (μm)	o/p	Pol.	Pump	Line	Thres.	Off.	Uncert.	Nu (1/cm)	Freq. (GHz)	Ref.
1040.	-3+		CO2	10.15R36	+4-		2.0E-2	9.6	288.	RA1
1000.	-3+		CO2	10.21R26	+3+		2.0E-2	10.1	301.	RA1
940.	-3-		CO2	10.89P46	+3		2.0E-2	10.7	321.	RA1
830.	-3+		CO2	9.59P24	+3+		2.0E-2	12.1	362.	RA1
710.	-2		CO2	9.54P18	+4-		2.0E-2	14.1	424.	RA1
700.	-3-		CO2	9.57P22	+4-		2.0E-2	14.3	429.	RA1
640.	-3+		CO2	10.46P06	+3+		2.0E-2	15.7	470.	RA1
634.471	0	P	CO2	9.55P20	+3-		2.0E-6	15.76116	472.5078	RA2CH3
601.897	-2+		CO2	10.79P38	+3		2.0E-6	16.61413	498.0791	RA2RA1
567.946	-2		CO2	10.55P16	+3		1.8E-6	17.60731	527.854	RA2RA1
540.	-4		CO2	10.37R04	+4-		2.0E-2	18.6	560.	RA1
530.	-3		CO2	9.52P16	+4-		2.0E-2	18.8	560.	RA1
520.	-3-		CO2	10.74P34	+4-		2.0E-2	19.3	580.	RA1
507.5840	0	N	CO2	10.61P22	+3+		1.6E-6	19.70117	590.626	RA2CH3
487.	-4-		CO2	9.47P10	+4-		2.0E-2	20.5	620.	RA1
445.	-3		CO2	10.26R18	+4-		2.0E-2	22.5	670.	RA1
442.1678	-2		CO2	10.55P16	+3		1.5E-6	22.61585	678.006	RA2
424.	-3+		CO2	10.20R28	+4-		2.0E-2	23.6	710.	RA1
385.9092	0-	N	CO2	10.61P22	+3		1.2E-6	25.91283	776.847	RA2CH3

Table 3.3.6.12
CF_2CH_2 (OR 1,1-$C_2H_2F_2$), 1,1-DIFLUOROETHYLENE

For line assignments see References DEH*, DH, SA*

Lambda (μm)	o/p	Pol.	Pump	Line	Thres.	Off.	Uncert.	Nu (1/cm)	Freq. (GHz)	Ref.
1020.	-1		CO2	10.53P14	+3		3.0E-2	9.8	294.	HO1
990.	-1	P	CO2	10.61P22			2.0E-3	10.10	302.8	DY1
890.	-1	P	CO2	10.61P22			2.0E-3	11.23	336.8	DY1
890.	-1	P	CO2	10.61P22			2.0E-3	11.24	336.8	DY1
880.	0		CO2	10.51P12	+3		3.0E-2	11.3	340.	HO1
770.	-3		CO2	10.49P10	+3+		2.0E-2	13.0	391.	RA2
764.	-1+	N	CO2	10.49P10	+3+		4.9E-3	13.09	392.	BE1
662.816	-1	P	CO2	10.63P24	+3+		2.2E-6	15.08715	452.3015	RA2DY1
570.	-1		CO2	10.63P24	+3		3.0E-2	17.6	530.	HO1
554.365	0	N	CO2	10.53P14	+3+		3.7E-6	18.03865	540.785	RA2DY1
520.	-2		CO2	10.55P16	+4-		2.0E-2	19.1	570.	RA1
464.3	-1	P	CO2	10.25R20			2.0E-3	21.54	646.	DY1
458.0	-1	P	CO2	10.70P30			2.0E-3	21.83	655.	DY1
407.2937	-1	P	CO2	10.53P14	+3+		1.3E-6	24.55231	736.060	RA2HO1
375.5449	0	P	CO2	10.51P12	+3+		1.2E-6	26.62797	798.287	RA2DY1
288.5	-1	P	CO2	10.51P12			2.0E-3	34.66	1039.	DY1

Table 3.3.6.13
CHFF (OR *cis*-1,2-$C_2H_2F_2$), *cis*-1,2-DIFLUOROETHYLENE

For line assignments see Reference DR1 (pump)

Lambda (μm)	o/p	Pol.	Pump	Line	Thres.	Off.	Uncert.	Nu (1/cm)	Freq. (GHz)	Ref.
310.8	+1-	N	CO2	9.68P34	+4		1.6E-3	32.18	965.	DR1
310.	0		CO2	10.16R34	+3+		1.0E-2	32.3	967.	AD1
307.5	+1	P	CO2	10.18R30	+4		1.6E-3	32.52	975.	DR1AD1
286.5	0+	N	CO2	9.62P28	+4		1.6E-3	34.90	1046.	DR1
262.	-1		CO2	10.22R24	+3+		1.0E-2	38.2	1140.	AD1

Lambda (μm)	o/p	Pol.	Pump	Line	Thres.	Off.	Uncert.	Nu (1/cm)	Freq. (GHz)	Ref.
260.1	+1	P	CO2	10.25R20	+4		1.8E-3	38.45	1153.	DR1AD1
242.6	+1-	P	CO2	10.29R14	+4		2.0E-3	41.22	1236.	DR1
231.0	+1-	N	CO2	10.14R38	+4		2.2E-3	43.29	1298.	DR1
228.	-1		CO2	9.55P20	+3+		1.0E-2	43.9	1310.	AD1
198.	-1+		CO2	10.37R04	+3+		1.0E-2	50.5	1510.	AD1
190.	-1		CO2	9.37R04	+3+		1.0E-2	52.6	1580.	AD1
185.	0		CO2	9.34R08	+3+		1.0E-2	54.1	1620.	AD1

Table 3.3.6.14
C_2H_3Br, VINYL BROMIDE, WITH ^{79}Br ^{81}Br MIXED 51%, 49%

For line assignments see Reference DY9

Lambda (μm)	o/p	Pol.	Pump	Line	Thres.	Off.	Uncert.	Nu (1/cm)	Freq. (GHz)	Ref.
1899.889	0-	P	CO2	10.59P20	+3+		1.0E-6	5.263465	157.7947	DY9
1614.888	-1+	P	CO2	10.65P26	+3+		1.0E-6	6.192381	185.6429	DY9
1394.063	0-	P	CO2	10.25R20	+3+		1.0E-6	7.173279	215.0495	DY9
1383.882	-1+	P	CO2	10.63P24	+3+		1.0E-6	7.226049	216.6315	DY9
1247.594	-1+	P	CO2	10.30R12	+3+		1.0E-6	8.015428	240.2965	DY9
990.6303	-1-	P	CO2	10.37R04	+3		1.0E-6	10.09458	302.6280	DY9
989.1904	0-	N	CO2	10.55P16	+3+		1.0E-6	10.10928	303.0685	DY9
985.8588	-1	N	CO2	10.38R02	+3		1.0E-6	10.14344	304.0927	DY9
963.4873	+1-	N	CO2	10.49P10	+3+		1.0E-6	10.37896	311.1535	DY9
936.1590	-1+	N	CO2	10.17P32	+3+		1.0E-6	10.68195	320.2367	DY9
934.2230	-1	P	CO2	9.62P28	+3+		1.0E-6	10.70408	320.9003	DY9
900.1338	-1	N	CO2	10.26R18	+3+		1.0E-6	11.10946	333.0532	DY9
853.4380	-1+	N	CO2	10.49P10	+3+		1.0E-6	11.71731	351.2762	DY9
826.9443	-2+	N	CO2	10.61P22	+3+		1.0E-6	12.09271	362.5304	DY9
784.2681	-1-	P	CO2	10.63P24	+3+		1.0E-6	12.75074	382.2576	DY9
780.1330	-1+	P	CO2	10.29P14	+3+		1.0E-6	12.81833	384.2838	DY9
741.1149	-2+	P	CO2	10.59P20	+3+		1.0E-6	13.49318	404.5155	DY9
724.1399	-1+	P	CO2	10.53P14	+3+		1.0E-6	13.80949	413.9980	DY9
712.	-2-	P	CO2	10.32R10	+3+		2.0E-3	14.04	421.1	DY9
707.2210	-1-	P	CO2	10.22R24	+3+		1.0E-6	14.13985	423.9021	DY9

Table 3.3.6.14 (continued)
C_2H_3Br, VINYL BROMIDE, WITH ^{79}Br ^{81}Br MIXED 51%, 49%

For line assignments see Reference DY9

Lambda (μm)	o/p	Pol.	Pump	Line	Thres.	Off.	Uncert.	Nu (1/cm)	Freq. (GHz)	Ref.
693.1396	0	P	CO2	10.27R16	+3+		1.0E-6	14.42711	432.5138	DY9
680.5414	0+	P	CO2	10.27R16	+3+		1.0E-6	14.69418	440.5205	DY9
649.4255	0+	P	CO2	10.57P18	+3+		1.0E-6	15.39823	461.6272	DY9
646.	-2+	N	CO2	10.21R26	+3+		2.0E-3	15.48	464.1	DY9
635.3548	0	N	CO2	10.21R26	+3+		1.0E-6	15.73924	471.8505	DY9
624.0958	0-	P	CO2	10.26R18	+3+		1.0E-6	16.02318	480.3629	DY9
618.4462	-1	P	CO2	10.18R30	+3+		1.0E-6	16.16956	484.7511	DY9
594.7286	-1-	N	CO2	10.72P32	+3+		1.0E-6	16.81439	504.0828	DY9
553.6962	-1+	N	CO2	10.81P40	+3		1.0E-6	18.06044	541.4385	DY9
528.4965	-1+	N	CO2	10.13R40	+3		1.0E-6	18.92160	567.2553	DY9
506.	-1	P	CO2	10.14R38	+3		2.0E-3	19.76	592.	DY9
490.0829	0-	P	CO2	10.55P16	+3+		1.0E-6	20.40471	611.7178	DY9
482.9615	-1+	N	CO2	10.65P26	+3+		1.0E-6	20.70558	620.7378	DY9
445.0	-1+	N	CO2	10.61P22	+3+		2.0E-3	22.47	674.	DY9
443.5	-1+	N	CO2	10.63P24	+3+		2.0E-3	22.55	676.	DY9
438.5069	-1+	P	CO2	10.67P28	+3+		1.0E-6	22.80466	683.6665	DY9
427.0	-2-	P	CO2	10.22R24	+3+		2.0E-3	23.42	702.	DY9
424.0	-1+	P	CO2	10.59P20	+3+		2.0E-3	23.58	707.	DY9
419.0	-2+	P	CO2	10.17R32	+3+		2.0E-3	23.87	715.	DY9
416.0	-1	N	CO2	10.23R22	+3+		2.0E-3	24.04	721.	DY9
411.0	-2+	P	CO2	10.21R26	+3+		2.0E-3	24.33	729.	DY9
396.0	-2+	P	CO2	9.68P34	+3+		2.0E-3	25.25	757.	DY9
370.0	-2	P	CO2	10.67P28	+3+		2.0E-3	27.03	810.	DY9
356.0	-1-	N	CO2	10.25R20	+3+		2.0E-3	28.09	842.	DY9
283.0	-1-	P	CO2	10.25R20	+3+		2.0E-3	35.34	1059.	DY9

Table 3.3.7.1
CH_3NH_2, METHYLAMINE

For line assignments see Reference LA3

Lambda (μm)	o/p	Pol.	Pump	Line	Thres.	Off.	Uncert.	Nu (1/cm)	Freq. (GHz)	Ref.
351.	-1-	P	CO2	9.79P46	+4-		2.0E-2	28.5	850.	LA3
349.	-1	P	CO2	10.26R18	+4-		2.0E-2	28.7	860.	LA3
314.8469	-2+		CO2	9.37R04	+3		1.5E-6	31.76147	952.185	RA2RA1
288.	-2		CO2	9.37R04	+3+		2.0E-2	34.7	1040.	RA1
283.	-1	N	CO2	9.79P46	+3+		2.0E-2	35.3	1060.	LA3
271.	-2+	N	CO2	9.32R12	+4-		2.0E-2	36.9	1110.	LA3
270.	-1-	N	CO2	9.28R18	+4-		2.0E-2	37.0	1110.	LA3
254.	-1	P	CO2	10.13R40	+3+		2.0E-2	39.4	1180.	LA3
251.3		P	CO2	9.59P24			2.0E-3	39.79	1193.	DY1
251.	-1-	N	CO2	9.26R22	+4		2.0E-2	39.8	1190.	LA3
246.	-1-	P	CO2	10.51P12	+4		2.0E-2	40.7	1220.	LA3
221.	-2+	P	CO2	10.25R20	+4		2.0E-2	45.2	1360.	LA3
220.	-2+	P	CO2	9.66P32	+4		2.0E-2	45.5	1360.	LA3
218.0		P	CO2	9.59P24			2.0E-3	45.87	1375.	DY1
198.0		P	CO2	9.59P24			2.0E-3	50.5	1514.	DY1
185.	-1-	N	CO2	9.31R14	+4-		2.0E-2	54.	1620.	LA3
180.	-4		CO2	9.79P46	+3+		2.0E-2	56.	1670.	RA1
179.	-2+	N	CO2	9.57P22	+4-		2.0E-2	56.	1670.	LA3
178.	-1	N	CO2	10.17R32	+3+		2.0E-2	56.	1680.	LA3
178.	-1	P	CO2	9.32R12	+3+		2.0E-2	56.	1680.	LA3
169.	-1	P	CO2	9.26R22	+4-		2.0E-2	59.	1770.	LA3
166.	-2+	N	CO2	9.66P32	+4-		2.0E-2	60.	1810.	LA3
165.	-2+		CO2	10.63P24	+4+		2.0E-2	61.	1820.	LA3
165.	-1	P	CO2	9.28R18	+3+		2.0E-2	61.	1820.	LA3
147.84469	+1	N	CO2	9.59P24	+3+		2.0E-7	67.63855	2027.7526	RA2DY1
147.	-2+	P	CO2	10.15R36	+4-		2.0E-2	68.	2040.	LA3
146.	-2+	P	CO2	9.34R08	+4-		2.0E-2	68.	2050.	LA3
145.	-1-	P	CO2	9.31R14	+4-		2.0E-2	69.	2070.	LA3
142.	-2+	N	CO2	10.25R20	+4-		2.0E-2	70.	2110.	LA3
128.	-1-	P	CO2	10.30R12	+4-		2.0E-2	78.	2340.	LA3

Table 3.3.7.1 (continued)
CH_3NH_2, METHYLAMINE

For line assignments see Reference LA3

Lambda (μm)	o/p	Pol.	Pump	Line	Thres.	Off.	Uncert.	Nu (1/cm)	Freq. (GHz)	Ref.
126.	-2+	N	CO2	10.35R06	+4-		2.0E-2	79.	2380.	LA3
120.	-2+	P	CO2	10.63P24	+4		2.0E-2	83.	2500.	LA3
119.	-2+	N	CO2	9.46P08	+4-		2.0E-2	84.	2520.	LA3
116.	-1-	P	CO2	9.77P44	+3+		2.0E-2	86.	2580.	LA3
100.	-2+	P	CO2	9.31R14	+4-		2.0E-2	100.	3000.	LA3

Table 3.3.7.2
CH_3CCH, METHYL ACETYLENE

For line assignments see Reference CH5 (pump)

Lambda (μm)	o/p	Pol.	Pump	Line	Thres.	Off.	Uncert.	Nu (1/cm)	Freq. (GHz)	Ref.
1174.87	0	P	CO2	10.86P44	+3+	-15	4.2E-5	8.5116	255.17	CH5
1097.11	0-	P	CO2	9.46P08	+4	-20	8.9E-5	9.1149	273.26	CH6
675.3	0-	P	CO2	9.73P40	+4+	0	1.5E-4	14.808	443.95	CH6
647.89	+1	P	CO2	10.53P14		0	7.6E-5	15.435	462.72	CH5
583.77	-2-	P	CO2	9.55P20	+4	-15	1.7E-4	17.130	513.55	CH6
566.4	0-	P	CO2	9.54P18	+5-	+25	1.7E-4	17.654	529.26	CH6
531.1	-2-	P	CO2	9.44P06	+4	0	1.9E-4	18.830	564.5	CH6
516.77	-2-	P	CO2	9.32R12	+4+	+35	1.9E-4	19.351	580.1	CH6
428.87	-2	P	CO2	9.18R38	+4	+35	2.3E-4	23.317	699.0	CH6

Table 3.3.7.3
CH_2CHCN, VINYL CYANIDE

Lambda (μm)	o/p	Pol.	Pump	Line	Thres.	Off.	Uncert.	Nu (1/cm)	Freq. (GHz)	Ref.
1180.	-3-		CO2	10.14R38	+3+		2.0E-2	8.4	253.	RA1
1160.	-4-		CO2	10.65P26	+4-		2.0E-2	8.7	259.	RA1
940.	-4		CO2	10.68P28	+4-		2.0E-2	10.6	319.	RA1
910.	-4		CO2	10.30R12	+4		2.0E-2	11.0	329.	RA1
830.	-4		CO2	10.26R18	+4-		2.0E-2	12.1	362.	RA1
790.	-4		CO2	10.13R40	+4-		2.0E-2	12.6	378.	RA1
780.	-4		CO2	10.12R42	+3+		2.0E-2	12.9	387.	RA1
740.	-4-		CO2	10.55P16	+4-		2.0E-2	13.6	406.	RA1
720.	-4-		CO2	10.84P42	+4-		2.0E-2	13.9	415.	RA1
630.	-4		CO2	10.35R06	+4-		2.0E-2	15.8	475.	RA1
620.	-4		CO2	10.30R12	+4		2.0E-2	16.1	481.	RA1
587.	-1	P	CO2	10.59P20			2.0E-3	17.05	511.	DY1
584.	-1	N	CO2	10.51P12			2.0E-3	17.12	513.	DY1
580.	-3		CO2	10.29R14	+3+		2.0E-2	17.3	520.	RA1
574.	-1	N	CO2	10.27R16			2.0E-3	17.41	522.	DY1
550.	-1	P	CO2	10.53P14			2.0E-3	18.18	545.	DY1
500.	-4-		CO2	9.32R12	+4		2.0E-2	19.9	600.	RA1
489.	-4-		CO2	10.48P08	+4-		2.0E-2	20.4	610.	RA1
270.6	-1	N	CO2	10.63P24			2.0E-3	36.95	1108.	DY1

Table 3.3.8.1
CH_3CH_2F (OR C_2H_5F), ETHYL FLUORIDE

Lambda (μm)	o/p	Pol.	Pump	Line	Thres.	Off.	Uncert.	Nu (1/cm)	Freq. (GHz)	Ref.
1521.376	-2		CO2	9.47P10	+4-		2.4E-6	6.57300	197.0535	RA2RA1
1440.	-3		CO2	9.46P08	+3+		2.0E-2	6.9	208.	RA2
1070.	-3+		CO2	9.33R10	+3+		2.0E-2	9.4	280.	RA1
1005.230	-2		CO2	9.62P28	+3+		1.6E-6	9.94797	298.2327	RA2RA1
851.9	-3+		CO2	9.64P30	+3+		1.2E-4	11.738	351.91	RA1/WA

Table 3.3.8.1 (continued)
CH_3CH_2F (OR C_2H_5F), ETHYL FLUORIDE

Lambda (μm)	o/p	Pol.	Pump	Line	Thres.	Off.	Uncert.	Nu (1/cm)	Freq. (GHz)	Ref.
660.	-3		CO2	9.29R16	+4-		2.0E-2	15.2	454.	RA2
620.40	-3		CO2	9.57P22	+3+		1.6E-4	16.119	483.22	RA1/WA
593.506	-2		CO2	9.69P36	+3		2.0E-6	16.84904	505.121	RA2RA1
519.075	-2		CO2	9.37R04	+3+		2.5E-6	19.26503	577.551	RA2RA1
502.2623	-2		CO2	9.25R24	+3		1.6E-6	19.90991	596.884	RA2RA1
486.	-2-		CO2	9.25R24	+3+		2.0E-2	20.6	620.	RA1
452.	-3		CO2	9.26R22	+3+		2.0E-2	22.1	660.	RA2
405.5044	-2		CO2	9.22R30	+3		1.3E-6	24.66064	739.308	RA2RA1
404.	-3-		CO2	9.68P34	+4-		2.4E-3	24.75	742.	RA1/WA
336.7	-3+		CO2	9.29R16	+4-		3.0E-4	29.700	890.4	RA1/WA
264.7	-2		CO2	9.54P18	+3+		3.8E-4	37.78	1132.6	RA1/WA

Table 3.3.8.2
CH_3CH_2Cl (OR C_2H_5Cl), ETHYL CHLORIDE

Lambda (μm)	o/p	Pol.	Pump	Line	Thres.	Off.	Uncert.	Nu (1/cm)	Freq. (GHz)	Ref.
1720.	-1		CO2	10.20R28	+4		1.0E-2	5.81	174.	JEN
1400.			CO2	10.14R38	+4		1.0E-2	7.14	214.	JEN
1350.	-2		CO2	10.18R30	+4		1.0E-2	7.41	222.	JEN
900.	-1		CO2	10.18R30	+4		1.0E-2	11.1	333.	JEN

Table 3.3.8.3
CH_3CHF_2 (OR $C_2H_4F_2$), 1,1-DIFLUOROETHANE

Lambda (μm)	o/p	Pol.	Pump	Line	Thres.	Off.	Uncert.	Nu (1/cm)	Freq. (GHz)	Ref.
530.	-1		CO2	10.59P20	+3		3.0E-2	18.8	560.	HO1
460.	-1		CO2	10.59P20	+3		3.0E-2	21.8	650.	HO1

Table 3.3.9.1
CH_3CH_2OH (OR C_2H_5OH), ETHYL ALCOHOL

Lambda(μm)	o/p	Pol.	Pump	Line	Thres.	Off.	Uncert.	Nu (1/cm)	Freq. (GHz)	Ref.
396.	-2		CO2	9.66P32	+4		1.0E-2	25.3	757.	JEN

Table 3.3.12.1
$(H_2CO)_3$, TRIOXANE (CYCLIC TRIMER OF FORMALDEHYDE)

Lambda(μm)	o/p	Pol.	Pump	Line	Thres.	Off.	Uncert.	Nu (1/cm)	Freq. (GHz)	Ref.
948.9247	-1-		CO2	9.25R24	+3		1.0E-6	10.53824	315.9286	DA2
891.	-1-		CO2	9.27R20	+3		4.9E-3	11.22	336.	DA2
890.	-1-		CO2	9.24R26	+3		4.9E-3	11.24	337.	DA2
815.	-1-		CO2	9.66P32	+3		4.9E-3	12.27	368.	DA2
750.	-1-		CO2	10.57P18	+3		4.9E-3	13.33	400.	DA2
712.	-1-		CO2	9.21R32	+3		4.9E-3	14.04	421.	DA2
696.	-1-		CO2	9.29R16	+3		4.9E-3	14.37	431.	DA2
680.	-1-		CO2	10.74P34	+3		4.9E-3	14.71	441.	DA2
619.	-1-		CO2	10.23R22	+3		4.9E-3	16.16	484.	DA2
512.	-1-		CO2	10.81P40	+3		4.9E-3	19.53	586.	DA2
460.	-1-		CO2	9.27R22	+3		4.9E-3	21.7	652.	DA2
433.	-1-		CO2	10.86P44	+3		4.9E-3	23.1	692.	DA2
384.	-1-		CO2	9.22R30	+3		4.9E-3	26.0	781.	DA2

CODE FOR WAVELENGTH TABLE

No. of Entries	Gas code	Table number; gas symbol, name
7	AR	Table 3.3.1.3: Ar, argon
4	B10T	Table 3.3.4.7: $^{10}BCl_3$, boron trichloride with ^{10}B isotope
4	B11T	Table 3.3.4.8: $^{11}BCl_3$, boron trichloride with ^{11}B isotope
	CO2A	Pump gas: $C^{18}O_2$, CO_2 with ^{18}O isotope
	CO2B	Pump gas: $^{13}CO_2$, CO_2 with ^{13}C isotope
	CO2C	Pump gas: $^{13}C^{18}O_2$, CO_2 with ^{13}C and ^{18}O isotopes
	CO2S	Pump gas: CO_2 sequence-band line (pump)
12	CDF	Table 3.3.6.13: CHFF (or *cis*-1,2-$C_2H_2F_2$), *cis* 1,2-difluoroethylene
5	DCN	Table 3.3.3.4: DCN, deuterium cyanide
16	DF	Table 3.3.6.12: CF_2CH_2 (or 1,1-$C_2H_2F_2$), 1,1-difluoroethylene
2	DFE	Table 3.3.8.3: CH_3CHF_2 (or $C_2H_4F_2$), 1,1-difluoroethane
48	DFM	Table 3.3.5.11: CH_2F_2, difluoromethane
47	DMA	Table 3.3.6.5: CH_3OD, O-deutero methanol
10	DMB	Table 3.3.5.17: CD_3Br, fully deuterated methyl bromide
20	DMC	Table 3.3.5.10: CD_3Cl, fully deuterated methyl chloride
1	DME	Table 3.3.6.8: CD_3OD, fully deuterated methanol
7	DMEC	Table 3.3.5.13: CD_2Cl_2, fully deuterated methylene chloride
32	DMI	Table 3.3.5.20: CD_3I, fully deuterated methyl iodide
2	DMIA	Table 3.3.5.21: $^{13}CD_3I$, fully deuterated methyl iodide with ^{13}C isotope
6	D2O	Table 3.3.3.2: D_2O, deuterium oxide (heavy water)
1	EA	Table 3.3.9.1: CH_3CH_2OH (or C_2H_5OH), ethyl alcohol
4	EC	Table 3.3.8.2: CH_3CH_2Cl (or C_2H_5Cl), ethyl chloride
16	EF	Table 3.3.8.1: CH_3CH_2F (or C_2H_5F), ethyl fluoride
47	FA	Table 3.3.5.4: HCOOH, formic acid
64	FAA	Table 3.3.5.6: HCOOD, O-deutero formic acid (formic acid "A")
20	FAB	Table 3.3.5.7: DCOOH, C-deutero formic acid (formic acid "B")
57	FAC	Table 3.3.5.8: DCOOD, di-deutero formic acid (formic acid "C")
6	FA13	Table 3.3.5.5: $H^{13}COOH$, formic acid with ^{13}C isotope
2	FOA	Table 3.3.4.5: HDCO, mono-deutero formaldehyde (formaldehyde "A")
13	FOB	Table 3.3.4.6: D_2CO, di-deutero formaldehyde (formaldehyde "B")
11	F12	Table 3.3.5.18: CF_2Cl_2, di-fluoro di-chloro methane (fluorocarbon-12)
10	F13	Table 3.3.5.22: CF_3Br, bromo 3-fluoromethane (fluorocarbon-1,3)
7	HCN	Table 3.3.3.3: HCN, hydrogen cyanide
2	HE	Table 3.3.1.1: He, helium
12	H2O	Table 3.3.3.1: H_2O, water
122	MA	Table 3.3.6.2: CH_3OH, methanol (methyl alcohol)
9	MAC	Table 3.3.7.2: CH_3CCH, methyl acetylene
110	MAD	Table 3.3.6.7: CD_3OH, C-trideutero methanol
68	MAD1	Table 3.3.6.4: CH_2DOH, C-deutero methanol
11	MAD2	Table 3.3.6.6: CHD_2OH, C-dideutero methanol
35	MAM	Table 3.3.7.1: CH_3NH_2, methylamine
40	MA13	Table 3.3.6.3: $^{13}CH_3OH$, methanol with ^{13}C isotope
9	MBA	Table 3.3.5.14: $CH_3{}^{79}Br$, methyl bromide with ^{79}Br
7	MBB	Table 3.3.5.15: $CH_3{}^{81}Br$, methyl bromide with ^{81}Br
20	MBR	Table 3.3.5.16: CH_3Br, methyl bromide with mixed ^{79}Br and ^{81}Br
20	MC	Table 3.3.5.9: CH_3Cl, methyl chloride
29	MCY	Table 3.3.6.9: CH_3CN, methyl cyanide (acetonitrile)
1	MEC	Table 3.3.5.12: CH_2Cl_2, methylene chloride
6	MF	Table 3.3.5.1: CH_3F, methyl fluoride
15	MFD	Table 3.3.5.3: CD_3F, fully deuterated methyl fluoride
1	MF13	Table 3.3.5.2: $^{13}CH_3F$, methyl fluoride with ^{13}C isotope
21	MI	Table 3.3.5.19: CH_3I, methyl iodide
23	MME	Table 3.3.6.10: CH_3SH, methyl mercaptan
1	ND3	Table 3.3.4.4: ND_3, fully deuterated ammonia
45	NE	Table 3.3.1.2: Ne, neon
6	NH2D	Table 3.3.4.3: NH_2D, monodeutero ammonia
105	NH3	Table 3.3.4.1: NH_3, ammonia

4	NH3A	Table 3.3.4.2: $^{15}NH_3$, ammonia with ^{15}N isotope
29	N2H4	Table 3.3.6.1: N_2H_4, hydrazine
1	OCS	Table 3.3.3.5: OCS, carbonyl sulfide
2	SO2	Table 3.3.3.6: SO_2, sulfur dioxide
13	TRI	Table 3.3.12.1: $(H_2CO)_3$, trioxane (cyclic trimer of formaldehyde)
45	VB	Table 3.3.6.14: C_2H_3Br, vinyl bromide, with ^{79}Br, ^{81}Br mixed 51%, 49%
19	VC	Table 3.3.6.11: CH_2CHCl (or H_2CCl), vinyl chloride
19	VCY	Table 3.3.7.3: CH_2CHCN, vinyl cyanide
3	XE	Table 3.3.1.4: Xe, xenon

REFERENCES

AD1 — **Amos, K. B. and Davis, J. A.**, Additional CW far-infrared laser lines from CO_2 laser pumped *cis* 1,2 $C_2H_2F_2$, *IEEE J. Quantum Electron.*, QE-16 (5), 574-575, 1980.

AR1* — **Arimondo, E. and Inguscio, M.**, The rotation-vibration constants of the $^{12}CH_3F$ ν_3 band, *J. Mol. Spectrosc.*, 75, 81-86, 1979.

AR2* — **Arimondo, E. and Inguscio, A.**, Assignments of laser lines in optically pumped CH_3CN, *Digest for the 4th Annu. Conf. on Infrared and Millimeter Waves,* IEEE Cat. No. 79 CH 1384-7 MTT, IEEE, Piscataway, N. J., 1979, 211-212; *Int. J. Infrared Millimeter Waves,* 1(3), 437—458, 1980.

BA* — **Baskakov, O. I., Dyubko, S. F., Moskienko, M. V., and Fesenko, L. D.**, Identification of active transitions in a formic acid vapor laser, *Sov. J. Quantum Electron.*, 7(4), 445—449, 1977.

BEN1 — **Benedict, W. S., Pollack, M. A., and Tomlinson, W. J., III,** The water-vapor laser, *IEEE J. Quantum Electron.*, QE-5 (2), 108—124, 1969.
Note: Detailed summaries of data on all observed laser lines in $H_2{}^{16}O$, $H_2{}^{18}O$, and $D_2{}^{16}O$ and their identifications are presented.

BE1 — **Bean, B. L. and Perkowitz, S.**, Complete frequency coverage for submillimetre laser spectroscopy with optically pumped CH_3OH, CH_3OD, CD_3OD and CH_2CF_2, *Opt. Lett.*, 1 (6), 202—204, 1977.

BL — **Blaney, T. G., Bradley, C. C., Edwards, G. J., and Knight, D. J. E.**, Absolute frequency measurement of a Lamb-dip stabilised water vapour laser oscillating at 10.7 THz (28 μm), *Phys. Lett. A*, 43 (5), 471—472, 1973.
Note: The full result (restricted here by the printing format) was (10 718 068.71 ± 0.03) MHz.

BL1 — **Blaney, T. G., Knight, D. J. E., and Murray-Lloyd, E.**, Frequency measurements of some optically-pumped laser lines in CH_3OD, *Opt. Commun.*, 25 (2), 176—178, 1978.
Note: 295- and 396-μm line powers from KON2; polarizations differ from KON2.

BL2 — **Blaney, T. G., Cross, N. R., Knight, D. J. E., Edwards, G. J., and Pearce, P. R.,** Frequency measurement at 4.25 THz (70.5 μm) using a Josephson harmonic mixer and phase-lock techniques, *J. Phys. D*, 13 (8), 1365-1370, 1980.

BV1 — **Belland, P., Veron, D., and Whitbourn, B.**, Scaling laws for cw 337-μm HCN waveguide lasers, *Appl. Opt.*, 15 (12), 3047—3053, 1976.

BV2 — **Belland, P. and Veron, D.**, Amplifying medium characteristics in optimised 190 μm/195 μm DCN waveguide lasers, *IEEE J. Quantum Electron.*, QE-16 (8), 885—889, 1980.

CH2 — **Chang, T. Y. and Bridges, T. J.**, Laser action at 452, 496 and 541 μm in optically pumped CH_3F, *Opt. Commun.*, 1 (9), 423—426, 1970.

CH3 — **Chang, T. Y., Bridges, T. J., and Burkhardt, E. J.**, cw submillimeter laser action in optically pumped methyl fluoride, methyl alcohol and vinyl chloride gases, *Appl. Phys. Lett.*, 17 (6), 249—251, 1970.

CH4 — **Chang, T. Y., Bridges, T. J., and Burkhardt, E. J.**, cw laser action at 81.5 and 263.4 μm in optically pumped ammonia gas, *Appl. Phys. Lett.*, 17 (9), 357—358, 1970.

CH5 — **Chang, T. Y. and McGee, J. D.,** Millimeter and submillimeter wave laser action in symmetric top molecules optically pumped via parallel absorption bands, *Appl. Phys. Lett.,* 19 (4), 103—105, 1971.
Note: 150 μsec chopped pump (200 W); lines stronger than 0.1 mW taken as potentially CW and listed.

CH6 — **Chang, T. Y. and McGee, J. D.,** Millimeter and submillimeter wave laser action in symmetric top molecules optically pumped via perpendicular absorption bands, *IEEE J. Quantum Electron.,* QE-12 (1), 62—65, 1976.
Note: 150 μsec chopped pump (200 W); all lines listed.

DAN — **Daneu, V., Hocker, L. O., Javan, A., Ramachandra-Rao, D., Szoke, A., and Zernike, F.,** Accurate laser wavelength measurements in the infrared and far infrared using a Michelson interferometer, *Phys. Lett. A,* 29 (6), 319—320, 1969.

DA1 — **Dangoisse, D., Deldalle, A., Splingard, J-P., and Bellet, J.,** Mésure précise des emissions continues du laser submillimétrique a acide formique, *C.R. Acad. Sci., Ser. B,* 283, 115—118, 1976.
Note: Where DY8 is a second reference, it is also a frequency measurement.

DA2 — **Dangoisse, D., Deldalle, A., Splingard, J-P., and Bellet, J.,** CW optically pumped laser action in D_2CO, HDCO and $(H_2CO)_3$, *IEEE J. Quantum Electron.,* QE-13 (9), 730—731, 1977
Notes: (1) 9.66P32 pump lines in doubt; (2) power levels and wavelength accuracy taken as for DE1.

DA3 — **Dangoisse, D., Willemot, E., Deldalle, A., and Bellet, J.,** Assignment of the HCOOH cw-submillimeter laser, *Opt. Commun.,* 28 (1), 111-116, 1979.
Note: Collation of formic acid assignments.

DA4* — **Dangoisse, D., Duterage, B., and Glorieux, P.,** Assignment of laser lines in optically pumped submillimetre lasers: HDCO, D_2CO, *IEEE J. Quantum Electron.,* QE-16 (3), 296—300, 1980.

DER* — **Deroche, J-C,** Assignment of submillimeter laser lines in methyl chloride, *J. Mol. Spectrosc.,* 69, 19—24, 1978.

DE1 — **Deldalle, A., Dangoisse, D., Splingard, J-P., and Bellet, J.,** Accurate measurements of cw optically pumped FIR laser lines of formic acid molecule and its isotopic species $H^{13}COOH$, HCOOD and DCOOD, *Opt. Commun.,* 22 (3), 333—336, 1977.
Notes: (1) $H^{13}COOH$ 480-μm line pump line (10.09R46) is in doubt; (2) these authors assign the 693788.5-MHz line to HCOOH against HCOOD given by DY8.

DG — **Danielewicz, E. J., Galantowicz, T. A., Foote, F. B., Reel, R. D., and Hodges, D. T.,** High performance at new FIR wavelengths from optically pumped CH_2F_2, *Opt. Lett.,* 4 (9), 280—282, 1979.

DGH* — **Duxbury, G., Gamble, T. J., and Herman, H.,** Assignments of optically pumped laser lines of 1,1 difluoroethylene, *IEEE Trans. Microwave Theory Tech.,* MTT-22 (12), 1108—1109, 1974.

DH — **Duxbury, G. and Herman, H.,** Optically pumped millimetre lasers, *J. Phys. B,* 11 (5), 935—949, 1978.

DR1 — **Danielewicz, E. J., Reel, R. D., and Hodges, D. T.,** New far-infrared CW optically pumped *cis*-$C_2H_2F_2$ laser, *IEEE J. Quantum Electron.,* QE-16 (4), 402—405, 1980.
Note: Lasing threshold taken conventionally as 10 W.

DW1 — **Danielewicz, E. J. and Weiss, C. O.,** Far infrared emission from $^{15}NH_3$ optically pumped by a CW sequence band CO_2 laser, *IEEE J. Quantum Electron.,* QE-14 (4), 222—223, 1978.
Note: This isotopic species was used in the natural abundance (0.37%) in research grade NH_3.

DW2 — **Danielewicz, E. J. and Weiss, C. O.,** New CW far-infrared laser lines from CO_2 laser-pumped CD_3OH, *IEEE J. Quantum Electron.,* QE-14 (7), 458—459, 1978.

DW3 — **Danielewicz, E. J. and Weiss, C. O.,** New efficient CW far-infrared optically pumped CH_2F_2 laser, *IEEE J. Quantum Electron.,* QE-14 (10), 705—707, 1978.
Note: Power levels are scaled from DW2, with 'VS' now 10 mW since 44 mW is given here for the 165.9 μm line.

DW4 — **Danielewicz, E. J. and Weiss, C. O.,** New cw far-infrared D_2O, $^{12}CH_3F$ and $^{14}NH_3$ laser lines, *Opt. Commun.,* 27(1), 98—100, 1978.

DW5* — **Danielewicz, E. J.,** Molecular parameters determining the performance of CW optically pumped FIR lasers, *Digest for the 4th Annu. Conf. on Infrared and Millimeter Waves,* IEEE Cat. No. 79 CH 1384-7 MTT, IEEE, Piscataway, N.J., 1979, 203—204.

DY — **Dyubko, S. F., Svich, V. A., and Valitov, R. A.,** SO_2 submillimetre laser generating at wavelengths 0.141 and 0.193 mm, *JETP Lett.,* 7 (11), 320, 1968.

DY1 — **Dyubko, S. F., Svich, V. A., and Fesenko, L. D.,** Submillimeter-band gas laser pumped by a CO_2 laser, *JETP Lett.,* 16 (11), 418—419, 1972.

DY2 — **Dyubko, S. F., Svich, V. A., and Fesenko, L. D.,** Submillimeter CH_3OH and CH_3OD lasers with optical pumping, *Sov. Phys. Tech. Phys.,* 18 (8), 1121, 1974.

DY3 — **Dyubko, S. F., Svich, V. A., and Fesenko, L. D.,** *Zh. Prikl. Spectrosk. (USSR),* 20 (4), 718—719, 1974; English translation: Stimulated emission of submillimeter waves by hydrazine, excited by a CO_2 laser, *J. Appl. Spectrosc.,* 20 (4), 545—546, 1974.

DY4 — **Dyubko, S. F., Svich, V. A., and Fesenko, L. D.,** Submillimeter laser emission of CH_3I molecules excited by CO_2, *Opt. Spectrosc.,* 37 (1), 118, 1974.

DY5 — **Dyubko, S. F., Svich, V. A., and Fesenko, L. D.,** Submillimeter laser using formic acid vapor pumped with carbon dioxide laser radiation, *Sov. J. Quantum Electron.,* 3 (5), 446, 1974.
Note: R-branch pump lines in doubt — see DY8.

DY6 — **Dyubko, S. F., Fesenko, L. D., Baskakov, O. I., and Svich, V. A.,** Use of CD_3I, CH_3I and CD_3Cl molecules as active substances for submillimeter lasers with optical pumping, *Zh. Prikl. Spectrosk. (USSR),* 23 (2) 317—320, 1975; English translation in *J. Appl. Spectrosc.,* 23, 1114—1116, 1975.
Notes: (1) Accurate wavelengths are from frequency; (2) ten units of power are taken as approximately 1 mW; (3) CO_2 power taken as 7 W as in DY5.

DY7 — **Dyubko, S. F., Svich, V. A., and Fesenko, L. D.,** An Experimental study of the radiation spectrum of submillimeter laser on CD_3OH molecules, *Izv. Vuz. Radiofiz. (USSR),* 18 (10), 1434—1437, 1975; English translation in *Radiophys. Quantum Electron.,* 18(10), 1058—1060, 1975.

DY8 — **Dyubko, S. F., Svich, A. V., and Fesenko, L. D.,** Submillimeter HCOOH, DCOOH, HCOOD, and DCOOD laser, *Sov. Phys. Tech. Phys.,* 20 (11), 1536—1538, 1975; Translation of: *Zh. Tekh. Fiz. (USSR),* 45, 2458—2461, 1975.
Note: For identification of active transitions see BA*.

DY9 — **Dyubko, S. F., Efimenko, M. N., Svich, V. A., and Fesenko, L. D.,** Stimulated emission of radiation from optically pumped vinyl bromide molecules, *Sov. J. Quantum Electron.,* 6 (5), 600—601, 1976.
Notes (for DY8 and DY9): (1) conventional pump threshold values of "+3+", approximately 2 W, have been used for all DY results, except for ~4 extreme lines in DY9 given "+3", approximately 1 W; (2) the submillimeter powers are interpolated from the indications of detector calibration given.

ED — **Edwards, G. J.,** National Physical Laboratory, Teddington, Middlesex, U.K., private communication, 1978.
Note: Wavelength checks on seven lines to ±0.05%.

EP1 — **Epton, P. J., Wilson, W. L., Jr., Tittel, F. K., and Rabson, R. A.,** Frequency measurement of the formic acid laser 311-μm line, *Appl. Opt.,* 18 (11), 1704—1705, 1979.
Note: Measured from HCN 311-μm line (964 313.4 MHz).

EV1 — **Evenson, K. M., Wells, J. S., Matarrese, L. M., and Elwell, J. B.,** Absolute frequency measurements of the 28- and 78-μm CW water vapor laser lines, *Appl. Phys. Lett.,* 16 (4), 159—162, 1970.

FE1 — **Fetterman, H. R., Schlossberg, H. R., and Parker, C. D.,** cw submillimeter laser generation in optically pumped Stark-tuned NH_3, *Appl. Phys. Lett.,* 23 (12), 684—686, 1973.
Notes: (1) The pumped transitions are Stark shifted to the pump frequency; (2) the wavelength accuracy is estimated, not given.

GR* — **Graner, G.,** Assignment of submillimeter laser lines in CH_3I, *Opt. Commun.,* 14 (1), 67—69, 1975.

GW1 — **Grinda, M. and Weiss, C. O.,** New far infrared laser lines from CD_3OH, *Opt. Commun.,* 26 (1), 91, 1978.
Note: Power levels taken as for DW2.

HA — **Hard, T. M.,** Sulfur dioxide submillimeter laser, *Appl. Phys. Lett.,* 14 (4), 130, 1969.

HEN* — **Henningsen, J. O.,** Assignment of laser lines in optically pumped CH_3OH, *IEEE J. Quantum Electron.,* QE-13 (6), 435—441, 1977.

HEN1— **Henningsen, J. O.,** New FIR laser lines from optically pumped CH_3OH: measurements and assignments, *IEEE J. Quantum Electron.,* QE-14 (12), 958—962, 1978.

HEN2 — **Henningsen, J. O. and Petersen, J. C.,** Observation and assignment of far-infrared laser lines from optically pumped $C^{13}H_3OH$, *Infrared Phys.,* 18 (5 and 6), 475—479, 1978.
Notes: (1)^{13}C species was 90% pure; (2)lines given as "particularly strong" taken as 1 mW; (3)threshold conventionally taken as 10 W, (25 W pump).

HEN3*— **Henningsen, J. O.,** Spectroscopy of the lasing CH_3OH molecule, *Digest for the 4th Annu. Conf. on Infrared and Millimeter Waves,* IEEE Cat. No. 79 CH 1384-7 MTT, IEEE, Piscataway, N.J., 1979, 197—198.

HEN4*— **Henningsen, J. O., Petersen, J. C., Petersen, F. R., Jennings, D. A., and Evenson, K. M.,** High resolution spectroscopy of vibrationally excited $^{13}CH_3OH$ by frequency measurement of FIR laser emission, *J. Mol. Spectrosc.,* 77 (2), 298—309, 1979.

HEN5*— **Henningsen, J. O.**, Stark effect in the CO_2 laser pumped CH_3OH far infrared laser as a technique for high resolution infrared spectroscopy, *J. Mol. Spectrosc.*, 83(1), 70—93, 1980.

HEN6*— **Henningsen, J. O.**, Improved molecular constants and empirical corrections for the torsional ground state of the C-O stretch fundamental of CH_3OH, *J. Mol. Spectrosc.*, 85(2), 282—300, 1981.

HE1 — **Heppner, J., Weiss, C. O., and Plainchamp, P.**, Far infrared gain measurements in optically pumped CH_3OH, *Opt. Commun.*, 23 (3), 381—384, 1977, (see footnote on page 383).
Notes: (1) Sign of pump is offset from NPL work; (2) lambda accuracy is estimated as 0.1%.

HOC1 — **Hocker, L. O., Javan, A., Ramachandra-Rao, D., Frenkel, L., and Sullivan, T.**, Absolute frequency measurement and spectroscopy of gas laser transitions in the far infrared, *Appl. Phys. Lett.*, 10 (5), 147—149, 1967.

HOC2— **Hocker, L. O. and Javan, A.**, Absolute frequency measurements on new cw HCN submillimeter laser lines, *Phys. Lett. A*, 25 (7), 489—490, 1967.

HOC3 — **Hocker, L. O. and Javan, A.**, Absolute frequency measurements of new cw DCN submillimeter laser lines, *Appl. Phys. Lett.*, 12 (4), 124—125, 1968.

HOC4 — **Hocker, L. O., Ramachandra-Rao, D., and Javan, A.**, Absolute frequency measurement of the 190 μm and 194 μm gas laser transitions, *Phys. Lett. A*, 24 (12), 690—691, 1967.

HO1 — **Hodges, D. T., Reel, R. D., and Barker, D. H.**, Low-threshold CW submillimeter- and millimeter-wave laser action in CO_2-laser-pumped $C_2H_4F_2$, $C_2H_2F_2$, and CH_3OH, *IEEE J. Quantum Electron.*, QE-9 (12), 1159—1160, 1973.

HP1 — **Herman, H. and Prewer, B. E.**, New FIR laser lines from optically pumped methanol analogues, *Appl. Phys.*, 19, 241—242, 1979.

HU* — **Hubner, G., Hassler, J. C., Coleman, P. D., and Steenbeckeliers, G.**, Assignments of the far-infrared SO_2 laser lines, *Appl. Phys. Lett.*, 18 (11), 511—513, 1971.

IN1 — **Inguscio, M., Moretti, A., and Strumia, F.**, New laser lines from optically-pumped CH_3OH: measurements and assignments, *Opt. Commun.*, 32 (1), 87—90, 1980.
Note: Conventional threshold value of 20 W assigned.

IN2 — **Inguscio, M., Moretti, A., and Strumia, F.**, IR-FIR Stark spectroscopy of CH_3OH around the 9-P(34) CO_2 laser line, *Digest for the 4th Annu. Conf. on Infrared and Millimeter Waves*, IEEE Cat. No. 79 CH 1384-7 MTT, IEEE, Piscataway, N. J., 1979, 205—206; **Bionducci, G., Inguscio, M., Moretti, A., and Strumia, F.**, Design of FIR molecular lasers frequency tunable by Stark effect: electric breakdown of CH_3OH, CH_3F, CH_3I, and CH_3CN, *Infrared Phys.*, 19, 297—308, 1979.

JEN — **Jennings, D. A., Evenson, K. M., and Jimenez, J. J.**, New CO_2 pumped CW far infrared laser lines, *IEEE J. Quantum Electron.*, QE-11 (8), 637, 1975.

KAR — **Karlov, N. V., Petrov, Yu. B., Prokhorov, A. M., and Stel'makh, O. M.**, Laser based on boron trichloride, *JETP Lett.*, 8, 12—14, 1968.

KON — **Kon, S., Hagiwara, E., Yano, T., and Hirose, H.**, Far infrared laser action in optically pumped CD_3OD, *Jpn. J. Appl. Phys.*, 14 (5), 731—732, 1975.

KON2 — **Kon, S., Yano, T., Hagiwara, E., and Hirose, H.**, Far infrared laser action in optically pumped CH_3OD, *Jpn. J. Appl. Phys.*, 14(11), 1861—1862, 1975.

KW1 — **Kramer, G. and Weiss, C. O.**, Frequencies of some optically pumped submillimetre laser lines, *Appl. Phys.*, 10, 187—188, 1976.

LA1 — **Landsberg, B. M.**, Optically pumped CW submillimetre emission lines from methyl mercaptan CH_3SH, *IEEE J. Quantum Electron.*, QE-16 (6), 684—685, 1980.

LA2— **Landsberg, B. M.**, New optically pumped CW submillimetre emission lines from OCS, CH_3OH and CH_3OD, *IEEE J. Quantum Electron.*, QE-16 (7), 704—706, 1980.
Note: (1) Output powers have been interpolated, but thresholds are as given by the author, (2) the OCS line is tentatively assigned to ^{17}OCS.

LA3 — **Landsberg, B. M.**, New CW FIR Laser lines From optically pumped ammonia Analogues, preprint June 1980; *Appl. Phys.*, 23, 127—130, 1980.
Notes: (1) ND_2H and NH_2D lines were obtained from NH_3/ND_3 mixtures; (2) assignment of the ND_3 line (to ND_3) is tentative; (3) FIR power, and wavelength accuracy are as for LA1 and LA2.

LA4 — **Landsberg, B. M.**, private communications, June 1980; New CW optically pumped FIR emissions in HCOOH, D_2CO and CD_3Br *Appl. Phys.*, 23, 345—348, 1980.

LD — **Lund, M. W. and Davis, J. A.**, New CW far-infrared laser lines from CO_2 laser-pumped CH_3OD, *IEEE J. Quantum Electron.*, QE-15 (7), 537—538, 1979.
Note: Polarizations of 134- and 100.8-μm lines disagree with those of LA2 (N against P) so that E is shown.

LEV — **Levine, J. S. and Javan, A.**, Far infrared continuous-wave laser oscillation in pure helium, *Appl. Phys. Lett.*, 14 (11), 348—350, 1969.

LO1 — **Lourtioz, J-M., Pontnau, J., Morillon-Chapey, M., and Deroche, J-C.**, Submillimetre Laser Action of CW Optically Pumped CF_2Cl_2 (fluorocarbon 12), preprint received 7 July 1980; *Int. J. Infrared Millimeter Waves,* 2(1), 49—63, 1981.
Note: Pump threshold taken conventionally as half the CO_2 laser power.

LO2 — **Lourtioz, J-M.**, private communication, July 1980; Lourtioz, J—M., Pontnau, J., and Meyer, C., Optically pumped CW CF_3Br FIR laser New emission lines and tentative assignments, *Int. J. Infrared Millimeter Waves,* 2(3), 525—532, 1981.
Note: Conventional pump threshold of 10 W assigned.

LM1 — **Lide, D. R. and Maki, A. G.**, On the explanation of the so-called CN laser, *Appl. Phys. Lett.,* 11(2), 62—64, 1967.

MA1 — **Mathias, L. E. S., Crocker, A., and Wills, M. S.**, Spectroscopic measurements of the laser emission from discharges in compounds of hydrogen, carbon and nitrogen, *IEEE J. Quantum Electron.,* QE-4 (4), 205—208, 1968.

MA2 — **Mathias, L. E. S., Crocker, A., and Wills, M. S.**, Pulsed laser emission from Helium at 95 μm, *IEEE J. Quantum Electron.,* QE-3 (4), 170, 1967.

MF1 — **Muller, W. M. and Flesher, G. T.**, Continuous-wave submillimeter oscillation in discharges containing C, N and H or D, *Appl. Phys. Lett.,* 10 (3), 93—94, 1967.

MK*— **Maki, A.G.**, Assignment of some DCN and HCN laser lines, *Appl. Phys. Lett.,* 12 (4) 122—124, 1968.

MK1*— **Maki, A. G.**, Further assignments for the far-infrared laser transitions of HCN and $HC^{15}N$, *J. Appl. Phys.,* 49 (1), 7—11, 1978.

MK2*— **Maki, A. G., Olson, W. B., and Sams, R. L.**, HCN rotational-vibrational energy levels and intensity anomalies determined from infrared measurements, *J. Mol. Spectrosc.,* 36(3), 433-447, 1970.

MO — **Moskienko, M. V. and Dyubko, S. F.**, Identification of generation lines of submillimeter methyl bromide and acetonitrile molecule laser, *Izv. Vuz. Radiofiz. (USSR),* 21 (7), 951—960, 1978; English translation in *Radiophys. Quantum Electron.,* 21(7), 668—675, 1978.

NH1 — **Ni, Y. C. and Heppner, J.**, New cw laser lines from CO_2 laser pumped CH_3OD, *Opt. Commun.,* 32(3), 459—460, 1980.
Notes: (1) Pump threshold is taken conventionally as 6 W, half the power given as entering the cavity, (2) far-IR power calibration is taken as 1, 0.1, and 0.01 mW for 'S', 'M' and 'W', comparing with LD and LA2.

PAT3 — **Patel, C. K. N.**, Gas lasers, in *Lasers; a Series of Advances,* Vol. 2, Levine, A. K., Ed., Marcel Dekker, New York, 1968.

PAT4 — **Patel, C. K. N., Faust, W. L., McFarlane, R. A., and Garrett, C. G. B.**, Laser action up to 57.335 μm in gaseous discharges (Ne, He - Ne), *Appl. Phys. Lett.,* 4(1), 18—19, 1964.

PAT5 — **Patel, C. K. N., Faust, W. L., McFarlane, R. A., and Garrett, C. G. B.**, CW optical maser action up to 133 μm (0.133 mm) in Neon discharges, *Proc. IEEE,* 52 (6), 713, 1964.

PET — **Petersen, F. R., Evensen, K. M., Jennings, D. A., Wells, J. S., Goto, K., and Jimenez, J. J.**, Far infrared frequency synthesis with stabilized CO_2 lasers: accurate measurements of the water vapor and methyl alcohol laser frequencies, *IEEE J. Quantum Electron.,* QE-11 (10), 838—843, 1975.

PE2 — **Petersen, F. R.**, National Bureau of Standards, Boulder, Colo., private communications, November 1978 and March 1979. The CH_3OH data is published in Petersen, F. R., Evenson, K. M., Jennings, D. A., and Scalabrin, A., New frequency measurements and laser lines of optically pumped $^{12}CH_3OH$, *IEEE J. Quantum Electron.,* QE-16 (3), 319—323, 1980; the $^{13}CH_3OH$ data is largely given in HEN4*; the CD_3OH data may be unpublished.

PE3 — **Petersen, F. R., Scalabrin, A., and Evenson, K. M.**, Frequencies of cw FIR laser lines from optically pumped CH_2F_2, *Int. J. Infrared Millimeter Waves,* 1 (1), 111—115, 1980.

POL1 — **Pollack, M. A., Frenkel, L., and Sullivan, T.**, Absolute frequency measurement of the 220 μm water vapor laser transition, *Phys. Lett. A,* 26 (8), 381—382, 1968.

POL2 — **Pollack, M. A., Bridges, T. J., and Tomlinson, W. J.**, Competitive and cascade coupling between transitions in the cw water vapor laser, *Appl. Phys. Lett.,* 10 (9), 253—256, 1967.

PO1 — **Pontnau, J., Lourtioz, J-M, and Meyer, C.**, Submillimetre laser action of CW optically pumped CF_3Br, *IEEE J. Quantum Electron.,* QE-15 (10), 1088—1090, 1979.
Note: The assignments to the Br isotope are tentative.

RA1 — **Radford, H. E.**, New CW lines from a submillimeter waveguide laser, *IEEE J. Quantum Electron.,* QE-11 (5), 213—214, 1975.

RA2 — **Radford, H. E.**, private communication, August 1976.
Note: New frequency measurements on 41 lines (waveguide laser); of these, some (Fabry-Perot) lines of wavelength between 0.1 and 0.7 mm are published in Radford, H. E., Petersen, F. R., Jennings, D. A., and Mucha, J. A., Heterodyne measurements of submillimeter laser spectrometer frequencies, *IEEE J. Quantum Electron.,* QE-13, 92—94, 1977.
Note: Power levels are taken from the second reference given or assumed −2 or −3.

RE — **Reid, J. and Oka, T.,** Direct observation of velocity-tuned multiphonon processes in the laser cavity, *Phys. Rev. Lett.,* 38 (2), 67—70, 1977.

RE1 — **Redon, M., Gastaud, C., and Fourrier, M.,** New CW far-infrared lasing in $^{14}NH_3$ using Stark tuning, *IEEE J. Quantum Electron.,* QE-15 (6), 412—414, 1979.

RE2 — **Redon, M., Gastaud, C., and Fourrier, M.,** Far-infrared emission in NH_3 using "forbidden" transitions pumped by a CO_2 laser, *Opt. Commun.,* 30 (1), 95—98, 1979.

RE3 — **Redon, M., Gastaud, C., and Fourrier, M.,** Far-infrared emissions in ammonia by infrared pumping using a N_2O laser, *Infrared Phys.,* 20, 93—98, 1980.

Notes: (1) Calculated wavelengths of accuracy ±0.03 cm^{-1} (RE5) are listed, since the transitions are assigned and confirmed; far-IR wavelength measurements have since been made to ±0.3% (RE5); (2) the Stark field in NH_3 shifts mainly the lower level of the pump-absorbing transition and has only a small effect (not measured) in tuning the FIR-emitting transition; (3) the FIR emission is polarized normal to the Stark field because the Stark cell acts as a waveguide; (4) the "pump offset" listed, e.g., "20.2", represents the Stark field in kilovolts per centimeter for the pump polarization shown (P or N with respect to the FIR); (5) when "E" is shown for the polarization, the Stark fields for both P- and N- pumping lie within ±5% of the value given; (6) improved Stark field measurements as in RE4 have been incorporated here (RE5); (7) a conventional pump threshold of 1 W has been assigned (N_2O-pumped lines).

RE4 — **Redon, M., Gastaud, C., and Fourrier, M.,** New CW FIR laser lines obtained in ammonia pumped by a CO_2 laser, using the Stark tuning method, *Int. J. Infrared Millimeter Waves,* 1 (1), 95—100, 1980.

Notes: (1) See notes for RE3; (2) a conventional pump threshold of 10 W is assigned (CO_2-pumped lines).

RE5 — **Redon, M.,** Universitė Pierre et Marie Curie, Paris, private communications, May 7, 1980 (summary of new lines) and May 16, 1980 (improved Stark fields for N_2O-pumped lines).

SA* — **Sattler, J. P., Worchesky, T. L. Tobin, M. S., Ritter, K. J., Daley, T. W., and Lafferty, W. J.,** Submillimeter-wave emission assignments for 1,1 difluoroethylene, *Int. J. Infrared Millimeter Waves,* 1 (1), 127—138, 1980.

SC — **Scalabrin, A. and Evenson, K. M.,** Additional cw FIR laser lines from optically pumped CH_2F_2, *Opt. Lett.,* 4 (9), 277—279, 1979.

Notes: (1) Pump threshold is taken conventionally as half the optimum power; (2) a horizontal dash for polarization is here taken as "E".

SC2 — **Scalabrin, A., Petersen, F. R., Evenson, K. M., and Jennings, D. A.,** Optically pumped cw CH_2DOH FIR laser: new lines and frequency measurements, *Int. J. Infrared Millimeter Waves,* 1 (1), 117—126, 1980.

Notes: (1) The 308.3- and 509.4-μm lines are additional to those published (from authors' preprint); (2) the pump threshold is taken conventionally as half the listed CO_2 laser power.

STE — **Steenbeckeliers, G. and Bellet, J.,** New interpretation of the far-infrared SO_2 laser spectrum, *J. Appl. Phys.,* 46 (6), 2620—2626, 1975.

TA1 — **Tanaka, A., Tanimoto, A., Murata, N., Yamanaka, M., and Yoshinaga, H.,** Optically pumped far infrared and millimeter wave waveguide lasers, *Jpn. J. Appl. Phys.,* 13 (9), 1491—1492, 1974.

TA2 — **Tanaka, A., Yamanaka, M., and Yoshinaga, H.,** New far-infrared laser lines from CO_2 laser pumped CH_3OH gas by using a copper waveguide cavity, *IEEE J. Quantum Electron.,* QE-11 (10), 853—854, 1975.

TO1 — **Tobin, M. S., Sattler, J. P., and Wood, G. L.,** CD_3F optically-pumped near millimeter laser, *Digest for the 4th Annu. Conf. on Infrared and Millimeter Waves,* IEEE Cat. No. 79 CH 1384-7 MTT, IEEE, Piscataway, N. J., 1979, 209—210; Optically-pumped CD_3F submillimeter wave laser, *Opt. Lett.,* 4 (11), 384—386, 1979.

Notes: (1) A conventional threshold of 10 W is assigned; (2) all lines are assumed CW, although all but the strongest three were observed with chopped pump radiation.

T3 — **Tanaka, A., Tanimoto, A., Murata, N., Yamanaka, M., and Yoshinaga, H.,** CW efficient optically pumped far-infrared waveguide NH_3 lasers, *Opt. Commun.,* 22 (1), 17—21, 1977.

Note: Appears as second/third reference, giving improved CW power out and pump threshold.

WA — **Wagner, R. J., Zelano, A. J., and Ngai, L. H.,** New submillimeter laser lines in optically pumped gas molecules, *Opt. Commun.,* 8 (1), 46—47, 1973.

Note: Used for wavelength values, although laser was pulsed.

WE1 — **Weiss, C. O., Grinda, M., and Siemsen, K.,** FIR laser lines of CH_3OH pumped by CO_2 laser sequence lines, *IEEE J. Quantum Electron.,* QE-13 (11), 892, 1977.

Notes: (1) Pump lines are CO2 sequence lines (CO2S), see Siemsen and Whitford, *Opt. Commun.,* 22 (1), 11—16, 1977; (2) uncertainty of wavelength not given but estimated at 1 μm or 0.5%; (3) formal threshold of 2 W assigned.

WO1 — **Wood, R. A., Davis, B. W., Vass, A., and Pidgeon, C. R.,** Application of an isotopically enriched $^{13}C^{16}O_2$ laser to an optically-pumped FIR laser, Preprint April 1980; *Opt. Lett.*, 5(4), 153—154, 1980.

WO2 — **Wood, R. A., Vass, A., Pidgeon, C. R., Colles, M. J., and Norris, B.,** High power FIR lasing in $^{15}NH_3$ optically pumped with an isotopically enriched CO_2 laser, *Opt. Commun.*, 33 (1) 89—90, 1980.

W1* — **Worchesky, T. L.,** Assignments of methyl alcohol submillimeter laser transitions, *Opt. Lett.*, 3 (12), 232—234, 1978.

ZD — **Ziegler, G. and Durr, U.,** Submillimeter laser action of CW optically pumped CD_2Cl_2,CH_2DOH and CHD_2OH, *IEEE J. Quantum Electron.*, QE-14 (10), 708, 1978.
Notes: (1) Gas purity about 96%; (2) conventional threshold of 2 W assigned.

Section 4
Table of Laser Wavelengths

4. TABLE OF WAVELENGTHS OF GAS LASERS

Table 4.1
GAS LASERS ARRANGED IN ORDER OF WAVELENGTH

In this table all gas lasers in the precedins sections are arranged in order of increasing wavelength. In the final column are listed table numbers where assignments of transitions, isotopes used, special notes, and references to laser action can be found.

Wavelengths cited may be from laser measurements, spectroscopic measurements, or calculated values. The introduction to the relevant chapters should be consulted for a discussion of the determination and accuracy of the wavelengths. Wavelengths lines in vacuum are given in italics.

Wavelength (μm)	Element or compound	Table No.
0.10982	para H_2	3.1.2.12a
0.11020	H_2	3.1.2.12a
0.11134	D_2	3.1.2.12a
0.11152	para H_2	3.1.2.12a
0.11189	H_2	3.1.2.12a
0.11377	D_2	3.1.2.12a
0.11386	HD	3.1.2.12a
0.11415	HD	3.1.2.12a
0.11446	para H_2	3.1.2.12a
0.11476	D_2	3.1.2.12a
0.11520	HD	3.1.2.12a
0.11565	D_2	3.1.2.12a
0.11584	D_2	3.1.2.12a
0.11600	para H_2	3.1.2.12a
0.11613	H_2	3.1.2.12a
0.11639	H_2	3.1.2.12a
0.11662	H_2	3.1.2.12a
0.11746	para H_2	3.1.2.12a
0.11758	H_2	3.1.2.12a
0.11781	HD	3.1.2.12a
0.10982	para H_2	3.1.2.12a
0.11805	H_2	3.1.2.12a
0.11881	D_2	3.1.2.12a
0.11893	H_2	3.1.2.12a
0.11900	HD	3.1.2.12a
0.11901	D_2	3.1.2.12a
0.11928	HD	3.1.2.12a
0.11975	D_2	3.1.2.12a
0.11994	D_2	3.1.2.12a
0.12010	HD	3.1.2.12a
0.12054	para H_2	3.1.2.12a
0.12064	D_2	3.1.2.12a
0.12067	H_2	3.1.2.12a
0.12093	H_2	3.1.2.12a
0.12113	HD	3.1.2.12a
0.12173	H_2	3.1.2.12a
0.12177	para H_2	3.1.2.12a
0.12189	H_2	3.1.2.12a
0.12214	H_2	3.1.2.12a
0.12236	H_2	3.1.2.12a
0.12280	D_2	3.1.2.12a
0.12284	HD	3.1.2.12a
0.12287	para H_2	3.1.2.12a
0.12299	H_2	3.1.2.12a
0.12323	H_2	3.1.2.12a
0.12356	D_2	3.1.2.12a
0.12383	para H_2	3.1.2.12a
0.12394	H_2	3.1.2.12a
0.12416	para H_2	3.1.2.12a
0.12417	H_2	3.1.2.12a
0.12424	D_2	3.1.2.12a
0.12441	D_2	3.1.2.12a
0.12457	HD	3.1.2.12b
0.12483	D_2	3.1.2.12a
0.12500	D_2	3.1.2.12a
0.12520	para H_2	3.1.2.12a
0.12528	HD	3.1.2.12b
0.12533	D_2	3.1.2.12a
0.1261	Ar_2	3.1.1.1
0.12795	para H_2	3.1.2.12b
0.13033	HD	3.1.2.12b
0.13036	D_2	3.1.2.12b
0.13386	para H_2	3.1.2.12b
0.13423	H_2	3.1.2.12b
0.13459	D_2	3.1.2.12b
0.13551	HD	3.1.2.12b
0.13598	para H_2	3.1.2.12b
0.13680	para H_2	3.1.2.12b
0.13888	D_2	3.1.2.12b
0.13990	para H_2	3.1.2.12b

Table 4.1 (continued)
GAS LASERS ARRANGED IN ORDER OF WAVELENGTH

Wavelength (μm)	Element or compound	Table No.	Wavelength (μm)	Element or compound	Table No.
0.14020	H_2	3.1.2.12b	0.15743	HD	3.1.2.12b
0.14075	para H_2	3.1.2.12b	0.15758	D_2	3.1.2.12b
0.14077	HD	3.1.2.12b	0.15759	F_2	3.1.2.11
0.14187	H_2	3.1.2.12b	0.15774	H_2	3.1.2.12b
0.14322	D_2	3.1.2.12b	0.15777	H_2	3.1.2.12b
0.14326	para H_2	3.1.2.12b	0.15777	para H_2	3.1.2.12b
0.14362	H_2	3.1.2.12b	0.15792	H_2	3.1.2.12b
0.14376	para H_2	3.1.2.12b	0.15800	para H_2	3.1.2.12b
0.14406	para H_2	3.1.2.12b	0.15800	H_2	3.1.2.12b
0.14409	H_2	3.1.2.12b	0.15801	HD	3.1.2.12b
0.1457	Kr_2	3.1.2.19	0.15777	para H_2	3.1.2.12b
0.14602	para H_2	3.1.2.12b	0.15792	H_2	3.1.2.12b
0.14638	H_2	3.1.2.12b	0.15800	para H_2	3.1.2.12b
0.14641	para H_2	3.1.2.12b	0.15800	H_2	3.1.2.12b
0.14670	H_2	3.1.2.12b	0.15001	HD	3.1.2.12b
0.14684	para H_2	3.1.2.12b	0.15808	H_2	3.1.2.12b
0.14865	H_2	3.1.2.12b	0.15809	HD	3.1.2.12b
0.14884	HD	3.1.2.12b	0.15811	para H_2	3.1.2.12b
0.14917	para H_2	3.1.2.12b	0.15814	para H_2	3.1.2.12b
0.14942	H_2	3.1.2.12b	0.15819	HD	3.1.2.12b
0.14952	H_2	3.1.2.12b	0.15825	HD	3.1.2.12b
0.15136	HD	3.1.2.12b	0.15831	HD	3.1.2.12b
0.15157	para H_2	3.1.2.12b	0.15863	D_2	3.1.2.12b
0.15199	para H_2	3.1.2.12b	0.15864	D_2	3.1.2.12b
0.15233	H_2	3.1.2.12b	0.15867	D_2	3.1.2.12b
0.15299	HD	3.1.2.12b	0.15869	D_2	3.1.2.12b
0.15349	para H_2	3.1.2.12b	0.15871	D_2	3.1.2.12b
0.15449	H_2	3.1.2.12b	0.15872	D_2	3.1.2.12b
0.15482	C^{3+}	2.7.1	0.15890	para H_2	3.1.2.12b
0.15501	para H_2	3.1.2.12b	0.15890	H_2	3.1.2.12b
0.15509	C^{3+}	2.7.1	0.15898	D_2	3.1.2.12b
0.15534	H_2	3.1.2.12b	0.15913	D_2	3.1.2.12b
0.15620	HD	3.1.2.12b	0.15913	H_2	3.1.2.12b
0.15690	para H_2	3.1.2.12b	0.15914	D_2	3.1.2.12b
0.15655	H_2	3.1.2.12b	0.15923	D_2	3.1.2.12b
0.15663	H_2	3.1.2.12b	0.15926	D_2	3.1.2.12b
0.15671	F_2	3.1.2.11	0.15934	H_2	3.1.2.12b
0.15673	H_2	3.1.2.12b	0.15934	para H_2	3.1.2.12b
0.15675	para H_2	3.1.2.12b	0.15938	HD	3.1.2.12b
0.15713	HD	3.1.2.12b	0.15955	HD	3.1.2.12b
0.15720	H_2	3.1.2.12b	0.15961	H_2	3.1.2.12b
0.15727	HD	3.1.2.12b	0.15974	HD	3.1.2.12b
0.1574	F_2	3.1.2.11	0.15993	para H_2	3.1.2.12b
0.15743	H_2	3.1.2.12b	0.16009	D_2	3.1.2.12b
0.15743	para H_2	3.1.2.12b	0.16021	D_2	3.1.2.12b

Table 4.1 (continued)
GAS LASERS ARRANGED IN ORDER OF WAVELENGTH

Wavelength (μm)	Element or compound	Table No.
0.16023	HD	3.1.2.12b
0.16024	para H_2	3.1.2.12b
0.16035	D_2	3.1.2.12b
0.16037	HD	3.1.2.12b
0.16046	HD	3.1.2.12b
0.16049	H_2	3.1.2.12b
0.16052	HD	3.1.2.12b
0.16057	HD	3.1.2.12b
0.16058	D_2	3.1.2.12b
0.16059	para H_2	3.1.2.12b
0.16059	H_2	3.1.2.12b
0.16062	H_2	3.1.2.12b
0.16062	para H_2	3.1.2.12b
0.16065	HD	3.1.2.12b
0.16065	D_2	3.1.2.12b
0.16067	HD	3.1.2.12b
0.16068	D_2	3.1.2.12b
0.16069	HD	3.1.2.12b
0.16075	HD	3.1.2.12b
0.16075	H_2	3.1.2.12b
0.16077	HD	3.1.2.12b
0.16079	HD	3.1.2.12b
0.16083	HD	3.1.2.12b
0.16083	H_2	3.1.2.12b
0.16083	para H_2	3.1.2.12b
0.16084	H_2	3.1.2.12b
0.16085	D_2	3.1.2.12b
0.16090	H_2	3.1.2.12b
0.16091	HD	3.1.2.12b
0.16096	D_2	3.1.2.12b
0.16103	HD	3.1.2.12b
0.16103	H_2	3.1.2.12b
0.16107	D_2	3.1.2.12b
0.16108	D_2	3.1.2.12b
0.16109	para H_2	3.1.2.12b
0.16113	HD	3.1.2.12b
0.16115	D_2	3.1.2.12b
0.16117	D_2	3.1.2.12b
0.16117	para H_2	3.1.2.12b
0.16117	H_2	3.1.2.12b
0.16120	D_2	3.1.2.12b
0.16124	D_2	3.1.2.12b
0.16126	D_2	3.1.2.12b
0.16132	D_2	3.1.2.12b
0.16132	para H_2	3.1.2.12b
0.16132	H_2	3.1.2.12b
0.16141	D_2	3.1.2.12b
0.16148	H_2	3.1.2.12b
0.16149	para H_2	3.1.2.12b
0.16165	H_2	3.1.2.12b
0.16166	D_2	3.1.2.12b
0.16395	H_2	3.1.2.12b
0.16415	H_2	3.1.2.12b
0.16429	para H_2	3.1.2.12b
0.16444	H_2	3.1.2.12b
0.16460	para H_2	3.1.2.12b
0.16900	ArCl	3.1.12
0.17000	Xe_2	2.30
0.17500	ArCl	3.1.1.2
0.175641	Kr^{3+}	2.11.3
0.181085	Cu	3.1.2.10b
0.183243	Kr^{4+}	2.11.3
0.184343	Ar^{4+}	2.11.2
0.187831	CO	3.1.2.10b
0.189784	CO	3.1.2.10b
0.1933	ArF	3.1.2.3
0.195006	CO	3.1.2.10b
0.195027	Kr^{3+}	2.11.3
0.196808	Kr^{3+}	2.11.3
0.197013	CO	3.1.2.10b
0.201842	Ne^{3+}	2.11.1
0.202219	Ne^{3+}	2.11.1
0.205108	Kr^{3+}	2.11.3
0.206530	Ne^{3+}	2.11.1
0.211398	Ar^{3+}	2.11.2
0.217770	Ne^{2+}	2.11.1
0.218086	Ne^{2+}	2.11.1
0.219192	Kr^{3+}	2.11.2
0.22200	KrCl	3.1.2.17
0.223244	Xe^{3+}	2.11.4
0.22434	Ag^{+}	2.4.2
0.224884	Ar^{3+}	2.11.2
0.225464	Kr^{3+}	2.11.3
0.22640	Au^{+}	2.4.3
0.22657	Ne^{4+}	2.11.1
0.22776	Ag^{+}	2.4.2
0.228579	Ne^{3+}	2.11.1
0.231536	Xe^{3+}	2.11.4
0.233848	Kr^{3+}	2.11.3
0.235255	Ne^{3+}	2.11.1

Table 4.1 (continued)
GAS LASERS ARRANGED IN ORDER OF WAVELENGTH

Wavelength (μm)	Element or compound	Table No.
0.235798	Ne^{3+}	2.11.1
0.236246	Br^{3+}	2.10.3
0.237320	Ne^{3+}	2.11.1
0.241784	Kr^{3+}	2.11.3
0.24700	CO^{+}	3.1.2.10c
0.247340	Ne^{2+}	2.11.1
0.247739	Xe^{2+}	2.11.4
0.24840	KrF	3.1.2.18
0.24858	Cu^{+}	2.4.1
0.24910	KrF	3.1.2.18
0.25065	Cu^{+}	2.4.1
0.251330	Ar^{3+}	2.11.3
0.252666	Xe^{3+}	2.11.4
0.25292	Cu^{+}	2.4.1
0.25335	Au^{+}	2.4.3
0.25800	Cl_2	3.1.2.7
0.258125	Br^{3+}	2.10.3
0.25906	Cu^{+}	2.4.1
0.25990	Cu^{+}	2.4.1
0.26003	Cu^{+}	2.4.1
0.260998	Ne^{2+}	2.11.1
0.26134	Ne^{2+}	2.11.1
0.26164	Au^{+}	2.4.3
0.262138	Ar^{3+}	2.11.2
0.264882	Ar^{3+}	2.11.2
0.263269	Cl^{2+}	2.10.2
0.263896	S_2^{+}	2.9.2
0.264936	Kr^{3+}	2.11.3
0.266440	Kr^{3+}	2.11.3
0.267792	Ne^{2+}	2.11.1
0.267869	Ne^{2+}	2.11.1
0.269194	Xe^{3+}	2.11.4
0.27007	Cu^{+}	2.4.1
0.27031	Cu^{+}	2.4.1
0.27220	Cu^{+}	2.4.1
0.274138	Kr^{3+}	2.11.3
0.275388	Ar^{2+}	2.11.2
0.275959	F_2^{+}	2.10.1
0.2777634	Ne^{2+}	2.11.1
0.27615	O^{4+}	2.9.1
0.278762	Br^{2+}	2.10.3
0.28180	XeBr	3.1.2.26
0.28222	Au^{+}	2.4.3
0.282608	F_3^{+}	2.10.1
0.28472	Au^{+}	2.4.3
0.28500	ClF	3.1.2.8
0.285537	Ar^{2+}	2.11.2
0.286673	Ne^{2+}	2.11.1
0.286422	Ar^{2+}	2.11.2
0.28935	Au^{+}	2.4.3
0.291292	Ar^{3+}	2.11.2
0.29150	Br_2	3.1.2.6
0.29181	Au^{+}	2.4.3
0.292623	Ar^{3+}	2.11.2
0.298389	O_2^{+}	2.9.1
0.299951	Fe	1.8.1
0.3002642	Ar^{2+}	2.11.2
0.301618	Fe	1.8.1
0.30240	Ar^{2+}	2.11.2
0.303164	Fe	1.8.1
0.304043	Fe	1.8.1
0.304715	O_2^{+}	2.9.1
0.3049704	Kr^{2+}	2.11.3
0.30548	Ar^{2+}	2.11.2
0.306346	O_3^{+}	2.9.1
0.30650	XeCl	3.1.2.27
0.30700	XeCl	3.1.2.27
0.30730	XeCl	3.1.2.27
0.30792	XeCl	3.1.2.27
0.307974	Xe^{2+}	2.11.4
0.30817	XeCl	3.1.2.27
0.30843	XeCl	3.1.2.27
0.312150	F_2^{+}	2.10.1
0.31220	Au	1.1.7
0.312436	Kr^{2+}	2.11.3
0.317418	F_2^{+}	2.10.1
0.31806	Au^{+}	2.4.2
0.319142	Cl^{2+}	2.10.2
0.320247	F^{+}	2.10.1
0.323951	Kr^{2+}	2.11.3
0.324692	Xe^{3+}	2.11.4
0.32500	CD^{+}	2.5.2
0.330596	Xe^{3+}	2.11.4
0.331975	Ne^{+}	2.11.1
0.332375	Ne^{+}	2.11.1
0.332486	S_2^{+}	2.9.2
0.33292	Ne^{+}	2.11.1
0.333087	Xe^{3+}	2.11.4

Table 4.1 (continued)
GAS LASERS ARRANGED IN ORDER OF WAVELENGTH

Wavelength (μm)	Element or compound	Table No.	Wavelength (μm)	Element or compound	Table No.
0.33311	Ne^{2+}	2.11.1	0.3370716	N_2	3.1.2.21a
0.333621	Ar^{2+}	2.11.2	0.3370731	N_2	3.1.2.21a
0.334479	Ar^{2+}	2.11.2	0.3370757	N_2	3.1.2.21a
0.334545	Ne^{+}	2.11.1	0.3370762	N_2	3.1.2.21a
0.334776	P^{3+}	2.8.2	0.3370787	N_2	3.1.2.21a
0.334974	Xe^{2+}	2.11.4	0.3370803	N_2	3.1.2.21a
0.335852	Ar^{2+}	2.11.2	0.3370821	N_2	3.1.2.21a
0.3364903	N_2	3.1.2.21a	0.3370843	N_2	3.1.2.21a
0.3365474	N_2	3.1.2.21a	0.3370924	N_2	3.1.2.21a
0.3365537	N_2	3.1.2.21a	0.3370941	N_2	3.1.2.21a
0.3366106	N_2	3.1.2.21a	0.3370990	N_2	3.1.2.21a
0.3366211	N_2	3.1.2.21a	0.3371042	N_2	3.1.2.21a
0.3366682	N_2	3.1.2.21a	0.3371082	N_2	3.1.2.21a
0.3366911	N_2	3.1.2.21a	0.3371120	N_2	3.1.2.21a
0.3367218	N_2	3.1.2.21a	0.3371129	N_2	3.1.2.21a
0.336732	N_2^{+}	2.8.1	0.3371141	N_2	3.1.2.21a
0.3368432	N_2	3.1.2.21a	0.3371147	N_2	3.1.2.21a
0.3368917	N_2	3.1.2.21a	0.3371179	N_2	3.1.2.21a
0.3369250	N_2	3.1.2.21a	0.3371271	N_2	3.1.2.21a
0.3369361	N_2	3.1.2.21a	0.3371312	N_2	3.1.2.21a
0.3369502	N_2	3.1.2.21a	0.3371371	N_2	3.1.2.21a
0.3369542	N_2	3.1.2.21a	0.3371398	N_2	3.1.2.21a
0.3369555	N_2	3.1.2.21a	0.3371427	N_2	3.1.2.21a
0.3369575	N_2	3.1.2.21a	0.3371433	N_2	3.1.2.21a
0.3369760	N_2	3.1.2.21a	0.33750	Kr^{2+}	2.11.3
0.3369838	N_2	3.1.2.21a	0.337826	Ne^{+}	2.11.1
0.3369852	N_2	3.1.2.21a	0.338134	O^{3+}	2.9.1
0.3369898	N_2	3.1.2.21a	0.338554	O^{3+}	2.9.1
0.3370081	N_2	3.1.2.21a	0.334280	Ne^{+}	2.11.1
0.3370121	N_2	3.1.2.21a	0.339286	Cl^{2+}	2.10.2
0.3370138	N_2	3.1.2.21a	0.33434	Ne^{+}	2.11.1
0.3370161	N_2	3.1.2.21a	0.339344	Cl^{2+}	2.10.2
0.3370169	N_2	3.1.2.21a	0.34200	I_2	3.1.2.16a
0.3370297	N_2	3.1.2.21a	0.34230	I_2	3.1.2.16a
0.3370316	N_2	3.1.2.21a	0.34240	I_2	3.1.2.16a
0.3370360	N_2	3.1.2.21a	0.34280	I_2	3.1.2.16a
0.3370374	N_2	3.1.2.21a	0.345132	B^{+}	2.6.1
0.3370434	N_2	3.1.2.21a	0.345425	Xe^{2+}	2.11.4
0.3370472	N_2	3.1.2.21a	0.347876	N^{3+}	2.8.1
0.3370529	N_2	3.1.2.21a	0.348302	N^{3+}	2.8.1
0.3370551	N_2	3.1.2.21a	0.348331	Xe^{3+}	2.11.4
0.3370559	N_2	3.1.2.21a	0.3488	XeF	3.1.2.28a
0.3370623	N_2	3.1.2.21a	0.349733	S_2^{+}	2.9.2
0.3370663	N_2	3.1.2.21a	0.350742	Kr^{2+}	2.11.4
0.3370682	N_2	3.1.2.21a	0.35111	XeF	3.1.2.28a

Table 4.1 (continued)
GAS LASERS ARRANGED IN ORDER OF WAVELENGTH

Wavelength (μm)	Element or compound	Table No.	Wavelength (μm)	Element or compound	Table No.
0.351112	Ar^{2+}	2.11.2	0.37738	O^{2+}	2.9.1
0.351415	Ar^{2+}	2.11.2	0.377572	Tl	1.3.4
0.353002	Cl^{2+}	2.10.2	0.378099	Xe^{2+}	2.11.4
0.35310	XeF	3.1.2.28a	0.379528	Ar^{2+}	2.11.2
0.35400	XeF	3.1.2.28a	0.380329	Xe^{3+}	2.11.4
0.354231	Xe^{2+}	2.11.4	0.38410	Xe^{2+}	2.11.4
0.356063	Cl^{2+}	2.10.2	0.385826	Ar^{2+}	2.11.2
0.356420	Kr^{2+}	2.11.3	0.39140	N_2^+	2.8.1
0.3575980	N_2	3.1.2.21a	0.39540	CO^+	3.1.2.10d
0.3576194	N_2	3.1.2.21a	0.397301	Xe^{3+}	2.11.4
0.3576650	N_2	3.1.2.21a	0.398399	Hg^+	2.5.3
0.3576320	N_2	3.1.2.21a	0.39930	Xe^{2+}	2.11.4
0.3576571	N_2	3.1.2.21a	0.399499	N^+	2.8.1
0.3576613	N_2	3.1.2.21a	0.402478	F^+	2.10.1
0.3576776	N_2	3.1.2.21a	0.4032987	Ga	1.3.2
0.3576899	N_2	3.1.2.21a	0.40450	K	1.1.3
0.35769	Ar^+	2.11.2	0.40470	K	1.1.3
0.3576955	N_2	3.1.2.21a	0.40499	Xe^{2+}	2.11.4
0.35866	Al^+	2.6.3	0.405779	Pd	1.4.5
0.35740	Al^+	2.6.3	0.406048	Xe^{2+}	2.11.4
0.360210	Cl^{2+}	2.10.2	0.40620	Pb	1.4.5
0.361283	Cl^{2+}	2.10.2	0.406736	Kr^{2+}	2.11.3
0.362289	Cl^{2+}	2.10.2	0.4078	Sr^+	2.1.4
0.363786	Ar^{2+}	2.11.2	0.40886	Ar	2.11.2
0.363954	Pb	1.4.5	0.40862	Ag^+	2.4.2
0.364548	Xe^{3+}	2.11.4	0.40889	Si^{3+}	2.7.2
0.36500	Hg	1.2.7	0.40970	N_2^+	2.8.1
0.365-0.570	S_2	3.1.2.24a	0.4101745	In	1.3.3
0.36600	Hg	1.2.7	0.410336	N_2^+	2.8.1
0.366920	Xe^{2+}	2.11.4	0.4113138	Kr^{2+}	2.11.3
0.37052	Ar^{2+}	2.11.2	0.413250	Cl^+	2.10.2
0.37060	Ca^+	2.1.3	0.41453	Xe^{2+}	2.11.4
0.370935	S_2^+	2.9.2	0.414660	Ar^{2+}	2.11.2
0.37130	Ne^+	2.11.1	0.415445	Kr^{2+}	2.11.3
0.372044	Cl^{2+}	2.10.2	0.4162	Sr^+	2.1.4
0.372711	O^+	2.9.1	0.417181	Kr^{2+}	2.11.3
0.37370	Ca^+	2.1.3	0.41720	Ga	1.3.2
0.374573	Xe^{2+}	2.11.4	0.418292	Ar^{2+}	2.11.2
0.374877	Cl^{2+}	2.10.2	0.42100	Rb	1.1.4
0.374947	O^+	2.9.1	0.42100	CO^+	3.1.2.10d
0.375468	O^{2+}	2.9.1	0.421405	Xe^{2+}	2.11.4
0.37572	O^{2+}	2.9.1	0.422225	P^{2+}	2.8.2
0.375989	O^{2+}	2.9.1	0.422651	Kr^{2+}	2.11.3
0.375994	Xe^{3+}	2.11.4	0.424026	Xe^{2+}	2.11.4
0.377134	Kr^+	2.11.3	0.427260	Xe^{2+}	2.11.4

Table 4.1 (continued)
GAS LASERS ARRANGED IN ORDER OF WAVELENGTH

Wavelength (μm)	Element or compound	Table No.
0.42781	N_2^+	3.1.2.22
0.428592	Xe^{2+}	2.11.4
0.429633	Xe^+	2.11.4
0.43000	Kr_2F	3.1.3.1
0.43050	Sr^+	2.1.4
0.430575	Xe^{3+}	2.11.4
0.431780	Kr^+	2.11.3
0.43400	H	1.1.1
0.434738	O^+	2.9.1
0.43500	$TbAl_3Cl_{12}$	3.1.4.3
0.435126	O^+	2.9.1
0.435835	Hg	1.2.7
0.437073	Ar^+	2.11.2
0.43836	Ar^+	2.11.2
0.43861	Kr^+	2.11.3
0.44130	Xe^{2+}	2.11.4
0.441439	O^+	2.9.1
0.44156	Cu^+	2.5.2
0.441697	O^+	2.9.1
0.44300	HgI	3.1.2.15
0.443422	Xe^{2+}	2.11.4
0.444328	Kr^{2+}	2.11.3
0.44500	HgI	3.1.2.15
0.44680	Se^+	2.9.3
0.44812	Ar^+	2.11.2
0.448855	I^+	2.10.4
0.45035	Xe^{2+}	2.11.4
0.45066	Cu^+	2.4.1
0.451089	N_2^+	2.8.1
0.45110	In	1.3.3
0.451486	N_2^+	2.8.1
0.45290	Fe	1.8.1
0.453379	I^{2+}	2.10.4
0.454504	Ar^+	2.11.2
0.455259	Si^{2+}	2.7.2
0.45550	Cs	1.1.5
0.455592	Cu^+	2.4.1
0.45563	Cu^+	2.4.1
0.455879	Xe^{3+}	2.11.4
0.45607	Bi^{2+}	2.8.5
0.456784	Si^{2+}	2.7.2
0.457720	Kr^+	2.11.3
0.457936	Ar^+	2.11.2
0.45830	Kr^+	2.11.3
0.460302	Xe^+	2.11.4
0.46046	Se^+	2.9.3
0.460552	O^+	2.9.1
0.460957	Ar^+	2.11.2
0.461528	Kr^+	2.11.3
0.46191	Se^+	2.9.3
0.461917	Kr^+	2.11.3
0.46210	N^+	2.8.1
0.463051	N^+	2.8.1
0.463392	Kr^+	2.11.3
0.46439	N^+	2.8.1
0.464740	C^{2+}	2.7.1
0.464740	Xe^{3+}	2.11.4
0.46486	Se^+	2.9.3
0.464908	O^+	2.9.1
0.465021	O^{2+}	2.7.1
0.465016	Kr^+	2.11.3
0.465025	Xe^{2+}	2.11.4
0.465795	Ar^+	2.11.2
0.46732	Cu^+	2.4.1
0.467373	Xe^{2+}	2.11.4
0.467440	I^{2+}	2.10.4
0.46750	Be^+	2.1.1
0.46756	I^+	2.10.4
0.468045	Kr^+	2.11.3
0.468050	In^+	2.6.3
0.46829	Cu^+	2.4.1
0.468357	Xe^{2+}	2.11.4
0.46941	Kr^+	2.11.3
0.47090	N_2^+	2.22
0.47103	Kr^{2+}	2.11.3
0.47185	Se^+	2.9.3
0.47220	Bi	1.5.5
0.472357	Xe^{2+}	2.11.4
0.472689	Ar^+	2.11.2
0.47370	Tl^+	2.6.4
0.474040	Cl^+	2.10.2
0.47406	Se^+	2.9.3
0.474266	Br^+	2.10.3
0.474894	Xe^{2+}	2.11.4
0.47545	Kr^{2+}	2.11.3
0.476244	Kr^{2+}	2.11.3
0.47641	Se^+	2.9.3
0.476488	Ar^+	2.11.2
0.47651	Se^+	2.9.3
0.476571	Kr^+	2.11.3

Table 4.1 (continued)
GAS LASERS ARRANGED IN ORDER OF WAVELENGTH

Wavelength (μm)	Element or compound	Table No.
0.476874	Cl^+	2.10.2
0.478134	Cl^+	2.10.2
0.47884	Ag^+	2.4.2
0.47945	Xe^{2+}	2.11.4
0.47963	Kr^+	2.11.3
0.47970	Hg^{2+}	2.5.3
0.48251	Kr^+	2.11.3
0.48300	XeF	3.1.2.28b
0.48406	Se^+	2.9.3
0.48433	Te^+	2.9.4
0.484550	Se^+	2.9.3
0.484666	Kr^+	2.11.3
0.48558	Cu^+	2.4.1
0.48600	XeF	3.1.2.28b
0.48610	H	1.1.1
0.48620	Xe^+	2.11.4
0.486948	Xe^{2+}	2.11.4
0.487986	Ar^+	2.11.2
0.48820	Cd^+	2.5.1
0.48870	Xe^+	2.11.4
0.488906	Ar^+	2.11.2
0.48884122	Se	1.6.3
0.489668	Cl^+	2.10.2
0.490473	Cl^+	2.10.2
0.49097	Cu^+	2.4.1
0.49116	Zn^+	2.5.1
0.491766	Cl^+	2.10.2
0.49250	Zn^+	2.5.1
0.492560	S^+	2.9.2
0.49317	Cu^+	2.4.1
0.493467	I^{2+}	2.10.4
0.4954148	Xe^{3+}	2.11.4
0.49650	Ar^+	2.11.2
0.496508	Xe^+	2.11.4
0.497271	Xe^+	2.11.4
0.49761	Se^+	2.9.3
0.49810	Tl^+	2.6.4
0.49867	I^+	2.10.4
0.49902	HgBr	3.1.2.13
0.49925	Ar^{2+}	2.11.2
0.500780	Xe^{3+}	2.11.4
0.50133	Cu^+	2.4.1
0.501424	S^+	2.9.2
0.501639	N^+	2.8.1
0.50164	Kr^{2+}	2.11.3
0.501717	Ar^+	2.11.2
0.50180	HgBr	3.1.2.13
0.50200	Te^+	2.9.4
0.50219	Cu^+	2.4.1
0.50220	Kr^+	2.11.3
0.5023	HgBr	3.1.2.13
0.50259	Cd^+	2.5.1
0.50260	HgBr	3.1.2.13
0.50272	Ag^+	2.4.2
0.503262	S^+	2.9.2
0.50375	Kr^+	2.11.3
0.50390	HgBr	3.1.2.13
0.50420	HgBr	3.1.2.13
0.504489	Xe^+	2.11.4
0.50460	HgBr	3.1.2.13
0.50521	Cu^+	2.4.1
0.505463	Br^+	2.10.3
0.50605	Cu^+	2.4.1
0.506210	Ar^+	2.11.2
0.50687	Se^+	2.9.3
0.507830	Cl^+	2.10.2
0.50790	Tl^+	2.6.4
0.50961	Se^+	2.9.3
0.510310	Cl^+	2.10.2
0.510554	Cu	1.1.6
0.51260	Kr^+	2.11.3
0.51315	Ge^+	2.7.3
0.514180	Ar^+	2.11.2
0.51419	Se^+	2.9.3
0.514533	Ar^+	2.11.2
0.51457	C^+	2.7.1
0.51520	Tl^+	2.6.4
0.515704	Xe^{3+}	2.11.4
0.515908	Xe^{3+}	2.11.4
0.516032	S^+	2.9.2
0.51760	Se^+	2.9.3
0.51784	Ge^+	2.7.3
0.517959	CO	3.1.3.2
0.51800	Xe_2Cl	3.1.3.2
0.518238	Br^+	2.10.3
0.520832	Kr^+	2.11.3
0.52100	Hg^{2+}	2.5.3
0.52143	I^+	2.10.4
0.52163	I^+	2.10.4
0.521790	Cl^+	2.10.2

Table 4.1 (continued)
GAS LASERS ARRANGED IN ORDER OF WAVELENGTH

Wavelength (μm)	Element or compound	Table No.
0.52182	Kr^+	2.11.3
0.521962	S^+	2.9.2
0.522130	Cl^+	2.10.2
0.52234	Xe^{2+}	2.11.4
0.52276	Se^+	2.9.3
0.52280	N_2^+	3.1.2.22
0.523826	Br^+	2.10.3
0.523889	Xe^{2+}	2.11.4
0.52450	Na_2	3.1.2.23a
0.52526	Se^+	2.9.3
0.52532	Se^+	2.9.3
0.52564	Te^+	2.9.4
0.52565	Xe^{3+}	2.11.4
0.526017	Xe^{3+}	2.11.4
0.526043	Xe^+	2.11.4
0.52615	Xe^+	2.11.4
0.52633	Na_2	3.1.2.23a
0.52713	Se^+	2.9.3
0.52720	Be^+	2.11
0.52740	Na_2	3.1.2.23a
0.52870	Ar^+	2.11.2
0.52982	Na_2	3.1.2.23a
0.52995	Na_2	3.1.2.23a
0.53055	Se^+	2.9.3
0.530868	Kr^+	2.11.3
0.53140	Xe^+	2.11.4
0.532088	S^+	2.9.2
0.53210	Na_2	3.1.2.23a
0.533203	Br^+	2.10.3
0.53330	Na_2	3.1.2.23a
0.53370	Cd^+	2.5.2
0.53380	Cd^+	2.5.2
0.53383	Na_2	3.1.2.23a
0.53400	Na_2	3.1.2.23a
0.534106	Mn	1.7.6
0.53415	Na_2	3.1.2.23b
0.53428	Na_2	3.1.2.23b
0.534331	Xe^{3+}	2.11.4
0.534334	Xe^{3+}	2.11.4
0.53450	Na_2	3.1.2.23b
0.534583	S^+	2.4.2
0.53490	Na_2	3.1.2.23a
0.53500	Ca	1.2.2
0.53503	Tl	1.3.4
0.535292	Xe^{3+}	2.11.4
0.53670	Xe^{2+}	2.11.4
0.53690	Na_2	3.1.2.23a
0.53721	Pb^+	2.7.5
0.53700	Na_2	3.1.2.23a
0.53710	Na_2	3.1.2.23a
0.53760	Na_2	2.1.2.23a
0.53760	XeO	3.1.2.29
0.53781	Na_2	3.1.2.23a
0.53850	Na_2	3.1.2.23b
0.53851	As^+	2.8.3
0.53863	Na_2	3.1.2.23b
0.5389	Na_2	3.1.2.23a
0.539215	Cl^+	2.10.2
0.539460	Xe^{3+}	2.11.4
0.54009	Xe^{2+}	2.11.4
0.54024	Na_2	3.1.2.23a
0.54075	I^+	2.10.4
0.54106	Na_2	3.1.2.23a
0.541311	Na_2	3.1.2.23a
0.54135	Xe^{2+}	2.11.4
0.541916	Xe^+	2.11.4
0.542036	Mn	1.7.6
0.542874	S^+	2.9.2
0.543287	S^+	2.9.2
0.5442	XeO	3.1.2.29
0.5443	I_2	3.1.2.16b
0.54469	Na_2	3.1.2.23a
0.54480	Na_2	3.1.2.23a
0.54498	Tl^+	2.9.4
0.545	$TbAl_3Cl_{12}$	3.1.4.3
0.545294	Eu	1.8.4
0.54538	S^+	2.9.2
0.54540	Na_2	3.1.2.23a
0.54540	Te^+	2.9.4
0.54546	Xe^{2+}	2.11.4
0.54610	Hg	1.2.7
0.54670	Na_2	3.1.2.23a
0.54690	Na_2	3.1.2.23a
0.547064	Mn	1.7.6
0.547374	S^+	2.9.2
0.54793	Tl^+	2.9.4
0.54800	Na_2	3.1.2.23a
0.5481345	Mn	1.7.6
0.54850	Na_2	3.1.2.23a
0.54916	Na_2	3.1.2.23a

Table 4.1 (continued)
GAS LASERS ARRANGED IN ORDER OF WAVELENGTH

Wavelength (μm)	Element or compound	Table No.	Wavelength (μm)	Element or compound	Table No.
0.54968	As^+	2.8.3	0.56053	CO	3.1.2.10a
0.54970	As^+	2.8.3	0.56086	Pb^+	2.7.5
0.54990	Na_2	3.1.2.23a	0.56228	Se^+	2.9.3
0.549931	Xe^{3+}	2.11.4	0.56250	I^+	2.10.4
0.55015	Kr^{2+}	2.11.3	0.56260	Te_2	3.1.2.25
0.55022	Ar^{2+}	2.11.2	0.5638	Te_2	3.1.2.25
0.550990	S^+	2.9.2	0.564012	S^+	2.9.2
0.55160	HgCl	3.1.2.14	0.56405	Te^+	2.9.4
0.551677	Mn	1.7.6	0.56420	Te_2	3.1.2.25
0.55228	Se^+	2.9.3	0.56430	Te_2	3.1.2.25
0.5523	HgCl	3.1.2.14	0.56460	Te_2	3.1.2.25
0.55245	Xe^{2+}	2.11.4	0.56470	Te_2	3.1.2.25
0.553776	Mn	1.7.6	0.564716	S^+	2.9.2
0.5550	HgCl	3.1.2.14	0.56490	Te_2	3.1.2.25
0.5550	I_2	3.1.2.16b	0.56500	Te_2	3.1.2.25
0.55582	As^+	2.8.3	0.56520	As^+	2.8.2
0.556511	S^+	2.9.2	0.56590	Xe^+	2.11.4
0.5567	I_2	3.1.2.16b	0.56661	Te^+	2.9.4
0.55671	Se^+	2.9.3	0.566662	N^+	2.8.1
0.5571	Tl_2	3.1.2.25	0.567603	N^+	2.8.1
0.5575	Tl_2	3.1.2.25	0.56780	Hg^+	2.5.3
0.5576	HgCl	3.1.2.14	0.56782	I^+	2.10.4
0.557625	Tl^+	2.9.4	0.567953	N^+	2.8.1
0.557714	Cu	1.8.4	0.56800	I_2	3.1.2.16b
0.55780	Tl_2	3.1.2.25	0.568192	Kr^+	2.11.3
0.55781	KrO	3.1.2.20	0.56869	N^+	2.8.1
0.55788939	O	1.6.1	0.56960	Te_2	3.1.2.25
0.55790	Te_2	3.1.22.25	0.56970	I_2	3.1.2.16b
0.55800	ArO	3.1.2.4	0.56979	Se^+	2.9.3
0.55810	Na_2	3.1.2.23a	0.570024	Cu	1.1.6
0.55840	HgCl	3.1.2.14	0.5701	Te_2	3.1.2.25
0.55890	Sn^+	2.7.4	0.57085	Te^+	2.9.4
0.5590	HgCl	3.1.2.14	0.57110	Te_2	3.1.2.25
0.5591	Na_2	3.1.2.23a	0.57140	Te_2	3.1.2.25
0.5591	Se^+	2.9.3	0.57150	Te_2	3.1.2.25
0.55921	CO	3.1.2.10a	0.57190	Te_2	3.1.2.25
0.559235	Xe^{3+}	2.11.4	0.57192	Bi^+	2.8.5
0.559237	O^{2+}	2.9.1	0.57200	Te_2	3.1.2.25
0.55931	I^+	2.10.4	0.57210	Te_2	3.1.2.25
0.55949	CO	3.1.2.10a	0.57240	Te_2	3.1.2.25
0.55970	Na_2	3.1.2.23a	0.57270	Xe^+	2.11.4
0.55977	Kr^{2+}	2.11.3	0.57415	Te^+	2.9.4
0.5598	CO	3.1.2.10a	0.5745	I_2	3.1.2.16b
0.56019	CO	3.1.2.10a	0.57479	Se^+	2.9.3
0.56040	CO	3.1.2.10a	0.57510	Xe^+	2.11.4

Table 4.1 (continued)
GAS LASERS ARRANGED IN ORDER OF WAVELENGTH

Wavelength (μm)	Element or compound	Table No.	Wavelength (μm)	Element or compound	Table No.
0.57534	Kr^+	2.11.3	0.59050	I_2	3.1.2.16b
0.57557	Te^+	2.9.4	0.59240	Te_2	3.1.2.25
0.57607	I^+	2.10.4	0.59270	Te_2	3.1.2.25
0.57640	I_2	3.1.2.16b	0.59290	Bi_2	3.1.2.5
0.57649	Te^+	2.9.4	0.59340	Te_2	3.1.2.25
0.57660	Te_2	3.1.2.25	0.59353	$Kr^{+,2+}$	2.11.3
0.57670	Te_2	3.1.2.25	0.59360	Te_2	3.1.2.25
0.576959	Hg	1.2.7	0.59365	Te^+	2.9.4
0.57730	Te_2	3.1.2.25	0.59490	Tl^+	2.6.4
0.57740	Te_2	3.1.2.25	0.595567	Xe^{3+}	2.11.4
0.57800	Te_2	3.1.2.25	0.59680	I_2	3.1.2.16b
0.57820	Cu	1.1.6	0.597112	Xe^+	2.11.4
0.57830	Te_2	3.1.2.25	0.59723	Te^+	2.9.4
0.57840	Te_2	3.1.2.25	0.59743	Te^+	2.9.4
0.57850	Te_2	3.1.2.25	0.60020	Te_2	3.1.2.25
0.57870	Te_2	3.1.2.25	0.60040	Te_2	3.1.2.25
0.57890	Te_2	3.1.2.25	0.60050	Te_2	3.1.2.25
0.57900	Te_2	3.1.2.25	0.60080	Te_2	3.1.2.25
0.579065	Hg	1.2.7	0.60090	Te_2	3.1.2.25
0.57930	Te_2	3.1.2.25	0.60149	Te^+	2.9.4
0.57940	Te_2	3.1.2.25	0.60210	Zn^+	2.5.1
0.57970	Te_2	3.1.2.25	0.602427	P^+	2.8.2
0.57980	Te_2	3.1.2.25	0.60250	I_2	3.1.2.16b
0.57987	Sr^+	2.7.4	0.603419	P^+	2.8.2
0.58150	I_2	3.1.2.16b	0.60376	$Kr^{2+,3+}$	2.11.3
0.581935	S^+	2.9.2	0.604322	P^+	2.8.2
0.58300	I_2	3.1.2.16b	0.60480	I_2	3.1.2.16b
0.58380	As^+	2.8.3	0.60563	Se^+	2.9.3
0.58410	Te_2	3.1.2.25	0.60500	Eu	1.8.4
0.58428	Se^+	2.9.3	0.60646	CO	3.1.2.10a
0.58490	Te_2	3.1.2.25	0.60651	Se^+	2.9.3
0.58510	Te^+	2.9.4	0.60674	CO	3.1.2.10a
0.58570	Te_2	3.1.2.25	0.60690	I^+	2.10.4
0.58590	Te_2	3.1.2.25	0.60699	CO	3.1.2.10a
0.58600	Ca	1.2.2	0.60720	Kr	2.11.3
0.58650	Te_2	3.1.2.25	0.60722	CO	3.1.2.10a
0.58660	Na	1.1.2	0.60742	CO	3.1.2.10a
0.58670	Se^+	2.9.3	0.60759	CO	3.1.2.10a
0.58690	Te_2	3.1.2.25	0.60820	Te_2	3.1.2.25
0.58700	Te_2	3.1.2.25	0.60824	Te^+	2.9.4
0.58740	Te_2	3.1.2.25	0.60830	Te_2	3.1.2.25
0.58800	I_2	3.1.2.16b	0.60850	Te_2	3.1.2.25
0.58900	Na	1.1.2	0.60870	Te_2	3.1.2.25
0.589330	Xe^+	2.11.4	0.608504	P^+	2.8.2
0.58940	Zn^+	2.5.1	0.60890	Te_2	3.1.2.25

Table 4.1 (continued)
GAS LASERS ARRANGED IN ORDER OF WAVELENGTH

Wavelength (μm)	Element or compound	Table No.
0.60940	Xe^+	2.11.4
0.609474	Cl^+	2.10.2
0.610121	Se^+	2.9.3
0.61028	Zn^+	2.5.1
0.61101	I_2	3.1.2.16b
0.611756	Br^+	2.10.3
0.61197087	Ne	1.9.2
0.61274	I^+	2.10.4
0.61300	Sb^+	2.8.4
0.6141720	Ba^+	2.1.5
0.61431	Ne	1.9.2
0.61500	Hg^+	2.5.3
0.61600	Bi_2	3.1.2.5
0.61620	Te_2	3.1.2.25
0.61650	Te_2	3.1.2.25
0.616574	P^+	2.8.2
0.61680	Te_2	3.1.2.25
0.616878	Br^+	2.10.3
0.61688	Kr^+	2.11.3
0.61700	Te_2	3.1.2.25
0.61702	As^+	2.8.3
0.61750	I_2	3.1.2.16b
0.61752	I_2	3.1.2.16b
0.617619	Xe^{2+}	2.11.4
0.617730	I_2	3.1.2.16b
0.617900	I_2	3.1.2.16b
0.617970	I_2	3.1.2.16b
0.618245	I_2	3.1.2.16b
0.618325	I_2	3.1.2.16b
0.618490	I_2	3.1.2.16b
0.618580	I_2	3.1.2.16b
0.61980	I_2	3.1.2.16b
0.62040	Te_2	3.1.2.25
0.62049	I^+	2.10.4
0.62140	Zn^+	2.5.1
0.62304	Te^+	2.9.4
0.62389	Xe^{2+}	2.11.4
0.62390	Bi_2	3.1.2.5
0.6239651	F	1.7.1
0.624545	Tl^+	2.9.4
0.62580	I_2	3.1.2.16b
0.62600	I_2	3.1.2.16b
0.62664950	Ne	1.9.2
0.62700	Xe^+	2.11.4
0.62780	Te_2	3.1.2.25
0.627818	Au	1.1.7
0.62866	Xe^{3+}	2.11.4
0.62870	Te_2	3.1.2.25
0.62880	Te_2	3.1.2.25
0.6293447	Ne	1.9.2
0.62950	Te_2	3.1.2.25
0.63000	Bi_2	3.1.2.5
0.63103	Kr^{2+}	2.11.3
0.63126	Kr^+	2.11.3
0.63282	Ne	1.9.2
0.63300	I_2	3.1.2.16b
0.63390	Bi_2	3.1.2.5
0.63399	I^+	2.10.4
0.634318	Xe^{2+}	2.11.4
0.634724	Si^+	2.7.2
0.6348508	F	1.7.1
0.63497	Te^+	1.6.4
0.63518	Ne	1.8.2
0.63520	I_2	3.1.2.16b
0.63548	Cd^+	2.5.2
0.635737	I^+	2.10.4
0.63601	Cd^+	2.5.2
0.63710	Te_2	3.1.2.25
0.637148	Si^+	2.7.2
0.63790	Te_2	3.1.2.25
0.638075	Sr	1.2.3
0.63810	Te_2	3.1.2.25
0.63880	Te_2	3.1.2.25
0.64028455	Ne	1.9.2
0.64043	Ag^+	2.4.2
0.6413651	F	1.7.1
0.64140	Bi_2	3.1.2.5
0.64170	Kr^+	2.11.3
0.64220	Bi_2	3.1.2.5
0.64439	Se^+	2.9.3
0.644981	Ca	1.2.2
0.64530	Sn^+	2.7.4
0.64710	Kr^+	2.11.3
0.64770	Te_2	3.1.2.25
0.64826	N^+	2.8.1
0.64828	Ar^+	2.11.2
0.64840	Te_2	3.1.2.25
0.648897	I^+	2.10.4
0.64900	I_2	3.1.2.16b
0.64901	Se^+	2.9.3

Table 4.1 (continued)
GAS LASERS ARRANGED IN ORDER OF WAVELENGTH

Wavelength (μm)	Element or compound	Table No.
0.649690	Ba^+	2.1.5
0.65016	Hg^{2+}	2.5.3
0.65100	Kr^+	2.11.3
0.65110	I_2	3.1.2.16b
0.65118	As^+	2.8.3
0.65162	I^+	2.10.4
0.65285	Xe^+	2.11.4
0.65346	Se^+	2.9.3
0.65610	Te_2	3.1.2.25
0.65700	Kr^+	2.11.3
0.65740	Te_2	3.1.2.25
0.65760	Bi_2	3.1.2.5
0.65780	C^+	2.7.1
0.657903	Sn	1.4.8
0.65810	Te_2	3.1.2.5
0.65820	Bi_2	3.1.2.5
0.65850	Te^+	2.9.4
0.65853	I^+	2.10.4
0.65920	I_2	3.1.2.16b
0.66013	CO	3.1.2.10a
0.66028	$Kr^{+,2+}$	2.11.3
0.66030	Bi_2	3.1.2.5
0.66049	CO	3.1.2.10a
0.66082	CO	3.1.2.10a
0.66109	CO	3.1.2.10a
0.66133	CO	3.1.2.10a
0.66153	CO	3.1.2.10a
0.66225	I^+	2.10.4
0.66402	O^+	2.9.1
0.66450	Eu^+	2.1.2.1
0.66450	I_2	3.1.2.16b
0.66462	Te^+	2.9.4
0.66500	Bi_2	3.1.2.5
0.66601	Pb^+	2.7.5
0.666694	O^+	2.9.1
0.667193	Si^+	2.7.2
0.667193	P	1.5.2
0.667227	I^+	2.10.4
0.66765	Te^+	2.9.4
0.66943	Xe^+	2.11.4
0.66995	Xe^{3+}	2.11.4
0.67020	Xe^+	2.11.4
0.672138	O^+	2.9.1
0.67300	Ar^+	2.11.2
0.67630	I_2	3.1.2.16b

Wavelength (μm)	Element or compound	Table No.
0.676457	Kr^+	2.11.3
0.67836	C^+	2.7.1
0.68090	Bi_2	3.1.2.5
0.68252	I^+	2.10.4
0.68440	Sn^+	2.7.4
0.68611	Ar^+	2.11.2
0.687096	Kr^+	2.11.3
0.68853	Te^+	2.9.4
0.69040	I^+	2.10.4
0.691998	Al^+	2.6.2
0.69360	I_2	3.1.2.16b
0.69510	Tl^+	2.6.4
0.696635	F	1.7.1
0.69960	Sb^+	2.8.4
0.70060	Bi_2	3.1.2.5
0.70130	Bi_2	3.1.2.5
0.70330	I^+	2.10.4
0.703745	F	1.7.1
0.70392	Te^+	2.9.4
0.704209	Al^+	2.6.2
0.70564	Al^+	2.6.2
0.706342	Se^+	2.9.3
0.7067124	He	1.9.1
0.70691661	Ar	1.9.3
0.7067162	He	1.9.1
0.70723	Xe^+	2.11.4
0.71025	As^+	2.8.3
0.71140	I_2	3.1.2.16b
0.712033	Ba	1.2.4
0.712788	F	1.7.1
0.7133974	Cd	1.2.6
0.71390	I^+	2.10.4
0.714894	Xe^{2+}	2.11.4
0.720237	F	1.7.1
0.72369	CO^+	2.5.2
0.72560	Cu^+	2.4.1
0.72843	Cd^+	2.5.2
0.72920	Bi_2	3.1.2.5
0.73010	Bi_2	3.1.2.5
0.73068569	Ne	1.9.2
0.7309033	F	1.7.1
0.73350	Bi_2	3.1.2.5
0.73460	Hg^+	2.5.3
0.734804	Ar^+	2.11.2
0.73640	Bi_2	3.1.2.5

Table 4.1 (continued)
GAS LASERS ARRANGED IN ORDER OF WAVELENGTH

Wavelength (μm)	Element or compound	Table No.	Wavelength (μm)	Element or compound	Table No.
0.73660	Bi_2	3.1.2.5	0.7497524	N_2	3.1.2.21b
0.73760	Bi_2	3.1.2.5	0.7497728	N_2	3.1.2.21b
0.739124	Se^+	2.9.3	0.7498898	N_2	3.1.2.21b
0.73980	Bi_2	3.1.2.5	0.7499013	N_2	3.1.2.21b
0.7398688	F	1.7.1	0.7499327	N_2	3.1.2.21b
0.74002	Cu^+	2.4.1	0.7499593	N_2	3.1.2.21b
0.74045	Cu^+	2.4.1	0.7499825	N_2	3.1.2.21b
0.74080	Bi_2	3.1.2.5	0.7500071	N_2	3.1.2.21b
0.74181	Hg^+	2.5.3	0.7500640	N_2	3.1.2.21b
0.74257	F	1.7.1	0.7500734	N_2	3.1.2.21b
0.74356	Kr^+	2.11.3	0.7501050	N_2	3.1.2.21b
0.74390	Cu^+	2.4.1	0.7501295	N_2	3.1.2.21b
0.74390	Bi_2	3.1.2.5	0.7501404	N_2	3.1.2.21b
0.74680	Bi_2	3.1.2.5	0.7501553	N_2	3.1.2.21b
0.74710	Bi_2	3.1.2.5	0.7502134	N_2	3.1.2.21b
0.747149	Al^+	2.6.2	0.7502729	N_2	3.1.2.21b
0.74750	Bi_2	3.1.2.5	0.7502768	N_2	3.1.2.21b
0.74783	Zn^+	2.5.1	0.7503035	N_2	3.1.2.21b
0.74820	Bi_2	3.1.2.5	0.7503371	N_2	3.1.2.21b
0.7482187	N_2	3.1.2.21b	0.7503418	N_2	3.1.2.21b
0.74827	F	1.7.1	0.7503642	N_2	3.1.2.21b
0.7485941	N_2	3.1.2.21b	0.7503669	N_2	3.1.2.21b
0.7486135	N_2	3.1.2.21b	0.7503697	N_2	3.1.2.21b
0.7486253	N_2	3.1.2.21b	0.7503838	N_2	3.1.2.21b
0.7486413	N_2	3.1.2.21b	0.7503960	N_2	3.1.2.21b
0.7487409	N_2	3.1.2.21b	0.7503994	N_2	3.1.2.21b
0.7488046	N_2	3.1.2.21b	0.7504106	N_2	3.1.2.21b
0.7488246	N_2	3.1.2.21b	0.7504160	N_2	3.1.2.21b
0.7489107	N_2	3.1.2.21b	0.7504184	N_2	3.1.2.21b
0.748914	F	1.7.1	0.7504274	N_2	3.1.2.21b
0.7489626	N_2	3.1.2.21b	0.7504598	N_2	3.1.2.21b
0.7489809	N_2	3.1.2.21b	0.7504768	N_2	3.1.2.21b
0.7490096	N_2	3.1.2.21b	0.750508	Ar^+	2.11.2
0.7490317	N_2	3.1.2.21b	*0.75059341*	Ar	1.9.3
0.4791510	N_2	3.1.2.21b	0.7505113	N_2	3.1.2.21b
0.7491705	N_2	3.1.2.21b	0.7505710	N_2	3.1.2.21b
0.7492379	N_2	3.1.2.21b	0.7505903	N_2	3.1.2.21b
0.7493082	N_2	3.1.2.21b	0.7506063	N_2	3.1.2.21b
0.7493716	N_2	3.1.2.21b	0.7506356	N_2	3.1.2.21b
0.7493910	N_2	3.1.2.21b	0.7508145	N_2	3.1.2.21b
0.7495086	N_2	3.1.2.21b	0.7509890	N_2	3.1.2.21b
0.7495465	N_2	3.1.2.21b	0.7510133	N_2	3.1.2.21b
0.7495660	N_2	3.1.2.21b	0.7510923	N_2	3.1.2.21b
0.7496024	N_2	3.1.2.21b	0.7511592	N_2	3.1.2.21b
0.7497256	N_2	3.1.2.21b	0.7512799	N_2	3.1.2.21b

Table 4.1 (continued)
GAS LASERS ARRANGED IN ORDER OF WAVELENGTH

Wavelength (μm)	Element or compound	Table No.	Wavelength (μm)	Element or compound	Table No.
0.7513003	N_2	3.1.2.21b	0.7622959	N_2	3.1.2.21b
0.7513569	N_2	3.1.2.21b	0.7623256	N_2	3.1.2.21b
0.7514357	N_2	3.1.2.21b	0.7623311	N_2	3.1.2.21b
0.75150	F	1.7.1	0.7623582	N_2	3.1.2.21b
0.7515079	N_2	3.1.2.21b	0.7623686	N_2	3.1.2.21b
0.7515446	N_2	3.1.2.21b	0.7623916	N_2	3.1.2.21b
0.7515650	N_2	3.1.2.21b	0.7624220	N_2	3.1.2.21b
0.7517728	N_2	3.1.2.21b	0.7624690	N_2	3.1.2.21b
0.7518013	N_2	3.1.2.21b	0.7624924	N_2	3.1.2.21b
0.75255	Kr^+	2.11.3	0.7625115	N_2	3.1.2.21b
0.75430	Bi_2	3.1.2.5	0.7625445	N_2	3.1.2.21b
0.75510	Bi_2	3.1.2.5	0.7625709	N_2	3.1.2.21b
0.7552235	F	1.7.1	0.7625770	N_2	3.1.2.21b
0.75560	Au^+	2.4.3	0.7625812	N_2	3.1.2.21b
0.7574329	N_2	3.1.2.21b	0.7625906	N_2	3.1.2.21b
0.7586439	N_2	3.1.2.21b	0.7626007	N_2	3.1.2.21b
0.75875	Zn^+	2.5.1	0.7626044	N_2	3.1.2.21b
0.7587693	N_2	3.1.2.21b	0.7626114	N_2	3.1.2.21b
0.75987	Bi^{2+}	2.8.5	0.7626180	N_2	3.1.2.21b
0.76004	Au^+	2.4.3	0.7626207	N_2	3.1.2.21b
0.76030	Kr	1.9.4	0.7626360	N_2	3.1.2.21b
0.7603477	N_2	3.1.2.21b	0.7626560	N_2	3.1.2.21b
0.7606374	N_2	3.1.2.21b	0.7626700	N_2	3.1.2.21b
0.7607626	N_2	3.1.2.21b	0.7626749	N_2	3.1.2.21b
0.7608801	N_2	3.1.2.21b	0.7626826	N_2	3.1.2.21b
0.7609853	N_2	3.1.2.21b	0.7628854	N_2	3.1.2.21b
0.7610759	N_2	3.1.2.21b	0.7629102	N_2	3.1.2.21b
0.7611082	N_2	3.1.2.21b	0.7630305	N_2	3.1.2.21b
0.761118	Zn^+	2.5.1	0.7631880	N_2	3.1.2.21b
0.7611514	N_2	3.1.2.21b	0.7632446	N_2	3.1.2.21b
0.7612105	N_2	3.1.2.21b	0.7633348	N_2	3.1.2.21b
0.7612526	N_2	3.1.2.21b	0.7633985	N_2	3.1.2.21b
0.7613260	N_2	3.1.2.21b	0.7634546	N_2	3.1.2.21b
0.7615347	N_2	3.1.2.21b	0.7634779	N_2	3.1.2.21b
0.7616994	N_2	3.1.2.21b	0.7635474	N_2	3.1.2.21b
0.7617357	N_2	3.1.2.21b	0.7636126	N_2	3.1.2.21b
0.76185	I^+	2.10.4	0.7636904	N_2	3.1.2.21b
0.76190	Xe^+	2.11.4	0.7637586	N_2	3.1.2.21b
0.7619288	N_2	3.1.2.21b	0.7638274	N_2	3.1.2.21b
0.7621029	Rb	1.1.4	0.7639571	N_2	3.1.2.21b
0.7620844	N_2	3.1.2.21b	0.7639715	N_2	3.1.2.21b
0.7620943	N_2	3.1.2.21b	0.7640383	N_2	3.1.2.21b
0.7621161	N_2	3.1.2.21b	0.7640794	N_2	3.1.2.21b
0.7622235	N_2	3.1.2.21b	0.7641929	N_2	3.1.2.21b
0.7622565	N_2	3.1.2.21b	0.7642478	N_2	3.1.2.21b

Table 4.1 (continued)
GAS LASERS ARRANGED IN ORDER OF WAVELENGTH

Wavelength (μm)	Element or compound	Table No.
0.7644612	N_2	3.1.2.21b
0.76674	Cu^+	2.4.1
0.76650	K	1.1.3
0.76749	Se^+	2.9.3
0.76990	K	1.1.3
0.772436	Se^+	2.9.3
0.7726542	S	1.6.2
0.77325	Zn^+	2.5.1
0.77358	I^+	2.10.4
0.77387	Cu^+	2.4.1
0.7743859	N_2	3.1.2.21b
0.7752354	N_2	3.1.2.21b
0.7753652	N_2	3.1.2.21b
0.775470	F	1.7.1
0.7757	Zn^+	2.5.1
0.7761570	Rb	1.1.4
0.77700379	Se	1.6.3
0.77789	Cu^+	2.4.1
0.77962	Se^+	2.9.3
0.78000	Rb	1.1.4
0.780022	F	1.7.1
0.78016	Te^+	2.9.4
0.78053	Cu^+	2.4.1
0.78078	Cu^+	2.4.1
0.78260	Cu^+	2.4.1
0.78280	Xe^+	2.11.4
0.78393	Se^+	2.9.3
0.78453	Cu^+	2.4.1
0.78460	P^+	2.8.2
0.78493	Na_2	3.1.2.23b
0.78569	Na_2	3.1.2.23b
0.78960	Cu^+	2.4.1
0.78974	Na_2	3.1.2.23b
0.78979	Na_2	3.1.2.23b
0.79027	Cu^+	2.4.1
0.79178	Na_2	3.1.2.23b
0.79214	Te^+	2.9.4
0.79295	Na_2	3.1.2.23b
0.79314	Kr^+	2.11.3
0.79370	Na_2	3.1.2.23b
0.79448	Cu^+	2.4.1
0.79450	Hg^+	2.5.3
0.79470	Rb	1.1.4
0.796070	Ni^+	2.3.1
0.79740	Na_2	3.1.2.23b
0.797480	Ni^+	2.3.1
0.79766	Na_2	3.1.2.23b
0.79882	Cu^+	2.4.1
0.79890	Xe^+	2.11.4
0.79909	Na_2	3.1.2.23b
0.79930	Kr^+	2.11.3
0.79966	Na_2	3.1.2.23b
0.80054	Ag^+	2.4.2
0.80084	Na_2	3.1.2.23b
0.80365	Na_2	3.1.2.23b
0.80393	Na_2	3.1.2.23b
0.80445	Na_2	3.1.2.23b
0.80537	Na_2	3.1.2.23b
0.80561	Na_2	3.1.2.23b
0.80669	Cd^+	2.5.2
0.80692	Bi^{2+}	2.8.5
0.80694	Na_2	3.1.2.23b
0.80805	Na_2	3.1.2.23b
0.808858	Cu^+	2.4.1
0.80960	Cu^+	2.4.1
0.810433	Kr	1.9.4
0.81440	I_2	3.1.2.16b
0.81702	I^+	2.10.4
0.819228	Cu^+	2.4.1
0.823162	Xe^+	2.11.4
0.82539	I^+	2.10.4
0.82545	Ag^+	2.4.2
0.82630	Ag^+	2.4.2
0.82729	Au^+	2.4.3
0.82770	Cu^+	2.4.1
0.82798	O_2	3.1.2.12c
0.82803	Kr^+	2.11.3
0.828321	Cu^+	2.4.1
0.83089	Se^+	2.9.3
0.83248	Ag^+	2.4.2
0.83300	Xe^+	2.11.4
0.8335149	C	1.4.1
0.83390	Kr^+	2.11.3
0.83519	H_2	3.1.2.12c
0.83580	I_2	3.1.2.16b
0.83795	Ag^+	2.4.2
0.83890	Cd^+	2.5.2
0.84035	Ag^+	2.4.2
0.840919	Xe	1.9.5
0.84430	Xe^+	2.11.4

Table 4.1 (continued)
GAS LASERS ARRANGED IN ORDER OF WAVELENGTH

Wavelength (μm)	Element or compound	Table No.	Wavelength (μm)	Element or compound	Table No.
0.844628	U	1.6.1	0.8707478	N_2	3.1.2.21b
0.844638	U	1.6.1	0.8710118	N_2	3.1.2.21b
0.844672	U	1.6.1	0.8710273	N_2	3.1.2.21b
0.844680	U	1.6.1	0.8712956	N_2	3.1.2.21b
0.84634	Ne	1.9.2	0.8713533	N_2	3.1.2.21b
0.84730	Kr^+	2.11.3	0.87140	Xe^+	2.11.4
0.851104	Cu^+	2.4.1	0.8715519	N_2	3.1.2.21b
0.85210	Cs	1.1.5	0.8716718	N_2	3.1.2.21b
0.85309	Cd^+	2.5.2	0.8717377	N_2	3.1.2.21b
0.854180	Ca^+	2.1.3	0.8717970	N_2	3.1.2.21b
0.85470	Hg^+	2.5.3	0.8718571	N_2	3.1.2.21b
0.85690	Xe^{2+}	2.11.4	0.8718654	N_2	3.1.2.21b
0.85780	I_2	3.1.2.16b	0.8719537	N_2	3.1.2.21b
0.85790	I_2	3.1..16b	0.8719562	N_2	3.1.2.21b
0.85820	Xe^+	2.11.4	0.8719791	N_2	3.1.2.21b
0.85890	Kr^+	2.11.3	0.8720251	N_2	3.1.2.21b
0.85940	N	1.5.1	0.8720284	N_2	3.1.2.21b
0.85990	Au^+	2.4.3	0.8720306	N_2	3.1.2.21b
0.86044	Te^+	2.9.4	0.8720419	N_2	3.1.2.21b
0.86228	Hg^+	2.5.3	0.8720848	N_2	3.1.2.21b
0.86284	N	1.5.1	0.8721155	N_2	3.1.2.21b
0.86376895	Ne	1.9.2	0.8721327	N_2	3.1.2.21b
0.86520	Cd^+	2.5.2	0.8721718	N_2	3.1.2.21b
0.866200	Ca^+	2.1.3	0.8721971	N_2	3.1.2.21b
0.8669223	N_2	3.1.2.21b	0.8722007	N_2	3.1.2.21b
0.8671332	N_2	3.1.2.21b	0.8722220	N_2	3.1.2.21b
0.86770	Hg^+	2.5.3	0.8722341	N_2	3.1.2.21b
0.86819216	Ne	1.9.2	0.8722569	N_2	3.1.2.21b
0.86840	Cr^+	2.2.1	0.8722836	N_2	3.1.2.21b
0.86901	Kr^+	2.11.3	0.8723057	N_2	3.1.2.21b
0.8692580	N_2	3.1.2.21b	0.8726333	N_2	3.1.2.21b
0.8696366	N_2	3.1.2.21b	0.8728430	N_2	3.1.2.21b
0.8697945	N_2	3.1.2.21b	0.8730453	N_2	3.1.2.21b
0.8698263	N_2	3.1.2.21b	0.8732394	N_2	3.1.2.21b
0.8699397	N_2	3.1.2.21b	0.8734247	N_2	3.1.2.21b
0.87000	Sr^+	2.1.4	0.87343	Te^+	2.9.4
0.8700670	N_2	3.1.2.21b	0.8735995	N_2	3.1.2.21b
0.8700684	N_2	3.1.2.21b	0.8737644	N_2	3.1.2.21b
0.8701481	N_2	3.1.2.21b	0.8739162	N_2	3.1.2.21b
0.8701718	N_2	3.1.2.21b	0.8740559	N_2	3.1.2.21b
0.8702451	N_2	3.1.2.21b	0.8742917	N_2	3.1.2.21b
0.8702681	N_2	3.1.2.21b	0.87476	Ag^+	2.4.2
0.8703093	N_2	3.1.2.21b	0.87641	Cs	1.1.5
0.8703457	N_2	3.1.2.21b	*0.87740648*	Ne	1.9.2
0.8704549	N_2	3.1.2.21b	0.87730	Ag^+	2.4.2

Table 4.1 (continued)
GAS LASERS ARRANGED IN ORDER OF WAVELENGTH

Wavelength (μm)	Element or compound	Table No.
0.87800	Ar^+	2.11.2
0.88040	I_2	3.1.2.16b
0.880428	I^+	2.10.4
0.88060	I_2	3.1.2.16b
0.88130	I_2	3.1.2.16b
0.88228702	O	1.6.1
0.8845349	N_2	3.1.2.21b
0.8856271	N_2	3.1.2.21b
0.8858470	N_2	3.1.2.21b
0.88653	Ne	1.9.2
0.88676	Au^+	2.4.3
0.88774	I^+	2.10.4
0.88778	Cd^+	2.5.2
0.88787	H_2	3.1.2.12c
0.888051	N_2	3.1.2.21b
0.8884527	N_2	3.1.2.21b
0.8886204	N_2	3.1.2.21b
0.8886378	N_2	3.1.2.21b
0.8887756	N_2	3.1.2.21b
0.8889111	N_2	3.1.2.21b
0.8889738	N_2	3.1.2.21b
0.8890243	N_2	3.1.2.21b
0.8891133	N_2	3.1.2.21b
0.8891769	N_2	3.1.2.21b
0.8892149	N_2	3.1.2.21b
0.8892940	N_2	3.1.2.21b
0.8896001	N_2	3.1.2.21b
0.8898930	N_2	3.1.2.21b
0.8899078	N_2	3.1.2.21b
0.89013	H_2	3.1.2.12c
0.8902711	N_2	3.1.2.21b
0.8904419	N_2	3.1.2.21b
0.8906097	N_2	3.1.2.21b
0.8906649	N_2	3.1.2.21b
0.8906994	N_2	3.1.2.21b
0.8907920	N_2	3.1.2.21b
0.8908808	N_2	3.1.2.21b
0.8908878	N_2	3.1.2.21b
0.8909451	N_2	3.1.2.21b
0.8909527	N_2	3.1.2.21b
0.8909750	N_2	3.1.2.21b
0.8910132	N_2	3.1.2.21b
0.8910480	N_2	3.1.2.21b
0.8910612	N_2	3.1.2.21b
0.8911001	N_2	3.1.2.21b
0.8911280	N_2	3.1.2.21b
0.8911502	N_2	3.1.2.21b
0.8911538	N_2	3.1.2.21b
0.8911608	N_2	3.1.2.21b
0.8911898	N_2	3.1.2.21b
0.8912139	N_2	3.1.2.21b
0.8918033	N_2	3.1.2.21b
0.8920184	N_2	3.1.2.21b
0.8922249	N_2	3.1.2.21b
0.8924223	N_2	3.1.2.21b
0.8926099	N_2	3.1.2.21b
0.8927865	N_2	3.1.2.21b
0.89290	Kr	1.9.4
0.8929509	N_2	3.1.2.21b
0.8931019	N_2	3.1.2.21b
0.8933580	N_2	3.1.2.21b
0.89430	Cs	1.1.5
0.89719	Tl^+	2.9.4
0.89789	Kr^+	2.11.3
0.89910237	Ne	1.9.2
0.89982	Tl^+	2.9.4
0.90370	I_2	3.1.2.16b
0.90380	I_2	3.1.2.16b
0.904539	Xe	1.9.5
0.90455	N	1.5.1
0.90470	I_2	3.1.2.16b
0.90630	Xe^+	2.11.4
0.90600	I_2	3.1.2.16b
0.912297	Ar	1.9.3
0.91630	HD	3.1.2.12c
0.91723217	Cs	1.1.5
0.91870	N	1.5.1
0.921800	Mg^+	2.1.2
0.924400	Mg^+	2.1.2
0.92493	Se^+	2.9.3
0.92650	Xe^+	2.11.4
0.92740	I_2	3.1.2.16b
0.92760	I_2	3.1.2.16b
0.92870	Xe^+	2.11.4
0.92880	I_2	3.1.2.16b
0.92950	I_2	3.1.2.16b
0.93050	I_2	3.1.2.16b
0.93500	Tl^+	2.6.4
0.93779	Te^+	2.9.4
0.93921	N	1.5.1

Table 4.1 (continued)
GAS LASERS ARRANGED IN ORDER OF WAVELENGTH

Wavelength (μm)	Element or compound	Table No.	Wavelength (μm)	Element or compound	Table No.
0.93960	Hg^+	2.5.3	1.04172	I_2^+	2.10.4
0.9405729	C	1.4.1	*1.0455985*	As	1.5.3
0.94510	Cl	1.7.2	1.04550	S	1.6.2
0.94892838	Ne	1.9.2	1.04700	Ar	1.9.3
0.95160	I_2	3.1.2.16b	1.05340	I_2	3.1.2.16b
0.95326	D_2	3.1.2.12c	1.05680	N	1.5.1
0.95450	I_2	3.1.2.16b	1.05860	Hg^+	2.5.3
0.95550	I_2	3.1.2.16b	1.06000	I	1.7.4
0.965389	N_2	3.1.2.21b	1.06000	Nd Al_5Cl_{12}	3.1.4.1
0.965778	Ar	1.9.3	1.06000	$Nd(thd)_3$	3.1.4.2
0.965846	N_2	3.1.2.21b	*1.0612556*	Sn	1.4.4
0.966599	N_2	3.1.2.21b	*1.0617063*	As	1.5.3
0.96680709	Ne	1.9.2	1.06110	N	1.5.1
0.967270	N_2	3.1.2.21b	1.06200	Sn^+	2.7.4
0.967758	N_2	3.1.2.21b	*1.0623574*	Ne	1.9.2
0.967943	N_2	3.1.2.21b	1.06300	Xe^+	2.11.4
0.968061	N_2	3.1.2.21b	1.06380	S	1.6.2
0.969552	N_2	3.1.2.21b	1.06596	Kr^+	1.4.1
0.96970	Xe^+	2.11.4	1.06800	C	1.4.1
0.969879	N_2	3.1.2.21b	1.06910	C	1.4.1
0.97660	I_2	3.1.2.16b	1.0707333	C	1.4.1
0.979970	Xe	1.9.5	1.07400	Sn^+	2.7.4
0.98000	I	1.7.4	1.07750	I_2	3.1.2.16b
0.9838743	Cd	1.2.6	1.07880	I_2	3.1.2.16b
0.98980	Eu^+	2.12.1	*1.0801000*	Ne	1.9.2
0.99400	Ca^+	2.1.3	1.08600	S_2	3.1.2.24b
0.99547	Se^+	2.9.3	*1.0867911*	Cd	1.2.6
0.99630	I_2	3.1.2.16b	1.091450	Sr^+	2.1.4
0.99730	I_2	3.1.2.16b	1.09150	S_2	3.1.2.24b
1.00190	I_2	3.1.2.16b	1.091500	Mg^+	2.1.2
1.00200	Eu^+	2.12.1	1.09170	S_2	3.1.2.24b
1.00530	I_2	3.1.2.16b	1.09200	S_2	3.1.2.24b
1.00800	P	1.5.2	1.09230	Ar^+	2.11.2
1.01000	I	1.7.4	1.09410	S_2	3.1.2.24b
1.01060	Eu^+	2.12.1	1.09460	S_2	3.1.2.24b
1.01230	Cs	1.1.5	1.09520	Mg^+	2.1.2
1.02250	I_2	3.1.2.16b	1.09900	S_2	3.1.2.24b
1.02450	I_2	3.1.2.16b	1.09963	CN	3.1.2.9
1.02550	I_2	3.1.2.16b	1.09965	CN	3.1.2.9
1.02740	I_2	3.1.2.16b	1.09966	CN	3.1.2.9
1.0298238	Ne	1.9.2	1.09974	CN	3.1.2.9
1.03000	I	1.7.4	1.09987	CN	3.1.2.9
1.0324559	Yb	1.7.5	1.10000	S_2	3.1.2.24b
1.03350	Sr^+	2.1.4	1.10007	CN	3.1.2.9
1.04094	Se^+	2.9.3	1.10031	CN	3.1.2.9

Table 4.1 (continued)
GAS LASERS ARRANGED IN ORDER OF WAVELENGTH

Wavelength (μm)	Element or compound	Table No.	Wavelength (μm)	Element or compound	Table No.
1.10061	CN	3.1.2.9	1.13820	Na	1.1.2
1.10096	CN	3.1.2.9	*1.1393552*	Ne	1.9.2
1.10136	CN	3.1.2.9	1.14040	Na	1.1.2
1.10182	CN	3.1.2.9	*1.1412258*	Ne	1.9.2
1.10232	CN	3.1.2.9	1.14480	Ar	1.9.3
1.10288	CN	3.1.2.9	1.14530	I_2	3.1.2.16b
1.10348	CN	3.1.2.9	1.14582	Kr	1.9.4
1.10414	CN	3.1.2.9	1.14640	I_2	3.1.2.16b
1.10445	CN	3.1.2.9	1.14850	Cd	1.2.6
1.10485	CN	3.1.2.9	1.15020	I_2	3.1.2.16b
1.10521	CN	3.1.2.9	1.15100	I_2	3.1.2.16b
1.10560	CN	3.1.2.9	1.15150	I_2	3.1.2.16b
1.10603	CN	3.1.2.9	1.15220	I_2	3.1.2.16b
1.10641	CN	3.1.2.9	*1.1522595*	As	1.5.3
1.10660	I_2	3.1.2.16b	*1.1525900*	Ne	1.9.2
1.10689	CN	3.1.2.9	*1.1528174*	Ne	1.9.2
1.10726	CN	3.1.2.9	1.15290	I_2	3.1.2.16b
1.10730	I_2	3.1.2.16b	1.15540	Cd	1.2.6
1.10782	CN	3.1.2.9	1.15870	S_2	3.1.2.24b
1.10879	CN	3.1.2.9	1.15900	Pb^+	2.7.5
1.10981	CN	3.1.2.9	*1.1604712*	Ne	1.9.2
1.11090	CN	3.1.2.9	*1.1617260*	Ne	1.9.2
1.11200	CN	3.1.2.9	*1.1663677*	Cd	1.2.6
1.11321	CN	3.1.2.9	1.16980	I_2	3.1.2.16b
1.1146071	Ne	1.9.2	1.17000	K	1.1.3
1.1163455	P	1.5.2	1.17110	I_2	3.1.2.16b
1.11650	H_2	3.1.2.12c	1.17180	I_2	3.1.2.16b
1.11179812	Hg	1.2.7	1.17400	I_2	3.1.2.16b
1.1180588	Ne	1.9.2	*1.1745636*	Cd	1.2.6
1.1186470	P	1.5.2	1.17500	I_2	3.1.2.16b
1.12060	I_2	3.1.2.16b	1.17500	Tl^+	2.6.4
1.12140	I_2	3.1.2.16b	*1.1770013*	Ne	1.9.2
1.12160	I_2	3.1.2.16b	*1.1787698*	P	1.5.2
1.12250	H_2	3.1.2.12c	*1.1792270*	Ne	1.9.2
1.12260	I_2	3.1.2.16b	*1.1874246*	Cd	1.2.6
1.12300	Sr^+	2.1.4	*1.1984187*	Si	1.4.2
1.1247708	As	1.5.3	*1.1988192*	Ne	1.9.2
1.12550	I_2	3.1.2.16b	1.20340	Si	1.4.2
1.13030	Ba	1.2.4	1.2068179	Ne	1.9.2
1.13280	I_2	3.1.2.16b	1.20960	Be^+	2.11
1.13340	I_2	3.1.2.16b	1.21397	Ar	1.9.3
1.13470	I_2	3.1.2.16b	*1.2156374*	Ar	1.9.3
1.13480	I_2	3.1.2.16b	1.21700	I_2	3.1.2.16b
1.13500	I_2	3.1.2.16b	1.22220	Hg	1.2.7
1.13500	Tl^+	2.6.4	1.22460	Hg	1.2.7

Table 4.1 (continued)
GAS LASERS ARRANGED IN ORDER OF WAVELENGTH

Wavelength (μm)	Element or compound	Table No.
1.24028	Ar	1.9.3
1.24340	K	1.1.3
1.2462797	Ne	1.9.2
1.24780	Ba^+	2.1.5
1.25230	K	1.1.3
1.25450	Hg	1.2.7
1.255136	Yb	1.7.5
1.2561370	Pb	1.4.5
1.25878	Se^+	2.9.3
1.25880724	Ne	1.9.2
1.2598449	N_2	1.9.2
1.2692672	Ne	1.9.2
1.27022	Ar	1.9.3
1.27140	Yb^+	2.12.2
1.27400	I_2	3.1.2.16b
1.2773017	Ne	1.9.2
1.28700	I_2	3.1.2.16b
1.2890684	Ne	1.9.2
1.28998	Mn	1.7.6
1.2915545	Ne	1.9.2
1.29250	I_2	3.1.2.16b
1.29400	I_2	3.1.2.16b
1.2945989	As	1.5.3
1.29810	Hg	1.2.7
1.3058983	Tm	1.6.5
1.30100	I_2	3.1.2.16b
1.30200	I_2	3.1.2.16b
1.305363	Zn	1.2.5
1.30610	H_2	3.1.2.12c
1.30690	I_2	3.1.2.16b
1.30806	I_2	3.1.2.16b
1.3103722	Pb	1.4.5
1.3104227	Tm	1.6.5
1.315059	Zn	1.2.5
1.3152443	I	1.7.4
1.31530	I_2	3.1.2.16b
1.31660	H_2	3.1.2.12c
1.31775	Kr	1.9.4
1.31920	I_2	3.1.2.16b
1.32000	I_2	3.1.2.16b
1.32820	I_2	3.1.2.16b
1.32910	I_2	3.1.2.16b
1.32938	Mn	1.7.6
1.32950	Kr^+	2.11.3
1.33100	I_2	3.1.2.16b
1.33190	Mn	1.7.6
1.33240	I_2	3.1.2.16b
1.33330	I_2	3.1.2.16b
1.33490	I_2	3.1.2.16b
1.33800	I_2	3.1.2.16b
1.333700	Tm	1.6.5
1.34060	I_2	3.1.2.16b
1.34180	I_2	3.1.2.16b
1.34192	I_2	3.1.2.16b
1.34210	I_2	3.1.2.16b
1.34290	I_2	3.1.2.16b
1.34295	N	1.5.1
1.342996	In	1.3.3
1.34530	Yb^+	2.12.2
1.3476544	Ar	1.9.3
1.3574217	Hg	1.2.7
1.35819	N	1.5.1
1.35880	Cs	1.1.5
1.36000	Cs	1.1.5
1.36100	Eu^+	2.12.1
1.3612294	Sn	1.4.4
1.36225	Kr	1.9.4
1.36367	Mn	1.7.6
1.36550	Hg	1.2.7
1.365620	Xe	1.9.5
1.3677207	Hg	1.2.7
1.37580	Cs	1.1.5
1.37590	Ag^+	2.4.2
1.37660	Rb	1.1.4
1.38590	Cl	1.7.2
1.38642	Mn	1.7.6
1.38910	Cl	1.7.2
1.3954389	Hg	1.2.7
1.39680	Ni	1.8.2
1.39975	Mn	1.7.6
1.3982714	Cd	1.2.6
1.40200	S	1.6.2
1.40948	Ar	1.8.3
1.4124892	As	1.5.3
1.41830	CN	3.1.2.9
1.41849	CN	3.1.2.9
1.41876	CN	3.1.2.9
1.41911	CN	3.1.2.9
1.41954	CN	3.1.2.9
1.42005	CN	3.1.2.9

Table 4.1 (continued)
GAS LASERS ARRANGED IN ORDER OF WAVELENGTH

Wavelength (μm)	Element or compound	Table No.
1.42065	CN	3.1.2.9
1.42132	CN	3.1.2.9
1.42207	CN	3.1.2.9
1.42289	CN	3.1.2.9
1.42380	CN	3.1.2.9
1.42478	CN	3.1.2.9
1.4258622	As	1.5.3
1.42583	CN	3.1.2.9
1.42696	CN	3.1.2.9
1.42760	Ne	1.9.2
1.42808	CN	3.1.2.9
1.4283698	Yb	1.7.5
1.42945	CN	3.1.2.9
1.431625	In	
1.43040	Ne	1.9.2
1.43081	CN	3.1.2.9
1.43210	Ne	1.9.2
1.43300	Ne	1.9.2
1.4331602	Cd	1.2.6
1.4343722	Tm	1.6.5
1.43460	Ne	1.9.2
1.43680	Ne	1.9.2
1.441920	In	1.3.3
1.44269	Kr	1.9.4
1.4478302	Cd	1.2.6
1.4489080	Tm	1.6.5
1.45400	C	1.4.1
1.45400	N	1.5.1
1.4545941	I	1.7.4
1.455304	N	1.5.1
1.4553718	Ni	1.8.2
1.4629079	As	1.5.3
1.46900	Cs	1.1.5
1.47500	Rb	1.1.4
1.47648	Kr	1.9.4
1.47700	Eu^+	2.12.1
1.4793059	Yb	1.7.5
1.4848636	Ne	1.9.2
1.4873294	Ne	1.9.2
1.4876284	Ne	1.9.2
1.4892012	Ne	1.9.2
1.4903576	Ne	1.9.2
1.4940304	Ne	1.9.2
1.49660	Kr	1.9.4
1.4998810	Tm	1.6.5
1.50000	Ba	1.2.4
1.502499	Mg	1.2.1
1.50460	Ar	1.1.3
1.5234875	Ne	1.9.2
1.5290	Rb	1.1.4
1.529954	Hg	1.2.7
1.53300	Kr	1.9.4
1.5335134	Pb	1.4.5
1.54220	S	1.6.2
1.5533401	I	1.7.4
1.55500	Hg^+	2.5.3
1.5716351	P	1.5.2
1.58697	Cl	1.7.2
1.58830	Si	1.4.2
1.59820	Ag^+	2.4.2
1.6057666	Xe	1.8.5
1.61800	Ar	1.9.3
1.6383650	Tm	1.6.5
1.6404449	Cd	1.2.6
1.6407031	Ne	1.9.2
1.6437081	Cd	1.2.6
1.64640	Ag^+	2.4.2
1.6486189	Cd	1.2.6
1.648791	P	1.5.2
1.64980	Yb^+	2.12.2
1.65200	Ar	1.9.3
1.65430	S	1.6.3
1.675663	Tm	1.6.5
1.68533	Kr	1.9.4
1.68965	Kr	1.9.4
1.6924775	Hg	1.2.7
1.69360	Kr	1.9.4
1.69410	Ar	1.9.3
1.6946636	Hg	1.2.7
1.7077038	Hg	1.2.7
1.7114554	Hg	1.2.7
1.7166616	Ne	1.9.2
1.72020	Ag^+	2.4.2
1.7323684	Tm	1.6.5
1.7330499	Xe	1.9.5
1.7334185	Hg	1.3.7
1.7367231	Ga	1.3.2
1.7459155	Yb	1.7.5
1.74460	Ag^+	2.4.2
1.74800	Ag^+	2.4.2

Table 4.1 (continued)
GAS LASERS ARRANGED IN ORDER OF WAVELENGTH

Wavelength (μm)	Element or compound	Table No.
1.7500485	Eu	1.8.4
1.77100	Cu^+	2.4.1
1.78430	Kr	1.9.4
1.791905	Ar	1.9.3
1.798400	Yb	1.7.5
1.8053474	As	1.5.3
1.80570	Yb^+	2.12.2
1.8068806	As	1.5.3
1.8135329	Hg	1.2.7
1.81673	Kr	1.9.4
1.81850	Kr	1.9.4
1.8199685	Cu	1.1.6
1.82000	Ba	1.2.4
1.8215302	Ne	1.9.2
1.8234051	Cu	1.1.6
1.8258313	Ne	1.9.2
1.82760	Ne	1.9.2
1.82820	Ne	1.9.2
1.83040	Ne	1.9.2
1.83100	Zn^+	2.5.1
1.83600	DF	3.1.1.5b
1.8380629	Ag	1.1.7
1.84030	Ne	1.9.2
1.84100	Ag^+	2.4.2
1.84640	Ag^+	2.4.2
1.85400	DF	3.1.1.5b
1.85910	Ne	1.9.2
1.85970	Ne	1.9.2
1.86850	Ne	1.9.2
1.87240	Ag^+	2.4.2
1.8736732	In	1.3.3
1.87510	H	1.1.1
1.87960	Ag^+	2.4.2
1.89200	Ag^+	2.4.2
1.8943842	P	1.5.2
1.90122415	Ba	1.2.4
1.9123124	Cd	1.2.6
1.9124005	Sm	1.8.2
1.91560	Cu^+	2.4.1
1.9216572	Kr	1.9.4
1.92600	Cu	1.8.4
1.93701923	Ag	1.1.6
1.94800	Cu^+	2.4.1
1.95430	He	1.9.1
1.95740	Ne	1.9.2
1.958248	Ne	1.9.2
1.9589851	Tm	1.6.5
1.96100	Eu	1.8.4
1.97100	Cu^+	2.4.1
1.97160	Ag^+	2.4.2
1.9722834	Tm	1.6.5
1.973362	Br	1.7.3
1.9754647	As	1.5.3
1.97550	Cl	1.7.2
1.9814602	Ge	1.4.3
1.98240	Ag^+	2.4.2
1.9835173	Yb	1.7.5
1.98616	CN	3.1.2.9
1.98629	CN	3.1.2.9
1.98658	CN	3.1.2.9
1.98701	CN	3.1.2.9
1.98759	CN	3.1.2.9
1.98831	CN	3.1.2.9
1.98918	CN	3.1.2.9
1.99020	CN	3.1.2.9
1.99135	CN	3.1.2.9
1.99263	CN	3.1.2.9
1.99406	CN	3.1.2.9
1.9947227	Tm	1.6.5
2.00040	Cu^+	2.4.1
2.00090	CN	3.1.2.9
2.00310	CN	3.1.2.9
2.0041327	Yb	1.7.5
2.00550	CN	3.1.2.9
2.00800	CN	3.1.2.9
2.01070	CN	3.1.2.9
2.01350	CN	3.1.2.9
2.01640	CN	3.1.2.9
2.01960	CN	3.1.2.9
2.020602	Cl	1.4.3
2.02623	Xe	1.8.5
2.0282741	As	1.5.3
2.0355792	Ne	1.9.2
2.0359432	Ne	1.9.2
2.04820	Sm	1.8.3
2.05813	He	1.9.1
2.060755	He	1.9.1
2.06160	Ar	1.9.3
2.0655993	C	1.4.1
2.07940	Ag^+	2.4.2

Table 4.1 (continued)
GAS LASERS ARRANGED IN ORDER OF WAVELENGTH

Wavelength (μm)	Element or compound	Table No.
2.0962339	P	1.5.2
2.09860	Ar	1.9.3
2.1023345	Ne	1.9.2
2.10410	Ne	1.8.2
2.1059135	Tm	1.6.5
2.11650	Kr	1.9.4
2.1186997	Yb	1.7.5
2.13320	Ar	1.9.3
2.14800	Yb^+	2.12.2
2.15340	Ar	1.9.3
2.1573497	Ba	1.2.4
2.17080	Ne	1.9.2
2.19020	Kr	1.9.4
2.20000	Ar	1.9.3
2.24750	Kr	1.9.4
2.25300	Rb	1.1.4
2.27920	H_2O	3.2.2.5
2.2801247	S	1.6.2
2.28540	Br	1.7.3
2.29300	Rb	1.1.4
2.31339	Ar	1.9.3
2.31930	Xe	1.9.5
2.32540	Ba	1.2.4
2.3266649	Ne	1.9.2
2.34900	CO	3.2.1.2a
2.35110	Br	1.7.3
2.37830	CO	3.2.1.2a
2.3785794	In	1.3.3
2.3851957	Tm	1.6.5
2.3957953	Ne	1.8.2
2.3962995	Ne	1.9.2
2.39660	Ar	1.9.3
2.404160	Mg^+	2.1.2
2.40826	CO	3.2.1.2a
2.41380	HF	3.2.1.5a
2.412520	Mg^+	2.1.2
2.4162547	Ne	1.9.2
2.4225538	Ne	1.9.2
2.4256255	Ne	1.9.2
2.42600	Kr	1.9.4
2.43310	HF	3.2.1.5a
2.43630	S	1.6.3
2.43770	Yb^+	2.12.2
2.43810	CO	3.2.1.2a
2.4466627	As	1.5.3
2.44660	Cl	1.7.2
2.4480996	V	1.5.5
2.45380	HF	3.2.1.5a
2.47000	CO	3.2.1.2a
2.4764593	Ba	1.2.4
2.48759	HF	3.2.1.5a
2.48250	Xe	1.9.5
2.5014408	Ar	1.9.3
2.50250	CO	3.2.1.2a
2.5152702	Xe	1.9.5
2.52342	Kr	1.9.4
2.53500	CO	3.2.1.2a
2.5400115	Ne	1.9.2
2.54946	Ar	1.9.3
2.5512187	Ar	1.9.3
2.55080	HF	3.2.1.5a
2.55150	Ba	1.2.4
2.5531329	Ne	1.9.2
2.5634025	Ar	1.9.3
2.5668023	Ar	1.9.3
2.56993	CO	3.2.1.2a
2.57890	HF	3.2.1.5a
2.5818111	Eu	1.8.4
2.592300	Ba^+	2.1.5
2.59860	I	1.7.4
2.60271	CO	3.2.1.2a
2.60850	HF	3.2.1.5a
2.6266703	Kr	1.9.4
2.62690	Xe	2.11.4
2.6288137	Kr	1.9.4
2.63708	CO	3.2.1.2a
2.63480	HF	3.2.1.5a
2.65146	Xe	2.11.4
2.6513946	O	1.6.1
2.6550282	Ar	1.9.3
2.6608387	Xe	2.11.4
2.66680	HF	3.2.1.5a
2.6672615	Xe	2.11.4
2.67270	HF	3.2.1.5a
2.67500	CO	3.2.1.2a
2.6843026	Ar	1.9.3
2.68864	CO	3.2.1.2a
2.69142	CO	3.2.1.2a
2.69620	HF	3.2.1.5a
2.7006079	Sm	1.8.3

Table 4.1 (continued)
GAS LASERS ARRANGED IN ORDER OF WAVELENGTH

Wavelength (μm)	Element or compound	Table No.
2.70752	HF	3.2.1.5a
2.7135274	Br	1.7.3
2.71359	CO	3.2.1.2a
2.7152859	Ar	1.9.3
2.7181668	Eu	1.8.4
2.72619	CO	3.2.1.2a
2.7275	HF	3.2.1.5a
2.72903	CO	3.2.1.2a
2.73192	CO	3.2.1.2a
2.73486	CO	3.2.1.2a
2.7363805	Ar	1.9.3
2.73787	CO	3.2.1.2a
2.74093	CO	3.2.1.2a
2.74406	CO	3.2.1.2a
2.74410	HF	3.2.1.5a
2.74724	CO	3.2.1.2a
2.75048	CO	3.2.1.2a
2.75200	CO	3.2.1.2a
2.75378	CO	3.2.1.2a
2.75714	CO	3.2.1.2a
2.75720	I	1.7.4
2.7580982	Ne	1.9.2
2.76040	HF	3.2.1.5a
2.76055	CO	3.2.1.2a
2.76403	CO	3.2.1.2a
2.76469	CO	3.2.1.2a
2.76470	CO	3.2.1.2a
2.76757	CO	3.2.1.2a
2.76760	CO	3.2.1.2a
2.77052	CO	3.2.1.2a
2.77117	CO	3.2.1.2a
2.77659	CO	3.2.1.2a
2.77991	CO	3.2.1.2a
2.78260	HF	3.2.1.5a
2.7826380	Ne	1.92
2.78290	CO	3.2.1.2a
2.78614	CO	3.2.1.2a
2.78828	CO	3.2.1.2a
2.78945	CO	3.2.1.2a
2.790	Rb	1.1.4
2.79020	HF	3.2.1.5a
2.79030	CO	3.2.1.2a
2.79281	CO	3.2.1.2a
2.79530	HF	3.2.1.5a
2.79624	CO	3.2.1.2a
2.79973	CO	3.2.1.2a
2.80420	CO	3.2.1.2a
2.80710	CO	3.2.1.2a
2.80202417	Ar	1.9.3
2.80328	CO	3.2.1.2a
2.80415	CO	3.2.1.2a
2.80689	CO	3.2.1.2a
2.80710	CO	3.2.1.2a
2.81010	CO	3.2.1.2a
2.81057	CO	3.2.1.2a
2.81430	CO	3.2.1.2a
2.81811	CO	3.2.1.2a
2.82130	HF	3.2.1.5a
2.82197	CO	3.2.1.2a
2.82272	CO	3.2.1.2a
2.8245953	Ar	1.9.3
2.82590	CO	3.2.1.2a
2.82603	CO	3.2.1.2a
2.82800	CO	3.2.1.2a
2.82940	CO	3.2.1.2a
2.82989	CO	3.2.1.2a
2.8318	HF	3.2.1.5a
2.83284	CO	3.2.1.2a
2.83634	CO	3.2.1.2a
2.8375	Br	1.7.3
2.83990	CO	3.2.1.2a
2.84352	CO	3.2.1.2a
2.84460	CO	3.2.1.2a
2.84463	CO	3.2.1.2a
2.84721	CO	3.2.1.2a
2.84760	CO	3.2.1.2a
2.85069	CO	3.2.1.2a
2.85070	CO	3.2.1.2a
2.85096	CO	3.2.1.2a
2.85400	HF	3.2.1.5a
2.85478	CO	3.2.1.2a
2.85866	CO	3.2.1.2a
2.8590043	Xe	2.11.4
2.86134	Kr	1.9.4
2.8620231	Ar	1.9.3
2.86261	CO	3.2.1.2a
2.8656	Kr	1.9.4
2.86570	HF	3.2.1.5a
2.86662	CO	3.2.1.2a
2.87038	CO	3.2.1.2a

Table 4.1 (continued)
GAS LASERS ARRANGED IN ORDER OF WAVELENGTH

Wavelength (μm)	Element or compound	Table No.	Wavelength (μm)	Element or compound	Table No.
2.87060	HF	3.2.1.5a	*2.9539*	HF	3.2.1.5a
2.87070	CO	3.2.1.2a	*2.9549*	HF	3.2.1.5a
2.87389	CO	3.2.1.2a	*2.95553*	CO	3.2.1.2a
2.87746	CO	3.2.1.2a	*2.9573*	HF	3.2.1.5a
2.87822932	Ar	1.9.3	*2.95986*	CO	3.2.1.2a
2.88110	CO	3.2.1.2a	*2.9643*	HF	3.2.1.5a
2.8843088	Ar	1.9.3	2.9663	Sm	1.8.3
2.88480	CO	3.2.1.2a	*2.9676035*	Ne	1.9.2
2.88857	CO	3.2.1.2a	*2.969*	OH	3.2.1.7
2.88890	HF	3.2.1.5a	*2.97061*	CO	3.2.1.2a
2.88920	CO	3.2.1.2a	*2.97248*	CO	3.2.1.2a
2.88921	CO	3.2.1.2a	*2.9725*	CO	3.2.1.2a
2.89230	CO	3.2.1.2a	*2.97454*	CO	3.2.1.2a
2.89233	CO	3.2.1.2a	*2.97566*	CO	3.2.1.2a
2.89240	CO	3.2.1.2a	*2.9757*	CO	3.2.1.2a
2.89630	CO	3.2.1.2a	*2.97854*	Cl	3.2.1.2a
2.90027	CO	3.2.1.2a	*2.9789*	CO	3.2.1.2a
2.90430	CO	3.2.1.2a	*2.97891*	CO	3.2.1.2a
2.9059	Ba^+	2.1.5	*2.9796792*	Ar	1.9.3
2.90840	CO	3.2.1.2a	*2.9812503*	Ne	1.9.2
2.91030	HF	3.2.1.5a	*2.9813368*	As	1.5.3
2.91110	HF	3.2.1.5a	*2.98261*	CO	3.2.1.2a
2.9134037	Ar	1.9.3	*2.9844656*	Kr	1.9.4
2.91965	CO	3.2.1.2a	*2.98676*	CO	3.2.1.2a
2.92221	HF	3.2.1.5a	*2.9878091*	Kr	1.9.4
2.9230381	Ba	1.3.4	*2.9896*	HF	3.2.1.5a
2.92337	CO	3.2.1.2a	*2.99097*	CO	3.2.1.2a
2.92560	HF	3.2.1.5a	*2.9989*	HF	3.2.1.5a
2.92715	CO	3.2.1.2a	*2.99962*	CO	3.2.1.2a
2.9280662	Ar	1.9.3	*3.00405*	CO	3.2.1.2a
2.92875	CO	3.2.1.2a	*3.0051*	HF	3.2.1.5a
2.92880	CO	3.2.1.2a	*3.0064*	HF	3.2.1.5a
2.93099	CO	3.2.1.2a	*3.00856*	CO	3.2.1.2a
2.9317981	Cs	1.1.5	3.010	Cs	1.1.5
2.93187	Cs	1.1.5	*3.0111339*	Cs	1.1.5
2.9319	Cs	1.1.5	*3.0118377*	Sr	1.2.3
2.93491	Cs	1.1.5	*3.01313*	CO	3.2.1.2a
2.935	OH	3.2.1.7	*3.01738*	CO	3.2.1.2a
2.93505	CO	3.2.1.2a	*3.01740*	CO	3.2.1.2a
2.9351	CO	3.2.1.2a	*3.0206*	CO	3.2.1.2a
2.93890	CO	3.2.1.2a	*3.02063*	CO	3.2.1.2a
2.94295	CO	3.2.1.2a	*3.0264*	HF	3.2.1.5a
2.9455858	Ne	1.9.2	*3.0267787*	Ne	1.9.2
2.94707	CO	3.2.1.2a	*3.02750*	CO	3.2.1.2a
2.95127	CO	3.2.1.2a	*3.0275836*	Ne	1.9.2

Table 4.1 (continued)
GAS LASERS ARRANGED IN ORDER OF WAVELENGTH

Wavelength (μm)	Element or compound	Table No.
3.03173	CO	3.2.1.2a
3.036	I	1.7.4
3.03605	CO	3.2.1.2a
3.04043	CO	3.2.1.2a
3.04489	CO	3.2.1.2a
3.046207	Ar	1.9.3
3.0461	HF	3.2.1.5a
3.0482	HF	3.2.1.5a
3.04942	CO	3.2.1.2a
3.0536574	Kr	1.9.4
3.05403	CO	3.2.1.2a
3.0582	HF	3.2.1.5a
3.05872	CO	3.2.1.2a
3.06348	CO	3.2.1.2a
3.0652	HF	3.2.1.5a
3.0664	Kr	1.9.4
3.0664	Cl	1.7.2
3.0668	CO	3.2.1.2a
3.06682	CO	3.2.1.2a
3.0670208	Sr	1.2.3
3.06831	CO	3.2.1.2a
3.0720016	Ne	1.9.2
3.07323	CO	3.2.1.2a
3.07793	CO	3.2.1.2a
3.078	OH	3.2.1.7
3.08234	CO	3.2.1.2a
3.08682	CO	3.2.1.2a
3.09138	CO	3.2.1.2a
3.0935	HF	3.2.1.5a
3.0958	HF	3.2.1.5a
3.09602	CO	3.2.1.2a
3.0961401	Cs	1.1.5
3.0982	HF	3.2.1.5a
3.0996226	Ar	1.9.3
3.10074	CO	3.2.1.2a
3.10554	CO	3.2.1.2a
3.1069	Ar	1.9.3
3.11041	CO	3.2.1.2a
3.1125	HF	3.2.1.5a
3.115	OH	3.2.1.7
3.11536	CO	3.2.1.2a
3.12039	CO	3.2.1.2a
3.12551	CU	3.2.1.2a
3.1333028	Ar	1.9.3
3.13448	CO	3.2.1.2a
3.1345761	Ar	1.9.3
3.1350	HF	3.2.1.5a
3.13915	CO	3.2.1.2a
3.1411	HF	3.2.1.5a
3.1415224	K	1.1.3
3.14390	CO	3.2.1.2a
3.1450	HF	3.2.1.5a
3.14873	CO	3.2.1.2a
3.14940	HF	3.2.1.5a
3.1514572	Kr	1.9.4
3.15364	CO	3.2.1.2a
3.157	OH	3.2.1.7
3.15863	CO	3.2.1.2a
3.1601267	K	1.1.3
3.16371	CO	3.2.1.2a
3.1640	HF	3.2.1.5a
3.1653653	Te	1.6.4
3.16886	CO	3.2.1.2a
3.1696	HF	3.2.1.5a
3.1739	HF	3.2.1.5a
3.17410	CO	3.2.1.2a
3.1748096	Pb	1.4.5
3.18347	CO	3.2.1.2a
3.18826	CO	3.2.1.2a
3.1912	HF	3.2.1.5a
3.19312	CO	3.2.1.2a
3.19806	CO	3.2.1.2a
3.2029	HF	3.2.1.5a
3.20309	CO	3.2.1.2a
3.2050778	Cs	1.1.5
3.20821	CO	3.2.1.2a
3.21340	CO	3.2.1.2a
3.2150	HF	3.2.1.5a
3.21868	CO	3.2.1.2a
3.22405	CO	3.2.1.2a
3.22930	HF	3.2.1.5a
3.23400	OH	3.2.1.7
3.2363	I	1.7.4
3.24380	HF	3.2.1.5a
3.25850	HF	3.2.1.5a
3.26030	HF	3.2.1.5a
3.2739	Xe	2.11.4
3.27400	OH	3.2.1.7
3.28900	Cd^+	2.5.2
3.29200	HF	3.2.1.5a

Table 4.1 (continued)
GAS LASERS ARRANGED IN ORDER OF WAVELENGTH

Wavelength (μm)	Element or compound	Table No.
3.29463	N_2	3.1.2.21c
3.29910	HF	3.2.1.5a
3.30149	N_2	3.1.2.21c
3.3044	HF	3.2.1.5a
3.30734	N_2	3.1.2.21c
3.30989	N_2	3.1.2.21c
3.3099055	Xe	1.9.5
3.31221	N_2	3.1.2.21c
3.31426	N_2	3.1.2.21c
3.31667	N_2	3.1.2.21c
3.31760	N_2	3.1.2.21c
3.3182141	Ne	1.9.2
3.31889	N_2	3.1.2.21c
3.32060	HF	3.2.1.5a
3.32069	N_2	3.1.2.21c
3.33500	HF	3.2.1.5a
3.3341754	Ne	1.9.2
3.34010	Kr	1.9.4
3.34679986	Kr	1.9.4
3.3510469	Ne	1.9.2
3.3520466	Ne	1.9.2
3.36660	Xe	1.9.5
3.37720	HF	3.2.1.5a
3.3813942	Ne	1.9.2
3.3849653	Ne	1.9.2
3.389503	S	1.6.2
3.3912244	Ne	1.9.2
3.3922345	Ne	1.9.2
3.4023945	Xe	1.9.5
3.407422	C	1.4.1
3.428	COS	3.2.2.2a
3.4296	I	1.7.4
3.4480843	Ne	1.9.2
3.45184	N_2	3.1.2.21c
3.45832	N_2	3.1.2.21c
3.46114	N_2	3.1.2.21c
3.46368	N_2	3.1.2.21c
3.4654	Sm	1.8.3
3.46596	N_2	3.1.2.21c
3.46795	N_2	3.1.2.21c
3.46967	N_2	3.1.2.21c
3.47109	N_2	3.1.2.21c
3.4729495	Ne	1.9.2
3.4882957	Kr	1.9.4
3.4894892	Kr	1.9.4
3.4909362	Cs	1.1.5
3.4933	DF	3.2.1.5b
3.5070	Xe	1.9.5
3.5117661	C	1.4.1
3.5214	DF	3.2.1.5b
3.5361	Sm	1.8.3
3.543052	Cl	1.7.2
3.5507	DF	3.2.1.5b
3.57278	HCl	3.2.1.4a
3.58110	DF	3.2.1.5b
3.5894556	Ne	1.9.2
3.601	B	1.3.1
3.60262	HCl	3.2.1.4a
3.6128	DF	3.2.1.5b
3.6140628	Cs	1.1.5
3.6164638	Ne	1.9.2
3.6219081	Xe	1.9.5
3.62349	N_2	3.1.2.21d
3.62614	N_2	3.1.2.21d
3.62910	N_2	3.1.2.21d
3.631236	Ar	1.9.3
3.63367	HCl	3.2.1.4a
3.63619	HCl	3.2.1.4a
3.63630	DF	3.2.1.5b
3.64313	N_2	3.1.2.21d
3.64472	N_2	3.1.2.21d
3.645	DF	3.2.1.5b
3.64662	N_2	3.1.2.21d
3.64883	N_2	3.1.2.21d
3.65138	N_2	3.1.2.21d
3.6518315	Xe	1.9.5
3.65424	N_2	3.1.2.21d
3.65745	N_2	3.1.2.21d
3.66095	N_2	3.1.2.21d
3.66483	N_2	3.1.2.21d
3.66599	HCl	3.2.1.4a
3.6665	DF	3.2.1.5b
3.66848	HCl	3.2.1.4a
3.6789254	Mg	1.2.1
3.6798859	Xe	1.9.5
3.6798	DF	3.2.1.5b
3.6825364	Mg	1.2.1
3.6858866	Xe	1.9.5
3.6899	N_2	3.1.2.21b
3.6983	DF	3.2.1.5b

Table 4.1 (continued)
GAS LASERS ARRANGED IN ORDER OF WAVELENGTH

Wavelength (μm)	Element or compound	Table No.	Wavelength (μm)	Element or compound	Table No.
3.69958	HCl	3.2.1.4a	3.8704	DF	3.2.1.5b
3.7013512	Ar	1.9.3	3.8757	DF	3.2.1.5b
3.70207	HCl	3.2.1.4a	3.8768	HCl	3.2.1.4a
3.7071	HCl	3.2.1.4a	3.8816	DF	3.2.1.5b
3.7086023	Ar	1.9.3	3.8840	HCl	3.2.1.4a
3.7097	HCl	3.2.1.4a	3.8864	HCl	3.2.1.4a
3.7143477	Ar	1.9.3	3.8903	DF	3.2.1.5b
3.7155	DF	3.2.1.5b	3.8950221	Xe	1.9.5
3.7310	DF	3.2.1.5b	3.9128	DF	3.2.1.5b
3.73450	HCl	3.2.1.4a	3.9155	DF	3.2.1.5b
3.7351	N_2	3.1.2.21b	3.9145	DF	3.2.1.5b
3.73698	HCl	3.2.1.4a	3.9149	HCl	3.2.1.4a
3.7383	HCl	3.2.1.4a	3.9181	HCl	3.2.1.4a
3.7408	HCl	3.2.1.4a	3.9205	HCl	3.2.1.4a
3.7520	DF	3.2.1.5b	3.9272	DF	3.2.1.5b
3.7563	DF	3.2.1.5b	3.93	Hg	1.2.7
3.7651	DF	3.2.1.5b	3.9487	DF	3.2.1.5b
3.7710	HCl	3.2.1.4a	3.9536	HCl	3.2.1.4a
3.77079	HCl	3.2.1.4a	3.956	Kr	1.9.4
3.77350	HCl	3.2.1.4a	3.9560	HCl	3.2.1.4a
3.7746325	Ne	1.9.2	3.9565	DF	3.2.1.5b
3.7742128	Kr	1.9.4	3.9572	DF	3.2.1.5b
3.7834	N_2	3.1.2.21d	3.9589222	Ba	1.2.4
3.7878	DF	3.2.1.5b	3.9654	DF	3.2.1.5b
3.7902	DF	3.2.1.5b	3.9806	Ne	1.9.2
3.7942	N	1.5.1	3.9843	DF	3.2.1.5b
3.79660	Cl	1.7.2	3.9909	HCl	3.2.1.4a
3.8007	DF	3.2.1.5b	3.9966035	Xe	1.9.5
3.8050	HCl	3.2.1.4a	3.9995	DF	3.2.1.5b
3.8074	HCl	3.2.1.4a	4.0032	DF	3.2.1.5b
3.80847	HCl	3.2.1.4a	4.0054	DF	3.2.1.5b
3.8135916	Te	1.3.4	4.0059	HCl	3.2.1.4a
3.8154	N	1.5.1	4.0079678	Ba	1.2.4
3.8206	DF	3.2.1.5b	4.0170	HBr	3.2.1.3a
3.8298	DF	3.2.1.5b	4.0176	HBr	3.2.1.3a
3.8375	DF	3.2.1.5b	4.0207278	Xe	1.9.5
3.8401	HCl	3.2.1.4a	4.0212	DF	3.2.1.5b
3.8425	HCl	3.2.1.4a	4.0295	HCl	3.2.1.4a
3.85	HCN	3.2.2.4	4.0404	HCl	3.2.1.4a
3.8501	DF	3.2.1.5b	4.0434	DF	3.2.1.5b
3.8509	HCl	3.2.1.4a	4.0464	DF	3.2.1.5b
3.8536	HCl	3.2.1.4a	4.0470	HBr	3.2.1.3a
3.8547	DF	3.2.1.5b	4.0475	HBr	3.2.1.3a
3.86573	Mg	1.2.1	4.0491	DF	3.2.1.5b
3.8696535	Xe	1.9.5	4.0594	DF	3.2.1.5b

Table 4.1 (continued)
GAS LASERS ARRANGED IN ORDER OF WAVELENGTH

Wavelength (μm)	Element or compound	Table No.
4.068	Kr	1.9.4
4.0764	HCl	3.2.1.4a
4.0783	HBr	3.2.1.3a
4.0788	HBr	3.2.1.3a
4.0892	DF	3.2.1.5b
4.0895	DF	3.2.1.5b
4.1107	HBr	3.2.1.3a
4.1112	HBr	3.2.1.3a
4.1140	HCl	3.2.1.4a
4.1337	DF	3.2.1.5b
4.1368	Sm	1.8.3
4.1369	DF	3.2.1.5b
4.1442	HBr	3.2.1.3a
4.1448	HBr	3.2.1.3a
4.1526299	Xe	1.9.5
4.1526711	Kr	1.9.4
4.1653	HBr	3.2.1.3a
4.1658	HBr	3.2.1.3a
4.1796	HBr	3.2.1.3a
4.1798	DF	3.2.1.5b
4.1820	X	1.9.5
4.1862	DF	3.2.1.5b
4.1970	HBr	3.2.1.3a
4.1975	HBr	3.2.1.3a
4.2013276	Mg	1.2.1
4.2044098	Ar	1.9.3
4.2181082	Cs	1.1.5
4.218950	Ne	1.9.2
4.2295	HBr	3.2.1.5a
4.2633	HBr	3.2.1.5a
4.2639	HBr	3.2.1.5a
4.2988	HBr	3.2.1.5a
4.2994	HBr	3.2.1.5a
4.31311	CO_2	3.2.2.1a
4.314	CO_2	3.2.2.1b
4.31626	CO_2	3.2.2.1a
4.31787	CO_2	3.2.2.1a
4.3203	CO_2	3.2.2.1a
4.32110	CO_2	3.2.2.1a
4.3213904	Eu	1.8.4
4.32271	CO_2	3.2.2.1a
4.32439	CO_2	3.2.2.1a
4.3249	CO_2	3.2.2.1a
4.3250	HBr	3.2.1.3a
4.3255	HBr	3.2.1.3a
4.32601	CO_2	3.2.2.1a
4.3276	CO_2	3.2.2.1a
4.32772	CO_2	3.2.2.1a
4.3285152	Ba	1.2.4
4.32937	CO_2	3.2.2.1
4.33100	CO_2	3.2.2.1a
4.3321362	I	1.7.4
4.33453	CO_2	3.2.2.1a
4.3354	HBr	3.2.1.3a
4.33623	CO_2	3.2.2.1a
4.33801	CO_2	3.2.2.1a
4.33974	CO_2	3.2.2.1a
4.340	CO_2	3.2.2.1b
4.34153	CO_2	3.2.2.1a
4.34331	CO_2	3.2.2.1a
4.346	CO_2	3.2.2.1c
4.35059	CO_2	3.2.2.1a
4.3513	CO_2	3.2.2.1a
4.3540	CO_2	3.2.2.1b
4.3549	CO_2	3.2.2.1a
4.3579	HBr	3.2.1.3a
4.3580	CO_2	3.2.2.1a
4.3585	HBr	3.2.1.3a
4.3612	CO_2	3.2.2.1a
4.3638859	Mg	1.2.1
4.3644	CO_2	3.2.2.1a
4.3677	CO_2	3.2.2.1a
4.371	CO_2	3.2.2.1c
4.3711	CO_2	3.2.2.1a
4.3745	CO_2	3.2.2.1a
4.3747938	Kr	1.9.4
4.376612	Kr	1.9.4
4.3770	CO_2	3.2.2.1c
4.3779	CO_2	3.2.2.1a
4.3814	CO_2	3.2.2.1a
4.382	CO_2	3.2.2.1c
4.3849	CO_2	3.2.2.1a
4.385	CO_2	3.2.2.1c
4.3920	CO_2	3.2.2.1c
4.3925	HBr	3.2.1.3a
4.3931	HBr	3.2.1.3a
4.398	CO_2	3.2.2.1c
4.4281	HBr	3.2.1.3a
4.4307	HBr	3.2.1.3a
4.4652	HBr	3.2.1.3a

Table 4.1 (continued)
GAS LASERS ARRANGED IN ORDER OF WAVELENGTH

Wavelength (μm)	Element or compound	Table No.	Wavelength (μm)	Element or compound	Table No.
4.4658	HBr	3.2.1.3a	*4.82395*	CO	3.2.1.2a
4.5041	HBr	3.2.1.3a	*4.83361*	CO	3.2.1.2a
4.5047	HBr	3.2.1.3a	*4.83371*	CO	3.2.1.2a
4.5330	HBr	3.2.1.3a	*4.84354*	CO	3.2.1.2a
4.5335	HBr	3.2.1.3a	*4.84403*	CO	3.2.1.2a
4.5393674	Xe	1.9.5	*4.85359*	CO	3.2.1.2a
4.5615027	O	1.6.1	*4.85457*	CO	3.2.1.2a
4.5678667	Xe	1.9.5	*4.85623*	CO	3.2.1.2a
4.5691	HBr	3.2.1.3a	*4.8564726*	I	1.7.4
4.5696	HBr	3.2.1.3a	*4.8629144*	I	1.7.4
4.5706441	Xe	1.9.5	*4.86370*	CO	3.2.1.2a
4.5851	N_2O	3.2.2.6c	*4.86520*	CO	3.2.1.2a
4.6	N_2O	3.2.2.6d	*4.86560*	Sm	1.2.3
4.60535	He	1.9.1	*4.87394*	CO	3.2.1.2a
4.60567	He	1.9.1	*4.87549*	CO	3.2.1.2a
4.6070	HBr	3.2.1.3a	*4.8773393*	Kr	1.9.4
4.6076	HBr	3.2.1.3a	*4.8831334*	Kr	1.9.4
4.6157396	Sn	1.4.4	*4.88429*	CO	3.2.1.2a
4.6109078	Xe	1.9.5	*4.88530*	CO	3.2.1.2a
4.6189	N_2O	3.2.2.6c	*4.89477*	CO	3.2.1.2a
4.6204	N_2O	3.2.2.6b	*4.89522*	CO	3.2.1.2a
4.6463	HBr	3.2.1.3a	*4.90527*	CO	3.2.1.2a
4.6467	HBr	3.2.1.3a	*4.90537*	CO	3.2.1.2a
4.6699795	Ba	1.2.4	*4.90868*	CO	3.2.1.2a
4.6812	N_2O	3.2.2.6b	*4.9160256*	Ar	1.9.3
4.6948356	Eu	1.8.4	*4.91543*	CO	3.2.1.2a
4.7151680	Ar	1.9.3	*4.91609*	CO	3.2.1.2a
4.7169143	Ba	1.2.4	*4.91849*	CO	3.2.1.2a
4.7184144	Ba	1.2.4	*4.9207077*	Ar	1.9.3
4.745130	CO	3.2.1.2a	*4.92623*	CO	3.2.1.2a
4.75450	CO	3.2.1.2a	*4.93614*	CO	3.2.1.2a
4.76398	CO	3.2.1.2a	*4.93808*	CO	3.2.1.2a
4.77	H_2O	3.2.2.5	*4.94667*	CO	3.2.1.2a
4.77350	CO	3.2.1.2a	*4.94805*	CO	3.2.1.2a
4.77689	CO	3.2.1.2a	*4.95733*	CO	3.2.1.2a
4.78330	CO	3.2.1.2a	*4.95815*	CO	3.2.1.2a
4.78608	CO	3.2.1.2a	*4.96277*	CO	3.2.1.2a
4.79312	CO	3.2.1.2a	*4.96812*	CO	3.2.1.2a
4.79537	CO	3.2.1.2a	*4.96837*	CO	3.2.1.2a
4.8021974	Yb	1.7.5	*4.97243*	CO	3.2.1.2a
4.80307	CO	3.2.1.2a	*4.97871*	CO	3.2.1.2a
4.80479	CO	3.2.1.2a	*4.97903*	CO	3.2.1.2a
4.81313	CO	3.2.1.2a	*4.98220*	CO	3.2.1.2a
4.81431	CO	3.2.1.2a	*4.98918*	CO	3.2.1.2a
4.82331	CO	3.2.1.2a	*4.99008*	CO	3.2.1.2a

Table 4.1 (continued)
GAS LASERS ARRANGED IN ORDER OF WAVELENGTH

Wavelength (μm)	Element or compound	Table No.	Wavelength (μm)	Element or compound	Table No.
4.99210	CO	3.2.1.2a	5.07581	CO	3.2.1.2a
4.99978	CO	3.2.1.2a	5.07799	CO	3.2.1.2a
4.9996952	Kr	1.9.4	5.07890	CN	3.2.1.1
5.00125	CO	3.2.1.2a	5.08110	DCl	3.2.1.4b
5.00212	CO	3.2.1.2a	5.08314	CO	3.2.1.2a
5.00321	DCl	3.2.1.4b	5.08380	CN	3.2.1.1
5.00980	DCl	3.2.1.4b	5.08417	CO	3.2.1.2a
5.01050	CO	3.2.1.2a	5.08600	Zn^+	2.5.1
5.01227	CO	3.2.1.2a	5.08686	CO	3.2.1.2a
5.01255	CO	3.2.1.2a	5.08844	CO	3.2.1.2a
5.01794	CO	3.2.1.2a	5.08960	CN	3.2.1.1
5.02135	CO	3.2.1.2a	5.09200	CO	3.2.1.2a
5.02259	CO	3.2.1.2a	5.09390	CN	3.2.1.1
5.0235584	Ar	1.9.3	5.09400	CO	3.2.1.2a
5.02398	CO	3.2.1.2a	5.09539	CO	3.2.1.2a
5.0243255	Xe	1.9.5	5.09683	CO	3.2.1.2a
5.02764	CO	3.2.1.2a	5.09803	CO	3.2.1.2a
5.0322846	Ba	1.2.4	5.09902	CO	3.2.1.2a
5.03233	CO	3.2.1.2a	5.10000	DCl	3.2.1.4b
5.03294	CO	3.2.1.2a	5.10043	CO	3.2.1.2a
5.0348	DCl	3.2.1.4b	5.10300	Ne	1.9.2
5.03554	CO	3.2.1.2a	5.10402	CO	3.2.1.2a
5.03720	CN	3.2.1.1	5.10420	CN	3.2.1.1
5.03745	CO	3.2.1.2a	5.10441	CO	3.2.1.2a
5.04120	DCl	3.2.1.4b	5.10490	CO	3.2.1.2a
5.04344	CO	3.2.1.2a	5.10670	DCl	3.2.1.4b
5.04346	CO	3.2.1.2a	5.1072522	Tl	1.3.4
5.0445	DCl	3.2.1.4b	5.10777	CO	3.2.1.2a
5.04724	CO	3.2.1.2a	5.10898	CO	3.2.1.2a
5.04740	CN	3.2.1.1	5.10934	CO	3.2.1.2a
5.04740	CO	3.2.1.2a	5.10974	CO	3.2.1.2a
5.05140	DCl	3.2.1.4b	5.11180	DCl	3.2.1.4b
5.05412	CO	3.2.1.2a	5.11212	CO	3.2.1.2a
5.05468	CO	3.2.1.2a	5.11413	CO	3.2.1.2a
5.05747	CO	3.2.1.2a	5.1146	CN	3.2.1.1
5.05780	CN	3.2.1.1	5.11994	CO	3.2.1.2a
5.05907	CO	3.2.1.2a	5.12029	CO	3.2.1.2a
5.06490	CO	3.2.1.2a	5.12058	CO	3.2.1.2a
5.0660871	Eu	1.8.4	5.1216467	Ar	1.9.3
5.06605	CO	3.2.1.2a	5.12200	DCl	3.2.1.4b
5.06670	DCl	3.2.1.4b	5.12275	CO	3.2.1.2a
5.06766	CO	3.2.1.2a	5.12438	CO	3.2.1.2a
5.06830	CN	3.2.1.1	5.12510	CN	3.2.1.1
5.071040	CO	3.2.1.2a	5.12789	CO	3.2.1.2a
5.07430	DCl	3.2.1.4b	5.12973	CO	3.2.1.2a

Table 4.1 (continued)
GAS LASERS ARRANGED IN ORDER OF WAVELENGTH

Wavelength (μm)	Element or compound	Table No.	Wavelength (μm)	Element or compound	Table No.
5.1311509	Kr	1.9.4	*5.18360*	CN	3.2.1.1
5.13156	CO	3.2.1.2a	*5.18644*	CO	3.2.1.2a
5.13237	CO	3.2.1.2a	*5.18790*	DCl	3.2.1.4b
5.13295	CO	3.2.1.2a	*5.18846*	CO	3.2.1.2a
5.13410	DCl	3.2.1.4b	*5.18862*	CO	3.2.1.2a
5.13470	CO	3.2.1.2a	*5.18987*	CO	3.2.1.2a
5.13580	CN	3.2.1.1	*5.19234*	CO	3.2.1.2a
5.13596	CO	3.2.1.2a	*5.19289*	CO	3.2.1.2a
5.13630	DCl	3.2.1.4b	*5.19339*	CO	3.2.1.2a
5.13682	CO	3.2.1.2a	*5.19450*	CN	3.2.1.1
5.14220	CN	3.2.1.1	*5.19979*	CO	3.2.1.2a
5.14267	CO	3.2.1.2a	*5.20025*	CO	3.2.1.2a
5.14310	DCl	3.2.1.4b	*5.20046*	CO	3.2.1.2a
5.14403	CO	3.2.1.2a	*5.20200*	DCl	3.2.1.4b
5.14408	CO	3.2.1.2a	*5.20345*	CO	3.2.1.2a
5.14526	CO	3.2.1.2a	*5.20475*	CO	3.2.1.2a
5.14660	CN	3.2.1.1	*5.20540*	CN	3.2.1.1
5.15110	DCl	3.2.1.4b	*5.20765*	CO	3.2.1.2a
5.15137	CO	3.2.1.2a	*5.20900*	DCl	3.2.1.4b
5.15197	CO	3.2.1.2a	*5.21110*	CO	3.2.1.2a
5.15240	CN	3.2.1.1	*5.21132*	CO	3.2.1.2a
5.15391	CO	3.2.1.2a	*5.21180*	DCl	3.2.1.4b
5.15590	CO	3.2.1.2a	*5.21219*	CO	3.2.1.2a
5.15750	CN	3.2.1.1	*5.21414*	CO	3.2.1.2a
5.15594	CO	3.2.1.2a	*5.21497*	CO	3.2.1.2a
5.15800	DCl	3.2.1.4b	*5.21640*	CN	3.2.1.1
5.16200	CO	3.2.1.2a	*5.21732*	CO	3.2.1.2a
5.16270	CN	3.2.1.1	*5.21860*	DCl	3.2.1.4b
5.16529	CO	3.2.1.2a	*5.22139*	CO	3.2.1.2a
5.16640	CO	3.2.1.2a	*5.22242*	CO	3.2.1.2a
5.16667	CO	3.2.1.2a	*5.22256*	CO	3.2.1.2a
5.16793	CO	3.2.1.2a	*5.22426*	CO	3.2.1.2a
5.16860	CN	3.2.1.1	*5.22497*	CO	3.2.1.2a
5.1711388	Ne	1.9.2	*5.22760*	CN	3.2.1.1
5.17216	CO	3.2.1.2a	*5.22998*	CO	3.2.1.2a
5.17320	CN	3.2.1.1	*5.23002*	CO	3.2.1.2a
5.17410	CO	3.2.1.2a	*5.23160*	CO	3.2.1.2a
5.17600	DCl	3.2.1.4b	*5.23370*	CN	3.2.1.1
5.17681	CO	3.2.1.2a	*5.23414*	CO	3.2.1.2a
5.17758	CO	3.2.1.2a	*5.23594*	CO	3.2.1.2a
5.17961	CO	3.2.1.2a	*5.23648*	CO	3.2.1.2a
5.18006	CO	3.2.1.2a	*5.23768*	CO	3.2.1.2a
5.18110	DCl	3.2.1.4b	*5.23890*	CN	3.2.1.1.
5.18192	CO	3.2.1.2a	*5.24195*	CO	3.2.1.2a
5.18245	CO	3.2.1.2a	*5.24287*	CO	3.2.1.2a

Table 4.1 (continued)
GAS LASERS ARRANGED IN ORDER OF WAVELENGTH

Wavelength (μm)	Element or compound	Table No.	Wavelength (μm)	Element or compound	Table No.
5.24350	CO	3.2.1.2a	*5.3097*	DCl	3.2.1.4b
5.24430	CN	3.2.1.1.	*5.3112*	CN	3.2.1.1
5.24550	CO	3.2.1.2a	*5.31284*	CO	3.2.1.2a
5.24587	CO	3.2.1.2a	*5.31342*	CO	3.2.1.2a
5.24704	CO	3.2.1.2a	*5.31451*	CO	3.2.1.2a
5.24884	CO	3.2.1.2a	*5.31660*	CO	3.2.1.2a
5.25030	CN	3.2.1.1	*5.31818*	CO	3.2.1.2a
5.25030	DCl	3.2.1.4b	*5.31926*	CO	3.2.1.2a
5.25243	CO	3.2.1.2a	*5.3229*	CN	3.2.1.1
5.25510	CO	3.2.1.2a	*5.32409*	CO	3.2.1.2a
5.25587	CO	3.2.1.2a	*5.32440*	DCl	3.2.1.4b
5.25774	CO	3.2.1.2a	*5.32582*	CO	3.2.1.2b
5.25828	CO	3.2.1.2a	*5.3258685*	Ne	1.9.2
5.26135	CO	3.2.1.2a	*5.32609*	CO	3.2.1.2a
5.26304	CO	3.2.1.2a	*5.32843*	CO	3.2.1.2b
5.2661	CN	3.2.1.1	*5.32669*	CO	3.2.1.2a
5.26902	CO	3.2.1.2a	*5.32963*	CO	3.2.1.2a
5.26966	CO	3.2.1.2a	*5.3297517*	Bi	1.5.5
5.26976	CO	3.2.1.2a	*5.33201*	CO	3.2.1.2a
5.27379	CO	3.2.1.2a	*5.33489*	CO	3.2.1.2a
5.27400	CO	3.2.1.2a	*5.33490*	CO	3.2.1.2a
5.2760	DCl	3.2.1.4b	*5.33738*	CO	3.2.1.2b
5.2771	CN	3.2.1.1	*5.33932*	CO	3.2.1.2b
5.28118	CO	3.2.1.2a	*5.33949*	CO	3.2.1.2a
5.28192	CO	3.2.1.2a	*5.34094*	CO	3.2.1.2a
5.2825643	Eu	1.8.4	*5.34122*	CO	3.2.1.2a
5.28224	CO	3.2.1.2a	*5.34326*	CO	3.2.1.2a
5.2829	DCl	3.2.1.4b	*5.34430*	DCl	3.2.1.4b
5.28468	CO	3.2.1.2a	*5.34564*	CO	3.2.1.2a
5.2879276	As	1.5.3	*5.35032*	CO	3.2.1.2a
5.28680	CO	3.2.1.2a	*5.35172*	CO	3.2.1.2a
5.2884	CN	3.2.1.1	*5.35295*	CO	3.2.1.2a
5.29250	CO	3.2.1.2a	*5.35465*	CO	3.2.1.2a
5.29285	CO	3.2.1.2a	*5.35490*	CO	3.2.1.2a
5.29420	CO	3.2.1.2a	*5.3562*	DCl	3.2.1.4b
5.29571	CO	3.2.1.2a	*5.3566973*	Xe	1.9.5
5.2977	CN	3.2.1.1	*5.35693*	CO	3.2.1.2a
5.2999768	Kr	1.9.4	*5.35794*	CO	3.2.1.2a
5.29974	CO	3.2.1.2a	*5.36151*	CO	3.2.1.2b
5.3019553	Kr	1.9.4	*5.36219*	CO	3.2.1.2a
5.30289	CO	3.2.1.2a	*5.36290*	DCl	3.2.1.4b
5.30328	CO	3.2.1.2b	*5.36484*	CO	3.2.1.2a
5.30465	CO	3.2.1.2a	*5.36534*	CO	3.2.1.2a
5.30667	CO	3.2.1.2a	*5.36566*	CO	3.2.1.2a
5.30687	CO	3.2.1.2a	*5.36816*	CO	3.2.1.2a

Table 4.1 (continued)
GAS LASERS ARRANGED IN ORDER OF WAVELENGTH

Wavelength (μm)	Element or compound	Table No.	Wavelength (μm)	Element or compound	Table No.
5.36985	CO	3.2.1.2a	*5.41452*	CO	3.2.1.2a
5.37114	CO	3.2.1.2a	*5.41470*	CO	3.2.1.2a
5.37280	CO	3.2.1.2b	*5.41603*	CO	3.2.1.2a
5.37448	CO	3.2.1.2a	*5.41752*	CO	3.2.1.2a
5.37592	CO	3.2.1.2a	*5.41864*	CO	3.2.1.2a
5.37686	CO	3.2.1.2a	*5.41965*	CO	3.2.1.2a
5.37702	CO	3.2.1.2b	*5.42341*	CO	3.2.1.2a
5.37765	CO	3.2.1.2a	*5.42425*	CO	3.2.1.2a
5.37790	CO	3.2.1.2a	*5.42646*	CO	3.2.1.2a
5.37855	CO	3.2.1.2a	*5.42648*	CO	3.2.1.2a
5.37953	CO	3.2.1.2a	*5.42724*	CO	3.2.1.2a
5.37990	DCl	3.2.1.4b	*5.42950*	DCl	3.2.1.4b
5.38108	CO	3.2.1.2a	*5.4306800*	Eu	1.8.4
5.38423	CO	3.2.1.2b	*5.43082*	CO	3.2.1.2a
5.38449	CO	3.2.1.2a	*5.43093*	CO	3.2.1.2a
5.38559	CO	3.2.1.2a	*5.43095*	CO	3.2.1.2a
5.38664	CO	3.2.1.2a	*5.43226*	CO	3.2.1.2b
5.38690	CO	3.2.1.2a	*5.43393*	CO	3.2.1.2a
5.38781	CO	3.2.1.2a	*5.43598*	CO	3.2.1.2a
5.38784	CO	3.2.1.2b	*5.43778*	CO	3.2.1.2b
5.38890	DCl	3.2.1.4b	*5.43858*	CO	3.2.1.2a
5.38904	CO	3.2.1.2a	*5.43924*	CO	3.2.1.2a
5.38914277	Ar	1.9.3	*5.43998*	CO	3.2.1.2a
5.39138	CO	3.2.1.2a	*5.44239*	CO	3.2.1.2a
5.39365	CO	3.2.1.2a	*5.44396*	CO	3.2.1.2b
5.39467	CO	3.2.1.2a	*5.44427*	CO	3.2.1.2a
5.39560	DCl	3.2.1.4b	*5.44485*	CO	3.2.1.2a
5.39580	CO	3.2.1.2b	*5.44637*	CO	3.2.1.2b
5.39603	CO	3.2.1.2a	*5.45075*	CO	3.2.1.2a
5.39750	CO	3.2.1.2a	*5.45084*	CO	3.2.1.2a
5.39800	CO	3.2.1.2a	*5.45217*	CO	3.2.1.2a
5.39879	CO	3.2.1.2b	*5.45397*	CO	3.2.1.2a
5.40136	CO	3.2.1.2a	*5.45429*	CO	3.2.1.2a
5.40166	CO	3.2.1.2a	*5.45560*	CO	3.2.1.2b
5.40185	CO	3.2.1.2a	*5.45770*	DCl	3.2.1.4b
5.40271	CO	3.2.1.2a	*5.45788*	CO	3.2.1.2a
5.40437	CO	3.2.1.2a	*5.45967*	CO	3.2.1.2b
5.4048094	Ne	1.9.2	*5.46167*	CO	3.2.1.2a
5.40530	CO	3.2.1.2a	*5.46230*	CO	3.2.1.2a
5.40751	CO	3.2.1.2b	*5.46350*	CO	3.2.1.2a
5.40851	CO	3.2.1.2a	*5.46526*	CO	3.2.1.2a
5.40678	CO	3.2.1.2a	*5.46571*	CO	3.2.1.2a
5.40988	CO	3.2.1.2b	*5.46738*	CO	3.2.1.2b
5.41018	CO	3.2.1.2a	*5.4676586*	Ar	1.9.3
5.41384	CO	3.2.1.2a	*5.4677812*	Ar	1.9.3

Table 4.1 (continued)
GAS LASERS ARRANGED IN ORDER OF WAVELENGTH

Wavelength (μm)	Element or compound	Table No.	Wavelength (μm)	Element or compound	Table No.
5.47054	CO	3.2.1.2a	*5.53856*	CO	3.2.1.2a
5.47082	CO	3.2.1.2b	*5.53929*	CO	3.2.1.2a
5.47166	CO	3.2.1.2a	*5.54066*	CO	3.2.1.2b
5.47273	CO	3.2.1.2a	*5.54095*	CO	3.2.1.2a
5.4749269	Xe	1.9.5	*5.54221*	CO	3.2.1.2a
5.47581	CO	3.2.1.2a	*5.54230*	DCl	3.2.1.4b
5.47759	CO	3.2.1.2a	*5.54457*	CO	3.2.1.2b
5.47850	CO	3.2.1.2a	*5.547327*	Ca	1.2.2
5.47913	CO	3.2.1.2b	*5.54712*	CO	3.2.1.2a
5.47980	Ba	1.2.4	*5.54983*	CO	3.2.1.2a
5.48210	CO	3.2.1.2b	*5.55209*	CO	3.2.1.2a
5.48394	CO	3.2.1.2a	*5.55280*	CO	3.2.1.2b
5.48559	CO	3.2.1.2a	*5.55432*	CO	3.2.1.2a
5.49190	CO	3.2.1.2a	*5.55446*	CO	3.2.1.2a
5.49350	DCl	3.2.1.4b	*5.55606*	CO	3.2.1.2b
5.49353	CO	3.2.1.2b	*5.56125*	CO	3.2.1.2a
5.49529	CO	3.2.1.2a	*5.56133*	CO	3.2.1.2a
5.498705	I	1.7.4	5.56360	Ba	1.2.4
5.50000	NO	3.2.1.6	*5.56506*	CO	3.2.1.2a
5.50001	CO	3.2.1.2b	*5.56659*	CO	3.2.1.2a
5.50100	Xe	1.9.5	*5.56770*	CO	3.2.1.2b
5.50139	CO	3.2.1.2a	*5.56870*	DBr	3.2.1.3b
5.50181	CO	3.2.1.2a	*5.5699805*	Kr	1.9.4
5.50358	CO	3.2.1.2b	*5.5704*	DBr	3.2.1.3b
5.50510	CO	3.2.1.2b	*5.57282*	CO	3.2.1.2a
5.50546	CO	3.2.1.2a	*5.5754726*	Xe	1.9.5
5.50680	CO	3.2.1.2a	*5.57473*	CO	3.2.1.2b
5.50840	DCl	3.2.1.4b	*5.57760*	DCl	3.2.1.4b
5.51094	CO	3.2.1.2b	*5.5779*	DBr	3.2.1.3b
5.51415	CO	3.2.1.2a	*5.57818*	CO	3.2.1.2a
5.51442	CO	3.2.1.2a	*5.57901*	CO	3.2.1.2a
5.51500	Ne	1.9.2	*5.57948*	CO	3.2.1.2a
5.51641	CO	3.2.1.2a	*5.58454*	CO	3.2.1.2a
5.51681	CO	3.2.1.2b	*5.58586*	CO	3.2.1.2b
5.51845	CO	3.2.1.2a	*5.5862769*	Kr	1.9.4
5.51919	CO	3.2.1.2a	*5.59140*	CO	3.2.1.2b
5.52200	CO	3.2.1.2b	*5.59146*	CO	3.2.1.2a
5.52664	CO	3.2.1.2a	*5.59159*	CO	3.2.1.2a
5.52744	CO	3.2.1.2a	*5.59489*	CO	3.2.1.2a
5.52761	CO	3.2.1.2a	*5.5983205*	C	1.4.1
5.52866	CO	3.2.1.2b	*5.59642*	CO	3.2.1.2a
5.53026	CO	3.2.1.2a	*5.59713*	CO	3.2.1.2b
5.53040	DCl	3.2.1.4b	*5.6034328*	Xe	1.9.5
5.53307	CO	3.2.1.2a	*5.60348*	CO	3.2.1.2b
5.53322	CO	3.2.1.2b	*5.60433*	CO	3.2.1.2a

Table 4.1 (continued)
GAS LASERS ARRANGED IN ORDER OF WAVELENGTH

Wavelength (μm)	Element or compound	Table No.	Wavelength (μm)	Element or compound	Table No.
5.60491	CO	3.2.1.2a	*5.69823*	CO	3.2.1.2a
5.60607	CO	3.2.1.2a	*5.70000*	HCOOS	3.2.4.5
5.60845	CO	3.2.1.2a	*5.70104*	CO	3.2.1.2a
5.60854	CO	3.2.1.2b	*5.70516*	CO	3.2.1.2b
5.61370	DCl	3.2.1.4b	*5.7067951*	Ne	1.9.2
5.61722	CO	3.2.1.2a	*5.71043*	CO	3.2.1.2a
5.61740	CO	3.2.1.2a	*5.71100*	DBr	3.2.1.3b
5.62010	CO	3.2.1.2b	*5.71300*	DBr	3.2.1.3b
5.62063	CO	3.2.1.2a	*5.71363*	CO	3.2.1.2a
5.62760	DBr	3.2.1.3b	*5.72147*	CO	3.2.1.2b
5.62807	CO	3.2.1.2b	*5.72179*	CO	3.2.1.2a
5.62888	CO	3.2.1.2a	*5.72392*	CO	3.2.1.2a
5.62910	DBr	3.2.1.3b	*5.72638*	CO	3.2.1.2a
5.63028	CO	3.2.1.2a	*5.72969*	CO	3.2.1.2b
5.6305126	Kr	1.9.4	*5.73084*	CO	3.2.1.2b
5.63229	CO	3.2.1.2a	*5.73369*	CO	3.2.1.2b
5.63298	CO	3.2.1.2a	*5.73381*	CO	3.2.1.2a
5.64350	CO	3.2.1.2b	*5.73757*	CO	3.2.1.2a
5.64549	CO	3.2.1.2a	*5.73929*	CO	3.2.1.2a
5.65127	CO	3.2.1.2a	*5.74100*	DBr	3.2.1.3b
5.65230	CO	3.2.1.2a	*5.74122*	CO	3.2.1.2a
5.65688	CO	3.2.1.2a	*5.74300*	DBr	3.2.1.3b
5.65815	CO	3.2.1.2a	*5.74599*	CO	3.2.1.2a
5.6582	DBr	3.2.1.3b	*5.74607*	CO	3.2.1.2b
5.6597	DBr	3.2.1.3b	*5.75139*	CO	3.2.1.2a
5.66260	CO	3.2.1.2b	*5.75238*	CO	3.2.1.2a
5.66425	CO	3.2.1.2a	*5.75291*	CO	3.2.1.2a
5.66609	CO	3.2.1.2b	*5.754965*	Ga	1.3.2
5.6667374	Ne	1.9.2	*5.75778*	CO	3.2.1.2a
5.66781	CO	3.2.1.2b	*5.75833*	CO	3.2.1.1.2a
5.67043	CO	3.2.1.2a	*5.75861*	CO	3.2.1.2a
5.67097	CO	3.2.1.2a	*5.76538*	CO	3.2.1.2a
5.67408	CO	3.2.1.2b	*5.76562*	CO	3.2.1.2a
5.67635	CO	3.2.1.2a	*5.77083*	CO	3.2.1.2a
5.68011	CO	3.2.1.2a	*5.77130*	CO	3.2.1.2a
5.68396	CO	3.2.1.2a	*5.772*	DBr	3.2.1.3b
5.68415	CO	3.2.1.2a	*5.7722389*	Eu	1.8.4
5.68570	CO	3.2.1.2b	*5.773*	DBr	3.2.1.3b
5.68669	CO	3.2.1.2a	*5.7773913*	Ne	1.9.2
5.68862	CO	3.2.1.2a	*5.77675*	CO	3.2.1.2b
5.6896	DBr	3.2.1.3b	*5.77904*	CO	3.2.1.2a
5.6911	DBr	3.2.1.3b	*5.77955*	CO	3.2.1.2a
5.69711	CO	3.2.1.2a	*5.78115*	CO	3.2.1.2a
5.69747	CO	3.2.1.2b	*5.78350*	CO	3.2.1.2a
5.69803	CO	3.2.1.2a	*5.78414*	CO	3.2.1.2a

Table 4.1 (continued)
GAS LASERS ARRANGED IN ORDER OF WAVELENGTH

Wavelength (μm)	Element or compound	Table No.	Wavelength (μm)	Element or compound	Table No.
5.78689	CO	3.2.1.2b	*5.8620*	DBr	3.2.1.3b
5.79262	CO	3.2.1.2b	*5.8626*	CO	3.2.1.2a
5.79308	CO	3.2.1.2a	*5.86306*	CO	3.2.1.2a
5.79389	CO	3.2.1.2a	*5.8648*	Hg	1.2.7
5.79634	CO	3.2.1.2a	*5.86508*	CO	3.2.1.2b
5.79715	CO	3.2.1.2b	*5.86623*	CO	3.2.1.2a
5.79845	CO	3.2.1.2b	*5.8665236*	Ar	1.9.3
5.80119	CO	3.2.1.2b	*5.86810*	CO	3.2.1.2a
5.802	DBr	3.2.1.3b	*5.87037*	CO	3.2.1.2b
5.803	DBr	3.2.1.3b	*5.8706*	NO	3.2.1.6
5.8037601	Ar	1.9.3	*5.87693*	CO	3.2.1.2a
5.804	DBr	3.2.1.3b	*5.87838*	CO	3.2.1.2a
5.8049	DBr	3.2.1.3b	*5.8789*	NO	3.2.1.6
5.80516	CO	3.2.1.2a	*5.8858082*	Ne	1.9.2
5.80638	CO	3.2.1.2a	5.8899	Ba	1.2.4
5.80841	CO	3.2.1.2a	*5.89071*	CO	3.2.1.2a
5.80934	CO	3.2.1.2a	*5.89096*	CO	3.2.1.2a
5.81005	CO	3.2.1.2b	*5.89244*	CO	3.2.1.2b
5.81032	CO	3.2.1.2b	*5.8928*	DBr	3.2.1.3b
5.81365	CO	3.2.1.2b	*5.8944*	DBr	3.2.1.3b
5.81742	CO	3.2.1.2a	*5.89446*	CO	3.2.1.2a
5.82031	CO	3.2.1.2a	*5.90320*	CO	3.2.1.2a
5.82252	CO	3.2.1.2a	*5.9036*	NO	3.2.1.6
5.82627	CO	3.2.1.2b	*5.9044*	CO	3.2.1.2b
5.82983	CO	3.2.1.2a	*5.9047*	CO	3.2.1.2b
5.832	DBr	3.2.1.3b	*5.90789*	CO	3.2.1.2a
5.83371	CO	3.2.1.2b	*5.9083*	NO	3.2.1.6
5.834	DBr	3.2.1.3b	*5.913*	Xe	1.9.5
5.83441	CO	3.2.1.2a	*5.91587*	CO	3.2.1.2a
5.83586	CO	3.2.1.2a	*5.91958*	CO	3.2.1.2a
5.836	DBr	3.2.1.3b	*5.92149*	CO	3.2.1.2b
5.837	DBr	3.2.1.3b	*5.92151*	CO	3.2.1.2a
5.83905	CO	3.2.1.2b	*5.92460*	DBr	3.2.1.3b
5.84242	CO	3.2.1.2a	*5.9261*	DBr	3.2.1.3b
5.84577	CO	3.2.1.2b	*5.92871*	CO	3.2.1.2a
5.8477292	Ar	1.9.3	*5.92882*	CO	3.2.1.2b
5.8462	NO	3.2.1.6	*5.93530*	CO	3.2.1.2a
5.84870	CO	3.2.1.2a	*5.94230*	NO	3.2.1.6
5.84937	CO	3.2.1.2a	*5.9495478*	Eu	1.8.4
5.85198	CO	3.2.1.2b	*5.94803*	CO	3.2.1.2a
5.85423	CO	3.2.1.2a	*5.94927*	CO	3.2.1.2a
5.8549	NO	3.2.1.6	*5.95354*	CO	3.2.1.2a
5.85517	CO	3.2.1.2a	*5.95390*	CO	3.2.1.2b
5.85799	CO	3.2.1.2b	*5.9546*	NO	3.2.1.6
5.85840	NO	3.2.1.6	*5.95490*	CO	3.2.1.2a

Table 4.1 (continued)
GAS LASERS ARRANGED IN ORDER OF WAVELENGTH

Wavelength (μm)	Element or compound	Table No.
5.9550	NO	3.2.1.6
5.9578742	Ne	1.9.2
5.9573	DBr	3.2.1.3b
5.9590	DBr	3.2.1.3b
5.96155	CO	3.2.1.2b
5.9632	NO	3.2.1.6
5.96342	CO	3.2.1.2a
5.96484	CO	3.2.1.2b
5.96594	CO	3.2.1.2a
5.96668	CO	3.2.1.2a
5.9673	NO	3.2.1.6
5.96826	CO	3.2.1.2a
5.97525	CO	3.2.1.2b
5.97560	NO	3.2.1.6
5.97688	CO	3.2.1.2b
5.97775	CO	3.2.1.2a
5.97851	CO	3.2.1.2a
5.97962	CO	3.2.1.2b
5.9799	NO	3.2.1.6
5.9830082	O	1.6.1
5.9833948	Hg	1.2.7
5.98180	CO	3.2.1.2a
5.9882	NO	3.2.1.6
5.99273	CO	3.2.1.2b
5.9931	NO	3.2.1.6
5.99551	CO	3.2.1.2a
6.00100	NO	3.2.1.6
6.00145	CO	3.2.1.2b
6.00418	CO	3.2.1.2a
6.00540	NO	3.2.1.6
6.00602	CO	3.2.1.2b
6.00941	CO	3.2.1.2a
6.01727	CO	3.2.1.2a
6.0192	NO	3.2.1.6
6.01945	CO	3.2.1.2b
6.0209	DBr	3.2.1.3b
6.0225	DBr	3.2.1.3b
6.02348	CO	3.2.1.2a
6.02668	CO	3.2.1.2b
6.0267	NO	3.2.1.6
6.03055	CO	3.2.1.2a
6.0324	NO	3.2.1.6
6.03308	CO	3.2.1.2b
6.03774	CO	3.2.1.2a
6.0386	NO	3.2.1.6
6.03954	CO	3.2.1.2b
6.0402	NO	3.2.1.6
6.0419	CO	3.2.1.6
6.04319	CO	3.2.1.2a
6.0529	DBr	3.2.1.3b
6.0530155	Ar	1.9.3
6.0543	NO	3.2.1.6
6.0544	DBr	3.2.1.3b
6.05583	CO	3.2.1.2a
6.05793	CO	3.2.1.2a
6.0592473	Eu	1.8.4
6.0628	NO	3.2.1.6
6.06351	CO	3.2.1.2b
6.06578	CO	3.2.1.2b
6.0673	NO	3.2.1.6
6.06866	CO	3.2.1.2a
6.07145	CO	3.2.1.2a
6.07915	CO	3.2.1.2a
6.0801	NO	3.2.1.6
6.08166	CO	3.2.1.2a
6.08545	CO	3.2.1.2a
6.0858	DBr	3.2.1.3b
6.0873	CO	3.2.1.2a
6.0884	NO	3.2.1.6
6.08856	CO	3.2.1.2b
6.09269	CO	3.2.1.2b
6.0934	NO	3.2.1.6
6.09484	CO	3.2.1.2a
6.09964	CO	3.2.1.2a
6.10133	CO	3.2.1.2b
6.1015	NO	3.2.1.6
6.10821	CO	3.2.1.2a
6.10999	CO	3.2.1.2a
6.11401	CO	3.2.1.2a
6.1200	DBr	3.2.1.3b
6.12031	CO	3.2.1.2a
6.1204	NO	3.2.1.6
6.1216	DBr	3.2.1.3b
6.12176	CO	3.2.1.2a
6.12253	CO	3.2.1.2a
6.12740	CO	3.2.1.2b
6.12857	CO	3.2.1.2a
6.1324	Xe	1.9.5
6.13525	CO	3.2.1.2a
6.13549	CO	3.2.1.2a

Table 4.1 (continued)
GAS LASERS ARRANGED IN ORDER OF WAVELENGTH

Wavelength (μm)	Element or compound	Table No.
6.13987	CO	3.2.1.2b
6.14069	CO	3.2.1.2b
6.1417	NO	3.2.1.6
6.1477551	Ga	1.3.2
6.14816	CO	3.2.1.2a
6.14941	CO	3.2.1.2a
6.15238	CO	3.2.1.2b
6.1538	NO	3.2.1.6
6.15415	CO	3.2.1.2b
6.1546	NO	3.2.1.6
6.1546	DBr	3.2.1.3b
6.1562	DBr	3.2.1.3b
6.1576	NO	3.2.1.6
6.16124	CO	3.2.1.2a
6.16507	CO	3.2.1.2b
6.1663	NO	3.2.1.6
6.16779	CO	3.2.1.2b
6.17451	CO	3.2.1.2a
6.17792	CO	3.2.1.2b
6.1792	NO	3.2.1.6
6.18161	CO	3.2.1.2b
6.1838	NO	3.2.1.6
6.18797	CO	3.2.1.2a
6.1903	DBr	3.2.1.3b
6.19095	CO	3.2.1.2b
6.1918	DBr	3.2.1.3b
6.1921	NO	3.2.1.6
6.19561	CO	3.2.1.2b
6.1972	NO	3.2.1.6
6.1973	NO	3.2.1.6
6.20161	CO	3.2.1.6
6.20407	CO	3.2.1.6
6.20415	CO	3.2.1.2.6
6.2055	NO	3.2.1.6
6.20978	CO	3.2.1.2b
6.2110	NO	3.2.1.6
6.21544	CO	3.2.1.2a
6.21666	CO	3.2.1.2a
6.21753	CO	3.2.1.2b
6.21910	NO	3.2.1.6
6.2237	DBr	3.2.1.3b
6.2249	NO	3.2.1.6
6.2272	DBr	3.2.1.3b
6.2289	CO	3.2.1.2a
6.22985	CO	3.2.1.2a
6.23084	CO	3.2.1.2b
6.23109	CO	3.2.1.2b
6.2381	NO	3.2.1.6
6.24302	CO	3.2.1.2a
6.24359	CO	3.2.1.2b
6.24367	CO	3.2.1.2a
6.24483	CO	3.2.1.2b
6.2511	NO	3.2.1.6
6.25637	CO	3.2.1.2a
6.25653	CO	3.2.1.2b
6.2566	DBr	3.2.1.3b
6.2581	DBr	3.2.1.3b
6.25875	CO	3.2.1.2b
6.2602	NO	3.2.1.6
6.2645	NO	3.2.1.2b
6.26964	CO	3.2.1.2b
6.26992	CO	3.2.1.2a
6.27285	CO	3.2.1.2b
6.2778	NO	3.2.1.6
6.28293	CO	3.2.1.2b
6.28365	CO	3.2.1.2a
6.2865	NO	3.2.1.6
6.291	DBr	3.2.1.3b
6.2913	NO	3.2.1.6
6.2916	CO	3.2.1.2a
6.292	CO	3.2.1.2a
6.2932	CO	3.2.1.2a
6.29640	CO	3.2.1.2b
6.29758	CO	3.2.1.2a
6.2998	NO	3.2.1.6
6.30076	CO	3.2.1.2a
6.30160	CO	3.2.1.2b
6.3051	NO	3.2.1.6
6.31005	CO	3.2.1.2b
6.3120378	Xe	1.9.5
6.31171	CO	3.2.1.2a
6.3136	NO	3.2.1.6
6.3154334	Xe	1.9.5
6.31382	CO	3.2.1.2a
6.3191	NO	3.2.1.6
6.32388	CO	3.2.1.2b
6.32602	CO	3.2.1.2a
6.32717	CO	3.2.1.2a
6.3274	NO	3.2.1.6
6.3279	DBr	3.2.1.3b

Table 4.1 (continued)
GAS LASERS ARRANGED IN ORDER OF WAVELENGTH

Wavelength (μm)	Element or compound	Table No.
6.3294	DBr	3.2.1.3b
6.3336	NO	3.2.1.6
6.33723	CO	3.2.1.2b
6.33790	CO	3.2.1.2b
6.34051	CO	3.2.1.2a
6.34054	CO	3.2.1.2a
6.35043	CO	3.2.1.2b
6.35211	CO	3.2.1.2b
6.35415	CO	3.2.1.2a
6.35525	CO	3.2.1.2a
6.36380	CO	3.2.1.2b
6.36650	CO	3.2.1.2b
6.3687777	Se	1.6.3
6.36798	CO	3.2.1.2a
6.37017	CO	3.2.1.2a
6.37640	NO	3.2.1.6
6.37136	CO	3.2.1.2b
6.38108	CO	3.2.1.2b
6.38201	CO	3.2.1.2a
6.3894	NO	3.2.1.6
6.39623	CO	3.2.1.2a
6.3980	NO	3.2.1.6
6.64008	CO	3.2.1.2a
6.6422525	K	1.1.3
6.64031	NO	3.2.1.6
6.40504	CO	3.2.1.2b
6.41351	CO	3.2.1.2a
6.42528	CO	3.2.1.2a
6.42620	NO	3.2.1.6
6.42704	CO	3.2.1.2a
6.43210	NO	3.2.1.6
6.44011	CO	3.2.1.2a
6.44077	CO	3.2.1.2a
6.44688	CO	3.2.1.2b
6.44797	CO	3.2.1.2b
6.45460	Ba	1.2.4
6.45469	CO	3.2.1.2a
6.45515	CO	3.2.1.2a
6.4566366	Sr	1.2.3
6.4575288	K	1.1.3
6.46051	CO	3.2.1.2b
6.46266	CO	3.2.1.2b
6.46882	CO	3.2.1.2a
6.47039	CO	3.2.1.2a
6.47435	CO	3.2.1.2b
6.477439	Hg	1.2.7
6.48315	CO	3.2.1.2a
6.48838	CO	3.2.1.2b
6.50245	CO	3.2.1.2a
6.50260	CO	3.2.1.2b
6.51242	CO	3.2.1.2a
6.51606	CO	3.2.1.2a
6.51702	CO	3.2.1.2b
6.52736	CO	3.2.1.2a
6.52988	CO	3.2.1.2a
6.53162	CO	3.2.1.2b
6.54252	CO	3.2.1.2a
6.54390	CO	3.2.1.2a
6.55788	CO	3.2.1.2a
6.55812	CO	3.2.1.2a
6.55908	CO	3.2.1.2b
6.56144	CO	3.2.1.2b
6.58720	CO	3.2.1.2a
6.60205	CO	3.2.1.2a
6.60285	CO	3.2.1.2b
6.60774	CO	3.2.1.2a
6.61711	CO	3.2.1.2a
6.61757	CO	3.2.1.2b
6.62161	CO	3.2.1.2a
6.63572	CO	3.2.1.2a
6.64788	CO	3.2.1.2a
6.65005	CO	3.2.1.2a
6.60000	CS_2	3.2.2.3a
6.66458	CO	3.2.1.2a
6.67933	CO	3.2.1.2a
6.69429	CO	3.2.1.2a
6.70208	CO	3.2.1.2a
6.70948	CO	3.2.1.2a
6.71966	I	1.7.4
6.72487	CO	3.2.1.2a
6.73029	CO	3.2.1.2a
6.74049	CO	3.2.1.2a
6.74471	CO	3.2.1.2a
6.7463414	Ar	1.9.3
6.75634	CO	3.2.1.2a
6.75935	CO	3.2.1.2a
6.7631543	Te	1.6.4
6.77240	CO	3.2.1.2a
6.7788016	Ne	1.9.2
6.78929	CO	3.2.1.2a

Table 4.1 (continued)
GAS LASERS ARRANGED IN ORDER OF WAVELENGTH

Wavelength (μm)	Element or compound	Table No.	Wavelength (μm)	Element or compound	Table No.
6.80459	CO	3.2.1.2a	*7.1764192*	Pb	1.4.5
6.8175155	O	1.6.1	*7.1853791*	Cs	1.1.5
6.82011	CO	3.2.1.2a	*7.18877*	CO	3.2.1.2a
6.83583	CO	3.2.1.2a	*7.19966*	CO	3.2.1.2a
6.84226	CO	3.2.1.2a	*7.204*	H_2O	3.2.2.5
6.8487052	Fe	1.8.1	*7.2168093*	Ar	1.9.3
6.85183	CO	3.2.1.2a	*7.21544*	CO	3.2.1.2a
6.8598869	O	1.6.1	*7.22428*	CO	3.2.1.2a
6.86824	CO	3.2.1.2a	*7.240*	Xe	1.9.5
6.87197	CO	3.2.1.2a	*7.24192*	CO	3.2.1.2a
6.8745531	O	1.6.1	*7.24779*	CO	3.2.1.2a
6.88448	CO	3.2.1.2a	*7.28104*	CO	3.2.1.2a
6.8884755	Ne	1.9.2	*7.285*	H_2O	3.2.2.5
6.88716	CO	3.2.1.2a	*7.2932*	Ne	1.9.3
6.90000	CS_2	3.2.2.3b	*7.297*	H_2O	3.2.2.5
6.90258	CO	3.2.1.2a	*7.29807*	CO	3.2.1.2a
6.9035028	I	1.7.4	*7.31071*	CO	3.2.1.2a
6.91823	CO	3.2.1.2a	*7.31536*	CO	3.2.1.2a
6.93411	CO	3.2.1.2a	*7.3168038*	Xe	1.9.5
6.9448712	Ar	1.9.3	*7.3228367*	Ne	1.9.2
6.94284	CO	3.2.1.2a	*7.32661*	CO	3.2.1.2a
6.9428548	Ar	1.9.3	*7.33291*	CO	3.2.1.2a
6.95022	CO	3.2.1.2a	*7.34277*	CO	3.2.1.2a
6.96656	CO	3.2.1.2a	*7.35071*	CO	3.2.1.2a
6.97276	CO	3.2.1.2a	*7.35917*	CO	3.2.1.2a
6.9876797	Ne	1.9.2	*7.3625232*	Kr	1.9.4
6.98807	CO	3.2.1.2a	*7.36878*	CO	3.2.1.2a
7.00361	CO	3.2.1.2a	*7.37584*	CO	3.2.1.2a
7.01938	CO	3.2.1.2a	*7.390*	H_2O	3.2.2.5
7.03539	CO	3.2.1.2a	*7.39276*	CO	3.2.1.2a
7.05164	CO	3.2.1.2a	*7.405131*	Ne	1.9.2
7.0580595	Kr	1.9.4	*7.40795*	Co	3.2.1.2a
7.06813	CO	3.2.1.2a	*7.4222794*	Ne	1.9.2
7.07876	CO	3.2.1.2a	*7.425*	H_2O	3.2.2.5
7.08486	CO	3.2.1.2a	*7.42739*	CO	3.2.1.2a
7.09210	CO	3.2.1.2a	*7.4313142*	Xe	1.9.5
7.093	H_2O	3.2.2.5	*7.44156*	CO	3.2.1.2a
7.098	Ne	1.9.2	*7.44511*	CO	3.2.1.2a
7.01083	CO	3.2.1.2a	*7.453*	H_2O	3.2.2.5
7.10784	CO	3.2.1.2a	*7.45785*	CO	3.2.1.2a
7.11905	CO	3.2.1.2a	*7.4588*	H_2O	3.2.2.5
7.12374	CO	3.2.1.2a	*7.4699904*	Ne	1.9.2
7.13988	CO	3.2.1.2a	*7.47439*	CO	3.2.1.2a
7.15626	CO	3.2.1.2a	*7.4799887*	Ne	1.9.2
7.17289	CO	3.2.1.2a	*7.49120*	CO	3.2.1.2a

Table 4.1 (continued)
GAS LASERS ARRANGED IN ORDER OF WAVELENGTH

Wavelength (μm)	Element or compound	Table No.
7.4995237	Ne	1.9.2
7.50827	CO	3.2.1.2a
7.52561	CO	3.2.1.2a
7.5313000	Ne	1.9.2
7.543	H_2O	3.2.2.5
7.54322	CO	3.2.1.2a
7.579798	Ne	1.9.2
7.5885028	Ne	1.9.2
7.590	H_2O	3.2.2.5
7.59362	CO	3.2.1.2a
7.5966	H_2O	3.2.2.5
7.61057	CO	3.2.1.2a
7.6163805	Ne	1.9.2
7.62780	CO	3.2.1.2a
7.6458386	Ne	1.9.2
7.64529	CO	3.2.1.2a
7.6511521	Ne	1.9.2
7.66307	CO	3.2.1.2a
7.68112	CO	3.2.1.2a
7.6926284	Ne	1.9.2
7.70	Ne	1.9.2
7.7090	H_2O	3.2.2.5
7.73422	CO	3.2.1.2a
7.7408259	Ne	1.9.2
7.740	H_2O	3.2.2.5
7.75159	CO	3.2.1.2a
7.7655499	Ne	1.9.2
7.76925	CO	3.2.1.2a
7.7815561	Ne	1.9.2
7.78719	CO	3.2.1.2a
7.8001903	Ar	1.9.3
7.8026187	Ar	1.9.3
7.80542	CO	3.2.1.2a
7.8065353	Ar	1.9.3
7.8071691	Te	1.6.4
7.82394	CO	3.2.1.2a
7.8369230	Ne	1.9.2
7.8716418	Ne	1.9.2
7.87994	CO	3.2.1.2a
7.89776	CO	3.2.1.2a
7.8953393	K	1.1.3
7.9020	SiH_4	3.2.4.7
7.91587	CO	3.2.1.2a
7.9220	SiH_4	3.2.4.7
7.9423392	Pb	1.4.5
7.9490	SiH_4	3.2.4.7
7.95298	CO	3.2.1.2a
7.9548	SiH_4	3.2.4.7
7.9690	SiH_4	3.2.4.7
7.9846694	Ne	1.9.2
7.9920	SiH_4	3.2.4.7
8.008892	Ne	1.9.2
8.03112	CO	3.2.1.2a
8.0341	C_2H_2	3.2.3.2a
8.0352	C_2H_2	3.2.3.2a
8.0356	C_2H_2	3.2.3.2a
8.0378	C_2H_2	3.2.3.2a
8.0380	C_2H_2	3.2.3.2a
8.0402	C_2H_2	3.2.3.2a
8.0409	C_2H_2	3.2.3.2a
8.0445	C_2H_2	3.2.3.[illegible]
8.04941	CO	3.2.3[illegible]a
8.0621949	Ne	1.9.2
8.06800	CO	3.2.1.2a
8.08689	CO	3.2.1.2a
8.10609	CO	3.2.1.2a
8.116	Ne	1.9.2
8.1482	N_2	3.1.2.21c
8.1736588	Ne	1.9.2
8.18427	N_2	3.1.2.21c
8.189	COS	3.2.2.2a
8.191	COS	3.2.2.2a
8.194	COS	3.2.2.2a
8.196	COS	3.2.2.2a
8.198	COS	3.2.2.2a
8.200	COS	3.2.2.2a
8.203	COS	3.2.2.2a
8.205	COS	3.2.2.2a
8.20692	CO	3.2.1.2a
8.207	COS	3.2.2.2a
8.210	COS	3.2.2.2a
8.2102	N_2	3.1.2.21c
8.212	COS	3.2.2.2a
8.215	COS	3.2.2.2a
8.217	COS	3.2.2.2a
8.219	COS	3.2.2.2a
8.222	COS	3.2.2.2a
8.22601	COS	3.2.2.2a
8.224	COS	3.2.2.2a
8.227	COS	3.2.2.2a

Table 4.1 (continued)
GAS LASERS ARRANGED IN ORDER OF WAVELENGTH

Wavelength (μm)	Element or compound	Table No.	Wavelength (μm)	Element or compound	Table No.
8.229	COS	3.2.2.2a	*8.3870*	COS	3.2.2.2a
8.232	COS	3.2.2.2a	*8.3900*	COS	3.2.2.2a
8.234	COS	3.2.2.2a	*8.3930*	COS	3.2.2.2a
8.2368	COS	3.2.2.2a	*8.3962*	COS	3.2.2.2a
8.237	COS	3.2.2.2a	*8.3999*	COS	3.2.2.2a
8.2416	COS	3.2.2.2a	*8.4024*	COS	3.2.2.2a
8.2439	COS	3.2.2.2a	*8.4055*	COS	3.2.2.2a
8.24541	CO	3.2.1.2a	*8.4085*	COS	3.2.2.2a
8.247	COS	3.2.2.2a	*8.4117*	COS	3.2.2.2a
8.249	COS	3.2.2.2a	*8.4146*	COS	3.2.2.2a
8.2543	COS	3.2.2.2a	*8.4178*	COS	3.2.2.2a
8.2571	COS	3.2.2.2a	*8.4213*	COS	3.2.2.2a
8.2595	COS	3.2.2.2a	*8.4243*	COS	3.2.2.2a
8.2623	COS	3.2.2.2a	*8.428*	COS	3.2.2.2a
8.2645	COS	3.2.2.2a	*8.431*	COS	3.2.2.2a
8.26514	CO	3.2.1.2a	*8.434*	COS	3.2.2.2a
8.2673	COS	3.2.2.2a	*8.437*	COS	3.2.2.2a
8.270	COS	3.2.2.2a	*8.441*	COS	3.2.2.2a
8.273	COS	3.2.2.2a	*8.444*	COS	3.2.2.2a
8.275	COS	3.2.2.2a	*8.447*	COS	3.2.2.2a
8.278	COS	3.2.2.2a	*8.451*	COS	3.2.2.2a
8.281	COS	3.2.2.2a	*8.454*	COS	3.2.2.2a
8.283	COS	3.2.2.2a	*8.457*	COS	3.2.2.2a
8.286	COS	3.2.2.2a	*8.48*	HCN	3.2.2.4
8.300	C_2H_2	3.2.3.2a	*8.4927853*	Fe	1.8.1
8.328	COS	3.2.2.2a	*8.529294*	He	1.9.1
8.330	COS	3.2.2.2a	*8.6*	COS	3.2.2.2b
8.336	COS	3.2.2.2a	*8.6414054*	Ne	1.9.2
8.3371447	Ne	1.9.2	*8.6554154*	Ne	1.9.2
8.342	COS	3.2.2.2a	*8.98767*	CO_2	3.2.2.1c
8.345	COS	3.2.2.2a	*8.99495*	CO_2	3.2.2.1c
8.348	COS	3.2.2.2a	*9.00238*	CO_2	3.2.2.1c
8.3496011	Ne	1.9.2	*9.0067086*	Xe	1.9.5
8.351	COS	3.2.2.2a	*9.00998*	CO_2	3.2.2.1c
8.354	COS	3.2.2.2a	*9.01775*	CO_2	3.2.2.1c
8.357	COS	3.2.2.2a	*9.0195724*	I	1.7.4
8.360	COS	3.2.2.2a	*9.02568*	CO_2	3.2.2.1c
8.3625	COS	3.2.2.2a	*9.03378*	CO_2	3.2.2.1c
8.3654	COS	3.2.2.2a	*9.04205*	CO_2	3.2.2.1c
8.3685	COS	3.2.2.2a	*9.05050*	CO_2	3.2.2.1c
8.3715	COS	3.2.2.2a	*9.05911*	CO_2	3.2.2.1c
8.3746	COS	3.2.2.2a	*9.06790*	CO_2	3.2.2.1c
8.3779	COS	3.2.2.2a	*9.07506*	CO_2	3.2.2.1c
8.3809	COS	3.2.2.2a	*9.07687*	CO_2	3.2.2.1c
8.3839	COS	3.2.2.2a	*9.08317*	CO_2	3.2.2.1c

Table 4.1 (continued)
GAS LASERS ARRANGED IN ORDER OF WAVELENGTH

Wavelength (μm)	Element or compound	Table No.	Wavelength (μm)	Element or compound	Table No.
9.08601	CO_2	3.2.2.1c	*9.19456*	CO_2	3.2.2.1c
9.0894630	Ne	1.9.2	*9.19583*	CO_2	3.2.2.1b
9.09145	CO_2	3.2.2.1c	*9.19865*	CO_2	3.2.2.1c
9.09349	CO_2	3.2.2.1a	*9.20073*	CO_2	3.2.2.1a
9.09533	CO_2	3.2.2.1c	*9.20101*	CO_2	3.2.2.1b
9.09976	CO_2	3.2.2.1a	*9.20506*	CO_2	3.2.2.1c
9.09991	CO_2	3.2.2.1c	*9.20624*	CO_2	3.2.2.1b
9.10484	CO_2	3.2.2.1c	*9.20917*	CO_2	3.2.2.1a
9.10623	CO_2	3.2.2.1a	*9.21003*	CO_2	3.2.2.1a
9.10855	CO_2	3.2.2.1c	*9.21009*	CO_2	3.2.2.1a
9.11291	CO_2	3.2.2.1a	*9.21152*	CO_2	3.2.2.1b
9.11452	CO_2	3.2.2.1c	*9.21575*	CO_2	3.2.2.1c
9.11737	CO_2	3.2.2.1c	*9.21685*	CO_2	3.2.2.1b
9.11979	CO_2	3.2.2.1a	*9.21777*	CO_2	3.2.2.1a
9.12438	CO_2	3.2.2.1c	*9.21969*	CO_2	3.2.2.1a
9.12637	CO_2	3.2.2.1c	*9.22240*	CO_2	3.2.2.1b
9.12689	CO_2	3.2.2.1a	*9.22662*	CO_2	3.2.2.1a
9.13430	CO_2	3.2.2.1a	*9.22664*	CO_2	3.2.2.1c
9.13443	CO_2	3.2.2.1c	*9.22769*	CO_2	3.2.2.1b
9.13784	CO_2	3.2.2.1b	*9.22953*	CO_2	3.2.2.1a
9.14173	CO_2	3.2.2.1a	*9.23319*	CO_2	3.2.2.1b
9.14239	CO_2	3.2.2.1b	*9.23570*	CO_2	3.2.2.1a
9.14467	CO_2	3.2.2.1c	*9.23773*	CO_2	3.2.2.1c
9.14492	CO_2	3.2.2.1c	*9.23874*	CO_2	3.2.2.1b
9.14699	CO_2	3.2.2.1b	*9.23931*	CO_2	3.2.2.1c
9.14948	CO_2	3.2.2.1a	*9.23961*	CO_2	3.2.2.1a
9.15164	CO_2	3.2.2.1b	*9.24435*	CO_2	3.2.2.1b
9.15447	CO_2	3.2.2.1c	*9.24503*	CO_2	3.2.2.1a
9.15634	CO_2	3.2.2.1b	*9.24995*	CO_2	3.2.2.1a
9.15745	CO_2	3.2.2.1a	*9.25002*	CO_2	3.2.2.1b
9.16110	CO_2	3.2.2.1b	*9.25137*	CO_2	3.2.2.1c
9.16421	CO_2	3.2.2.1c	*9.25461*	CO_2	3.2.2.1a
9.16570	CO_2	3.2.2.1c	*9.25574*	CO_2	3.2.2.1b
9.17407	CO_2	3.2.2.1a	*9.26053*	CO_2	3.2.2.1a
9.17114	CO_2	3.2.2.1c	*9.26362*	CO_2	3.2.2.1c
9.17567	CO_2	3.2.2.1b	*9.26444*	CO_2	3.2.2.1a
9.17649	CO_2	3.2.2.1c	*9.27136*	CO_2	3.2.2.1a
9.179196	K	1.1.3	*9.27452*	CO_2	3.2.2.1a
9.18063	CO_2	3.2.2.1b	*9.27607*	CO_2	3.2.2.1c
9.18273	CO_2	3.2.2.1a	*9.28244*	CO_2	3.2.2.1a
9.18425	CO_2	3.2.2.1c	*9.28485*	CO_2	3.2.2.1a
9.18565	CO_2	3.2.2.1b	*9.28873*	CO_2	3.2.2.1c
9.18748	CO_2	3.2.2.1c	*9.28379*	CO_2	3.2.2.1a
9.19071	CO_2	3.2.2.1b	*9.29545*	CO_2	3.2.2.1a
9.19161	CO_2	3.2.2.1a	*9.3*	NH_3	3.2.3.4a

Table 4.1 (continued)
GAS LASERS ARRANGED IN ORDER OF WAVELENGTH

Wavelength (μm)	Element or compound	Table No.	Wavelength (μm)	Element or compound	Table No.
9.30158	CO_2	3.2.2.1c	9.44669	CO_2	3.2.2.1b
9.30539	CO_2	3.2.2.1a	9.45007	CO_2	3.2.2.1c
9.30630	CO_2	3.2.2.1a	9.45055	CO_2	3.2.2.1a
9.31464	CO_2	3.2.2.1c	9.45416	CO_2	3.2.2.1b
9.31725	CO_2	3.2.2.1a	9.45661	CO_2	3.2.2.1c
9.31742	CO_2	3.2.2.1a	9.45805	CO_2	3.2.2.1a
9.32725	CO_2	3.2.2.1c	9.46170	CO_2	3.2.2.1b
9.32790	CO_2	3.2.2.1c	9.46479	CO_2	3.2.2.1c
9.32880	CO_2	3.2.2.1a	9.46485	CO_2	3.2.2.1a
9.32937	CO_2	3.2.2.1a	9.46931	CO_2	3.2.2.1b
9.34006	CO_2	3.2.2.1c	9.47196	CO_2	3.2.2.1c
9.34045	CO_2	3.2.2.1a	9.47305	CO_2	3.2.2.1a
9.34137	CO_2	3.2.2.1c	9.4747	H_2O	3.2.2.5
9.34176	CO_2	3.2.2.1a	9.47697	CO_2	3.2.2.1b
9.35237	CO_2	3.2.2.1a	9.47943	CO_2	3.2.2.1a
9.35307	CO_2	3.2.2.1a	9.47972	CO_2	3.2.2.1c
9.35441	CO_2	3.2.2.1a	9.48469	CO_2	3.2.2.1b
9.35504	CO_2	3.2.2.1c	9.48754	CO_2	3.2.2.1c
9.36456	CO_2	3.2.2.1a	9.48835	CO_2	3.2.2.1a
9.36629	CO_2	3.2.2.1a	9.49248	CO_2	3.2.2.1b
9.36734	CO_2	3.2.2.1a	9.49431	CO_2	3.2.2.1a
9.36892	CO_2	3.2.2.1c	9.49487	CO_2	3.2.2.1c
9.37702	CO_2	3.2.2.1a	9.49996	CO_2	3.2.2.1e
9.37972	CO_2	3.2.2.1a	9.50033	CO_2	3.2.2.1b
9.38053	CO_2	3.2.2.1a	9.50333	CO_2	3.2.2.1c
9.38301	CO_2	3.2.2.1c	9.50394	CO_2	3.2.2.1a
9.38976	CO_2	3.2.2.1a	9.50824	CO_2	3.2.2.1b
9.39336	CO_2	3.2.2.1c	9.50826	CO_2	3.2.2.1e
9.39380	H_2O	3.2.2.5	9.50948	CO_2	3.2.2.1a
9.39400	CO_2	3.2.2.1a	9.51024	CO_2	3.2.2.1c
9.39730	CO_2	3.2.2.1c	9.51621	CO_2	3.2.2.1b
9.40277	CO_2	3.2.2.1a	9.51676	CO_2	3.2.2.1e
9.40722	CO_2	3.2.2.1c	9.51933	CO_2	3.2.2.1c
9.41181	CO_2	3.2.2.1c	9.51981	CO_2	3.2.2.1a
9.41472	CO_2	3.2.2.1a	9.52425	CO_2	3.2.2.1b
9.41738	CO_2	3.2.2.1b	9.52494	CO_2	3.2.2.1a
9.42129	CO_2	3.2.2.1c	9.52547	CO_2	3.2.2.1e
9.42462	CO_2	3.2.2.1b	9.52583	CO_2	3.2.2.1c
9.42653	CO_2	3.2.2.1c	9.53235	CO_2	3.2.2.1b
9.42889	CO_2	3.2.2.1a	9.53435	CO_2	3.2.2.1e
9.43191	CO_2	3.2.2.1b	9.53556	CO_2	3.2.2.1c
9.43557	CO_2	3.2.2.1c	9.53597	CO_2	3.2.2.1a
9.43937	CO_2	3.2.2.1b	9.54051	CO_2	3.2.2.1b
9.44148	CO_2	3.2.2.1c	9.54070	CO_2	3.2.2.1a
9.44333	CO_2	3.2.2.1a	9.54165	CO_2	3.2.2.1a

Table 4.1 (continued) GAS LASERS ARRANGED IN ORDER OF WAVELENGTH

Wavelength (μm)	Element or compound	Table No.	Wavelength (μm)	Element or compound	Table No.
9.54348	CO_2	3.2.2.1c	*9.65235*	CO_2	3.2.2.1d
9.54873	CO_2	3.2.2.1b	*9.65531*	CO_2	3.2.2.1c
9.55200	CO_2	3.2.2.1c	*9.65590*	CO_2	3.2.2.1a
9.55243	CO_2	3.2.2.1a	*9.65742*	CO_2	3.2.2.1a
9.55280	CO_2	3.2.2.1e	*9.65773*	CO_2	3.2.2.1e
9.55676	CO_2	3.2.2.1a	*9.66335*	CO_2	3.2.2.1d
9.55702	CO_2	3.2.2.1b	*9.66944*	CO_2	3.2.2.1e
9.55768	CO_2	3.2.2.1c	*9.67463*	CO_2	3.2.2.1d
9.56232	CO_2	3.2.2.1e	*9.67597*	CO_2	3.2.2.1a
9.56537	CO_2	3.2.2.1b	*9.67700*	CO_2	3.2.2.1a
9.56740	H_2O	3.2.2.5	*9.68138*	CO_2	3.2.2.1e
9.56866	CO_2	3.2.2.1c	*9.68622*	CO_2	3.2.2.1d
9.56918	CO_2	3.2.2.1a	*9.69355*	CO_2	3.2.2.1e
9.57206	CO_2	3.2.2.1e	*9.69483*	CO_2	3.2.2.1a
9.57312	CO_2	3.2.2.1a	*9.69622*	CO_2	3.2.2.1a
9.57379	CO_2	3.2.2.1b	*9.69610*	CO_2	3.2.2.1d
9.57394	CO_2	3.2.2.1c	*9.70000*	NH_3	3.2.3.4a
9.58227	CO_2	3.2.2.1b	*9.70020*	Xe	1.9.5
9.58555	CO_2	3.2.2.1c	*9.70596*	Xe	3.2.2.1e
9.58623	CO_2	3.2.2.1a	*9.71029*	Xe	3.2.2.1d
9.58979	CO_2	3.2.2.1a	*9.71400*	Xe	3.2.2.1a
9.59081	CO_2	3.2.2.1b	*9.71505*	Xe	3.2.2.1a
9.59216	CO_2	3.2.2.1e	*9.72275*	Xe	3.2.2.1d
9.59942	CO_2	3.2.2.1b	*9.73348*	Xe	3.2.2.1a
9.60000	NH_3	3.2.3.4a	*9.73419*	Xe	3.2.2.1a
9.60169	CO_2	3.2.2.1d	*9.73558*	CO_2	3.2.2.1d
9.60254	CO_2	3.2.2.1e	*9.74870*	CO_2	3.2.2.1d
9.60265	CO_2	3.2.2.1c	*9.75000*	CH_3F	3.2.4.3
9.60357	CO_2	3.2.2.1a	*9.75326*	CO_2	3.2.2.1a
9.60675	CO_2	3.2.2.1a	*9.75365*	CO_2	3.2.2.1a
9.60809	CO_2	3.2.2.1b	*9.76212*	CO_2	3.2.2.1d
9.61126	CO_2	3.2.2.1d	*9.76467*	CO_2	3.2.2.1e
9.61314	CO_2	3.2.2.1e	*9.77336*	CO_2	3.2.2.1a
9.61998	CO_2	3.2.2.1c	*9.77586*	CO_2	3.2.2.1d
9.62110	CO_2	3.2.2.1d	*9.77838*	CO_2	3.2.2.1e
9.62122	CO_2	3.2.2.1d	*9.78992*	CO_2	3.2.2.1d
9.62395	CO_2	3.2.2.1a	*9.79233*	CO_2	3.2.2.1e
9.62403	CO_2	3.2.2.1e	*9.79353*	CO_2	3.2.2.1a
9.63123	CO_2	3.2.2.1a	*9.79377*	CO_2	3.2.2.1a
9.63499	CO_2	3.2.2.1d	*9.80652*	CO_2	3.2.2.1e
9.63754	CO_2	3.2.2.1c	*9.81395*	CO_2	3.2.2.1a
9.63917	CO_2	3.2.2.1a	*9.81450*	CO_2	3.2.2.1a
9.64161	CO_2	3.2.2.1a	*9.82095*	CO_2	3.2.2.1e
9.64165	CO_2	3.2.2.1d	*9.83554*	CO_2	3.2.2.1a
9.64625	CO_2	3.2.2.1e	*9.83564*	CO_2	3.2.2.1e

Table 4.1 (continued)
GAS LASERS ARRANGED IN ORDER OF WAVELENGTH

Wavelength (μm)	Element or compound	Table No.	Wavelength (μm)	Element or compound	Table No.
9.85057	CO_2	3.2.2.1e	*~10*	COF_2	3.2.3.3
9.85630	CO_2	3.2.2.1a	*10.00202*	CO_2	3.2.2.1f
9.86575	CO_2	3.2.2.1e	*10.00218*	CO_2	3.2.2.1f
9.87304	CO_2	3.2.2.1d	*10.00910*	CO_2	3.2.2.1g
9.87656	CO_2	3.2.2.1a	*10.01992*	CO_2	3.2.2.1g
9.88117	CO_2	3.2.2.1e	*10.01415*	CO_2	3.2.2.1f
9.88923	CO_2	3.2.2.1d	*10.02591*	CO_2	3.2.2.1a
9.89074	CO_2	3.2.2.1g	*10.02645*	CO_2	3.2.2.1f
9.89843	CO_2	3.2.2.1g	*10.03100*	CO_2	3.2.2.1g
9.89686	CO_2	3.2.2.1e	*10.03108*	CO_2	3.2.2.1d
9.90057	CO_2	3.2.2.1a	*10.03152*	CO_2	3.2.2.1e
9.90576	CO_2	3.2.2.1d	*10.03347*	CO_2	3.2.2.1a
9.90636	CO_2	3.2.2.1g	*10.03908*	CO_2	3.2.2.1f
9.91279	CO_2	3.2.2.1e	*10.04132*	CO_2	3.2.2.1a
9.91451	CO_2	3.2.2.1g	*10.04234*	CO_2	3.2.2.1g
9.91788	CO_2	3.2.2.1f	*10.04940*	CO_2	3.2.2.1a
9.92262	CO_2	3.2.2.1d	*10.04953*	CO_2	3.2.2.1e
9.92289	CO_2	3.2.2.1a,g	*10.05039*	CO_2	3.2.2.1d
9.92733	CO_2	3.2.2.1f	*10.05205*	CO_2	3.2.2.1d
9.92898	CO_2	3.2.2.1e	*10.05393*	CO_2	3.2.2.1d
9.93151	CO_2	3.2.2.1g	*10.05793*	CO_2	3.2.2.1a
9.93709	CO_2	3.2.2.1f	*10.063422*	Ne	1.9.2
9.93983	CO_2	3.2.2.1d	*10.06536*	CO_2	3.2.2.1f
9.94542	CO_2	3.2.2.1e	*10.06579*	CO_2	3.2.2.1g
9.94552	CO_2	3.2.2.1a	*10.06668*	CO_2	3.2.2.1a
9.94715	CO_2	3.2.2.1f	*10.07006*	CO_2	3.2.2.1d
9.94945	CO_2	3.2.2.1g	*10.07572*	CO_2	3.2.2.1a
9.95738	CO_2	3.2.2.1d	*10.07792*	CO_2	3.2.2.1g
9.95753	CO_2	3.2.2.1f	*10.07900*	CO_2	3.2.2.1f
9.95877	CO_2	3.2.2.1g	*10.08328*	CO_2	3.2.2.1c
9.96213	CO_2	3.2.2.1e	*10.08506*	CO_2	3.2.2.1a
9.96821	CO_2	3.2.2.1f	*10.08633*	CO_2	3.2.2.1e
9.96835	CO_2	3.2.2.1g	*10.08761*	CO_2	3.2.2.1b
9.96848	CO_2	3.2.2.1a	*10.09009*	CO_2	3.2.2.1d
9.97528	CO_2	3.2.2.1d	*10.09031*	CO_2	3.2.2.1g
9.97816	CO_2	3.2.2.1g	*10.09229*	CO_2	3.2.2.1c
9.97909	CO_2	3.2.2.1e	*10.0930*	CO_2	3.2.2.1f
9.97922	CO_2	3.2.2.1b	*10.09469*	CO_2	3.2.2.1a
9.98394	CO_2	3.2.2.1b	*10.09704*	CO_2	3.2.2.1d
9.98823	CO_2	3.2.2.1g	*10.10186*	CO_2	3.2.2.1b
9.99054	CO_2	3.2.2.1b	*10.10297*	CO_2	3.2.2.1g
9.99177	CO_2	3.2.2.1a	*10.10435*	CO_2	3.2.2.1c
9.99353	CO_2	3.2.2.1d	*10.10462*	CO_2	3.2.2.1a
9.99630	CO_2	3.2.2.1e	*10.10512*	CO_2	3.2.2.1e
9.99853	CO_2	3.2.2.1g	*10.10676*	CO_2	3.2.2.1b

Table 4.1 (continued)
GAS LASERS ARRANGED IN ORDER OF WAVELENGTH

Wavelength (μm)	Element or compound	Table No.	Wavelength (μm)	Element or compound	Table No.
10.10734	CO_2	3.2.2.1f	*10.18921*	CO_2	3.2.2.1b
10.11048	CO_2	3.2.2.1d	*10.19105*	CO_2	3.2.2.1a
10.11295	CO_2	3.2.2.1c	*10.19265*	CO_2	3.2.2.1c
10.11484	CO_2	3.2.2.1a	*10.19485*	CO_2	3.2.2.1a
10.11591	CO_2	3.2.2.1g	*10.19531*	CO_2	3.2.2.1b
10.11678	CO_2	3.2.2.1f	*10.20149*	CO_2	3.2.2.1b
10.12186	CO_2	3.2.2.1c	*10.20203*	CO_2	3.2.2.1c
10.12190	CO_2	3.2.2.1b	*10.20280*	CO_2	3.2.2.1a
10.12203	CO_2	3.2.2.1f	*10.20398*	CO_2	3.2.2.1c
10.12419	CO_2	3.2.2.1e	*10.20676*	CO_2	3.2.2.1g
10.12535	CO_2	3.2.2.1a	*10.20715*	CO_2	3.2.2.1a
10.12615	CO_2	3.2.2.1a	*10.20774*	CO_2	3.2.2.1b
10.12709	CO_2	3.2.2.1b	*10.21315*	CO_2	3.2.2.1c
10.12912	CO_2	3.2.2.1g	*10.21407*	CO_2	3.2.2.1b
10.13107	CO_2	3.2.2.1c	*10.21501*	CO_2	3.2.2.1a
10.13123	CO_2	3.2.2.1d	*10.21562*	CO_2	3.2.2.1c
10.13237	CO_2	3.2.2.1b	*10.22001*	CO_2	3.2.2.1g
10.13616	CO_2	3.2.2.1a	*10.22048*	CO_2	3.2.2.1b
10.13624	CO_2	3.2.2.1a	*10.22180*	CO_2	3.2.2.1g
10.13776	CO_2	3.2.2.1b	*10.22696*	CO_2	3.2.2.1b
10.14058	CO_2	3.2.2.1c	*10.22756*	CO_2	3.2.2.1a
10.14260	CO_2	3.2.2.1g	*10.23317*	CO_2	3.2.2.1c
10.14313	CO_2	3.2.2.1b	*10.23631*	CO_2	3.2.2.1c
10.14352	CO_2	3.2.2.1e	*10.23713*	CO_2	3.2.2.1g
10.14662	CO_2	3.2.2.1a	*10.23961*	CO_2	3.2.2.1c
10.14725	CO_2	3.2.2.1a	*10.24013*	CO_2	3.2.2.1a
10.14863	CO_2	3.2.2.1b	*10.24370*	CO_2	3.2.2.1f
10.15039	CO_2	3.2.2.1c	*10.24663*	CO_2	3.2.2.1a
10.15235	CO_2	3.2.2.1d	*10.24834*	CO_2	3.2.2.1c
10.15420	CO_2	3.2.2.1b	*10.25238*	CO_2	3.2.2.1c
10.15637	CO_2	3.2.2.1g	*10.25276*	CO_2	3.2.2.1g
10.15750	CO_2	3.2.2.1a	*10.25314*	CO_2	3.2.2.1a
10.15865	CO_2	3.2.2.1a	*10.26039*	CO_2	3.2.2.1a
10.15985	CO_2	3.2.2.1b	*10.26069*	CO_2	3.2.2.1c
10.16050	CO_2	3.2.2.1c	*10.26525*	CO_2	3.2.2.1c
10.16557	CO_2	3.2.2.1b	*10.26643*	CO_2	3.2.2.1a
10.16626	CO_2	3.2.2.1a	*10.26867*	CO_2	3.2.2.1g
10.17033	CO_2	3.2.2.1a	*10.27334*	CO_2	3.2.2.1c
10.17091	CO_2	3.2.2.1c	*10.27445*	CO_2	3.2.2.1a
10.17385	CO_2	3.2.2.1d	*10.27844*	CO_2	3.2.2.1c
10.17724	CO_2	3.2.2.1b	*10.27967*	CO_2	3.2.2.1f
10.17951	CO_2	3.2.2.1a	*10.28002*	CO_2	3.2.2.1a
10.18163	CO_2	3.2.2.1c	*10.28488*	CO_2	3.2.2.1g
10.18231	CO_2	3.2.2.1a	*10.28880*	CO_2	3.2.2.1a
10.18391	CO_2	3.2.2.1b	*10.29195*	CO_2	3.2.2.1c

Table 4.1 (continued)
GAS LASERS ARRANGED IN ORDER OF WAVELENGTH

Wavelength (μm)	Element or compound	Table No.	Wavelength (μm)	Element or compound	Table No.
10.29391	CO_2	3.2.2.1a	*10.39681*	CO_2	3.2.2.1f
10.29822	CO_2	3.2.2.1f	*10.39790*	CO_2	3.2.2.1a
10.30139	CO_2	3.2.2.1g	*10.39955*	CO_2	3.2.2.1a
10.30243	CO_2	3.2.2.1a	*10.39977*	N_2O	3.2.2.6a
10.30347	CO_2	3.2.2.1a	*10.40312*	O	1.6.1
10.30577	CO_2	3.2.2.1c	*10.40354*	CO_2	3.2.2.1c
10.30819	CO_2	3.2.2.1a	*10.40477*	CO_2	3.2.2.1a
10.30830	N_2O	3.2.2.6a	*10.40676*	CO_2	3.2.2.1g
10.31565	N_2O	3.2.2.6a	*10.40770*	N_2O	3.2.2.6a
10.31616	CO_2	3.2.2.1a	*10.41105*	N_2O	3.2.2.6a
10.31716	CO_2	3.2.2.1f	*10.41571*	N_2O	3.2.2.6a
10.31819	CO_2	3.2.2.1g	*10.41587*	CO_2	3.2.2.1a
10.31843	CO_2	3.2.2.1a	*10.41687*	N_2O	3.2.2.6a
10.32258	CO_2	3.2.2.1a	*10.41771*	CO_2	3.2.2.1f
10.32305	N_2O	3.2.2.6a	*10.41982*	CO_2	3.2.2.1c
10.33018	CO_2	3.2.2.1a	*10.42327*	CO_2	3.2.2.1a
10.33050	N_2O	3.2.2.6a	*10.42376*	N_2O	3.2.2.6a
10.33370	CO_2	3.2.2.1a	*10.42539*	CO_2	3.2.2.1g
10.33530	CO_2	3.2.2.1g	*10.42674*	N_2O	3.2.2.6a
10.33649	CO_2	3.2.2.1f	*10.43179*	CO_2	3.2.2.1b
10.33737	CO_2	3.2.2.1a	*10.43186*	N_2O	3.2.2.6a
10.33800	N_2O	3.2.2.6a	*10.43466*	N_2O	3.2.2.6a
10.34451	CO_2	3.2.2.1a	*10.43644*	CO_2	3.2.2.1c
10.34555	N_2O	3.2.2.6a	*10.43901*	CO_2	3.2.2.1f
10.34928	CO_2	3.2.2.6a	*10.44001*	N_2O	3.2.2.6a
10.35246	CO_2	3.2.2.6a	*10.44051*	CO_2	3.2.2.1b
10.35271	CO_2	3.2.2.6g	*10.44059*	CO_2	3.2.2.1a
10.35314	N_2O	3.2.2.6a	*10.44178*	CO_2	3.2.2.1c
10.35620	CO_2	3.2.2.1f	*10.44263*	N_2O	3.2.2.6a
10.35912	CO_2	3.2.2.1a	*10.44433*	CO_2	3.2.2.1g
10.36079	N_2O	3.2.2.6a	*10.44821*	N_2O	3.2.2.6a
10.36518	CO_2	3.2.2.1a	*10.45065*	N_2O	3.2.2.6a
10.36785	CO_2	3.2.2.1a	*10.451505*	Te	1.3.4
10.36848	N_2O	3.2.2.6a	*10.45341*	CO_2	3.2.2.1c
10.37042	CO_2	3.2.2.1g	*10.45460*	N_2O	3.2.2.6a
10.37404	CO_2	3.2.2.1a	*10.45803*	CO_2	3.2.2.1a
10.37623	N_2O	3.2.2.6a	*10.45823*	CO_2	3.2.2.1b,a
10.37631	CO_2	3.2.2.1f	*10.45848*	CO_2	3.2.2.1c
10.38138	CO_2	3.2.2.1a	*10.45871*	N_2O	3.2.2.6a
10.38354	CO_2	3.2.2.1a	*10.46072*	CO_2	3.2.2.1f
10.38403	N_2O	3.2.2.6a	*10.46359*	CO_2	3.2.2.1g
10.38759	CO_2	3.2.2.1c	*10.46476*	N_2O	3.2.2.6a
10.38843	CO_2	3.2.2.1g	*10.46684*	N_2O	3.2.2.6a
10.38926	CO_2	3.2.2.1a	*10.46721*	CO_2	3.2.2.6b
10.39187	N_2O	3.2.2.6a	*10.47072*	CO_2	3.2.2.6c

Table 4.1 (continued)
GAS LASERS ARRANGED IN ORDER OF WAVELENGTH

Wavelength (μm)	Element or compound	Table No.	Wavelength (μm)	Element or compound	Table No.
10.47312	N_2O	3.2.2.6a	*10.53306*	N_2O	3.2.2.6a
10.47501	N_2O	3.2.2.6a	*10.53366*	N_2O	3.2.2.6a
10.47545	CO_2	3.2.2.1a	*10.53407*	CO_2	3.2.2.1g
10.47553	CO_2	3.2.2.1c	*10.54186*	N_2O	3.2.2.6a
10.47619	CO_2	3.2.2.1a	*10.54224*	N_2O	3.2.2.6a
10.47629	CO_2	3.2.2.1b	*10.54224*	CO_2	3.2.2.1b
10.48153	N_2O	3.2.2.6a	*10.54349*	CO_2	3.2.2.1c
10.48283	CO_2	3.2.2.1f	*10.54376*	CO_2	3.2.2.1g
10.48316	CO_2	3.2.2.1g	*10.54509*	CO_2	3.2.2.1e
10.48324	N_2O	3.2.2.6a	*10.54720*	CO_2	3.2.2.1c
10.48544	CO_2	3.2.2.1b	*10.54838*	CO_2	3.2.2.1a
10.48838	CO_2	3.2.2.1c	*10.55066*	N_2O	3.2.2.6a
10.48998	N_2O	3.2.2.6a	*10.55088*	N_2O	3.2.2.6a
10.49151	N_2O	3.2.2.6a	*10.55140*	CO_2	3.2.2.1a
10.49292	CO_2	3.2.2.1c	*10.55202*	CO_2	3.2.2.1b
10.49319	CO_2	3.2.2.1a	*10.55640*	CO_2	3.2.2.1e
10.49469	CO_2	3.2.2.1b	*10.55945*	N_2O	3.2.2.6a
10.49849	N_2O	3.2.2.6a	*10.56189*	CO_2	3.2.2.1b
10.49983	N_2O	3.2.2.6a	*10.56259*	CO_2	3.2.2.1c
10.50271	CO_2	3.2.2.1c	*10.56460*	CO_2	3.2.2.1g
10.50304	CO_2	3.2.2.1g	*10.56600*	CO_2	3.2.2.1c
10.50402	CO_2	3.2.2.1b	*10.56744*	CO_2	3.2.2.1a
10.50534	CO_2	3.2.2.1f	*10.56800*	CO_2	3.2.2.1e
10.50639	CO_2	3.2.2.1c	*10.56830*	N_2O	3.2.2.6a
10.50706	N_2O	3.2.2.6a	*10.56847*	N_2O	3.2.2.6a
10.50821	N_2O	3.2.2.6a	*10.57105*	CO_2	3.2.2.1a
10.51066	CO_2	3.2.2.1c	*10.57185*	CO_2	3.2.2.1b
10.51126	CO_2	3.2.2.1a	*10.57710*	N_2O	3.2.2.6a
10.51288	CO_2	3.2.2.1c	*10.57746*	N_2O	3.2.2.6a
10.51312	CO_2	3.2.2.1a	*10.57988*	CO_2	3.2.2.1e
10.51344	CO_2	3.2.2.1b	*10.58190*	CO_2	3.2.2.1b
10.51567	N_2O	3.2.2.6a	*10.58205*	CO_2	3.2.2.1c
10.51664	N_2O	3.2.2.6a	*10.58518*	CO_2	3.2.2.1c
10.52295	CO_2	3.2.2.1b	*10.58576*	CO_2	3.2.2.1g
10.52325	CO_2	3.2.2.1g	*10.58595*	N_2O	3.2.2.6a
10.52333	CO_2	3.2.2.1e	*10.58684*	CO_2	3.2.2.1a
10.52434	N_2O	3.2.2.6a	*10.59103*	CO_2	3.2.2.1a
10.52476	CO_2	3.2.2.1c	*10.59104*	CO_2	3.2.2.1a
10.52512	N_2O	3.2.2.6a	*10.59204*	CO_2	3.2.2.1b
10.52827	CO_2	3.2.2.1f	*10.59205*	CO_2	3.2.2.1c
10.52875	CO_2	3.2.2.1c	*10.59485*	N_2O	3.2.2.6a
10.52985	CO_2	3.2.2.1a	*10.59559*	N_2O	3.2.2.6a
10.53000	C_2H_4	3.2.5	*10.60000*	NH_3	3.2.3.4a
10.53209	CO_2	3.2.2.1a	*10.60063*	CO_2	3.2.2.1d
10.53255	CO_2	3.2.2.1b	*10.60168*	CO_2	3.2.2.1c

Table 4.1 (continued)
GAS LASERS ARRANGED IN ORDER OF WAVELENGTH

Wavelength (μm)	Element or compound	Table No.	Wavelength (μm)	Element or compound	Table No.
10.60228	CO_2	3.2.2.1b	*10.67459*	CO_2	3.2.2.1a
10.60380	N_2O	3.2.2.6a	*10.67659*	CO_2	3.2.2.1b
10.60451	CO_2	3.2.2.1c	*10.67921*	CO_2	3.2.2.1d
10.60471	CO_2	3.2.2.1c	*10.67953*	CO_2	3.2.2.1d
10.60474	N_2O	3.2.2.6a	*10.67988*	N_2O	3.2.2.6a
10.60658	CO_2	3.2.2.1a	*10.68503*	CO_2	3.2.2.1c
10.60724	CO_2	3.2.2.1g	*10.68540*	CO_2	3.2.2.1e
10.61139	CO_2	3.2.2.1a	*10.68565*	CO_2	3.2.2.1a
10.61261	CO_2	3.2.2.1b	*10.68759*	CO_2	3.2.2.1b
10.61281	N_2O	3.2.2.6a	*10.68904*	CO_2	3.2.2.1a
10.61310	CO_2	3.2.2.1a	*10.68953*	N_2O	3.2.2.6a
10.61394	N_2O	3.2.2.6a	*10.69263*	CO_2	3.2.2.1d
10.61726	CO_2	3.2.2.1e	*10.69363*	CO_2	3.2.2.1d
10.62187	N_2O	3.2.2.6a	*10.69639*	CO_2	3.2.2.1a
10.62209	CO_2	3.2.2.1c	*10.69869*	CO_2	3.2.2.1b
10.62304	CO_2	3.2.2.1b	*10.69923*	N_2O	3.2.2.6a
10.62319	N_2O	3.2.2.6a	*10.7*	NH_3	3.2.3.4a
10.62462	CO_2	3.2.2.1c	*10.70633*	CO_2	3.2.2.1d
10.62585	CO_2	3.2.2.1d	*10.70652*	CO_2	3.2.2.1a
10.62668	CO_2	3.2.2.1a	*10.70679*	CO_2	3.2.2.1c
10.62905	CO_2	3.2.2.1g	*10.70801*	CO_2	3.2.2.1d
10.63030	CO_2	3.2.2.1e	*10.70878*	N_2O	3.2.2.6a
10.63210	CO_2	3.2.2.1a	*10.70989*	CO_2	3.2.2.1b
10.63250	N_2O	3.2.2.6a	*10.71055*	CO_2	3.2.2.1a
10.63355	CO_2	3.2.2.1b	*10.71473*	CO_2	3.2.2.1c
10.63886	CO_2	3.2.2.1d	*10.71857*	CO_2	3.2.2.1a
10.64187	N_2O	3.2.2.6a	*10.71879*	N_2O	3.2.2.6a
10.64268	CO_2	3.2.2.1c	*10.72029*	CO_2	3.2.2.1d
10.64365	CO_2	3.2.2.1e	*10.72118*	CO_2	3.2.2.1b
10.64417	CO_2	3.2.2.1b	*10.72265*	CO_2	3.2.2.1a
10.64710	CO_2	3.2.2.1a	*10.72775*	CO_2	3.2.2.1a
10.65118	CO_2	3.2.2.1g	*10.72866*	N_2O	3.2.2.6a
10.65215	CO_2	3.2.2.1d	*10.73243*	CO_2	3.2.2.1a
10.65316	CO_2	3.2.2.1a	*10.73258*	CO_2	3.2.2.1b
10.65317	CO_2	3.2.2.1d	*10.73452*	CO_2	3.2.2.1d
10.65488	CO_2	3.2.2.1b	*10.73468*	N_2O	3.2.2.6a
10.65726	CO_2	3.2.2.1e	*10.73729*	NH_3	3.2.3.4a
10.66077	N_2O	3.2.2.6a	*10.73757*	CO_2	3.2.2.1d
10.66366	CO_2	3.2.2.1c	*10.73859*	N_2O	3.2.2.6a
10.66568	CO_2	3.2.2.1b	*10.74112*	CO_2	3.2.2.1a
10.66571	CO_2	3.2.2.1d	*10.74394*	NH_3	3.2.3.4a
10.66605	CO_2	3.2.2.1d	*10.74466*	N_2O	3.2.2.6a
10.66789	CO_2	3.2.2.1a	*10.74857*	N_2O	3.2.2.6a
10.67030	N_2O	3.2.2.6a	*10.74903*	CO_2	3.2.2.1d
10.67118	CO_2	3.2.2.1e	*10.74934*	CO_2	3.2.2.1a

Table 4.1 (continued)
GAS LASERS ARRANGED IN ORDER OF WAVELENGTH

Wavelength (μm)	Element or compound	Table No.	Wavelength (μm)	Element or compound	Table No.
10.75153	CO_2	3.2.2.1c	*10.83941*	CO_2	3.2.2.1a
10.75277	CO_2	3.2.2.1d	*10.84102*	N_2O	3.2.2.6a
10.75387	NH_3	3.2.3.4a	*10.84183*	CO_2	3.2.2.1d
10.75449	N_2O	3.2.2.6a	*10.84559*	N_2O	3.2.2.6a
10.75468	CO_2	3.2.2.1a	*10.84981*	CO_2	3.2.2.1d
10.75861	N_2O	3.2.2.6a	*10.85159*	N_2O	3.2.2.6a
10.76381	CO_2	3.2.2.1d	*10.85359*	CO_2	3.2.2.1e
10.76406	CO_2	3.2.2.1a	*10.85600*	N_2O	3.2.2.6a
10.76438	N_2O	3.2.2.6a	*10.85826*	CO_2	3.2.2.1d
10.76710	NH_3	3.2.3.4a	*10.85978*	CO_2	3.2.2.1a
10.76824	NH_3	3.2.3.4a	*10.86222*	N_2O	3.2.2.6a
10.76871	N_2O	3.2.2.6a	*10.86647*	N_2O	3.2.2.6a
10.77130	CO_2	3.2.2.1a	*10.86697*	CO_2	3.2.2.1d
10.77433	N_2O	3.2.2.6a	*10.87081*	CO_2	3.2.2.1e
10.77451	CO_2	3.2.2.6c	*10.87171*	CO_2	3.2.2.1a
10.77731	CO_2	3.2.2.6a	*10.87290*	N_2O	3.2.2.6a
10.77886	CO_2	3.2.2.6d	*10.87498*	CO_2	3.2.2.1d
10.77886	N_2O	3.2.2.6a	*10.87700*	N_2O	3.2.2.6a
10.78399	CO_2	3.2.2.1d	*10.88365*	N_2O	3.2.2.6a
10.78434	N_2O	3.2.2.6a	*10.88443*	CO_2	3.2.2.1d
10.78739	CO_2	3.2.2.1a	*10.88473*	CO_2	3.2.2.1a
10.78908	CO_2	3.2.2.1a	*10.88759*	N_2O	3.2.2.6a
10.78908	N_2O	3.2.2.6a	*10.88891*	CO_2	3.2.2.1e
10.79362	CO_2	3.2.2.1a	*10.89018*	CO_2	3.2.2.1a
10.79419	CO_2	3.2.2.1d	*10.89198*	CO_2	3.2.2.1d
10.79440	N_2O	3.2.2.6a	*10.89446*	N_2O	3.2.2.6a
10.79792	CO_2	3.2.2.1c	*10.89631*	CO_2	3.2.2.1a
10.79935	N_2O	3.2.2.6a	*10.89824*	N_2O	3.2.2.6a
10.80002	CO_2	3.2.2.1d	*10.90027*	CO_2	3.2.2.1g
10.80003	CO_2	3.2.2.1a	*10.90096*	CO_2	3.2.2.1g
10.80452	N_2O	3.2.2.6a	*10.90217*	CO_2	3.2.2.1d
10.80968	N_2O	3.2.2.6a	*10.90533*	N_2O	3.2.2.6a
10.80979	CO_2	3.2.2.1d	*10.90733*	CO_2	3.2.2.1e
10.81111	CO_2	3.2.2.1a	*10.90895*	N_2O	3.2.2.6a
10.81470	N_2O	3.2.2.6a	*10.90927*	CO_2	3.2.2.1d
10.81634	CO_2	3.2.2.1a,d	*10.91010*	CO_2	3.2.2.1a
10.82007	N_2O	3.2.2.6a	*10.91123*	CO_2	3.2.2.1g
10.82372	CO_2	3.2.2.1a	*10.91626*	N_2O	3.2.2.6a
10.82494	N_2O	3.2.2.6a	*10.91972*	N_2O	3.2.2.6a
10.82567	CO_2	3.2.2.1d	*10.92133*	CO_2	3.2.2.1a
10.83052	N_2O	3.2.2.6a	*10.92147*	CO_2	3.2.2.1a
10.83293	CO_2	3.2.2.1d	*10.92246*	CO_2	3.2.2.1g
10.83527	CO_2	3.2.2.1a	*10.92609*	CO_2	3.2.2.1e
10.83523	N_2O	3.2.2.6a	*10.92735*	N_2O	3.2.2.6a
10.83559	CO_2	3.2.2.1e	*10.93055*	N_2O	3.2.2.6a

Table 4.1 (continued)
GAS LASERS ARRANGED IN ORDER OF WAVELENGTH

Wavelength (μm)	Element or compound	Table No.	Wavelength (μm)	Element or compound	Table No.
10.93071	CO_2	3.2.2.1a	*11.02472*	CO_2	3.2.2.1d
10.93396	CO_2	3.2.2.1g	*11.02496*	CO_2	3.2.2.1d
10.93590	CO_2	3.2.2.1a	*11.02873*	CO_2	3.2.2.1d
10.93830	N_2O	3.2.2.6a	*11.02898*	N_2O	3.2.2.6a
10.94144	N_2O	3.2.2.6a	*11.02974*	CO_2	3.2.2.1a
10.94235	CO_2	3.2.2.1a	*11.03544*	CO_2	3.2.2.1g
10.94518	CO_2	3.2.2.1e	*11.03850*	CO_2	3.2.2.1a
10.94572	CO_2	3.2.2.1g	*11.04061*	N_2O	3.2.2.6a
10.94676	CO_2	3.2.2.1a	*11.04354*	CO_2	3.2.2.1a
10.94942	CO_2	3.2.2.1a	*11.04471*	CO_2	3.2.2.1d
10.95149	CO_2	3.2.2.1a	*11.04578*	CO_2	3.2.2.1e
10.95240	N_2O	3.2.2.6a	*11.04823*	CO_2	3.2.2.1d
10.95774	CO_2	3.2.2.1g	*11.04932*	CO_2	3.2.2.1g
10.96059	N_2O	3.2.2.1a	*11.05259*	CO_2	3.2.2.1a
10.96214	CO_2	3.2.2.1a	*11.0610*	CO_2	3.2.2.1a
10.96341	N_2O	3.2.2.6a	*11.06347*	CO_2	3.2.2.1g
10.96460	CO_2	3.2.2.1e	*11.06501*	CO_2	3.2.2.1d
10.97002	CO_2	3.2.2.1g	*11.06696*	CO_2	3.2.2.1e
10.97183	N_2O	3.2.2.6a	*11.06803*	CO_2	3.2.2.1d
10.97261	CO_2	3.2.2.1a	*11.07160*	CO_2	3.2.2.1a
10.97449	N_2O	3.2.2.6a	*11.0762*	CO_2	3.2.2.1a
10.98000	C_2H_2	3.2.5	*11.07789*	CO_2	3.2.2.1g
10.981641	Ne	1.9.2	*11.08363*	CO_2	3.2.2.1a
10.98258	CO_2	3.2.2.1g	*11.08563*	CO_2	3.2.2.1d
10.98314	N_2O	3.2.2.6a	*11.08813*	CO_2	3.2.2.1d
10.98438	CO_2	3.2.2.1e	*11.08851*	CO_2	3.2.2.1e
10.98527	CO_2	3.2.2.1a	*11.09258*	CO_2	3.2.2.1g
10.98563	N_2O	3.2.2.6a	*11.0999*	CO_2	3.2.2.1a
10.98566	CO_2	3.2.2.1d	*11.10014*	CO_2	3.2.2.1a
10.98882	CO_2	3.2.2.1a	*11.10656*	CO_2	3.2.2.1d
10.9944	CO_2	3.2.2.1a	*11.10699*	CO_2	3.2.2.1f
10.99450	N_2O	3.2.2.6a	*11.1073*	CO_2	3.2.2.1a
10.99539	CO_2	3.2.2.1g	*11.10754*	CO_2	3.2.2.1g
11.00449	CO_2	3.2.2.1e	*11.10855*	CO_2	3.2.2.1d
11.00503	CO_2	3.2.2. 1d	*11.11042*	CO_2	3.2.2.1e
11.00593	N_2O	3.2.2.6a	*11.12278*	CO_2	3.2.2.1g
11.00731	CO_2	3.2.2.1a	*11.1238*	CO_2	3.2.2.1a
11.00847	CO_2	3.2.2.1g	*11.12782*	CO_2	3.2.2.1d
11.00954	CO_2	3.2.2.1d	*11.12915*	CO_2	3.2.2.1a
11.01000	NH_3	3.2.3.4a	*11.12928*	CO_2	3.2.2.1d
11.01080	NH_3	3.2.3.4a	*11.1309*	CO_2	3.2.2.1a
11.01593	CO_2	3.2.2.1a	*11.13270*	CO_2	3.2.2.1e
11.01595	CO_2	3.2.2.1a	*11.13520*	CO_2	3.2.2.1d
11.01743	N_2O	3.2.2.6a	*11.13829*	CO_2	3.2.2.1a
11.02182	CO_2	3.2.2.1g	*11.14940*	CO_2	3.2.2.1d

Table 4.1 (continued)
GAS LASERS ARRANGED IN ORDER OF WAVELENGTH

Wavelength (μm)	Element or compound	Table No.	Wavelength (μm)	Element or compound	Table No.
11.14968	CO_2	3.2.2.1f	*11.35462*	CO_2	3.2.2.1d
11.15033	CO_2	3.2.2.1d	*11.35602*	CO_2	3.2.2.1g
11.15407	CO_2	3.2.2.1g	*11.35885*	CO_2	3.2.2.1d
11.15537	CO_2	3.2.2.1e	*11.36170*	CO_2	3.2.2.1f
11.15867	CO_2	3.2.2.1a	*11.37542*	CO_2	3.2.2.1g
11.16443	CO_2	3.2.2.1f	*11.37905*	CO_2	3.2.2.1d
11.17131	CO_2	3.2.2.1d	*11.38390*	CO_2	3.2.2.1d
11.17170	CO_2	3.2.2.1d	*11.39512*	CO_2	3.2.2.1g
11.17841	CO_2	3.2.2.1e	*11.39833*	CO_2	3.2.2.1f
11.17943	CO_2	3.2.2.1f	*11.40384*	CO_2	3.2.2.1d
11.18868	CO_2	3.2.2.1a	*11.40932*	CO_2	3.2.2.1d
11.19339	CO_2	3.2.2.1d	*11.41514*	CO_2	3.2.2.1g
11.19354	CO_2	3.2.2.1d	*11.41705*	CO_2	3.2.2.1f
11.19468	CO_2	3.2.2.1f	*11.42899*	CO_2	3.2.2.1d
11.20185	CO_2	3.2.2.1d	*11.43511*	CO_2	3.2.2.1d
11.20879	NH_3	3.2.3.4a	*11.43546*	CO_2	3.2.2.1g
11.20968	CO_2	3.2.2.1f	*11.43605*	CO_2	3.2.2.1f
11.21020	CO_2	3.2.2.1f	*11.45452*	CO_2	3.2.2.1d
11.21541	CO_2	3.2.2.1f	*11.45533*	CO_2	3.2.2.1f
11.21612	CO_2	3.2.2.1d	*11.45610*	CO_2	3.2.2.1g
11.22568	CO_2	3.2.2.1f	*11.46044*	NH_3	3.2.3.4a
11.22598	CO_2	3.2.2.1f	*11.46129*	CO_2	3.2.2.1d
11.23776	CO_2	3.2.2.1e	*11.47135*	NH_3	3.2.3.4a
11.23903	CO_2	3.2.2.1d	*11.47490*	CO_2	3.2.2.1f
11.24209	CO_2	3.2.2.1f	*11.476*	CS_2	3.2.2.3a
11.24991	CO_2	3.2.2.1e	*11.47706*	CO_2	3.2.2.1g
11.25832	CO_2	3.2.2.1f	*11.48042*	CO_2	3.2.2.1d
11.26045	CO_2	3.2.2.1d	*11.4823*	CS_2	3.2.2.3a
11.26100	NH_3	3.2.3.4a	*11.48786*	CO_2	3.2.2.1d
11.26229	CO_2	3.2.2.1d	*11.4893*	CS_2	3.2.2.3a
11.26590	CO_2	3.2.2.1d	*11.49474*	CO_2	3.2.2.1f
11.27488	CO_2	3.2.2.1f	*11.4962*	CS_2	3.2.2.3a
11.28347	CO_2	3.2.2.1d	*11.49835*	CO_2	3.2.2.1g
11.29171	CO_2	3.2.2.1f	*11.5031*	CS_2	3.2.2.3a
11.298683	Xe	1.9.5	*11.50671*	CO_2	3.2.2.1d
11.29963	CO_2	3.2.2.1g	*11.5099*	CS_2	3.2.2.3a
11.30684	CO_2	3.2.2.1d	*11.5166*	CS_2	3.2.2.3a
11.30881	CO_2	3.2.2.1d	*11.51996*	CO_2	3.2.2.1g
11.30986	CO_2	3.2.2.1d	*11.52074*	NH_3	3.2.3.4a
11.31813	CO_2	3.2.2.1g	*11.5237*	CS_2	3.2.2.3a
11.32617	CO_2	3.2.2.1f	*11.52446*	NH_3	3.2.3.4a
11.33056	CO_2	3.2.2.1d	*11.5307*	CS_2	3.2.2.3a
11.33418	CO_2	3.2.2.1d	*11.5376*	CS_2	3.2.2.3a
11.33693	CO_2	3.2.2.1g	*11.54189*	CO_2	3.2.2.1g
11.34380	CO_2	3.2.2.1f	*11.5446*	CS_2	3.2.2.3a

Table 4.1 (continued)
GAS LASERS ARRANGED IN ORDER OF WAVELENGTH

Wavelength (μm)	Element or compound	Table No.	Wavelength (μm)	Element or compound	Table No.
11.553	CS_2	3.2.2.3a	*12.036591*	Sb	1.5.4
11.560	CS_2	3.2.2.3a	*12.03872*	NH_3	3.2.3.4a
11.56417	CO_2	3.2.2.1g	*12.04925*	CO_2	3.2.2.1f
11.568	CS_2	3.2.2.3a	*12.07694*	CO_2	3.2.2.1f
11.582	CS_2	3.2.2.3a	*12.10499*	CO_2	3.2.2.1f
11.58689	CO_2	3.2.2.1f	*12.11418*	NH_3	3.2.3.4a
11.5962	CS_2	3.2.2.3a	*12.13342*	CO_2	3.2.2.1f
11.60907	CO_2	3.2.2.1f	*12.140522*	Ar	1.9.3
11.63081	CO_2	3.2.2.1f	*12.15575*	NH_3	3.2.3.4a
11.63303	CO_2	3.2.2.1g	*12.18444*	NH_3	3.2.3.4a
11.65286	CO_2	3.2.2.1f	*12.214*	CS_2	3.2.2.3b
11.65669	CO_2	3.2.2.1g	*12.237*	CS_2	3.2.2.3b
11.67521	CO_2	3.2.2.1f	*12.24521*	NH_3	3.2.3.4a
11.68070	CO_2	3.2.2.1g	*12.24700*	CS_2	3.2.3b
11.69787	CO_2	3.2.2.1f	*12.24911*	NH_3	3.2.3.4a
11.70507	CO_2	3.2.2.1g	*12.26105*	NH_3	3.2.3.4a
11.71208	NH_3	3.2.3.4a	*12.266358*	Xe	1.9.5
11.71582	NH_3	3.2.3.4a	*12.28136*	NH_3	3.2.3.4a
11.72084	CO_2	3.2.2.1f	*12.31072*	NH_3	3.2.3.4a
11.72712	NH_3	3.2.3.4a	*12.35002*	NH_3	3.2.3.4a
11.72981	CO_2	3.2.2.1g	*12.37821*	NH_3	3.2.3.4a
11.74413	CO_2	3.2.2.1f	*12.38425*	NH_3	3.2.3.4a
11.74637	NH_3	3.2.3.4a	*12.39560*	NH_3	3.2.3.4a
11.76773	CO_2	3.2.2.1f	*12.52781*	NH_3	3.2.3.4a
11.79165	CO_2	3.2.2.1f	*12.53999*	NH_3	3.2.3.4a
11.796	NH_3	3.2.3.4a	*12.56068*	NH_3	3.2.3.4a
11.80167	NH_3	3.2.3.4a	*12.568814*	K	1.1.3
11.81589	CO_2	3.2.2.1f	*12.59059*	NH_3	3.2.3.4a
11.83	H_2O	3.2.2.5	*12.63063*	NH_3	3.2.3.4a
11.84046	CO_2	3.2.2.1f	*12.68213*	NH_3	3.2.3.4a
11.860553	Ne	1.9.2	*12.69716*	NH_3	3.2.3.4a
11.86536	CO_2	3.2.2.1f	*12.80959*	NH_3	3.2.3.4a
11.89060	CO_2	3.2.2.1f	*12.81145*	NH_3	3.2.3.4a
11.902069	Ne	1.9.2	*12.81532*	NH_3	3.2.3.4a
11.91617	CO_2	3.2.2.1f	*12.82765*	NH_3	3.2.3.4a
11.94209	CO_2	3.2.2.1f	*12.835306*	Ne	1.9.2
11.959	CS_2	3.2.2.3b	*12.85*	HCN	3.2.2.4
11.96000	H_2O	3.2.2.5	*12.84863*	NH_3	3.2.3.4a
11.96835	CO_2	3.2.2.1f	*12.87890*	NH_3	3.2.3.4a
11.97859	NH_3	3.2.3.4a	*12.917304*	Xe	1.9.5
11.983	CS_2	3.2.2.3b	*12.91746*	NH_3	3.2.3.4a
11.99025	NH_3	3.2.3.4a	*12.97163*	NH_3	3.2.3.4a
11.99496	CO_2	3.2.2.1f	*12.98366*	NH_3	3.2.3.4a
12.01008	NH_3	3.2.3.4a	*13.03715*	NH_3	3.2.3.4a
12.02192	CO_2	3.2.2.1f	*13.11233*	NH_3	3.2.3.4a

Table 4.1 (continued)
GAS LASERS ARRANGED IN ORDER OF WAVELENGTH

Wavelength (μm)	Element or compound	Table No.
13.12477	NH_3	3.2.3.4a
13.144	CO_2	3.2.2.1a
13.14593	NH_3	3.2.3.4a
13.154	CO_2	3.2.2.1a
13.159	CO_2	3.2.2.1a
13.17643	NH_3	3.2.3.4a
13.188714	Cd	1.2.6
13.21725	NH_3	3.2.3.4a
13.23390	NH_3	3.2.3.4a
13.26978	NH_3	3.2.3.4a
13.33580	NH_3	3.2.3.4a
13.411	NH_3	3.2.3.4a
13.540	CF_3I	3.2 .4.2
13.541	CO_2	3.2.2.1a
13.570	CF_3I	3.2.4.2
13.57749	NH_3	3.2.3.4a
13.63	CF_3I	3.2.4.2
13.65533	NH_3	3.2.3.4a
13.72555	NH_3	3.2.3.4a
13.72608	NH_3	3.2.3.4a
13.758432	Ne	1.9.2
13.87	CO_2	3.2.2.1a
14.1	CO_2	3.2.2.1a
14.1681	CO_2	3.2.2.1a
14.2001	CO_2	3.2.2.1a
14.2325	CO_2	3.2.2.1a
14.3	NH_3	3.2.3.4b
14.58202	Cd	1.2.6
14.78	NH_3	3.2.3.4a
14.8	NH_3	3.2.3.4b
14.930	Ne	1.9.2
15.037067	Ar	1.9.3
15.04	NH_3	3.2.3.4a
15.08	NH_3	3.2.3.4a
15.19	NSF	3.2.2.8
15.2	NH_3	3.2.3.4a
15.26	NSF	3.2.2.8
15.29	CF_4	3.2.4.1b
15.306	CF_4	3.2.4.1b
15.32	CF_4	3.2.4.1b
15.34	NSF	3.2.2.8
15.39	CF_4	3.2.4.1b
15.42	NSF	3.2.2.8
15.42	CF_4	3.2.4.1b
15.44	CF_4	3.2.4.1b
15.45	CF_4	3.2.4.1b
15.46	CF_4	3.2.4.1b
15.46	CF_4	3.2.4.1b
15.47	CF_4	3.2.4.1b
15.47	NH_3	3.2.3.4a
15.47	NSF	3.2.2.8
15.48	CF_4	3.2.4.1a
15.49	CF_4	3.2.4.1a
15.51	CF_4	3.2.4.1c
15.53	CF_4	3.2.4.1b
15.54	CF_4	3.2.4.1b,c
15.547	CF_4	3.2.4.1a
15.55	CF_4	3.2.4.1c
15.57	CF_4	3.2.4.1a
15.58	CF_4	3.2.4.1a,c
15.59359	CO_2	3.2.2.1a
15.60	CF_4	3.2.4.1a,c
15.607	CF_4	3.2.4.1a
15.61	CF_4	3.2.4.1c
15.61	NSF	3.2.2.8
15.62	CF_4	3.2.4.1b,c
15.63106	CO_2	3.2.2.1a
15.64	CF_4	3.2.4.1c
15.65	CF_4	3.2.4.1c
15.66877	CO_2	3.2.2.1a
15.7	NH_3	3.2.3.4b
15.70670	CO_2	3.2.2.1a
15.71	CF_4	3.2.4.1a,c
15.711	CH_3CCH	3.2.6.1
15.716	CH_3CCH	3.2.6.1
15.721	CH_3CCH	3.2.6.1
15.74	CF_4	3.2.4.1a,b
15.74488	CO_2	3.2.2.1a
15.75	CF_4	3.2.4.1b
15.78	CF_4	3.2.4.1b
15.78148	CO_2	3.2.2.1a
15.78330	CO_2	32.2.1a
15.79	CF_4	3.2.4.1c
15.80	CF_4	3.2.4.1b
15.82	CF_4	3.2.4.1b,c
15.82196	CO_2	3.2.2.1a
15.83	CF_4	3.2.4.1a
15.838	NSF	3.2.2.8
15.84	CF_4	3.2.4.1a
15.844	CF_4	3.2.4.1a

Table 4.1 (continued)
GAS LASERS ARRANGED IN ORDER OF WAVELENGTH

Wavelength (μm)	Element or compound	Table No.	Wavelength (μm)	Element or compound	Table No.
15.845	CF_4	3.2.4.1a	*16.19082*	CO_2	3.2.2.1a
15.847	CF_4	3.2.4.1a	*16.19278*	CO_2	3.2.2.1a
15.85	CF_4	3.2.4.1a	*16.19494*	CO_2	3.2.2.1a
15.86086	CO_2	3.2.2.1a	*16.19724*	CO_2	3.2.2.1a
15.87260	NH_3	3.2.3.4a	*16.20*	CF_4	3.2.4.1b
15.87758	NH_3	3.2.3.4a	*16.21*	CF_4	3.2.4.1a,c
15.89	CF_4	3.2.4.1b	*16.22*	CF_4	3.2.4.1c
15.89	NSF	3.2.2.8	*16.23*	CF_4	3.2.4.1b
15.905	SF_6	3.2.6.2	*16.24*	CF_4	3.2.4.1a
15.91	CF_4	3.2.4.1b	*16.245*	CF_4	3.2.4.1a
15.91292	NH_3	3.2.3.4a	*16.25*	CF_4	3.2.4.1a,b,c
15.92	CF_4	3.2.4.1b	*16.259*	CF_4	3.2.4.1a
15.93	CF_4	3.2.4.1b	*16.26*	CF_4	3.2.4.1b
15.93935	CO_2	3.2.2.1a	*16.27*	CF_4	3.2.4.1a,b
15.94	CF_4	3.2.4.1b	*16.279*	CF_4	3.2.4.1a
15.94637	NH_3	3.2.3.4a	*16.28*	CF_4	3.2.4.1c
15.95	CF_4	3.2.4.1c	*16.29*	CF_4	3.2.4.1b
15.968063	K	1.1.3	*16.30*	CF_4	3.2.4.1b,c
15.97896	CO_2	3.2.2.1a	*16.32*	CF_4	3.2.4.1b
15.98	NSF	3.2.2.8	*16.32*	$FClO_3$	3.2.4.4
15.99	CF_4	3.2.4.1c	*16.33*	CF_4	3.2.4.1c
16.0	NH_3	3.2.3.4b	*16.34*	$FClO_3$	3.2.4.4
16.00	CF_4	3.2.4.1a,b	*16.34*	CF_4	3.2.4.1a
16.02	CF_4	3.2.4.1a	*16.35*	$FClO_3$	3.2.4.4
16.03	NSF	3.2.2.8	*16.35*	CF_4	3.2.4.1c
16.06	CF_4	3.2.4.1a	*16.36*	CF_4	3.2.4.1b
16.07	CF_4	3.2.4.1a	*16.36699*	CO_2	3.2.2.1a
16.08	CF_4	3.2.4.1c	*16.39*	CH_3CCH	3.2.6.1
16.09	CF_4	3.2.4.1a,b	*16.4*	NOCl	3.2.2.7
16.10	CF_4	3.2.4.1b	*16.40*	CF_4	3.2.4.1a
16.11	CF_4	3.2.4.1a,b	*16.40907*	CO_2	3.2.2.1a
16.121	CH_3CCH	3.2.6.1	*16.42*	CH_3CCH	3.2.6.1
16.14	CF_4	3.2.4.1c	*16.44*	CF_4	3.2.4.1c
16.15	CF_4	3.2.4.1b,c	*16.45*	$FClO_3$	3.2.4.4
16.15	NSF	3.2.2.8	*16.45148*	CO_2	3.2.2.1a
16.16	CF_4	3.2.4.1b	*16.48*	CF_4	3.2.4.1b
16.18	CF_4	3.2.4.1a	*16.49*	$FClO_3$	3.2.4.4
16.18191	CO_2	3.2.2.1a	*16.49408*	CO_2	3.2.2.1a
16.18267	CO_2	3.2.2.1a	*16.50*	$FClO_3$	3.2.4.4
16.18364	CO_2	3.2.2.1a	*16.52*	$FClO_3$	3.2.4.4
16.18474	CO_2	3.2.2.1a	*16.52*	NOCl	3.2.2.7
16.18605	CO_2	3.2.2.1a	*16.53696*	CO_2	3.2.2.1a
16.18744	CO_2	3.2.2.1a	*16.56*	$FClO_3$	3.2.4.4
16.18906	CO_2	3.2.2.1a	*16.57*	NOCl	3.2.2.7
16.19	NSF	3.2.2.8	*16.58012*	CO_2	3.2.2.1a

Table 4.1 (continued)
GAS LASERS ARRANGED IN ORDER OF WAVELENGTH

Wavelength (μm)	Element or compound	Table No.	Wavelength (μm)	Element or compound	Table No.
16.586	CO_2	3.2.2.1a	17.639	CO_2	3.2.2.1c
16.596	CO_2	3.2.2.1b	17.665	C_2D_2	3.2.3.2b
16.597	CO_2	3.2.2.1a	17.684	CO_2	3.2.2.1c
16.61	$FClO_3$	3.2.4.4	17.772	C_2D_2	3.2.3.2b
16.62350	CO_2	3.2.2.1a	17.730	CO_2	3.2.2.1c
16.638076	Ne	1.9.2	17.775	CO_2	3.2.2.1c
16.66	$FClO_3$	3.2.4.4	17.778	C_2D_2	3.2.3.2b
16.66719	CO_2	3.2.2.1a	17.8	NH_3	3.2.3.4b
16.667472	Ne	1.9.2	17.803541	Ne	1.92
16.69	NOCl	3.2.2.7	17.821	CO_2	3.2.2. 1c
16.7	NOCl	3.2.2.7	17.835	C_2D_2	3.2.3.2b
16.71112	CO_2	3.2.2.1a	17.839876	Ne	1.9.2
16.73	$FClO_3$	3.2.4.4	17.888255	Ne	1.9.2
16.75	$FClO_3$	3.2.4.4	19.915	CO_2	3.2.2.1c
16.75	NOCl	3.2.2.7	17.962	CO_2	3.2.2.1c
16.75533	CO_2	3.2.2.1a	18.010	CO_2	3.2.2.1c
16.77	$FClO_3$	3.2.4.4	18.053	CO_2	3.2.2.1c
16.79	$FClO_3$	3.2.4.4	18.21	NH_3	3.2.3.4a
16.79983	CO_2	3.2.2.1a	18.3	BCl_3	3.2.3.1
16.82	$FClO_3$	3.2.4.4	18.5067	Ne	1.9.2
16.86	NOCl	3.2.2.7	18.505324	Xe	1.9.5
16.89361	Ne	1.9.2	18.67	C_2D_2	3.2.3.2b
16.9	NOCl	3.2.2.7	18.79	C_2D_2	3.2.3.2b
16.927	CO_2	3.2.2.1b	18.8	BCl_3	3.2.3.1
16.93	$FClO_3$	3.2.4.4	18.84	C_2D_2	3.2.3.2b
16.946194	Ne	1.9.2	18.85	C_2D_2	3.2.3.2b
16.97	CO_2	3.2.2.1b	18.92674	NH_3	3.2.3.4a
16.99	NOCl	3.2.2.7	18.96	C_2D_2	3.2.3.2b
17.00	$FClO_3$	3.2.4.4	18.97	C_2D_2	3.2.3.2b
17.023	CO_2	3.2.2.1a	18.983	COS	3.2.2.2a
17.029	CO_2	3.2.2.1a	19.03	C_2D_2	3.2.3.2b
17.036	CO_2	3.2.2.1a	19.057	COS	3.2.2.2a
17.048	CO_2	3.2.2.1a	19.081	C_2D_2	3.2.3.2b
17.157279	Ne	1.9.2	19.1	BCl_3	3.2.3.1
17.188155	Ne	1.9.2	19.13	C_2D_2	3.2.3.2b
17.280	CO_2	3.2.2.1c	19.20	C_2D_2	3.2.3.2b
17.370	CO_2	3.2.2.1a	19.27	C_2D_2	3.2.3.2b
17.376	CO_2	3.2.2.1a	19.4	BCl_3	3.2.3.1
17.390	CO_2	3.2.2.1a	19.55019	NH_3	3.2.3.4a
17.45	C_2D_2	3.2.3.2b	19.67	C_2D_2	3.2.3.2b
17.463	CO_2	3.2.2.1c	19.94700	C_2D_2	3.2.3.2b
17.498	C_2D_2	3.2.3.2b	20.01	C_2D_2	3.2.3.2b
17.56	C_2D_2	3.2.3.2b	20.010	C_2D_2	3.2.3.2b
17.596	C_2CO_2	3.2.2.1c	20.073	C_2D_2	3.2.3.2b
17.610	C_2D_2	3.2.3.2b	20.13	C_2D_2	3.2.3.2b

Table 4.1 (continued)
GAS LASERS ARRANGED IN ORDER OF WAVELENGTH

Wavelength (μm)	Element or compound	Table No.	Wavelength (μm)	Element or compound	Table No.
20.2	BCl_3	3.2.3.1	*34.552*	Ne	3.3.1.2,1.9.2
20.202	C_2D_2	3.2.3.2b	*34.679*	Ne	3.3.1.2,1.9.2
20.267	C_2D_2	3.2.3.2b	*34.79000*	CH_3OH	3.3.6.3
20.332	C_2D_2	3.2.3.2b	*34.8*	CD_3OH	3.3.6.7
20.38798	NH_3	3.2.3.4a	*35.0*	CD_3OD	3.3.6.8
20.44	C_2D_2	3.2.3.2b	*35.602*	Ne	3.3.1.2,1.9.2
20.478165	Ne	1.9.2	*37.0*	CH_3OH	3.3.6.2
20.6	BCl_3	3.2.3.1	*37.231*	Ne	3.3.1.2,1.9.2
21.05409	NH_3	3.2.3.4a	*37.6*	CD_3OH	3.3.6.7
21.471	NH_3	3.2.3.4a	*37.85421*	CH_3OH	3.3.6.2
21.750951	Ne	1.9.2	*39.92423*	CH_3OH	3.3.6.2
22.4	BCl_3	3.2.3.1	*40.1*	CD_3OH	3.3.6.7
22.542	NH_3	3.2.3.4a	*41.0*	CD_3OD	3.3.6.8
22.563	NH_3	3.2.3.4a	*41.25*	CD_3OH	3.3.6.7
22.71	NH_3	3.2.3.4a	*41.35487*	CD_3OH	3.3.6.7
22.836211	Ne	1.9.2	*41.40*	CD_3OH	3.3.6.7
23.0	BCl_3	3.2.3.1	*41.741*	Ne	3.3.1.2,1.9.2
23.675	NH_3	3.2.3.4a	*41.90*	CH_3OH	3.3.6.3
23.86	NH_3	3.2.3.4a	*42.15908*	CH_3OH	3.3.6.2
24.78	SiF_4	3.2.4.6	*42.31*	CH_3OH	3.3.6.2
24.918	NH_3	3.2.3. 4a	*45.50*	CH_2DOH	3.3.6.4
25.12	NH_3	3.2.3.4a	*43.1*	CH_3OH	3.3.6.2
25.21617	Ne	1.9.2	*43.47*	CH_3OH	3.3.6.2
25.31	SiF_4	3.2.4.6	*43.69729*	CD_3OH	3.3.6.7
25.36	SiF_4	3.2.4.6	*44.0*	CH_2DOH	3.3.6.4
25.40	SiF_4	3.2.4.6	*46.7*	CH_3OD	3.3.6.5
25.67	SiF_4	3.2.4.6	*47.244*	H_2O	3.3.3.1
25.68	SiF_4	3.2.4.6	*47.46315*	H_2O	3.3.3.1
25.77	SiF_4	3.2.4.6	*47.687*	H_2O	3.3.3.1
25.79	SiF_4	3.2.4.6	*47.91*	NH_3	3.3.4.1
26.01	SiF_4	3.2.4.6	*49.8*	CD_3OH	3.3.6.7
26.14	SiF_4	3.2.4.6	*50.705*	Ne	3.3.1.2,1.9.2
26.282	NH_3	3.2.3.4a	*52.030*	NH_3	3.3.4.1
26.943974	Ar	1.9.3	*52.425*	Ne	3.3.1.2,1.9.2
28.052222	Ne	1.9.2	*52.800*	NH_3	3.3.4.1
30.69	NH_3	3.2.3.4a	*53.486*	Ne	3.3.1.2,1.9.2
31.47	NH_3	3.2.3.4a	*53.500*	CH_3OH	3.3.6.2
31.552211	Ne	1.9.2	*53.820*	CD_3OH	3.3.6.7
31.930315	Ne	1.9.2	*54.019*	Ne	3.3.1.2,1.9.2
31.951	NH_3	3.2.3.4a	*54.117*	Ne	3.3.1.2,1.9.2
32.105777	Ne	1.9.2	*54.460*	NH_3	3.3.4.1
32.13	NH_3	3.2.3.4a	*55.088*	H_2O	3.3.3.1
32.518950	Ne	1.9.2	*55.37004*	CH_3OH	3.3.6.2,1.9.2
33.029	H_2O	3.3.3.1	*55.537*	Ne	3.3.1.2
33.815657	Ne	1.9.2	*55.56*	CD_3OH	3.3.6.7

Table 4.1 (continued)
GAS LASERS ARRANGED IN ORDER OF WAVELENGTH

Wavelength (μm)	Element or compound	Table No.	Wavelength (μm)	Element or compound	Table No.
57.0	CH_3OD	3.3.6.5	*84.2789*	D_2O	3.3.3.2
57.040	NH_3	3.3.4.1	*85.047*	Ne	3.3.1.2,1.9.2
57.355	Ne	3.3.1.2,1.9.2	*85.3179*	CH_3OH	3.3.6.3
58.010	NH_3	3.3.4.1	*85.60093*	CH_3OH	3.3.6.2
59.37	NH_3	3.3.4.1	*85.79*	CH_3OH	3.3.6.3
60.10	CD_3OH	3.3.6.7	*86.*	NH_2D	3.3.4.3
60.17327	CH_3OH	3.3.6.2	*86.11179*	CH_3OH	3.3.6.3
61.61330	CH_3OH	3.3.6.2	*86.23938*	CH_3OH	3.3.6.2
62.96597	CH_3OH	3.3.6.2	*86.4*	CD_3OH	3.3.6.7
63.09639	CH_3OH	3.3.6.3	*86.96200*	Ne	3.3.1.2,1.9.2
63.36954	CH_3OH	3.3.6.2	*87.*	ND_3	3.3.4.4
64.50	NH_3	3.3.4.1	*87.09*	NH_3	3.3.4.1
65.51	NH_3	3.3.4.1	*87.10*	CH_2DOH	3.3.6.4
67.24	NH_3	3.3.4.1	*87.4*	NH_3	3.3.4.1
67.47941	CD_3OH	3.3.6.7	*87.90*	CH_2DOH	3.3.6.4
67.49536	CH_3OH	3.3.6.2	*87.90*	CH_3OH	3.3.6.3
67.75	NH_3	3.3.4.1	*88.06*	NH_3	3.3.4.1
68.32900	Ne	3.3.1.2,1.9.2	*88.471*	Ne	3.3.1.2,1.9.2
69.50	CH_3OD	3.3.6.5	*89.6*	CH_3OD	3.3.6.5
69.67956	CH_3OH	3.3.6.2	*89.859*	Ne	3.3.1.2,1.9.2
70.3	CH_3OD	3.3.6.5	*90.40*	CH_2DOH	3.3.6.4
70.51163	CH_3OH	3.3.6.2	*90.93*	NH_3	3.3.4.1
71.0	CD_3OH	3.3.6.7	*92.54392*	CH_3OH	3.3.6.2
72.108	Ne	3.3.1.2,1.9.2	*92.66428*	CH_3OH	3.3.6.2
72.74778	D_2O	3.3.3.2	*92.88*	NH_3	3.3.4.2
73.30642	CH_3OH	3.3.6.2	*93.02*	Ne	3.3.1.2,1.9.2
73.401	H_2O	3.3.3.1	*94.45*	NH_3	3.3.4.1
75.561687	Xe	1.9.5	*94.5*	D_2O	3.3.3.2
76.1	CD_3OH	3.3.6.7	*95.55106*	CH_2F_2	3.3.5.11
77.	NH_2D	3.3.4.3	*95.788*	He	3.3.1.1,1.9.1
77.40565	CH_3OH	3.3.6.2	*96.52239*	CH_3OH	3.3.6.2
77.48939	CH_3OH	3.3.6.3	*96.67*	NH_3	3.3.4.1
77.90489	CH_3OH	3.3.6.2	*97.51854*	CH_3OH	3.3.6.2
78.0	CD_3OD	3.3.6.8	*98*	CH_3OH	3.3.6.2
78.27	NH_3	3.3.4.1	*100*	CH_3NH_2	3.3.7.1
78.44333	H_2O	3.3.3.1	*100.00*	CH_2DOH	3.3.6.4
79.09101	H_2O	3.3.3.1	*100.80647*	CH_3OH	3.3.6.2
79.62	NH_3	3.3.4.1	*101.6*	CH_3OD	3.3.6.5
80.0	CH_3OD	3.3.6.5	*102.02349*	CH_2DOH	3.3.6.4
80.0	CH_3OH	3.3.6.2	*102.6*	CD_3OH	3.3.6.7
80.6	CH_3OH	3.3.6.2	*103.12463*	CH_3OD	3.3.6.5
81.48	NH_3	3.3.4.1	*103.48079*	CH_3OH	3.3.6.3
81.500	NH_3	3.3.4.1	*103.58629*	CH_3OH	3.3.6.3
81.5571	CD_3OH	3.3.6.7	*105.14719*	CH_3OH	3.3.6.3
81.9	CH_3OD	3.3.6.5	*105.35*	NH_3	3.3.4.1

Table 4.1 (continued)
GAS LASERS ARRANGED IN ORDER OF WAVELENGTH

Wavelength (μm)	Element or compound	Table No.	Wavelength (μm)	Element or compound	Table No.
105.51827	CH_2F_2	3.3.5.11	124.4	Ne	3.3.1.2,1.9.2
106	CH_3OD	3.3.6.5	124.43170	CH_2DOH	3.3.6.4
106.07	Ne	3.3.1.2,1.9.2	126	CH_3NH_2	3.3.7.1
107.72019	D_2O	3.3.3.2	126.1	Ne	3.3.1.2,1.9.2
108	NH_2D	3.3.4.3	127	CH_3SH	3.3.6.10
108.66842	CD_3OH	3.3.6.7	128	CH_3NH_2	3.3.7.1
108.81775	CH_2DOH	3.3.6.4	128	CH_3SH	3.3.6.10
108.94124	CH_2DOH	3.3.6.4	128.0	CH_3OD	3.3.6.5
109.29579	CH_2F_2	3.3.5.11	128.62900	HCN	3.3.3.3
110	CH_3OD	3.3.6.5	128.7	CD_3OH	3.3.6.7
110.43238	CH_3OH	3.3.6.3	129.5497	CH_3OH	3.3.6.2
110.7	CH_3OD	3.3.6.5	131.56276	CH_3OH	3.3.6.7
111.9	NH_3	3.3.4.2	132.8	Ne	3.3.1.2,1.9.2
112.22	NH_3	3.3.4.1	133.11960	CH_3OH	3.3.6.7
112.3	CD_3OH	3.3.6.7	133.9	HCOOH	3.3.5.4
112.53224	CH_2DOH	3.3.6.4	133.99765	CH_2F_2	3.3.5.11
112.6	D_2O	3.3.3.2	134	CH_3OD	3.3.6.5
113	NH_2D	3.3.4.3	135	CH_3OD	3.3.6.5
113.73188	CH_3OH	3.3.6.2	135.17175	CH_2DOH	3.3.6.4
113.8	CH_3OD	3.3.6.5	135.17256	CH_2DOH	3.3.6.4
114.29	NH_3	3.3.4.1	135.26932	CH_2F_2	3.3.5.11
115.32	H_2O	3.3.3.1	135.83350	CH_2DOH	3.3.6.4
115.82318	CH_3OH	3.3.6.3	136	CH_3OD	3.3.6.5
116	CH_3NH_2	3.3.7.1	136.62721	CD_3OH	3.3.6.7
116	CH_3SH	3.3.6.10	137	CH_3OD	3.3.6.5
116.27	NH_3	3.3.4.1	140.30	CH_2DOH	3.3.6.4
117	CH_3SH	3.3.6.10	140.89	SO_2	3.3.3.6
117.08507	CH_2DOH	3.3.6.4	141	CH_3OD	3.3.6.5
117.22707	CH_3OD	3.3.6.5	142	CH_3NH_2	3.3.7.1
117.72748	CH_2F_2	3.3.5.11	144.11787	CD_3OH	3.3.6.7
117.95948	CH_3OH	3.3.6.2	145	CH_3NH_2	3.3.7.1
118.01308	CH_3OH	3.3.6.3	145.5	CH_3OH	3.3.6.2
118.5910	H_2O	3.3.3.1	145.6	CH_3OD	3.3.6.5
118.83409	CH_3OH	3.3.6.2	145.66171	CH_3OD	3.3.6.5
119	CH_3NH_2	3.3.7.1	146	CH_3NH_2	3.3.7.1
119	CD_3OD	3.3.6.8	146.09738	CH_3OH	3.3.6.3
119.02	NH_3	3.3.4.1	147	CH_3NH_2	3.3.7.1
120	CH_3NH_2	3.3.7.1	147	CH_3SH	3.3.6.10
121	CH_3OH	3.3.6.2	147.15000	NH_3	3.3.4.1
121.20	CH_3OH	3.3.6.3	147.84469	CH_3NH_2	3.3.7.1
122.46551	CH_2F_2	3.3.5.11	147.97	CH_3OH	3.3.6.3
122.46581	CH_2F_2	3.3.5.11	148.59041	CH_3OH	3.3.6.3
123.26	CH_3OH	3.3.6.3	149.27228	CH_3OH	3.3.6.3
124	CH_3SH	3.3.6.10	149.38792	CH_2DOH	3.3.6.4
124	NH_2D	3.3.4.3	149.61284	CH_2DOH	3.3.6.4

Table 4.1 (continued)
GAS LASERS ARRANGED IN ORDER OF WAVELENGTH

Wavelength (μm)	Element or compound	Table No.	Wavelength (μm)	Element or compound	Table No.
150	CD_3OD	3.3.6.8	*171.75758*	CH_3OH	3.3.6.3
150.57167	CH_2DOH	3.3.6.4	*171.8*	CH_3DOH	3.3.6.4
150.81629	CH_2DOH	3.3.6.4	*172.8*	CD_3F	3.3.5.3
151.25369	CH_3OH	3.3.6.2	*172.84620*	CH_2DOH	3.3.6.4
151.49	NH_3	3.3.4.1	*176*	CH_3OH	3.3.6.2
152	CH_3OH	3.3.6.2	*178*	CH_3NH_2	3.3.7.1
152.07569	CH_3OH	3.3.6 .3	*178*	CH_3OH	3.3.6.2
152.7	CH_2DOH	3.3.6.4	*179*	CH_3NH_2	3.3.7.1
152.9	NH_3	3.3.4.2	*179.0*	CD_3OH	3.3.6.7
155.28	NH_3	3.3.4.1	*179.0*	CHD_2OH	3.3.6.6
155.6	CD_3F	3.3.5.3	*179.0*	CH_3OD	3.3.6.5
157.92848	CH_3OH	3.3.6.3	*179.72791*	CH_3OH	3.3.6.2
158.0	CD_3OH	3.3.6.7	*180*	CH_3NH_2	3.3.7.1
158.51348	CH_2F_2	3.3.5.11	*180.4*	CH_3OH	3.3.6.2
158.96020	CH_2F_2	3.3.5.11	*180.74051*	CD_3OH	3.3.6.7
159.2	CH_3OH	3.3.6.2	*181.92643*	N_2H_4	3.3.6.1
159.21794	CH_2DOH	3.3.6.4	*182*	CH_3OD	3.3.6.5
159.67569	CH_3OH	3.3.6.2	*182.10000*	CH_3OD	3.3.6.5
161	CH_3SH	3.3.6.10	*182.10000*	CH_2DOH	3.3.6.4
162.70	CH_2DOH	3.3.6.4	*182.56629*	CH_3OH	3.3.6.7
163.03353	CH_3OH	3.3.6.2	*183.62132*	CH_2DOH	3.3.6.4
164.0	CH_3OH	3.3.6.2	*184*	CD_3OD	3.3.6.8
164.50760	CH_3OH	3.3.6.2	*184.0*	CD_3OH	3.3.6.7
164.56421	CH_3OH	3.3.6.2	*184.30590*	CH_2F_2	3.3.5.11
164.60038	CH_3OH	3.3.6.2	*185*	CH_3NH_2	3.3.7.1
164.69747	CH_3OH	3.3.6.2	*185*	CHFF	3.3.6.13
164.74645	CH_2DOH	3.3.6.4	*185*	CH_3SH	3.3.6.10
164.78320	CH_3OH	3.3.6.2	*185.50040*	CH_3OH	3.3.6.2
165	CH_3NH_2	3.3.7.1	*185.9*	CH_3OH	3.3.6.2
165	CD_3OD	3.3.6.8	*186*	CH_3OD	3.3.6.5
165.0	CHD_2OH	3.3.6.6	*186.04219*	CH_3OH	3.3.6.2
166	CH_3NH_2	3.3.7.1	*188.41111*	CH_2DOH	3.3.6.4
166.63105	CH_2F_2	3.3.5.11	*188.42390*	CD_3OH	3.3.6.7
166.67665	CH_2F_2	3.3.5.11	*189.30000*	CH_2DOH	3.3.6.4
167.35235	CH_2DOH	3.3.6.4	*189.9490*	DCN	3.3.3.4
167.54117	CH_2DOH	3.3.6.4	*190*	CHFF	3.3.6.13
167.58700	CH_3OH	3.3.6.2	*190.0090*	DCN	3.3.3.4
168.0	CHD_2OH	3.3.6.6	*190.72590*	CH_3OH	3.3.6.2
168.1	CH_3OD	3.3.6.5	*191.5*	CH_3OH	3.3.6.2
168.84	CH_3OH	3.3.6.3	*191.61960*	CH_3OH	3.3.6.2
169	CH_3NH_2	3.3.7.1	*191.84803*	CH_2F_2	3.3.5.11
169	CH_3OD	3.3.6.5	*191.9*	CD_3OH	3.3.6.7
170.57637	CH_3OH	3.3.6.2	*192.72*	SO_2	3.3.3.6
171.3	CH_3OH	3.3.6.2	*192.78*	CH_3F	3.3.5.1
171.67	D_2O	3.3.3.2	*192.9072*	N_2H_4	3.3.6.1

Table 4.1 (continued)
GAS LASERS ARRANGED IN ORDER OF WAVELENGTH

Wavelength (μm)	Element or compound	Table No.
193.14158	CH_3OH	3.3.6.2
193.90445	CH_2F_2	3.3.5.11
194.06320	CH_3OH	3.3.6.2
194.44761	CH_2F_2	3.3.5.11
194.7027	DCN	3.3.3.4
194.7644	DCN	3.3.3.4
195.0	HDCO	3.3.4.5
195.49558	CH_2DOH	3.3.6.4
196.0	HDCO	3.3.4.5
196.10	CH_2DOH	3.3.6.4
196.5	HCOOH	3.3.5.4
198.0	CH_3NH_2	3.3.7.1
198	CHFF	3.3.6.13
198.66433	CH_3OH	3.3.6.2
200.0	CH_2DOH	3.3.6.4
200.0	CD_3F	3.3.5.3
201	CD_3OH	3.3.6.7
201.5	CD_3F	3.3.5.3
202.40	CH_3OH	3.3.6.2
202.46490	CH_2F_2	3.3.5.11
203.63580	CH_3OH	3.3.6.3
204.38720	DCN	3.3.3.4
205	CH_3SH	3.3.6.10
205	CH_3OH	3.3.6.2
206.0	CD_3F	3.3.5.3
206.68740	CH_2DOH	3.3.6.4
206.90	CH_3OH	3.3.6.2
208	CH_3OH	3.3.6.2
208.4121	CH_3OH	3.3.6.3
209.9302	CH_3OH	3.3.6.2
211.00	HCN	3.3.3.3
211.2629	CH_3OH	3.3.6.2
211.3148	CH_3OH	3.3.6.2
212	CH_3OD	3.3.6.5
212.5	CH_2DOH	3.3.6.4
213.4625	CH_3OH	3.3.6.2
214.35	CH_3OH	3.3.6.2
214.5791	CH_2F_2	3.3.5.11
215.0	NH_3	3.3.4.1
215.0812	CD_3OH	3.3.6.7
215.37246	CH_3OD	3.3.6.5
216.3	He	3.3.1.1,1.9.1
216.8	CH_2DOH	3.3.6.4
218.0	CH_3NH_2	3.3.7.1
218.0	CH_2DOH	3.3.6.4
218.0	DCOOD	3.3.5.8
218.0	NH_3	3.3.4.2
218.22	CH_3OH	3.3.6.2
218.3	NH_3	3.3.4.1
219.0960	CH_2DOH	3.3.6.4
220	CD_3OH	3.3.6.7
220	CH_3NH_2	3.3.7.1
220.22790	H_2O	3.3.3.1
221	CH_3NH_2	3.3.7.1
222	CD_3OH	3.3.6.7
223	CD_3OH	3.3.6.7
223.50	CH_3OH	3.3.6.2
223.9	NH_3	3.3.4.1
224	CH_3SH	3.3.6.10
224	CH_3OD	3.3.6.5
224	CD_3Cl	3.3.5.10
224.22560	CH_2DOH	3.3.6.4
225	CH_3OD	3.3.6.5
225.1	NH_3	3.3.4.1
225.5159	CH_3OH	3.3.6.2
226.2974	CH_2DOH	3.3.6.4
227.15	CH_3Cl	3.3.5.9
227.6570	CH_2F_2	3.3.5.11
228	CHFF	3.3.6.13
229	CD_3OD	3.3.6.8
229	CH_3OD	3.3.6.5
230.10590	CH_2F_2	3.3.5.11
231.0	CHFF	3.3.6.13
232	CD_3OH	3.3.6.7
232.7	CH_3OH	3.3.6.2
232.7884	CH_3OH	3.3.6.2
232.93906	CH_3OH	3.3.6.2
233	D_2CO	3.3.4.6
233.91570	N_2H_4	3.3.6.1
234	CH_3SH	3.3.6.10
234	CH_3OD	3.3.6.5
234.0	N_2H_4	3.3.6.1
235.6541	CH_2F_2	3.3.5.11
236	CD_3OH	3.3.6.7
236.25	CH_3Cl	3.3.5.9
236.5303	CH_3OH	3.3.6.3
236.5915	CH_2F_2	3.3.5.11
236.6008	CH_2F_2	3.3.5.11
237.5230	CH_3OH	3.3.6.3
237.60	CH_3OH	3.3.6.2

Table 4.1 (continued)
GAS LASERS ARRANGED IN ORDER OF WAVELENGTH

Wavelength (μm)	Element or compound	Table No.
238	CHD_2OH	3.3.6.6
238	CH_3OD	3.3.6.5
238	CD_3OH	3.3.6.7
238.5227	CH_3OH	3.3.6.3
240.0	HCOOD	3.3.5.6
240.98	CH_3Cl	3.3.5.9
241	CH_3OD	3.3.6.5
241.2	DCOOD	3.3.5.8
242.47270	CH_3OH	3.3.6.2
242.5	CH_3OH	3.3.6.2
242.6	CHFF	3.3.6.13
242.79	CH_3OH	3.3.6.2
244	D_2CO	3.3.4.6
245	CD_3Cl	3.3.5.10
245	D_2CO	3.3.4.6
245.0	CH_3Br	3.3.5.16
246	CH_3NH_2	3.3.7.1
246	CD_3Cl	3.3.5.10
246.5	N_2H_4	3.3.6.1
247.3	CD_3F	3.3.5.3
247.5	CD_3F	3.3.5.3
248.1220	CH_2DOH	3.3.6.4
249	CD_2Cl_2	3.3.5.13
249	CD_3Cl	3.3.5.10
249.7204	CH_2DOH	3.3.6.4
250.10	NH_3	3.3.4.1
250.50	N_2H_4	3.3.6.1
250.7813	CH_3OH	3.3.6.2
251	CH_3NH_2	3.3.7.1
251	CH_3OH	3.3.6.2
251.1398	CH_3OH	3.3.6.2
251.3	CH_3NH_2	3.3.7.1
251.4324	CH_3OH	3.3.6.2
251.91	CH_3F	3.3.5.1
253.5530	CH_3OH	3.3.6.2
253.60	CH_3OH	3.3.6.2
253.7196	CD_3OH	3.3.6.7
254	CH_3NH_2	3.3.7.1
254	CD_2Cl_2	3.3.5.13
254.5	HCOOH	3.3.5.4
255	CD_3OD	3.3.6.8
256	D_2CO	3.3.4.6
256.0270	CH_2F_2	3.3.5.11
257.1	NH_3	3.3.4.1
258	CH_2Cl_2	3.3.5.12
258.4356	CD_3OH	3.3.6.7
260	CHD_2OH	3.3.6.6
260	HCOOH	3.3.5.5
260.1	CHFF	3.3.6.13
261	CH_3OH	3.3.6.2
261.03	CH_3Cl	3.3.5.9
261.7292	CH_2F_2	3.3.5.11
262	CHFF	3.3.6.13
262	CH_3SH	3.3.6.10
262.0	N_2H_4	3.3.6.1
263.40	NH_3	3.3.4.1
263.70	CH_3OH	3.3.6.2
264.1	CH_3Br	3.3.5.14
264.5359	CH_3OH	3.3.6.2
264.7	C_2H_5F	3.3.8.1
264.8014	N_2H_4	3.3.6.1
265	CD_3OH	3.3.6.7
265.0	N_2H_4	3.3.6.1
265	CD_3F	3.35.3
265.1	DCOOH	3.3.5.7
266	CD_3OH	3.3.6.7
266.1	DCOOD	3.3.5.8
266.73520	CH_2DOH	3.3.6.4
267	CD_3OH	3.3.6.7
267.4432	CH_3OH	3.3.6.2
268	CD_3OH	3.3.6.7
268.5722	CH_3OH	3.3.6.3
268.8	NH_3	3.3.4.1
270	CH_3NH_2	3.3.7.1
270	CH_3OH	3.3.6.2
270.0055	CH_2F_2	3.3.4.11
270.6000	CH_2CHCN	3.3.7.3
271	CH_3NH_2	3.3.7.1
271.3	CH_3Cl	3.3.5.9
271.5	N_2H_4	3.3.6.1
272	CD_3I	3.3.5.20
272.0	DCOOH	3.3.5.7
272.2516	CH_2DOH	3.3.6.4
272.3389	CH_2F_2	3.3.5.11
273.0037	CH_2DOH	3.3.6.4
273.4	NH_3	3.3.4.1
274	CH_3OH	3.3.6.2
275.00	CH_3CL	3.3.5.9
275.09	CH_3Cl	3.3.5.9
276.1	DCOOD	3.3.5.8

Table 4.1 (continued)
GAS LASERS ARRANGED IN ORDER OF WAVELENGTH

Wavelength (μm)	Element or compound	Table No.	Wavelength (μm)	Element or compound	Table No.
276.7157	CD_3OH	3.3.6.7	*293.8217*	CH_3OH	3.3.6.2
276.8	NH_3	3.3.4.1	*293.9015*	CH_2F_2	3.3.5.11
277	CD_3OH	3.3.6.7	*294*	D_2CO	3.3.4.6
278	CD_3OH	3.3.6.7	*294.3*	CH_3Br	3.3.5.16
278.5	HCOOH	3.3.5.4	*294.81098*	CH_3OD	3.3.6.5
278.8048	CH_3OH	3.3.6.2	*295.3967*	CH_2DOH	3.3.6.4
279	D_2 CO	3.3.4.6	*295.6394*	CH_2DOH	3.3.6.4
279.3	NH_3	3.3.4.1	*297*	CD_3OH	3.3.6.7
279.4	CH_3OD	3.3.6.5	*297*	CD_3Br	3.3.5.17
279.8	CH_3Br	3.3.5.16	*298*	CH_3SH	3.3.6.10
280	CH_3OD	3.3.6.5	*298.0*	DCOOD	3.3.5.8
280.2183	CH_3OH	3.3.6.3	*298.2910*	CH_2F_2	3.3.5.11
280.2397	CH_3OH	3.3.6.3	*299*	CD_3OD	3.3.6.8
280.9341	CH_3OH	3.3.6.2	*299*	CD_3OH	3.3.6.7
281.18	CH_3CN	3.3.6.9	*300.0*	CH_2DOH	3.3.6.4
281.7	CH_3Cl	3.3.5.9	*301*	CD_3I	3.3.5.20
281.98	CH_3CN	3.3.6.9	*301.2754*	N_2H_4	3.3.6.1
283	CH_3NH_2	3.3.7.1	*301.3*	NH_3	3.3.4.1
283.0	C_2H_3Br	3.3.6.14	*301.9943*	CH_3OH	3.3.6.2
283.1	DCOOD	3.3.5.8	*302.2781*	HCOOH	3.3.5.4
286	CH_3OH	3.3.6.2	*304.1*	HCOOD	3.3.5.6
286.1974	CD_3OH	3.3.6.7	*304.0832*	DCOOD	3.3.5.8
286.5	CHFF	3.3.6.13	*305.72611*	CH_3OD	3.3.6.5
286.7241	CD_3OH	3.3.6.7	*306.3*	NH_3	3.3.4.1
286.8	CH_3Cl	3.3.5.9	*307.5*	CHFF	3.3.6.13
286.9	CH_3CN	3.3.6.9	*307.65*	CH_3Cl	3.3.5.9
287.3076	CD_3OH	3.3.6.7	*307.78*	CH_3OH	3.3.6.3
287.6672	CH_2F_2	3.3.5.11	*308.0405*	CH_2DOH	3.3.6.4
288	CH_3NH_2	3.3.7.1	*308.2957*	CH_2DOH	3.3.6.4
288	CD_3Cl	3.3.5.10	*309*	CD_3OH	3.3.6.7
288.5	CF_2CH_2	3.3.6.12	*309.5*	HCOOH	3.3.5.4
288.5	NH_3	3.3.4.1	*309.5*	NH_3	3.3.4.1
289.4	NH_3	3.3.4.1	*309.7140*	HCN	3.3.3.3
289.4999	CH_2F_2	3.3.5.11	*310*	CHFF	3.3.6.13
290	CD_3OH	3.3.6.7	*310*	CD_3OH	3.3.6.7
290	CD_3Br	3.3.5.17	*310*	DCOOD	3.3.5.8
290	NH_3	3.3.4.1	*310.8*	CHFF	3.3.6.13
290.4	NH_3	3.3.4.1	*310.8870*	HCN	3.3.3.3
290.62	CH_3OH	3.3.6.2	*311.07*	CH_3Br	3.3.5.16
291.0	NH_3	3.3.4.1	*311.0747*	N_2H_4	3.3.6.1
291.61	CH_3OH	3.3.6.3	*311.10*	CH_3Br	3.3.5.16
291.9	HCOOD	3.3.5.6	*311.2*	CH_3OH	3.3.6.2
292.1415	CH_3OH	3.3.6.2	*311.20*	CH_3Br	3.3.5.15
292.50	CH_3OH	3.3.6.2	*311.21*	CH_3Br	3.3.5.16
293.6480	CD_3Cl	3.3.5.10	*311.554*	HCOOH	3.3.5.4

Table 4.1 (continued)
GAS LASERS ARRANGED IN ORDER OF WAVELENGTH

Wavelength (μm)	Element or compound	Table No.
312	CD_3OD	3.3.6.8
312.0	DCOOH	3.3.5.7
313	HCOOH	3.3.5.5
314.8469	CH_3NH_2	3.3.7.1
316	CH_3SH	3.3.6.10
318	CD_3Cl	3.3.5.10
319	CH_3SH	3.3.6.10
319.9	HCOOH	3.3.5.4
320.0	CH_3OD	3.3.6.5
320	D_2CO	3.3.4.6
321	CD_3OH	3.3.6.7
322.4522	CH_2DOH	3.3.6.4
323.1	DCOOD	3.3.5.8
323.3	CD_3F	3.3.5.3
324	CH_3SH	3.3.6.10
324	D_2CO	3.3.4.6
324.1	HCOOD	3.3.5.6
325.17	CH_3OH	3.3.6.3
325.2	DCOOD	3.3.5.8
325.9	HCOOD	3.3.5.6
326.4230	CH_2F_2	3.3.5.11
327.0	N_2H_4	3.3.6.1
328.4570	DCOOH	3.3.5.7
330	CH_3OD	3.3.6.5
331.5	N_2H_4	3.3.6.1
331.6694	N_2H_4	3.3.6.1
332.6034	CH_3OH	3.3.6.3
332.86	CH_3Br	3.3.5.16
333.15	CH_3Br	3.3.5.16
334.0	CH_3Cl	3.3.5.9
335.1831	HCN	3.3.3.3
335.7087	DCOOD	3.3.5.8
336.0000	CD_3OH	3.3.6.7
336	N_2H_4	3.3.6.1
336.2461	CH_2DOH	3.3.6.4
336.3	HCOOH	3.3.5.4
336.5578	HCN	3.3.3.3
336.6	CD_3F	3.3.5.3
336.7	C_2H_5F	3.3.8.1
338.9638	$C^{13}H_3OH$	3.3.6.3
339	CD_3OD	3.3.6.8
339.9	HCOOD	3.3.5.6
340.3566	CH_2DOH	3.3.6.4
341	CH_3SH	3.3.6.10
341	CD_3Br	3.3.5.17
341	D_2CO	3.3.4.6
341.8	DCOOH	3.3.5.7
342	CD_2Cl_2	3.3.5.13
346	CD_3OH	3.3.6.7
346	CHD_2OH	3.3.6.6
346	D_2CO	3.3.4.6
346.3	CH_3CN	3.3.6.9
346.48750	CH_3OH	3.3.6.2
347.0	HCOOD	3.3.5.6
349	CH_3NH_2	3.3.7.1
349.0	CD_3F	3.3.5.3
349.3	CH_3Cl	3.3.5.9
350	CD_3OH	3.3.6.7
350.2	DCOOD	3.3.5.8
351	CH_3NH_2	3.3.7.1
351	CH_3SH	3.3.6.10
351	CD_3OH	3.3.6.7
351.0	HCOOD	3.3.5.6
351.9	DCOOD	3.3.5.8
351.9	HCOOD	3.3.5.6
352	CD_3OH	3.3.6.7
352.5	CH_3OD	3.3.6.5
352.7	CH_3Br	3.3.5.16
353	CD_3OH	3.3.6.7
353.1	HCOOD	3.3.5.6
354	CD_3OD	3.3.6.8
355	CHD_2OH	3.3.6.6
355.1261	CH_2F_2	3.3.5.11
355.2	HCOOD	3.3.5.6
356.0	C_2H_3Br	3.3.6.14
356.0	HCOOD	3.3.5.6
358.2	HCOOD	3.3.5.6
358.9	$C^{13}H_3OH$	3.3.6.3
359.9	HCOOH	3.3.5.4
361.2	HCOOD	3.3.5.6
362.1	DCOOH	3.3.5.7
363	CHD_2OH	3.3.6.6
365.2	DCOOH	3.3.5.7
366.9	DCOOD	3.3.5.8
367	CD_3Br	3.3.5.17
368.4	CD_3F	3.3.5.3
368.862	N_2H_4	3.3.6.1
369.1137	CH_3OH	3.3.6.2
369.9678	HCOOD	3.3.5.6
370.0	C_2H_3Br	3.3.6.14

Table 4.1 (continued)
GAS LASERS ARRANGED IN ORDER OF WAVELENGTH

Wavelength (μm)	Element or compound	Table No.	Wavelength (μm)	Element or compound	Table No.
370	CH_3SH	3.3.6.10	*395.7124*	HCOOD	3.3.5.6
370	CD_3OH	3.3.6.7	*396*	C_2H_5OH	3.3.9.1
372.0	HCOOD	3.3.5.6	*396*	C_2H_3Br	3.3.6.14
372.5	N_2H_4	3.3.6.1	*396.0*	DCOOD	3.3.5.8
372.5282	HCN	3.3.3.3	*396.00*	CH_2DOH	3.3.6.4
372.68	CH_3F	3.3.5.1	*397.1*	DCOOD	3.3.5.8
372.87	CH_3CN	3.3.6.9	*398*	CD_3OH	3.3.6.7
373.0	N_2H_4	3.3.6.1	*398.1*	HCOOD	3.3.5.6
373.8	HCOOD	3.3.5.6	*403*	CH_3SH	3.3.6.10
374.0861	CH_2OOH	3.3.6.4	*404*	C_2H_5F	3.3.8.1
375	NH_3	3.3.4.2	*404.0*	HCOOH	3.3.5.4
375.5449	CF_2CH_2	3.3.6.12	*404.7*	NH_3	3.3.4.1
377.45	CH_3I	3.3.5.19	*405.0*	HCOOH	3.3.5.4
378.4	OCS	3.3.3.5	*405.5044*	C_2H_5F	3.3.8.1
378.57	CH_3Cl	3.3.5.9	*405.5848*	HCOOH	3.3.5.4
379	CH_3SH	3.3.6.10	*406*	CD_3OD	3.3.6.8
380.02	CH_3Br	3.3.5.16	*407*	CD_3OH	3.3.6.7
380.5654	DCOOD	3.3.5.8	*407.2937*	CF_2CH_2	3.3.6.12
381.9956	CH_2F_2	3.3.5.11	*407.7*	CH_3Br	3.3.5.16
382.6392	CH_2F_2	3.3.5.11	*409*	CD_3OH	3.3.6.7
383.2845	CD_3Cl	3.3.5.10	*410*	CD_3OH	3.3.6.7
384	$(H_2CO)_3$	3.3.12.1	*411.0*	C_2H_3Br	3.3.6.14
384	CH_3SH	3.3.6.10	*411.2*	HCOOD	3.3.5.6
384.7	CD_3F	3.3.5.3	*412*	CD_3OH	3.3.6.7
385	CD_3OH	3.3.6.7	*414*	CD_3OD	3.3.6.8
385.9092	CH_2CHCl	3.3.6.11	*414.1*	DCOOD	3.3.5.8
386	CD_3OH	3.3.6.7	*414.98*	CH_3Br	3.3.5.14
386.3392	CH_3OH	3.3.6.2	*415.2*	DCOOD	3.3.5.8
386.4	CH_3CN	3.3.6.9	*416.0*	C_2H_3Br	3.3.6.14
387.3	CH_3CN	3.3.6.9	*416.5223*	CH_3OH	3.3.6.2
387.5591	CH_2DOH	3.3.6.4	*417.0*	HCOOD	3.3.5.6
387.8	HCOOD	3.3.5.6	*417*	CH_3OD	3.3.6.5
388.4	CH_3CN	3.3.6.9	*418.0827*	CH_3OH	3.3.6.2
389.9070	DCOOD	3.3.5.8	*418.1*	HCOOH	3.3.5.4
390	CD_3I	3.3.5.20	*418.2703*	CH_2F_2	3.3.5.11
390.1	CH_3OH	3.3.6.2	*418.3*	CH_3Br	3.3.5.16
390.5	CH_3I	3.3.5.19	*418.6129*	HCOOH	3.3.5.4
391.6886	HCOOD	3.3.5.6	*418.71180*	CD_3OH	3.3.6.7
392.0687	CH_3OH	3.3.6.2	*419.0*	C_2H_3Br	3.3.6.14
392.48	CH_3I	3.3.5.19	*420.3911*	HCOOH	3.3.5.4
392.9	HCOOD	3.3.5.6	*421*	CD_3OH	3.3.6.7
393.6311	HCOOH	3.3.5.4	*422*	CD_3OH	3.3.6.7
394.7009	CH_2F_2	3.3.5.11	*422.1512*	CH_2DOH	3.3.6.4
395.0	HCOOD	3.3.5.6	*422.8*	CH_3Br	3.3.5.16
395.1488	DCOOD	3.3.5.8	*424.0*	C_2H_3Br	3.3.6.14

Table 4.1 (continued)
GAS LASERS ARRANGED IN ORDER OF WAVELENGTH

Wavelength (μm)	Element or compound	Table No.
424	CH_2CHCl	3.3.6.11
425.2	DCOOD	3.3.5.8
426	CHD_2OH	3.3.6.6
427.0	C_2H_3Br	3.3.6.14
427.04	CH_3CN	3.3.6.9
427.2	CH_2DOH	3.3.6.4
428	CD_3Br	3.3.5.17
428.87	CH_3CCH	3.3.7.2
429.6898	HCOOD	3.3.5.6
430	CD_3Br	3.3.5.17
430.4380	HCOOD	3.3.5.6
430.55	CH_3CN	3.3.6.9
432.1094	HCOOH	3.3.5.4
432.6313	HCOOH	3.3.5.4
432.6665	HCOOH	3.3.5.4
433	$(H_2CO)_3$	3.3.12.1
433.1036	CD_3I	3.3.5.20
433.2	DCOOH	3.3.5.7
433.2	HCOOD	3.3.5.6
433.2353	DCOOH	3.3.5.7
434.9514	CH_2F_2	3.3.5.11
435	CD_3OH	3.3.6.7
435.7718	N_2H_4	3.3.6.1
437.4510	HCOOH	3.3.5.4
438.5069	C_2H_3Br	3.3.6.14
440	CD_3Br	3.3.5.17
441.2	CH_3CN	3.3.6.9
442.1678	CH_2CHCl	3.3.6.11
442.8	DCOOD	3.3.5.8
443.2645	CD_3Cl	3.3.5.10
443.5	C_2H_3Br	3.3.6.14
444.3862	CD_3I	3.3.5.20
444.8	HCOOH	3.3.5.4
445.0	C_2H_3Br	3.3.6.14
445	CH_2CHCl	3.3.6.11
445.8996	HCOOH	3.3.5.4
446.5054	HCOOH	3.3.5.4
446.8	HCOOD	3.3.5.6
446.8730	HCOOH	3.3.5.4
447.1421	CH_3I	3.3.5.19
448.5335	HCOOH	3.3.5.5
449.7997	CD_3Cl	3.3.5.10
450.1	HCOOD	3.3.5.6
450.9799	HCOOD	3.3.5.6
451.4754	CH_2DOH	3.3.6.4
452	C_2H_5F	3.3.8.1
452.2	DCOOD	3.3.5.8
452.40	CH_2DOH	3.3.6.4
453.3974	CH_3CN	3.3.6.9
455	CD_3OH	3.3.6.7
456	CH_3SH	3.3.6.10
457.3	CH_3I	3.3.5.19
457.3410	DCOOD	3.3.5.8
458.0	CF_2CH_2	3.3.6.12
458.5229	HCOOH	3.3.5.4
459.2	CH_3I	3.3.5.19
460	CH_3CHF_2	3.3.8.3
460	$(H_2CO)_3$	3.3.12.1
460.5619	CD_3I	3.3.5.20
461.0718	N_2H_4	3.3.6.1
461.2	CH_3Cl	3.3.5.9
461.2610	HCOOD	3.3.5.6
461.3847	CH_3OH	3.3.6.3
464.3	CF_2CH_2	3.3.6.12
464.4123	CH_2F_2	3.3.5.11
464.7567	CD_3Cl	3.3.5.10
466.25	CH_2CN	3.3.6.9
466.5461	DCOOH	3.3.5.7
468.2359	CH_3DOH	3.3.6.4
469	CD_2Cl_2	3.3.5.13
469.0233	CH_3OH	3.3.6.2
469.2	DCOOD	3.3.5.8
472	CD_3OH	3.3.6.7
472.1	HCOOD	3.3.5.6
472.9	HCOOD	3.3.5.6
477.4	HCOOD	3.3.5.6
477.9	CH_3I	3.3.5.19
478.9	DCOOD	3.3.5.8
479.9040	DCOOH	3.3.5.7
480	CD_3OH	3.3.6.7
480	HCOOH	3.3.5.5
480.0	CH_3CN	3.3.6.9
480.3101	CD_3Cl	3.3.5.10
482.9615	C_2H_3Br	3.3.6.14
483	CD_3OH	3.3.6.7
483	CHD_2OH	3.3.6.6
483.5	N_2H_4	3.3.6.1
486	C_2H_5F	3.3.8.1
486.1	CH_3OH	3.3.6.2
487	CH_2CHCl	3.3.6.11

Table 4.1 (continued)
GAS LASERS ARRANGED IN ORDER OF WAVELENGTH

Wavelength (μm)	Element or compound	Table No.	Wavelength (μm)	Element or compound	Table No.
487.2260	CD_3I	3.3.5.20	*525.32*	CH_3I	3.3.5.19
489	CH_2CHCN	3.3.7.3	*526.4856*	DCOOD	3.3.5.8
490.0829	C_2H_3Br	3.3.6.14	*527.2146*	DCOOD	3.3.5.8
490.3909	CD_3I	3.3.5.20	*527.8730*	N_2H_4	3.3.6.1
491.8906	DCOOD	3.3.5.8	*528.4965*	C_2H_3Br	3.3.6.14
493.1562	HCOOD	3.3.5.6	*529.3*	CH_3I	3.3.5.19
494.6461	CH_3CN	3.3.6.9	*530*	CD_3Br	3.3.5.17
495	CD_3OD	3.3.6.8	*530*	CH_2CHCl	3.3.6.11
495	CD_3OH	3.3.6.7	*530*	CH_3CHF_2	3.3.8.3
495	CH_3OH	3.3.6.2	*531*	HCOOD	3.3.5.6
496.072	CH_3F	3.3.5.1	*531.1*	CH_3Br	3.3.5.16
496.1009	CH_3F	3.3.5.1	*531.1*	CH_3CCH	3.3.7.2
498	CD_3OH	3.3.6.7	*533.655*	N_2H_4	3.3.6.1
498.0	CH_3OD	3.3.6.5	*533.6783*	HCOOH	3.3.5.4
498.0	HCOOD	3.3.5.6	*533.7006*	HCOOH	3.3.5.4
500	CH_2CHCN	3.3.7.3	*535*	HCOOH	3.3.5.4
502.2623	C_2H_5F	3.3.8.1	*540*	CH_2CHCl	3.3.6.11
503.0567	CH_2F_2	3.3.5.11	*540*	CD_3I	3.3.5.20
506	C_2H_3Br	3.3.6.14	*540.9864*	CH_2F_2	3.3.5.11
507.5840	CH_2CHCl	3.3.6.11	*542.99*	CH_3I	3.3.5.19
508	CD_3OH	3.3.6.7	*545.21*	CH_3Br	3.3.5.15
508	DCOOD	3.3.5.8	*545.39*	CH_3Br	3.3.5.16
508.4	CH_3I	3.3.5.19	*550*	CH_2CHCN	3.3.7.3
508.5	CH_3Br	3.3.5.16	*550*	CD_3Br	3.3.5.17
508.7911	DCOOD	3.3.5.8	*551*	CD_3OH	3.3.6.7
509.3717	CH_2DOH	3.3.6.4	*553*	CD_3OH	3.3.6.7
510.2	CH_3CN	3.3.6.9	*553.6962*	C_2H_3Br	3.3.6.14
511.4451	CH_2F_2	3.3.5.11	*554*	CD_3OH	3.3.6.7
511.9	CH_3Cl	3.3.5.9	*554.365*	CF_2CH_2	3.3.6.12
512	$(H_2CO)_3$	3.3.12.1	*556.876*	CD_3I	3.3.5.20
513.0022	HCOOH	3.3.5.4	*560*	CD_3Br	3.3.5.17
513.0157	HCOOH	3.3.5.4	*561.2939*	DCOOD	3.3.5.8
514.9507	DCOOD	3.3.5.8	*561.4*	CH_3CN	3.3.6.9
515.1695	HCOOH	3.3.5.4	*564.7*	CH_3Br	3.3.5.15
516.77	CH_3CCH	3.3.7.2	*566.4*	CH_3CCH	3.3.7.2
517	CD_3OH	3.3.6.7	*567.1065*	HCOOD	3.3.5.6
517.33	CH_3I	3.3.5.19	*567.5316*	CH_2F_2	3.3.5.11
518	CHD_2OH	3.3.6.6	*567.8683*	DCOOD	3.3.5.8
519.075	C_2H_5F	3.3.8.1	*567.946*	CH_2CHCl	3.3.6.11
519.303	CD_3Cl	3.3.5.10	*568.8*	CH_3Cl	3.3.5.9
520	CH_2CHCl	3.3.6.11	*569.477*	CD_3I	3.3.5.20
520	CD_2Cl_2	3.3.5.13	*570*	CF_2CH_2	3.3.6.12
520	CF_2CH_2	3.3.6.12	*570.5687*	CH_3OH	3.3.6.2
523.0914	CH_2DOH	3.3.6.4	*574*	CH_2CHCN	3.3.7.3
523.406	CD_3I	3.3.5.20	*576.17*	CH_3I	3.3.5.19

Table 4.1 (continued)
GAS LASERS ARRANGED IN ORDER OF WAVELENGTH

Wavelength (μm)	Element or compound	Table No.
578.90	CH_3I	3.3.5.19
580	CH_2CHCN	3.3.7.3
580.3872	HCOOH	3.3.5.4
580.8010	HCOOH	3.3.5.4
582.5536	HCOOD	3.3.5.6
583	CD_3OH	3.3.6.7
583.77	CH_3CCH	3.3.7.2
583.87	CH_3I	3.3.5.19
584	CH_2CHCN	3.3.7.3
585.72	CH_3Br	3.3.5.15
587	CH_2CHCN	3.3.7.3
588.0276	CH_2F_2	3.3.5.11
590	HCOOD	3.3.5.6
591.6157	DCOOD	3.3.5.8
593	DCOOD	3.3.5.8
593.506	C_2H_5F	3.3.8.1
594	HCOOD	3.3.5.6
594.7286	C_2H_3Br	3.3.6.14
599	CD_3OH	3.3.6.7
599.550	CD_3I	3.3.5.20
601.897	CH_2CHCl	3.3.6.11
602.4870	CH_3OH	3.3.6.2
614.110	CD_3I	3.3.5.20
614.2851	CH_3OH	3.3.6.2
614.3	CF_2Cl_2	3.3.5.18
616.3351	CH_2DOH	3.3.6.4
618.4462	C_2H_3Br	3.3.6.14
619	$(H_2CO)_3$	3.3.12.1
620	CH_2CHCN	3.3.7.3
620.40	C_2H_5F	3.3.8.1
624.0958	C_2H_3Br	3.3.6.14
624.4301	CH_3OH	3.3.6.2
629.8442	CH_3OH	3.3.6.3
630	CH_2CHCN	3.3.7.3
630.1661	HCOOD	3.3.5.6
631.0000	CD_2Cl_2	3.3.5.13
631.9	CH_3Br	3.3.5.16
632.0	CH_3Br	3.3.5.16
634.471	CH_2CHCl	3.3.6.11
635.3548	C_2H_3Br	3.3.6.14
638.4	CF_2Cl_2	3.3.5.18
639.1282	DCOOH	3.3.5.7
640	CH_2CHCl	3.3.6.11
640	CD_3I	3.3.5.20
642.5999	CH_2F_2	3.3.5.11
644	CD_3I	3.3.5.20
645	DCOOD	3.3.5.8
646	C_2H_3Br	3.3.6.14
646	CD_3OH	3.3.6.7
647.3485	DCOOH	3.3.5.7
647.89	CH_3CCH	3.3.7.2
648	CD_3OH	3.3.6.7
649.4255	C_2H_3Br	3.3.6.14
652.6800	CH_3CN	3.3.6.9
657	HCOOD	3.3.5.6
657.2391	CH_2F_2	3.3.5.11
658.5300	CH_3Br	3.3.5.15
660	C_2H_5F	3.3.8.1
660	HCOOD	3.3.5.6
660.582	CD_3I	3.3.5.20
660.70	CH_3Br	3.3.5.16
662.816	CF_2CH_2	3.3.6.12
666	DCOOD	3.3.5.8
667.232	CD_3I	3.3.5.20
668	HCOOD	3.3.5.6
669.531	HCOOH	3.3.5.4
670.094	CD_3I	3.3.5.20
670.114	CD_3I	3.3.5.20
671	CH_3I	3.3.5.19
675.3	CH_3CCH	3.3.7.2
680	$(H_2CO)_3$	3.3.12.1
680	CD_3OH	3.3.6.7
680.5414	C_2H_3Br	3.3.6.14
682.6	CH_2DOH	3.3.6.4
684.7	CF_2Cl_2	3.3.5.18
685	CD_3OH	3.3.6.7
689.9981	HCOOD	3.3.5.6
690	CD_3I	3.3.5.21
691.119	CD_3I	3.3.5.20
692	HCOOD	3.3.5.6
693.1396	C_2H_3Br	3.3.6.14
694	CH_3OH	3.3.6.2
694.1893	CH_3OH	3.3.6.2
695	CD_3OH	3.3.6.7
695.3499	CH_3OH	3.3.6.2
695.6720	HCOOD	3.3.5.6
696	$(H_2CO)_3$	3.3.12.1
697.4552	DCOOH	3.3.5.7
698.555	CD_3Cl	3.3.5.10
699.4226	CH_3OH	3.3.6.2

Table 4.1 (continued)
GAS LASERS ARRANGED IN ORDER OF WAVELENGTH

Wavelength (μm)	Element or compound	Table No.
700	CH_2CHCl	3.3.6.11
702	CD_3OH	3.3.6.7
703	CD_3OH	3.3.6.7
704.53	CH_3CN	3.3.6.9
705	HCOOH	3.3.5.4
707.2210	C_2H_3Br	3.3.6.14
710	CH_2CHCl	3.3.6.11
710	DCOOH	3.3.5.7
711	CD_3OH	3.3.6.7
712	$(H_2CO)_3$	3.3.12.1
712	C_2H_3Br	3.3.6.14
713.1056	DCOOH	3.3.5.7
713.72	CH_3CN	3.3.6.9
715.4	CH_3Br	3.3.5.14
719.3	CH_3I	3.3.5.19
720	CH_2CHCN	3.3.7.3
721	N_2H_4	3.3.6.1
722	CD_3OH	3.3.6.7
724.1399	C_2H_3Br	3.3.6.14
724.9203	CH_2F_2	3.3.5.11
726.9203	DCOOD	3.3.5.8
727.9491	HCOOD	3.3.5.6
730.3230	CD_3I	3.3.5.20
733.0000	HCOOD	3.3.5.6
733.5740	D_2CO	3.3.4.6
734.1616	N_2H_4	3.3.6.1
734.2620	CD_3I	3.3.5.20
735.130	CD_3Cl	3.3.5.10
737	DCOOD	3.3.5.8
740	CH_2CHCN	3.3.7.3
741.1149	C_2H_3Br	3.3.6.14
741.62	CH_3CN	3.3. 6.9
742.572	HCOOH	3.3.5.4
744.050	HCOOH	3.3.5.4
745	CD_3OH	3.3.6.7
745	CD_3I	3.3.5.20
749.29	CH_3Br	3.3.5.14
749.36	CH_3Br	3.3.5.14
750	$(H_2CO)_3$	3.3.12.1
750	CH_3CN	3.3.6.9
751.4	CF_2Cl_2	3.3.5.18
752.6808	D_2CO	3.3.4.6
752.7485	DCOOH	3.3.5.7
760	CD_3OH	3.3.6.7
760	HCOOH	3.3.5.4
761.7617	DCOOD	3.3.5.8
762.50	CH_2DOH	3.3.6.4
764	CF_2CH_2	3.3.6.12
765.2	CF_2Cl_2	3.3.5.18
770	CF_2CH_2	3.3.6.12
744	CD_3OH	3.3.6.7
779.8744	DCOOD	3.3.5.8
780	CH_2CHCN	3.3.7.3
780.1330	C_2H_3Br	3.3.6.14
784.2681	C_2H_3Br	3.3.6.14
786.1617	HCOOH	3.3.5.4
786.9419	HCOOH	3.3.5.4
788.4820	CD_3I	3.3.5.20
788.9192	HCOOH	3.3.5.5
789.4203	DCOOD	3.3.5.8
789.8396	HCOOH	3.3.5.4
790	CH_2CHCN	3.3.7.3
792	CD_3Cl	3.3.5.10
795	N_2H_4	3.3.6.1
795	DCOOD	3.3.5.8
802	N_2H_4	3.3.6.1
806	CD_3I	3.3.5.21
812	DCOOD	3.3.5.8
813.7572	HCOOD	3.3.5.6
815	$(H_2CO)_3$	3.3.12.1
819	HCOOD	3.3.5.6
824	CF_3Br	3.3.5.22
826	HCOOD	3.3.5.6
826.9443	C_2H_3Br	3.3.6.14
829	CD_2Cl_2	3.3.5.13
830	CH_2CHCN	3.3.7.3
830	CH_2CHCl	3.3.6.11
831.13	CH_3Br	3.3.5.15
835	DCOOD	3.3.5.8
843.2369	DCOOD	3.3.5.8
851.9	C_2H_5F	3.3.8.1
853.4380	C_2H_3Br	3.3.6.14
854.4	CH_3CN	3.3.6.9
858.7	CF_2Cl_2	3.3.5.18
869	CD_3OD	3.3.5.13
870.80	CH_3Cl	3.3.5.9
871.5850	CD_3OH	3.3.6.7
877.5481	DCOOD	3.3.5.8
880	CF_2CH_2	3.3.6.12
883	CF_3Br	3.3.5.22

Table 4.1 (continued)
GAS LASERS ARRANGED IN ORDER OF WAVELENGTH

Wavelength (μm)	Element or compound	Table No.	Wavelength (μm)	Element or compound	Table No.
883.598	CD_3Cl	3.3.5.10	*1020*	CF_2CH_2	3.3.6.12
890	$(H_2CO)_3$	3.3.12.1	*1025*	CF_2Cl_2	3.3.5.18
890	CF_2CH_2	3.3.6.12	*1030.378*	HCOOH	3.3.5.5
891	$(H_2CO)_3$	3.3.12.1	*1040*	CH_2CHCl	3.3.6.11
895	CD_3I	3.3.5.20	*1040*	CF_3Br	3.3.5.22
900	C_2H_5Cl	3.3.8.2	*1047.579*	DCOOH	3.3.5.7
900.1338	C_2H_3Br	3.3.6.14	*1063.29*	CH_3I	3.3.5.19
910	CH_2CHCN	3.3.7.3	*1070*	C_2H_5F	3.3.8.1
918.610	CD_3I	3.3.5.20	*1070.231*	DCOOD	3.3.5.8
919.9355	HCOOD	3.3.5.6	*1083*	CF_3Br	3.3.5.22
925.5	CH_3Br	3.3.5.14	*1086.89*	CH_3CN	3.3.6.9
926.2087	HCOOD	3.3.5.6	*1097.11*	CH_3CCH	3.3.7.2
927.9814	DCOOD	3.3.5.8	*1099.544*	CD_3I	3.3.5.20
930	HCOOH	3.3.5.4	*1100*	CD_3OH	3.3.6.7
934.2230	C_2H_3Br	3.3.6.14	*1146*	CD_3OH	3.3.6.7
935.0095	DCOOD	3.3.5.8	*1151*	CF_3Br	3.3.5.22
936.1590	C_2H_3Br	3.3.6.14	*1157.318*	HCOOD	3.3.5.6
936.6023	DCOOD	3.3.5.8	*1158*	DCOOD	3.3.5.8
940	CH_2CHCl	3.3.6.11	*1160*	CH_2CHCN	3.3.7.3
940	CH_2CHCN	3.3.7.3	*1161.676*	HCOOD	3.3.5.6
944.0	CH_3Cl	3.3.5.9	*1164*	CF_2Cl_2	3.3.5.18
948.9247	$(H_2CO)_3$	3.3.12.1	*1164.8*	CH_3CN	3.3.6.9
953.880	CD_3I	3.3.5.20	*1174.87*	CH_3CCH	3.3.7.2
958.2500	CH_3Cl	3.3.5.9	*1180*	CH_2CHCN	3.3.7.3
963.4873	C_2H_3Br	3.3.6.14	*1205*	CF_2Cl_2	3.3.5.18
964	CH_3I	3.3.5.19	*1213.362*	HCOOH	3.3.5.4
968	CD_3OH	3.3.6.7	*1221.79*	CH_3F	3.3.5.2
971.8064	DCOOH	3.3.5.7	*1223.658*	CH_3OH	3.3.6.2
980	CF_2Cl_2	3.3.5.18	*1237.966*	DCOOH	3.3.5.7
981.709	CD_3I	3.3.5.20	*1239.480*	CD_3Cl	3.3.5.10
985.8588	C_2H_3Br	3.3.6.14	*1247.594*	C_2H_3Br	3.3.6.14
986.3125	HCOOD	3.3.5.6	*1253.738*	CH_3I	3.3.5.19
989.1904	C_2H_3Br	3.3.6.14	*1281.649*	DCOOD	3.3.5.8
990	CF_2CH_2	3.3.6.12	*1290*	CD_3OH	3.3.6.7
990.51	CH_3Br	3.3.5.16	*1310.4*	CH_3Br	3.3.5.14
990.6303	C_2H_3Br	3.3.6.14	*1350*	C_2H_5Cl	3.3.8.2
992	CH_3F	3.3.5.1	*1351.8*	CH_3CN	3.3.6.9
998.5140	DCOOD	3.3.5.8	*1383.882*	C_2H_3Br	3.3.6.14
1000	CH_2 CHCl	3.3.6.11	*1394.063*	C_2H_3Br	3.3.6.14
1005.230	C_2H_5F	3.3.8.1	*1400*	C_2H_5Cl	3.3.8.2
1005.348	CD_3I	3.3.5.20	*1440*	C_2H_5F	3.3.8.1
1007	N_2H_4	3.3.6.1	*1448.0958*	CH_2F_2	3.3.5.11
1009.409	DCOOD	3.3.5.8	*1450*	CD_3F	3.3.5.3
1014.9	CH_3CN	3.3.6.9	*1490*	CD_3F	3.3.5.3
1016.3	CH_3CN	3.3.6.9	*1521.376*	C_2H_5F	3.3.8.1

Table 4.1 (continued)
GAS LASERS ARRANGED IN ORDER OF WAVELENGTH

Wavelength (μm)	Element or compound	Table No.	Wavelength (μm)	Element or compound	Table No.
1530	CF_3Br	3.3.5.22	*1886.87*	CH_3Cl	3.3.5.9
1541.75	HCOOD	3.3.5.6	*1890*	CF_3Br	3.3.5.22
1549.505	CD_3I	3.3.5.20	*1899.889*	C_2H_3Br	3.3.6.14
1550.	CF_3Br	3.3.5.22	*1965.34*	CH_3Br	3.3.5.14
1572.64	CH_3Br	3.3.5.14	*1990.757*	CD_3Cl	3.3.5.10
1614.888	C_2H_3Br	3.3.6.14	*2140*	CF_3Br	3.3.5.22
1692	CF_3Br	3.3.5.22	*2650*	CH_3Br	3.3.5.15
1720	C_2H_5Cl	3.3.8.2			
1730.833	HCOOD	3.3.5.6			
1814.37	CH_3CN	3.3.6.9			

Index

INDEX

A

Afterglow, 101, 117, 118
Ag, 13, 118, 190, 191, 202—203
energy level diagram of, 175
Al, 208
Alkali metals, 7, 120
Amplified spontaneous emission (ASE) mode, 4
Ar, 81—87, 103, 105, 111, 118, 122, 227—229, 244, 245, 430
energy level diagram of, 181—183
excitation mechanisms in, 193—195
operating parameters of, 196—197
performance of, 195—197
Ar_2, 276
Arc-formation, 103
ArCl, 276
ArArO, 277
Ar-O_2, 110
ASE, see Amplified spontaneous emission
Asymmetric top molecules, 318—320
Au, 13-14, 190, 191, 204
energy level diagram of, 176
Average electron temperature, 101

B

B, 24, 208
^{10}B, 123
Ba, 16—17, 104, 199
BCl_3, 319
wavelength of, 392
$^{10}BCl_3$, 439
$^{11}BCl_3$, 439
Be, 198
Bi, 36, 104, 118, 215
Bi_2, 277—278
Br, 46—47, 104, 121, 223
Br_2, 278

C

C, 27, 28, 105, 210
Ca, 15, 104, 118, 120, 198
Cascade pumping, 129
radiative, 116
Cd, 18—20, 103, 118, 190—192, 206, 239
energy level diagram of, 178
Cd_3Br, 454
C_2D_2, 318
wavelength of, 393—394
CD_3Cl, 449
CD_2Cl_2, 452
CD_3F, 440
CD_3I, 140, 456—457
$^{13}CD_3I$, 457
CD_3OD, 472—473
CD_3OH, 469—472
Ce, 11, 104, 118
CF_4, 320, 321
$C^{12}F_4$, 398—399
$C^{13}F_4$, 399—400
$C^{14}F_4$, 400—401
C_2F_5I, 140
C_2F_7I, 140, 141, 142
CF_3Br, 458
CF_2CH_2, 476
CF_2Cl_2, 455
CF_3I, 120, 140, 319
wavelength of, 401
C_2H_2, 318
wavelength of, 392
C_2H_4, 320
wavelength of, 403
C_2H_5Cl, 482
C_2H_5F, 481—482
C_2H_5OH, 483
Charge-exchange, 239
CH_3Br, 453—454
$CH_3{}^{79}Br$, 452
$CH_3{}^{81}Br$, 453
C_2H_3Br, 477—478
CH_3CCH, 480
wavelength of, 403
CH_2CHCl, 475
CH_3CH_2Cl, 482
CH_2CHCN, 481
CH_3CHF_2, 482
CH_3CH_2F, 481—482
CH_3CH_2OH, 483
CH_3Cl, 448
CH_2Cl_2, 451
CH_3CN, 473
CH_2DOH, 463—464
CHD_2OH, 468
Chemical pumping, 104, 120—122
$C_2H_4F_2$, 482
1,1-$C_2H_2F_2$, 476
cis-1,2-$C_2H_2F_2$,476
CH_2F_2, 450—451
CH_3F, 319, 439
wavelength of, 401
$^{13}CH_3F$, 440
CHFF, 476
CH_3I, 455—456
CH_3NH_2, 479—480
CH_3OD, 466—468
CH_3OH, 414, 415, 459—463
$^{13}CH_3OH$, 464—466
CH_3SH, 474
Cl, 44—45, 222
Cl_2, 278
ClF, 278
CN, 278, 323
wavelength of, 278—280, 323
CO, 104, 313
wavelength of, 280—281

CO_2, 118, 314, 315, 413—415
vibrational modes of, 318
CO_2 pump, 417
$C^{12}O^{16}$, 324—342
$C^{12}O^{16}_2$, 360—368
$C^{12}O^{16}O^{18}$, 368—370
$C^{12}O^{18}_2$, 371—374
$C^{13}O^{16}$, 343—347
$C^{13}O^{16}_2$, 374—377
$C^{13}O^{18}_2$, 377—379
$C^{14}O^{16}$, 379—381
$C^{14}O^{18}_2$, 381—383
COF_2, 320
wavelength of, 394
Collisions
deactivation of, 143
electron, 194
Penning, 239
quenching of, 138
resonant, 105
Column, positive, 100
Conservation of spin, see also Wigner spin rule, 105, 106
COS, 314
$C^{12}OS$, 383—385
$C^{13}OS$, 385
Cr, 199
CS_2, 314
$C^{12}S_2$, 385—386
$C^{13}S_2$, 386
Cu, 11—12, 113, 134—139, 189, 191, 200
energy level diagram of, 174
CuAcAc, 135
CuBr, 135, 136
$Cu(C_5H_7O_2)_2$, 135
CuCl, 135, 136
CuI, 135, 136
$Cu(NC_3)_2$, 135
Cu_2O, 135
CW optically pumped operation, 142

D

D_2, 282, 285—286
DBr, 349—350
DC-excitation, 100—101, 130
DCl, 352—353
DCN, 432
D_2CO, 438
DCOOD, 446—447
DCOOH, 445
Deactivation of collisions, 143
Density of electrons, 138
DF wavelength, 355—357
Diatomic electronic transition, 276—306
Diatomic molecules, 313—314
Diatomic vibrational transition, 323—359
Diffusion loss, 101
Dimers, 120
Direct electron impact,122, 133, 134
Direct nuclear pumping (DNP), 104, 105, 122
Direct optical pumping, 104, 118—119
Discharge pumping, 421
glow, 411
Discharges
DC-excited, 100—101
double, 103
glow, 100
hollow-cathode, 101, 191
pin, 103
positive-column, 100, 191
pulsed, 101—102
RF-excited, 100—101
short rise-time pulsed, 102
transversely excited, 102—103
Dissociation-excitation double-pulse mode, 138
Dissociative-excitation transfer (atom-molecule), 107—116
Dissociative-recombination process, 101
DNP, see Direct nuclear pumping
D_2O, 415, 431
Double discharge, 103
Double-pulse excitation, 135, 138
Drive-out, 101

E

$\delta E\infty$, 105
Electric field, 101
Electron collision, 194
Electron concentration, 101
Electron density, 138
Electron excitation, 103
Electronic impact, 313
Electronic transition, 273—312
Electron impact excitation, 102, 113, 129, 131, 193, 318
direct, 122, 133, 134
Electron temperature, 103, 129, 131, 132, 138
average, 101
Energy level diagrams, 173-187, 412-416
Eu, 54—56, 104, 238
Excitation, see also Pumping, 100
Ar, 241
collosions, 189, 193, 239
double-pulse, 135
electron impact, see Electron impact excitation
pulsed electron-beam, 103
recombining plasmas and, 103—104
selective, 111
Excitation transfer, 105—107
dissociative (atom-molecule), 107—116

F

F, 42—44, 132—133, 221
F_2, 281
Far infrared lasers, 411—419, 421
$FClO_3$, 401—402
Fe, 53-53, 118
Five-atomic vibrational transition, 398—403

Flat-plate lasers, 125
Four-atomic molecules, 314—318
Four-atomic vibrational transition, 392—398
Frequency measurements, 416
Frequency standards, 142—145

G

Ga, 25, 104
GaI_3, 120
Gas, 421
  temperature of, 128
Gas lasers
  far-IR, 421
  ionized, 171—269
  neutral, 3
G, 28, 211
Glow discharge, 100, 101, 411

H

H, 6, 103, 313
H_2, 282—286
HBr, 347—348
H_2CCl, 475
HCD, see Hollow-cathode discharges
HCl, 350—351
HCN, 314, 411, 432
  energy level diagram of, 412
  wavelength of, 386—387
$(H_2CO)_3$, 483
HCOOD, 443—445
HCOOH, 320, 441—442
  wavelength of, 402
$H^{13}COOH$, 442
HD, 283—286
HDCO, 438
He, 55, 57, 106, 118, 126, 239, 240, 428
  operating parameters for, 243
  performance of, 243
^3He, 123
Heating, 313
$He\text{-}F_2$, 132
He-HF, 133
$^3He\text{-}Kr$, 122
He-Ne, 105, 116, 123-128, 144, 145
$^3He\text{-}Ne$, 122
$He\text{-}NF_3$, 132, 133
He-Xe, 353—355
Hg, 21—24, 104, 118, 120, 190—192, 207, 240
  energy level diagram of, 179
HgBr, 287
HgCl, 287
HgI, 288
H_2O, 320, 411, 413, 430—431
  energy level diagram of, 413
  wavelength of, 387
Hollow-cathode discharges (HCD), 101, 191
Hollow-cathode excited lasers, 124—125

I

I, 47—49, 104, 120, 141—144, 192—193, 224—225
  photodissociation of, 139—142
I_2, 288—291
Impact excitation, see Electron impact excitation
Infrared lasers, 411—419
InI, 120
Inversion, 314
Ionized gas lasers, 171—269
Ion laser transitions, 101
Irradiation, 140

K

K, 8—9, 104
Kr, 88—92, 103, 105, 111, 118, 122, 129, 230—233, 251
  energy level diagram of, 184, 185
  performance of, 197
Kr_2, 292
KrCl, 291
KrF, 291—292
Kr_2F, 307
KrO, 292—293

L

Lasing gas, 421
Length standards, 142—145
Lifetime ratios, 129
Linear triatomic molecules, 314—318
Longitudinal electric field, 101
Loss of diffusion, 101

M

Measurements, 416
Metals, 7, 120
Metal-vapor ion laser transitions, 101
Mg, 14—15, 118, 198
Mn, 51—52

N

N, 31—33, 105, 212—213
N_2, 293—301
Na, 7
Na_2, 301—302
Na-H, 117
Nd, 118
$NdAl_3Cl_{12}$, 307
$NdAl_4Cl_{15}$, 307
$Nd(thd)_3$, 307
ND_3, 437

Ne, 58—80, 107, 111, 118, 122, 128, 226—227, 411, 412, 428—429
energy level diagram of, 180, 412
performance of, 197
Near-IR Xe lasers, 128—132
Negative glow, 101
Ne-O_2, 110
Neutral laser transitions, 3—100
Neutral resonance, 102
NH_3, 319, 414, 433—436
$N^{15}H_3$, 319, 397
$^{15}NH_3$, 437
N_2H_4, 458—459
$N^{14}H_3$, 394—397
NH_2D, 437
Ni, 53, 199
$N^{14}N^{15}O$, 391
$N^{15}N^{14}O$, 391
NO, 313
wavelength of, 357—359
N_2O, 314
$N_2{}^{14}O$, 388—319
$N^{15}{}_2O$, 391
NOCl, 391
NSF, 391—392
Nuclear pumping, 104, 122—123

O

O, 37—38, 117, 120, 215—216
O_2, 111
OCS, 432
OH, 313, 414
wavelength of, 359
Optical pumping, 142, 413, 417, 421
direct, 104, 118—119
photodissociative, 104, 119—120, 139—142

P

P, 33—34, 213
Partial inversion, 314
Passively stabilized lasers, 142
Pb, 29-30, 118, 212
PCD, see Positive-column discharges
pD, see Tube-diameter product
Penning collisions, 239
Perfluoroalkyliodides, 120
Performance, 191—193, 254, 255
Ar, 195, 197, 244
Kr, 197, 251
Ne, 197
Xe, 197, 258
Photodissociative optical pumping, 104, 119—120, 139—142
Photo-preionization, 103
Pin discharges, 103
Plasmas, 103—104
POPOP, 308
Polyatomic electronic transition, 307—308
Positive-column discharges (PCD), 100, 191
Precision frequency measurements, 416
Pulsed electron-beam excitation, 103
Pulse discharges, 101—102
Pumping, see also Excitation
cascade, 116, 129
chemical, 104, 120—122
CO_2, 417
direct nuclear, 104, 105, 122
direct optical, 104, 118—119
discharge, see Discharge pumping
glow discharge, 411
nuclear, 104—105, 122—123
optical, 104, 118—120, 142, 413, 417, 421
photodissociative optical, 104, 119—120
radiative cascade, 116
recombination, 116—118

Q

Quantum number, 313
Quenching, 138

R

Radiation trapping, 111
Radiative cascade pumping, 116
Rb, 9—10, 104
Recombination, 101, 129, 139, 142
Recombination pumping, 116—118
Recombining plasma excitation, 103—104
Red F, 132—133
Resonance
neutral, 102
transition of, 113
trapping of, 136
Resonant collision, 105
RF-excitation, 100—101, 130
Rise-time pulsed discharges, 102
Rogowski profile, 103

S

S, 38—39, 104, 217
S_2, 302
Samarium, 54
SAW spectral analyzer, 418
Sb, 35-36, 214
Se, 40, 104, 192, 218—219
operating parameters for, 243
performance of, 243
Selective excitation, 111
Self-terminating, 102, 113-114, 134, 135, 142
transitions, 114
Seven-atomic vibrational transition, 403—404
SF_6, 320
wavelength of, 404
Short rise-time pulsed discharges, 102
Si, 27, 210

SiF_4, 320
SiH_4, 320
wavelength of, 402—403
Six-atomic vibrational transition, 403
Sm, 54
Sn, 29, 118, 211
SO_2, 433
Spectral analyzer, 418
Spin conservation, 105, 106
Spontaneous emission, 4
Sputtering, 102
Sr, 16, 199
Stabilized lasers, 142
Standards
frequency, 142—145
length, 142—145
Symmetric top molecules, 318—320

T

$TbAl_3Cl_{12}$, 308
Te, 40, 220—221
Te_2, 303—304
TEM_{∞} mode, 128
Temperature
average electron, 101
electron, 101, 103, 129, 131, 132, 138
gas, 128
Th, 26, 209
Thermal drive-out, 101
Thermal heating, 313
TlHg, 120
Tm, 41—42
Transfer
dissociative excitation (atom-molecule), 107—116
excitation, 105—107
Transitions
electronic, 273—312
metal-vapor ion, 101
neutral, 3—100
resonance of, 113
unidentified, 99—100
vibrational, 313—409
Tranversely excited discharges, 102—103
Trapping
radiation, 111
resonance, 136
Triatomic electronic transition, 307
Triatomic molecules, linear, 314—318
Triatomic vibrational transition, 360—391
Tube-diameter product (pD), 101

U

^{235}U, 122, 123
$^{235}UF_6$, 123
Unidentified laser transitions, 99—100
UV irradiation, 140
UV laser performance, 254—255

V

V, 37
Vacuum-UV photo-preionization, 103
Vibrational quantum number, 313
Vibrational states, 313
Vibrational transition lasers, 313—409
diatomic, 323—359
five-atomic, 398—403
four-atomic, 392—398
seven-atomic, 403—404
six-atomic, 403
triatomic, 360—391
Volume-excited structures, 103

W

Waveguide lasers, 125
Wavelengths, table of, 495—568
Wigner spin rule, 105, 106

X

Xe, 92—98, 103, 111, 118, 122, 128—132, 233—237, 430
energy level diagram of, 186, 187
performance of, 197—198, 258
Xe_2, 306
Xe-Ar, 129
XeBr, 304
XeCl, 305
Xe_2Cl, 307
XeF, 305—306
Xe-He, 128—132
XeO, 306

Y

Yb, 50, 238

Z

Zn, 18, 118, 190, 191, 205
energy level diagram of, 177